ADVANCES IN CHEMICAL PHYSICS

VOLUME LXXXIII

Advances in
CHEMICAL PHYSICS

Edited by

I. PRIGOGINE

University of Brussels
Brussels, Belgium
and
University of Texas
Austin, Texas

and

STUART A. RICE

Department of Chemistry
and
The James Franck Institute
The University of Chicago
Chicago, Illinois

VOLUME LXXXIII

AN INTERSCIENCE® PUBLICATION
JOHN WILEY & SONS, INC.
NEW YORK • CHICHESTER • BRISBANE • TORONTO • SINGAPORE

In recognition of the importance of preserving what has
been written, it is a policy of John Wiley & Sons, Inc. to
have books of enduring value published in the United
States printed on acid-free paper, and we exert our best
efforts to that end.

An Interscience® Publication

Library of Congress Cataloging Number: 58-9935

ISBN 0-471-54018-8

Printed in the United States of America

10 9 8 7 6 5 4 3 2 1

CONTRIBUTORS TO VOLUME LXXXIII

A. C. ALBRECHT, Department of Chemistry, Cornell University, Ithaca, New York

JEREMY K. BURDETT, Department of Chemistry, The James Franck Institute, and the NSF Center for Superconductivity, The University of Chicago, Chicago, Illinois

JEFFREY A. CINA, Department of Chemistry and The James Franck Institute, The University of Chicago, Chicago, Illinois

NOEL A. CLARK, Condensed Matter Laboratory, Department of Physics, University of Colorado, Boulder, Colorado

W. T. COFFEY, School of Engineering, Department of Microelectronics and Electrical Engineering, Trinity College, Dublin, Ireland

P. J. CREGG, School of Engineering, Department of Microelectronics and Electrical Engineering, Trinity College, Dublin, Ireland

JACK H. FREED, Baker Laboratory of Chemistry, Cornell University, Ithaca, New York

KARL F. FREED, The James Franck Institute and the Department of Chemistry, The University of Chicago, Chicago, Illinois

MATTHEW A. GLASER, Condensed Matter Laboratory, Department of Physics, University of Colorado, Boulder, Colorado

VINCENT HURTUBISE, The James Franck Institute and the Department of Chemistry, The University of Chicago, Chicago, Illinois

YU. P. KALMYKOV, The Institute of Radioengineering and Electronics, Russian Academy of Sciences, Moscow, Russia

DUCKHWAN LEE, Department of Chemistry, Sogang University, Seoul, Korea

ANTONINO POLIMENO, Department of Physical Chemistry, University of Padua, Padua, Italy

VÍCTOR ROMERO-ROCHÍN, Instituto de Fisica, Universidad Nacional Autonoma de Mexico, Mexico

TIMOTHY J. SMITH, JR., Department of Chemistry and The James Franck Institute, The University of Chicago, Chicago, Illinois

INTRODUCTION

Few of us can any longer keep up with the flood of scientific literature, even in specialized subfields. Any attempt to do more and be broadly educated with respect to a large domain of science has the appearance of tilting at windmills. Yet the synthesis of ideas drawn from different subjects into new, powerful, general concepts is as valuable as ever, and the desire to remain educated persists in all scientists. This series, *Advances in Chemical Physics*, is devoted to helping the reader obtain general information about a wide variety of topics in chemical physics, a field which we interpret very broadly. Our intent is to have experts present comprehensive analyses of subjects of interest and to encourage the expression of individual points of view. We hope that this approach to the presentation of an overview of a subject will both stimulate new research and serve as a personalized learning text for beginners in a field.

I. PRIGOGINE
STUART A. RICE

CONTENTS

ADVANCES IN CHEMICAL PHYSICS

VOLUME LXXXIII

TIME-RESOLVED OPTICAL TESTS FOR ELECTRONIC GEOMETRIC PHASE DEVELOPMENT

JEFFREY A. CINA* and TIMOTHY J. SMITH, JR.

Department of Chemistry and the James Franck Institute, The University of Chicago, Chicago, Illinois

VÍCTOR ROMERO-ROCHÍN

Instituto de Física, Universidad Nacional Autónoma de México, México, D.F.

CONTENTS

* Camille and Henry Dreyfus Teacher-Scholar.

Advances in Chemical Physics, Volume LXXXIII, Edited by I. Prigogine and Stuart A. Rice.
ISBN 0-471-54018-8 © 1993 John Wiley & Sons, Inc.

I. INTRODUCTION

The occurrence of electronic geometric phase factors in time-dependent states of molecules is a subtle consequence of the Born–Oppenheimer method. Based on the fortunate separation in timescales between electronic and nuclear motions, the Born–Oppenheimer treatment provides the familiar and useful vision of molecular dynamics in which the nuclear motion proceeds, sometimes in a more or less classical fashion, on adiabatic electronic potential energy surfaces. In the presence of a degeneracy or near degeneracy between electronic potential energy surfaces at some nuclear configuration, it is also possible for the Born–Oppenheimer separation to give rise to an induced adiabatic vector potential in the nuclear Hamiltonian [1, 2]. In some cases, the adiabatic vector potential is the source of an effective magnetic field which exerts a force on the nuclei. The adiabatic vector potential can also be the source of an electronic geometric Berry phase, which appears in the time-dependent molecular wave function as a line integral of the adiabatic vector potential along the path or paths of nuclear motion. As with any quantum mechanical phase effect, electronic geometric phase development can only be observed in an interference experiment. The purpose of this article is to summarize our own recent theoretical studies on the accessibility of electronic geometric phases to measurement in a certain class of time-resolved interference experiments, termed wave packet interferometry experiments. Geometric phases in general and molecular geometric phases in particular have both been the subject of recent review articles [3–6], so we do not attempt to survey the field as a whole.

The wave packet interferometry experiments proposed and investigated here depend on the newly developed capability of producing sequences of ultrashort optical pulses among which there are fixed and actively stabilized optical phase relationships [7, 8]. When a molecular electronic transition is driven by a resonant pair of optical-phase-controlled light pulses, the excited state wave packets produced by the two pulses bear a definite quantum mechanical phase relationship and are therefore subject to interference with each other [8, 9]. In the situation considered here, of small transition amplitude and pulse durations short on a vibrational timescale, the interference is between two copies of the ground state nuclear wave function, one of which has propagated for the inter-pulse delay time under the excited state nuclear Hamiltonian, the other of which has developed under the ground state Hamiltonian. The quantum interference can be constructive or destructive, leading respectively to increased or decreased excited state population relative to the separate effects of the two pulses. Since the magnitude and sign of the interference signal for a given inter-pulse delay depend on the time development of the electronically excited wave packet produced by the initial pulse and the reference (ground state) wave packet transferred by the delayed pulse, the signal is sensitive to the overall quantum phase development of the two packets, including any geometric contributions.

We can sketch the basic idea of the class of experiments considered by adopting an idealized model, due to Longuet-Higgins and co-workers, of a molecule exhibiting an adiabatic electronic sign change. The adiabatic electronic sign change is the simplest example of a molecular Berry phase. The model consists of three "electronic states" of a charged particle on a ring and the ring, representing the nuclear degrees of freedom, sustains elliptical deformations. A nodeless electronic ground state is energetically well separated from the other states. But the adiabatic electronic excited states, with nodal points connected by lines parallel or perpendicular to the instantaneous elliptical distortion, become degenerate at the circular ring configuration [10].

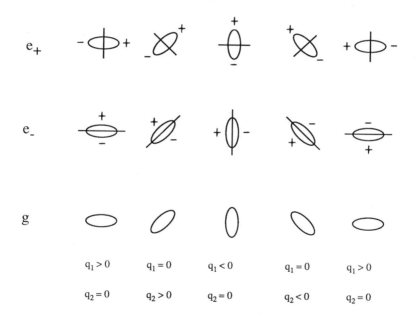

A cyclic change in direction of the elliptical deformation, describing a closed path in the (q_1, q_2) plane, is an example of molecular pseudo-rotation. In the excited adiabatic electronic states, a complete pseudorotation leads to a sign change of the electronic wave function [11], in the ground state it does not. To observe the sign change experimentally, one would use a sequence of nonresonant pulses to drive coherent pseudo-rotation of the molecule in its electronic ground state by an impulsive Raman process. The first pulse of a resonant in-phase-locked pulse pair transfers a small nuclear amplitude of the pseudorotating system to an excited state, say $|e_-\rangle$, (this choice is governed by the polarization of the resonant pulses). The delayed pulse transfers to the excited state an additional amplitude which may interfere constructively, destructively, or not at all with the propagated initial amplitude. Let us assume for the sake of illustration that the pseudorotational period is the same in the ground and excited states and consider inter-pulse delays corresponding to whole numbers of the pseudorotational periods. The initial propagated amplitude carries the sign-changing excited electronic wave function, while the reference amplitude generated by the delayed pulse carries the nodeless electronic ground state. Thus, the sense of interference will depend on whether the system completes an odd or even number of complete molecular pseudorotations during the inter-pulse delay. The resulting dependence of the excited state population on odd versus even numbers of nuclear excursions around the point of electronic degeneracy is the evidence we seek for electronic Berry phase development in the excited state.

We should note that Loss, Goldbart, and Balatsky, and Loss and Goldbart have considered a system analogous to the Longuet-Higgins model of a Jahn-Teller molecule [12]. They have described a geometrical phase resulting from the quantum orbital motion of electrons in a mesoscopic normal metal ring subject to an inhomogeneous magnetic field and have shown that the Berry's phase leads to persistent equilibrium charge and spin currents. In addition, certain molecular manifestations of the nonabelian generalizations of the geometric phase have been the subject of a recent investigation [13].

The experiments proposed here, and wave packet interferometry experiments in general, should be distinguished from recent suggestions for temporal control of chemical reaction dynamics. Tannor and Rice, and Tannor, Kosloff, and Rice have considered temporally separated optical pulses of different colors interacting with molecular systems in pump-dump experiments [14]. They have shown that by controlling the duration of propagation on the excited state electronic potential energy surface, corresponding to the delay between the two pulses, selectivity of products formed on the ground state electronic potential energy surface can in principle be achieved. In addition, a greater degree of control can be gained by modifying the phase structure of the individual pump and dump pulses. The effect of temporally controlled transfer of nuclear probability density among different electronic potential energy surfaces on the chemical reaction rate has in fact been studied in a recent experiment [15].

Another strategy for optical control of chemical reaction dynamics has been investigated by Brumer and Shapiro. The simplest version of their scheme relies on quantum mechanical interference between one and three photon absorption pathways to a given final continuum state [16]. Thus a fixed relative optical phase between continuous wave light sources of different colors must be maintained. Some essential features of Brumer and Shapiro's proposed method have been implemented in recent experimental work [17].

This article is organized in the following way. Section II presents a model Hamiltonian consisting of three electronic states and a pair of nuclear modes. The forms of the adiabatic electronic states, their energies and the induced vector potential are specified. In Section III, time-dependent perturbation theory is used to obtain an expression for the interference signal produced when a system executing coherent molecular pseudorotation in its ground electronic state undergoes an electronic transition driven by a resonant pair of in-phase-locked light pulses. In order to specify matrix elements of the electronic dipole moment operator, we adopt the Longuet-Higgins model values. We state the conditions under which excited state motion is adiabatic. As an aid to interpretation, we adapt the well studied locally quadratic Hamiltonian approach to Gaussian wave packet dynamics [18] to incorporate the effects of a locally linear vector potential. The equations of motion are found for the Gaussian parameters. A simple expression is obtained for the interference signal under the assumption that the locally quadratic Hamiltonian gives an adequate treatment of the nuclear motion in the excited state. The calculated signal in the locally quadratic theory is shown to be independent of the arbitrary choice of phases in the adiabatic electronic states (i.e., the approximate method is gauge invariant). In Section IV, interferograms are calculated and discussed. The interference signal is calculated in the case of weak electronic-nuclear coupling and large amplitude pseudorotation, using the locally quadratic Hamiltonian approach. The weak coupling-large amplitude case clearly manifests the adiabatic electronic sign change. We check the resiliency of this result by direct numerical calculation. We also investigate the spherical average necessary to predict the interference signal of an ensemble of randomly oriented molecules. For fairly short delays, the interferogram in the strongly coupled system at intermediate amplitude also proves amenable to calculation with the locally quadratic theory, despite the instability of the method in this case. The strong coupling interferogram for arbitrary time delay is obtained by direct numerical diagonalization of the Hamiltonian and interpreted with numerically calculated excited state wave packets and by comparison to the locally quadratic result. There are three appendices. Appendix A discusses the preparation of coherent pseudo-rotation by nonresonant impulsive Raman excitation and its dependence on molecular orientation. Appendices B and C list some details pertaining to the locally quadratic theory.

II. HAMILTONIAN AND INDUCED VECTOR POTENTIAL

Our model system will be a molecule with three electronic levels. The electronic ground state is taken to be nondegenerate. The two excited electronic states are Jahn–Teller active, being degenerate at the symmetrical nuclear configuration and coupled by distortion of the molecule away from the symmetrical configuration. The symmetric shape corresponds to values $q_1 = q_2 = 0$ of the two internal coordinates. We suppress all other vibrational modes. In the experiments suggested here, the most significant effects of spatial orientation will be in determining the nature of the time-dependent vibrational state prepared by a sequence of light pulses with specified spatial polarizations. We therefore treat the molecule as having a specific fixed orientation, as in a low temperature crystal. It may prove advantageous to perform the experiments in a jet, so we later discuss an orientational average.

In the electronic ground state, $|g\rangle$, the nuclear Hamiltonian,

$$H_g = \frac{\mathbf{p}^2}{2M} + \frac{M\omega_g^2}{2} \mathbf{q}^2 \qquad (2.1)$$

is a two dimensional harmonic oscillator, with $\mathbf{q} = (q_1, q_2)$ and $\mathbf{p} = (p_1, p_2)$. The Hamiltonian in the electronic excited manifold is given by

$$H_e = \epsilon + \frac{\mathbf{p}^2}{2M} + \frac{M\omega_e^2}{2} \mathbf{q}^2 + \kappa q_1 \sigma_1 + \kappa q_2 \sigma_2 \qquad (2.2)$$

where the constant, $\kappa = k\sqrt{\hbar M \omega_e^3}$, defines the strength of the electronic-nuclear coupling and k is the dimensionless vibronic coupling parameter introduced by Longuet-Higgins and co-workers [10, 19–21]. The Pauli matrices in Eq. (2.2) are constructed from the nuclear coordinate independent electronic states, $|+\rangle$ and $|-\rangle$, in the forms

$$\sigma_1 = |+\rangle\langle-| + |-\rangle\langle+|$$
$$\sigma_2 = i|-\rangle\langle+| - i|+\rangle\langle-| \qquad (2.3)$$
$$\sigma_3 = |+\rangle\langle+| - |-\rangle\langle-|$$

An external magnetic field or spin-orbit coupling would add a term $z\sigma_3$, with z a constant, to the excited state Hamiltonian (2.2) [1(c), 22]. The strategy outlined below for measuring the adiabatic electronic sign change will be equally well applicable to the complex Berry phase factors occurring in the presence of a σ_3 term.

The development of electronic geometric phase factors is governed by an adiabatic vector potential induced in the nuclear kinetic energy when we extend the Born–Oppenheimer separation to the degenerate pair of states [1, 23–25]. To see how the induced vector potential appears, we consider the family of transformations which diagonalize the excited state electronic coupling in the form

$$\kappa(q_1\sigma_1 + q_2\sigma_2) = \kappa q U(\mathbf{q})\sigma_3 U^\dagger(\mathbf{q}) , \qquad (2.4)$$

so that it is expressed as a rotation of the diagonal Pauli matrix dependent on the nuclear coordinate operators. The unitary operator in Eq. (2.4) can be written as a product,

$$U(\mathbf{q}) = U_\perp(\mathbf{q})U_\parallel(\mathbf{q}) \qquad (2.5)$$

where

$$U_\perp(\mathbf{q}) = e^{-i(\phi/2)\sigma_3}e^{-i(\pi/4)\sigma_2}e^{i(\phi/2)\sigma_3} , \qquad (2.6)$$

with $q_1 = q \cos \phi$ and $q_2 = q \sin \phi$, and

$$U_\parallel(\mathbf{q}) = e^{if(\mathbf{q})\sigma_3}e^{ig(\mathbf{q})} \qquad (2.7)$$

$f(\mathbf{q})$ and $g(\mathbf{q})$ in (2.7) are arbitrary single-valued functions. The eigenkets of the operator (2.4) are the adiabatic electronic states,

$$|e_\pm(\mathbf{q})\rangle = U(\mathbf{q})|\pm\rangle \qquad (2.8)$$

whose energies, including the harmonic term from Eq. (2.2), correspond to the electronic potential energy surfaces

$$\epsilon_\pm(\mathbf{q}) = \epsilon \pm \kappa q + \frac{M\omega_e^2}{2}\, \mathbf{q}^2 \qquad (2.9)$$

The presence of a conical degeneracy at $\mathbf{q} = 0$ between the upper and lower linear Jahn–Teller surfaces of Eq. (2.9) accounts for the occurrence of nontrivial electronic geometric phase factors in this system [1, 2, 26].

Applying the unitary operator (2.5) to the molecular Hamiltonian for the excited state manifold allows an easy identification of the adiabatic and nonadiabatic parts. We write the Hamiltonian (2.2) as

$$H_e = U(\mathbf{q}) \mathcal{H}_e U^\dagger(\mathbf{q}) \tag{2.10}$$

where

$$\mathcal{H}_e = \epsilon + \frac{1}{2M} \left(\mathbf{p} + \mathbf{A}(\mathbf{q}) \right)^2 + \frac{M\omega_e^2}{2} \mathbf{q}^2 + \kappa q \sigma_3 \tag{2.11}$$

The vector potential in Eq. (2.11) is given by

$$\mathbf{A} = \frac{\hbar}{i} U^\dagger(\mathbf{q}) \nabla_q U(\mathbf{q})$$

$$\mathbf{A} = \mathbf{A}_0 + \mathbf{A}_1 \sigma_1 + \mathbf{A}_2 \sigma_2 + \mathbf{A}_3 \sigma_3 \tag{2.12}$$

The coefficients are

$$\mathbf{A}_0 = \hbar \left(\frac{\partial g}{\partial q_1}, \frac{\partial g}{\partial q_2} \right) \tag{2.13}$$

$$\mathbf{A}_1 = \frac{\hbar}{2} \left(\frac{\partial \phi}{\partial q_1}, \frac{\partial \phi}{\partial q_2} \right) \cos(\phi + 2f) \tag{2.14}$$

$$\mathbf{A}_2 = \frac{\hbar}{2} \left(\frac{\partial \phi}{\partial q_1}, \frac{\partial \phi}{\partial q_2} \right) \sin(\phi + 2f) \tag{2.15}$$

and $\mathbf{A}_3 = \mathbf{A}_{3\perp} + \mathbf{A}_{3\parallel}$, with a transverse (divergence-free) part

$$\mathbf{A}_{3\perp} = \frac{\hbar}{2} \left(\frac{\partial \phi}{\partial q_1}, \frac{\partial \phi}{\partial q_2} \right) \tag{2.16}$$

and a longitudinal (curl-free) part

$$\mathbf{A}_{3\parallel} = \hbar \left(\frac{\partial f}{\partial q_1}, \frac{\partial f}{\partial q_2} \right) \tag{2.17}$$

(see, for example, [27]).

We can then decompose $\mathcal{H}_e$ into adiabatic (electronically diagonal) and nonadiabatic (electronically off-diagonal) parts; i.e., $\mathscr{H}_e = \mathscr{H}_{ad} + \mathscr{H}_{nad}$, with

$$\mathscr{H}_{ad} = \epsilon + \frac{1}{2M} (\mathbf{p} + \mathbf{A}_0 + \sigma_3 \mathbf{A}_3)^2 + \frac{M\omega_e^2}{2} \mathbf{q}^2 + \kappa q \sigma_3 + \frac{\mathbf{A}_1^2 + \mathbf{A}_2^2}{2M}$$
(2.18)

$$\mathscr{H}_{nad} = (\sigma_1 \mathbf{A}_1 + \sigma_2 \mathbf{A}_2) \cdot \frac{\mathbf{p}}{2M} + \frac{\mathbf{p}}{2M} \cdot (\sigma_1 \mathbf{A}_1 + \sigma_2 \mathbf{A}_2)$$
(2.19)

The transverse part of the induced adiabatic vector potential, (2.16), which appears in the kinetic energy of the adiabatic nuclear Hamiltonian, (2.18), governs electronic geometric phase development [25, 28]. The two dimensional curl of $\mathbf{A}_{3\perp}$ yields a solenoidal "magnetic field" of the form $\pi \hbar \delta(q_1) \delta(q_2)$ (in the general case of $z \neq 0$, it yields a monopolar field). The $\mathbf{A}_1^2 + \mathbf{A}_2^2 \sim q^{-2}$ term is an "electric gauge potential" giving rise to an inverse cubed force directed away from the point of degeneracy [29].

III. INTERFERENCE SIGNAL

A. General Expression

The model Hamiltonian of Section II captures some of the essential features of the electronic and vibrational structure of polyatomic molecules, like benzene and *sym*-triazine, that have both nondegenerate and degenerate, Jahn–Teller active, electronic levels. In this section interference experiments are described which will be sensitive to the geometric phase development accompanying adiabatic nuclear motion on either of the electronic potential energy surfaces in the Jahn–Teller pair.

The system is to be prepared in a coherently pseudorotating nuclear wave packet in the ground electronic state. Specifically, the wave packet is a moving displaced vibrational ground state of the two-dimensional oscillator with the form

$$\langle \mathbf{q} | \bar{\psi}(t) \rangle = \exp \left[\frac{i}{\hbar} (\bar{\alpha}(q_1 - \bar{q}_{1t})^2 + \bar{\alpha}(q_2 - \bar{q}_{2t})^2 + \bar{p}_1(q_1 - \bar{q}_{1t}) \right.$$
$$\left. + \bar{p}_2(q_2 - \bar{q}_{2t}) + \bar{\gamma}) \right]$$
(3.1)

In Eq. (3.1), $\bar{\alpha} = iM\omega_g/2$, and the other barred quantities are the usual time-dependent parameters. The formation of the coherent state (3.1) by nonresonant optical impulsive excitation [30] and the dependence of the initial values of the parameters on pulse sequence and molecular orientation are discussed in Appendix A.

The coherently pseudorotating molecule then interacts with a phase-locked pair of electronically resonant light pulses. The laser electric field, polarized in the **I** direction of a space-fixed cartesian frame, gives an interaction of the form

$$V(t) = -\hat{\mu} \cdot \mathbf{I}(E(t) + E(t - t_d)) \cos(\Omega t) \tag{3.2}$$

where $\hat{\mu}$ is the electronic dipole moment operator, connecting $|g\rangle$ with $|\pm\rangle$, and $E(t)$ is the field envelope. Since both pulses in (3.2) have the same carrier wave, they are said to be phase-locked at the frequency Ω. The pulses are short compared to the vibrational timescale and hence will transfer copies of the moving wave packet to one or both of the excited state electronic potential energy surfaces. It is now experimentally possible to produce ultrashort laser pulse-pairs phase-locked in phase (as in (3.2)), in quadrature, or out of phase at the carrier frequency or at any other component frequency [8].

A straightforward first order perturbation theory calculation gives the probability that the interaction (3.2) will leave the system electronically excited. The excited state population can be measured experimentally by monitoring fluorescence or photoionization. Of particular interest is the cross term coming from interaction with the fields of both pulses, which is given by

$$P^{\text{int}}(t_d) = \frac{2}{\hbar^2} \text{Re} \int_{-\infty}^{\infty} dt'' \int_{-\infty}^{\infty} dt' E(t'' - t_d) \cos(\Omega t'') E(t') \cos(\Omega t')$$

$$\times \langle \bar{\psi}(0)|e^{iH_g t''/\hbar} \langle g|\hat{\mu} \cdot \mathbf{I} e^{-iH_e(t''-t')/\hbar} \hat{\mu} \cdot \mathbf{I}|g\rangle e^{-iH_g t'/\hbar}|\bar{\psi}(0)\rangle . \tag{3.3}$$

The quantity (3.3) represents the effect of quantum mechanical interference between the two excited state amplitudes generated by the pulse pair from the moving ground state wave packet. The amplitude prepared by the initial pulse propagates under H_e during the interpulse delay; it can acquire geometric phases and, in general, undergo complicated nonadiabatic dynamics [31–33]. A reference wave packet, which has undergone comparatively simple motion under H_g, is transferred to the excited state by the delayed pulse. It is assumed that no radiative or material dephasing processes disrupt the interference of the wave packets on the timescale of the interpulse delay.

It is possible to evaluate Eq. (3.3) for arbitrary pulse shapes using the well known numerical diagonalization of the Jahn–Teller Hamiltonian [10]. However, geometric phase development will be most clearly manifest under simplifying experimental conditions which make possible an illuminating approximate treatment of the interference signal. If we specialize to Gaussian pulses

$$E(t) = E_0 e^{-t^2/2\tau^2} \rightarrow E_0 \tau \sqrt{2\pi} \delta(t) \tag{3.4}$$

which approach delta functions on a vibrational timescale, and make a rotating wave approximation, Eq. (3.3) reduces to

$$P^{\text{int}}(t_d) = \frac{\pi E_0^2 \tau^2}{\hbar^2} \text{Re}[e^{i\Omega t_d} \langle \bar{\psi}(0)|e^{iH_g t_d/\hbar} \langle g|\hat{\mu} \cdot \mathbf{I} e^{-iH_e t_d/\hbar} \hat{\mu} \cdot \mathbf{I}|g\rangle|\bar{\psi}(0)\rangle]. \tag{3.5}$$

The quantity in square brackets in (3.5) resembles the overlap kernel in the time-dependent theory of the continuous-wave absorption spectrum [34], but here involves the nonstationary ground state wave packet $|\bar{\psi}(0)\rangle$ rather than a stationary vibrational wave function. The interference signal in the impulsive limit directly measures the overlap between pseudo-rotating wave packets propagated in the ground and excited states for a time t_d.

B. Particular Example

We need the specific form of the matrix elements of the electronic dipole moment operator as functions of the nuclear coordinates in order to calculate interferograms using (3.3) or (3.5). To illustrate ideas, we adopt a simple realization of the Hamiltonian (2.1) and (2.2), due to Longuet-Higgins and co-workers [10], which provides closed-form expressions for the $\mathbf{q}$ dependence of the matrix elements of μ. Longuet-Higgins' realization of a Jahn–Teller active system treats the motion of a single electron on a continuous ring which is subject to elliptical distortions. It therefore mimics the Jahn–Teller effect in a monocyclic aromatic molecule.

We focus on the electronic ground state of the charged particle, $\langle \xi|g\rangle = (2\pi)^{-1/2}$, and the first, degenerate, pair of excited states, $\langle \xi|\pm\rangle = (2\pi)^{-1/2} e^{\pm i\xi}$. The dipole moment operator,

$$\hat{\mu} = \mu(\mathbf{i} \cos \xi + \mathbf{j} \sin \xi) \tag{3.6}$$

has transition elements [35]

$$\langle \pm|\hat{\mu}|g\rangle = \frac{\mu}{2} (\mathbf{i} \mp i\mathbf{j}) \tag{3.7}$$

The unit vectors in (3.6) and (3.7) belong to a molecule-fixed cartesian frame $(\mathbf{i}, \mathbf{j}, \mathbf{k})$ in which $\mathbf{k}$ is normal to the molecular plane. The ring distortions are all combinations of two orthogonal elliptical deformations, q_1 and q_2, having their major axes at a $45°$ angle. The major axis of the q_1 mode defines the $\mathbf{i}$ direction in the molecule-fixed frame and the q_2 distortion is along the $(2)^{-1/2}(\mathbf{i} + \mathbf{j})$ direction.

The adiabatic electronic states of Eq. (2.8) follow the ring distortion. Their transition elements with the ground state follow from (3.7) and are given by

$$\langle e_\pm(\mathbf{q})|\hat{\mu}|g\rangle = \frac{\mu}{2\sqrt{2}}\, e^{-i(g(\mathbf{q})\pm f(\mathbf{q}))}(\mathbf{i}(1 \pm e^{\mp i\phi}) + i\mathbf{j}(\mp 1 + e^{\mp i\phi})) \tag{3.8}$$

Notice that the transitions to $|e_+(\mathbf{q})\rangle$ and $|e_-(\mathbf{q})\rangle$ are parallel and perpendicular, respectively, with respect to the major axis of the ellipse (for instance, a field along $\mathbf{i}$ prepares the state $|+\rangle + |-\rangle \cong |e_+(\phi = 0)\rangle$ or $|e_-(\phi = \pi)\rangle$). To calculate the interferogram, we must also specify the relative orientations of the molecule-fixed $(\mathbf{i}, \mathbf{j}, \mathbf{k})$ and space-fixed $(\mathbf{I}, \mathbf{J}, \mathbf{K})$ axes. As is explained further in Appendix A, we use the freedom in defining q_1 and q_2 to ensure that the $\mathbf{ik}$-plane includes the polarization direction $\mathbf{I}$ of the phase-locked pulse pair.

C. Adiabatic Approximation

Since we are interested in electronic phase effects accompanying adiabatic nuclear excursions, and are considering motion near a conical intersection, we must specify the conditions under which adiabaticity obtains. A wave packet moving on one of the excited state potential surfaces will not undergo nonadiabatic transition to the other provided it is further from $q = 0$ than its width and that

$$\frac{\hbar^2}{4\kappa^2 q_t^6}(q_t^2\dot{q}_t^2 - (\mathbf{q}_t \cdot \dot{\mathbf{q}}_t)^2) = \frac{\hbar^2\dot{\phi}_t^2}{4\kappa^2 q_t^2} \ll 1 \tag{3.9}$$

where $\mathbf{q}_t$ and $\dot{\mathbf{q}}_t$ are the average position and velocity [25]. Since $\dot{\phi}_t$ is the instantaneous pseudorotational frequency, Eq. (3.9) is the usual requirement that the local electronic splitting exceeds the frequency of nuclear motion. A ground state wave packet, $|\bar{\psi}(0)\rangle$, obeying (3.9) will initially evolve adiabatically following an impulsive transition to the excited state surfaces. Adiabatic motion will continue until the trajectory of the wave packet or its spreading give significant amplitude for $q < \hbar\omega_e/2\kappa$.

To introduce the adiabatic approximation in the expression (3.5) for the interferogram, we use the excited state Hamiltonian in the form (2.10) and propagate under the adiabatic Hamiltonian (2.18) while neglecting the nonadiabatic contribution (2.19). Thus, we take

$$e^{-iH_e t_d/\hbar} \cong U(\mathbf{q})[|+\rangle e^{-i\mathscr{H}_{ad}^+ t_d/\hbar}\langle +| + |-\rangle e^{-i\mathscr{H}_{ad}^- t_d/\hbar}\langle -|]U^\dagger(\mathbf{q})$$

(3.10)

in which $\mathscr{H}_{ad}^\pm$ are given by (2.18) with σ_3 replaced by ± 1. The quantity (3.10) operates on the state

$$\hat{\mu} \cdot \mathbf{I}|g\rangle|\bar{\psi}(0)\rangle = (\mathbf{i} \cdot \mathbf{I})\frac{\mu}{2}(|+\rangle + |-\rangle)|\bar{\psi}(0)\rangle$$

(3.11)

in the interferogram expression (we have used $j \cdot I = 0$). We may center the initial wave packet around a negative value of q_1 (i.e., $\phi \approx \pi$) so that (3.11) is a pseudorotating wave packet in the lower electronic excited state (cf. following (3.8)). The first term in square brackets in (3.10), which governs adiabatic motion on the upper electronic potential energy surface, essentially vanishes in operation on (3.11); polarization perpendicular to the instantaneous long axis gives little amplitude on the upper surface. The $\mathscr{H}_{ad}^+$ term would also be discriminated against if the light pulses had spectral width less than $2\kappa q/\hbar$ and the center frequency were near resonance with the lower excited state surface. Combining the results of this paragraph with Eq. (3.5) gives an interference signal

$$P^{int}(t_d) = \pi \left(\frac{E_0 \mu \tau(\mathbf{i} \cdot \mathbf{I})}{2\hbar}\right)^2 \text{Re}\{\sim\}$$

$$\{\sim\} = \tfrac{1}{2}e^{i\Omega t_d}\langle \bar{\psi}(0)|e^{iH_g t_d/\hbar}(1 - e^{-i\phi})e^{i(g-f)}e^{-i\mathscr{H}_{ad}^- t_d/\hbar}e^{-i(g-f)}$$
$$\times (1 - e^{i\phi})|\bar{\psi}(0)\rangle$$

(3.12)

in the adiabatic approximation. The initial condition on the location of the ground state wave packet assumed in (3.12) might appear difficult to produce except in an oriented sample, such as a low temperature crystal. In Appendix A, we show that in fact large amplitude excitation along the q_1-mode can in principle be achieved for all but a small fraction of the molecules that will interact with the polarized resonant pulse pair in a randomly oriented sample.

D. Gaussian Wave Packets

We have found, by direct numerical propagation under H_e, that initially Gaussian pseudorotating wave packets often remain fairly well localized over several periods of motion. Even a wave packet excited into the trough of a *sym*-triazine-like system with $k \sim 2$ executes more than a complete pseudorotation before breaking up into several pieces (see further, below). The persistence of localized wave packets and the resulting prominent peaks in the calculated interference signal suggest that at least the important central portions of excited state wave packets may be treated as Gaussians with time-dependent width, momentum, coordinate and phase parameters. Such a treatment, which employs a simple analytic form for the wave function, will also be helpful in identifying geometric phase effects in the predicted interference signal.

We therefore adapt the locally quadratic Hamiltonian treatment of Gaussian wave packets, pioneered by Heller [18], to a system with an induced adiabatic vector potential. The locally quadratic theory replaces the anharmonic time-independent nuclear Hamiltonian by a time-dependent Hamiltonian which is taken to be of second order about the instantaneous center of the wave packet. Since the nuclear wave packet continually evolves under an effective harmonic Hamiltonian, an initially Gaussian wave form remains Gaussian. The treatment yields equations of motion for the wave function parameters that can be solved numerically [36–38]. The locally quadratic Hamiltonian includes a second order expansion of the scalar potential, consisting of the last three terms in Eq. (2.18), which we write as

$$v(\mathbf{q}) = v + v_{q_1}(q_1 - q_{1t}) + v_{q_2}(q_2 - q_{2t}) + v_{q_1 q_2}(q_1 - q_{1t})(q_2 - q_{2t})$$
$$+ \tfrac{1}{2} v_{q_1 q_1}(q_1 - q_{1t})^2 + \tfrac{1}{2} v_{q_2 q_2}(q_2 - q_{2t})^2 \qquad (3.13)$$

Expressions for the various coefficients are listed in Appendix B.

The presence of an induced adiabatic vector potential in the kinetic energy of (2.18) raises the question of how to properly handle the expansion of $\mathbf{A}_0(\mathbf{q}) + \sigma_3 \mathbf{A}_3(\mathbf{q})$ about the center of the wave packet. This question can be posed and resolved, as follows, by recalling the definition (2.12) of that quantity in terms of $U(\mathbf{q})$. The interferogram expression (3.12) calls for propagation of wave packets under $\mathcal{H}_{ad}^-$. But the Gaussian wave packet $|\psi(0)\rangle$ appears there preceded by exponential factors depending on $\mathbf{q} - \mathbf{q}_t$ to all orders. There is no need to approximate the factor $\exp(-ig(\mathbf{q}) + if(\mathbf{q}))$; it merely cancels the longitudinal contributions (2.13) and (2.17) to the vector potential in (2.18). The other factor, $(1 - \exp(i\phi))$, comes from the electronic matrix elements of $U_\perp^\dagger(\mathbf{q})$ involved in the operation of (3.10) on (3.11). This factor accounts for a slight distortion of the wave packet upon excitation to the lower Jahn–Teller surface, which occurs despite the arbitrarily short pulse duration. To remain consistent with the Gaussian approximation, we must truncate the expansion of $\phi(\mathbf{q})$ in the exponent at second order, writing

$$\phi(\mathbf{q}) = \phi_t + \phi_{q_1}(q_1 - q_{1t}) + \phi_{q_2}(q_2 - q_{2t}) + \phi_{q_1 q_2}(q_1 - q_{1t})(q_2 - q_{2t})$$
$$+ \tfrac{1}{2}\phi_{q_1 q_1}(q_1 - q_{1t})^2 + \tfrac{1}{2}\phi_{q_2 q_2}(q_2 - q_{2t})^2 \tag{3.14}$$

In more general terms, consistent application of the Gaussian approximation requires use of the quadratic expansion (3.14) in the unitary operator $U_\perp(\mathbf{q})$ given by (2.6). Hence, we obtain a *linear* approximation to the transverse vector potential (2.16) of the form

$$\mathbf{A}_{3\perp}(\mathbf{q}) \cong (a + a_{q_1}(q_1 - q_{1t}) + a_{q_2}(q_2 - q_{2t}), b + b_{q_1}(q_1 - q_{1t})$$
$$+ b_{q_2}(q_2 - q_{2t})) \tag{3.15}$$

which in turn entails a locally uniform approximation to the field derived from the induced vector potential. The coefficients in the expansions (3.14) and (3.15) are given for reference in Appendix B [39].

Determination of the time-dependent nuclear wave function now becomes a straightforward exercise in the locally quadratic theory. We need to propagate Gaussian wave packets of the form

$$\langle \mathbf{q}|\bar{\psi}(t)\rangle = \exp\left[\frac{i}{\hbar}\left(\alpha(q_1 - q_{1t})^2 + \beta(q_2 - q_{2t})^2 + 2\delta(q_1 - q_{1t})(q_2 - q_{2t})\right.\right.$$
$$\left.\left. + p_1(q_1 - q_{1t}) + p_2(q_2 - q_{2t}) + \gamma\right)\right] \tag{3.16}$$

under the locally quadratic version of the Hamiltonian $\mathcal{H}_{\mathrm{ad}}^-$ with the longitudinal parts of the vector potential removed from the kinetic energy operator (i.e., the Hamiltonian of Eq. (2.18) with σ_3 replaced by -1 and both $\mathbf{A}_0$ and $\mathbf{A}_{3\parallel}$ set to zero). Equations of motion for the parameters in (3.16) are obtained in the usual way, by substituting that function in the time-dependent Schrödinger equation and equating coefficients of the various powers of $(q_1 - q_{1t})$ and $(q_2 - q_{2t})$. We write $\alpha = \alpha_r + i\alpha_i$, etcetera for the complex parameters and obtain the following equations of motion after some algebra:

$$\dot{q}_{1t} = \frac{1}{M}(p_1 - a) \tag{3.17}$$

$$\dot{q}_{2t} = \frac{1}{M}(p_2 - b) \tag{3.18}$$

$$\dot{p}_{1t} = -v_{q_1} + \frac{1}{M}(p_1 - a)a_{q_1} + \frac{1}{M}(p_2 - b)b_{q_1} \tag{3.19}$$

$$\dot{p}_{2t} = -v_{q_2} + \frac{1}{M}(p_2 - b)b_{q_2} + \frac{1}{M}(p_1 - a)a_{q_2} \tag{3.20}$$

$$\dot{\gamma}_r = p_1 \dot{q}_{1t} + p_2 \dot{q}_{2t} - \frac{1}{2M} (p_1 - a)^2 - \frac{1}{2M} (p_2 - b)^2 - v - \frac{\hbar}{M} (\alpha_i + \beta_i) \tag{3.21a}$$

$$= \frac{M}{2} (\dot{q}_{1t}^2 + \dot{q}_{2t}^2) + a\dot{q}_{1t} + b\dot{q}_{2t} - v - \frac{\hbar}{M} (\alpha_i + \beta_i) \tag{3.21b}$$

$$\dot{\gamma}_i = \frac{\hbar}{M} (\alpha_r + \beta_r) \tag{3.22}$$

$$\dot{\alpha}_i = -\frac{4}{M} (\alpha_r \alpha_i + \delta_r \delta_i) + \frac{2}{M} (a_{q_1} \alpha_i + b_{q_1} \delta_i) \tag{3.23}$$

$$\dot{\alpha}_r = \frac{2}{M} (\alpha_i^2 - \alpha_r^2 + \delta_i^2 - \delta_r^2) + \frac{2}{M} (a_{q_1} \alpha_r + b_{q_1} \delta_r) - \tfrac{1}{2} v_{q_1 q_1} \tag{3.24}$$

$$\dot{\beta}_i = -\frac{4}{M} (\beta_r \beta_i + \delta_r \delta_i) + \frac{2}{M} (b_{q_2} \beta_i + a_{q_2} \delta_i) \tag{3.25}$$

$$\dot{\beta}_r = \frac{2}{M} (\beta_i^2 - \beta_r^2 + \delta_i^2 - \delta_r^2) + \frac{2}{M} (b_{q_2} \beta_r + a_{q_2} \delta_r) - \tfrac{1}{2} v_{q_2 q_2} \tag{3.26}$$

$$\dot{\delta}_i = -\frac{2}{M} (\alpha_i + \beta_i)\delta_r - \frac{2}{M} (\alpha_r + \beta_r)\delta_i + \frac{1}{M} (a_{q_2} \alpha_i + b_{q_1} \beta_i) \tag{3.27}$$

$$\dot{\delta}_r = -\frac{2}{M} (\alpha_r + \beta_r)\delta_r + \frac{2}{M} (\alpha_i + \beta_i)\delta_i + \frac{1}{M} (a_{q_2} \alpha_r + b_{q_1} \beta_r) - \tfrac{1}{2} v_{q_1 q_2} \tag{3.28}$$

To evaluate the interference signal (3.12) in the Gaussian approximation, one needs to solve the twelve equations of motion (3.17) through (3.28) twice. In one instance, the initial conditions are the parameters of the ground state wave packet $\langle q|\psi(0)\rangle$ given in (3.1). In the second instance, the initial values of some parameters are slightly modified by incorporating the coefficients of the expansion (3.14) in the multiplying factor $\exp(i\phi(q))$. The two sets of initial conditions are given explicitly in Appendix C.

Equations (3.17)–(3.20) are the usual classical Hamilton's equations for a particle in the vector potential $(a(q), b(q))$. Equation (3.21b) for the phase of the wave function involves the Lagrangian for such a particle plus a quantum correction. The equation of motion for the imaginary part of the overall phase, (3.22), is redundant in the sense that it merely ensures continued adherence to the normalization condition

$$e^{-\gamma_i/\hbar} = [4(\alpha_i \beta_i - \delta_i^2)/\pi^2 \hbar^2]^{1/4} \tag{3.29}$$

In practice it is useful to solve (3.22), despite its redundancy, since it provides a check on the accuracy of numerical integration.

The magnitude and sign of the interference signal will be sensitive to geometric phase development through its effect on the classical action $\gamma_r + \int dt(\hbar/M)(\alpha_i + \beta_i)$ (see Eq. (3.21b)). In particular, the line integral

$$\int_0^{t_d} dt(a\dot{q}_{1t} + b\dot{q}_{2t}) = \int_C (dq_{1t}a + dq_{2t}b) \qquad (3.30)$$

where C is the path taken by $\mathbf{q}_t$, is the electronic Berry phase at the center of the wave packet.

In general, one might consider the influence of the induced adiabatic vector potential on the mean values of the coordinates, $\mathbf{q}_t$; these of course directly affect the value of the interference signal. In the $z = 0$ case to which we have specialized, however, the transverse vector potential $\mathbf{A}_{3\perp}$ has vanishing curl except at the origin and therefore does not exert a Lorentz force. For that reason, the effect of $\mathbf{A}_{3\perp}$ in the $z = 0$ case has been likened to the Aharonov–Bohm effect [1]. In contrast, the repulsive inverse-cubed force arising from the $(\mathbf{A}_1^2 + \mathbf{A}_2^2)/2M$ term will always affect $\mathbf{q}_t$ through its presence in v_{q_1} and v_{q_2}.

We have dropped several terms involving products of two of the quantities a_{q_1}, a_{q_2}, b_{q_1} and b_{q_2} from Eqs. (3.24), (3.26) and (3.28). We must omit these terms because there are other terms of similar magnitude which would come, like those omitted, from quadratic corrections to $\mathbf{A}_{3\perp}^2(\mathbf{q}_t)/2M$ in the Hamiltonian (2.18), had the expansion of $\mathbf{A}_{3\perp}$ not been stopped at first order. Order of magnitude arguments suggest that those quadratic corrections will be of little consequence under the condition assumed here, that q_t is somewhat greater than the width of the wave packet. Numerical solutions of the parameter equations of motion, to be described below, were found to be sensibly independent of whether these terms, and similarly incomplete quadratic corrections to $(\mathbf{A}_1^2 + \mathbf{A}_2^2)/2M$, were retained (see Appendix B).

The semi-classical equations of motion obtained above involve only the transverse adiabatic vector potential which is, by definition, independent of the choice of gauge functions $f(\mathbf{q})$ and $g(\mathbf{q})$. The $(\mathbf{A}_1^2 + \mathbf{A}_2^2)/2M$ term in the potential is also independent of those two arbitrary functions. The locally quadratic approach to Gaussian dynamics therefore gives physically equivalent results for any choice of $f(\mathbf{q})$ and $g(\mathbf{q})$. The finding that the locally quadratic Hamiltonian approach developed here is strictly invariant with respect to choice of phases of the adiabatic electronic eigenstates supersedes the approximate discussion of gauge invariance given earlier by Romero-Rochin and Cina [25] (see also [40]).

Calculation of the interference signal (3.12) from the ground and excited state wave packets (3.1) and (3.16) finally requires the evaluation of four Gaussian integrals whose general form,

$$Q \equiv \int_{-\infty}^{\infty} dx \int_{-\infty}^{\infty} dy \exp\left[\frac{i}{\hbar}\left(ax^2 + by^2 + 2dxy + mx + ny + g\right)\right]$$

$$= \frac{\pi\hbar}{(d^2 - ab)^{1/2}} \exp\left[\frac{i}{\hbar}\left(g + \frac{m^2 b + n^2 a - 2mnd}{4(d^2 - ab)}\right)\right] \qquad (3.31)$$

with

$$a = \alpha - \bar{\alpha}^* \qquad (3.32)$$

$$b = \beta - \bar{\beta}^* \qquad (3.33)$$

$$d = \delta - \bar{\delta}^* \qquad (3.34)$$

$$m = p_1 - \bar{p}_1 - 2\alpha(q_{1t} - \bar{q}_{1t}) - 2\delta(q_{2t} - \bar{q}_{2t}) \qquad (3.35)$$

$$n = p_2 - \bar{p}_2 - 2\beta(q_{2t} - \bar{q}_{2t}) - 2\delta(q_{1t} - \bar{q}_{1t}) \qquad (3.36)$$

$$g = \gamma - \bar{\gamma}^* - p_1(q_{1t} - \bar{q}_{1t}) - p_2(q_{2t} - \bar{q}_{2t})$$
$$+ \alpha(q_{1t} - \bar{q}_{1t})^2 + \beta(q_{2t} - \bar{q}_{2t})^2 + 2\delta(q_{1t} - \bar{q}_{1t})(q_{2t} - \bar{q}_{2t}) \qquad (3.37)$$

is given here for reference. Equations (3.32)–(3.37) are found by moving the origin of integration to the mean position of the ground state reference wave packet. In general, all six parameters (a, b, d, m, n, and g) will be complex.

There has been some work on semi-classical quantization of the linear Jahn–Teller Hamiltonian (2.2) [41, 42]. The quantization scheme which bears the closest relation to the present wave packet treatment involved the calculation of classical trajectories while slowly turning on the nonadiabatic interaction [41(a)]. The main emphasis of that work was the development of a method for obtaining energy levels in molecules with nonadiabatic dynamics which might then be applied to larger multimode systems.

IV. CALCULATED INTERFEROGRAMS

A. Weak Coupling, Large Amplitude Case

We can illustrate the sensitivity of the interference signal to electronic geometric phase development in the Jahn–Teller active electronic state by examining an idealized limiting case. In the case of large amplitude pseudorotation and weak electronic-nuclear coupling, the predicted signal takes a simple form. To see this, we seek the limiting situation in which the nuclear motion in the excited state is nearly identical to the harmonic ground state motion despite the force coming from electronic-nuclear coupling (i.e., the contributions $\kappa q_{1t}/q_t$ and $\kappa q_{2t}/q_t$ to the right-hand sides of (3.19) and (3.20) respectively). The force from electronic-nuclear coupling will change the coordinate by an amount of order $\Delta q \sim 2\pi(\Delta p)/M\omega \sim (2\pi/\omega)^2\kappa/M = (2\pi)^2 k\sqrt{\hbar/M\omega}$ over a vibrational period (we assume $\omega_g = \omega_e = \omega$). The requirement that this change in coordinate be less than the width of the wave packet, $\sqrt{\hbar/2M\omega}$, leads to the weak coupling condition, $k \ll 1$. The stipulation of large amplitude ensures adiabaticity according to (3.9), a stringent requirement when k is small. Under the large amplitude condition, it is safe to ignore the dispersion in ϕ in the interferogram expression (3.12). Taking $\bar{\phi}_0 = \pi$, the signal is given by

$$P^{\text{int}}(t_d) = \frac{\pi E_0^2 \mu^2 \tau^2}{4\hbar^2} (\mathbf{i} \cdot \mathbf{l})^2 \operatorname{Re}[e^{i\Omega t_d}(1 - e^{-i\bar{\phi} t_d})Q] . \tag{4.1}$$

In the preceding expression, Q is of the form (3.40), with

$$a = iM\omega \tag{4.2a}$$

$$b = iM\omega \tag{4.2b}$$

$$d = n = m = 0 \tag{4.2c}$$

$$g = -\epsilon t_d + \int_0^{t_d} dt \kappa q_t + \int_0^{t_d} dt(a\dot{q}_{1t} + b\dot{q}_{2t}) + i\hbar \ln(\pi\hbar/M\omega) \tag{4.2d}$$

The differences in momentum, position, α, β, and δ between the excited state wave packet and the ground state reference wave packet are taken to be negligibly small in the weak coupling-large amplitude limit.

If the molecule is oriented in the **IJ**-plane, the pseudorotation will be circular (see Appendix A); the ground state angle variable is $\bar{\phi}_{t_d} = \omega t_d + \pi$. A choice of locked frequency on vertical resonance, $\hbar\Omega = \epsilon - \kappa\bar{q}_0$, cancels the first two terms on the right-hand side of (4.2d), which together constitute the electronic dynamical phase. In the case of circular pseudorotation, the electronic geometric phase is given simply by

$$\frac{1}{\hbar} \int_0^{t_d} dt(a\dot{q}_{1t} + b\dot{q}_{2t}) = \frac{\omega t_d}{2} \tag{4.3}$$

and the interferogram reduces to [43]

$$P^{\text{int}}(t_d) = \frac{\pi E_0^2 \mu^2 \tau^2}{2\hbar^2} \cos\left(\frac{\omega t_d}{2}\right) \tag{4.4}$$

Equation (4.4) shows that under the limiting conditions of large amplitude, weak coupling and circular pseudorotation, the time dependence of the in-phase interference signal reduces to an oscillation with twice the pseudorotational period. The ground and excited state wave packets differ only in their phase factors, and the differences in dynamical phase are canceled by the choice of locked frequency on vertical resonance. The electronic geometric phase, which develops in the excited state but not the ground state, and amounts to a sign change after each complete pseudorotation, is the only remaining phase difference. Thus, destructive interference is observed after odd numbers of pseudorotations and constructive interference after even numbers of pseudorotations.

Equations (4.1) and (4.4) predict vanishing interference for delay times halfway between complete pseudorotations, despite the fact that the ground and excited state packets are still spatially overlapping. This absence of interference is a consequence of the perpendicular and parallel selection rules mentioned below Eq. (3.8); as a result of them, the initial and reference wave packets go to different excited adiabatic surfaces for t_d halfway between complete cycles.

As mentioned earlier, it is possible to calculate the interference signal directly from (3.5) without resorting to adiabatic or locally quadratic approximations, by diagonalizing the Longuet-Higgins model Hamiltonian with the standard procedure [10]. Figure 1 shows the exact interfer-

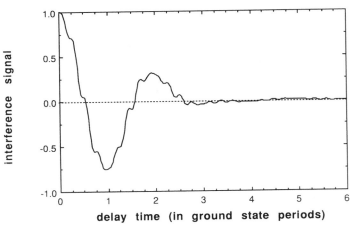

Figure 1. Exact interference signal, P^{int}, for the case of weak coupling, $k = 0.2$, and large amplitude pseudorotation, $\bar{q}_{10} = -10\sqrt{\hbar/M\omega}$. The system is oriented in the **IJ**-plane and the locked laser frequency is on vertical resonance. The time delay, t_d is measured in units of the ground state pseudorotational period in this and all subsequent figures.

ence signal obtained in that way, for a case of fairly weak coupling, $k = 0.2$, and amplitude, $\bar{q}_{10} = -10\sqrt{\hbar/M\omega}$, large enough that adiabaticity should be fairly well obeyed. The molecule is taken to be oriented in the **IJ**-plane so the pseudorotation is circular, and again $\omega_g = \omega_e = \omega$. Figure 1 is to be compared with Fig. 2, which shows the same signal calculated in the adiabatic and locally quadratic Hamiltonian approximations. Several conclusions can be drawn from the two figures. The oscillation of the interference signal at twice the ground state pseudorotational period evidently survives significant relaxation of the asymptotic weak coupling-large amplitude limit. That the Gaussian treatment allows us to definitively ascribe the sign changes in Fig. 2 to electronic geometric phase development is demonstrated by the Fig. 3 which graphs the phase difference $g/\hbar + \Omega t_d$ (with $\Omega = (\epsilon - \kappa\bar{q}_0)/\hbar$) as a function of time. The phase difference accumulates at a rate close to π radians per pseudorotation and is therefore dominated by the geometrical contribution.

The values of the time-dependent Gaussian parameters shown in Fig. 4 indicate that the wave packet does not spread significantly over several pseudorotational periods. We conclude that the damping of the interference signal under the conditions of Figs. 1 and 2 is more a consequence of trajectory differences between the ground and excited states than a result of wave packet spreading. The high frequency wiggles in Fig. 1 occur at a frequency $2\kappa q/\hbar \approx 4\omega$ equal to the local splitting between the two excited

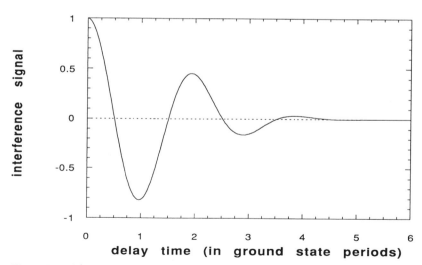

Figure 2. The same signal as in Fig. 1, calculated in the adiabatic and locally quadratic approximations.

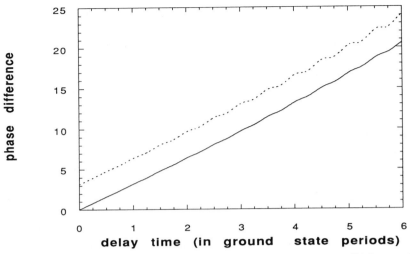

Figure 3. The phase difference between each of the two Gaussian contributions to the excited state wave packet and the ground state wave packets plus Ωt_d. The solid line gives the phase difference for the initially undistorted wave packet and the dashed line gives the phase difference for the distorted packet.

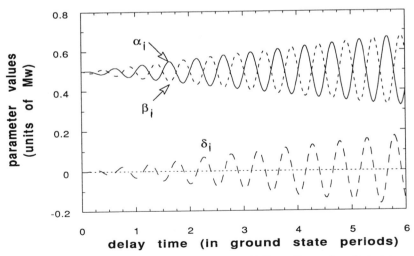

Figure 4. Gaussian width parameters for the initially undistorted excited state wave packet as a function of propagation time.

state surfaces. They arise from the small amplitude for initial excitation to the upper surface of the Jahn–Teller pair, which is separated from the phase-locked frequency Ω (vertically resonant with the lower surface) by the local splitting. The absence of the rapid wiggles from Fig. 2 does not represent a failure of the locally quadratic treatment per se, but rather results from the neglect of the upper excited state surface in Eq. (3.12) (i.e., the omission of the first term in square brackets in (3.10)).

Since the locally quadratic theory replaces the exact time-independent nuclear Hamiltonian with an effective time-dependent Hamiltonian, it is not true in general that the approximate treatment conserves energy. A calculation of the expectation value of the full molecular Hamiltonian in the approximate excited state wave function showed that, in the case considered here, the energy is conserved within 0.05% over six periods of motion. It is also of interest to check the overlap between the exact excited state amplitude and the approximate excited state wave function propagated under the locally quadratic Hamiltonian. In the present case, the real part of that overlap was found to decrease slowly, falling from 1.0 to 0.71 in two ground state periods. The real part of the overlap remains positive until crossing over to negative values just beyond four pseudo-rotational periods. The magnitude squared of the overlap decreases by 3.3% in six periods.

If phase-sensitive experiments of the kind suggested here are to be performed in the vapor phase or in a supersonic jet, it will be necessary to consider the spherically averaged interference signal appropriate to a randomly oriented sample. The spherical orientational distribution reduces to an effective cylindrical distribution, since molecules in the **JK**-plane have their electronic transition moments perpendicular to the **I** polarization of the resonant pulse pair. Figure 5 shows the exact interferogram for the $k = 0.2$ system oriented in the **IK**-plane. Again, the ground state vibration prepared by the nonresonant impulsive Raman sequence (Appendix A) has initial amplitude $\bar{q}_{10} = -10\sqrt{\hbar/M\omega}$. For molecules oriented in the **IK**-plane, the q_2 mode has no initial momentum, and the excited state wave packet vibrates along q_1, making nonadiabatic crossings to the upper Jahn–Teller surface at the conical intersection. The nonzero interference signals seen in Fig. 5 at half-odd multiples of $\gamma\pi/\omega$ are direct measures of the nonadiabatic transition amplitude, since the parallel selection rule places the reference wave packet on the upper surface at those times.

Figure 6 shows the interferogram for a randomly oriented collection of $k = 0.2$ molecules. Although the interferogram does not vanish upon spherical averaging, the oscillation of the interference signal at twice the pseudorotational period resulting from the adiabatic geometric sign

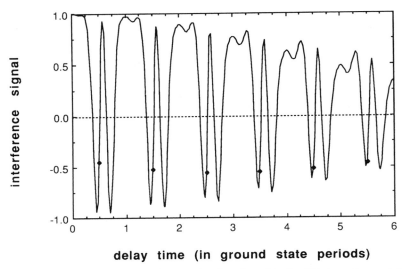

Figure 5. Same as Fig. 1, but for a molecule oriented in the **IK**-plane. Large dots at half-odd multiples of the ground state period mark the interference signal due to nonadiabatic transition amplitude in the excited state.

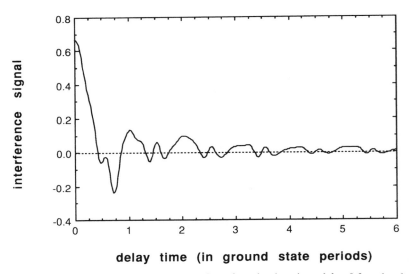

Figure 6. Interference signal for a collection of randomly oriented $k = 0.2$ molecules.

change is no longer immediately apparent. Nevertheless, the averaged interferogram clearly resembles a superposition of Fig. 1, which shows the sign change, and Fig. 5, which does not. The interference signal in Fig. 6 is on the same scale as Figs. 1 and 5. It takes a values less than one at zero delay time because the planes of some of the molecules make a nonzero angle with the $\mathbf{I}$ polarization direction (i.e., because of the same $(\mathbf{i} \cdot \mathbf{I})^2$ prefactor in (3.12) that renders the orientational distribution effectively cylindrical).

B.   Strong Coupling, Intermediate Amplitude Case

It is of some interest to determine the interference signal in the strong electronic-nuclear coupling case, where, in addition to acquiring electronic geometric phases, the excited state wave packet has a trajectory very different from that of the ground state packet. Since optical impulsive excitation of the ground state pseudorotation is driven by electronic-nuclear coupling in the excited state [30] (see also Eq. (A.7)), strong coupling will facilitate the preparation of large amplitude motion. Figure 7 shows the interferogram for a system with $k = 2.1$, an electronic-nuclear coupling constant equal to the experimental value for the 3s $^1E'$ Rydberg state of *sym*-triazine [44]. The initial amplitude of the circularly pseudo-

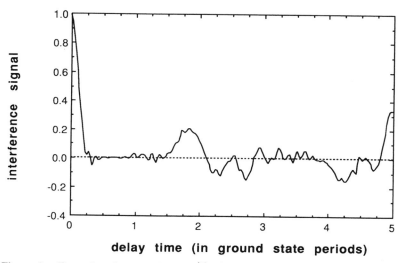

Figure 7. Exact interference signal, P^{int}, for the case of strong electronic-nuclear coupling, $k = 2.1$, and intermediate amplitude pseudorotation, $\bar{q}_{10} = -\kappa/M\omega^2 = -k\sqrt{\hbar/M\omega}$. The system is oriented in the **IJ**-plane and the locked laser frequency is on vertical resonance.

rotating ground state wave packet is $\bar{q}_{10} = \kappa/M\omega^2$, so excitation is into the trough of the Jahn–Teller system. The locked frequency Ω is vertically resonant. The probability density on the lower Jahn–Teller surface, corresponding to the wave packet prepared by the initial laser pulse, is shown for several propagation times in Fig. 8. That figure is helpful in understanding the form of the strong coupling interferogram. Pseudo-rotation proceeds more slowly in the excited state than in the ground state; after the initial coincidence, there is no significant interference until the ground state wave packet "laps" the excited state packet near $t_d = 2(2\pi/\omega)$. That same delay happens to be the time after which the wave packet begins to break up into several pieces. Subsequently, when the delay time equals an integer number of ground state periods, and the perpendicular selection rule places the reference wave packet on the lower Jahn–Teller surface, it almost invariably interferes significantly with the propagated initial amplitude. But there is little discernible

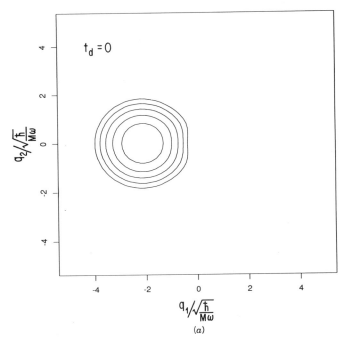

(a)

Figure 8. The square of the nuclear amplitude on the lower adiabatic Jahn–Teller surface at several propagation times in the $k = 2.1$ system. The motion proceeds in a counterclockwise direction. The wave packet does not become delocalized until after it has completed one pseudorotation in the excited state at $t_d \approx 2(2\pi/\omega)$.

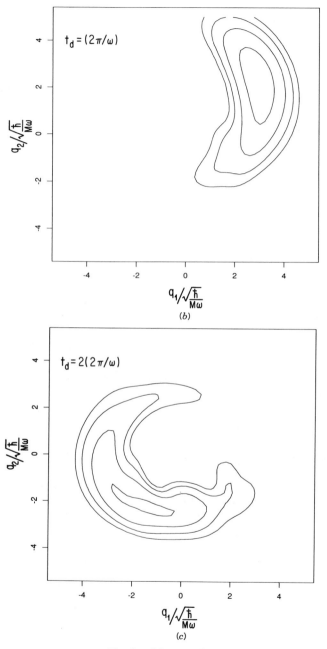

Fig. 8. (*Continued*).

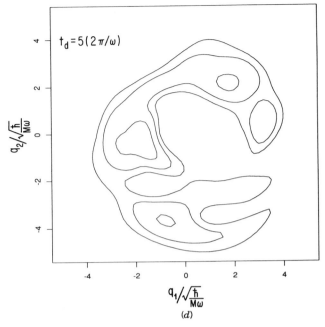

Fig. 8. (*Continued*).

regularity to the sense, constructive versus destructive, of the long-time interference signal.

The interference signal of Fig. 7 will obviously defy treatment with any form of single- or double-Gaussian wave packet at times longer than a few ground state periods. Nevertheless, the approximate interferogram of Fig. 9, calculated in the locally quadratic approximation, reproduces the initial decay and shows an interference signal near two ground state periods (about one excited state pseudorotation) that resembles the exact result. After one pseudorotation, the centers of the two Gaussians making up the excited state wave packet have each acquired geometrical phases of about π, but the aperiodic nature of their trajectories, shown in Fig. 10, leads to complicated time dependence in the dynamical phase. The phase functions, $g\hbar + \Omega t_d$, for the two Gaussians are shown in Fig. 11. The spikes in Fig. 11 come from the terms in Eq. (3.37) depending on the coordinate differences between the ground and excited state wave packets. Those differences are expected to be small when interference is most pronounced.

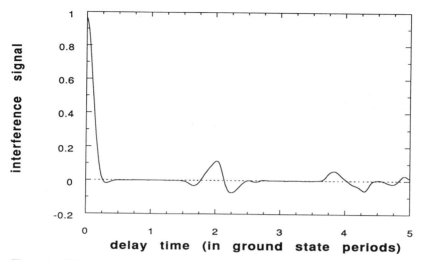

Figure 9. The same signal as in Fig. 6, calculated in the adiabatic and locally quadratic approximations.

The strong coupling, intermediate amplitude case rapidly becomes unstable in the locally quadratic treatment. The width parameters for the initially undistorted Gaussian are graphed in Fig. 12. They take exceedingly small values (corresponding to large spatial widths) at regular intervals. These unphysically large rms widths result from parametric resonance in the equations of motion (3.23), (3.25) and (3.27) for the imaginary parts of α, β and δ [45]. As a result of overestimating the mean-squared variations of position and momentum, and also allowing amplitude to enter the region of the conical intersection on one adiabatic surface only, the Gaussian wave packet exhibits rapid growth in the expectation value of the energy. The average value of the molecular Hamiltonian in the double-Gaussian excited state wave function increases by a factor of 1.2 within one ground state period and becomes 5.3 times its initial value in two periods. The real part of the overlap between the exact and approximate excited state wave functions is shown in Fig. 13. While decreasing steadily, the real part of the overlap remains positive well beyond the time of the first recurrence at $t_d \approx 2(2\pi/\omega)$. The magnitude squared of the overlap (not shown) decreases from 0.93 to 0.10 in two ground state periods.

We must conclude that the Gaussian wave packet provides a poor representation of the global features of the excited state wave function in the strong coupling case. However, the locally quadratic Hamiltonian treatment evidently describes the central region of the excited state wave packet, while it remains localized, accurately enough to recover the form of the interference signal through the first recurrence.

It is possible that a slight improvement in the treatment of the nuclear motion, based on the time-dependent variational principle, will accurately predict the interference signal on the short timescale necessary to observe geometric phase development, without suffering the instabilities of the locally quadratic method [36, 37]. Such an improvement may come at the cost of describing the excited state wave function as a superposition of

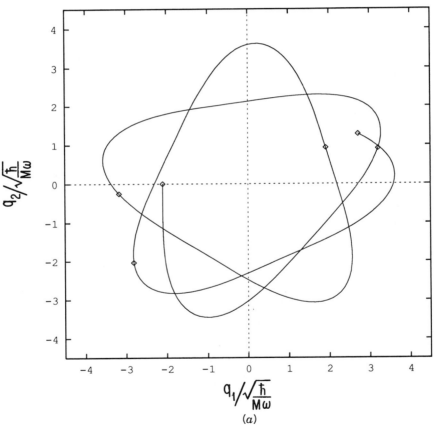

(a)

Figure 10. Trajectories of the undistorted (a) and distorted (b) contributions to the excited state wave functions. Tick marks on the trajectories are at intervals of $(2\pi/\omega)$.

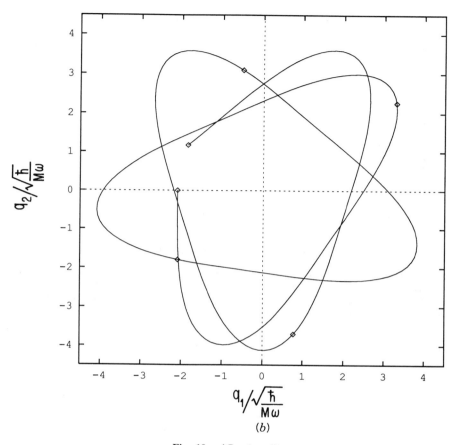

Fig. 10. (*Continued*).

many, rather than just two, Gaussian wave packets. The time-dependent
Hartree approach to wave packet propagation, developed by Coalson and
co-workers [46], might also be of use here, but a similar cautionary note
applies. What is needed is a form for the wave function flexible enough to
yield an accurate interferogram on the timescale of a few pseudorotations, yet simple enough to be helpful in discerning the contribution
of geometric phase development to the signal.

In this connection, it is of interest to note that when nonadiabatic
effects are small, as in strongly coupled systems at low energy or
substantial vibronic angular momentum, a semi-classical quantization
scheme beginning with the adiabatic nuclear Hamiltonian yields fairly

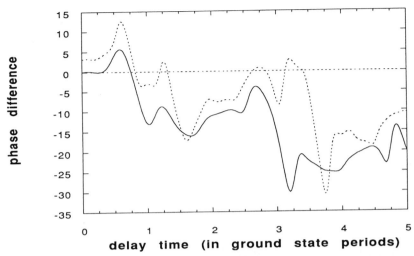

Figure 11. The phase difference between each of the two Gaussian contributions to the excited state wave packet and the ground state wave packets plus Ωt_d. The solid line gives the phase difference for the initially undistorted wave packet and the dashed line gives the phase difference for the distorted packet.

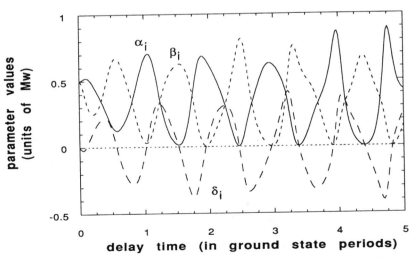

Figure 12. Width parameters for the initially undistorted Gaussian contribution to the excited state nuclear wave function.

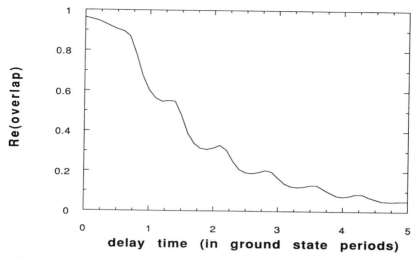

Figure 13. Real part of the overlap between the exact excited state nuclear wave function and the approximate wave function of the locally quadratic theory.

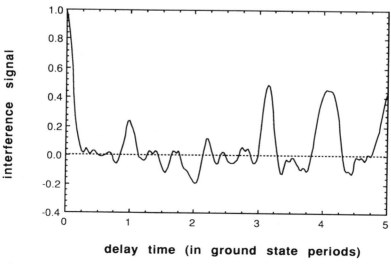

Figure 14. Exact interferogram for a $k = 2.1$ system oriented in the **IK**-plane.

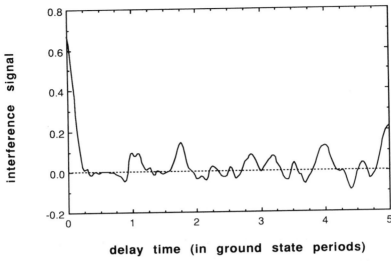

Figure 15. Spherically averaged interferogram for the $k = 2.1$ system.

accurate energies [41a]. Since the case of strong coupling and intermediate amplitude considered above involves a nonstationary superposition of eigenfunctions of intermediate vibronic angular momentum, it would be worth investigating whether the results of such a method can provide information useful in interpreting the interferogram in that case.

Figure 14 gives the interference signal for a system oriented in the **IK**-plane. The spherically averaged interferogram is shown in Fig. 15. Again, many peaks in the latter can be identified with peaks in the limiting cases of Figs. 7 and 14.

V. CONCLUDING REMARKS

We hope that the studies summarized above will motivate time-resolved interference experiments on Jahn–Teller systems. Half-odd quantum numbers for molecular pseudorotation have been reported in the continuous wave spectra of Na_3 [47], benzene [48] and *sym*-triazine [44]. The half-odd quantum numbers are strong evidence for the presence of geometric phase factors. Benzene and *sym*-triazine have nondegenerate ground states and would therefore be the most obvious choices for phase sensitive measurements of the kind we propose. The $3s\,^1E'$ Rydberg state of *sym*-triazine has the simplifying feature of exhibiting Jahn–Teller activity in only one doubly degenerate mode, namely the ν_6 ring bending mode.

The hardest part of the experiments will likely be the preparation of coherent pseudorotation of large amplitude (see Appendix A). One preliminary experiment could be to apply a train of short nonresonant pulses, spaced at the ground state vibrational period, to a diatomic molecule such as iodine and test whether the interference signal produced with a subsequent phase locked pulse-pair is detectably altered.

Since the ground state of the model system studied here is a two-dimensional isotropic harmonic oscillator, one might imagine preparing initial *stationary* states of well-defined vibrational (pseudorotational) angular momentum, rather than the coherently vibrating states considered above. If such states can be prepared, wave packet-like behavior might still be observed in the interference signal, for the following reason. The perpendicular and parallel selection rules have the consequence that the negative-q_1 and positive-q_1 halves of the annular ground state wave function are transferred to the lower and upper adiabatic states, respectively. If the center frequency of the resonant pulses were tuned to the lower surface, and the spectral pulse width did not exceed the local electronic splitting, the positive-q_1 side of the wave function would fail to make the transition. As a result, the annulus would become a crescent in the excited state.

APPENDIX A: DEPENDENCE OF OPTICAL IMPULSIVE EXCITATION OF PSEUDOROTATION ON MOLECULAR ORIENTATION

The molecule-fixed axes, $(\mathbf{i}, \mathbf{j}, \mathbf{k})$, are related to the space-fixed $(\mathbf{I}, \mathbf{J}, \mathbf{K})$ axes by Euler angles (Θ, Φ, Γ). We define the angles so that (Θ, Φ) gives the spatial direction of the vector $\mathbf{k}$, which is normal to the plane of the molecule. Moreover, the value of Γ is fixed by making the requirement that $\mathbf{I}$ have no $\mathbf{j}$ component (see following Eq. (3.8)). Thus, we have

$$\begin{pmatrix} \mathbf{I} \\ \mathbf{J} \\ \mathbf{K} \end{pmatrix} =$$

$$\begin{pmatrix} \cos\Theta\cos\Phi\cos\Gamma - \sin\Phi\sin\Gamma & -\cos\Theta\cos\Phi\sin\Gamma - \sin\Phi\cos\Gamma & \sin\Theta\cos\Phi \\ \cos\Theta\sin\Phi\cos\Gamma + \cos\Phi\sin\Gamma & -\cos\Theta\sin\Phi\sin\Gamma + \cos\Phi\cos\Gamma & \sin\Theta\sin\Phi \\ -\sin\Theta\cos\Gamma & \sin\Theta\sin\Gamma & \cos\Theta \end{pmatrix}$$

$$\times \begin{pmatrix} \mathbf{i} \\ \mathbf{j} \\ \mathbf{k} \end{pmatrix} \tag{A.1}$$

where $\Gamma = -\tan^{-1}(\sin\Phi/\cos\Theta\cos\Phi)$.

We wish to describe the preparation of a nuclear wave packet corresponding to coherent molecular pseudorotation in the electronic ground state by a sequence of nonresonant light pulses [30]. The system is driven by a pair of vibrationally abrupt pulses

$$\mathbf{E}(t) = f(t - t')\mathbf{I} \cos \chi t + f(t - t'') \frac{\mathbf{I} + \mathbf{J}}{\sqrt{2}} \cos \chi t \qquad (A.2)$$

separated by one quarter of a pseudorotational period and polarized at 45° (we choose $t' = -5\pi/2\omega$ and $t'' = -2\pi/\omega$). The frequency χ is far below electronic resonance. The delayed pulse could be phase-shifted by a random angle, as there is no requirement of phase coherence between the two pulses. The field (A.2) prepares the state $|\Psi(0)\rangle = |g\rangle|\psi(0)\rangle$, where the wave packet, $\langle \mathbf{q}|\psi(0)\rangle$, is of the form (3.1) with

$$\bar{q}_{10} = \frac{\pi_0}{M\omega_g}((\mathbf{I} \cdot \mathbf{j})^2 - (\mathbf{I} \cdot \mathbf{i})^2) = -\frac{\pi_0}{M\omega_g}(\mathbf{I} \cdot \mathbf{i})^2 \qquad (A.3)$$

$$\bar{q}_{20} = -\frac{\pi_0}{M\omega_g} 2(\mathbf{I} \cdot \mathbf{i})(\mathbf{I} \cdot \mathbf{j}) = 0 \qquad (A.4)$$

$$\bar{p}_1 = \frac{\pi_0}{2}((\mathbf{I} \cdot \mathbf{j} + \mathbf{J} \cdot \mathbf{j})^2 - (\mathbf{I} \cdot \mathbf{i} + \mathbf{J} \cdot \mathbf{i})^2) \qquad (A.5)$$

$$\bar{p}_2 = -\pi_0(\mathbf{I} \cdot \mathbf{i} + \mathbf{J} \cdot \mathbf{i})(\mathbf{I} \cdot \mathbf{j} + \mathbf{J} \cdot \mathbf{j}) \qquad (A.6)$$

The momentum impulse π_0 is related to the pulse energy by

$$\pi_0 = \frac{\kappa\mu^2}{\epsilon^2} \int dt f^2(t) \qquad (A.7)$$

The inability of the initial, I-polarized, pulse to couple to the q_2-mode and the periodic nature of the ground state motion lead to the vanishing of $\bar{q}_{20}$ seen in Eq. (A.4). Therefore, regardless of molecular orientation, there is nonzero vibrational displacement only along the q_1-mode at $t = 0$. By writing Eq. (A.3) explicitly as $\bar{q}_{10} = -\pi_0/M\omega_g(1 - \sin^2\Theta \cos^2\Phi)$, we see that the range of orientations within which the magnitude of $\bar{q}_{10}$ is less than a certain value, say $\sqrt{\hbar/2M\omega_e}$, decreases with increasing pulse energy. In practice, the nonresonant pulse pair will be replaced with a sequence of such pairs, spaced at the ground state pseudorotational period, in order to drive the pseudorotation to large amplitude and discriminate against other low frequency modes [30, 49–52].

APPENDIX B. QUADRATIC EXPANSION COEFFICIENTS

Here we give explicit expressions for the coefficients in the power series expansions of the potential, the azimuthal angle, and the vector potential. The coefficients in the quadratic expansion, (3.13), of the potential are

$$v = \frac{M\omega_e^2}{2} q_t^2 - \kappa q_t + \frac{\hbar^2}{8Mq_t^2} \tag{B.1}$$

$$v_{q_1} = M\omega_e^2 q_{1t} - \kappa \frac{q_{1t}}{q_t} - \frac{\hbar^2 q_{1t}}{4Mq_t^4} \tag{B.2}$$

$$v_{q_2} = M\omega_e^2 q_{2t} - \kappa \frac{q_{2t}}{q_t} - \frac{\hbar^2 q_{2t}}{4Mq_t^4} \tag{B.3}$$

$$v_{q_1q_1} = M\omega_e^2 - \kappa \frac{q_{2t}^2}{q_t^3} \tag{B.4}$$

$$v_{q_1q_2} = \kappa \frac{q_{1t}q_{2t}}{q_t^3} \tag{B.5}$$

$$v_{q_2q_2} = M\omega_e^2 - \kappa \frac{q_{1t}^2}{q_t^3} \tag{B.6}$$

The terms proportional to $\hbar^2$ in (B.2) and (B.3) are the contributions of the repulsive $(\mathbf{A}_1^2 + \mathbf{A}_2^2)/2M \sim q^{-2}$ term to v_{q_1} and v_{q_2}, respectively [29]. Terms of order $\hbar^2/Mq_t^4$, representing quadratic corrections to that portion of the potential, have been dropped from (B.4) and (B.6) for the same reason that terms of the same form were dropped from Eqs. (3.24), (3.26) and (3.28).

The coefficients in the quadratic expansion, (3.14), of the azimuthal angle, $\phi(\mathbf{q})$, have the explicit forms

$$\phi_t = \tan^{-1}\left(\frac{q_{2t}}{q_{1t}}\right) \tag{B.7}$$

$$\phi_{q_1} = -\frac{q_{2t}}{q_t^2} \tag{B.8}$$

$$\phi_{q_2} = \frac{q_{1t}}{q_t^2} \tag{B.9}$$

$$\phi_{q_1q_1} = \frac{2q_{1t}q_{2t}}{q_t^4} \tag{B.10}$$

$$\phi_{q_1 q_2} = \frac{q_{2t}^2 - q_{1t}^2}{q_t^4} \tag{B.11}$$

$$\phi_{q_2 q_2} = -\frac{2 q_{1t} q_{2t}}{q_t^4} \tag{B.12}$$

The coefficients in the linear expansion, (3.15), of the transverse adiabatic vector potential, $\mathbf{A}_{3\perp}(\mathbf{q})$, are given by

$$a = -\frac{\hbar q_{2t}}{2 q_t^2} \tag{B.13}$$

$$a_{q1} = \frac{\hbar q_{1t} q_{2t}}{q_t^4} \tag{B.14}$$

$$a_{q2} = \frac{\hbar (q_{2t}^2 - q_{1t}^2)}{2 q_t^4} \tag{B.15}$$

$$b = \frac{\hbar q_{1t}}{2 q_t^2} \tag{B.16}$$

$$b_{q1} = \frac{\hbar (q_{2t}^2 - q_{1t}^2)}{2 q_t^4} \tag{B.17}$$

$$b_{q2} = -\frac{\hbar q_{1t} q_{2t}}{q_t^4} \tag{B.18}$$

APPENDIX C: INITIAL CONDITIONS

Here we list the two sets of initial conditions used for the parameter equations of motion in calculating the interference signal (3.12) in the Gaussian approximation. One Gaussian has initial conditions corresponding directly to the $t = 0$ values of the parameters of the ground state wave packet:

$$q_{10} = \bar{q}_{10} < 0, \qquad q_{20} = \bar{q}_{20} = 0 \tag{C.1}$$

$$p_1 = \bar{p}_1, \qquad p_2 = \bar{p}_2 \tag{C.2}$$

$$\alpha_r(0) = 0, \qquad \alpha_i(0) = \frac{M \omega_g}{2} \tag{C.3}$$

$$\beta_r(0) = 0, \qquad \beta_i(0) = \frac{M \omega_g}{2} \tag{C.4}$$

$$\delta_r(0) = 0, \qquad \delta_i(0) = 0 \tag{C.5}$$

$$\gamma_r(0) = 0\,, \qquad \gamma_i(0) = \frac{\hbar}{2}\ln\left(\frac{\pi\hbar}{M\omega_g}\right) \qquad (C.6)$$

Some of the parameters of the second Gaussian are slightly shifted due to the inclusion of the coefficients, (B.7)–(B.12), of the quadratic expansion of the azimuthal angle variable in $\exp[i\phi(\mathbf{q})]$:

$$q_{10} = \bar{q}_{10} < 0\,, \qquad q_{20} = \bar{q}_{20} = 0 \qquad (C.7)$$

$$p_1 = \bar{p}_1\,, \qquad p_2 = \bar{p}_2 + \frac{\hbar}{\bar{q}_{10}} \qquad (C.8)$$

$$\alpha_r(0) = 0\,, \qquad \alpha_i(0) = \frac{M\omega_g}{2} \qquad (C.9)$$

$$\beta_r(0) = 0\,, \qquad \beta_i(0) = \frac{M\omega_g}{2} \qquad (C.10)$$

$$\delta_r(0) = -\frac{\hbar}{2\bar{q}_{10}^{2}}\,, \qquad \delta_r(0) = 0 \qquad (C.11)$$

$$\gamma_r(0) = \hbar\pi\,, \qquad \gamma_i(0) = \frac{\hbar}{2}\ln\left(\frac{\pi\hbar}{M\omega_g}\right) \qquad (C.12)$$

The reference wave packet, propagated under the ground state Hamiltonian and then transferred to the excited state by the delayed pulse, also consists of a sum of two Gaussians. The first Gaussian has parameters corresponding directly to those of the ground state wave packet at $t = t_d$. Some of the parameters of the second Gaussian are shifted due to the inclusion of the coefficients of a quadratic expansion, about the center of the ground state packet, of the azimuthal angle variable in $\exp[-i\phi(\mathbf{q})]$. We do not bother to write these parameters out explicitly as the shifts are analogous to those in (C.7)–(C.12) and involve the coefficients (B.7)–(B.12) evaluated at $t = t_d$.

Note added in proof. Very recently, A. Furlan, M. J. Riley, and S. Leutwyler [*J. Chem. Phys.* **96**, 7306 (1992)] have reported a resonant two-photon ionization spectrum of the $S_1 \leftarrow S_0$ vibronic transitions of supersonically cooled triptycene. The excited state exhibits $E' \otimes e'$ Jahn–Teller activity with linear and quadratic vibronic coupling in a single, degenerate, benzene wagging mode.

Acknowledgments

We have benefitted from numerous discussions of wave packet interferometry with its inventors, N. F. Scherer, R. J. Carlson, A. J. Ruggiero, M. Du, L. D. Ziegler, G. R. Fleming and S. A. Rice. One of us (J. A. C.) thanks R. D. Coalson for helpful advice on the locally quadratic approach to Gaussian dynamics. This work was paid for by a grant from the National Science Foundation. Acknowledgment is made to the donors of the Petroleum Research Fund, administered by the ACS, for partial support of this research. T. J. S. Jr. was supported in part by a Department of Education National Needs Fellowship. J.A.C. acknowledges the support of The Camille and Henry Dreyfus Teacher-Scholar Award Program.

References

1. (a) C. A. Mead and D. G. Truhlar, *J. Chem. Phys.* **70**, 2284 (1979); (b) C. A. Mead, *Chem. Phys.* **49**, 23 (1980); (c) **49**, 33 (1980).

2. M. V. Berry, *Proc. R. Soc. London Ser. A* **392**, 45 (1984).

3. C. A. Mead, "The Geometric Phase in Molecular Systems," *Rev. Mod. Phys.* **64**, 51 (1992).

4. A. Shapere and F. Wilczek (eds.) *Geometric Phases in Physics* (World Scientific, Singapore, 1989).

5. J. W. Zwanziger, M. Koenig, and A. Pines, *Annu. Rev. Phys. Chem.* **41**, 601 (1990).

6. A. Bohm, L. J. Boya, and B. Kendrick, *Phys. Rev. A* **43**, 1206 (1991).

7. N. F. Scherer, A. J. Ruggiero, M. Du, and G. R. Fleming, *J. Chem. Phys.* **93**, 856 (1990).

8. (a) N. F. Scherer, R. J. Carlson, A. Matro, M. Du, A. J. Ruggiero, V. Romero-Rochin, J. A. Cina, G. R. Fleming, and S. A. Rice, *J. Chem. Phys.* **95**, 1487 (1991); (b) N. F. Scherer, A. Matro, R. J. Carlson, M. Du, L. D. Ziegler, J. A. Cina, and G. R. Fleming, *J. Chem. Phys.* **196**, 4180 (1992).

9. (a) H. Metiu and V. Engel, *J. Opt. Soc. Am.* **B7**, 1709 (1990); (b) V. Engel and H. Metiu, "Interference Effects in the One Photon Dissociation of a Molecule with Two Laser Pulses," unpublished manuscript; (c) R. Bavli, V. Engel, and H. Metiu, *J. Chem. Phys.* **96** 2600 (1992).

10. H. C. Longuet-Higgins, U. Öpik, M. H. L. Pryce, and R. A. Sack, *Proc. R. Soc. London Ser. A* **244**, 1 (1958).

11. The adiabatic electronic states pictured here are constrained to be real and hence double-valued. This is equivalent to incorporating the electronic Berry phase into the definition of the adiabatic states and is done here only for illustration (compare Eq. (2.8)).

12. (a) D. Loss, P. Goldbart, and A. V. Balataky, *Phys. Rev. Lett.* **65**, 1655 (1990); (b) D. Loss and P. Goldbart, "Persistent Current from Berry's Phase in Mesoscopic Systems," preprint (June, 1991).

13. H. L. Davis, Ph.D. dissertation, The University of Chicago (1989).

14. (a) D. J. Tannor and S. A. Rice, *J. Chem. Phys.* **83**, 5013 (1985); (b) D. J. Tannor, R. Kosloff, and S. A. Rice, *J. Chem. Phys.* **85**, 5805 (1986); (c) D. J. Tannor and S. A. Rice, *Adv. Chem. Phys.* **70**, 441 (1988).

15. E. D. Potter, J. L. Herreck, S. Pedersen, Q. Liu, and A. H. Zewail, *Nature* **335**, 66 (1992).

16. (a) P. Brumer and M Shapiro, *Faraday Discuss. Chem. Soc.* **82**, 177 (1986); (b) M. Shapiro, J. W. Hepburn, and P. Brumer, *Chem. Phys. Lett.* **149**, 451 (1988); (c) C.K. Chan, P. Brumer, and M. Shapiro, *J. Chem. Phys.* **94**, 2688 (1991).

17. (a) S.M. Park, S. Lu, and R. J. Gordon, *J. Chem. Phys.* **94**, 8622 (1991). (b) S. -D. Liu, S. M. Park, Y. Xie, and R. J. Gordon, *J. Chem. Phys.* **96**, 6613 (1992).

18. E. J. Heller, *J. Chem Phys.* **62**, 1544 (1975).

19. H. C. Longuet-Higgins, *Adv. Spectrosc.* **2**, 429 (1961).

20. G. Herzberg and H. C. Longuet-Higgins, *Discuss. Faraday Soc.* **35**, 77 (1963).

21. H. C. Longuet-Higgins, *Proc. R. Soc. London Ser A* **344**, 147 (1975).

22. A. J. Stone, *Proc. R. Soc. London Ser. A* **351**, 141 (1976).

23. J. Moody, A. Shapere, and F. Wilczek, *Phys. Rev. Lett.* **56**, 893 (1986).

24. M. Stone, *Phys. Rev. D* **33**, 1191 (1986).

25. V. Romero-Rochin and J. A. Cina, *J. Chem. Phys.* **91**, 6103 (1989).

26. R. Jackiw, *Comments At. Mol. Phys.* **21**, 71 (1988).

27. A. Messiah, *Quantum Mechanics*, vol. II (North-Holland, Amsterdam, 1976), p. 1017.

28. H. Kuratsuji and S. Iida, *Prog. Theor. Phys.* **74**, 439 (1985).

29. M. V. Berry and R. Lim, *J. Phys. A: Math. Gen.* **23**, L655 (1990).

30. J. A Cina and V. Romero-Rochin, *J. Chem. Phys.* **93**, 3844 (1990).

31. For discussions of nonadiabatic nuclear dynamics in Jahn–Teller systems from several different points of view, see: C. S. Sloane and R. Silbey, *J. Chem. Phys.* **56**, 6031 (1972); and refs. 32 and 33.

32. H. Köppel, W. Domcke, and L. S. Cederbaum, *Adv. Chem. Phys.* **57**, 59 (1984).

33. I. B. Bersuker and V. Z. Polinger, in *Advances in Quantum Chemistry*, vol. 15, P.-O. Lowdin (ed.) (Academic Press, New York, 1982), p. 85.

34. (a) E. J. Heller, *J. Chem. Phys.* **68**, 2066 (1978); (b) *Acc. Chem. Res.*, **14**, 368 (1981).

35. The expression (3.7) for the dipole matrix elements differs by a factor of $\frac{1}{2}$ from that used in refs. 25, 30 and 43.

36. (a) R. Heather and H. Metiu, *Chem. Phys. Lett.* **118**, 558 (1985); (b) *J. Chem. Phys.* **84**, 3250 (1986).

37. R. D. Coalson and M. Karplus, *J. Chem. Phys.* **93**, 3919 (1990).

38. R. G. Littlejohn, *Phys. Rep.* **138**, 193 (1986).

39. There is a parallel between the wave packet-based semi-classical treatment of nuclear motion developed here and the Thomas–Fermi theory of an electronic system in a slowly varying vector potential. In the semi-classical electronic theory as well as here, one naturally arrives at a locally linear approximation to the scalar-potential-derived forces and a locally uniform approximation to the magnetic force derived from the vector potential. See, R. A. Harris and J. A. Cina, *J. Chem. Phys.* **79**, 1381 (1983); C. J. Grayce and R. A. Harris, *Molec. Phys.* **71**, 1 (1990).

40. T. Pacher, C. A. Mead, L. S. Cederbaum, and H. Köppel, *J. Chem. Phys.* **91**, 7057 (1989).

41. (a) J. W. Zwanziger, E. R. Grant, and G. S. Ezra, *J. Chem. Phys.* **85**, 2089 (1986); (b) J. W. Zwanziger and E. R. Grant, *J. Chem. Phys.* **90**, 2357 (1989).

42. H.-D. Meyer and W. H. Miller, J. Chem. Phys. **70**, 3214 (1979).

43. J. A. Cina, *Phys. Rev. Lett.* **66**, 1146 (1991).

44. R. L. Whetten, K. S. Haber, and E. R. Grant, *J. Chem. Phys.* **84**, 1270 (1986).

45. L. D. Landau and E. M. Lifshitz, *Mechanics*, 3rd edition (Pergamon Press, Oxford, 1976), p. 80.

46. M. Messina and R. D. Coalson, *J. Chem. Phys.* **90**, 4015 (1989).

47. G. Delacrétaz, E. R. Grant, R. L. Whetten, L. Wöste, and J. W. Zwanziger, *Phys. Rev. Lett.* **56**, 2598 (1986).

48. R. L. Whetten and E. R. Grant, *J. Chem. Phys.* **84**, 654 (1986).

49. (a) A. M. Weiner, D. E. Leaird, G. P. Wiederrecht, and K. A. Nelson, *Science* **247**, 1317 (1990); *J. Opt. Soc. Am. B* **8**, 1264 (1991); (b) Y.-X. Yan, E. B. Gamble, Jr., and K. A. Nelson, *J. Chem. Phys.* **83**, 5391 (1985).

50. J. Chesnoy and A. Mohktari, *Phys. Rev. A* **38**, 3566 (1988).

51. Y. J. Yan and S. Mukamel, *J. Chem. Phys.* **94**, 997 (1991).

52. A. M. Walsh and R. F. Loring, *J. Chem. Phys.* **93**, 7566 (1990).

ON GLOBAL ENERGY CONSERVATION IN NONLINEAR LIGHT–MATTER INTERACTION: THE NONLINEAR SPECTROSCOPIES, ACTIVE AND PASSIVE

DUCKHWAN LEE* AND A. C. ALBRECHT

Department of Chemistry, Cornell University, Ithaca, New York

CONTENTS

---

* Permanent Address: Department of Chemistry, Sogang University, Seoul, Korea.

Advances in Chemical Physics, Volume LXXXIII, Edited by I. Prigogine and Stuart A. Rice.
ISBN 0-471-54018-8 © 1993 John Wiley & Sons, Inc.

ABSTRACT

In the nonlinear interaction of light with matter, global energy conservation is an issue of spectroscopic significance. It is examined here in detail. Energy conservation must be achieved by balancing the cycle-averaged rate of net energy transfer between the fields and the medium with the cycle-averaged rate of energy change in the medium as derived by the change in the population among its eigenstates. In the absence of any material resonances with the radiation fields, no net exchange of energy between the fields and the medium is possible, and the total energy among the fields and within the medium must be separately conserved. In such *passive* processes, matter serves catalytically to alter the state of the radiation fields. In the presence of one or more resonances, there must be net energy transfer between the fields and the medium. The energy change in the medium must appear as change in the population among its eigenstates. The population change can be readily understood for *active spectroscopies* in which light–matter interaction is directly in quadrature form in the incident fields. However, it is shown how transitions among the eigenstates are also possible through interactions which are not directly in quadrature form in the incident fields, corresponding to *passive* processes having resonances—the *passive spectroscopies*. Whatever the specific nature of the light–matter interaction, the signal field can recognize only the frequency and the absolute phase of the (nonlinear) induced polarization produced in the medium. Global energy conservation for the combined system of the field and the medium is explicitly demonstrated for resonant passive processes, such as CARS and CSRS, up to third order in the fields, by employing the diagrammatic technique. Finally, certain ambiguities inherent in the diagrammatic techniques are resolved.

I. INTRODUCTION

The nonlinear interaction of light with matter is useful both as an optical method for generating new radiation fields and as a spectroscopic means for probing the quantum-mechanical structure of molecules [1–5]. Light–matter interactions can be formally classified [5, 6] as either *active* or *passive processes* and for electric field based interactions with ordinary molecules (electric dipole approximation), both may be described in terms of the familiar nonlinear electrical susceptibilities. The nonlinear electrical susceptibility represents the material response to incident CW radiation and its microscopic quantum-mechanical formalism can be found directly by diagrammatic techniques based on the perturbative density matrix approach including dephasing effects in their fast-modulation limit [7]. Since time-independent (DC) fields can only induce a stationary polarization of the medium, thus altering its eigenstates, it is convenient to regard them as a part of the *dark* Hamiltonian, particularly when dealing with optical spectroscopies.

In *active processes* such as the one- or multiphoton absorption and emission spectroscopies, the interactions are always in *quadrature* form. The "absorptive" and "emissive" Fourier components of the radiation field appear directly in conjugate pairs, and at least one material resonance with the radiation must exist, that between the initial and final eigenstates. Energy is exchanged between light and matter in a manner that survives cycle-averaging. On the other hand, in *passive processes*, the interactions are not always in quadrature form. Furthermore, coherence in the material response and the phase (or momentum) matching condition for the radiation fields are essential. In *pure passive processes*, those without any resonance, the energy exchanged between light and matter does not survive cycle-averaging and matter serves only catalytically to alter the nature of the radiation. Material resonances are not necessary (though they can be useful) for generating new radiation fields. Harmonic generation and the so-called parametric wave mixing processes belong to such passive processes. Although *active processes* can appear only at odd order in the perturbative expansion of the induced polarization, *passive processes* may appear at any order. At odd order, there exists a subgroup of nonresonant passive processes that is maximally in quadrature at the polarization level. Let such processes be referred to as *nonlinear dispersion* since they are the nonlinear analog of linear dispersion (for the odd frequency that is not in quadrature).

Light–matter interactions, whether active or passive, may at the same time be classified as *resonant* (partially or fully) and *nonresonant* depending on the presence or absence of material resonances with the radiation fields at some point(s) during the evolution of the field perturbed system. As noted, all active processes must contain at least one resonance, while passive processes are usually (though not always) nonresonant. In the complete absence of resonance, the *permutation symmetry* of the electrical susceptibilities [3, 8, 9] permits an independent definition of thermodynamic systems; matter and radiation. There is no net energy exchange between the medium and the fields. The total energy of the radiation and of the medium is separately conserved. Thus, as demonstrated by the Manley–Rowe relations [8], the total energy of *photons* created in any given nonresonant passive process must be equal to the total energy of the *photons* destroyed. Furthermore, the momenta of photons must be conserved, as given by the phase matching condition. At the same time, the medium is left completely unchanged and thus plays a purely passive (but catalytic) role in the interaction.

For processes containing intermediate and/or final resonances, however, a net exchange of energy between the fields and the medium becomes possible, the *active processes* being the most obvious example. Now, the permutation symmetry of the electrical susceptibilities breaks down, and the radiation and the medium are no longer independent thermodynamic systems because of the continuous net exchange of energy between them. In *active processes*, with or without intermediate resonances, global energy conservation within the composite radiation/matter system requires that the net energy gained or lost by all *photons* must equal the energy lost or gained by the material as expressed by transitions between the initial and final eigenstates in medium. Furthermore, the interaction is directly in quadrature form, regarded as essential for inducing population changes in the medium by the action of photons. Thus in *active processes* energy conservation is achieved by balancing the cycle-averaged rate of total energy transfer per unit volume, W_F, between the radiation and the medium, with the cycle-averaged rate of energy change, W_M, in the unit volume of the medium as derived by the change in the population among its eigenstates.

Passive processes that are of spectroscopic interest of course require material resonances. For example, nondegenerate four-wave mixing processes such as CARS ($\omega_4 = \omega_1 - \omega_2 + \omega_3$ with ω_1, $\omega_3 > \omega_2$) and CSRS ($\omega_4 = \omega_1 - \omega_2 + \omega_3$ with ω_1, $\omega_3 < \omega_2$) seek out a Raman-type resonance, $\omega_R^0 = |\omega_1 - \omega_2|$ [10]. In these coherent Raman processes, the total energy of *photons* appears to be conserved just as in nonresonant passive processes, that is $\hbar\omega_1 + \hbar\omega_3 = \hbar\omega_2 + \hbar\omega_4$. Nevertheless, the lack of permutation symmetry in the electrical susceptibility, because of the material resonance, suggests that at the same time a continuous net energy exchange between the field and the medium has taken place. In fact, it has already been pointed out [10, 11] that in some cases the diagrammatic description of such processes apparently leaves the medium *excited* as a result of interactions of this type although the interacting fields are not directly in quadrature. Thus, global energy conservation appears to fail in resonant passive processes in which the medium appears to suffer a population change ($W_M \neq 0$) while at the same time the energy remains conserved within the radiation fields ($W_F = 0$). This contradiction must be resolved and in any case the creation of population change in the medium in the absence of quadrature among the radiation fields is a novel event that requires further consideration.

It is the purpose of this paper to examine the energy conservation requirement for nonlinear light–matter interactions, in general, and for passive processes that contain resonances, in particular. The semi-classical approach, in which the radiation fields satisfy the classical Maxwell's equations, is used to define W_F in terms of the nonlinear electrical susceptibilities. In the *fully nonresonant passive process* corresponding to $(s + 1)$-wave mixing, it has already been shown that the *absolute phase* of the $(s + 1)$th radiation field generated through the nonlinear interaction must be shifted by $\pi/2$ (*out-of-phase*) from the *absolute phase* of the induced polarization produced at the sth order [3, 8]. On the other hand, in *active processes* we shall see that there can be no phase shift (*in-phase*) of the $(s + 1)$th field from the sth order polarization, due to the quadrature form of interaction at that level. By analyzing the wave equation for the radiation field, it will be shown that for *resonant passive processes* the $(s + 1)$th radiation field generated in an $(s + 1)$-wave mixing processes contains both *in-phase* and *out-of-phase* components with respect to the polarization produced at the sth order. In such cases, a net energy flow between the radiation and the matter becomes possible through the in-phase component even in the absence of quadrature among the fields, and energy is not conserved separately within the radiation and the material subsystems.

As mentioned [6, 11, 12], the diagrammatic techniques based on the perturbative density matrix approach are useful for obtaining microscopic quantum-mechanical expressions for the nonlinear electrical susceptibilities. It will be shown that the diagrammatic technique can also be used to calculate W_M. Global energy conservation in resonant nonlinear light–matter interaction is confirmed upon finding that the W_F obtained from Maxwell's equations is identical to the W_M obtained from the density matrix approach. It will be demonstrated that the diagrammatic technique that uses the familiar energy ladder diagrams [6] provides additional insight into nonlinear optical processes. Finally, we shall see how a net population change in the medium need not always require immediate quadrature in the radiation fields.

In Section II, the wave equation obtained from Maxwell's equations is analyzed. All incident fields are taken as CW and material resonances are explicitly included by generalizing to complex wave vectors and complex electrical susceptibilities. It will be shown that the difference between the *absolute phase* of the signal wave, the $(s + 1)$th wave, and the *absolute phase* of the induced polarization produced by s incident waves is dependent upon the nature of the light–matter interaction as well as the presence or absence of material resonances. In Section III, W_F is defined in terms of the nonlinear electrical susceptibility and the difference of the absolute phases. In Section IV, the density matrix approach and the diagrammatic technique for obtaining the quantum-mechanical expressions for electrical susceptibilities are briefly summarized. It is shown how the W_M determined from the diagrammatic technique is identical to W_F, thus satisfying global energy conservation in nonlinear light–matter interaction. By way of illustration, energy conservation is demonstrated explicitly in Section V for several nonlinear processes up to third order ($s = 3$).

II. THE CLASSICAL WAVE EQUATION FOR RADIATION FIELDS IN NONLINEAR MEDIA

In this section, the linear ($s = 1$) and nonlinear ($s > 1$) wave equations for the electric field of the signal wave are analyzed in detail. The difference in the absolute phase of the signal field and that of the induced polarization is examined in the presence of material resonances which generate both complex wave vectors and complex electrical susceptibilities.

Maxwell's equations can be combined to give the wave equation for the electric field, $\mathbf{E}(\mathbf{r}, t)$, at point $\mathbf{r}$ and time t, in an electrically neutral and nonmagnetic medium in the absence of external current [3, 8, 13–15],

$$-\nabla^2 \mathbf{E}(\mathbf{r}, t) + (1/c^2)\frac{\partial^2 \mathbf{E}(\mathbf{r}, t)}{\partial t^2} = -(4\pi/c^2)\frac{\partial^2 \mathbf{P}(\mathbf{r}, t)}{\partial t^2} \qquad (2.1)$$

Here, the vector relation, $\nabla \times \nabla \times \mathbf{E}(\mathbf{r}, t) = \nabla(\nabla \cdot \mathbf{E}(\mathbf{r}, t)) - \nabla^2 \mathbf{E}(\mathbf{r}, t)$, and Gauss's law, $\nabla \cdot \mathbf{E}(\mathbf{r}, t) = 0$, have been used. $\mathbf{P}(\mathbf{r}, t)$ is the macroscopic polarization induced in the medium at position $\mathbf{r}$ and at time t by the electric field. In the perturbative approach, we have

$$\mathbf{P}(\mathbf{r}, t) = \sum_{s=1} \mathbf{P}^{(s)}(\mathbf{r}, t) \qquad (2.2)$$

where $\mathbf{P}^{(s)}(\mathbf{r}, t)$ is the sth order induced polarization arising from s incident fields. (Practically speaking, nonlinear induced polarizations ($s > 1$) are important only when the incident fields are sufficiently strong to produce a detectable ($s + 1$)th wave).

The general solution for Eq. (2.1) can be written in the plane wave form,

$$\mathbf{E}(\mathbf{r}, t) = \tfrac{1}{2} \sum_l \{\mathbf{E}_l(\mathbf{r}; \phi_l) \exp[i(\mathbf{k}_l \cdot \mathbf{r} - \omega_l t)] + \text{c.c.}\}$$

$$= \tfrac{1}{2} \sum_l \mathbf{E}_l(\mathbf{r})\{\exp[i(\mathbf{k}_l \cdot \mathbf{r} - \omega_l t + \phi_l(\mathbf{r}))] + \text{c.c.}\} \qquad (2.3)$$

where $\mathbf{k}_l$ and ω_l are the wave vector and the frequency (always positive) for the lth Fourier component of $\mathbf{E}(\mathbf{r}, t)$. And, $\phi_l(\mathbf{r})$ is called the *absolute phase* which is taken as real. The summation is over all Fourier components and Cartesian polarization unit vectors, $\hat{e}_l$, present in the medium.

For CW fields, the field amplitude, $E_l(r)$, is time-independent. In general, the wave vector k_l is complex,

$$k_l = k'_l + ik''_l . \tag{2.4}$$

For a homogeneous plane wave [14, 15], the directions of k'_l and k''_l are the same. (In anisotropic crystals [3, 8, 14], k_l^2 depends on the polarization direction of the field and then k_l is not always perpendicular to the polarization unit vector $\hat{e}_l$).

Although complex notation is often used for the sake of convenience, the electric field, $E(r, t)$, is a classically measurable property and thus must be real, as given in Eq. (2.3). Its direct measurement at optical frequencies, however, is not practicable because of the slow response time of electric field detectors [16]. Instead, intensity (quadrature) detectors using the photoelectric effect are usually employed to record the cycle-averaged energy flux, $(c/8\pi)|E(r, t)|^2$ [17], or photons.

With complex plane wave notation, the radiation field is formally divided into the "absorptive" component (with ω_l, k_l, and ϕ_l) and the "emissive" component (with $-\omega_l$, $-k_l^*$, and $-\phi_l$), corresponding to the first and the c.c. terms, respectively, in Eq. (2.3). For the *active processes*, which appear only at odd orders ($s = $ odd) in perturbation, the absorptive and emissive Fourier components are always present in pairs and the frequency of the signal field (the $(s + 1)$th wave), ω_j, is identical to the one unpaired component remaining among the s (odd) incident fields, $\{\omega_i\}$. In general, $E_l(r; \phi_l)$ is a complex vector satisfying $E_l(r; \phi_l)^* = E_l(r; -\phi_l)$ due to the reality of $E(r, t)$. On the other hand, $E_l(r) = \mathcal{E}_l(r)\hat{e}_l$ is *real* and positive definite. There are two independent Cartesian polarization unit vectors $\hat{e}_l$ for each k_l. The absolute phases for the two Cartesian polarization components are the same for linearly polarized light and they are different for elliptically polarized light [19].

The sth order induced polarization $P^{(s)}(r, t)$ can likewise be written in Fourier expansion form [20],

$$P^{(s)}(r, t) = \tfrac{1}{2} \sum_p \{P_p^{(s)}(r) \exp[i(k_p \cdot r - \omega_p t + \phi_p(r))] + \text{c.c.}\} \tag{2.5}$$

where the summation is over all Fourier components and polarization unit vectors. $P^{(s)}(r, t)$ is real, and $P_p^{(s)}(r)$ induced by the s incident fields can be written in terms of the electrical susceptibility $\chi^{(s)}(\omega_p) \equiv \chi^{(s)}(\omega_p = \xi_1\omega_1 + \xi_2\omega_2 + \cdots + \xi_s\omega_s)$ (an $(s + 1)$th rank tensor), as

$$P_p^{(s)}(r) = \chi^{(s)}(\omega_p = \xi_1\omega_1 + \xi_2\omega_2 + \cdots + \xi_s\omega_s)E_1(r)E_2(r)\cdots E_s(r) \tag{2.6}$$

where

$$\omega_p = \sum_{i=1}^{s} \xi_i \omega_i \tag{2.7a}$$

$$\mathbf{k}_p = \sum_{i=1}^{s} \{\xi_i \mathbf{k}_i' + i\mathbf{k}_i''\} \tag{2.7b}$$

and the *absolute phase* of the polarization at ω_p is given by

$$\phi_p = \sum_{i=1}^{s} \xi_i \phi_i \tag{2.7c}$$

Here, ξ_i takes on the value of $+1$ for an absorptive Fourier component (with ω_i and $\mathbf{k}_i$), or -1 for an emissive Fourier component (with $-\omega_i$ and $-\mathbf{k}_i^*$). Thus, the set $\{\xi_i\}$ specifies a particular combination of phasing of the incident fields as they polarize the medium. Whenever there is a material resonance, the electrical susceptibility becomes complex. The reality condition for $\mathbf{P}^{(s)}(\mathbf{r}, t)$, however, requires that $\chi^{(s)}(\omega_p = \xi_1\omega_1 + \xi_2\omega_2 + \cdots + \xi_s\omega_s)^* = \chi^{(s)}(-\omega_p = -\xi_1\omega_1 - \xi_2\omega_2 - \cdots - \xi_s\omega_s)$.

A. Linear and Nonlinear Wave Equations

The general wave equation, Eq. (2.1), is now combined with the plane wave forms of Eqs. (2.3) and (2.5) to find the linear and nonlinear wave equations for the signal field at ω_j. Specifically, we derive in this section the expression for the difference between the absolute phase of the signal field and that of the induced polarization, a parameter that proves to be of central importance in guaranteeing global energy conservation. In order to simplify the notation, we consider here only isotropic media such as gases, liquid fluids, or cubic crystals. However, the results obtained are quite general and also must be valid for light–matter interactions in anisotropic media.

When the nonlinear induced polarization is negligible, $\mathbf{P}_p^{(s)}(\mathbf{r}) = 0$ for all $s > 1$, Eq. (2.1) can be combined with Eqs. (2.3) and (2.5) to give the linear wave equation for the signal field at ω_j,

$$\nabla^2 \mathbf{E}_j(\mathbf{r}; \phi_j) + 2ik_j \nabla_j \mathbf{E}_j(\mathbf{r}; \phi_j) = 0 \tag{2.8}$$

Here,

$$\nabla_j = (1/k_j)\mathbf{k}_j \cdot \nabla \tag{2.9a}$$

$$k_j^2 = (\omega_j/c)^2 \{1 + 4\pi\chi^{(1)}(\omega_j = \omega_j)\} \tag{2.9b}$$

in which ∇_j is the derivative along the direction of wave vector $\mathbf{k}_j$.

The first term on the left hand side of Eq. (2.8) is usually neglected through the slowly varying amplitude approximation (SVAA) [3, 8]. Then, with $\mathbf{E}_j(\mathbf{r}; \phi_j) = \mathbf{E}_j(\mathbf{r}) \exp[i\phi_j(\mathbf{r})]$, Eq. (2.8) becomes

$$2k_j[i\nabla_j\mathbf{E}_j(\mathbf{r}) - \mathbf{E}_j(\mathbf{r})\nabla_j\phi_j(\mathbf{r})] = 0 \tag{2.10}$$

Both $\mathbf{E}_j(\mathbf{r})$ and $\phi_j(\mathbf{r})$ are real, and the real and imaginary parts of Eq. (2.10) have to be separately satisfied. Thus, both the field amplitude, $\mathscr{E}_j(\mathbf{r})$ or $\mathbf{E}_j(\mathbf{r})$, and the absolute phase, $\phi_j(\mathbf{r})$, must not vary with $\mathbf{r}$. On the other hand, $\mathbf{E}(\mathbf{r}, t)$ will decrease exponentially with $\mathbf{r}$ when $\mathbf{k}_j$ is complex due to the presence of material resonance.

Next consider the case where the nonlinear induced polarization at sth order ($s > 1$) at the frequency ω_p (given by Eq. (2.7a)) is *not* negligible. The wave equation for the signal field at $\omega_j = \omega_p$, Eq. (2.1) with $\mathbf{P}_p^{(s)}(\mathbf{r})$ given by Eq. (2.6) becomes in the SVAA

$$2k_j[i\nabla_j\mathbf{E}_j(\mathbf{r}) - \mathbf{E}_j(\mathbf{r})\nabla_j\phi_j(\mathbf{r})] = 4\pi(\omega_j/c)^2\chi^{(s)}(\omega_j = \xi_1\omega_1 + \xi_2\omega_2 + \cdots + \xi_s\omega_s)$$
$$\times \mathbf{E}_1(\mathbf{r})\mathbf{E}_2(\mathbf{r})\cdots\mathbf{E}_s(\mathbf{r}) \exp[-i(\Delta\mathbf{k}'_j \cdot \mathbf{r} + \Delta\phi_j(\mathbf{r})) + \Delta\mathbf{k}''_j \cdot \mathbf{r}] \tag{2.11}$$

where $\Delta\mathbf{k}'_j$ and $\Delta\mathbf{k}''_j$ are respectively defined as

$$\Delta\mathbf{k}'_j \equiv \mathbf{k}'_j - \mathbf{k}'_p = \mathbf{k}'_j - \sum_{i=1}^{s} \xi_i\mathbf{k}'_i \tag{2.12a}$$

$$\Delta\mathbf{k}''_j \equiv \mathbf{k}''_j - \mathbf{k}''_p = \mathbf{k}''_j - \sum_{i=1}^{s} \mathbf{k}''_i \tag{2.12b}$$

And, in Eq. (2.11),

$$\Delta\phi_j(\mathbf{r}) \equiv \phi_j(\mathbf{r}) - \phi_p(\mathbf{r}) = \phi_j(\mathbf{r}) - \sum_{i=1}^{s} \xi_i\phi_i(\mathbf{r}) \tag{2.12c}$$

represents the *difference* between the *absolute phase* of the signal field at ω_j and the *absolute phase* of the sth order polarization at $\omega_p = \omega_j$. The SVAA for Eq. (2.11) is justified when the nonlinear effect is not so strong as to significantly alter the transverse profiles of the incident fields as they move through the medium.

Once the signal field at ω_j is generated in the medium through the sth order interaction, it can act in turn as an incident field to produce the fields at $\{\omega_i\}$ also at the sth order. Thus, such an induced polarization, $\mathbf{P}_i^{(s)}(\mathbf{r})$ at ω_i, is determined by $\chi^{(s)}(\omega_i)$ with

$$\omega_i = \xi_i\left\{\omega_j - \sum_{l \neq i}^{s} \xi_l\omega_l\right\} \tag{2.13}$$

The wave equation for $\mathbf{E}_i(\mathbf{r})$ has the same form as Eq. (2.11) and the $\Delta \mathbf{k}'_i$ and $\Delta \phi_i(\mathbf{r})$ that would be contained in it are given as

$$\Delta \mathbf{k}'_i \equiv \mathbf{k}'_i - \xi_i \left\{ \mathbf{k}'_j - \sum_{l \neq i}^s \xi_l \mathbf{k}'_l \right\} = -\xi_i \Delta \mathbf{k}'_j \qquad (2.14a)$$

and

$$\Delta \phi_i(\mathbf{r}) \equiv \phi_i(\mathbf{r}) - \xi_i \left\{ \phi_j(\mathbf{r}) - \sum_{l \neq i}^s \xi_l \phi_l(\mathbf{r}) \right\} = -\xi_i \Delta \phi_j(\mathbf{r}) \qquad (2.14b)$$

where $\xi_i^2 = 1$ is used.

Now, as shown in Appendix A, Eq. (2.11) can be used to derive the equations for $\mathscr{E}_j(\mathbf{r})$ and $\Delta \phi_j(\mathbf{r})$,

$$\nabla_j \mathscr{E}_j(\mathbf{r}) = \frac{2\pi(\omega_j/c)^2}{k_j'^2 + k_j''^2} \left[A_j(\mathbf{r}) \sin \Delta_j(\mathbf{r}) - B_j(\mathbf{r}) \cos \Delta_j(\mathbf{r}) \right] \qquad (2.15a)$$

$$\nabla_j \Delta \phi_j(\mathbf{r}) = \frac{A_j(\mathbf{r}) \cos \Delta_j(\mathbf{r}) + B_j(\mathbf{r}) \sin \Delta_j(\mathbf{r})}{A_j(\mathbf{r}) \sin \Delta_j(\mathbf{r}) - B_j(\mathbf{r}) \cos \Delta_j(\mathbf{r})} \nabla_j \{\ln \mathscr{E}_j(\mathbf{r})\}$$

$$+ \sum_{i=1}^s \frac{\xi_i A_i(\mathbf{r}) \cos \Delta_j(\mathbf{r}) - B_i(\mathbf{r}) \sin \Delta_j(\mathbf{r})}{\xi_i A_i(\mathbf{r}) \sin \Delta_j(\mathbf{r}) + B_i(\mathbf{r}) \cos \Delta_j(\mathbf{r})} \nabla_j \{\ln \mathscr{E}_i(\mathbf{r})\} \qquad (2.15b)$$

Here, $\Delta_j(\mathbf{r}) \equiv \Delta \mathbf{k}'_j \cdot \mathbf{r} + \Delta \phi_j(\mathbf{r})$, and $A_j(\mathbf{r})$ and $B_j(\mathbf{r})$ happen to be the real and imaginary parts of $k_j^* \{\hat{e}_j \cdot \mathbf{P}_j^{(s)}(\mathbf{r})\} \exp[\Delta \mathbf{k}''_j \cdot \mathbf{r}]$, respectively, which are given as

$$A_j(\mathbf{r}) = [k'_j \zeta'_j + k''_j \zeta''_j] \mathscr{E}_1(\mathbf{r}) \mathscr{E}_2(\mathbf{r}) \cdots \mathscr{E}_s(\mathbf{r}) \exp[\Delta \mathbf{k}''_j \cdot \mathbf{r}] \qquad (2.16a)$$

$$B_j(\mathbf{r}) = [k'_j \zeta''_j - k''_j \zeta'_j] \mathscr{E}_1(\mathbf{r}) \mathscr{E}_2(\mathbf{r}) \cdots \mathscr{E}_s(\mathbf{r}) \exp[\Delta \mathbf{k}''_j \cdot \mathbf{r}] \qquad (2.16b)$$

Here, a (scalar) Cartesian component of the electrical susceptibility, $\zeta_j = \zeta'_j + i\zeta''_j$, has been introduced for convenience. It is defined by

$$\zeta_j = \hat{e}_j \cdot \chi^{(s)}(\omega_j = \xi_1 \omega_1 + \xi_2 \omega_2 + \cdots + \xi_s \omega_s) \cdot \hat{e}_1 \hat{e}_2 \cdots \hat{e}_s \qquad (2.17)$$

$A_i(\mathbf{r})$ and $B_i(\mathbf{r})$ are similarly defined in terms of $\mathbf{k}'_i$, $\mathbf{k}''_i$, $\Delta \mathbf{k}''_i$, and $\chi^{(s)}(\omega_i)$, and by replacing the amplitude factors in Eqs. (2.16a) and (2.16b) appropriately (i.e. $\mathscr{E}_i(\mathbf{r}) \to \mathscr{E}_j(\mathbf{r})$).

According to Eq. (2.15a), when there is signal field/polarization momentum mismatch, $\mathbf{k}'_j \neq \mathbf{k}'_p$, the signal field $\mathscr{E}_j(\mathbf{r})$ will oscillate along the direction of $\mathbf{k}_j$ with the period of the coherence length, $l_c = 2\pi/\Delta k'_j$, provided $\Delta\phi_j(\mathbf{r})$ does not depend on $\mathbf{r}$ too strongly [3, 8]. The signal field can build up coherently from the polarization over macroscopic distances only when the phase matching condition is satisfied, $\Delta\mathbf{k}'_j = 0$. The phase matching condition is evident already in the equation for the field amplitude, $E_j(\mathbf{r})$, Eq. (2.11). It represents the interference between the free wave (the general solution of the homogeneous equation, $\chi^{(s)}(\omega_j) = 0$) and the bound wave (the particular solution of the inhomogeneous equation, $\chi^{(s)}(\omega_j) \neq 0$), both at ω_j. In general, these two waves experience different dispersion in the medium through their different wave vectors, $\mathbf{k}_j$ and $\mathbf{k}_p$. For passive processes with intermediate resonances, the phase matching condition is given by $\Delta\mathbf{k}'_j = 0$ while $\Delta\mathbf{k}''_j$ does not usually vanish.

B.   The Difference in Absolute Phases, $\Delta\phi_j(\mathbf{r})$

As can be seen from Eq. (2.15a), $\Delta\phi_j(\mathbf{r})$ is crucial for determining the amplitude of the signal field, $\mathscr{E}_j(\mathbf{r})$, in the medium. It is also an important factor for determining W_F and W_M, as will be shown below. In this section, we examine $\Delta\phi_j(\mathbf{r})$ for various nonlinear light–matter interactions.

From now on the phase matching condition is assumed, thus $\Delta\mathbf{k}'_j = 0$ and $\Delta_j(\mathbf{r}) = \Delta\phi_j(\mathbf{r})$. This immediately applies to all active processes (naturally phase-matched) and all (resonant and nonresonant) passive processes that are phase-matched. Equation (2.15b) can be significantly simplified under certain circumstances.

When the light-matter interaction is directly in quadrature form, s is odd and at most only $\frac{1}{2}(s - 1)$ different frequencies are incident since each field, but one, appears with both *absorptive* and *emissive* components. Thus the absolute phase of only one field survives the sum in Eq. (2.12c). The odd field is the Fourier component at ω_j so that finally $\Delta\phi_j(\mathbf{r}) \equiv \phi_j(\mathbf{r}) - \phi_j(\mathbf{r}) = 0$. That is,

$$\Delta\phi_j(\mathbf{r}) = 0 \qquad (2.18)$$

Such interactions include all *active processes* and all *nonresonant nonlinear dispersions*. Thus, in such spectroscopies, the signal field must always remain *in-phase* with the induced polarization and $\mathscr{E}_j(\mathbf{r})$ is determined from Eq. (2.15a) with $\Delta_j(\mathbf{r}) = 0$.

Nonresonant passive processes in which the light–matter interactions are not directly in quadrature (not fully conjugately paired) have been discussed by Amstrong et al. [8] and other [3]. In this case, all wave vectors and electrical susceptibilities are real. From (2.16b), it can be seen that $B_j(\mathbf{r}) = 0$ and thus $B_i(\mathbf{r}) = 0$ for all i. Then, Eq. (2.15b) reduces to

$$\nabla_j \Delta \phi_j(\mathbf{r}) = \frac{\cos \Delta \phi_j(\mathbf{r})}{\sin \Delta \phi_j(\mathbf{r})} \nabla_j \ln[\mathscr{E}_j(\mathbf{r}) \mathscr{E}_1^{\xi_1}(\mathbf{r}) \cdots \mathscr{E}_s^{\xi_s}(\mathbf{r})] \qquad (2.19)$$

This equation, when integrated on both sides from the incident boundary $(\mathbf{r} = 0)$ to $\mathbf{r}$ in the medium along the direction of $\mathbf{k}_j$, results in the following invariance:

$$\mathscr{E}_j(r) \mathscr{E}_1^{\xi_1}(\mathbf{r}) \mathscr{E}_2^{\xi_2}(\mathbf{r}) \cdots \mathscr{E}_s^{\xi_s}(\mathbf{r}) \cos \Delta \phi_j(\mathbf{r})$$
$$= \mathscr{E}_j(0) \mathscr{E}_1^{\xi_1}(0) \mathscr{E}_2^{\xi_2}(0) \cdots \mathscr{E}_s^{\xi_s}(0) \cos \Delta \phi_j(0) \qquad (2.20)$$

Thus, if there is no signal field at the incident boundary, $\mathscr{E}_j(0) = 0$, the above invariance requires that (for $-\pi \le \Delta \phi_j(\mathbf{r}) \le \pi$)

$$\Delta \phi_j(\mathbf{r}) = \pm \tfrac{1}{2} \pi \qquad (2.21)$$

for all $\mathbf{r}$ where $\mathscr{E}_i(\mathbf{r}) \ne 0$ and $\mathscr{E}_j(\mathbf{r}) \ne 0$. That is, the signal field generated in the medium through such a *nonresonant passive process* must contain only a component that is *out-of-phase* with respect to the induced polarization. The sign of $\Delta \phi_j$ is fixed under the constraint that $\mathscr{E}_j(\mathbf{r})$ remains positive definite everywhere inside the medium. The amplitude of the signal field is determined by Eq. (2.15a) coupled with similar equations for the incident field amplitudes. Even when a weak signal field exists at the incident boundary, $\Delta \phi_j(\mathbf{r})$ must approach $\pm \tfrac{1}{2} \pi$ as the signal field becomes amplified since the right hand side of Eq. (2.20) is finite while the left hand side increases with $\mathscr{E}_j(\mathbf{r})$.

For *resonant passive processes*, the situation is more complicated. Now, Eq. (2.15b) must be solved simultaneously with the equations for the amplitudes of the signal and the incident fields as given in Eq. (2.15a). When incident fields can drive material resonances, their principle source of depletion is through the imaginary part of their wave vector, $\mathbf{k}_i''$. Their additional depletion through higher order events can often be neglected. Then, $\mathscr{E}_i(\mathbf{r})$ remains constant and Eq. (2.15b) now leads to the following constraint:

$$[A_j^0 \cos \Delta \phi_j(\mathbf{r}) + B_j^0 \sin \Delta \phi_j(\mathbf{r})] \mathscr{E}_j(\mathbf{r})$$
$$= [A_j^0 \cos \Delta \phi_j(0) + B_j^0 \sin \Delta \phi_j(0)] \mathscr{E}_j(0) \qquad (2.22)$$

where

$$A^0_j = k'_j \zeta'_j + k''_j \zeta''_j \tag{2.23a}$$

$$B^0_j = k'_j \zeta''_j - k''_j \zeta'_j \tag{2.23b}$$

In the absence of any signal field at the incident boundary, $\mathscr{E}_j(0) = 0$, Eq. (2.22) constrains $\Delta\phi_j(\mathbf{r})$ to be constant throughout the medium with

$$\tan \Delta\phi_j(\mathbf{r}) = \frac{k'_j \zeta'_j + k''_j \zeta''_j}{k''_j \zeta'_j - k'_j \zeta''_j} \tag{2.24a}$$

Furthermore, in nearly degenerate four-wave mixing processes without electronic resonances like ordinary CARS and CSRS, the material resonance is in the form of $|\omega_1 - \omega_2| = \omega_R$; no incident field is individually in resonance. In such cases, all wave vectors are real while the electrical susceptibilities are complex. Then, Eq. (2.24a) is further simplified to

$$\tan \Delta\phi_j(\mathbf{r}) = -\zeta'_j / \zeta''_j \tag{2.24b}$$

It can be seen that Eq. (2.24b) reduces to $\Delta\phi_j(\mathbf{r}) = \pm\frac{1}{2}\pi$ (Eq. (2.21)) for *nonresonant passive processes* in which $\zeta''_j = 0$. On the other hand, $\Delta\phi_j(\mathbf{r})$ approaches 0 as $|\zeta''_j|$ increases. That is, in the absence of any resonance, the generated signal field contains only an out-of-phase component with respect to the polarization (Eq. (2.21)). But the in-phase component increases as the incident fields are tuned to some material resonance. More generally, the signal field in any resonant $(s + 1)$-wave passive process contains both *in-phase* and *out-of-phase* components with respect to the induced polarization, whose ratio is determined by the relative magnitudes of the real and imaginary parts of the electrical susceptibility and the wave vectors.

When either Eq. (2.24a) or Eq. (2.24b) is satisfied, Eq. (2.15a) becomes

$$\nabla_j \mathscr{E}_j(\mathbf{r}) = 2\pi(\omega_j/c)^2 \frac{|\zeta'_j + i\zeta''_j|}{|k'_j + ik''_j|} \mathscr{E}_1(0)\mathscr{E}_2(0) \cdots \mathscr{E}_s(0) \exp[\Delta\mathbf{k}''_j \cdot \mathbf{r}] \tag{2.25}$$

Here, the depletion of the incident fields is taken to be only by linear processes. Thus $\mathscr{E}_i(\mathbf{r}) = \mathscr{E}_i(0)$ for all $i = 1, 2, \ldots, s$. The amplitude of the signal field is proportional to the magnitude of the electrical susceptibility, $|\zeta'_j + i\zeta''_j|$, at ω_j. In obtaining Eq. (2.25), the ambiguity in the sign of $\cos\{\Delta\phi_j(\mathbf{r})\} = \pm[1 + \tan^2\{\Delta\phi_j(\mathbf{r})\}]^{1/2}$ has been fixed so that the signal field builds up along the direction of $\mathbf{k}_j$.

In this section, we have shown how $\Delta\phi_j(\mathbf{r})$ is determined according to the nature of the light–matter interactions and the presence or absence of material resonances. For interactions of the quadrature type, the *active processes* or the *nonresonant passive dispersions*, the signal field is always *in-phase* with respect to the induced polarization, namely $\Delta\phi_j(\mathbf{r}) = 0$. For all remaining *passive interactions*, where the light–matter interactions are not directly in quadrature in the fields, the signal field must be *out-of-phase*, $\Delta\phi_j(\mathbf{r}) = \pm\frac{1}{2}\pi$, in the absence of material resonance, but it must contain both *in-phase* and *out-of-phase* components in the presence of resonance $(0 \le |\Delta\phi_j(\mathbf{r})| < \frac{1}{2}\pi)$.

III. ENERGY EXCHANGE RATES: THE MACROSCOPIC EXPRESSION FOR W_F

Next we determine the cycle-averaged rate of total energy transfer per unit volume, W_F, between the radiation fields and the medium in terms of the electrical susceptibilities and the all important $\Delta\phi_j(\mathbf{r})$ just examined in detail. Global energy conservation for resonant light–matter interactions will be confirmed by noting how the W_F, derived here, is equivalent to W_M, the cycle-averaged rate of energy change in unit volume of the medium, to be discussed in Section IV.

The cycle-averaged rate of work done by the CW field in unit volume of the medium at $\mathbf{r}$ is given by [1, 3, 8, 9, 13]

$$W_F(\mathbf{r}) = \frac{1}{T}\int_0^T \mathbf{E}(\mathbf{r}, t)\,\frac{\partial\mathbf{P}(\mathbf{r}, t)}{\partial t}\,dt \tag{3.1}$$

where $\mathbf{E}(\mathbf{r}, t)$ and $\mathbf{P}(\mathbf{r}, t)$ together satisfy the wave equation (Eq. (2.1)) and T represents a time sufficient to encompass many oscillations of $\mathbf{E}(\mathbf{r}, t)$ and $\mathbf{P}(\mathbf{r}, t)$. The energy conservation law for the radiation field, as given by Poynting's theorem [13], follows directly from Eq. (3.1) and Maxwell's equations. According to Poynting's theorem, the time rate of change of the field energy within unit volume at $\mathbf{r}$ *plus* the energy flux out of the volume at $\mathbf{r}$ through the boundary surface of the unit volume is equal to the negative of $W_F(\mathbf{r})$. For global energy conservation in the composite field/matter system, however, an energy change in the medium must balance any change in the total energy of the radiation.

If the plane wave forms for $\mathbf{E}(\mathbf{r}, t)$ and $\mathbf{P}^{(s)}(\mathbf{r}, t)$ in Eqs. (2.3) and (2.5) are introduced into Eq. (3.1), the cycle-averaged total energy exchange rate per unit volume between the radiation field and the medium through any sth order light–matter interaction $(s \geq 1)$ is given as

$$W_F^{(s)}(\mathbf{r}) = \sum_l W_F^{(s)}(\mathbf{r}; \omega_l) \tag{3.2}$$

where

$$W_F^{(s)}(\mathbf{r}; \omega_l) = \sum_l \tfrac{1}{2}\omega_l \, \mathrm{Im}\{\tilde{\mathbf{E}}_l(\mathbf{r}) \cdot \tilde{\mathbf{P}}_p^{(s)}(\mathbf{r}) \exp[-i\Delta_l(\mathbf{r})]\} \tag{3.3}$$

in which the summation is over all (positive) frequency components of the field present in the medium, including the signal field $(l \equiv j)$ and all of the incident fields $(l \equiv i,\ 1 \leq i \leq s)$. And, as before, $\Delta_l(\mathbf{r}) = \Delta\mathbf{k}_l' \cdot \mathbf{r} + \Delta\phi_l(\mathbf{r})$. The imaginary part of the wave vector in Eqs. (2.3) and (2.5) is absorbed into the amplitudes, $\tilde{\mathbf{E}}_l(\mathbf{r}) = \mathbf{E}_l(\mathbf{r})\exp[-\mathbf{k}_l'' \cdot \mathbf{r}]$ and $\tilde{\mathbf{P}}_p^{(s)}(\mathbf{r}) = \mathbf{P}_p^{(s)}(\mathbf{r})\exp[-\mathbf{k}_p'' \cdot \mathbf{r}]$. Here, $\Delta\mathbf{k}_l'$ and $\Delta\phi_l(\mathbf{r})$ are given by Eq. (2.12) for the signal field $(l = j)$ at ω_j and they are given by Eq. (2.14) for the incident fields $(l = i)$ at ω_i. $W_F^{(s)}(\mathbf{r})$ is positive if energy is transferred from the field to the medium as a result of the interaction; it is negative if energy is transferred from the medium to the field [6]. It is noted here that the Fourier components of $\mathbf{E}(\mathbf{r}, t)$ and $\mathbf{P}(\mathbf{r}, t)$ must be at the same frequency in order to survive the cycle-average integral in Eq. (3.1).

$W_F^{(s)}(\mathbf{r}; \omega_j)$ represents the cycle-averaged energy exchange rate per unit volume between the medium and the signal field at ω_j that arises from the s-fold interaction of the incident fields with the medium at position $\mathbf{r}$. By using Eqs. (2.6), (2.17), and (3.3), we can express $W_F^{(s)}(\mathbf{r}; \omega_j)$ in terms of the electrical susceptibility

$$W_F^{(s)}(\mathbf{r}; \omega_j) = \tfrac{1}{2}\omega_j \tilde{K}(\mathbf{r}) \, \mathrm{Im}\{\zeta_j \exp[-i\Delta_j(\mathbf{r})]\}$$
$$= \tfrac{1}{2}\omega_j \tilde{K}(\mathbf{r})[\zeta_j'' \cos \Delta_j(\mathbf{r}) - \zeta_j' \sin \Delta_j(\mathbf{r})] \tag{3.4}$$

where

$$\tilde{K}(\mathbf{r}) = \tilde{\mathscr{E}}_j(\mathbf{r})\tilde{\mathscr{E}}_1(\mathbf{r})\tilde{\mathscr{E}}_2(\mathbf{r}) \cdots \tilde{\mathscr{E}}_s(\mathbf{r}) \tag{3.5}$$

which is real, and in which $\tilde{\mathscr{E}}_i(\mathbf{r}) = \mathscr{E}_i(\mathbf{r})\exp[-\mathbf{k}_i'' \cdot \mathbf{r}]$. The real and imaginary parts of the scalar component, ζ_j, of the electrical susceptibility have been defined in Eq. (2.17).

Equations (3.2) and (3.4) apply to both resonant and nonresonant processes. For passive processes, first of all, the phase matching condition may not always be satisfied. For the phase mismatched case, $\Delta\mathbf{k}_j' \neq 0$, it is evident from Eq. (3.4) that $W_F^{(s)}(\mathbf{r}; \omega_j)$ oscillates along $\mathbf{r}$ with the period equal to the coherence length, l_c. Whenever $W_F^{(s)}(\mathbf{r}; \omega_j) < 0$, the signal field is being generated through the nonlinear interaction at point $\mathbf{r}$ in the medium. Whenever $W_F^{(s)}(\mathbf{r}; \omega_j) > 0$, the signal field is being destroyed through the interaction. Thus, there can be no significant buildup of the signal field inside the medium. This is consistent with the discussion presented in Section II.A within the context of the equation for the signal field amplitude $\mathcal{E}_j(\mathbf{r})$.

The sth order cycle-averaged rate of total energy transfer per unit volume, $W_F^{(s)}(\mathbf{r})$, between the medium and all fields present in the medium is obtained, as given in Eq. (3.2), by adding the contributions from all Fourier components of the field including the signal field at ω_j and the incident fields at $\{\omega_i\}$. Thus, when the phase matching condition is satisfied, $\Delta \mathbf{k}'_j = 0$, Eq. (3.2) can be written as

$$W_F^{(s)}(\mathbf{r}) = \tfrac{1}{2} \omega_j \tilde{K}(\mathbf{r})[\zeta''_j \cos \Delta\phi_j(\mathbf{r}) - \zeta'_j \sin \Delta\phi_j(\mathbf{r})]$$
$$+ \sum_{i=1}^{s} \tfrac{1}{2} \omega_i \tilde{K}(\mathbf{r})[\zeta''_i \cos \Delta\phi_j(\mathbf{r}) + \zeta'_i \xi_i \sin \Delta\phi_j(\mathbf{r})] \qquad (3.6)$$

where Eq. (2.14b) has been used for the second line to express $\Delta\phi_i(\mathbf{r})$ in terms of $\Delta\phi_j(\mathbf{r})$. The first part of Eq. (3.6) represents the cycle-averaged rate of energy exchange per unit volume between the signal field and the medium. The second part denotes the contribution from all remaining incident fields.

For *fully nonresonant passive processes*, all wave vectors and electrical susceptibilities are real. Furthermore, there exists *permutation symmetry* among the electrical susceptibilities [3, 8, 9] such that

$$\zeta'_j = \zeta'_i, \quad \text{for all } i = 1, 2, \ldots, s \qquad (3.7)$$

and $K(\mathbf{r}) = \mathcal{E}_j(\mathbf{r}) \mathcal{E}_1(\mathbf{r}) \mathcal{E}_2(\mathbf{r}) \cdots \mathcal{E}_s(\mathbf{r})$. From these relations along with Eq. (2.14), the following Manley–Rowe relations [8] are readily recognized:

$$\frac{W_F^{(s)}(\mathbf{r}; \omega_j)}{\hbar \omega_j} = -\frac{W_F^{(s)}(\mathbf{r}; \omega_i)}{\xi_i \hbar \omega_i}, \quad \text{for all } i = 1, 2, \ldots, s \qquad (3.8)$$

In nonresonant passive dispersions, in which the light–matter interactions are maximally in quadrature at the polarization level, but there is no material resonance, no energy transfer between the medium and any Fourier component of the field is possible because $\Delta\phi_j(\mathbf{r}) = 0$ (See Eq. (3.4) with $\zeta''_j = 0$). Equation (3.8), however, remains valid for such processes.

It is evident now from Eqs. (3.6) and (3.7), and the condition, $\omega_j = \omega_p \equiv \xi_1\omega_1 + \xi_2\omega_2 + \cdots + \xi_s\omega_s$ (Eq. (2.7a)), that in the absence of resonance $W_F^{(s)}(\mathbf{r})$ vanishes for all $\mathbf{r}$, indicating that (as expected) for *nonresonant passive processes* there is *no energy exchange* between the medium and the radiation fields present in the medium. Here, as we have noted, the medium simply plays a purely catalytic role for the transfer of energy among the incident fields and the signal field. The signal field is generated at the expense of the incident fields, not of the medium. In such cases, the medium in the presence of radiation can be defined as an independent thermodynamic system [2, 3]. That is, for *nonresonant passive processes*, the energy conservation within the radiation and within the medium must be separately satisfied ($W_F^{(s)}(\mathbf{r}) = W_M^{(s)}(\mathbf{r}) = 0$). Only for this case, the condition, $\omega_j = \omega_p$, can be taken as the separate energy conservation condition for the radiation field; the total energy of the *photons* created (the signal field and those incident fields with $\xi_i = -1$) is the same as the sum of the energies of the *photons* destroyed (those incident fields with $\xi_i = +1$). It can be also seen from Eq. (3.4) that the maximum energy transfer to the signal field through *nonresonant passive processes* occurs when $\Delta\phi_j(\mathbf{r}) = \pm\frac{1}{2}\pi$ (Eq. (2.21)) [1, 3, 8].

For *all resonant processes*, *active* and *passive*, Eq. (3.4) remains valid. However the permutation symmetry (Eq. (3.7)), among the electrical susceptibilities, and the Manley–Rowe relations (Eq. (3.8)) break down in the presence of resonance. Now, $W_F^{(s)}(\mathbf{r})$ in Eq. (3.6) does not vanish in general, requiring that a net exchange of energy takes place between the radiation and the medium, not only in active processes, but also in passive processes having resonance. (Thus, $W_F^{(s)}(\mathbf{r}) \neq 0$). The combined system of radiation and the medium is no longer separable into uncoupled thermodynamic systems. Now, since the energy of the radiation is not separately conserved, the condition, $\omega_j = \omega_p$, can no longer be regarded as the energy conservation condition for the radiation. It simply continues to specify the frequency of the signal field, whether it is being generated from the induced polarization or is already present and is interacting with the induced polarization.

As discussed in Section II.B, for *active processes*, conjugate pairing of fields exists at the signal level so that the phase matching condition is automatically satisfied and $\Delta_j(\mathbf{r}) = 0$ for all $\mathbf{r}$. In this case, Eq. (3.4) becomes

$$W_F^{(s)}(\mathbf{r}; \omega_j) = \tfrac{1}{2}\omega_j\tilde{K}(\mathbf{r})\zeta_j'' \tag{3.9}$$

where $\tilde{K}(\mathbf{r})$ becomes a product of $\frac{1}{2}(s + 1)$ intensities involved in an sth order active process.

Whenever $W_F^{(s)}(\mathbf{r}) \neq 0$, energy change in the medium is essential and it can only appear as a population change among its quantum eigenstates, as is discussed next. A population change in the medium through active processes is easily understood from the quadrature form of the light–matter interaction. Here, the cycle-averaged rate of the population change is proportional to the intensity of the incident fields or $|\mathbf{E}_i|^2 = \mathbf{E}_i^*\mathbf{E}_i$, and one or more *photons* is said to be either absorbed or emitted by the medium through such quadrature form of interaction. Thus, active processes are classified by the number of *photons* involved or the number of quadrature interactions. In fact, Eq. (3.9) can be used to determine the expression for the *hyper cross-section* that characterizes an r-photon ($r = \frac{1}{2}(s + 1)$) *active process*, as shown explicitly by Lee and Albrecht [6].

For *resonant passive processes*, Eq. (3.6) still remains valid. However, due to the presence of material resonances, $W_F^{(s)}(\mathbf{r})$ does not vanish in general, just as for active processes. Again, $W_F^{(s)}(\mathbf{r}) \neq 0$ can be achieved only if there is a corresponding energy change in the medium through a population change, despite the fact that the light–matter interaction is not in general of quadrature form at the polarization level. It will be shown in the next section how the population among eigenstates of the medium can indeed be altered by such a nonquadrature form of interaction.

IV. DENSITY MATRIX APPROACH, THE DIAGRAMMATIC TECHNIQUE, AND THE MICROSCOPIC EXPRESSIONS FOR $W_F^{(s)}(\mathbf{r})$ AND $W_M^{(s)}(\mathbf{r})$

The density matrix approach to nonlinear light–matter interaction is briefly summarized by way of introducing the corresponding diagrammatic technique based on the familiar energy ladder diagrams to develop the microscopic quantum-mechanical expressions for the nonlinear electrical susceptibilities. It is also shown how the diagrammatic technique can be used to find the quantum-mechanical expression for the cycle-averaged rate of population change in unit volume of the medium, W_M.

Depending on the nature of the radiation/matter interaction—whether it is active or passive and whether resonance is present or absent for the latter—the Hilbert space of the medium is partitioned into subspaces belonging to the individual molecular components (termed the *chromophore* space) whose eigenstates are made explicit. The remaining portion of the Hilbert space is assigned to the nonradiative part of the bath. Its eigenstates, not made explicit, are reduced to stochastic dynamics. As such they are responsible for the nonradiative source of damping. For example, in the *active* spectroscopies the *chromophore* often corresponds to an appropriate subspace of a spectroscopically active solute molecule dissolved in a spectroscopically inactive solvent whose Hilbert space is fully assigned to the bath. But in *passive processes*, which are coherent, both solute and solvent in this same system have subspaces (resonant or not) that must be treated explicitly and they each contribute to the signal. The system is now *bichromophoric*. Here we shall avoid the notational complications of multichromophoric systems and choose to treat the system as consisting of a single chromophore having a number density, N. In practice, the definition must depend on the nature of the interaction and, in addition, we must be prepared to generalize to the multichromophore context. In any case, whenever the *chromophore* is in resonance with the radiation, the electrical susceptibility becomes complex due to the presence of the bath.

At the electric-dipole approximation (EDA) and in the fast-modulation limit of the chromophore–bath interaction, the molecular density matrix element, $\rho_{mn}(t)$, in the *dark* molecular basis, $\{|m\rangle\}$, satisfies the master equation [6],

$$i\hbar \, \partial\rho_{mn}(t)/\partial t = \hbar\omega_{mn}\rho_{mn}(t) + \langle m|[-\vec{\mu} \cdot \mathbf{E}(t), \rho(t)]|n\rangle \qquad (4.1)$$

where $\omega_{mn} = \omega^0_{mn} - i\gamma_{mn}$ with $\omega^0_{mn} = (\epsilon_m - \epsilon_n)/\hbar$. The *dark* eigenstate $|m\rangle$ is an eigenstate of the *dark* Hamiltonian $\mathbf{H}^0$ with eigenvalue ϵ_m. γ_{mn} is the phenomenological damping constant between the dark states m and n. $\mathbf{E}(t)$ and $\vec{\mu}$ are the electric field strength at the chromophore and the electric dipole operator, respectively. As previously stated, any time-independent DC field present in the medium is conveniently included in $\mathbf{H}^0$. It is emphasized that at the EDA, reference to the propagation property of the radiation field is completely absent at the chromophore site. The propagation property, important to the coherent (passive) spectroscopies, is incorporated at the EDA by attaching the appropriate retardation factor, $\exp[i(\mathbf{k} \cdot \mathbf{r} + \phi(\mathbf{r}))]$, as a phase factor to the matrix element of the commutator in Eq. (4.1) [8]. Here, $\mathbf{r}$ is a coarse position vector in the medium not to be available in the spatial integral over the local intra-chromophore space.

The solution to Eq. (4.1) is usually found by employing the perturbative procedure whose sth order solution, $\rho^{(s)}_{mn}(t)$, is given in iterative form as

$$\rho^{(s)}_{mn}(\mathbf{r}, t) = (1/i\hbar) \exp[-i\omega_{mn}t] \sum_i \tilde{\mathscr{E}}_i(\mathbf{r}) \exp[i(\mathbf{k}'_i \cdot \mathbf{r} + \phi_i)]$$

$$\times \int_{-\infty}^{t} \langle m|[-\vec{\mu} \cdot \hat{e}_i, \rho^{(s-1)}(\mathbf{r}, t')]|n\rangle \exp[i(\omega_{mn} - \xi_i\omega_i)t'] \, dt' \qquad (4.2)$$

where the field at frequency ω_i interacts with the chromophore either to change the bra to $\langle m|$ (with the ket in $|n\rangle$) or the ket to $|n\rangle$ (with the bra in $\langle m|$). It can be readily recognized from Eq. (4.2) that $\rho_{mn}^{(s)}(\mathbf{r}, t)$ can also be expressed in the Fourier expansion,

$$\rho_{mn}^{(s)}(\mathbf{r}, t) = \tfrac{1}{2} \sum_p \{\rho_{mn}^{(s)}(\mathbf{r}; \omega_p) \exp[i(\mathbf{k}_p' \cdot \mathbf{r} - \omega_p t + \phi_p(\mathbf{r}))] + \text{c.c.}\} \quad (4.3)$$

where $\mathbf{k}_p$, ω_p and $\phi_p(\mathbf{r})$ have been given in Eqs. (2.7). Successive applications of the time integral in Eq. (4.2) produces the usual time-ordered integrals or the Dyson series, and gives $\rho_{mn}^{(s)}(\mathbf{r}; \omega_p)$. The diagrammatic techniques [12, 21], however, help in understanding nonlinear light–matter interaction by demonstrating the importance of the time-ordering of the interactions, as first pointed out by Lynch and Bloembergen [22]. It also provides a simple method for writing down microscopic expressions for the nonlinear electrical susceptibilities. The time-evolution diagram in its energy ladder form [6] clearly displays the changes that take place in the quantum states of the medium. On the other hand, we have noted how when describing *passive processes* certain diagrams appear to call for population changes among the chromophore eigenstates, when it would seem that none should occur. It turns out that diagrams that show no population change are also misleading. This problem will be discussed and resolved below.

A. Diagrammatic Technique for Electrical Susceptibility

Once the time-dependent density matrix is determined (Eq. (4.2) is solved), the quantum-mechanical expression for the sth order electrical susceptibility expression $\chi^{(s)}(\omega_p = \xi_1\omega_1 + \xi_2\omega_2 + \cdots + \xi_s\omega_s)$ is easily obtained. Thus, the sth order induced polarization $\mathbf{P}^{(s)}(\mathbf{r}, t)$ is given as

$$\mathbf{P}^{(s)}(\mathbf{r}, t) = N \operatorname{Tr}\{\vec{\mu}\rho^{(s)}(\mathbf{r}, t)\} \quad (4.4)$$

where N is the number density of chromophores in the medium. The sth order electrical susceptibility $\chi^{(s)}(\omega_p)$ is obtained by extracting the Fourier component of $\exp[-i\omega_p t]$ from both sides of Eqs. (2.6) and (4.4). By using the scalar form of the sth order electrical susceptibility as introduced through Eq. (2.17), we can obtain

$$\zeta_j\tilde{K}(\mathbf{r}) = N\tilde{\mathscr{E}}_j(\mathbf{r}) \sum_m \sum_n \langle m|\vec{\mu} \cdot \hat{e}_j|n\rangle \rho_{mn}^{(s)}(\mathbf{r}; \omega_p) \quad (4.5)$$

where $\tilde{K}(\mathbf{r})$ is given in Eq. (3.5) and $\omega_j = \omega_p$. Equation (4.5) can be used to find the algebraic expression for ζ_j.

Yee and Gustafson [12] first suggested a diagrammatic technique that uses double Feynman diagrams. Here, the time-evolution of the chromophore states is represented by vertices on two vertical time axes, one describing the bra ($\langle m|$) evolution and the other the ket ($|m\rangle$) evolution. Such double Feynman diagrams [11, 12, 21] clearly emphasize the importance of the relative time ordering of the bra and ket state evolution. However, material resonances (or their absence) at various steps in the time-evolution are often not clear from such diagrams.

The time-evolution of chromophore states can also be represented diagrammatically using the familiar energy ladder diagrams [6] in which solid horizontal lines are used for real eigenstates and dashed horizontal lines for virtual states. Examples will be given shortly. In general, however, the polarized state of the chromophore created by the successive interaction with each field is denoted, with time increasing from left to right, by successively placed vertical *solid* or *dashed* arrows, each representing one of the s steps of the field–matter interaction and each connecting a pair of horizontal lines (variously solid or dashed). The length of a given vertical arrow is proportional to the frequency of the interacting field at that step. A solid arrow indicates a change in the ket condition and a dashed arrow indicates a change in the bra condition. The absorptive component ($+\omega_i$; $\xi_i = +1$) of the ith incident field is represented as upward for the solid arrow and as downward for the dashed arrow. For the emissive component ($-\omega_i$; $\xi_i = -1$), the solid arrow becomes downward and the dashed arrow upward.

Each type of arrow (solid or dashed) in its first appearance in the time sequence must begin from the initial state, usually (though not necessarily) taken as the ground state, g. In a given diagram, upon tracing the subset of all solid arrows during the subsequent time-evolution, one must find head-to-tail continuity. The same is true for all dashed arrows. Whatever the specific nature of a given time-evolution diagram, the induced polarization of the chromophore in an sth order process is entirely determined by the bra ($\langle m|$) reached at the head of the last appearing (most recent) dashed arrow and by the ket ($|n\rangle$) at the head of the last appearing solid arrow in the diagram (necessarily, one of these must appear at the sth step). The corresponding sth order polarized state of the chromophore is represented by a *wavy* line at the end of the time sequence in the diagram to indicate the off-diagonal (in general) coherence that is set up at sth order. The length of the wavy line must be proportional to the frequency of the signal field, $\omega_j = \omega_p \equiv \xi_1\omega_1 + \xi_2\omega_2 + \cdots + \xi_s\omega_s$. The time evolution of individual steps in an sth order process using the energy ladder diagram is entirely equivalent to that displayed in the double Feynman diagrams with the added advantage that resonance conditions are clearly displayed.

By way of illustration, one of the diagrams for the third order electrical susceptibility $(s = 3)$, $\chi^{(3)}(\omega_4 = \omega_1 - \omega_2 + \omega_3)$, is depicted in Fig. 1. Here, the chromophore is initially in its ground state (both bra and ket in state g). After interacting first with the field at ω_1, the ket state becomes a virtual ket state, $|m\rangle$, which is in turn changed to a real ket state, $|f\rangle$, by the interaction with the field at ω_2. At the third order, the field at ω_3 interacts with the bra state, still in its ground state, $\langle g|$, taking it into the virtual bra state, $\langle n|$. Thus in this diagram, after interacting with the fields ω_1, ω_2, and ω_3 at the third order, the chromophore is left polarized across the real ket state, $|f\rangle$, and the virtual bra state, $\langle n|$. This final polarized state of the chromophore is represented by the wavy line between $|f\rangle$ and $\langle n|$. It might appear from this diagram that the bra state of the chromophore should collapse to the real state, $\langle f|$, after interacting with the signal field at $\omega_4 = \omega_1 - \omega_2 + \omega_3$, indicating an apparent overall change of population of the chromophore into state f. In this diagram, the resonance at $\omega_1 - \omega_2 = \omega_{fg}^0$ is clearly evident. However, as will be seen shortly, the outcome of such a polarization event depends on the difference between the absolute phase of the polarization and the absolute phase of the signal field that is generated from it, or is already present in the medium. It turns out that the discharge of the polarization might, alternatively, leave the chromophore unexcited in its ground state, g. The diagram is at best ambiguous as to the outcome.

In general when s different Fourier components of the field interact with the chromophore, and there is no laboratory ordering of their timing (they are all CW), there are $s!$ distinctive ways of arranging in time the s arrows of different length. Furthermore, since each arrow can be either solid or dashed, there are altogether $2^s s!$ time-evolution diagrams whose "sum" or "superposition" represents $\rho_{mn}^{(s)}(\mathbf{r}; \omega_p)$. To obtain the quantum-mechanical expression for the contribution to $\rho_{mn}^{(s)}(\mathbf{r}; \omega_p)$ from the process described by a given time-evolution diagram, each solid or dashed arrow must be identified with a transition moment and an energy denominator. Whenever a given arrow head touches a solid horizontal line, a material

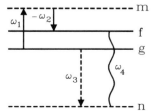

Figure 1. An example of time-evolution diagrams using the energy ladder diagram.

resonance is possible depending on the previous history. However, whenever an arrow head reaches a dashed horizontal line, a virtual transition at that step is indicated regardless of the previous history. Each virtual state, represented by a dashed horizontal line, must involve a sum over the complete set of dark eigenstates. On the other hand, the sum can be limited as usual to just the contribution from the two resonant states which provide a pole that dominates the sum.

Similar rules can be applied to find the expression for the sth order electrical susceptibility. Here, the wavy line in a diagram is identified with the transition moment exposed in Eq. (4.5). Each of $2^s s!$ time-evolution diagrams corresponds to a single term in the susceptibility which contains $(s + 1)$ transition moments and s energy denominators. Such rules can be found in [6] and are briefly summarized here in Appendix B.

The diagrammatically obtained electrical susceptibility expressions can then be used to find the microscopic expression for $W_F^{(s)}(\mathbf{r})$ as given through Eq. (3.6).

B. Rate of Population Change and the Expression for $W_M^{(s)}$

The diagrammatic technique explained above is also useful for finding the quantum-mechanical expression for the cycle-averaged rate of population change in the chromophore as a result of nonlinear light–matter interaction. The population of a given chromophore eigenstate is represented by its diagonal density matrix element.

Whatever the specific nature of the light–matter interaction, the state of polarization of the chromophore after s interactions with Fourier components of the radiation is simply that found at the sth order, as represented by the wavy line in time-evolution diagrams. Quite generally, after s field-chromophore interactions, the state of polarization falls naturally into three distinct classes. The chromophore may be polarized across two virtual states, across a real and a virtual state, or across two real states. That is, the m and n in $\rho_{mn}^{(s)}(\mathbf{r}, t)$ may be (A) both virtual, (B) one real and one virtual, and (C) both real states. To maintain this generality, we indicate in Fig. 2 the outcome of each of these "most recent" appearing events by partial solid and dashed arrows for any sth order process. The three classes of polarization are evident: Class A (Fig. 2(A)), Class B (Figs. 2(B$_1$) and 2(B$_2$)), and Class C (Fig. 2(C)).

For a Class A polarization, the chromophore will radiate a signal field at ω_p, and both bra and ket states return to the ground state. However, when either or both of the final bra and ket states are real (Classes B and C), there can be two outcomes. The chromophore may either radiate a signal field at ω_p or interact with a signal field already present at ω_p. In the latter case, the chromophore may be left populated in an eigenstate

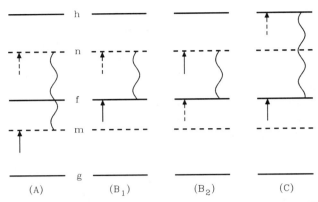

Figure 2. Possible polarized states after s interactions with the fields. The short solid and dashed arrows represent the most recent interactions of the fields with the ket and the bra states, respectively, in the evolution at sth order. Here, the relative time-ordering of the short solid and dashed arrows is entirely ignored. The wavy line emphasizes the polarized state of the chromophore. The solid horizontal lines, g, f, and h, represent real molecular states, while the dashed horizontal lines, m and n, represent nonresonant virtual states.

different from the initial one. The diagrams are ambiguous as to the outcome. For example, if upon discharging the sth order polarization the ket state $|f\rangle$ in Fig. 2(B$_1$) were to return to the initial ground state $|g\rangle$ by radiating a signal field at ω_p, the population of the chromophore in the initial ground state, $\rho_{gg}^{(s+1)}(\mathbf{r}, t)$, would remain unchanged. Alternatively, if the polarization were driven to collapse in the presence of an appropriate signal field at ω_p, the bra state is changed to $\langle f|$, thus causing a real change in the population of state f, $\rho_{ff}^{(s+1)}(\mathbf{r}, t) \neq 0$. After collapsing, the polarized state in Fig. 2(C) may leave the chromophore either in excited states, f or h, or in the ground state, g. The actual fate of polarized states of Classes B and C depends on the difference, $\Delta\phi_j(\mathbf{r})$, between the absolute phase of the polarization and the absolute phase of the signal field present at position $\mathbf{r}$ in the medium, as will be discussed shortly in more detail.

The cycle-averaged rate of population change (per unit volume) of a state f, due to the interaction with s incident fields, $\{\omega_1, \omega_2, \ldots, \omega_s\}$, and a signal field at $\omega_j = \omega_p \equiv \xi_1\omega_1 + \xi_2\omega_2 + \cdots + \xi_s\omega_s$, can be defined as

$$\Gamma_f^{(s)}(\mathbf{r}) = (N/T) \int_0^T \left\{ -(1/i\hbar) \sum_n [\langle f|\vec{\mu} \cdot \mathbf{E}(\mathbf{r}, t)|n\rangle \rho_{nf}^{(s)}(\mathbf{r}, t) \right.$$

$$\left. - \rho_{fn}^{(s)}(\mathbf{r}, t)\langle n|\vec{\mu} \cdot \mathbf{E}(\mathbf{r}, t)|f\rangle] \right\} dt \qquad (4.6)$$

Here, $\mathbf{E}(\mathbf{r}, t)$ is the total field present in the medium including the signal and all incident fields. Then, Eqs. (2.3) and (4.3) can be used to reduce the cycle-average integral in Eq. (4.6) to

$$
\Gamma_f^{(s)}(\mathbf{r}) = (1/2\hbar) \sum_l \tilde{\mathscr{E}}_l(\mathbf{r}) \, \mathrm{Im} \Bigg\{ N \sum_n [\langle f | \vec{\mu} \cdot \hat{e}_l | n \rangle \rho_{nf}^{(s)}(\mathbf{r}; \omega_l)
$$

$$
- \rho_{fn}^{(s)}(\mathbf{r}; \omega_l) \langle n | \vec{\mu} \cdot \hat{e}_l | f \rangle] \exp[-i\{\Delta \mathbf{k}_l' \cdot \mathbf{r} + \Delta \phi_l(\mathbf{r})\}] \Bigg\}
$$

$$(4.7)$$

where the summation on l is over all fields present in the medium, the signal field ($l \equiv j$) and the incident fields ($l \equiv i$, $1 \le i \le s$). The first term represents the time evolution in which, after s interactions with the fields, the ket state of the chromophore becomes $|f\rangle$ at the sth step, while the bra state is in a virtual state $\langle n|$, as shown in Fig. 2(B_1). Then at the $(s + 1)$th step, the chromophore interacts with the field at ω_l to change the bra state to $\langle f|$, causing a population change in f, ($\rho_{ff}^{(s+1)}(\mathbf{r}, \omega_l) \neq 0$). For the second term, on the other hand, the bra state reaches $\langle f|$ at the sth step as shown in Fig. 2(B_2) and at the $(s + 1)$th step the field at ω_l brings the ket state to $|f\rangle$. It is clear that only those time-orderings in which either the ket or bra state reaches the state f at the sth order (the last step before the wavy line) as shown in Figs. 2(B_1), (B_2), and (C), can contribute to $\Gamma_f^{(s)}(\mathbf{r})$.

By comparing Eq. (4.7) with Eq. (4.5), it can be readily recognized that

$$
\Gamma_f^{(s)}(\mathbf{r}) = (1/2\hbar) \tilde{K}(\mathbf{r}) \sum_{\mathcal{Q}} \sum_l \mathrm{Im} \Big\{ [\zeta_{l,\mathrm{I}} - \zeta_{l,\mathrm{II}}] \exp[-i\{\Delta \mathbf{k}_l' \cdot \mathbf{r} + \Delta \phi_l(\mathbf{r})\}] \Big\}
$$

$$
= (1/2\hbar) \tilde{K}(\mathbf{r}) \sum_{\mathcal{Q}} \sum_l \Big\{ [\zeta_{l,\mathrm{I}}'' - \zeta_{l,\mathrm{II}}''] \cos\{\Delta \mathbf{k}_l' \cdot \mathbf{r} + \Delta \phi_l(\mathbf{r})\}
$$

$$
- [\zeta_{l,\mathrm{I}}' - \zeta_{l,\mathrm{II}}'] \sin\{\Delta \mathbf{k}_l' \cdot \mathbf{r} + \Delta \phi_l(\mathbf{r})\} \Big\}
$$

$$(4.8)$$

where $\mathcal{Q}$ represents all possible time-orderings that can contribute to $\Gamma_f^{(s)}(\mathbf{r})$, and, $\zeta_{l,\mathrm{I}}$ is a term in the electrical susceptibility $\chi^{(s)}(\omega_l)$, expressed as a scalar Cartesian component given in Eq. (2.17), corresponding to a diagram in which the ket state becomes $|f\rangle$ after s interactions with fields, as shown in Fig. 2(B_1). (We shall call this a "Type I" term.) Likewise $\zeta_{l,\mathrm{II}}$ is a term ("Type II") corresponding to a diagram in which the bra state becomes $\langle f|$ after s interactions as shown in Fig. 2(B_2). The diagram in Fig. 2(C) belongs to Type I for the resonant state f and to Type II for the resonant state h. Apparently, $\zeta_{l,\mathrm{I}}$ and $\zeta_{l,\mathrm{II}}$ can be obtained by the same diagrammatic rules used to find the electrical susceptibility expressions.

Equation (4.8) provides additional insight into the wavy line used in the time-evolution diagrams. As defined above, the wavy line represents the polarization induced in the medium by s interactions with the incident fields. This polarization must oscillate in time with frequency $\omega_p = \xi_1\omega_1 + \xi_2\omega_2 + \cdots + \xi_s\omega_s$ and carry the absolute phase factor $\phi_p = \xi_1\phi_1 + \xi_2\phi_2 + \cdots + \xi_s\phi_s$. When both bra and ket states are in virtual states (Class A) as in Fig. 2(A), there can be no cycle-averaged population change in any of the chromophore states; the virtual bra and ket states must return to their initial ground state by radiating a field at ω_p. However, when after s field-chromophore interactions one or both of the bra and ket states is in a real chromophore state (Classes B and C) as in Figs. 2(B_1), (B_2), and (C), the ambiguity of the wavy line suggests two possibilities in general. Suppose ζ_l is pure imaginary ($\zeta_l' = 0$). Then, according to Eq. (4.8), there can be real population change if there is a field at $\omega_l = \omega_p$ such that $\Delta\phi_l \equiv \phi_l - \phi_p \neq \pm\frac{1}{2}\pi$ (not out-of-phase). That is, any in-phase component of the field at ω_l present in the medium will change the bra state from a virtual state $\langle n|$ to the real state $\langle f|$ in Figs. 2(B_1) and (C), resulting in population change in the chromophore. On the other hand, the creation of purely out-of-phase component of the field at ω_l from the collapse of the induced polarization is accompanied by a return of the ket state in $|f\rangle$ to the ground state $|g\rangle$ (see Section II.B) and the chromophore suffers no population change. The condition of $\omega_j = \omega_p$ ($\equiv \xi_1\omega_1 + \xi_2\omega_2 + \cdots + \xi_s\omega_s$) is necessary for both cases.

A change in the population of chromophore states is usually regarded as requiring a *quadrature* form of the interaction with the fields, as in active processes. Such a transition is *quadratic* in the incident field amplitudes since the rate of transition is proportional to the intensities of the incident fields. However, Eq. (4.8) indicates that an appropriate nonquadrature form of interaction can also change the population. The signal field in the medium sees only the status of the polarization induced in the medium; the details of how the polarization has been created are not important, other than assigning to it an absolute phase.

The cycle-averaged rate of energy change in unit volume of the medium, $W_M^{(s)}(\mathbf{r})$, can be obtained from Eq. (4.8) simply by multiplying by $\hbar\omega_{fg}$,

$$W_M^{(s)}(\mathbf{r}) = \sum_f \hbar\omega_{fg}\Gamma_f^{(s)}(\mathbf{r}) \tag{4.9}$$

It is noted that Eq. (4.9) is valid even when the chromophore is initially not in its ground state, g, since the trace of the density matrix is conserved; thus

$$\Gamma_g^{(s)}(\mathbf{r}) = -\sum_{f \neq g} \Gamma_f^{(s)}(\mathbf{r}) \tag{4.10}$$

We have shown here that the quantum-mechanical expressions for both $W_F^{(s)}(\mathbf{r})$ and $W_M^{(s)}(\mathbf{r})$ can be found from the time-evolution diagrams for nonlinear light–matter interactions. It is also shown that a population change among the chromophore eigenstates does not always require immediate quadrature form of the interactions. Population changes can take place prior to any quadrature detection of photons.

For global energy conservation for the composite field/matter system, any change in the total energy of the fields must accompany the corresponding change in the total energy of the medium. In other words, the condition for global energy conservation requires that

$$W_F^{(s)}(\mathbf{r}) = W_M^{(s)}(\mathbf{r}) \tag{4.11}$$

For *nonresonant passive processes*, both $W_F^{(s)}(\mathbf{r})$ and $W_M^{(s)}(\mathbf{r})$ vanish separately and global energy conservation in Eq. (4.11) is trivial, $W_F^{(s)}(\mathbf{r}) = W_M^{(s)}(\mathbf{r}) = 0$. However, for *resonant processes*, *active* and *passive*, neither $W_F^{(s)}(\mathbf{r})$ nor $W_M^{(s)}(\mathbf{r})$ vanishes in general and Eq. (4.11) must be confirmed by examining the quantum-mechanical expressions for $W_F^{(s)}(\mathbf{r})$ and $W_M^{(s)}(\mathbf{r})$ as is done in the next section for processes up to third order.

V. ENERGY CONSERVATION FOR RESONANT PROCESSES

The global energy conservation condition, Eq. (4.11), is explicitly demonstrated for resonant processes up to third order ($s \leq 3$), particularly for resonant passive processes, at exact resonance, where population change can be achieved at a nonquadrature level by the fields. The phase matching condition is assumed, $\Delta \mathbf{k}_j' = 0$. As before, the material resonance at the one-photon level is taken into account by the complex wave vector, $\mathbf{k}_i = \mathbf{k}_i' + i\mathbf{k}_i''$, whose imaginary part is absorbed into the amplitude, $\tilde{\mathscr{E}}_i(\mathbf{r}) = \mathscr{E}_i(\mathbf{r}) \exp[-\mathbf{k}_i'' \cdot \mathbf{r}]$. The electrical susceptibility is expressed in terms of the scalar Cartesian component, ζ_j, as given in Eq. (2.17).

A. First Order Processes

The first order light–matter interaction described by the electrical susceptibility, $\chi^{(1)}(\omega_1 = \omega_1)$, may be viewed as quadrature in its simplest form. The signal field at the first order oscillates at the same frequency as the incident field and the phase matching condition is trivial. As shown in Section II, the signal field at this lowest order is always in-phase with the induced polarization, $\Delta\phi_1(\mathbf{r}) = 0$, regardless of the presence or absence of the material resonance.

In the absence of resonance, $\omega_1 \neq \omega_{fg}^0$, there can be no net energy transfer between the field and the medium. The first order nonresonant interaction, often known as *linear dispersion*, determines the characteristics of linear light propagation in the medium through Eq. (2.9b). This can be regarded as the first order analog of the higher order contribution to dispersion corresponding to passive process appearing only at odd orders in light–matter interactions that are maximally in quadrature at the polarization level.

When a material resonance is present, $\omega_1 = \omega_{fg}^0$, as shown in the diagram in Fig. 3 for example, there can be cycle-averaged energy transfer between the medium and the field. The induced polarization is of Class C in Fig. 2, in which the induced polarization is set up between two real states, the excited ket state, $|f\rangle$, and the ground bra state, $\langle g|$. Figure 3 can represent one-photon absorption, the most common *active process*. For this case, Eq. (3.2) becomes, with Eq. (3.9),

$$W_F^{(1)}(\mathbf{r}) = \tfrac{1}{2}\omega_1 \tilde{\mathscr{E}}_1(\mathbf{r})^2 \zeta_1'' \tag{5.1a}$$

At the same time, since the ket state is in the real state, $|f\rangle$, after the first interaction (before the wavy line), the diagram in Fig. 3 belongs to Type I of Eq. (4.8). Then, Eq. (4.9) becomes

$$W_M^{(1)}(\mathbf{r}) = \tfrac{1}{2}\omega_{fg}^0 \tilde{\mathscr{E}}_1(\mathbf{r})^2 \zeta_1'' \tag{5.1b}$$

since $\Delta\phi_1 = 0$. Energy conservation for the combined system of the medium and light (Eq. (4.11)) is evident from Eqs. (5.1a) and (5.1b). At exact resonance, $\omega_{fg}^0 = \omega_1$ and $W_F^{(1)}(\mathbf{r}) = W_M^{(1)}(\mathbf{r})$. In other words, the rate of energy lost by the field is the same as the rate of energy gained by the medium as indicated by a population change among its eigenstates.

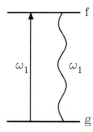

Figure 3. First order active process corresponding to one-photon absorption with the resonance condition, $\omega_{fg}^0 = \omega_1$.

Where the medium is initially in an excited eigenstate, the resonant active process in this two-level problem is one-photon emission, or fluorescence. In this case, the resonant state would lie below the initial state. The field gains energy through the interaction while the medium loses energy.

B. Second Order Processes

At second order in optical fields ($s = 2$), only nonquadrature *passive processes*, harmonic generation and parametric three-wave mixing, are possible. In fully nondegenerate second order processes, three frequency components of the field at ω_1, ω_2, and ω_3 are present in the medium. Without loss of generality, we can take $\xi_1 = \xi_2 = 1$ in $\omega_1 = \xi_1 \omega_2 + \xi_2 \omega_3$ and $\omega_2 > \omega_3$. For a given field acting as a signal field, ω_j ($j = 1$, 2, or 3), the remaining two fields play the role of the incident fields that induce the polarization at ω_p ($= \omega_j$). For example, if $\omega_p = \omega_2 + \omega_3$, then the nondegenerate second order process corresponds to a sum frequency generation in which the fields at ω_2 and ω_3 induce the second order polarization at $\omega_1 = \omega_2 + \omega_3$. If ω_j is chosen as either $\omega_2 = \omega_1 - \omega_3$ ($\xi_1 = -\xi_3 = 1$) or $\omega_3 = \omega_1 - \omega_2$ ($\xi_1 = -\xi_2 = 1$), then the second order process corresponds to difference frequency generation or parametric down-conversion. According to Eq. (2.14b), $\Delta\phi_1(\mathbf{r}) = -\Delta\phi_2(\mathbf{r}) = -\Delta\phi_3(\mathbf{r})$. The second order induced polarization produced in the medium must contain all three frequency components and the action of all three fields must be considered when exploring global energy conservation.

For each signal field, ω_j ($j = 1$, 2, or 3), there are 8 ($= 2^s \, s!$ with $s = 2$) time-evolution diagrams. Figure 4 shows eight such diagrams for $\omega_j = \omega_1$ (and $\xi_1 = \xi_2 = 1$). Here, only the initial ground state is represented by the solid horizontal line and the virtual states are suppressed for the sake of clarity in the diagrams. If there are material resonances, the solid horizontal lines must be inserted at appropriate positions in the diagrams to represent the real eigenstates. For each of the diagrams in Fig. 4, the corresponding microscopic electrical susceptibility expression can be found by the diagrammatic procedure summarized in Appendix B. These expressions for the second order electrical susceptibility in the Cartesian component form (Eq. (2.17)) are listed in Table I where the sums over intermediate virtual states m and n, are implicit.

The time-evolution diagrams for the signal field at ω_1 in Fig. 4 are designated by a permutation notation, $\mathfrak{Q}_h$, where $\mathfrak{Q}$ takes on a lower case letter (a, b, c, ...). The numeric subscript h, which identifies the relative time orderings of the interactions in the ket and bra spaces, can be either suppressed or take on the value 1 or 2 as explained below. Here, $\mathfrak{Q}$ ($= a$, b, c, d, e, f) identifies the particular product of the three transition moments in the expression of $\chi^{(2)}(\omega_1)$ for a given diagram as

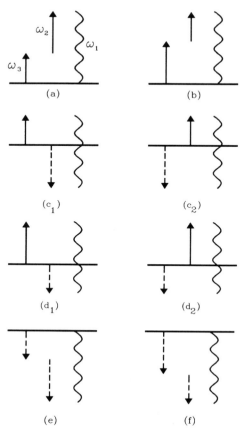

Figure 4. Second order time-evolution diagrams for $\omega_1 = \omega_2 + \omega_3$ ($\omega_2 > \omega_3$). The virtual states are suppressed and the corresponding susceptibilities are listed in the third column of Table I headed $\zeta_1(\mathfrak{Q}_h)$, $\mathfrak{Q}_h = a, b, c_1, c_2, d_1, d_2, e, f$.

listed in Table I. Since there are three different transition moments in $\chi^{(2)}(\omega_1)$, there must be six (=3!) distinct products of the three transition moments, each product represented by $\mathfrak{Q}$. Each transition moment in $\chi^{(2)}(\omega_1)$ indicates which two real and/or virtual states are connected by an arrow or wavy line in a diagram. Thus, $\mathfrak{Q}$ also identifies the vertical locations of the arrows and the wavy line in a diagram. When two arrows are of the same kind, both solid or both dashed, in a diagram (for example, $\mathfrak{Q} = a, b, e,$ and f), the time-ordering of the interaction is uniquely defined because the first arrow must begin from the ground

TABLE I
Second Order Electrical Susceptibilities[a]

Ω_h[b]	$(2\hbar^2/N)D_{\mathfrak{L}}$[c]	$\zeta_1(\Omega_h)$[d]	$\zeta_2(\Omega_h)$[e]	$\zeta_3(\Omega_h)$[e]
a_1		$D_a/(\omega_{mg}-\omega_3)(\omega_{ng}-\omega_1)$	$D_a/(\omega_{ng}-\omega_1)(\omega_{mn}+\omega_2)$	$D_a/(\omega_{ng}-\omega_1)(\omega_{mg}-\omega_3)$
a	$\langle g|\mu_1|n\rangle\langle n|\mu_2|m\rangle\langle m|\mu_3|g\rangle$			
a_2			$D_a/(\omega_{mg}^*-\omega_3)(\omega_{nm}-\omega_2)$	$D_a/(\omega_{ng}^*+\omega_3)(\omega_{mg}-\omega_3)$
b_1		$D_b/(\omega_{mg}-\omega_2)(\omega_{ng}-\omega_1)$	$D_b/(\omega_{ng}-\omega_1)(\omega_{mg}-\omega_2)$	$D_b/(\omega_{ng}-\omega_1)(\omega_{mn}^*+\omega_3)$
b	$\langle g|\mu_1|n\rangle\langle n|\mu_3|m\rangle\langle m|\mu_2|g\rangle$			
b_2				$D_b/(\omega_{mg}^*-\omega_2)(\omega_{nm}-\omega_3)$
c_1		$D_c/(\omega_{mg}-\omega_3)(\omega_{nm}^*+\omega_1)$	$D_c/(\omega_{mg}^*-\omega_3)(\omega_{ng}+\omega_2)$	$D_c/(\omega_{ng}+\omega_2)(\omega_{mg}-\omega_3)$
c	$\langle g|\mu_2|n\rangle\langle n|\mu_1|m\rangle\langle m|\mu_3|g\rangle$			
c_2		$D_c/(\omega_{ng}^*+\omega_2)(\omega_{mn}-\omega_1)$		
d_1		$D_d/(\omega_{mg}-\omega_2)(\omega_{nm}^*+\omega_1)$	$D_d/(\omega_{ng}+\omega_3)(\omega_{mg}-\omega_2)$	$D_d/(\omega_{mg}^*-\omega_2)(\omega_{ng}^*+\omega_3)$
d	$\langle g|\mu_3|n\rangle\langle n|\mu_1|m\rangle\langle m|\mu_2|g\rangle$			
d_2		$D_d/(\omega_{ng}^*+\omega_3)(\omega_{mn}-\omega_1)$		
e_1		$D_e/(\omega_{ng}^*+\omega_3)(\omega_{mg}^*+\omega_1)$	$D_e/(\omega_{ng}+\omega_3)(\omega_{mn}+\omega_2)$	$D_e/(\omega_{mg}^*+\omega_1)(\omega_{ng}^*+\omega_3)$
e	$\langle g|\mu_3|n\rangle\langle n|\mu_2|m\rangle\langle m|\mu_1|g\rangle$			
e_2			$D_e/(\omega_{mg}^*+\omega_1)(\omega_{nm}-\omega_2)$	
f_1		$D_f/(\omega_{ng}^*+\omega_2)(\omega_{mg}^*+\omega_1)$	$D_f/(\omega_{mg}^*+\omega_1)(\omega_{ng}+\omega_2)$	$D_f/(\omega_{ng}+\omega_2)(\omega_{mn}^*+\omega_3)$
f	$\langle g|\mu_2|n\rangle\langle n|\mu_3|m\rangle\langle m|\mu_1|g\rangle$			
f_2				$D_f/(\omega_{mg}^*+\omega_1)(\omega_{nm}-\omega_3)$

[a] $\omega_{ab} = \omega_{ab}^0 - i\gamma_{ab}$. The susceptibilities are expressed in the form of Eq. (2.17).

[b] See the text for the convention used for Ω_h.

[c] $\mu_i = \vec{\mu} \cdot \hat{e}_i$.

[d] The corresponding time-evolution diagrams are given in Fig. 4 with the same P.

[e] The time-evolution diagrams, not shown in this paper, can be readily deduced by the diagrammatic rule in Appendix B.

state, $|g\rangle$ for a solid arrow and $\langle g|$ for a dashed arrow. Since there is only one possible time-ordering for such interactions, the subscript h in $\mathfrak{Q}_h$ can be entirely suppressed for simplicity. On the other hand, there are two possible time-orderings of the interactions when the two arrows are of different kinds, one solid and one dashed, in a diagram (for example $\mathfrak{Q} = c$ and d). Here, both arrows must begin from the ground state, g, and the wavy line connects the two states at the heads of the two arrows. The time-orderings of the arrows in such *mixed interaction diagrams* are identified by the numerical subscript h in $\mathfrak{Q}_h$; $h = 1$ if the solid arrow appears first and $h = 2$ if the dashed arrow appears first. The electrical susceptibilities for such diagrams evidently share the same product of the transition moments. The mixed interaction diagrams (Figs. 4(c_1) and (c_2), and Figs. 4(d_1) and (d_2)) are responsible for dephasing or pressure induced extra resonances [24, 25].

In Table I, the second order electrical susceptibilities for the polarizations at ω_2 and ω_3 are also listed. The corresponding time-evolution diagrams can be easily deduced from the susceptibility expressions (or *vice versa*). The diagrams for polarizations at ω_2 and ω_3 can also be classified according to the permutation notation, $\mathfrak{Q}_h$ or $\mathfrak{Q}$, in the same fashion explained above for the diagrams in Fig. 4 in which the polarization is at ω_1.

When there is no material resonance, the imaginary part of the complex frequency, $\omega_{ab} = \omega_{ab}^0 - i\gamma_{ab}$, can be entirely ignored. Then, the sum of two susceptibility terms corresponding to the mixed interaction diagrams, $\zeta_1'(c_1) + \zeta_1'(c_2)$, for example, can be seen as identical to $\zeta_2'(c)$ and $\zeta_3'(c)$ since $\omega_{ng}^0 - \omega_{mg}^0 = \omega_{nm}^0$ and $\omega_1 = \omega_2 + \omega_3$. Thus, the permutation symmetry for the electrical susceptibilities in the absence of material resonance as given by Eq. (3.7) can be readily seen from Table I. As shown in Section II, in such a case the total energy of the fields and the medium is separately conserved ($W_F^{(2)}(\mathbf{r}) = W_M^{(2)}(\mathbf{r}) = 0$). The signal field is always out-of-phase ($\Delta\phi_j(\mathbf{r}) = \pm\frac{1}{2}\pi$) with respect to the induced polarization as given in Eq. (2.21). The signal field acquires energy only from the incident fields; the medium plays a purely catalytic role.

For resonant processes, resonant excited states must be added to the diagrams in Fig. 4 by use of solid horizontal lines. When any of the arrow heads in the diagrams touches a solid horizontal line, the corresponding electrical susceptibility term may contain the resonant energy denominators and display resonance enhancement as the resonance condition is approached. For example, if there is an excited state such that $\omega_{mg}^0 = \omega_3$, the diagram in Fig. 4(c_1) becomes resonant with the energy denominator $\omega_{mg}^0 - \omega_3$. The diagram in Fig. 4(c_2), however, remains nonresonant (See Table I). The diagrams that are resonant under various resonance conditions are identified in Table II. In general for resonant processes, only the dominant, resonant, electrical susceptibility terms need to be considered and the nonresonant susceptibility terms are often either entirely neglected by the rotating wave approximation [3, 6, 10] or are taken as a part of the total nonresonant background contribution of the medium.

As shown in Section IV, there can be population change among the resonant eigenstates through the nonquadrature passive processes. The diagrams contributing to $W_M^{(2)}(\mathbf{r})$ in Eq. (4.9) can be identified from Fig. 4 and they are classified in Table II according to the Types of Eq. (4.8).

Table II can be utilized to confirm global energy conservation for the composite field/matter system for various resonant three-wave mixing processes when the resonance condition is exactly satisfied. Thus, for singly resonant sum frequency generation with $\omega_{ng}^0 = \omega_1$, all ζ_j's are pure imaginary when the resonance condition is exactly satisfied. Thus, according to Eqs. (3.6) and (4.9),

$$W_F^{(2)}(\mathbf{r}) = \tfrac{1}{2}\tilde{K}(\mathbf{r})(\omega_1 + \omega_2 + \omega_3)\{\zeta_1''(a) + \zeta_1''(b)\} \cos \Delta\phi_1(\mathbf{r})$$

(5.2a)

and

$$W_M^{(2)}(\mathbf{r}) = \tfrac{1}{2}\tilde{K}(\mathbf{r})2\omega_{ng}^0\{\zeta_1''(a) + \zeta_1''(b)\} \cos \Delta\phi_1(\mathbf{r})$$ (5.2b)

since $\zeta_1''(a) = \zeta_2''(a_1) = \zeta_3''(a)$ and $\zeta_1''(b) = \zeta_2''(b) = \zeta_3''(b_1)$ which are clear from Table I and the auxiliary conditions in Table II. Here, $\tilde{K}(\mathbf{r}) = \mathscr{E}_1(\mathbf{r})\mathscr{E}_2(\mathbf{r})\mathscr{E}_3(\mathbf{r})$. Equation (4.11) is clearly satisfied when $\omega_1 = \omega_2 + \omega_3 = \omega_{ng}^0$, and only the in-phase component, $\Delta\phi_1(\mathbf{r}) = 0$, of the signal field contributes to the cycle-averaged energy exchange as discussed above.

For singly resonant difference frequency generation with $\omega_{mg}^0 = \omega_2$, for example, $\zeta_1''(b) = \zeta_2''(b) = -\zeta_3''(b_2)$ and $\zeta_1''(e) = \zeta_2''(e_1) = -\zeta_3''(e)$ at exact resonance. The diagram for $\zeta_3(e)$ is Type II for $W_M^{(2)}(\mathbf{r})$. The expressions for $W_F^{(2)}(\mathbf{r})$ and $W_M^{(2)}(\mathbf{r})$ can be derived and they are indeed equivalent at exact resonance, $\omega_{mg}^0 = \omega_2$,

$$W_M^{(2)}(\mathbf{r}) = W_F^{(2)}(\mathbf{r}) = \tilde{K}(\mathbf{r})\omega_{mg}^0\{\chi_1''(b) + \chi_1''(e)\} \cos \Delta\phi_1(\mathbf{r})$$ (5.3)

with $\tilde{K}(\mathbf{r}) = \mathscr{E}_1(\mathbf{r})\tilde{\mathscr{E}}_2(\mathbf{r})\mathscr{E}_3(\mathbf{r})$. A similar result can be obtained for $\omega_{mg}^0 = \omega_3$.

For doubly resonant second order passive processes, for example, $\omega_{mg}^0 = \omega_1$ and $\omega_{mg}^0 = \omega_3$, all susceptibilities are now pure real at exact resonance. Then, only the out-of-phase component $(\Delta\phi_j(\mathbf{r}) = \pm\tfrac{1}{2}\pi)$ of the signal field contributes to the energy exchange. From Table I,

TABLE II

Second Order Time-Evolution Diagrams Contributing to $W_F^{(2)}(\mathbf{r})$ and $W_M^{(2)}(\mathbf{r})$

Resonance Conditions	Diagrams for $W_F^{(2)}(\mathbf{r})$			Diagrams for $W_M^{(2)}(\mathbf{r})$[a]			Auxiliary Conditions[b]
	ω_1	ω_2	ω_3	ω_1	ω_2	ω_3	
$\omega_{ng}^0 = \omega_1$	a, b	a_1, b	a, b_1	$a[A], b[A]$	$a_1[A]$	$b_1[A]$	$\omega_{mg}^0 - \omega_3 = \omega_{mn}^0 + \omega_2$
$\omega_{mg}^0 = \omega_2$	b, d_1	b, d	b_2, d	$d_1[A]$	$b[A], d[A]$	$b_2[B]$	$\omega_{mg}^0 - \omega_1 = \omega_{nm}^0 - \omega_3$
$\omega_{mg}^0 = \omega_3$	a, c_1	a_2, c	a, c	$c_1[A]$	$a_2[B]$	$a[A], c[C]$	$\omega_{ng}^0 - \omega_1 = \omega_{nm}^0 - \omega_2$
$\omega_{ng}^0 = \omega_1$	b	b	b_1, b_2	$b[C]$	$b[D]$	$b_1[E], b_2[E]$	$\omega_{nm}^0 = \omega_3$
$\omega_{mg}^0 = \omega_2$							
$\omega_{mg}^0 = \omega_1$	a	a_1, a_2	a	$a[C]$	$a_1[E], a_2[E]$	$a[D]$	$\omega_{nm}^0 = \omega_2$
$\omega_{mg}^0 = \omega_3$							

[a] The type of contribution to Eq. (4.8) is shown in brackets: [A] = Type I; [B] = Type II; [C] = Type I for n; [D] = Type I for m; [E] = Type I for n and Type II for m.

[b] Derived from the resonance condition and $\omega_{ng}^0 = \omega_{nm}^0 + \omega_{mg}^0$ and $\omega_1 = \omega_2 + \omega_3$.

$\zeta'_1(b) = \zeta'_2(b) \neq \zeta'_3(b_1) \neq \zeta'_3(b_2)$ at exact resonance. Again, the equivalence of $W_F^{(2)}(\mathbf{r})$ and $W_M^{(2)}(\mathbf{r})$ at exact resonance can be readily confirmed as

$$W_M^{(2)}(\mathbf{r}) = W_F^{(2)}(\mathbf{r})$$

$$= \tfrac{1}{2}\tilde{K}(\mathbf{r})\omega_{mg}^0\{-\zeta'_1(b) + \zeta'_3(b_1) + \zeta'_3(b_2)\}\sin\Delta\phi_1(\mathbf{r}) \quad (5.4)$$

where the diagrams corresponding to $\zeta'_3(b_1)$ and $\zeta'_3(b_2)$ can change the populations of both resonant states, m and n. Here, $\tilde{K}(\mathbf{r}) = \tilde{\mathscr{E}}_1(\mathbf{r})\tilde{\mathscr{E}}_2(\mathbf{r})\tilde{\mathscr{E}}_3(\mathbf{r})$.

It can be readily seen from Eqs. (5.2), (5.3), and (5.4), that global energy conservation as given in Eq. (4.11) is achieved in all second order (singly or doubly) *resonant passive processes* when the resonance conditions are exactly satisfied.

C. Third Order Processes

At third order ($s = 3$), both *active* and *passive processes* are possible. Without loss of generality all *active processes* at third order are described by the electrical susceptibility $\chi^{(3)}(\omega_1 = \omega_1 + \omega_2 - \omega_2)$ with a resonance condition between the initial and final states. Here, the light–matter interaction is in quadrature and the energy change through population change in the medium can be readily understood. For example, $\omega_1 + \omega_2 = \omega_{fg}^0$ for two-photon absorption and $|\omega_1 - \omega_2| = \omega_{fg}^0$ for Raman scattering. For resonant secondary radiation (RSR, or fully resonant Raman scattering with $|\omega_1 - \omega_2| = \omega_{fg}^0$ and $\omega_2 = \omega_{mg}^0$), there has been extensive discussion on the distinction between the resonant Raman scattering (RRS) and the hot luminescence (HL) components [26, 27]. The diagrammatic technique has been shown to be useful in identifying the RRS and HL components in RSR.

Among *passive processes* at third order, four-wave mixing (4WM) processes described by susceptibilities like $\chi^{(3)}(\omega_4 = \omega_1 - \omega_2 + \omega_3)$ are often of particular interest. Here, three incident fields at ω_1, ω_2, and ω_3, for example, produce an induced polarization in the medium at $\omega_4 = \omega_1 - \omega_2 + \omega_3$ ($\xi_1 = -\xi_2 = \xi_3 = 1$). Although *passive processes* do not usually require material resonances, the enhanced 4WM signal due to the presence of a Raman-type resonance,

$$\omega_1 - \omega_2 = \omega_4 - \omega_3 = \omega_{fg}^0 \quad (5.5)$$

can be used to probe the resonant state, f, thus making resonant 4WM a useful spectroscopic tool [10]. For fully nondegenerate 4WM processes, we can assume $\omega_1 > \omega_2$ and $\omega_4 > \omega_3$ without loss of generality. Among 48 time-evolution diagrams for each signal field, only the four diagrams shown in Figs. 5 and 6 recognize the Raman-type resonance. (The virtual states are again suppressed in Figs. 5 and 6.) For the reasons evident in the last two steps of the interactions in the diagrams, the resonant 4WM is called CARS when the signal field is ω_4 or ω_1, and it is CSRS when the field at ω_2 or ω_3 is detected. The corresponding susceptibility expressions are listed in Tables III and IV. The diagrams and the corresponding electrical susceptibilities are again classified by the permutation notation

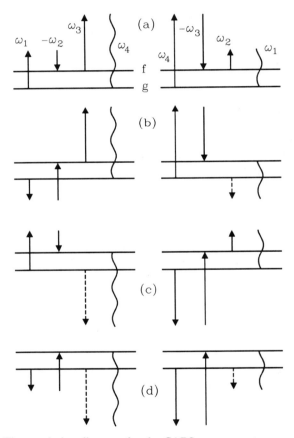

Figure 5. Time-evolution diagrams for the CARS processes at $\omega_4 = \omega_1 - \omega_2 + \omega_3$ and $\omega_1 = \omega_4 - \omega_3 + \omega_2$ ($\omega_4 > \omega_1 > \omega_3 > \omega_2$). The resonant state f, satisfying the Raman-type resonance, $\omega_{fg}^0 = \omega_1 - \omega_2$, is included and the virtual states are suppressed. The corresponding susceptibilities are listed as $\zeta_4(\mathfrak{Q})$ and $\zeta_1(\mathfrak{Q})$, respectively, in Table III, $\mathfrak{Q} = a, b, c, d$.

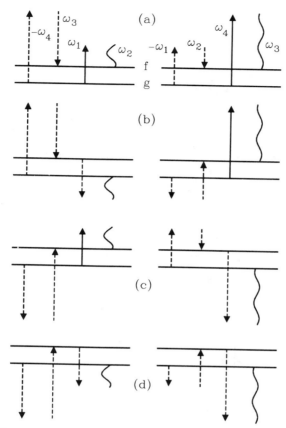

Figure 6. Time-evolution diagrams for the CSRS processes at $\omega_2 = \omega_3 - \omega_4 + \omega_1$ and $\omega_3 = \omega_2 - \omega_1 + \omega_4$ ($\omega_4 > \omega_1 > \omega_3 > \omega_2$), satisfying the Raman-type resonance, $\omega_{fg}^0 = \omega_4 - \omega_1$. The corresponding susceptibilities are listed as $\zeta_2(\Omega)$ and $\zeta_3(\Omega)$, respectively, in Table III, $\Omega = a, b, c, d$.

Ω. (The numerical subscript h in Ω_h is dropped here for simplicity.) As already mentioned, some of the diagrams appear to leave the medium excited, after the interactions. This is the case, for example, in Figs. 5(c) and (d) for the signal detected at ω_4.

In the CARS experiment, the incident fields at ω_1, ω_2, and ω_3 (with wave vectors, $\mathbf{k}_1$, $\mathbf{k}_2$, and $\mathbf{k}_3$), are spatially overlapped in the medium and the signal field at ω_4 is generated from the medium along the direction determined by the phase matching condition, $\mathbf{k}_4' = \mathbf{k}_1' - \mathbf{k}_2' + \mathbf{k}_3'$. According to Eq. (2.15a), the signal field strength, $\mathscr{E}_4(\mathbf{r})$, is determined by the

TABLE III

Third Order Electrical Susceptibilities Responsible for the CARS Diagrams in Fig. 5[a]

ζ	$(4\hbar^3 N)D_\zeta$	$\zeta_4(\Omega)$	$\zeta_1(\Omega)$
a	$\langle g|\mu_4|n\rangle\langle n|\mu_3|f\rangle\langle f|\mu_2|m\rangle\langle m|\mu_1|g\rangle$	$D_a/(\omega_{mg}-\omega_1)(\omega_{fg}-\omega_1+\omega_2)(\omega_{ng}-\omega_4)$	$D_a/(\omega_{ng}-\omega_4)(\omega_{fg}-\omega_4+\omega_3)(\omega_{mg}-\omega_1)$
b	$\langle g|\mu_4|n\rangle\langle n|\mu_3|f\rangle\langle f|\mu_1|m\rangle\langle m|\mu_2|g\rangle$	$D_b/(\omega_{mg}^*+\omega_2)(\omega_{fg}-\omega_1+\omega_2)(\omega_{ng}-\omega_4)$	$D_b/(\omega_{ng}-\omega_4)(\omega_{fg}-\omega_4+\omega_3)(\omega_{mf}^*+\omega_1)$
c	$\langle g|\mu_3|n\rangle\langle n|\mu_4|f\rangle\langle f|\mu_2|m\rangle\langle m|\mu_1|g\rangle$	$D_c/(\omega_{mg}-\omega_1)(\omega_{fg}-\omega_1+\omega_2)(\omega_{nf}^*+\omega_4)$	$D_c/(\omega_{ng}-\omega_4)(\omega_{fg}-\omega_4+\omega_3)(\omega_{mg}-\omega_1)$
d	$\langle g|\mu_3|n\rangle\langle n|\mu_4|f\rangle\langle f|\mu_1|m\rangle\langle m|\mu_2|g\rangle$	$D_d/(\omega_{mg}^*+\omega_2)(\omega_{fg}-\omega_1+\omega_2)(\omega_{nf}^*+\omega_4)$	$D_d/(\omega_{ng}-\omega_4)(\omega_{fg}-\omega_4+\omega_3)(\omega_{mf}^*+\omega_1)$

$^a\omega_{ab}=\omega_{ab}^0+i\gamma_{ab}$. The susceptibilities are expressed in the form of Eq. (2.17) and $\mu_j=\vec{\mu}\cdot\hat{e}_j$.

TABLE IV

Third Order Electrical Susceptibilities Responsible for the CSRS Diagrams in Fig. 6[a]

ζ	$(4\hbar^3 N)D_\zeta$	$\zeta_2(\Omega)$	$\zeta_3(\Omega)$
a	$\langle g|\mu_4|n\rangle\langle n|\mu_3|f\rangle\langle f|\mu_2|m\rangle\langle m|\mu_1|g\rangle$	$D_a/(\omega_{ng}^*-\omega_4)(\omega_{fg}^*-\omega_4+\omega_3)(\omega_{mf}-\omega_2)$	$D_a/(\omega_{mg}^*-\omega_1)(\omega_{fg}^*-\omega_1+\omega_2)(\omega_{nf}-\omega_3)$
b	$\langle g|\mu_4|n\rangle\langle n|\mu_3|f\rangle\langle f|\mu_1|m\rangle\langle m|\mu_2|g\rangle$	$D_b/(\omega_{ng}^*-\omega_4)(\omega_{fg}^*-\omega_4+\omega_3)(\omega_{mg}+\omega_2)$	$D_b/(\omega_{mg}^*-\omega_1)(\omega_{fg}^*-\omega_1+\omega_2)(\omega_{nf}-\omega_3)$
c	$\langle g|\mu_3|n\rangle\langle n|\mu_4|f\rangle\langle f|\mu_2|m\rangle\langle m|\mu_1|g\rangle$	$D_c/(\omega_{ng}+\omega_3)(\omega_{fg}^*-\omega_4+\omega_3)(\omega_{mf}-\omega_2)$	$D_c/(\omega_{mg}^*-\omega_1)(\omega_{fg}^*-\omega_1+\omega_2)(\omega_{ng}+\omega_3)$
d	$\langle g|\mu_3|n\rangle\langle n|\mu_4|f\rangle\langle f|\mu_1|m\rangle\langle m|\mu_2|g\rangle$	$D_d/(\omega_{ng}+\omega_3)(\omega_{fg}^*-\omega_4+\omega_3)(\omega_{mg}+\omega_2)$	$D_d/(\omega_{mg}^*-\omega_1)(\omega_{fg}^*-\omega_1+\omega_2)(\omega_{ng}+\omega_3)$

$^a\omega_{ab}=\omega_{ab}^0-i\gamma_{ab}$. The susceptibilities are expressed in the form of Eq. (2.17) and $\mu_j=\vec{\mu}\cdot\hat{e}_j$.

susceptibility, $\chi^{(3)}(\omega_4)$, and $\Delta\phi_4(\mathbf{r})$ is also given in terms of $\chi^{(3)}(\omega_4)$ as shown in Eq. (2.24). Thus, only four CARS resonant diagrams for ω_4 in Fig. 5, which determine $\chi^{(3)}(\omega_4)$, need to be considered for the CARS signal at ω_4. Likewise, the CSRS resonant diagrams in Fig. 6 are important in determining the CSRS signal at ω_2 or ω_3.

On the other hand, as indicated by the summation over all Fourier components present in the medium in Eqs. (3.6) and (4.8), all 16 CARS/CSRS resonant diagrams in Figs. 5 and 6 must be considered to establish global energy conservation in the combined system of the field and the medium. However, in order to determine $\Delta\phi_j(\mathbf{r})$ and $\Delta\phi_i(\mathbf{r})$ in Eqs. (3.6) and (4.8), a set of four resonant diagrams corresponding to $\chi^{(3)}(\omega_j)$, as given by Eqs. (2.24) and (2.14b), needs to be considered.

For electronically nonresonant (ordinary) CARS/CSRS, all electrical susceptibilities become pure imaginary when the Raman-type resonance condition in Eq. (5.5) is satisfied. Then, the following relations between the susceptibilities,

$$\zeta_4''(\mathfrak{Q}) = \zeta_1''(\mathfrak{Q}) = -\zeta_2''(\mathfrak{Q}) = -\zeta_3''(\mathfrak{Q}) \tag{5.6}$$

can be readily seen for all $\mathfrak{Q}$ $(=a, b, c,$ and $d)$ from Tables III and IV. As before, the expressions for the rates of energy transfer, $W_F^{(3)}(\mathbf{r})$ and $W_M^{(3)}(\mathbf{r})$, can be found and both of them are given as

$$W_F^{(3)}(\mathbf{r}) = W_M^{(3)}(\mathbf{r})$$
$$= \omega_{fg}^0 K(\mathbf{r})\{\zeta_4''(a) + \zeta_4''(b) + \zeta_4''(c) + \zeta_4''(d)\}\cos\Delta\phi_4 \tag{5.7}$$

with $K(\mathbf{r}) = \mathscr{E}_1(\mathbf{r})\mathscr{E}_2(\mathbf{r})\mathscr{E}_3(\mathbf{r})\mathscr{E}_4(\mathbf{r})$. Global energy conservation (Eq. (4.11)) is again evident. As long as $\Delta\phi_4 \neq \frac{1}{2}\pi$, there will be some population change in the medium.

For electronically resonant CARS/CSRS, where the resonance conditions, $\omega_{mg}^0 = \omega_1 (\omega_{mf}^0 = \omega_2)$ and $\omega_{ng}^0 = \omega_4 (\omega_{nf}^0 = \omega_3)$, are also satisfied in addition to the Raman-type resonance in Eq. (5.5), two diagrams in Figs. 5(a) and 6(a) recognize all such resonances. Then, the induced polarization (wavy line) for the CARS in Fig. 5(a) is set up between the ground and one of the resonant states. On the other hand, for the CSRS in Fig. 6(a), the polarization is generated across two excited states, each initially unpopulated. CARS and CSRS are known to exhibit different line shapes in the presence of inhomogeneous broadening [28]. The CARS diagrams in Fig. 5(a) are of Type I for the state m or n; there can be no population change in the state f through such time-evolution. The CSRS diagrams in

Fig. 6(a) are of Type I for the state m or n, but they are of Type II for the state f. Again, the equivalence of $W_F^{(3)}(\mathbf{r})$ and $W_M^{(3)}(\mathbf{r})$ at exact resonance as given in Eq. (4.11) can be verified since both $W_F^{(3)}(\mathbf{r})$ and $W_M^{(3)}(\mathbf{r})$ are given as

$$
W_F^{(3)}(\mathbf{r}) = W_M^{(3)}(\mathbf{r})
$$

$$
= \tfrac{1}{2}\tilde{K}(\mathbf{r})\{[\omega_{mg}^0 + \omega_{ng}^0]\zeta_4''(a) + \omega_{mf}^0\zeta_2''(a) + \omega_{nf}^0\zeta_3''(a)\}\cos\Delta\phi_4
$$
(5.8)

where $\tilde{K}(\mathbf{r}) = \tilde{\mathscr{E}}_1(\mathbf{r})\,\tilde{\mathscr{E}}_2(\mathbf{r})\,\tilde{\mathscr{E}}_3(\mathbf{r})\,\tilde{\mathscr{E}}_4(\mathbf{r})$.

In this section, global energy conservation for the combined system of the field and the medium at the exact resonance is explicitly demonstrated for various *resonance passive processes* up to third order by deriving the microscopic expressions for $W_F^{(3)}(\mathbf{r})$ and $W_M^{(3)}(\mathbf{r})$ through the diagrammatic technique using energy ladder diagrams. It is shown here that there is finite energy transfer between the fields and the medium even when the light–matter interactions are not in direct quadrature form at the polarization level.

VI. CONCLUDING REMARKS

The wave equation derived from classical Maxwell's equations is important for determining the strength of the signal field generated through nonlinear light–matter interaction. It also provides useful information concerning the relationship between the absolute phase of the nonlinear induced polarization produced in the medium and the absolute phase of the generated signal field. The difference in absolute phases, $\Delta\phi_j$, is important for the rate of energy transfer between the medium and the radiation as well as for the rate of energy change in the medium arising from the population change, as shown in Eqs. (3.6) and (4.8). When the light–matter interactions are of direct quadrature form in the incident fields, as for all active processes as well as for those nonresonant passive processes describing linear and nonlinear dispersion, the signal field must remain in-phase ($\Delta\phi_j = 0$) with respect to the induced polarization. For nonquadrature light–matter interactions, $\Delta\phi_j$ depends on the presence or absence of some material resonance with the incident fields as shown in Eq. (2.24a). In the absence of any resonances, the signal field must be out-of-phase ($\Delta\phi_j = \pm\tfrac{1}{2}\pi$). However, in the presence of resonances, both in-phase and out-of-phase components are present in the signal field ($0 \le |\Delta\phi_j| < \tfrac{1}{2}\pi$). This is just the condition for a transfer of energy between the radiation and the medium that will survive cycle averaging.

When there is no incident field at the signal frequency (ω_j) and the signal wave vector ($\mathbf{k}_j$), $\Delta\phi_j$ will remain constant throughout the medium, as given by Eq. (2.24a). However, if such an incident field were present, $\Delta\phi_j$ will change through the medium from some predetermined value at the incident boundary to the value given by Eq. (2.24a), once the signal field generated in the medium is dominant.

In this paper, it has been explicitly demonstrated that, whenever there is material resonance with the incident fields, irrespective of the specific nature of the interactions, there is finite energy transfer between the fields and the medium that survives cycle-averaging. Global energy conservation for the composite field/matter system is achieved by balancing the change in the total energy in the fields with the corresponding change in the energy of the medium. The frequency conserving condition, $\omega_j = \omega_p \equiv \xi_1\omega_1 + \xi_2\omega_2 + \cdots + \xi_s\omega_s$ is valid for all light–matter interactions, but it represents energy conservation among photons only when the total energy of the radiation field is conserved—*nonresonant passive processes*.

Global energy conservation is demonstrated in this paper only at the exact resonance between the CW fields and the chromophore states of the medium. When the fields are off-resonant (but within the bandwidth), global energy conservation can be demonstrated by treating the Hilbert space for the bath explicitly, not just as a source of stochastic dynamics. Now, the resonance condition has to be expressed in terms of the combined chromophore-bath eigenstates. Furthermore, when pulsed fields are used, the analysis in this paper can be extended by properly including the pulse envelope functions into the cycle-averaged W_F and W_M in Eqs. (3.1) and (4.6). Of course when considering short-time measurements, the time-energy uncertainty limitation on the global energy conservation statement may become significant.

Finally, apparent ambiguities in the diagrams have been resolved. Thus a given diagram, as conventionally constructed, conserves frequency, not energy. The algebraic sum of all arrows in an sth order diagram (with opposite signs for solid and dashed) correctly gives the frequency of the polarization at sth order. The diagram also correctly depicts the evolution of the density matrix up to sth order. However, since a diagram does not in general depict the outcome of the collapse of the polarization at the $(s + 1)$th step, it can correctly indicate neither the population change in the material nor photon conservation within the radiation after $s + 1$ events. As we have seen, the outcome of the collapsing polarization depends on the difference of the absolute phase of the signal field and that of the material polarization, a quantity that depends on experimental conditions naturally not available in a diagram.

Of course for the *nonresonant passive processes* the frequency conservation intrinsic to a diagram trivially denotes energy conservation among the photons, since with no material resonance, collapse of the polarization can only produce photons.

APPENDIX A

In this appendix, Eqs. (2.15a) and (2.15b) are derived. Equation (2.11) can be separated into the real and imaginary parts, from which the following equations can be readily obtained after taking the inner product with $\hat{e}_j$,

$$\nabla_j \mathscr{E}_j(\mathbf{r}) = \frac{2\pi(\omega_j/c)^2}{k_j'^2 + k_j''^2} \left[A_j(\mathbf{r}) \sin \Delta_j(\mathbf{r}) + B_j(\mathbf{r}) \cos \Delta_j(\mathbf{r}) \right]$$

(A.1a)

$$\mathscr{E}_j(\mathbf{r})\nabla_j \phi_j(\mathbf{r}) = \frac{2\pi(\omega_j/c)^2}{k_j'^2 + k_j''^2} \left[B_j(\mathbf{r}) \sin \Delta_j(\mathbf{r}) + A_j(\mathbf{r}) \cos \Delta_j(\mathbf{r}) \right]$$

(A.1b)

where $A_j(\mathbf{r})$ and $B_j(\mathbf{r})$ are the real and imaginary parts of $k_j^* \{\hat{e}_j \cdot \mathbf{P}_j^{(s)}(\mathbf{r})\}$ $\exp[\Delta \mathbf{k}_j'' \cdot \mathbf{r}]$, respectively, which are given in Eqs. (2.16) and (2.17), and $\Delta_j(\mathbf{r}) = \Delta \mathbf{k}_j' \cdot \mathbf{r} + \Delta \phi_j(\mathbf{r})$. Equation (A.1a) combined with the relation, $\nabla_j \{\ln \mathscr{E}_j(\mathbf{r})\} = \{1/\mathscr{E}_j(\mathbf{r})\}\nabla_j \mathscr{E}_j(\mathbf{r})$, can be used to rewrite Eq. (A.1b) as

$$\nabla_j \phi_j(\mathbf{r}) = \frac{A_j(\mathbf{r}) \cos \Delta_j(\mathbf{r}) + B_j(\mathbf{r}) \sin \Delta_j(\mathbf{r})}{A_j(\mathbf{r}) \sin \Delta_j(\mathbf{r}) - B_j(\mathbf{r}) \cos \Delta_j(\mathbf{r})} \nabla_j \ln \mathscr{E}_j(\mathbf{r}) \qquad (A.2)$$

Similar equations for the incident fields ($1 \le i \le s$) can be written

$$\nabla_i \phi_i(\mathbf{r}) = -\frac{A_i(\mathbf{r}) \cos \Delta_j(\mathbf{r}) - \xi_i B_i(\mathbf{r}) \sin \Delta_j(\mathbf{r})}{\xi_i A_i(\mathbf{r}) \sin \Delta_j(\mathbf{r}) + B_i(\mathbf{r}) \cos \Delta_j(\mathbf{r})} \nabla_i \ln \mathscr{E}_i(\mathbf{r}) \qquad (A.3)$$

where $\Delta_i(\mathbf{r})$ is replaced with $\Delta_j(\mathbf{r})$ by $\Delta_i(\mathbf{r}) = -\xi_i \Delta_j(\mathbf{r})$ (Eqs. (2.14a) and (2.14b)). Furthermore, $\nabla_i \phi_i(\mathbf{r}) = \nabla_j \phi_i(\mathbf{r})$ and $\nabla_i \{\ln \mathscr{E}_i(\mathbf{r})\} = \nabla_j \{\ln \mathscr{E}_i(\mathbf{r})\}$ for an isotropic medium. Now, Eq. (2.15b) readily follows from Eq. (2.12c) with Eqs. (A.2) and (A.3) and $\xi_i^2 = 1$.

APPENDIX B

The rules for obtaining the quantum-mechanical expression for nonlinear electrical susceptibility corresponding to a time-evolution diagram are briefly summarized. For simplicity, the initial bra and ket states are limited to the ground state g. The details of the diagrammatic technique can be found in Ref. 6.

To every straight arrow, solid or dashed, in a diagram, there belong one transition moment and one energy denominator. To the wavy line, there belongs only a transition moment and no energy denominator. More explicitly:

1. In general the transition moment $\langle m|\pm\vec{\mu}\cdot\hat{e}|n\rangle$ introduces the polarization vector $\hat{e}$ of the field active at a given step. For a solid arrow:

   a. the negative sign is used, and

   b. m is the new ket state (at the head of the arrow) and n is the old ket state (at the tail of the arrow).

   For a dashed arrow:

   a. the positive sign is used, and

   b. n is the new bra state (at the head of the arrow) and m is the old bra state (at the tail of the arrow).

   For the wavy line:

   a. the positive sign is used, and

   b. m and n are the bra and ket states linked by the wavy line.

2. In general the energy denominator is derived by first noting both the bra and ket states at the time corresponding to the head of the given arrow. This is called the "current" condition. The energy denominator expression for *that event* is made up of:

   a. the energy of the current bra state minus the energy of the current ket state, plus

   b. the algebraic sum of the energies of all photons that have been active up to and including the arrow under consideration, plus

   c. $i\hbar$ times the damping constant (when resonant) associated with the superposition state of the current ket and bra states.

   d. a wavy line has no associated energy denominator.

Thus, the sth order susceptibility expression for a given diagram is obtained by summing over the complete set of eigenstates, the product of (i) the $s+1$ transition moments from (1) above; (ii) the s energy denominators from (2) above; (iii) the number of molecules in unit volume, N; and (iv) the numerical factor 2^{1-s} (see [20]).

Acknowledgments

This work was supported in part by the Materials Science Center at Cornell University. D. L. thanks the Ministry of Education of Korea and the Korean Science and Engineering Foundation for their support.

References

1. N. Bloembergen, *Nonlinear Optics* (Benjamin, Reading, 1965).

2. Y. R. Shen, *The Principle of Nonlinear Optics* (Wiley, New York, 1984).

3. M. Schubert and B. Wilhelmi, *Nonlinear Optics and Quantum Electronics* (Wiley, New York, 1986).

4. P. N. Butcher and D. Cotter, *The Element of Nonlinear Optics* (Cambridge University Press, Cambridge, 1990).

5. B. S. Wherrett, in *Nonlinear Optics*, P. G. Harper and B. S. Wherrett (eds.) (Academic Press, London, 1977).

6. D. Lee and A. C. Albrecht, in *Advances in Infrared and Raman Spectroscopy*, R. J. Clark and R. E. Hester (eds.) (Wiley-Heydon, New York, 1985), vol. 12.

7. For example, R. Kubo, *Adv. Chem. Phys.* **15**, 101 (1969); R. Kubo, in *Fluctuation, Relaxation and Resonance in Magnetic Systems*, D. ter Haar (ed.) (Oliver and Boyd, Edinburgh, 1962).

8. J. A. Amstrong, N. Bloembergen, J. Ducuing, and P. S. Pershan, *Phys. Rev.* **127**, 1918 (1962).

9. Y. R. Shen, *Phys. Rev.* **167**, 818 (1968); P. S. Pershan, *Phys. Rev.* **130**, 919 (1983); N. Bloembergen and P. S. Pershan, *Phys. Rev.* **128**, 606 (1962).

10. M. A. Dugan and A. C. Albrecht, *Phys. Rev.* **A23**, 3877 (1991).

11. S. J. A. Druet, B. Attal, T. K. Gustafson, and J. P. Taran, *Phys. Rev.* **A18**, 1529 (1978).

12. T. K. Yee and T. K. Gustafson, *Phys. Rev.* **A18**, 1597 (1978).

13. J. D. Jackson, *Classical Electrodynamics*, 2nd edition (Wiley, New York, 1975).

14. M. Born and E. Wolf, *Principle of Optics*, 6th edition (Pergamon, Oxford, 1980).

15. L. D. Landau and E. M. Lifschitz, *Electrodynamics of Continuous Media*, 2nd edition (Pergamon, Oxford, 1984).

16. L. Mandel and E. Wolf, *Rev. Modern Phys.* **37**, 231 (1965).

17. The quantum-mechanical operator, $\hat{E}$, for the electric field is hermitian, consistent with the fact that $\mathbf{E}(\mathbf{r}, t)$ is a measurable property. The expectation value of $\hat{E}$ vanishes for pure number (or Fock) states of radiation. However, it survives for pure coherent (or Glauber) states of radiation. The quantum-mechanical operator for the intensity is given in terms of $\hat{E}^{(-)}\hat{E}^{(+)}$, or the number operator, $\hat{n} = \hat{a}^{+}\hat{a}$ (see [18]).

18. R. J. Glauber, *Phys. Rev.* **130**, 2529 (1963); **131**, 2766 (1963); W. H. Louisell, *Quantum Statistical Properties of Radiation* (Wiley, New York, 1973); B. W. Shore, *The Theory of Coherent Atomic Excitation* (Wiley, New York, 1990).

19. For elliptically polarized light, the complex polarization unit vectors are often more convenient. They are defined by combining the real Cartesian polarization unit vectors, $\hat{e}_l$, with the complex absolute phase part, $\exp[i\phi_l]$, in Eq. (2.3) (see [12]).

20. A factor of 2^{1-s} has been absorbed in the expression for the sth order electrical susceptibility [6].

21. S. J. A. Druet, J. P. E. Taran, and C. J. Borde, *J. Phys.* **40**, 819 (1979).

22. N. Bloembergen, H. Lotem, and R. T. Lynch, *Ind. J. Pure Appl. Phys.* **6**, 151 (1977).

23. J. T. Fourkas, R. Trebino, and M. D. Fayer, *J. Chem. Phys.*, in press.

24. G. Grynberg and P. R. Berman, *Phys. Rev.* **A43**, 3994 (1991) and refs. therein.

25. Y. R. Prior, A. R. Bogdan, M. Bagenais, and N. Bloembergen, *Phys. Rev. Lett.* **46**, 111 (1981).

26. Reference 6 and refs. therein.

27. H. Kono, Y. Nomura, and Y. Fujimura, *Adv. Chem. Phys.* **80**, 403 (1991). Here, the time-correlation function formalism is used to classify the RSR into the fluorescence-like, the Raman-like and the interference-like components. This classification corresponds to the nomenclature II in [6] and the interference-like component is not positive definite in general.

28. M. T. Riebe and J. C. Wright, *J. Chem. Phys.* **88**, 2981 (1988).

A MANY-BODY STOCHASTIC APPROACH TO ROTATIONAL MOTIONS IN LIQUIDS*

ANTONINO POLIMENO[†] and JACK H. FREED

Baker Laboratory of Chemistry, Cornell University, Ithaca, New York

CONTENTS

*Supported by NSF Grants CHE9004552 and DMR8901718.
[†]Permanent address: Department of Physical Chemistry, University of Padua, via Loredan 2, 35131 Padua, Italy.

Advances in Chemical Physics, Volume LXXXIII, Edited by I. Prigogine and Stuart A. Rice.
ISBN 0-471-54018-8 © 1993 John Wiley & Sons, Inc.

I. METHODOLOGY

A. Introduction

Classic Brownian motion has been widely applied in the past to the interpretation of experiments sensitive to rotational dynamics. ESR and NMR measurements of T_1 and T_2 for small paramagnetic probes have been interpreted on the basis of a simple Debye model, in which the rotating solute is considered a rigid Brownian rotator, such that the time scale of the rotational motion is much slower than that of the angular momentum relaxation and of *any other degree of freedom* in the liquid system. It is usually accepted that a fairly accurate description of the molecular dynamics is given by a Smoluchowski equation (or the equivalent Langevin equation), that can be solved analytically in the absence of external mean potentials.

Since the pioneering contribution of Debye [1], one-body Smoluchowski equations have provided a general framework for the study of dielectric relaxation in liquids, neutron scattering, and infrared spectroscopy. The basic hypothesis is that the solute degrees of freedom are the only "relevant" (i.e., slow when compared with the timescale of the experiment) variables in the system, and that the surrounding liquid medium behaves as a homogeneous bath whose internal degrees of freedom are rapidly relaxing. This simple picture has had many substantial refinements and improvements. Perrin [2], Sack [3], Fixman and Rider [4], Hubbard [5], McClung [6], Morita [7], and many others have contributed by including anisotropy and inertial effects and by studying detailed numerical solutions to classic Fokker–Planck–Kramers equations for the tumbling of a general top. Good agreement between the experimental data and theoretical predictions can often be obtained at moderate viscosities and pressures. Also, the influence of a mean potential of interaction has been extensively studied, since the original work of Favro [8].

However, when the experimental results associated with the molecular tumbling become more precise, as is often the case when magnetic resonance techniques are involved, the one-body approach become questionable, and a more sophisticated insight into the many-body nature of the liquid is required. Usually a simple Debye approach fails in interpreting molecular dynamics data obtained for liquids of "molecules which are highly anisotropic in shape, for example rod-like molecules, or molecules which interact via anisotropic forces, such as the case where hydrogen bonding occurs, or finally molecules which display high internal mobility like bulk polymers" [9]; in short whenever a Markovian description of the solute degrees of freedom is unacceptable, due to the effect of solvent degrees of freedom whose relaxation timescale is comparable to the solute correlation time. Substantial departures from predictions of Brownian motion theory are observed in extreme conditions, for example, when very low temperatures or very high viscosities, such as in supercooled organic fluids, are considered; or when there are strong interactions between the solute and the immediate solvent surroundings, such as in ordered liquid phases or highly polar liquids. ESR studies in ordered and isotropic fluids over a wide range of temperatures and pressures [10, 11], NMR data [12], highly viscous fluid studies [13–16], dielectric experiments performed in glassy liquids [17–22], far infrared spectroscopy of polar solvents [23] are only a few examples of studies that have been particularly sensitive to the inadequacies of stochastic single-body models.

In principle, the presence of slow stochastic torques directly affecting the solute reorientational motion can be dealt with in the framework of generalized stochastic Fokker–Planck equations including frequency-dependent frictional terms. However, the non-Markovian nature of the time evolution operator does not allow an easy treatment of this kind of model. Also, it may be difficult to justify the choice of frequency dependent terms on the basis of a sound physical model. One would like to take advantage of some knowledge of the physical system under investigation to set up a "relevant" time evolution operator that is more or less able to account for the main relaxation processes affecting the solute. One way this can be accomplished is by including collective degrees of freedom, which can, at least partially, account for the non-Markovian nature of the motion of the isolated probe.

Many theoretical models have been proposed in the past for including some "solvent" degrees of freedom, representing in a qualitative way the complex environment around the solute molecule. The "itinerant oscillator" model (IOM) developed by Coffey and co-workers [23–25] is an interesting attempt to improve on the limitations inherent in the one-body Debye approach. The molecule is considered to be coupled by a harmonic potential to a cage of solvent particles reorienting as a whole, and some calculations with a cosine potential have been attempted. The system "molecule + cage" reorients in a fixed plane and the additional solvent molecules are described merely as a source for a damping force (torque) affecting both the molecule and the cage. A bidimensional Langevin equation, or the corresponding linearized Fokker–Planck–Kramers equation, is used to calculate the usual correlation functions of interest, and dielectric relaxation and far infrared data are interpreted in terms of this model (and also compared with molecular dynamics simulations).

The itinerant oscillator model can be seen in the context of the more general "reduced model" theory due to Grigolini and co-workers [26–29]. Again, the main idea is to account for the complex behavior of the medium as a non-Markovian bath which affects the rotational (and/or translational) motion of the probe. This bath is thought of as added "virtual" degrees of freedom whose features simulate, in a multidimensional Langevin equation scheme, the "real" time dependent generalized Langevin equation,

We briefly note, at this point, the contribution of Zwan and Hynes [30–32] that is in line with these previous approaches. These authors consider a generalization of the IOM for a simple internal-dipole isomerization reaction in which the interaction with the rest of the solvent is implicitly split into a dissipative interaction (generating the usual damping terms, considered small by Zwan and Hynes) and long-distance interactions with "a pair of solvent outer dipoles". The picture is very schematic (again only linearized potentials are considered), but the concept of a *third interacting body* dynamically coupled to the probe and the "slow modes" previously defined, is interesting. Note that Zwan and Hynes use their initial multidimensional linear Langevin equation to obtain a generalized Langevin equation in a single reaction coordinate, which they solve with the aid of a Grote–Hynes approach (cf. a recent comparison with MD results [32]).

Finally, a comparison with the models developed in the past by Freed and co-workers is in order [28, 33–35]. With the objective of interpreting observed departures from simple Debye behavior in many liquid state ESR experiments, they considered two main physical models based on the characteristic correlation times of the stochastic torques acting on the probe, compared with that of the probe motion itself. In the so-called "fluctuating torques" (FT) models the probe can be seen as larger (and slower), or at least of comparable dimensions to the solvent molecules. Because of the rapid reorganization of the surroundings, only dissipative friction effects are exerted by the solvent on the probe. On the other hand, in the "slowly relaxing local structure" (SRLS) model, the probe can be seen as smaller (and faster) than the solvent "structure", whose motion about the probe is slow enough that the probe reorients relative to the instantaneous value of the intermolecular potentials. A rationalization of these models is achieved by Stillman and Freed [33], who are able to obtain, using arguments based on the stochastic Liouville approach, general augmented Fokker–Planck equations describing simple model cases. We note in passing the similar objectives of this stochastic Liouville approach and the reduced model theory of Grigolini.

Recently Kivelson and Miles [36] and Kivelson and Kivelson [37] have attempted to rationalize some of the physical observations concerning supercooled organic liquids [13–20] by adopting a many-body description. The reorientational relaxation of an asymmetric top is assumed to take place in a potential $V(\Omega - \Omega^*)$ where Ω are the Euler angles specifying the orientation of the top, while Ω^* are defined as an unspecified "equilibrium position" for the top in the mean field potential V. The so-called β relaxation is related to the diffusional motion within a potential well, whereas the so-called α relaxation is identified as a "random restructuring of the torsional potential", that is, of Ω^*, which can be considered as a function of some "slow environmental variable X" [37]. This model is, in spirit, very similar to the SRLS model of Freed and co-workers. Kivelson et al. rationalize the multiexponential form of the rotational correlation functions observed in many supercooled fluids in terms of a memory function approach; that is, the correlation function is expressed as a Mori continued fraction expansion truncated at the second term [36]. The second memory function is supposed to be a phenomenological biexponential function. In this simple way a qualitative description of the α, β and Poley relaxation processes is achieved, although the behavior of the librational signal is not very well explained if compared to the experimental evidence (a weak, temperature dependent signal is calculated). No real attempts at relating these considerations with microscopic or mesoscopic models is made by the authors; the model is proposed as an extension of the so-called "three-variable theory" [38].

To summarize, in complex liquids, where the bath cannot be considered as a simple collection of very fast modes which can be eliminated in the usual Markovian approximation, the spectrum of stochastic torques acting on the probe can be modeled in terms of virtual or "ghost" degrees of freedom, coupled to the molecular ones in a multidimensional Langevin or Fokker–Planck formalism. The new modes are able to simulate, in some qualitative way, the complex features of the real solvent (e.g., reduced model theory), and they can be interpreted in terms of a formal Mori expansion (e.g., a three variables theory), or they can be chosen with an intuitive physical meaning (FT/SRLS and IOM models). Generally speaking, an interaction potential must be introduced to describe the coupling between real and virtual modes, but second order interactions, mediated by other solvents modes, should also be considered in order to simulate dissipative contribution to the torques affecting the probe (Zwan and Hynes models).

Clearly, a general theory able to naturally include other solvent modes in order to simulate a dissipative solute dynamics is still lacking. Our aim is not so ambitious, and we believe that an effective working theory, based on a self-consistent set of hypotheses of *microscopic* nature is still far off. Nevertheless, a *mesoscopic* approach in which one is not limited to the one-body model, can be very fruitful in providing a fairly accurate description of the experimental data, provided that a clever choice of the reduced set of coordinates is made, and careful analytical and computational treatments of the improved model are attained. In this paper, it is our purpose to consider a description of rotational relaxation in the formal context of a many-body Fokker–Planck–Kramers equation (MFPKE). We shall devote Section I to the analysis of the formal properties of multivariate FPK operators, with particular emphasis on systematic procedures to eliminate the non-essential parts of the collective modes in order to obtain manageable models. Detailed computation of correlation functions is reserved for Section II. A preliminary account of our approach has recently been presented in two Letters which address the specific questions of (1) the Hubbard–Einstein relation in a mesoscopic context [39] and (2) bifurcations in the rotational relaxation of viscous liquids [40].

In Section I.B we discuss how to devise a general MFPKE to describe complex liquids. A three-body model will be presented as a description of a system in which at least two significant additional sets of solvent degrees of freedom are introduced. In Section I.C we show the relation between some of the previously cited approaches and particular cases of our model. In particular, augmented Fokker–Planck equations (AFPE) of Stillman and Freed are seen to be directly related to the MFPK formalism. Section I.D is devoted to the explicit study of a two-dimensional planar version of the three-body model of Section I.B. In Section I.E we consider the actual relation between AFPEs and MFPKEs in a test case. A summary is given in Section I.F. The projection procedure employed in the treatment of large MFPKEs is described in Appendices A and B.

B. A Many-Body Approach to Complex Fluids

A set of collective degrees of freedom representing, at least qualitatively, the main effects exerted by the complex medium in the immediate surroundings of the rotating solute, needs to be incorporated into the initial one-body description of the molecular dynamics. Following suggestions of many authors, we choose to think in terms of an instantaneous structure of the solvent molecules around the reorienting probe, a sort of loose "cage" that can be considered as a dynamical structure relaxing in the same time range as the solute rotational coordinates (i.e., it is a slowly relaxing local structure). Thus the relevant phase space is increased by three Euler angles for the orientation of the solvent local structure, and also by the three components of the corresponding angular momentum vector. The resulting two-body scheme is formally that of two interacting rigid tops; the first one being the solute molecule, the second one the average of the instantaneous orientations of the solvent molecules in the near environment of the probe.

The picture can be improved, if necessary, by adding faster solvent degrees of freedom, coupled both to the probe (the first body or body 1, from now on) and the solvent structure (the second body). That is, we suppose that the second body does not account for all of the effects exerted by the real environment, but only for the slowest ones, since "... motion of an individual molecule in a (ordered) fluid should be a complex process involving ... long-range (and slow) hydrodynamic effects to short-range (and fast) molecular couplings" [11]. Note that if we limit our analysis to the timescale of the reduced system solute + solvent structure (that we may well suppose to be orders of magnitude slower than the rest of the liquid system, except maybe in very viscous fluids), any faster mode will be seen as giving an additional frictional effect, after its elimination as an explicit degree of freedom by a projection procedure. Thus it will be possible to see that the introduction of a fast third (or additional) collective body interacting with the solute and the solvent structure can be considered as the approximate source of the fluctuating forces/torques invoked by Freed et al.

Although our primary interest is concerned with the study of the rotational dynamics of the solute, we may consider part or all of the additional solvent degrees of freedom as point vectors, or fields. An example of a fast translation-like mode coupled to a rotator is given by a stochastic polarization or "reaction field" in polar solvents [41]. Note that, at least as a starting point, we shall always include the conjugate momentum coordinates in the system phase space. That is, we shall always initially consider the multivariate Kramers equation including all the position *and* velocity degrees of freedom.

1. Many-Body Fokker–Planck–Kramers Equations

Let us suppose that the liquid system is described by a MFPKE in $N + 1$ rigid bodies (the solute, or body 1 and N rotational solvent modes or "bodies"), each characterized by inertia and friction tensors $\mathbf{I}_n$ and $\boldsymbol{\xi}_n$, a set of Euler angles $\boldsymbol{\Omega}_n$, and an angular momentum vector $\mathbf{L}_n$ ($n = 1, \ldots, N + 1$) *plus* K fields, each defined by a generalized mass tensor and friction tensor $\mathbf{M}_k$ and $\boldsymbol{\xi}_k$, a position vector $\mathbf{X}_k$ and the conjugate linear momentum $\mathbf{P}_k$ ($k = 1, \ldots, K$). The time evolution of the joint conditional probability $P(\boldsymbol{\Omega}^0, \mathbf{X}^0, \mathbf{L}^0, \mathbf{P}^0 | \boldsymbol{\Omega}, \mathbf{X}, \mathbf{L}, \mathbf{P}, t)$ (where $\boldsymbol{\Omega}$, $\mathbf{X}$, etc. stand for the collection of Euler angles, field coordinates etc.) for the system is governed by the multivariate Fokker–Planck–Kramers equation

$$\frac{\partial}{\partial t} P = -\hat{\Gamma} P \tag{1.1}$$

and the initial conditions are

$$P|_{t=0} = \prod_{n=1}^{N+1} \delta(\boldsymbol{\Omega}_n - \boldsymbol{\Omega}_n^0) \delta(\mathbf{L}_n - \mathbf{L}_n^0) \prod_{k=1}^{K} \delta(\mathbf{X}_k - \mathbf{X}_k^0) \delta(\mathbf{P}_k - \mathbf{P}_k^0) \tag{1.2}$$

where the FPKE partial differential operator is given by the sum of Kramers operators for each body and field

$$\hat{\Gamma} = \sum_{n=1}^{N+1} \hat{\Gamma}_n + \sum_{k=1}^{K} \hat{\Gamma}_k \tag{1.3}$$

The rotational operator for the nth body is defined according to Hwang and Freed [35] as

$$\hat{\Gamma}_n = i\mathbf{L}_n \mathbf{I}_n^{-1} \hat{\mathbf{J}}_n + \mathbf{T}_n \nabla_n - \hat{\mathbf{P}}_n \nabla_n - k_B T \nabla_n \boldsymbol{\xi}_n \left(\nabla_n + \frac{1}{k_B T} \mathbf{I}_n^{-1} \mathbf{L}_n \right) \tag{1.4}$$

The vector operator $\hat{\mathbf{J}}_n$ is the angular momentum operator for the nth body; note that the generator of infinitesimal rotation ($\mathbf{M}_n$) is simply proportional to $\hat{\mathbf{J}}_n$, that is, $\mathbf{M}_n = i\hat{\mathbf{J}}_n$; $\mathbf{T}_n$ is the torque acting on the nth body, which we take as generated from a general potential V depending on all the displacement coordinates of the system

$$\mathbf{T}_n = -i\hat{\mathbf{J}}_n V \tag{1.5}$$

Finally ∇_n is the gradient operator on the $\mathbf{L}_n$ subspace, while $\hat{\mathbf{P}}_n$ is a precessional term, whose Cartesian components in the molecular frame fixed on the nth body are given by

$$\hat{P}_{n_i} = \left(\frac{1}{I_{n_k}} - \frac{1}{I_{n_j}} \right) L_{n_j} L_{n_k} \epsilon_{ijk} \tag{1.6}$$

where I_{n_i} is a principal value of the inertia tensor $\mathbf{I}_n$ and ϵ_{ijk} is the Levi–Civita symbol.

The translation operator for field $\mathbf{X}_n$ is defined accordingly as the three-dimensional Kramers operator

$$\hat{\Gamma}_k = \mathbf{P}_k \mathbf{M}_k^{-1} \nabla_{\mathbf{X}_k} + \mathbf{F}_k \nabla_{\mathbf{P}_k} - k_B T \nabla_{\mathbf{P}_k} \boldsymbol{\xi}_k \left(\nabla_{\mathbf{P}_k} + \frac{1}{k_B T} \mathbf{M}_k^{-1} \mathbf{P}_k \right) \tag{1.7}$$

where $\mathbf{F}_k$ is the restoring force generated by the gradient operator $\nabla_{\mathbf{X}_k}$ on V

$$\mathbf{F}_k = -\nabla_{\mathbf{X}_k} V \tag{1.8}$$

and $\nabla_{\mathbf{P}_k}$ is the gradient operator on the subspace $\mathbf{P}_k$. In the following we will consider only isotropic space, and we will conveniently define all the vectors and vectors operators in a unique laboratory frame.

The potential function V still must be made explicit in order to complete the description of the system. A general multipole expansion in terms of first, second rank, etcetera interactions depending only on the relative orientation between each pair of bodies can be taken, as well as a multipole-field term (e.g., a dipole-field) for the pairwise interaction between each body and field. Finally each stochastic field is subjected to a harmonic potential, to parametrize in the most economical way the amplitude of the stochastic fluctuations. The complete potential is then

$$\frac{V}{k_B T} = \sum_{n=1}^{N+1} \sum_{n'=1}^{N+1} V_{nn'}(\boldsymbol{\Omega}_n - \boldsymbol{\Omega}_{n'}) - \sum_{n=1}^{N+1} \sum_{k=1}^{K} \mu_k \mathbf{X}_k \mathbf{u}_k + \frac{1}{2} \sum_{k=1}^{K} \mathbf{X}_k \boldsymbol{\Xi}_k^2 \mathbf{X}_k \tag{1.9}$$

where

$$V_{nn'}(\mathbf{\Omega}_n - \mathbf{\Omega}_{n'}) = \sum_{R_{nn'},P_{nn'},Q_{nn'}} v_{P_{nn'}Q_{nn'}}^{R_{nn'}} \mathscr{D}_{P_{nn'}Q_{nn'}}^{R_{nn'}*}(\mathbf{\Omega}_n - \mathbf{\Omega}_{n'})$$

(1.10)

where $\mathscr{D}_{P_{nn'}Q_{nn'}}^{R_{nn'}*}$ is the (adjoint of) the Wigner rotation function of rank $R_{nn'}$ and components $P_{nn'}$, $Q_{nn'}$. The dipolar coupling between each body and each field is expressed in terms of the inner product between the field $\mathbf{X}_n$ and a unit vector $\mathbf{u}_n$ fixed on the body (so that the quantity $\mu_n \mathbf{u}_n$ can be interpreted, if desired, as the dipole moment of the nth body); the (diagonal) matrix $\mathbf{\Xi}_n$ has elements which measure the amplitude of the fluctuations of the components of the field $\mathbf{X}_n$.

2. Three-Body Fokker–Planck–Kramers Equation

In the following paragraphs we shall apply the previous general formulas to a simplified description of a liquid system in which only three bodies are retained: the solute molecule (body 1), a slowly relaxing local structure or solvent cage (body 2), and a fast stochastic field ($\mathbf{X}$) as a source of fluctuating torque. Although this is a minimal description if compared to the general approach of the previous section, it should still represent a considerable improvement with respect to the usual one-body schemes, since it explicitly includes both a fast and a slow solvent mode.

The reduced Markovian phase space is now given by the Euler angles specifying the position of the solute rotator $\mathbf{\Omega}_1$ and the three components of the corresponding angular momentum vector $\mathbf{L}_1$, plus the analogous quantities $\mathbf{\Omega}_2$ and $\mathbf{L}_2$ for the solvent structure plus the fast field $\mathbf{X}$ and its conjugate linear momentum $\mathbf{P}$. The conditional probability for the system $P(\mathbf{\Omega}_1^0, \mathbf{\Omega}_2^0, \mathbf{X}^0, \mathbf{L}_1^0, \mathbf{L}_2^0, \mathbf{P}^0|\mathbf{\Omega}_1, \mathbf{\Omega}_2, \mathbf{X}, \mathbf{L}_1, \mathbf{L}_2, \mathbf{P}, t)$ is now driven by the MFPK operator

$$\hat{\Gamma} = \hat{\Gamma}_1 + \hat{\Gamma}_2 + \hat{\Gamma}_\mathbf{X}$$

(1.11)

where $\hat{\Gamma}_1$ and $\hat{\Gamma}_2$ are given by Eq. (1.4) and $\hat{\Gamma}_\mathbf{X}$ by Eq. (1.7). A further simplification will be introduced by considering an isotropic fluid composed of spherical top molecules (but with embedded dipoles, quadrupoles, etc.). Not much changes for molecules of cylindrical symmetry (i.e., symmetric tops). Thus all the inertial, mass and friction tensors for each body and the field will be treated as scalars. The precessional terms can be completely neglected, and all the suboperators can be written easily in a unique laboratory frame. The direct potential term between the solute and the solvent cage will include only first and second rank interactions, and they will be dependent only on the relative angle between $\mathbf{u}_1$ and $\mathbf{u}_2$ (see caption to Fig. 1)

$$\frac{V}{k_B T} = -v_1 P_1(\boldsymbol{\Omega}_2 - \boldsymbol{\Omega}_1) - v_2 P_2(\boldsymbol{\Omega}_2 - \boldsymbol{\Omega}_1) - (\mu_1 \mathbf{u}_1 + \mu_2 \mathbf{u}_2)\mathbf{X} + \frac{\Xi^2}{2}\mathbf{X}^2$$
(1.12)

here P_1 and P_2 are the Legendre polynomials of rank 1 and 2, respectively. Note that any direct dipole–dipole interaction between body 1 and body 2, is included in the first rank part (a minus sign has been extracted for future convenience from the first and second rank parameters).

A variety of interesting physical situations can now be obtained in the framework of the three-body model just defined, by carefully choosing the range of variation of the frictional parameters: ξ_1, the friction exerted by the rest of the solvent on body 1, ξ_2, the friction of body 2 and ξ_X, the friction on the field; and the energetic parameters v_1, v_2, μ_1, μ_2 (Ξ being renormalized to 1, cf. next section). For instance, one can consider the case of a fast solute interacting via a nematic-like interaction potential with a slow (large) solvent structure in the absence or presence of a fast

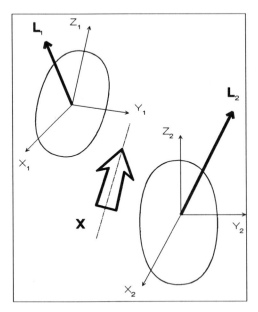

Figure 1. A three-body scheme for a complex liquid. Note that $\mathbf{u}_1$ and $\mathbf{u}_2$ are aligned respectively along the z_1 and z_2 axes.

fluctuating field ($v_1 = 0$, $v_2 \neq 0$, $\xi_2 \gg \xi_1$). Or one can choose the case in which only the interaction between the solute probe and the field is present, ignoring any local structure ($v_1 = v_2 = 0$, $\xi_2 = 0$). A planar Smoluchowski equivalent of this latter case was recently used for the interpretation of dielectric friction effects in polar isotropic liquids [41].

In many physical systems of experimental interest, it is usually possible to devise a reduced phase space of coordinates and/or momenta in which an accurate description is achievable. For instance, in a highly viscous fluid one may neglect all the momenta $\mathbf{L}_1$, $\mathbf{L}_2$ and $\mathbf{P}$ given their very fast relaxation with respect the time scale relaxation of the position coordinates $\mathbf{\Omega}_1$, $\mathbf{\Omega}_2$, $\mathbf{X}$. In many cases, the field vector (and its conjugate linear momentum) can be considered as a fast mode with respect to the rest of the system, so that both $\mathbf{X}$ and $\mathbf{P}$ can be projected out. One can also suppose that, although inertial effects are unimportant for the large solvent structure, that is, $\mathbf{L}_2$ is a fast coordinate, some inertia is still affecting the motion of body 1, so that $\mathbf{L}_1$ must be retained. If all the additional solvent degrees of freedom are eliminated, and only $\mathbf{\Omega}_1$ is left, the single body Smoluchowski equation is recovered.

C. Elimination of Fast Variables

Our purpose in this section is to obtain a simpler time evolution operator from the complete one of the previous section via a systematic elimination of any fast variables initially included in the system. In order to handle efficiently the algebra involved, with the smallest number of independent parameters, it is convenient to introduce from here on rescaled, dimensionless quantities (see Table I) and to "symmetrize" [42] the initial MFPK operator via the usual similarity transformation

$$\tilde{\Gamma} = P_{eq}^{-1/2} \hat{\Gamma} P_{eq}^{1/2} \tag{1.13}$$

where P_{eq} is the Boltzmann distribution function over the total energy (potential plus kinetic). It is the unique eigenfunction of zero eigenvalue of the unsymmetrized operator. The final symmetrized and rescaled time evolution operator is then written explicitly

$$\begin{aligned}
\tilde{\Gamma} = {}& \omega_1^s (i\mathbf{L}_1 \hat{\mathbf{J}}_1 + \mathbf{T}_1 \nabla_1) - \omega_1^c \exp(\mathbf{L}_1^2/4)\nabla_1 \exp(-\mathbf{L}_1^2/2)\nabla_1 \exp(\mathbf{L}_1^2/4) \\
&+ \omega_2^s (i\mathbf{L}_2 \hat{\mathbf{J}}_2 + \mathbf{T}_2 \nabla_2) - \omega_2^c \exp(\mathbf{L}_2^2/4)\nabla_2 \exp(-\mathbf{L}_2^2/2)\nabla_2 \exp(\mathbf{L}_2^2/4) \\
&+ \omega_X^s (\mathbf{P}\nabla_X + \mathbf{F}\nabla_P) - \omega_X^c \exp(\mathbf{P}^2/4)\nabla_P \exp(-\mathbf{P}^2/2)\nabla_P \exp(\mathbf{P}^2/4)
\end{aligned} \tag{1.14}$$

TABLE I
Rescaled Units and Parameters[a,b,c]

$$\tilde{\mathbf{L}}_{1,2} = (k_B T I_{1,2})^{-1/2} \mathbf{L}_{1,2}$$
$$\tilde{\mathbf{P}} = (k_B T m)^{-1/2} \mathbf{P}$$
$$\tilde{\mathbf{X}} = (k_B T)^{-1/2} \Xi \mathbf{X}$$
$$\tilde{\mu}_{1,2} = (k_B T)^{-1/2} \Xi^{-1} \mu_{1,2}$$
$$\omega_{1,2}^c = I_{1,2}^{-1} \xi_{1,2}$$
$$\omega_{\mathbf{x}}^c = m^{-1} \xi_{\mathbf{x}}$$
$$\omega_{1,2}^s = (k_B T)^{1/2} I_{1,2}^{-1/2}$$
$$\omega_{\mathbf{x}}^s = m^{-1/2} \Xi$$
$$\tilde{V} = (k_B T)^{-1} V$$

[a]Where the tilde symbol stands for rescaled units, and it is neglected throughout the text.

[b]Rescaled units are dimensionless except for the four ω terms, which are in angular frequency units.

[c]Subscripts 1, 2 imply the symbol for either body 1 or body 2.

while the rescaled potential (in $k_B T$ units according to Table I) is given by

$$V = -v_1 P_1(\mathbf{\Omega}_2 - \mathbf{\Omega}_1) - v_2 P_2(\mathbf{\Omega}_2 - \mathbf{\Omega}_1) - (\mu_1 \mathbf{u}_1 + \mu_2 \mathbf{u}_2)\mathbf{X} + \tfrac{1}{2}\mathbf{X}^2 \tag{1.15}$$

The *streaming* frequencies ω_1^s and ω_2^s in Eq. (1.14) are related to the inertial motions of body 1 and body 2, respectively (i.e., they are the inverses of the correlation times for the deterministic motion of the two bodies). The *collisional* frequencies ω_1^c and ω_2^c are a measure of the direct coupling with the stochastic environment, that is, of the dissipative contribution to the dynamics. An analogous interpretation may hold for the frequencies $\omega_{\mathbf{x}}^s$ and $\omega_{\mathbf{x}}^c$, related to the streaming and stochastic drift of the field.

1. Field Mode Projection

According to the previous section, we shall start by considering $\mathbf{X}$ and $\mathbf{P}$ as fast degrees of freedom, relaxing on a much more rapid timescale than the orientational coordinates and momenta of the solute and the solvent cage. Many different projection schemes are available to handle stochastic partial differential operators. Here we choose to adopt a slightly modified total time ordered cumulant (TTOC) expansion procedure, directly related to the well known resolvent approach. In order to make this chapter self-contained, we summarize the method in the Appendices and its application to the cases considered here and in the next section.

Given only that ω_X^s and ω_X^c are much larger than any other frequency in the system, one can easily eliminate both $\mathbf{X}$ and $\mathbf{P}$ in a simple step, via a projection based on the eigenfunctions of the monodimensional FPK operator for a single particle in a harmonic field [43]. Following the detailed scheme outlined in Appendix A, after projecting out the field and its momentum, one obtains the following MFPK operator in the remaining two bodies coordinates:

$$\tilde{\Gamma} = \omega_1^s(i\mathbf{L}_1\hat{\mathbf{J}}_1 + \mathbf{T}_1\nabla_1) - \exp(\mathbf{L}_1^2/4)\nabla_1\omega_1^c\exp(-\mathbf{L}_1^2/2)\nabla_1\exp(\mathbf{L}_1^2/4)$$

$$+ \omega_2^s(i\mathbf{L}_2\hat{\mathbf{J}}_2 + \mathbf{T}_2\nabla_2) - \exp(\mathbf{L}_2^2/4)\nabla_2\omega_2^c\exp(-\mathbf{L}_2^2/2)\nabla_2\exp(\mathbf{L}_2^2/4)$$

$$- \exp(\mathbf{L}_1^2/4)\nabla_1\omega_{12}^c\exp(-\mathbf{L}_1^2/4 - \mathbf{L}_2^2/4)\nabla_2\exp(\mathbf{L}_2^2/4)$$

$$- \exp(\mathbf{L}_2^2/4)\nabla_2\omega_{21}^c\exp(-\mathbf{L}_1^2/4 - \mathbf{L}_2^2/4)\nabla_1\exp(\mathbf{L}_1^2/4) \qquad (1.16)$$

One remaining effect from the projected fast field is given by the redefined two-body potential with respect to which the torques $\mathbf{T}_1$ and $\mathbf{T}_2$ are defined; the only modification is a redefined first rank potential parameter

$$V = -\mu_1 P_1(\Omega_2 - \Omega_1) - v_2 P_2(\Omega_2 - \Omega_1) \qquad (1.17)$$

$$v_1 \to v_1 + \mu_1\mu_2 \qquad (1.18)$$

and a constant term proportional to $\mu_1^2 + \mu_2^2$ that we neglect since it only affects the arbitrary zero of energy.

But the major contribution of the projected fast field to the resulting operator is given by a new frictional tensor (or collisional frequency tensor), which includes coupling terms between body 1 and 2 that are of a purely "dynamic" nature; that is, they do not affect the final equilibrium distribution. The collisional matrices, modified by the averaged action of the fast field, may be expressed in the following way:

$$\begin{pmatrix} \boldsymbol{\omega}_1^c & \boldsymbol{\omega}_{12}^c \\ \boldsymbol{\omega}_{21}^c & \boldsymbol{\omega}_2^c \end{pmatrix} = \begin{bmatrix} \omega_1^c\mathbf{1} - \omega_1\mathbf{U}_1^2 & -(\omega_1\omega_2)^{1/2}\mathbf{U}_1\mathbf{U}_2 \\ -(\omega_1\omega_2)^{1/2}\mathbf{U}_2\mathbf{U}_1 & \omega_2^c\mathbf{1} - \omega_2\mathbf{U}_2^2 \end{bmatrix} \qquad (1.19)$$

where ω_1 and ω_2 are proportional to the field collisional frequency ω_X^c

$$\omega_n = \mu_n^2 \frac{\omega_n^{s2}\omega_X^c}{\omega_X^{s2}} \qquad (1.20)$$

with $n = 1, 2,$; $\mathbf{U}_1$ and $\mathbf{U}_2$ are angular dependent 3×3 matrices defined as

$$\mathbf{U}_n \equiv -i\hat{\mathbf{J}}_n \otimes \mathbf{u}_n \qquad (1.21)$$

that is, the pqth Cartesian component of $\mathbf{U}_j$ is proportional to the result of the application of the p component of $\hat{\mathbf{J}}_I$ on the q component of the unitary vector $\mathbf{u}_J$. Note that the new collisional matrix is naturally a symmetric and positive definite matrix. If it is evaluated in the molecular frame fixed on body 1 (2), the diagonal block for body 1 (2) is a constant diagonal one, while the diagonal friction block for body 2 (1) and the coupling friction blocks are only dependent upon the relative orientation $\mathbf{\Omega}_2 - \mathbf{\Omega}_1$.

The effect of the new frictional term can be important whenever a strong initial coupling is supposed to exist between the solute and the fast mode. It is not difficult to show that a close relation exists between the frictional coupling terms of our MFPKE and the Stillman and Freed augmented Fokker–Planck equation (AFPE) in the case of a so-called "fluctuating torque" model. A close analogy between AFPE and MFPKE formalisms can be easily achieved if we consider the motion of the second body as completely diffusive. One can eliminate as a fast variable the angular momentum $\mathbf{L}_2$ from the previous two-body MFPKE (cf. Eq. (1.16)), following again a TTOC scheme (see Appendix A). A new hybrid (partly inertial and partly diffusive) time evolution operator is found for the system $(\mathbf{\Omega}_1, \mathbf{\Omega}_2, \mathbf{L}_1)$ whose form is given as

$$
\begin{aligned}
\tilde{\Gamma} = {} & \omega_1^s (i\mathbf{L}_1 \hat{\mathbf{J}}_1 + \mathbf{T}_1 \nabla_1) - \exp(\mathbf{L}_1^2/4)\nabla_1 \omega_1' \exp(-\mathbf{L}_1^2/2)\nabla_1 \exp(\mathbf{L}_1^2/4) \\
& - i \exp(\mathbf{L}_1^2/4)\nabla_1 \mathbf{f} \exp(-\mathbf{L}_1^2/4 - V/2)\hat{\mathbf{J}}_1 \exp(V/2) \\
& - i \exp(V/2)\hat{\mathbf{J}}_1 \mathbf{f}^{\mathrm{tr}} \exp(-\mathbf{L}_1^2/4 - V/2)\nabla_1 \exp(\mathbf{L}_1^2/4) \\
& + \exp(V/2)\hat{\mathbf{J}}_2 \mathbf{D}_2^0 \exp(-V)\hat{\mathbf{J}}_2 \exp(V/2)
\end{aligned}
\tag{1.22}
$$

with new angle dependent matrices that are defined in terms of ω_1^c, ω_2^c and ω_{12}^c

$$
\boldsymbol{\omega}_1' = \boldsymbol{\omega}_1^c - \boldsymbol{\omega}_{12}^c (\boldsymbol{\omega}_2^c)^{-1} \boldsymbol{\omega}_{21}^c
\tag{1.23}
$$

$$
\mathbf{f} = \omega_1^s \boldsymbol{\omega}_{12}^c (\boldsymbol{\omega}_2^c)^{-1}
\tag{1.24}
$$

$$
\mathbf{D}_2^0 = \omega_2^{s2} (\boldsymbol{\omega}_2^c)^{-1}
\tag{1.25}
$$

This is a two-body AFPE that is fully equivalent to those described by Stillman and Freed, including *both* a fluctuating torque effect (matrix $\mathbf{f}$) and a slowly relaxing local structure (interaction potential V); the equivalence of the two approaches will be further investigated in the next section for the case of a planar model.

If the momentum $\mathbf{L}_1$ itself is considered as a fast relaxing variable, that is, the motion of the solute is supposed to be completely diffusive, then it is possible to further reduce the phase space to only the rotational coordinates $\mathbf{\Omega}_1$ and $\mathbf{\Omega}_2$. The two-body Smoluchowski operator that is left after performing the TTOC projection is

$$\tilde{\Gamma} = \exp(V/2)\hat{\mathbf{J}}_1\mathbf{D}_1\exp(-V)\hat{\mathbf{J}}_1\exp(V/2) + \exp(V/2)\hat{\mathbf{J}}_1\mathbf{D}_{12}$$

$$\times \exp(-V)\hat{\mathbf{J}}_2\exp(V/2) + \exp(V/2)\hat{\mathbf{J}}_2\mathbf{D}_{21}$$

$$\times \exp(-V)\hat{\mathbf{J}}_2\exp(V/2) + \exp(V/2)\hat{\mathbf{J}}_2\mathbf{D}_2\exp(-V)\hat{\mathbf{J}}_2\exp(V/2)$$

$$(1.26)$$

and we can again write down the diffusive matrix blocks in terms of ω_1^c, ω_2^c, ω_{12}^c

$$\mathbf{D}_1 = \omega_1^{s2}[\boldsymbol{\omega}_1^c - \boldsymbol{\omega}_{12}^c(\boldsymbol{\omega}_2^c)^{-1}\boldsymbol{\omega}_{21}^c]^{-1} \tag{1.27}$$

$$\mathbf{D}_2 = \omega_2^{s2}[\boldsymbol{\omega}_2^c - \boldsymbol{\omega}_{21}^c(\boldsymbol{\omega}_1^c)^{-1}\boldsymbol{\omega}_{12}^c]^{-1} \tag{1.28}$$

$$\mathbf{D}_{12} = \mathbf{D}_{21}^{tr} = \omega_1^s\omega_2^s[\boldsymbol{\omega}_{21}^c - \boldsymbol{\omega}_2^c(\boldsymbol{\omega}_{12}^c)^{-1}\boldsymbol{\omega}_1^c]^{-1} \tag{1.29}$$

In glassy liquids or supercooled organic fluids the viscosity affecting all the positional and orientational variables is supposed to be rather large. We can then consider a third reduced equation, describing the coupled evolution of $\mathbf{\Omega}_1$, $\mathbf{\Omega}_2$, $\mathbf{X}$, after a straightforward elimination of all the momenta $\mathbf{L}_1$, $\mathbf{L}_2$ and $\mathbf{P}$ from Eq. (1.14). We then easily obtain a three-body Smoluchowski equation with a 9×9 diffusion matrix that is diagonal and constant

$$\tilde{\Gamma} = -D_\mathbf{X}\exp(V/2)\nabla_\mathbf{X}\exp(-V)\nabla_\mathbf{X}\exp(V/2) + D_1\exp(V/2)\hat{\mathbf{J}}_1$$

$$\times \exp(-V)\hat{\mathbf{J}}_1\exp(V/2) + D_2\exp(V/2)\hat{\mathbf{J}}_2\exp(-V)\hat{\mathbf{J}}_2\exp(V/2)$$

$$(1.30)$$

and where the diffusion coefficients are related to the initial collisional frequencies, that is,

$$D_\mathbf{X} = \omega_\mathbf{X}^{s2}/\omega_\mathbf{X}^c, \qquad D_1 = \omega_1^{s2}/\omega_1^c, \qquad D_2 = \omega_2^{s2}/\omega_2^c.$$

D. Planar Model

There are several reasons for considering planar equivalents of some of the above 3D-models. First of all, the heavy matrix notation employed in the previous section can be discarded, and the number of degrees of freedom for the complete system is reduced from 18 to 8 (two polar angles of rotation, their conjugate momenta, which are proportional to the angular velocities, and two in-plane components for the reaction field plus their conjugate momenta). The numerical treatment of the resulting MFPK equation is easier, and a comparison between different levels of complexity in the dynamical description can be made; that is, one could consider the explicit effects of the static and the dynamic interaction between the two rotators in detail. In this way one can obtain useful insight for predicting the behavior of the much more difficult three-dimensional case. Also, one can use the planar model in order to test approximate analytical treatments.

Planar models are also important for comparing our work to some of the previous theoretical studies along the same lines, for example, the planar augmented Fokker–Planck equation described by Stillman and Freed (see next section) and the itinerant oscillator model of Coffey and Evans.

1. Planar Dipoles in a Polar Fluid

Let us consider a system made of two planar dipoles, reorienting in the xy plane of the laboratory frame, and interacting with the components X_1, X_2 of a stochastic field lying in the same plane. Our starting equation, the planar equivalent of equation (1.14) is much simplified. All the frequency matrices are now scalars, the precessional terms are obviously not present and only one angular variable for each rotator has to be considered. The complete time evolution operator in a rescaled and symmetrized form is then given by

$$
\tilde{\Gamma} = \omega_1^s \left(L_1 \frac{\partial}{\partial \phi_1} - \frac{\partial V}{\partial \phi_1} \frac{\partial}{\partial L_1} \right) - \omega_1^c \exp(L_1^2/4) \frac{\partial}{\partial L_1} \exp(-L_1^2/2) \frac{\partial}{\partial L_1}
$$
$$
\times \exp(L_1^2/4)
$$
$$
+ \omega_2^s \left(L_2 \frac{\partial}{\partial \phi_2} - \frac{\partial V}{\partial \phi_2} \frac{\partial}{\partial L_2} \right) - \omega_2^c \exp(L_2^2/4) \frac{\partial}{\partial L_2} \exp(-L_2^2/2) \frac{\partial}{\partial L_2}
$$
$$
\times \exp(L_2^2/4)
$$
$$
+ \omega_x^s \left(P_1 \frac{\partial}{\partial X_1} - \frac{\partial V}{\partial X_1} \frac{\partial}{\partial P_1} \right) - \omega_x^c \exp(p_1^2/4) \frac{\partial}{\partial P_1} \exp(-p_1^2/2) \frac{\partial}{\partial P_1}
$$
$$
\times \exp(p_1^2/4)
$$
$$
+ \omega_x^s \left(P_2 \frac{\partial}{\partial X_2} - \frac{\partial V}{\partial X_2} \frac{\partial}{\partial P_2} \right) - \omega_x^c \exp(p_2^2/4) \frac{\partial}{\partial P_2} \exp(-p_2^2/2) \frac{\partial}{\partial P_2}
$$
$$
\times \exp(p_2^2/4) \tag{1.31}
$$

The potential function for the system is chosen to be

$$V = -v_1 \cos(\phi_1 - \phi_2) - v_2 \cos 2(\phi_1 - \phi_2)$$
$$- (\mu_1 \cos \phi_1 + \mu_2 \cos \phi_2)X_1 - (\mu_1 \sin \phi_1 + \mu_2 \sin \phi_2)X_2 + \tfrac{1}{2}X_1^2 + \tfrac{1}{2}X_2^2$$
$$(1.32)$$

We can now use our projection technique to recover averaged time evolution operators in which some of the system coordinates are considered as fast. An interesting case is given by the model in which the solvent polarization relaxes faster than the reorientational molecular modes, that is, the equivalent of equation (1.16). Note that now the matrices $\mathbf{U}_{1,2}$ (where the subscripts 1, 2 imply we are referring to both $\mathbf{U}_1$ and $\mathbf{U}_2$), are simply given by $(-\sin \phi_{1,2}, \cos \phi_{1,2})^{tr}$ and the resulting diagonal elements of the final friction matrix are constant, so

$$\tilde{\Gamma} = \omega_1^s \left(L_1 \frac{\partial}{\partial \phi_1} - \frac{\partial V}{\partial \phi_1} \frac{\partial}{\partial L_1} \right) - \omega_1 \exp(L_1^2/4) \frac{\partial}{\partial L_1} \exp(-L_1^2/2) \frac{\partial}{\partial L_1}$$
$$\times \exp(L_1^2/4)$$
$$+ \omega_2^s \left(L_2 \frac{\partial}{\partial \phi_2} - \frac{\partial V}{\partial \phi_2} \frac{\partial}{\partial L_2} \right) - \omega_2 \exp(L_2^2/4) \frac{\partial}{\partial L_2} \exp(-L_2^2/2) \frac{\partial}{\partial L_2}$$
$$\times \exp(L_2^2/4)$$
$$- \omega_{12} \exp(L_1^2/4) \frac{\partial}{\partial L_1} \exp(-L_1^2/4 - L_2^2/4) \frac{\partial}{\partial L_2} \exp(L_2^2/4)$$
$$- \omega_{21} \exp(L_2^2/4) \frac{\partial}{\partial L_2} \exp(-L_1^2/4 - L_2^2/4) \frac{\partial}{\partial L_1} \exp(L_1^2/4) \quad (1.33)$$

and now $\omega_{1,2}$ and ω_{12} are

$$\omega_1 = \omega_1^c + \frac{\mu_1^2 \omega_1^{s2}}{\omega_X^{s2}} \omega_X^c \quad (1.34)$$

$$\omega_2 = \omega_2^c + \frac{\mu_2^2 \omega_2^{s2}}{\omega_X^{s2}} \omega_X^c \quad (1.35)$$

$$\omega_{12} = \omega_{21} = \frac{\mu_1 \mu_2 \omega_1^s \omega_2^s}{\omega_X^{s2}} \cos(\phi_1 - \phi_2) \quad (1.36)$$

and the potential V is again the direct interaction between the two planar rotators, with a renormalized u_1. The diagonal terms of the friction matrix have the well known Nee–Zwanzig form for the friction exerted by a polar viscous fluid on a reorienting dipole (dielectric friction). This is not surprising, since our model considers for simplicity only a first rank interaction between the system and its environment. Note that the frictional coupling depends explicitly only on the relative orientation (in this planar model the difference angle between the absolute angles ϕ_1 and ϕ_2), as in the case of the three-dimensional model. If one neglects the frictional coupling terms, what is left is the IOM equation for two Brownian dipoles proposed by Coffey and Evans.

E. Augmented Fokker–Planck Equations and MFPKEs

The model proposed by Stillman and Freed (SF) in their 1980 paper [33] is very versatile. By choosing carefully (i) the *coupling forces* between molecule variables (x_1) and augmented ones (x_2), and (ii) the potential function in the final equilibrium distribution, one can easily recover a variety of mathematical forms, reflecting different physical cases. The SF procedure starts from considering a system coupled to a second one in a deterministic way (interaction potential); the latter, in the absence of any coupling is described by a FP operator. The first step to obtain a description of the full system is to write the stochastic Liouville equation (SLE), according to Kubo [44] and Freed [45]

$$\frac{\partial}{\partial t} P(\mathbf{x}_1, \mathbf{p}_1, \mathbf{x}_2, t) = -(\hat{\mathscr{L}}_1 + \hat{R}_2) P(\mathbf{x}_1, \mathbf{p}_1, \mathbf{x}_2, t) \qquad (1.37)$$

The Liouville operator $\hat{\mathscr{L}}_1$ contains a potential term depending on $\mathbf{x}_2$; the Fokker–Planck operator $\hat{R}_2$ is considered for the sake of simplicity merely diffusive (so that $\mathbf{p}_2$ does not enter into the calculation). The SLE is not rigorous, since it does not contain terms related to the back reaction of system 1 on system 2. That is, it does not tend to the correct equilibrium, zero eigenvalue, solution. Stillman and Freed "complete" it by requiring that a given equilibrium solution P_{eq} is recoverable. They accomplish this by modifying some reversible or irreversible drift terms, in a manner consistent with the Graham–Haken relations [46], which are based upon detailed balance, as well as with physical intuition. This finally leads to an augmented Fokker–Planck operator for the probability function. A number of points can now be highlighted. (1) The only physical aspects of the model are the interaction force $\mathbf{f}(\mathbf{x}_1, \mathbf{x}_2)$ in $\hat{\mathscr{L}}_1$ and the potential function $V(\mathbf{x}_1, \mathbf{x}_2)$ defining P_{eq}; (2) the result accounts for the back reaction of 1 on 2; (3) one can usually obtain an ALE (augmented Langevin equation) from the AFPE; (4) as long as sensible choices of $\mathbf{f}$ and V are made, SF are able to show that the basic FP equation can be recovered, in the limit when $\mathbf{x}_2$ or $\mathbf{p}_1$ become fast variables; (5) two main classes of models have been obtained: *fluctuating torque models* (only frictional effects are found), and *slowly relaxing local structure models* (no frictional effects, but a reorganization of the potential energy is found). Finally an AFPE can be generalized to contain spin-dependent terms, treating the spin Hamiltonian as a potential. Also, other fast modes can be added in a simple way as collisional operators in the AFPE. On the other hand, some aspects of the entire procedure are not well defined. One starts with a flawed formulation (i.e., the SLE does not obey detailed balance); the next step (i.e., the modification based on detailed balance conditions) is not uniquely defined and requires physical intuition. The MFPKE while initially more constraining, leads to a more precisely defined formulation. The relation between MFPKE and AFPE is better understood in the context of the general properties of Fokker–Planck operators, that are briefly reviewed in the next section.

1. Fokker–Planck Operators: The Graham–Haken Conditions

The general operator of a FP operator is

$$\hat{\Gamma} = \sum_i \frac{\partial}{\partial q_i} K_i - \sum_{ij} \frac{\partial_2}{\partial q_i \partial q_j} K_{ij} \tag{1.38}$$

where q_i are a set of general variables and K_{ij} is a symmetric tensor. Haken defines the irreversible and reversible drift coefficients as

$$D_i = \tfrac{1}{2}(K_i + \epsilon_i K_i) \tag{1.39}$$

$$J_i = \tfrac{1}{2}(K_i - \epsilon_i K_i) \tag{1.40}$$

where $\mathcal{T} q_i = \epsilon_i q_i$ ($\epsilon_i = \pm 1$), $\mathcal{T}$ the time reversal operator. In order that the FP has the stationary solution $P_{\mathrm{eq}} = \mathcal{N} \exp(-V)$ it follows that

$$K_{ij} = \epsilon_i \epsilon_j \mathcal{T} K_{ij} \tag{1.41}$$

$$D_i - \sum_j \frac{\partial K_{ij}}{\partial q_j} = -\sum_j K_{ij} \frac{\partial V}{\partial q_j} \tag{1.42}$$

$$\sum_i \left(\frac{\partial J_i}{\partial q_i} - J_i \frac{\partial V}{\partial q_i} \right) = 0 \tag{1.43}$$

(note that $\mathcal{T} V = V$). An alternative form of Eq. (1.38) may be obtained, in vector notation as

$$\hat{\Gamma} = \left(\frac{\partial}{\partial \mathbf{q}} \right) \mathbf{J} - \left(\frac{\partial}{\partial \mathbf{q}} \right) \mathbf{K} P_{\mathrm{eq}} \left(\frac{\partial}{\partial \mathbf{q}} \right) P_{\mathrm{eq}}^{-1} \tag{1.44}$$

In the following we shall write a general FP operator having the equilibrium solution P_{eq} in the form of Eq. (1.44). The vector $\mathbf{J}$ satisfies the following relations

$$\mathscr{T}\mathbf{J} = -\mathbf{J} \tag{1.45}$$

$$\left(\frac{\partial}{\partial \mathbf{q}}\right) \mathbf{J} P_{eq} = \mathbf{0} \tag{1.46}$$

When $\mathbf{J} = 0$ one recovers the so-called "potential condition", which means that the operator $\hat{\Gamma}$ has no reversible part.

2. Analysis of a Simple System According to the Stillman–Freed Procedure

We consider here for simplicity a one-dimensional system constructed from a generalized solute coordinate x_1 and its conjugate momentum p_1 coupled to a diffusive solvent coordinate x_2 via a potential $V = V_1(x_1) + V_2(x_2) + V_{int}(x_1, x_2)$. According to SF, the (renormalized and rescaled) stochastic Liouville operator is

$$\tilde{\Gamma} = \omega_1^s \left(p_1 \frac{\partial}{\partial x_1} - \frac{\partial V_1}{\partial x_1} \frac{\partial}{\partial p_1} - \frac{\partial V_{int}}{\partial x_1} \frac{\partial}{\partial p_1} \right) - \omega_1^c \exp(p_1^2/4) \frac{\partial}{\partial p_1}$$

$$\times \exp(-p_1^2/2) \frac{\partial}{\partial p_1} \exp(p_1^2/4)$$

$$- D_2 \exp(V_2/2) \frac{\partial}{\partial x_2} \exp(-V_2) \frac{\partial}{\partial x_2} \exp(V_2/2) \tag{1.47}$$

The SL operator is given simply by the sum of the FPK operator for subsystem 1 plus the Smoluchowski operator for subsystem 2. It is not complete, in the sense that it does not have a meaningful solution for $t \to +\infty$, which should be the equilibrium distribution. If we require that the total system tends to the Boltzmann distribution given by the total energy (including the interaction term V_{int})

$$P_{eq} \propto \exp[-(p_1^2/2 + V_1 + V_2 + V_{int})] \tag{1.48}$$

the slowly relaxing local structure model will be recovered. In this case SF modify the irreversible term in x_2 in a way that is equivalent to substituting V_2 with V in the Smoluchowski part of the operator

$$\tilde{\Gamma} = \omega_1^s \left(p_1 \frac{\partial}{\partial x_1} - \frac{\partial V}{\partial x_1} \frac{\partial}{\partial p_1} \right) - \omega_1^c \exp(p_1^2/4) \frac{\partial}{\partial p_1}$$

$$\times \exp(-p_1^2/2) \frac{\partial}{\partial p_1} \exp(p_1^2/4)$$

$$- D_2 \exp(V/2) \frac{\partial}{\partial x_2} \exp(-V) \frac{\partial}{\partial x_2} \exp(V/2) \tag{1.49}$$

For clarity the streaming term for subsystem 1 has been rewritten with respect to the total V; (obviously $\partial V_2/\partial x_1 = 0$). If the equilibrium is required to be independent of the interaction energy, that is,

$$P_{eq} \propto \exp[-(p_1^2/2 + V_1 + V_2)] \tag{1.50}$$

a fluctuating torque model is obtained, with an AFP operator written as

$$\tilde{\Gamma} = \omega_1^s \left(p_1 \frac{\partial}{\partial x_1} - \frac{\partial V}{\partial x_1} \frac{\partial}{\partial p_1} \right) - \omega_1^c \exp(p_1^2/4) \frac{\partial}{\partial p_1} \exp(-p_1^2/2) \frac{\partial}{\partial p_1}$$

$$\times \exp(p_1^2/4)$$

$$- \omega_1^s \left(\exp(V/2) \frac{\partial}{\partial x_2} f \exp(-V/2 - p_1^2/4) \frac{\partial}{\partial p_1} \exp(p_1^2/4) \right.$$

$$+ \exp(p_1^2/4) \frac{\partial}{\partial p_1} f \exp(-V/2 - p_1^2/4) \frac{\partial}{\partial x_2} \exp(V/2) \Big)$$

$$- D_2 \exp(V/2) \frac{\partial}{\partial x_2} \exp(-V) \frac{\partial}{\partial x_2} \exp(V/2) \tag{1.51}$$

where V is now simply $V_1 + V_2$, and the function f is defined by

$$f = \exp(V_2) \int dx_2 \exp(-V_2) \left(\frac{\partial V_{int}}{\partial x_1} \right) \tag{1.52}$$

These are essentially SF Eqs. (4.4) (SRLS case) and (2.36) (FT case).

3. MFPKE Approach

It is easy to show that the AFPEs obtained in the previous section can be recovered from a complete system (x_1, p_1, x_2, p_2). Let us consider a FPK operator defined with the potential $V(x_1, x_2)$ and the collisional matrix

$$\boldsymbol{\omega}^c = \begin{pmatrix} \omega_1^c & \omega_{int} \\ \omega_{int} & \omega_2^c \end{pmatrix} \tag{1.53}$$

where ω_{int} is a general function of x_1, x_2, which we shall see in the following is closely related to the function f used in the SF procedure. The total MFKP operator is

$$\tilde{\Gamma} = \omega_1^s \left(p_1 \frac{\partial}{\partial x_1} - \frac{\partial V}{\partial x_1} \frac{\partial}{\partial p_1} \right) - \omega_1^c \exp(p_1^2/4) \frac{\partial}{\partial p_1}$$

$$\times \exp(-p_1^2/2) \frac{\partial}{\partial p_1} \exp(p_1^2/4)$$

$$+ \omega_2^s \left(p_2 \frac{\partial}{\partial x_2} - \frac{\partial V}{\partial x_2} \frac{\partial}{\partial p_2} \right) - \omega_2^c \exp(p_2^2/4) \frac{\partial}{\partial p_2}$$

$$\times \exp(-p_2^2/2) \frac{\partial}{\partial p_2} \exp(p_2^2/4)$$

$$- \omega_1^c \exp(p_1^2/4) \frac{\partial}{\partial p_1} \exp(-p_1^2/2) \frac{\partial}{\partial p_1} \exp(p_1^2/4) - \omega_{int}$$

$$\times \exp(p_1^2/4) \frac{\partial}{\partial p_1} \exp(-p_1^2/4 - p_2^2/4) \frac{\partial}{\partial p_2} \exp(p_2^2/4)$$

$$- \omega_2^c \exp(p_2^2/4) \frac{\partial}{\partial p_2} \exp(-p_2^2/2) \frac{\partial}{\partial p_2} \exp(p_2^2/4) - \omega_{int}$$

$$\times \exp(p_2^2/4) \frac{\partial}{\partial p_2} \exp(-p_2^2/4 - p_1^2/4) \frac{\partial}{\partial p_1} \exp(p_1^2/4) \qquad (1.54)$$

Let us now consider the projected operator obtained when p_2 is a fast variable, so that subsystem 2 is diffusive. Following the TTOC procedure, a reduced MFKP operator is recovered that is given by

$$\tilde{\Gamma} = \omega_1^s \left(p_1 \frac{\partial}{\partial x_1} - \frac{\partial V}{\partial x_1} \frac{\partial}{\partial p_1} \right) + \omega_1' \hat{S}_1^+ \hat{S}_1^-$$

$$- \omega_1^s \left[\left(\frac{\partial}{\partial x_2} - \frac{1}{2} \frac{\partial V}{\partial x_2} \right) g \hat{S}_1^- + g \hat{S}_1^+ \left(\frac{\partial}{\partial x_2} - \frac{1}{2} \frac{\partial V}{\partial x_2} \right) \right]$$

$$- \frac{\omega_1^{s2}}{\omega_2^c} \exp(V/2) \frac{\partial}{\partial x_2} \exp(-V) \frac{\partial}{\partial x_2} \exp(V/2) \qquad (1.55)$$

where ω_1' and g are given by

$$\omega_1' = \omega_1^c - \frac{\omega_{int}^2}{\omega_2} \qquad (1.56)$$

$$g = \frac{\omega_{int}}{\omega_2^c} \qquad (1.57)$$

and the $\hat{S}_1^{\pm}$ are the lowering and raising operators ($p_1/2 \mp \partial/\partial p_1$). This reduced operator is a unified form for the cases treated by SF provided that one does not consider as an additive contribution the correction to ω_1^c. (This is due to the fact that the simple treatment of SF merely adds the collisional term in p_1 as a contribution of other unspecified "fast modes" without considering in detail any dependence of the friction coefficient for the first system). For instance, if ω_{int} is chosen to be zero, the SRLS case is recovered; while if V is given only by $V_1(x_1)$ and $V_2(x_2)$ and ω_{int} is not zero, the FT case is found (just identify the SF function f with the actual g, thus relating the roles of V_{int} in the SF approach and ω_{int} in the MFPK model through Eqs. (1.52) and (1.57)). From a purely algebraic point of view it is straightforward to understand why the AFPEs recovered by SF are so intimately related to a bidimensional MFPKE. In fact, it is clear that SF can obtain a model that is consistent with simple MFPKE provided that they modify, according to Haken's conditions, only the irreversible drift coefficients (vector $\mathbf{D}$) and the reversible drift coefficients (vector $\mathbf{J}$) without changing the assumed diffusion tensor (matrix $\mathbf{K}$). The initial system in the SF derivation is made by a Kramers subsystem (x_1, p_1) and by a diffusive one x_2

$$P_{\text{eq}} = \mathcal{N} \exp(-\tfrac{1}{2} p_1^2 - V_1 - V_2) \tag{1.58}$$

$$\mathbf{J} = \begin{pmatrix} \omega_1^s p_1 \\ -\omega_1^s \dfrac{\partial V_1}{\partial x_1} \\ 0 \end{pmatrix} \tag{1.59}$$

$$\mathbf{K} = \begin{pmatrix} 0 & 0 & 0 \\ 0 & \omega_1^c & 0 \\ 0 & 0 & \omega_2^c \end{pmatrix} \tag{1.60}$$

Here J_1 is associated with x_1, J_2 with p_1, J_3 with x_2. The SL approach requires that we modify $\mathbf{J}$ by adding a term to the partial derivative of V_{int} with respect to x_1

$$\mathbf{J} = \begin{pmatrix} \omega_1^s p_1 \\ -\omega_1^s \dfrac{\partial V_1}{\partial x_1} - \omega_s \dfrac{\partial V_{\text{int}}}{\partial x_1} \\ 0 \end{pmatrix} \tag{1.61}$$

This is the reactive force on the first system as a result of its interaction with the second system. In order to obtain a proper equation in the SRLS case, SF modify the irreversible term in x_2, that is, D_3. In the present notation this is merely equivalent to substituting Eq. (1.59) by Eq. (1.48). In the FT case SF modify J_3; that is, they add a term $-\omega_1^s p_1 f$ to the reversible drift coefficient in x_2, which was previously zero, and leave P_{eq} unmodified. In both cases these are the minimal modifications required to achieve detailed balance. No changes in the diffusion tensor elements are introduced, although such possibilities exist. This "minimum effort" choice yields equations derivable from a MFPKE in which the full set of variable (x_1, p_1, x_2, p_2) is considered.

F. Discussion and Summary of Methodology

In the final section of Section I we summarize our methodology and we discuss briefly some of the recent theoretical contributions of other authors, that we have found to be useful or complementary to our techniques.

1. Discussion

In the past 10 years or so, there have been a number of theoretical contributions to the fundamental problem of describing fluids in a mesoscopic context. If one wants to go beyond the usual Debye formulation, it is evident that the simplicity of one-body stochastic models must be abandoned. Stochastic models which are able to describe the dynamical behavior of a complex liquid (for instance, a highly viscous solution), exact their price in terms of a more involved formalism. One must be careful to achieve a balance between complexity in formulation and new information gained from the model. Often one can resort to a phenomenological model, which may or may not be the starting point for a more complete (and complicated) theoretical treatment.

Kivelson and co-workers [36, 37] have recently given some useful suggestions. Their models of liquids at high viscosity are "simple" and relatively easy to discuss: for instance, in [37] three different dynamical models are tested to predict some of the known properties of glassy liquids (a single body relaxing in a potential cage subjected to slow diffusion (a), or to a strong collision motion (b); or in the presence of torsional barriers (c)). Unfortunately, a purely qualitative discussion may be not sufficient to analyze "simple" models. It is necessary (i) to define exactly all the physical (and *mathematical*) hypotheses underlying a given model and then (ii) to treat it computationally in a rigorous way, in order to gain a complete understanding. In this chapter so far, we have attempted to clarify the first point, that is, we have described what we consider a useful methodology to define exactly the "equation of motions" of complex liquids. In Section II we consider the second point, and we present a systematic study of two- and three-body stochastic models, together with the description of the formal tools necessary to deal with the multidimensional Fokker–Planck operators in three dimensions.

We have chosen to encompass our methodology in the necessarily limited framework of rotational FPK operators for describing the solute molecule and the solvent cages (slow fluctuating solvent structures); with translational FPK operators for describing stochastic fields (fast fluctuating solvent structures). We are aware that a truly complete description should also include in a many-body stochastic view, the interaction between the rotational and translational degrees of freedom of the solute and/or of the solvent. In addition, one can use different formal approaches to obtain improved (i.e., many-body) kinetic equations for the orientational distribution of a solute molecule strongly interacting with the solvent. In this respect, Bagchi et al. [47] have recently provided an analysis for explaining the anomalous rotational behavior of glassy liquids by including the translational motions of the solvent molecules and the density fluctuations of the solvent in the Debye–Smoluchowski description, which is particularly interesting since it could provide links between mesoscopic stochastic theories and advanced microscopic and mode-mode coupling treatments. They obtain an integro-differential kinetic equation in the orientational distribution probability function of the solute, which is appropriate for highly viscous fluids only. No explicit mean field potential or inertial effects are included.

Finally rototranslational coupling has been investigated in two recent papers by Wey and Patey [48, 49], using the general approach of the Van Hove functions described within the Kerr approximation, which relates the rototranslational correlation function of the solute to the joint conditional probability in both the position and orientation of the molecule. This method is helpful in providing a physical and mathematical framework for rototranslational coupling in complex fluids. However, it requires as a starting point a well defined equation of motion for the conditional probability. Wey and Patey have tested only one-body stochastic equations (such as the Fick–Debye and the Berne–Pecora equations), which are necessarily restricted.

2. Summary

We have attempted to provide a general approach to build multi-dimensional stochastic operators of the Fokker–Planck–Kramers type, for describing the time evolution of an extended set of degrees of freedom in complex liquids. This set contains the orientation of a probe molecule (first body) and its conjugate angular momentum vector, plus similar coordinates for a collection of N bodies. Each of them is an additional *solvent body*. Also, a collection of K stochastic fields is introduced. The time evolution operator for the system of $6 \times (N + K + 1)$ degrees of freedom is given by a sum over rotational and translational FPK operators. The only source of coupling (at this stage) is given by a potential depending on the mutual orientations of each body and field.

For the case of two rotators and one stochastic field ($N = 1$ and $K = 1$), it has been shown (using a TTOC expansion procedure) how to eliminate, as fast variables, some of the original degrees of freedom (e.g., the stochastic field and its momentum) in order to obtain models which contain coupling terms just in the friction tensor of the rotators. The reduced two-body Fokker–Planck–Kramers (2BFPK) equation has been shown to be formally equivalent to the augmented Fokker–Planck equation described by Stillman and Freed [33]. In the planar case, that is, when both the probe and the solvent body are described as planar dipoles, and any residual frictional effect due to a fast field is neglected, one obtains the IOM equation of Coffey and Evans [23–25].

II. COMPUTATIONAL TREATMENT

A. Introduction

In the first section we have discussed a general methodology for the theoretical description of rotational dynamics of rigid solute molecules in complex solvents. Many-body Fokker–Planck–Kramers equations (MFPKE), including collective solvent degrees of freedom (either rotational ones, i.e., rigid bodies, or translational ones, i.e., vector fields), and their conjugate momenta, have been described as convenient tools to reproduce (or simulate) the complexity of an actual liquid system.

In Section II, we apply our stochastic models to physical systems of interest. Although the methodology was developed mainly to interpret complex features of ESR spectra over a wide range of temperatures, viscosities and solvent compositions, we believe that it could profitably be applied to many other experimental techniques, sensitive to rotational dynamics effects (such as dielectric relaxation, Raman and neutron scattering, NMR measurements) in liquids. Preliminary results on two- and three-body models, have been encouraging for the study of "slowly relaxing local structure" (SRLS) and "fluctuating torque" (FT) effects in isotropic liquids at moderate and high viscosities [39]; and for the interpretation of the bifurcation phenomenon in glassy and supercooled fluids [40]. Here we describe in detail the computational approach that is needed to treat many-body MFPK operators, provide extensive results on several rotational models, and discuss their application for interpreting liquid behavior.

In Section II.B we briefly review the usage of the complex symmetric Lanczos algorithm for treating MFPK operators, with particular attention to the problem of the choice of a suitable set of basis functions for a many-body problem. In Section II.C we consider the case of two spherical rotators in a highly damped fluid (Smoluchowski regime) as a first example of the application of angular momentum coupling techniques to Fokker–Planck operators (two-body Smoluchowski model, 2BSM). This approach is extended in Section II.D for studying a three-body system (two rotators plus one field), again in the overdamped regime (3BSM). Sections II.E. and II.F are devoted to the analysis of two-body models in the full phase space of rotational coordinates *and* momenta of the two rotators (two-body Kramers models, 2BKM), for a total of twelve degrees of freedom, all fully coupled together, at least in principle. Section II.G. contains a discussion of results concerning the various models. Rotational correlation functions and momentum correlation functions for body 1 are discussed, together with their dominant eigenvalues; a detailed analysis of the dominant eigenmodes of the system is given in each case.

Finally, a comparison of the MFPKE approach with molecular dynamics, ESR and stimulated light scattering experiments is contained in Section II.H. Detailed formulations of reduced matrix elements are given in Appendix C.

B. Computational Strategy

A powerful and general method for numerical solution of Fokker–Planck (FP) operators has been given by Moro and Freed [50]. It involves first establishing a complex symmetric matrix representation with a basis set of orthonormal functions, followed by a tridiagonalization procedure utilizing the Lanczos algorithm. The usage of the conjugate gradient algorithm as an alternative procedure to tridiagonalize the initial matrix has been considered by Vasavada et al. [51]. A thorough review of the usage of iterative algorithms for solving stochastic Liouville and FP equations has been provided by Schneider and Freed [42]. The interested reader can consult this reference for further details. In this section we will focus our attention on the optimization, for the many-body systems considered, of the matrix representation rather than on the detailed computational treatment of the matrix itself.

We start with the time-dependent conditional probability for the stochastic system $P(\mathbf{q}^0|\mathbf{q}, t)$, where $\mathbf{q}$ is a complete set of stochastic variables. The time evolution of P is governed by the Fokker–Planck–Kramers (FPK) equation [cf. Eq. (1.1)]:

$$\frac{\partial}{\partial t} P(\mathbf{q}^0|\mathbf{q}, t) = -\hat{\Gamma} P(\mathbf{q}^0|\mathbf{q}, t) \tag{2.1}$$

with the intial condition [cf. Eq. (1.2)]

$$P(\mathbf{q}^0|\mathbf{q}, 0) = \delta(\mathbf{q} - \mathbf{q}^0) \tag{2.2}$$

In the following, $\mathbf{q}$ will be the collection of rotational coordinates $\mathbf{\Omega}_1$,

$\Omega_2, \ldots, \Omega_{N+1}$ and fields $X_1, X_2, \ldots, X_K$ and of their conjugate momenta $L_1, L_2, \ldots, L_{N+1}$ and $P_1, P_2, \ldots, P_K$ (cf. Section I.B.1). The operator $\hat{\Gamma}$ is given as a sum of FPK operators, each of them defined in the (Ω_n, L_n) or (X_k, P_k) subspace. The total energy E of the system is given by the potential energy of interaction plus the total kinetic energy, and it defines the equilibrium distribution $P_{eq}(q)$, that is, the unique eigenfunction of $\hat{\Gamma}$ with a zero eigenvalue. Thus

$$E(\mathbf{q}) = U(\Omega_1, \Omega_2, \ldots, \Omega_{N+1}, X_1, X_2, \ldots, X_K)$$
$$+ \frac{1}{2} \sum_{n=1}^{N+1} L_n I_n^{-1} L_n + \frac{1}{2} \sum_{k=1}^{K} P_k M_k^{-1} P_k \qquad (2.3)$$

$$P_{eq}(\mathbf{q}) = \frac{\exp[-E(\mathbf{q})/k_B T]}{\langle \exp[-E(\mathbf{q})/k_B T] \rangle} \qquad (2.4)$$

where $\langle \; \rangle$ standard for the integration on the full phase-space of $\mathbf{q}$ coordinates and momenta. It is useful to apply a similarity transformation to $\hat{\Gamma}$, which renders it possible to obtain a complex symmetric matrix representation of the operator (or a real symmetric one, if $\hat{\Gamma}$ is Hermitean). The transformation is simply [cf. Eq. (1.13)]

$$\tilde{\Gamma} = P_{eq}^{-1/2} \hat{\Gamma} P_{eq}^{1/2} \qquad (2.5)$$

note that the "symmetrized" operator has the same eigenvalues as the unsymmetrized one, while the eigenfunctions are multiplied by $P_{eq}^{-1/2}$. Then by representing $\tilde{\Gamma}$ in a complete orthonormal set of basis functions that are invariant under the classical time reversal operation, a complex symmetric matrix representation is guaranteed [42].

1. Correlation Functions, Spectral Densities and Lanczos Algorithm

Usually we are interested in the (auto)correlation function $G(t)$ of an observable (i.e., a function of some stochastic coordinates). In the following we will consider either rotational correlation functions (i.e., involving the spherical harmonics $Y_{jm}(\Omega_1)$) or momentum correlation functions (i.e., involving the components of L_1) for the first rotator (body 1), identified as the solute molecule

$$G_{jm}^R(t) = \langle Y_{jm}(t) | Y_{jm}(0) \rangle = \langle Y_{jm}(\Omega_1) P_{eq}^{1/2} | \exp(-\tilde{\Gamma} t) | Y_{jm}(\Omega_1) P_{eq}^{1/2} \rangle \qquad (2.6)$$

$$G_m^J(t) = \langle L_m(t) | L_m(0) \rangle = \langle L_{1_m} P_{eq}^{1/2} | \exp(-\tilde{\Gamma} t) | L_{1_m} P_{eq}^{1/2} \rangle \qquad (2.7)$$

Instead of computing the correlation functions directly, one can take the Fourier–Laplace transforms, or spectral densities

$$J_{jm}^R(\omega) = \int_0^{+\infty} dt \exp(-i\omega t)\langle Y_{jm}(t)|Y_{jm}(0)\rangle$$

$$= \langle Y_{jm}(\mathbf{\Omega}_1)P_{eq}^{1/2}|(i\omega + \tilde{\Gamma})^{-1}|Y_{jm}(\mathbf{\Omega}_1)P_{eq}^{1/2}\rangle \qquad (2.8)$$

$$J_m^J(\omega) = \int_0^{+\infty} dt \exp(-i\omega t)\langle L_m(t)|L_m(0)\rangle$$

$$= \langle L_{1_m}P_{eq}^{1/2}|(i\omega + \tilde{\Gamma})^{-1}|L_{1_m}P_{eq}^{1/2}\rangle \qquad (2.9)$$

Following the procedure developed by Moro and Freed, one obtains a matrix representation of $\tilde{\Gamma}$ and a vector representation of the function $FP_{eq}^{1/2}$ (where F is the observable, for example, $Y_{jm}(\mathbf{\Omega}_1)$ or $\mathbf{L}_{1_i}$), utilizing an appropriate set of basis functions. Given the (complex) symmetric matrix Γ and the "right vector" $\mathbf{v}$ (formed from $FP_{eq}^{1/2}$), one is left with the evaluation of the resolvent, the generic form of which is

$$J(\omega) = \mathbf{v} \cdot (i\omega\mathbf{1} + \Gamma)^{-1} \cdot \mathbf{v} \qquad (2.10)$$

We shall set N be the dimension of the finite basis subset used to represent $\hat{\Gamma}$ and $\mathbf{v}$. The calculation can be performed with great efficiency using an iterative algorithm, such as the Lanczos algorithm, that transforms Γ into a tridiagonalized form. A continued fraction expansion is then obtained:

$$J^{(N,n)}(\omega) = \cfrac{1}{i\omega + \alpha_1 -} \cfrac{\beta_i^2}{i\omega + \alpha_2 -} \cfrac{\beta_2^2}{i\omega + \alpha_3 -} \cdots \cfrac{\beta_{n-2}^2}{i\omega + \alpha_{n-1} -} \cfrac{\beta_{n-1}^2}{i\omega + \alpha_n}$$
$$(2.11)$$

where n is the number of iterations (Lanczos steps) necessary to achieve convergence; usually $n \ll N$. The α_i coefficients are the diagonal elements of the tridiagonal complex symmetric matrix, whereas the β_i are the extradiagonal ones. Given the tridiagonal matrix, one can also calculate the eigenvalues λ_i associated with the spectral density, by means of an efficient diagonalization procedure for tridiagonal matrices (e.g., the QR algorithm).

In practice, although the entire procedure has been shown to be extremely effective in dealing with stochastic systems of 2–3 degrees of freedom (as well as in a stochastic Liouville equation with spin coordinates as well), its application to larger systems (with degrees of freedom ranging from 4 to as many as 12) is not so straightforward, because of a dramatic increase in computation time and memory space requirements, even if a powerful supercomputer is used. The bottleneck is usually the matrix dimension N, which can be very large. It is therefore of considerable importance to optimize the basis set utilized to represent the operator in order to minimize N.

C. Two-Body Smoluchowski Model

The model that we are going to consider in this section is given by two spherical rotators, simply called body 1 and body 2. Body 1 is the solute molecule, whereas body 2 is the instantaneous structure of solvent molecules in the immediate surroundings of the solute. The rest of the solvent is described as a homogeneous, isotropic and continuous viscous fluid. In the overdamped regime, the system is described by a Smoluchowski equation in the phase space $(\mathbf{\Omega}_1, \mathbf{\Omega}_2)$, where $\mathbf{\Omega}_1$ and $\mathbf{\Omega}_2$ are respectively the set of Euler angles specifying the orientation of a fixed frame on body 1 with respect the lab frame, and an analogous set for the orientation of a fixed frame on body 2.

In accordance with Table I, we will adopt from the beginning a dimensionless set of units. The symmetrized, rescaled time evolution operator for the model is then (compare with equation (1.30) for the three-body case)

$$\tilde{\Gamma} = D_1 P_{eq}^{-1/2} \hat{\mathbf{J}}_1 P_{eq} \hat{\mathbf{J}}_1 P_{eq}^{-1/2} + D_2 P_{eq}^{-1/2} \hat{\mathbf{J}}_2 P_{eq} \hat{\mathbf{J}}_2 P_{eq}^{-1/2} \qquad (2.12)$$

The equilibrium distribution function P_{eq} is defined according to Eq. (2.4); but the relevant part of the total energy is given just by the (rescaled to $k_B T$) potential energy function V (cf. Section I)

$$V(\mathbf{\Omega}_1, \mathbf{\Omega}_2) = -\sum_R v_R P_R(\mathbf{\Omega}_2 - \mathbf{\Omega}_1) \qquad (2.13)$$

where $P_R(\mathbf{\Omega})$ is the Legendre polynomial of rank R, and Eq. (2.13) implies that U depends only on the relative orientations of bodies 1 and 2 (the minus sign is only for convenience). We shall consider the expansion of Eq. (2.13) up to $R = 2$ (i.e., only first or second rank interactions are included). $\hat{\mathbf{J}}_1$ and $\hat{\mathbf{J}}_2$ are respectively the "angular momentum" operators for body 1 and for body 2 in the laboratory frame of reference. For future usage, we define also the total "angular momentum" operator of the system as

$$\hat{\mathbf{J}} = \hat{\mathbf{J}}_1 + \hat{\mathbf{J}}_2 \qquad (2.14)$$

and we rewrite $\tilde{\Gamma}$ in a more convenient form for the actual calculation of the matrix elements (although less elegant than Eq. (2.12)) as

$$\tilde{\Gamma} = D_1\hat{\mathbf{J}}_1^2 + D_2\hat{\mathbf{J}}_2^2 + D_1 G_1 + D_2 G_2 \tag{2.15}$$

where the functions G_m $(m = 1, 2)$ are defined

$$G_m = \frac{1}{4}\mathbf{T}_m^2 + \frac{i}{2}(\hat{\mathbf{J}}_m\mathbf{T}_m)_{\text{fun}} \tag{2.16}$$

and where $(\ )_{\text{fun}}$ indicates that what is contained within, acts as a function, not an operator. Also, $\mathbf{T}_m$ is the torque acting on body m due to V, that is,

$$\mathbf{T}_m = -i\hat{\mathbf{J}}_m V \tag{2.17}$$

1. Uncoupled and Coupled Basis Sets

A simple choice for a complete basis set of functions for obtaining a matrix representation Γ is the uncoupled set

$$|J_1 M_1 K_1; J_2 M_2 K_2\rangle = |J_1 M_1 K_1\rangle |J_2 M_2 K_2\rangle \tag{2.18}$$

where each function $|J_m M_m K_m\rangle$ is given by [53, 54]

$$|J_m M_m K_m\rangle = \frac{[J_m]^{1/2}}{(8\pi^2)^{1/2}} \mathscr{D}_{M_m K_m}^{J_m^*}(\Omega_m) \tag{2.19}$$

Hereafter we let $[J] = 2J + 1$. This is a complete orthonormal set given by the direct product of Wigner functions in the set of Euler angles Ω_1 and Ω_2. Note that since the phase space is six dimensional, we have six distinct quantum numbers to cope with. However, the potential V that we have chosen is independent of the azimuthal angles γ_1 and γ_2, and this is reflected in the fact that Γ will be diagonal in K_1 and K_2. In the following, the K_m quantum numbers will be discarded from any formula, if not otherwise specified, since only the matrix block with $K_1 = K_2 = 0$ will be of interest.

It is possible to further reduce the number of effective (nondiagonal) quantum numbers taking advantage of the spherical symmetry of the fluid to determine other "constants of the motion" (note however that all the following considerations also hold for molecules with cylindrical symmetry). Let us consider the tensorial properties of the functions and operators defined in the previous paragraph with respect the "total" angular momentum operator $\hat{\mathbf{J}}$ given by Eq. (2.14). Obviously, $\hat{\mathbf{J}}_1$ and $\hat{\mathbf{J}}_2$ are themselves first rank tensor (i.e., vector) operators. Furthermore, one may rewrite the Rth component of the potential

$$v_R P_R(\mathbf{\Omega}_1, \mathbf{\Omega}_2) = -V_R \sum_{S=-R}^{R} (-)^S |R - S0\rangle_1 |RS0\rangle_2 \qquad (2.20)$$

in a form clearly showing its nature as a zero rank tensor (scalar) with respect to $\hat{\mathbf{J}}$. Note that

$$v_R = \frac{V_R[R]}{8\pi^2} \qquad (2.21)$$

Since it is simply an exponential function of V, P_{eq} is also a scalar. It follows directly from Eq. (2.12) that $\tilde{\Gamma}$ itself is a scalar, as it must be to satisfy Eq. (2.1). One can also arrive at this result from Eq. (2.15), noting that $\mathbf{T}_1$ and $\mathbf{T}_2$ are vector operators. It is also easy to see that

$$\mathbf{T}_1 = -\mathbf{T}_2 = \mathbf{T} \qquad (2.22)$$

The vector $\mathbf{T}$ will simply be called the "torque" in the following, without specifying any index. Note also that $G_1 = G_2 = G$.

From these considerations, one concludes that the coupled basis set

$$|J_1 J_2 JM\rangle = \sum_{M_1 M_2} C(J_1 M_1 J_2 M_2 JM)|J_1 M_1\rangle |J_2 M_2\rangle \qquad (2.23)$$

where $C(J_1 M_1 J_2 M_2 JM)$ is a Clebsch–Gordan coefficient, is the most suitable set of basis functions for the present problem (K_1 and K_2 have been neglected). In fact, due to spherical symmetry, both J and M are "good" quantum numbers, that is, Γ is diagonal in them (note that this is still true for cylindrical spatial symmetry, while for the completely asymmetric case only J is a "good" quantum number).

The initial vector $\mathbf{v}$ must also be evaluated. Instead of computing directly the vector representation of the given rotational function (i.e., the spherical harmonic in $\mathbf{\Omega}_1$, which is an element of the uncoupled basis set), one can evaluate the *matrix* representation of the function, which we call $\mathbf{M}$, and then multiply it by the vector representation of $P_{eq}^{1/2}$, which we call $\mathbf{v}_0$, whose calculation is relatively easy utilizing the coupled basis set (see Appendix C.2). That is, let

$$\mathbf{v} = \mathbf{M}\mathbf{v}_0 \qquad (2.24)$$

Then

$$(\mathbf{v})_\Lambda = \langle \Lambda | F P_{eq}^{1/2} \rangle \qquad (2.25)$$

$$(\mathbf{v}_0)_\Lambda = \langle \Lambda | P_{eq}^{1/2} \rangle \qquad (2.26)$$

$$(\mathbf{M})_{\Lambda,\Lambda'} = \langle \Lambda | F | \Lambda' \rangle \qquad (2.27)$$

where Λ stands for the collection of quantum numbers, and the rotational observable F is simply the basis vector $|PQ0\rangle$, that is, the Qth component of the P rank rotational function in Ω_1. One can see by inspection that only the elements with $J = 0$, $M = 0$ are not zero in the vector $\mathbf{v}_0$, and that only the matrix block defined by the conditions $J' = 0$, $M' = 0$, $\Delta(JPJ')$, $M = Q$ has to be considered in $\mathbf{M}$ (Δ is the triangle condition); then the only nonzero elements of $\mathbf{v}$ are those for which $J = P$, $M = Q$. It follows that the only matrix block we need to compute in Γ satisfies the conditions $J = P$, $M = Q$.

2. Matrix Elements

A clear advantage of employing angular momentum coupling techniques is the possibility of using the Wigner–Eckart (WE) theorem to simplify the calculation of matrix elements in the coupled basis set [4]. In this two-body case, only two nondiagonal quantum numbers, J_1 and J_2 have to be considered. (In general, if $N + 1$ rotators are present, a generalized coupled basis set allows one to have $2N$ effective nondiagonal quantum numbers.)

Let us now consider $\tilde{\Gamma}$ given by Eq. (2.12). The terms proportional to $\hat{\mathbf{J}}_1^2$ and $\hat{\mathbf{J}}_2^2$ are diagonal; and we may write for the matrix element of Eq. (2.12)

$$\langle \Lambda | \tilde{\Gamma} | \Lambda' \rangle = [D_1 J_1(J_1 + 1) + D_2 J_2(J_2 + 1)]\delta_{\Lambda\Lambda'} + (D_1 + D_2)\langle \Lambda | G | \Lambda' \rangle$$

$$(2.28)$$

where the sets Λ and Λ' are characterized by $K_1 = K_1' = 0$, $K_2 = K_2' = 0$, $J = J' = P$, $M = M' = Q$. The matrix element of G is

$$\langle \Lambda | G | \Lambda' \rangle = \tfrac{1}{4} \langle \Lambda | \mathbf{T}^2 | \Lambda' \rangle = \frac{i}{2} \langle \Lambda | \hat{\mathbf{J}}_1 \cdot \mathbf{T} | \Lambda' \rangle \qquad (2.29)$$

that is, it is reduced to a sum of matrix elements of scalar products of operators of form $\hat{\mathbf{A}} \cdot \hat{\mathbf{B}}$. (For future convenience, we will call Γ_0 the complete matrix element *without the factor* $\delta_{JJ'}\delta_{MM'}$.)

In general one finds, from the WE theorem (weak form for noncommuting operators)

$$\langle \Lambda | \hat{\mathbf{A}} \cdot \hat{\mathbf{B}} | \Lambda' \rangle = [J]^{-1} \sum_{J_1'' J_2'' J''} (-)^{J+J''} (J_1 J_2 J \| \hat{\mathbf{A}} \| J_1'' J_2'' J'')$$

$$\times (J_1'' J_2'' J'' \| \hat{\mathbf{B}} \| J_1' J_2' J') \delta_{JJ'} \delta_{MM'} \qquad (2.30)$$

where $\hat{\mathbf{A}}$, $\hat{\mathbf{B}}$ are either $\hat{\mathbf{J}}_1$ or $\mathbf{T}$. The reduced matrix element of $\hat{\mathbf{J}}_1$ is given by (see [4])

$$(J_1 J_2 J \| \hat{\mathbf{J}}_1 \| J_1' J_2' J') = (-)^{J_1 + J_2 + J' + 1} [JJ']^{1/2} \begin{Bmatrix} J_1 & J & J_2 \\ J' & J_1 & 1 \end{Bmatrix}$$

$$\times [J_1 (J_1 + 1)(2J_1 + 1)]^{1/2} \delta_{J_1 J_1'} \delta_{J_2 J_2'} \qquad (2.31)$$

and the reduced matrix element of $\mathbf{T}$ is evaluated in Appendix C.1. The final matrix Γ is real symmetric. (Note that $[JJ'] \equiv [J][J']$).

The matrix element $(\mathbf{M})_{\Lambda\Lambda'}$ of Eq. (2.27) is easily computed from the WE theorem, and one obtains

$$\langle \Lambda | F | \Lambda' \rangle = \frac{[P]^{1/2}}{(8\pi^2)^{1/2}} [J_1]^{1/2} \begin{pmatrix} J_1 & P & J_1' \\ 0 & 0 & 0 \end{pmatrix} \delta_{J_2 J_2'} \delta_{J_1 J_2'} \qquad (2.32)$$

with $J = P$, $M = Q$, $J' = 0$, $M' = 0$ and $K_1 = K_1' = K_2 = K_2' = 0$. No explicit dependence on Q is present; that is, given the spherical symmetry, all the rotational correlation functions are independent of Q. The components of the vector $\mathbf{v}_0$ are calculated in Appendix C.2.

D. Three-Body Smoluchowski Model

A further elaboration in describing the rotational dynamics of a solute in a complex environment is obtained by increasing the number of interacting solvent modes included in the time evolution operator. Theoretically, one could consider a new set of collective degrees of freedom for each relaxation process that is relevant for the solute dynamics. In practice, computational problems soon arise. However, a three-body description can still be treated rather easily, and it is the subject of the present section. Instead of considering a third rotational set of coordinates, we have chosen to define the third "body" as a stochastic, vector-like *field* **X**. One can think of a polarization coordinate, or of the fluctuating solvent dipole moment interacting with the probe. We shall consider only first rank interaction between the solute body and the solvent structure with the stochastic field. The effect of a fast field, as a source of a "fluctuating torque" relaxation mechanism on the solute dynamics has already been partially explored. A summary of our computational results is presented in a later section. Here we deal with the formulation of the three-body model and its detailed mathematical treatment.

1. The Model

The symmetrized and rescaled time evolution operator for the system described by the set $(\mathbf{\Omega}_1, \mathbf{\Omega}_2, \mathbf{X})$ is simply defined adding to the two-body operator in Eq. (2.12) the translational Smoluchowski term for the field to obtain [cf. Eq. (1.30)]

$$\tilde{\Gamma} = D_1 P_{eq}^{-1/2} \hat{\mathbf{J}}_1 P_{eq} \hat{\mathbf{J}}_1 P_{eq}^{-1/2} + D_2 P_{eq}^{-1/2} \hat{\mathbf{J}}_2 P_{eq} \hat{\mathbf{J}}_2 P_{eq}^{-1/2}$$
$$- D_{\mathbf{X}} P_{eq}^{-1} \nabla_{\mathbf{X}} P_{eq} \nabla_{\mathbf{X}} P_{eq}^{-1/2} \qquad (2.33)$$

where $\nabla_{\mathbf{X}}$ is the gradient operator in the $\mathbf{X}$ subspace. The equilibrium distribution function is now defined with respect the following potential

$$V(\mathbf{\Omega}_1, \mathbf{\Omega}_2, \mathbf{X}) = V_0(\mathbf{\Omega}_1, \mathbf{\Omega}_2) - \mu_1 \mathbf{X} \mathbf{u}_1 - \mu_2 \mathbf{X} \mathbf{u}_2 + \tfrac{1}{2} \mathbf{X}^2 \qquad (2.34)$$

Note that the dimensionless units defined in Table I are used, so that the curvature along the $\mathbf{X}$ direction is renormalized to 1. Here U_0 is the two-body interaction potential defined in Eq. (2.13). The two terms linear in $\mathbf{X}$ are the "dipolar" interaction energy (with $\mathbf{u}_1$ and $\mathbf{u}_2$ two unit vectors, respectively, along the z-axis of the fixed frame for the solute and the solvent body, cf. Fig. 1). Finally a quadratic term in $\mathbf{X}$ has been added in order to confine the fluctuations of the stochastic field.

2. Matrix Representation

An efficient treatment of the time evolution operator defined in Eq. (2.33) can be achieved by performing a canonical transformation of coordinates acting on the field $\mathbf{X}$. We define the shifted vector $\mathbf{X} \rightarrow \mathbf{X} - \mu_1 \mathbf{u}_1 - \mu_2 \mathbf{u}_2$ as a new set of field coordinates. The potential is now decoupled

$$V(\mathbf{\Omega}_1, \mathbf{\Omega}_2, \mathbf{X}) = V_0(\mathbf{\Omega}_1, \mathbf{\Omega}_2) + \tfrac{1}{2} \mathbf{X}^2 \qquad (2.35)$$

Note, however that the first rank coefficient in V_0 is modified slightly as $v_1 \rightarrow v_1 + \mu_1 \mu_2$. Although the potential form is simplified, new terms arise in the operator itself. Skipping straightforward algebraic details, the following equation is obtained:

$$\tilde{\Gamma} = \tilde{\Gamma}_0 + D_X \hat{\mathbf{S}}^+ \hat{\mathbf{S}}^-$$

$$- D_1\mu_1(\tfrac{1}{2}\mathbf{T}_1 + i\hat{\mathbf{J}}_1)\mathbf{U}_1\hat{\mathbf{S}}^- + D_1\mu_1\hat{\mathbf{S}}^+\mathbf{U}_1(\tfrac{1}{2}\mathbf{T}_1 - i\hat{\mathbf{J}}_1)$$

$$- D_2\mu_2(\tfrac{1}{2}\mathbf{T}_2 + i\hat{\mathbf{J}}_2)\mathbf{U}_2\hat{\mathbf{S}}^- + D_2\mu_2\hat{\mathbf{S}}^+\mathbf{U}_2(\tfrac{1}{2}\mathbf{T}_2 - i\hat{\mathbf{J}}_2)$$

$$- D_1\mu_1^2\hat{\mathbf{S}}^+\mathbf{U}_1^2\hat{\mathbf{S}}^- - D_2\mu_2^2\hat{\mathbf{S}}^+\mathbf{U}_2^2\hat{\mathbf{S}}^- \tag{2.36}$$

where $\hat{\mathbf{S}}^\pm = \tfrac{1}{2}\mathbf{X} \mp \nabla_X$ are the lowering and raising (vector) operators for the three-dimensional harmonic oscillator $\mathbf{X}$. $\mathbf{T}_1$ and $\mathbf{T}_2$ are the torques for body 1 and 2, respectively, due to V_0 ($\mathbf{T}_1 = -\mathbf{T}_2 = \mathbf{T}$); finally $\mathbf{U}_1$ and $\mathbf{U}_2$ are 3×3 matrices defined (for $m = 1, 2$) as

$$\mathbf{U}_m = i(\hat{\mathbf{J}}_m \otimes u_m)_{\text{fun}} \tag{2.37}$$

$\tilde{\Gamma}_0$ is the two-body Smoluchowski operator given by Eq. (2.12).

We now have to treat a system of 9 degrees of freedom. It is possible to use techniques of angular momentum coupling that are analogous to those employed for the two-body case. We define the angular momentum operators

$$\hat{\mathbf{j}} = -i\mathbf{X} \times \nabla_X \tag{2.38}$$

$$\hat{\mathbf{J}}_T = \hat{\mathbf{j}} + \hat{\mathbf{J}} \tag{2.39}$$

and $\hat{\mathbf{J}}$ is defined according to Eq. (2.14). In the following we will sometimes call $\hat{\mathbf{j}}$ "little" angular momentum, $\hat{\mathbf{J}}$ "big" angular momentum and $\hat{\mathbf{J}}_T$ "total" angular momentum. We use a double coupling scheme to determine the most convenient basis set for the problem. We start from the uncoupled basis set

$$|J_1M_1K_1; J_2M_2K_2; njm\rangle = |J_1M_1K_1\rangle|J_2M_2K_2\rangle|njm\rangle \tag{2.40}$$

given by the direct product of the uncoupled two-body set with the functions $|njm\rangle$ defined in terms of the polar coordinates X, θ and ϕ for the field; that is,

$$|njm\rangle = |nj\rangle|jm\rangle \tag{2.41}$$

$$|nj\rangle = \left[\frac{n!}{2^{j+1}(j+n+1/2)!}\right]^{1/2} X^j \mathscr{L}_n^{(j+1/2)}(X^2/2)\exp(-X^2/4) \tag{2.42}$$

$$|jm\rangle = Y_{jm}(\phi, \theta) \tag{2.43}$$

where $\mathcal{L}_q^p$ is the pth order Laguerre polynomial of degree q [52]. As in the previous case, two quantum numbers, K_1 and K_2, are obviously diagonal. Note that the system is still spherically symmetric and the potential does not depend on the Euler angles γ_1 and γ_2. We shall neglect K_1 and K_2 in the following whenever possible, since only the matrix blocks with $K_1 = K_2 = 0$ will be computed.

We may proceed in our coupling scheme by first considering the coupling of $\hat{\mathbf{J}}_1$ and $\hat{\mathbf{J}}_2$ to give $\hat{\mathbf{J}}$,

$$|J_1 J_2 JM; njm\rangle = \sum_{M_1 M_2} C(J_1 M_1 J_2 M_2 JM)|J_1 M_1; J_2 M_2; njm\rangle \quad (2.44)$$

In this basis set only the two-body operator $\tilde{\Gamma}_0$ is diagonal with respect to J and M. A fully coupled basis set is then obtained by coupling together $\hat{\mathbf{j}}$ and $\hat{\mathbf{J}}$ to give $\hat{\mathbf{J}}_T$

$$|n J_1 J_2 J_j J_T M_T\rangle = \sum_{mM} C(JM jm J_T M_T)|J_1 J_2 JM; njm\rangle \quad (2.45)$$

J_T and M_T are "good" (i.e., diagonal) quantum numbers for $\hat{\Gamma}$. Note that from an initial nine-dimensional problem, we are left with a five-dimensional one. The relevant quantum numbers are n, J_1, J_2, J and j.

The calculation of the matrices Γ and $\mathbf{M}$, and the vector $\mathbf{v}_0$ can now proceed along the same lines as the previous section. The general vector element $(\mathbf{v}_0)_\Lambda$ is exactly the same one given by Eq. (C.10) in Appendix C.2, but the factor $\mathscr{F}$ in that equation is now $\delta_{J_T 0} \delta_{M_T 0} \delta_{n0} \delta_{j0} \delta_{m0}$. The matrix element $(\mathbf{M})_{\Lambda, \Lambda'}$ is simply

$$\langle \Lambda | F | \Lambda' \rangle = \frac{[P]^{1/2}}{(8\pi^2)^{1/2}} [J_1]^{1/2} \begin{pmatrix} J_1 & P & J_1' \\ 0 & 0 & 0 \end{pmatrix} \delta_{J_2 J_2'} \delta_{J_1 J_2'} \delta_{JP} \delta_{J'0} \delta_{nn'} \delta_{jj'} \quad (2.46)$$

with $J_T = P$, $M_T = Q$, $J_T' = 0$, $M_T' = 0$. It follows that the only matrix block of Γ that is needed is defined by $J_T = J_T' = P$ and $M_T = M_T' = Q$. The matrix Γ is obtained by a systematic usage of the WE theorem. We may write $\tilde{\Gamma}$ of Eq. (2.36) in the straightforward but convenient form

$$\tilde{\Gamma} = \tilde{\Gamma}_0 + [D_\mathbf{x} + \tfrac{2}{3}(D_1 \mu_1^2 + D_2 \mu_2^2)]\hat{S}^+ \hat{S}^-$$

$$+ D_1 \mu_1 (\hat{O}_1^\dagger + \hat{O}_1) + D_2 \mu_2 (\hat{O}_2^\dagger + \hat{O}_2)$$

$$+ D_1 \mu_1^2 \hat{S} : \hat{G}_1 + D_2 \mu_2^2 \hat{S} : \hat{G}_2 \quad (2.47)$$

where the double dot symbol means the scalar product of two second rank Cartesian tensors. Here $\tilde{\Gamma}_0$ is the two-body operator; $\hat{O}_1$ and $\hat{O}_2$ are vector operators, while $\hat{\mathbf{S}}$, $\hat{\mathbf{G}}_1$ and $\hat{\mathbf{G}}_2$ are *matrix* operators, introduced for their convenient tensor properties (cf. below). They are defined according to the following equations ($m = 1, 2$):

$$\hat{\mathbf{O}}_m = \hat{\mathbf{S}}^+ \mathbf{u}_m \times (\tfrac{1}{2}\mathbf{T}_m - i\hat{\mathbf{J}}_m) \tag{2.48}$$

$$\hat{\mathbf{S}} = \hat{\mathbf{S}}^+ \hat{\mathbf{S}}^{-tr} \tag{2.49}$$

$$\hat{\mathbf{G}}_m = -\mathbf{U}_m^2 - \tfrac{2}{3}\mathbf{1} \tag{2.50}$$

where we have systematically used a Cartesian notation for representing the various tensor products and the general property of the matrix $\mathbf{U}_m$ is given in terms of the unit matrices $\mathbf{u}_m$ by

$$\mathbf{U}_m \mathbf{r} = \mathbf{u}_m \times \mathbf{r} \tag{2.51}$$

where $\mathbf{r}$ is a generic vector.

We can now consider each term separately. The two-body operator $\tilde{\Gamma}_0$ has the same matrix representation in the two-body coupled basis set and in the present three-body coupled basis set. The next term in $\hat{\mathbf{S}}^+\hat{\mathbf{S}}^-$ is diagonal in the chosen basis set [4]. Then the matrix representation of these diagonal terms is

$$\Gamma_{\text{diagonal}} = \Gamma_0 \delta_{J_T J_T'} \delta_{M_T M_T'} \delta_{JJ'} \delta_{nn'} \delta_{jj'} + [D_{\mathbf{x}} + \tfrac{2}{3}(D_1 \mu_1^2 + D_2 \mu_2^2)](2n + j)\delta_{\Lambda\Lambda'} \tag{2.52}$$

The term with off-diagonal elements from $\hat{\mathbf{O}}_1$ is considered next. From the WE theorem (strong form), and using the equivalence between a first rank tensor product and the external product of two vectors, we obtain

$$\langle \Lambda | \hat{\mathbf{O}}_1^+ | \Lambda' \rangle = i(2)^{1/2}(-)^{1+j+J'+J_T} \begin{Bmatrix} J & j & J_T \\ j' & J' & 1 \end{Bmatrix}$$

$$\times (J_1 J_2 J \| [\mathbf{u}_1 \otimes (\tfrac{1}{2}\mathbf{T}_1 - i\hat{\mathbf{J}}_1)]^{(1)} \| J_1' J_2' J')$$

$$\times (nj \| \hat{\mathbf{S}}^+ \| n'j') \delta_{J_T J_T'} \delta_{M_T M_T'} \tag{2.53}$$

The reduced matrix element of $\hat{\mathbf{S}}^+$ is evaluated in Appendix C.3; the reduced matrix element involving $\mathbf{u}_1$ is straightforwardly evaluated using the general formula

$$(J_1 J_2 J \| [\hat{\mathbf{A}} \otimes \hat{\mathbf{B}}]^{(1)} \| J_1' J_2' J') = [1]^{1/2}(-)^{(J+1+J')} \sum_{J_1'' J_2'' J''} \begin{Bmatrix} 1 & 1 & 1 \\ J & J'' & J' \end{Bmatrix}$$

$$\times (J_1 J_2 J \| \hat{\mathbf{A}} \| J_1'' J_2'' J'')(J_1'' J_2'' J'' \| \hat{\mathbf{B}} \| J_1' J_2' J') \tag{2.54}$$

where $\hat{\mathbf{A}}$ and $\hat{\mathbf{B}}$ can be $\mathbf{T}_1$, $\hat{\mathbf{J}}_1$, $\mathbf{u}_1$. The reduced matrix element of the torque and of the angular momentum operator are given respectively in Appendix C.1 and in the previous section. The reduced matrix element of $\mathbf{u}_1$ is proportional to the reduced matrix element of the function $|100\rangle_1$ (see Appendix C.1)

$$(J_1 J_2 J \| \mathbf{u}_1 \| J_1' J_2' J') = (-)^{J_1 + J_2 + J' + 1} [JJ']^{1/2} \begin{Bmatrix} J_1 & J & J_2 \\ J' & J_1' & 1 \end{Bmatrix}$$

$$\times \frac{(8\pi^2)^{1/2}}{[1]^{1/2}} (J_1 \| 1 \| J_1') \qquad (2.55)$$

where $(J_1 \| 1 \| J_1')$ is the reduced matrix element of a first rank spherical harmonic. The matrix element proportional to $\hat{O}_1^\dagger$ is obtained by exchanging Λ and Λ' in the previous formulas. The matrix elements of $\hat{O}_2$ and $\hat{O}_2^\dagger$ are evaluated in a similar manner.

Finally the matrix elements of the mixed operators in $\hat{\mathbf{G}}_1$ and $\hat{\mathbf{G}}_2$ may be considered. Both $\hat{\mathbf{S}}$ and $\hat{\mathbf{G}}_1$ ($\hat{\mathbf{G}}_2$) are second rank spherical tensors. It follows that

$$\langle \Lambda | \hat{\mathbf{S}} : \hat{\mathbf{G}}_1 | \Lambda' \rangle = (-)^{J' + j + J_T} \begin{Bmatrix} J & j & J_T \\ j' & J' & 2 \end{Bmatrix}$$

$$\times (J_1 J_2 J \| \hat{\mathbf{G}}_1 \| J_1' J_2' J')(nj \| \hat{\mathbf{S}} \| n'j') \delta_{J_T J_T'} \delta_{M_T M_T'}$$

$$(2.56)$$

and the reduced matrix element of $\hat{\mathbf{S}}$ is given in Appendix C.3. The reduced matrix element of $\hat{\mathbf{G}}_1$ is proportional to the reduced matrix element of the function $|200\rangle_1$

$$(J_1 J_2 J \| \hat{\mathbf{G}}_1 \| J_1' J_2' J') = (-)^{J_1 + J_2 + J' + 2} [JJ']^{1/2} \begin{Bmatrix} J_1 & J & J_2 \\ J' & J_1' & 2 \end{Bmatrix}$$

$$\times \left(-\frac{2}{3} \frac{(8\pi^2)^{1/2}}{[2]^{1/2}} (J_1 \| 2 \| J_1') \right) \qquad (2.57)$$

The calculation of the matrix element of $\hat{\mathbf{S}} : \hat{\mathbf{G}}_2$ proceeds along the same lines. Note that, as was the case with the two-body problem, a real symmetric matrix is obtained.

E. Two-Body Kramers Model: Slowly Relaxing Local Structure

The next model considered in this work is Kramers description of a two-body system, that is, the generalization of model (a) in order to account for inertial effects. The time evolution operator is given by a

Fokker–Planck–Kramers rotational operator involving the rotational coordinates of bodies 1 and 2, and their conjugate angular momenta, $\mathbf{L}_1$ and $\mathbf{L}_2$. The final phase space to be considered is then twelve-dimensional. In practice, we will find that only eight effective (nondiagonal) quantum numbers need to be considered in a properly chosen coupled basis set of functions, for two spherical (or symmetric) rotators. Still, the matrices needed for computations have huge dimensions, and the numerical treatment is far from easy, especially when large potential couplings and/or low friction regimes are explored.

1. Slowly Relaxing Local Structure Model

Again we consider the symmetrized and rescaled time evolution operator, obtained by summation of the two rotational FPK operators for bodies 1 and 2, in the presence of the usual interaction potential. Since we suppose that both the bodies are spherical, no precessional terms are present [6] [cf. Eq. (1.14) for the three-body case]:

$$\tilde{\Gamma} = \omega_1^s(i\mathbf{L}_1\hat{\mathbf{J}}_1 + \mathbf{T}_1\nabla_1) - \omega_1^c \exp(\mathbf{L}_1^2/4)\nabla_1 \exp(-\mathbf{L}_1^2/2)\nabla_1 \exp(\mathbf{L}_1^2/4)$$
$$+ \omega_2^s(i\mathbf{L}_2\hat{\mathbf{J}}_2 + \mathbf{T}_2\nabla_2) - \omega_2^c \exp(\mathbf{L}_2^2/4)\nabla_2 \exp(-\mathbf{L}_2^2/2)\nabla_2 \exp(\mathbf{L}_2^2/4)$$
$$(2.58)$$

The same definition and properties of the torque vectors holds as in Eq. (2.17); ∇_1 and ∇_2 are the gradient operators acting respectively in the subspaces $\mathbf{L}_1$ and $\mathbf{L}_2$. The frequency parameter ω_1^s is the streaming frequency; it is the characteristic frequency for the deterministic motion of body 1 and it is inversely proportional to the square root of the moment of inertia I_1. ω_1^c is the *collisional* frequency of body 1, and it is a direct measure of the dissipative effect due to the solvent, since it is proportional to the friction exerted by the medium on the body. Analogous parameters ω_2^s and ω_2^c are defined for body 2. See Table I for the explicit definitions.

The equilibrium distribution function is defined with respect to the total energy of the system

$$E = V_0(\mathbf{\Omega}_1, \mathbf{\Omega}_2) + \tfrac{1}{2}\mathbf{L}_1^2 + \tfrac{1}{2}\mathbf{L}_2^2 \qquad (2.59)$$

including the interaction potential between the two bodies and the (rescaled) kinetic energy. The coupling between body 1 and body 2 is given only by the potential; no "hydrodynamic" interactions, that is, frictional coupling terms, are included. A situation close to models in which the solute (body 1) reorients in a potential resulting from a slowly relaxing solvent local structure (body 2) is then recovered, as was discussed previously.

2. Matrix representation

The numerical treatment is again based on the matrix representation of the operator on a coupled basis set of functions, followed by the application of the Lanczos algorithm. Following the same method used in the previous section, we define the two "little" angular momentum operators (one for each body) and the overall "little" angular momentum operator

$$\hat{\mathbf{j}}_1 = -i\mathbf{L}_1 \times \nabla_1 \tag{2.60}$$

$$\hat{\mathbf{j}}_2 = -i\mathbf{L}_2 \times \nabla_2 \tag{2.61}$$

$$\hat{\mathbf{j}} = \hat{\mathbf{j}}_1 + \hat{\mathbf{j}}_2 \tag{2.62}$$

and the total angular momentum operator

$$\hat{\mathbf{J}}_T = \hat{\mathbf{j}} + \hat{\mathbf{J}} \tag{2.63}$$

where $\hat{\mathbf{J}}$ is defined by Eq. (2.14). It is easy to see that $\tilde{\Gamma}$ is a scalar with respect to $\hat{\mathbf{J}}_T$. The initial uncoupled basis set is given by

$$\begin{aligned}
&|J_1 M_1 K_1; J_2 M_2 K_2; n_1 j_1 m_1; n_2 j_2 m_2\rangle \\
&= |J_1 M_1 K_1\rangle |J_2 M_2 K_2\rangle \times |n_1 j_1 m_1\rangle |n_2 j_2 m_2\rangle
\end{aligned} \tag{2.64}$$

where the functions $|n_1 j_1 m_1\rangle$ and $|n_2 j_2 m_2\rangle$ are defined with respect to the polar coordinates L_1, θ_1, ϕ_1 and L_2, θ_2, γ_2, respectively. As usual, the K_1 and K_2 quantum numbers are diagonal and will be neglected in the following. The coupling scheme involves the coupling of $\hat{\mathbf{j}}_1$ and $\hat{\mathbf{j}}_2$; then the coupling of $\hat{\mathbf{J}}_1$ and $\hat{\mathbf{J}}_2$; finally $\hat{\mathbf{j}}$ and $\hat{\mathbf{J}}$ are coupled together to give $\hat{\mathbf{J}}_T$.

$$\begin{aligned}
&|J_1 M_1; J_2 M_2; n_1 n_2 j_1 j_2 jm\rangle \\
&\qquad = \sum_{m_1 m_2} C(j_1 m_1 j_2 m_2 jm) |J_1 M_1; J_2 M_2; n_1 j_1 m_1; n_2 j_2 m_2\rangle
\end{aligned} \tag{2.65}$$

$$\begin{aligned}
|J_1 J_2 JM; n_1 n_2 j_1 j_2 jm\rangle &= \sum_{M_1 M_2} C(J_1 M_1 J_2 M_2 JM) \\
&\qquad \times |J_1 M_1; J_2 M_2; n_1 n_2 j_1 j_2 jm\rangle
\end{aligned} \tag{2.66}$$

$$\begin{aligned}
|n_1 n_2 j_1 j_2 J_1 J_2 j J J_T M_T\rangle &= \sum_{mM} C(jmJM J_T M_T) \\
&\qquad \times |J_1 J_2 JM; n_1 n_2 j_1 j_2 jm\rangle
\end{aligned} \tag{2.67}$$

The total angular momentum quantum numbers, J_T and M_T are diagonal, that is, "constants of motion".

The calculation of the matrix Γ is now a straightforward application of previous formulas. The vector representation of P_{eq} again has the elements defined in Appendix C.3, where $\mathcal{F}$ is now equal to $\delta_{J_T 0}\delta_{M_T 0}\delta_{j0}\delta_{j_1 0}\delta_{j_2 0}\delta_{n_1 0}\delta_{n_2 0}$. The matrix elements for the rotational correlation function $(F = Y_{PQ})$ $(\mathbf{M}^R)_{\Lambda,\Lambda'}$ are given by

$$
\langle \Lambda | F | \Lambda' \rangle = \frac{[P]^{1/2}}{(8\pi^2)^{1/2}} [J_1]^{1/2}
$$

$$
\times \begin{pmatrix} J_1 & P & J_1' \\ 0 & 0 & 0 \end{pmatrix} \delta_{J_2 J_2'}\delta_{J_1 J_2'}\delta_{JP}\delta_{J'0}\delta_{jj'}\delta_{n_1 n_1'}\delta_{j_1 j_1'}\delta_{n_2 n_2'}\delta_{j_2 j_2'}
$$

$$(2.68)$$

and $J_T = P$, $J_T' = 0$, $M_T = Q$, $M_T' = 0$, $K_1 = K_1' = K_2 = K_2' = 0$; whereas the matrix elements $(\mathbf{M}^J)_{\Lambda,\Lambda'}$ for the momentum correlation function (mth spherical component of the first rank tensor $\mathbf{L}_1$, $F = \mathbf{L}_{1_m}$) are given by

$$
\langle \Lambda | F | \Lambda' \rangle = (-)^{1+j_1+j_2+J}[1J]^{-1/2}[jj']^{1/2}\begin{Bmatrix} j_1 & j & j_2 \\ j' & j_1 & 1 \end{Bmatrix}
$$

$$
\times (n_1 j_1 \| \mathbf{L}_1 \| n_1' j_1')\delta_{J_1 J_1'}\delta_{J_2 J_1'}\delta_{JJ'}\delta_{Jj'}\delta_{n_2 n_2'}\delta_{j_2 j_2'} \quad (2.69)
$$

and $J_T = 1$, $J_T' = 0$, $M_T = m$, $M_T' = 0$, $K_1 = K_1' = K_2 = K_2' = 0$. We conclude this section by writing down the complete matrix element for the time evolution operator:

$$
\langle \Lambda | \tilde{\Gamma} | \Lambda' \rangle = \omega_1^s(-)^{J+J_T+j_1+j_2}[jj']^{1/2}\begin{Bmatrix} j & J & J_T \\ J' & j' & 1 \end{Bmatrix}\begin{Bmatrix} j_1 & j & j_2 \\ j' & j_1 & 1 \end{Bmatrix}
$$

$$
\times \delta_{J_T J_T'}\delta_{M_T M_T'}\delta_{j_2 j_2'}\delta_{n_2 n_2'}
$$

$$
\times \left[i(-)^{J_1+J_2+J'}[JJ']^{1/2}\begin{Bmatrix} J_1 & J & J_2 \\ J' & J_1' & 1 \end{Bmatrix}(n_1 j_1 \| \mathbf{L}_1 \| n_1' j_1') \right.
$$

$$
\times [J_1(J_1+1)(2J_1+1)]^{1/2}\delta_{J_1 J_1'}\delta_{J_2 J_2'} - (n_1 j_1 \| \nabla_1 \| n_1' j_1')
$$

$$
\times (J_1 J_2 J \| \mathbf{T} \| J_1' J_2' J') \Big]
$$

$$
+ (2n_1 + j_1)\omega_1^c\delta_{\Lambda,\Lambda'}
$$

$$
+ \omega_2^s(-)^{j+J_T+j_1+j_2'+j'}[jj']^{1/2}\begin{Bmatrix} j & J & J_T \\ J' & j' & 1 \end{Bmatrix}\begin{Bmatrix} j_2 & j & j_1 \\ j' & j_2 & 1 \end{Bmatrix}
$$

$$
\times \delta_{J_T J_T'}\delta_{M_T M_T'}\delta_{j_1 j_1'}\delta_{n_1 n_1'}
$$

$$\times \left[i(-)^{J_1 + J_2'} [JJ']^{1/2} \begin{Bmatrix} J_2 & J & J_1 \\ J' & J_2' & 1 \end{Bmatrix} (n_2 j_2 \| \mathbf{L}_2 \| n_2' j_2') \right.$$

$$\times [J_2(J_2 + 1)(2J_2 + 1)]^{1/2} \delta_{J_1 J_1'} \delta_{J_2 J_2'} + (n_2 j_2 \| \nabla_2 \| n_2' j_2')$$

$$\left. \times (J_1 J_2 J \| \mathbf{T} \| J_1' J_2' J') \right]$$

$$+ (2n_2 + j_2) \omega_2^c \delta_{\Lambda, \Lambda'} \tag{2.70}$$

Note that the final matrix is *complex symmetric*, since the operator is non-Hermitean.

F. Two-Body Kramers Model: Fluctuating Torques

As was discussed in Section I, if one considers a three-body Kramers model and projects out the third set of solvent coordinates (and the conjugate momenta), a MFPKE is found in the remaining coordinates, with a frictional coupling between the solute and the solvent cage. This is a system close to the fluctuating torque (FT) case discussed by Stillman and Freed, except that an explicit description of the momentum of the solvent cage is added and the structure of the (frictional) coupling is deduced from an analytic model, rather than chosen to satisfy conditions of detailed balance. Therefore a more precise model is obtained at the price of less freedom in choosing the physical parameters.

1. Fluctuating Torque

After projecting the fast variables $\mathbf{X}$, $\mathbf{P}$ what is left is a two-body Kramers operator having the form [i.e., Eq. (1.16)]

$$\tilde{\Gamma} = \omega_1^s (i\mathbf{L}_1 \hat{\mathbf{J}}_1 + \mathbf{T}_1 \nabla_1) - \exp(\mathbf{L}_1^2/4) \nabla_1 \omega_1^c \exp(-\mathbf{L}_1^2/2) \nabla_1 \exp(\mathbf{L}_1^2/4)$$

$$+ \omega_2^s (i\mathbf{L}_2 \hat{\mathbf{J}}_2 + \mathbf{T}_2 \nabla_2) - \exp(\mathbf{L}_2^2/4) \nabla_2 \omega_2^c \exp(-\mathbf{L}_2^2/2) \nabla_2 \exp(\mathbf{L}_2^2/4)$$

$$- \exp(\mathbf{L}_1^2/4) \nabla_1 \omega_{12}^c \exp(-\mathbf{L}_1^2/4 - \mathbf{L}_2^2/4) \nabla_2 \exp(\mathbf{L}_2^2/4)$$

$$- \exp(\mathbf{L}_2^2/4) \nabla_2 \omega_{21}^c \exp(-\mathbf{L}_1^2/4 - \mathbf{L}_2^2/4) \nabla_1 \exp(\mathbf{L}_1^2/4) \tag{2.71}$$

The streaming operator is substantially unchanged compared to Eq. (2.58) (except for an additional contribution to the first rank interaction potential). The collisional operator is defined in terms of an orientational dependent friction matrix (or "collisional frequency" matrix in the present dimensionless formulation) as

$$\begin{pmatrix} \omega_1^c & \omega_{12}^c \\ \omega_{21}^c & \omega_2^c \end{pmatrix} = \begin{bmatrix} \omega_1^c \mathbf{1} - \omega_1 \mathbf{U}_1^2 & -(\omega_1 \omega_2)^{1/2} \mathbf{U}_1 \mathbf{U}_2 \\ -(\omega_1 \omega_2)^{1/2} \mathbf{U}_2 \mathbf{U}_1 & \omega_2^c \mathbf{1} - \omega_2 \mathbf{U}_2^2 \end{bmatrix} \tag{2.72}$$

where $\omega_{1,2}$ are defined with respect $\mu_{1,2}$, $D_{\mathbf{X}}$ [see Eqs. (1.19)–(1.21)]. Obviously, all the new collisional terms retain the characteristic tensorial properties allowing the use of the same coupled basis set as in the previous section.

2. Matrix Elements

The calculation of the matrix $\mathbf{M}$ and the starting vector $\mathbf{v}$ proceeds exactly along the same lines discussed in the previous paragraph (since they depend only on the structure of P_{eq}). The matrix element $\mathbf{\Gamma}$ can be conveniently evaluated by Eq. (2.71) in the form

$$\tilde{\mathbf{\Gamma}} = \tilde{\mathbf{\Gamma}}_s + [\hat{S}_1^{(0)}\mathbf{1} + \hat{\mathbf{S}}_1^{(2)}] : \left[\left(\omega_1^c + \frac{2}{3}\,\omega_1\right)\mathbf{1} + \omega_1\hat{\mathbf{G}}_1\right]$$

$$+ [\hat{S}_2^{(0)}\mathbf{1} + \hat{\mathbf{S}}_2^{(2)}] : \left[\left(\omega_2^c + \frac{2}{3}\,\omega_2\right)\mathbf{1} + \omega_2\hat{\mathbf{G}}_2\right]$$

$$+ [\hat{S}_{12}^{(0)}\mathbf{1} + \hat{\mathbf{S}}_{12}^{(2)}] : \left[\frac{2}{3}(\omega_1\omega_2)^{1/2}\mathbf{u}_1\mathbf{u}_2\mathbf{1} + (\omega_1\omega_2)^{1/2}\hat{\mathbf{G}}_{12}\right]$$

$$+ [\hat{S}_{21}^{(0)}\mathbf{1} + \hat{\mathbf{S}}_{21}^{(2)}] : \left[\frac{2}{3}(\omega_1\omega_2)^{1/2}\mathbf{u}_1\mathbf{u}_2\mathbf{1} + (\omega_1\omega_2)^{1/2}\hat{\mathbf{G}}_{21}\right] \quad (2.73)$$

where

$$\hat{S}_i^{(0)} = \frac{1}{3}\,\hat{\mathbf{S}}_i^+ \cdot \hat{\mathbf{S}}_i^- \quad (2.74)$$

$$\hat{\mathbf{S}}_i^{(2)} = \hat{\mathbf{S}}_i^+ \otimes \hat{\mathbf{S}}_i^- - \hat{S}_i^{(0)}\mathbf{1} \quad (2.75)$$

$$\hat{S}_{ij}^{(0)} = \frac{1}{3}\,\hat{\mathbf{S}}_i^+ \cdot \hat{\mathbf{S}}_j^- \quad (2.76)$$

$$\hat{\mathbf{S}}_{ij}^{(2)} = \hat{\mathbf{S}}_i^+ \otimes \hat{\mathbf{S}}_j^- - \hat{S}_{ij}^{(0)}\mathbf{1} \quad (2.77)$$

with $i, j = 1, 2$ $(i \neq j)$. These are the zero and second rank irreducible tensors built from $\hat{\mathbf{S}}_i^{\pm}$, the raising and lowering operators in $\mathbf{L}_i$

$$\hat{\mathbf{S}}_i^{\pm} = \frac{1}{2}\,\mathbf{L}_i \mp \nabla_i \quad (2.78)$$

The collisional operator is obtained by taking the product with the zero and second rank irreducible tensors built from $\mathbf{U}_i$, that is,

$$\mathbf{U}_i^2 = -\frac{2}{3}\,\mathbf{1} - \hat{\mathbf{G}}_i \quad (2.79)$$

$$\mathbf{U}_i\mathbf{U}_j = -\frac{2}{3}\,\mathbf{u}_i\mathbf{u}_j - \hat{\mathbf{G}}_{ij} \quad (2.80)$$

The matrix element of the streaming operator $\tilde{\Gamma}_s$ is equal to the one evaluated in Section E (cf. the terms in ω_1^s and ω_2^s in Eq. (2.70)). The contribution of the collisional part is given by

$$\langle \Lambda | \tilde{\Gamma}_c | \Lambda' \rangle = s_1 + s_2 + s_{12} + s_{21} \qquad (2.81)$$

That is, it is a sum of matrix elements from the four terms that $\tilde{\Gamma}_c$ was split into in Eq. (2.73). Here only s_1 and s_{12} are written, since s_2 and s_{21} are obtained by permuting indices 1 and 2 (note that $J_T = J_T'$, $M_T = M_T'$, $K_1 = K_1' = 0$, $K_1 = K_1' = 0$).

$$s_1 = \left(\omega_1^c + \frac{2}{3}\,\omega_1 \right)(2n_1 + j_1)\delta_{\Lambda\Lambda'}$$

$$+ \frac{1}{3}\,\omega_1(-)^{j_1+j_2+J_1+J_2+J+J'+J_T}[jj'JJ'J_1J_1']^{1/2}\begin{pmatrix} J_1 & 2 & J_1' \\ 0 & 0 & 0 \end{pmatrix}\frac{(-)^{j_i}}{\begin{pmatrix} j_1 & 2 & j_1' \\ 0 & 0 & 0 \end{pmatrix}}$$

$$\times \begin{Bmatrix} j & J & J_T \\ J' & j' & 2 \end{Bmatrix}\begin{Bmatrix} j_1 & j & j_2 \\ j' & J_1' & 2 \end{Bmatrix}\begin{Bmatrix} J_1 & J & J_2 \\ J' & J_1' & 2 \end{Bmatrix}$$

$$\times [(2n_1 + j_1)\delta_{j_1j_1'}\delta_{n_1n_1'} - 3\langle n_1j_10|\hat{S}_{1_0}^+\hat{S}_{1_0}^-|n_1'j_1'0\rangle]\delta_{j_2j_2'}\delta_{n_2n_2'}\delta_{J_2J_2'} \qquad (2.82)$$

$$s_{12} = \frac{2}{3}\,(\omega_1\omega_2)^{1/2}(-)^{j_1+j_2+j+J_1'+J_1+J}[J_1J_1'J_2J_2']^{1/2}\begin{pmatrix} J_1 & 1 & J_1' \\ 0 & 0 & 0 \end{pmatrix}\begin{pmatrix} J_2 & 1 & J_2' \\ 0 & 0 & 0 \end{pmatrix}$$

$$\times \begin{Bmatrix} j_1 & j_2 & j \\ j_2' & j_1' & 1 \end{Bmatrix}\begin{Bmatrix} J_1 & J_2 & J \\ J_2' & J_1' & 1 \end{Bmatrix}(n_1j_1\|\hat{S}_1^+\|n_1'j_1')(n_2j_2\|\hat{S}_2^-\|n_2'j_2')\delta_{jj'}\delta_{JJ'}$$

$$- \frac{1}{6}\,(\omega_1\omega_2)^{1/2}(-)^{J+J'+J_T}\frac{\begin{Bmatrix} j & J & J_T \\ J' & j & 2 \end{Bmatrix}}{\begin{pmatrix} j & 2 & J' \\ 0 & 0 & 0 \end{pmatrix}\begin{pmatrix} j & 2 & j' \\ 0 & 0 & 0 \end{pmatrix}}$$

$$\times \left[3\langle n_1n_2j_1j_2j0|\hat{S}_{1_0}^+\hat{S}_{2_0}^-|n_1'n_2'j_1'j_2'j'0\rangle \right.$$

$$\left. - (-)^{j_1+j_2+j}\begin{Bmatrix} j_1 & j_2 & j \\ j_2' & j_1' & 1 \end{Bmatrix}(n_1j_1\|\hat{S}_1^+\|n_1'j_1')(n_2j_2\|\hat{S}_2^-\|n_2'j_2')\delta_{jj'} \right]$$

$$\times \left[3\langle J_1J_2J0|u_{1_0}u_{2_0}|J_1'J_2'J'0\rangle - (-)^{J_1+J_1'}[J'J_1J_1'J_2J_2']^{1/2} \right.$$

$$\left. \times \begin{Bmatrix} J_1 & J_2 & J \\ J_2' & J_1' & 1 \end{Bmatrix}\delta_{JJ'} \right] \qquad (2.83)$$

The reduced matrix elements are given by

$$\langle n_1j_10|\hat{S}_{1_0}^+\hat{S}_{1_0}^-|n_1'j_1'0\rangle = \sum_{n_1''j_1''}(-)^{j_1+j_1''}\begin{pmatrix} j_1 & 1 & j_1'' \\ 0 & 0 & 0 \end{pmatrix}\begin{pmatrix} j_1'' & 1 & j_1' \\ 0 & 0 & 0 \end{pmatrix}$$

$$\times (n_1j_1\|\hat{S}_1^+\|n_1''j_1'')(n_1''j_1''\|\hat{S}_1^-\|n_1'j_1') \qquad (2.84)$$

$$\langle n_1 n_2 j_1 j_2 j 0 | \hat{S}^+_{1_0} \hat{S}^-_{2_0} | n'_1 n'_2 j'_1 j'_2 j' 0 \rangle = (-)^{j-1+j_2+j+j'} [jj']^{1/2}$$

$$\times (n_1 j_1 \| \hat{S}^+_1 \| n'_1 j'_1)(n_2 j_2 \| \hat{S}^-_2 \| n'_2 j'_2)$$

$$\times \mathcal{S}(j_1 j_2 j J'_1 j'_2 j') \qquad (2.85)$$

$$\langle J_1 J_2 J 0 | u_{1_0} u_{2_0} | J'_1 J'_2 J' 0 \rangle = (-)^{J-J'} [JJ'J - 1 J'_1 J_2 J'_2]^{1/2}$$

$$\times \begin{pmatrix} J_1 & 1 & J'_1 \\ 0 & 0 & 0 \end{pmatrix} \begin{pmatrix} J_2 & 1 & J'_2 \\ 0 & 0 & 0 \end{pmatrix}$$

$$\times \mathcal{S}(J_1 J - 2 J J'_1 J'_2 J') \qquad (2.86)$$

$$\mathcal{S}(l_1 l_2 l l'_1 l'_2 l') = \sum_k \begin{pmatrix} l_1 & l & l_2 \\ k & 0 & -k \end{pmatrix} \begin{pmatrix} l'_1 & l' & l'_2 \\ k & 0 & -k \end{pmatrix}$$

$$\times \begin{pmatrix} l_1 & l & l'_1 \\ -k & 0 & k \end{pmatrix} \begin{pmatrix} l_2 & l & l'_2 \\ -k & 0 & k \end{pmatrix} \qquad (2.87)$$

G. Results

In this section, we discuss the numerical results we have obtained for the four different models discussed in the previous sections: the 2BSM (two-body Smoluchowski model) and its generalization given by the 3BSM (three-body Smoluchowski model) describe diffusional systems; whereas the 2BKM-SRLS (two-body Kramers model in the "slowly relaxing local structure" version) and the 2BKM-FT (two-body Kramers model in the "fluctuating torque" version) include the conjugate momentum vectors.

In discussing a many-body stochastic model one needs an overview of the time evolution behavior of the system over a significantly large range of parameters, in order to explore physical regimes of interest. Thus, in all cases, we have obtained results for both first and second rank orientational correlation functions for the first body (the solute), while varying the energetic and frictional parameters; for the inertial models, momentum correlation functions have also been computed.

A common feature of all the stochastic models considered here is the presence of several important decay times, usually at least as many as the number of stochastic coordinates included in the system, but even more are found under certain conditions. To display the multiexponential decay of a process one can use different representations. First of all, such evidence can be obtained by plotting the correlation function $G(t)$ versus t. Also a representation in the frequency domain by spectral densities $J(\omega)$ versus ω can be useful. Cole–Cole plots may also have a certain usefulness, but they do not give much more information. We have chosen to give only time domain representations here, largely for reasons of space. A few spectral densities are shown in our initial reports. If a more detailed description is required, the best way to proceed seems to be the analysis of the dominant kinetic constants (i.e., eigenvalues of the time evolution operator), contributing significantly to the decay process (see below). The correlation function $G(t)$ can be written in terms of the eigenvalues λ_i of the time evolution operator according to the following expansion:

$$G(t) = \sum_i w_i \exp(-\lambda_i t) \qquad (2.88)$$

where each eigenvalue λ_i has a weight w_i. In all the table entries we show the set of eigenvalues having weights larger than or equal to a cut-off value ϵ. A measure of the overall correlation time of the process (i.e., the best approximation to a single exponential decay constant) is given, calculated as the zero frequency value of the spectral density.

It is interesting to investigate the eigenvectors corresponding to the dominant eigenvalues in a few cases. From the explicit expansion over the basis functions used for the matrix representation of the operator, one can obtain insight into the kind of motion represented by the ith mode (e.g., one can decide if it is mainly the isolated motion of the first body or if the solvent degrees of freedom are involved). Also, it is possible to gain information on the truncation criteria with respect to the different quantum numbers. This is particularly useful in dealing with models with more than three relevant (i.e., nondiagonal) quantum numbers.

1. Computational Procedures

As pointed out above, the numerical algorithm with which we have chosen to evaluate the eigenvalues and eigenvectors of the many-body stochastic operators, and to compute the temporal decay of a given correlation function $G(t)$, consists of: (1) determining a suitable set of basis functions via standard angular momentum techniques; (2) obtaining the matrix representation Γ of the symmetrized operator, and the initial vector $\mathbf{v}$; (this vector is calculated as the product of the matrix representation $\mathbf{M}$ of the observable function and the vector $\mathbf{v}_0$, which represents $P_{eq}^{1/2}$, cf. Eq. (2.26)); (3) applying a real symmetric implementation of the Lanczos algorithm (for Hermitean operators) or a complex symmetric one (for non-Hermitean cases) to transform Γ into tridiagonal form $\mathbf{T}$; (4) obtaining from $\mathbf{T}$, by straightforward diagonalization of the eigenvalue spectrum, and computing the temporal decay of $G(t)$ (alternatively one can directly calculate the spectral density $J(\omega)$ using a continued fraction expansion [50, 42]); (5) determining the eigenvectors corresponding to some eigenvalues (see below).

In this subsection, we wish to clarify some technical details concerning the computational procedure. One of the most serious difficulties one has to deal with when considering a many-body operator is how to check the internal consistency of the expressions. After all, when considering a 2BKM one has to solve a partial differential equation with 12 variables (i.e., the dimension of the phase space), and this is by no means a straightforward task. First of all, one requires a test of the algebraic formulas that give Γ, v_0, and v. Even though the procedures are clear, and based on the systematic usage of the Wignert–Eckart theorem, the large numbers of degrees of freedom involved, means possible algebraic mistakes that may be hard to find. For this purpose, we have found it very useful to check our algebraic manipulations made by hand, using standard computer algebra software packages such as *Reduce* [55] and *Mathematica* [56]. We did not write complete programs to perform all the algebraic steps; rather we checked separate parts of the calculation.

Another very useful way of testing our results has been to use two independent routes to numerically evaluate v_0. The first route is reviewed in Appendix C. It consists of the direct evaluation of the vector elements in the coupled basis which largely involves numerical integration of the function $P_{eq}^{1/2}$. This direct approach is convenient for the case of rotational invariance which is characteristic of the physical systems we have studied. A second route has previously been recommended by Moro and Freed [50] and Schneider and Freed [42]. They consider the following expression

$$\lim_{s \to 0^+} [s\mathbf{I} + \Gamma]\mathbf{v}_0 = \mathbf{c} \qquad (2.89)$$

where $\mathbf{c}$ is an arbitrary vector with a component along $\mathbf{v}_0$. Equation (2.89) follows from the fact that $P_{eq}^{1/2}$ is the unique stationary solution of the symmetrized operator $\tilde{\Gamma}$ (i.e., the eigenvector of zero eigenvalue). One solves it for $\mathbf{v}_0$, by using some efficient algorithm for large linear systems (e.g., the conjugate gradient method). Note that the calculation of $\mathbf{v}_0$ by Eq. (2.89) involves the direct use of Γ. Since the formulation of Γ is algebraically the most challenging step, we regarded agreement of $\mathbf{v}_0$ obtained by both methods as largely a confirmation of a correctly expressed Γ (as well as a reliable $\mathbf{v}_0$). In all cases we succeeded with this test to within appropriate numerical round-off error.

When one is reasonably sure of the algebraic formulas and programs, it is still necessary to check the convergence of each calculation, both with respect to the number of basis functions used (i.e., the dimension N of Γ) and the number of Lanczos steps (i.e., the dimension n of $\mathbf{T}$). Although one can use sophisticated pruning procedures in order to minimize N [51], we have used the simple criterion of repeating the calculation by increasing both N and n until there is a relative variation less than δ in *all* the dominant eigenvalues (i.e., all the eigenvalues having a relative weight larger than or equal to ϵ). Usually δ and ϵ have both been chosen to equal 10^{-3}.

Finally, we discuss the procedure adopted to evaluate the eigenvector $\mathbf{v}_\lambda$ corresponding to a chosen eigenvalue λ, since that is not normally delivered by the Lanczos algorithm. We have followed the suggestion of Cullum and Willoughby [57]. First, we evaluated the eigenvector $\mathbf{v}'_\lambda$ of $\mathbf{T}$ in the basis of Lanczos vectors. This is an n-dimensional vector, which can be easily obtained by an expression similar to Eq. (2.89)

$$\lim_{s \to 0^+} [(s + \lambda)\mathbf{I} + \mathbf{T}]\mathbf{v}'_\lambda = \mathbf{c} \tag{2.90}$$

One can now evaluate the Ritz eigenvector $\mathbf{v}_\lambda$ (i.e., the eigenvector of $\boldsymbol{\Gamma}$ in terms of its components in the original basis set) by simply premultiplying $\mathbf{v}'_\lambda$ by the transformation matrix $\mathbf{S}$

$$\mathbf{T} = \mathbf{S}^{tr}\boldsymbol{\Gamma}\mathbf{S} \tag{2.91}$$

where $\mathbf{S}$ is the $n \times N$ matrix whose ith row is the ith Lanczos vector (within round-off errors, $\mathbf{S}^{tr}\mathbf{S}$ is the $n \times n$ unit matrix). This last procedure is usually done by repeating the Lanczos tridiagonalization, so there is no need to store the n Lanczos vectors.

2. Two-Body Smoluchowski Model

We start with the two-body Smoluchowski model (2BSM); the details of the formulation (matrix and starting vector) are discussed in Section II.C. A stochastic system made of two spherical rotators in a diffusive (Smoluchowski) regime has been used recently to interpret typical bifurcation phenomena of supercooled organic liquids [40]. In that work it was shown that the presence of a slow body coupled to the solute causes unusual decay behavior that is strongly dependent on the rank of the interaction potential.

In all the 2BSM calculations presented here, the diffusion coefficient D_1 equals 1, which defines the unit of frequency (inverse time); whereas the diffusion coefficient for the solvent, D_2 varied from 10 (very fast solvent relaxation) to 1, 0.1, 0.01 (very slow solvent relaxation). In the $D_2 = 10$ case, one finds that the reorientation of the solute is virtually independent of the solvent; a projection procedure could easily be adopted in this case to yield a one-body Smoluchowski equation for body 1 with perturbational corrections from body 2. The temporal decay of the first and second rank correlation functions is then typically monoexponential. When the solvent is relaxing slowly (i.e., D_2 is in the range 1–0.01), the effect of the large cage of the rapid motion of the probe becomes increasingly important. The decay of the correlation functions of both ranks is already different from that of a single exponential for $D_2 = 1$, and for $D_2 = 0.1$ the biexponential behavior is characterized by significantly different decay rates, since the separation of timescales for the two bodies is large.

An analysis of the different effects of a first rank versus a second rank interaction is instructive. A second rank potential between the bodies generates an apparent "strong collision" effect; that is, the motion of the first body in the potential field of the slow solvent body is dominated by the jump rate between the two equivalent minima. Only correlation functions of odd parity are sensitive to this jump motion, so that if an averaged unique correlation time is computed, this is significantly modified for the first rank case (i.e., τ_1) with respect to the free diffusional motion (i.e., no coupling) regime. Thus, for a fairly large range of parameters, the ratio τ_1/τ_2 is lower than 3 (which is the typical value for a purely diffusive description) and often very close to 1 (typical of a strong collision description). When the potential is first rank, there is no comparable "jump" motion, and the ratio τ_1/τ_2 is always equal to or larger than 3. In other words, a second rank interaction potential (i.e., $v_2 \neq 0$) causes the solute (usually the faster body) to reorient in the instantaneous cage induced by the solvent (the slower body) or to jump to the other potential minimum. In this way, a *two-body small-step diffusion* model can exhibit features that are typical of a *one-body-strong collision* description [58].

The numerical results are collected in Tables II and III (respectively, for first rank and second rank correlation functions for a first rank interaction potential) and Tables IV and V (second rank potential). Each entry is defined for a value of the potential parameter (v_1 or v_2) and a value of the diffusion coefficient of the second body (D_2). The column on the left contains the zero-frequency spectral density or autocorrelation time

$$\tau_{1,2} = \int_0^{+\infty} dt G_{1,2} = J_{1,2}(0) \qquad (2.92)$$

(note again that the subscript 1, 2 refers to both τ_1 and τ_2, etc.), whereas the column on the right contains the dominant eigenvalue(s) of the process: for each mode λ_i the corresponding weight ω_i is given in parentheses.

Let us consider Table II in detail. When the solvent body is fast ($D_2 = 10$), the only effect on the rotational correlation time τ_1 for an increasing tight interaction with the probe is a modest variation (going from 0.5 for $v_1 = 0$ to 0.53 for $v_1 = 4$). The solvent readjusts itself rapidly

TABLE II

2BSM: First Rank Correlation Times (Left Column), Dominant Eigenvalues (Right Column)[a] and Some of the Corresponding Eigenvectors[b]

v_1	D_2								
	10.0		1.0		0.1		0.01	
0.0	0.500	2.000 (1.000)	0.500	2.000 (1.000)	0.500	2.000 (1.000)	0.500	2.000 (1.000)
1.0	0.506	1.970 (0.997)	0.555	1.500 (0.650)	0.999	0.197 (0.118)	5.410	0.019 (0.096)
				2.865 (0.340)		2.190 (0.869)		2.160 (0.762)
2.0	0.519	1.919 (0.996)	0.665	1.251 (0.763)	1.980	0.191 (0.334)	14.97	0.019 (0.292)[2a]
				4.153 (0.274)		2.725 (0.631)		2.603 (0.571)[2b]
3.0	0.529	1.883 (0.996)	0.757	1.139 (0.831)	2.812	0.188 (0.504)	23.11	0.019 (0.456)
				5.829 (0.156)		3.515 (0.448)		3.281 (0.428)
4.0	0.534	1.863 (0.996)	0.815	1.089 (0.870)	3.380	0.186 (0.612)	28.74	0.019 (0.560)[2c]
				7.765 (0.119)		4.469 (0.343)		4.130 (0.356)[2d]

Eigenvalue 2a			Eigenvalue 2b			Eigenvalue 2c			Eigenvalue 2d		
$|c_i|^2$	J_1	J_2	$|c_i|^2$	J_1	J_2	$|c_i|^2$	J_1	J_2	$|c_i|^2$	J_1	J_2
0.761	0	1	0.003	0	1	0.481	0	1	0.512	1	0
0.077	1	0	0.705	1	0	0.142	1	0	0.245	1	2
0.146	1	2	0.237	1	2	0.275	1	2	0.129	2	1
0.006	2	1	0.036	2	1	0.037	2	1	0.081	2	3
0.008	2	3	0.016	2	3	0.053	2	3	0.017	3	2
			0.001	3	2	0.004	3	2	0.012	3	4
						0.005	3	4	0.004	4	3

[a] These are calculated for $D_1 = 1$ (which defines the frequency scale) for increasing first rank potential coupling. For each dominant eigenvalue the relative weight is given (in parentheses).

[b] J is constant and equal to 1.

TABLE III

2BSM: Second Rank Correlation Times (Left Column), Dominant Eigenvalues (Right Column)[a] and Some of the Corresponding Eigenvectors[b]

v_1	$D_2=10.0$		$D_2=1.0$		$D_2=0.1$		$D_2=0.01$	
0.0	0.167	6.000 (1.000)	0.167	6.000 (1.000)	0.167	6.000 (1.000)	0.167	6.000 (1.000)
1.0	0.168	5.910 (0.994)	0.175	3.805 (0.197)	0.187	0.591 (0.006)	0.244	0.059 (0.004)
				6.297 (0.464)		2.404 (0.067)		2.186 (0.062)
				6.632 (0.327)		6.186 (0.895)		6.163 (0.845)
2.0	0.171	5.756 (0.990)	0.196	3.532 (0.427)	0.269	0.575 (0.056)	0.838	0.059 (0.039)[3a]
				7.160 (0.376)		2.975 (0.175)		2.891 (0.028)[3b]
				8.194 (0.152)		6.738 (0.690)		6.625 (0.500)[3c]
3.0	0.175	5.650 (0.989)	0.217	3.353 (0.572)	0.404	0.565 (0.145)	2.027	0.059 (0.101)
				8.499 (0.319)		3.813 (0.281)		3.319 (0.237)
				10.45 (0.072)		7.647 (0.470)		7.364 (0.258)
4.0	0.177	5.591 (0.988)	0.234	3.249 (0.660)	0.550	0.559 (0.244)	3.422	0.059 (0.195)[3d]
				10.17 (0.268)		4.805 (0.333)		4.169 (0.301)[3e]
				13.34 (0.036)		8.898 (0.293)		8.350 (0.134)[3f]

Eigenvalue 3a			Eigenvalue 3b			Eigenvalue 3c			Eigenvalue 3d			Eigenvalue 3e			Eigenvalue 3f		
$|c_i|^2$	J_1	J_2	$|c_i|^2$	J_1	J_2	$|c_i|^2$	J_1	J_2	$|c_i|^2$	J_1	J_2	$|c_i|^2$	J_1	J_2	$|c_i|^2$	J_1	J_2
0.761	0	2	0.009	0	2	0.055	1	1	0.481	0	2	0.001	0	2	0.102	1	1
0.094	1	1	0.694	1	1	0.003	1	3	0.172	1	1	0.467	1	1	0.129	1	3
0.130	1	3	0.241	1	3	0.630	2	0	0.245	1	3	0.290	1	3	0.294	2	0
0.003	2	0	0.026	2	0	0.265	2	2	0.019	2	0	0.088	2	0	0.009	2	2
0.004	2	2	0.011	2	2	0.021	2	4	0.026	2	2	0.031	2	2	0.390	2	4
0.001	2	4	0.015	2	4	0.018	3	1	0.045	2	4	0.090	2	4	0.028	3	1
			0.001	3	1	0.005	3	3	0.002	3	3	0.014	3	1	0.035	3	5
									0.004	3	5	0.002	3	3	0.004	4	2
												0.012	3	5	0.006	4	6

[a] These are calculated for $D_1 = 1$ for increasing first rank potential coupling. For each dominant eigenvalue the relative weight is given (in parentheses).

[b] J is constant and equal to 2.

141

TABLE IV

2BSM: First Rank Correlation Times (Left Column), Dominant Eigenvalues (Right Column)[a] and Some of the Corresponding Eigenvectors[b]

v_2	D_2							
	10.0		1.0		0.1		0.01	
0.0	0.500	2.000 (1.000)	0.500	2.000 (1.000)	0.500	2.000 (1.000)	0.500	2.000 (1.000)
1.0	0.505	1.979 (0.999)	0.534	1.835 (0.973)	0.608	1.291 (0.637)	0.642	1.064 (0.494)
						3.108 (0.356)		2.811 (0.491)
2.0	0.518	1.926 (0.961)	0.641	1.487 (0.945)	1.068	0.693 (0.694)	1.353	0.495 (0.627)[4a]
				9.368 (0.045)		4.424 (0.275)		4.040 (0.337)[4b]
3.0	0.531	1.877 (0.998)	0.772	1.226 (0.937)	2.027	0.387 (0.694)	3.522	0.210 (0.731)
				12.39 (0.041)		6.305 (0.158)		5.774 (0.224)
4.0	0.540	1.849 (0.998)	0.869	1.097 (0.947)	3.263	0.256 (0.831)	9.167	0.087 (0.801)[4c]
				16.63 (0.031)		8.753 (0.102)		8.025 (0.143)[4d]

Eigenvalue 4a			Eigenvalue 4b			Eigenvalue 4c			Eigenvalue 4d		
$|c_i|^2$	J_1	J_2	$|c_i|^2$	J_1	J_2	$|c_i|^2$	J_1	J_2	$|c_i|^2$	J_1	J_2
0.321	1	0	0.640	1	0	0.263	1	0	0.610	1	0
0.620	1	2	0.331	1	2	0.516	1	2	0.310	1	2
0.025	3	2	0.015	3	2	0.086	3	2	0.035	3	2
0.034	3	4	0.012	3	3	0.112	3	3	0.027	3	4
						0.010	5	5	0.009	5	4
						0.012	5	5	0.006	5	6

[a] These are calculated for $D_1 = 1$ for increasing second rank potential coupling. For each dominant eigenvalue the relative weight is given (in parentheses).

[b] J is constant and equal to 1.

TABLE V

2BSM: Second Rank Correlation Times (Left Column), Dominant Eigenvalues (Right Column)[a] and Some of the Corresponding Eigenvectors[b]

v_2	D_2 = 10.0		D_2 = 1.0		D_2 = 0.1		D_2 = 0.01	
0.0	0.167	6.000 (1.000)	0.167	6.000 (1.000)	0.167	6.000 (1.000)	0.167	6.000 (1.000)
1.0	0.168	5.936 (0.998)	0.178	4.785 (0.595)	0.252	0.593 (0.059)	0.981	0.059 (0.050)
				7.567 (0.386)		5.939 (0.653)		5.565 (0.323)
2.0	0.172	5.777 (0.996)	0.209	3.899 (0.709)	0.506	7.327 (0.148)	3.402	6.042 (0.274)
				10.00 (0.268)		0.577 (0.230)		0.059 (0.190)[5a]
						6.548 (0.565)		6.043 (0.258)
3.0	0.176	5.631 (0.995)	0.246	3.414 (0.792)	0.809	8.100 (0.107)	6.303	6.679 (0.357)[5b]
				13.43 (0.174)		0.563 (0.412)		0.059 (0.369)
						7.949 (0.365)		7.198 (0.151)
4.0	0.179	5.546 (0.995)	0.273	3.192 (0.846)	1.050	9.470 (0.120)	8.648	8.019 (0.332)
						0.554 (0.561)		0.059 (0.511)[5c]
						10.06 (0.216)		9.996 (0.298)[5d]
						11.54 (0.138)		

Eigenvalue 5a

$|c_i|^2$	J_1	J_2
0.789	0	2
0.040	2	0
0.056	2	2
0.097	2	4
0.002	4	2
0.003	4	4
0.003	4	6

Eigenvalue 5b

$|c_i|^2$	J_1	J_2
0.002	0	2
0.339	2	0
0.085	2	2
0.550	2	4
0.008	4	2
0.013	4	6

Eigenvalue 5c

$|c_i|^2$	J_1	J_2
0.474	0	2
0.093	2	0
0.132	2	2
0.226	2	4
0.020	4	2
0.017	4	4
0.034	4	6
0.001	6	4
0.001	6	6

Eigenvalue 5d

$|c_i|^2$	J_1	J_2
0.006	0	2
0.325	2	0
0.085	2	2
0.480	2	4
0.038	4	2
0.002	4	4
0.057	4	6
0.043	6	4

[a] These are calculated for $D_1 = 1$ for increasing second rank potential coupling. For each dominant eigenvalue the relative weight is given (in parentheses).

[b] J is constant and equal to 2.

143

to the solute motion. For lower values of the diffusion coefficient of the solvent body, the decay of the correlation function is controlled by two dominant modes: one of them (the fast one) may be related to the rotational diffusion of the first body relative to the instantaneous orientation of the solvent body, and the other one to the free rotational diffusion of the solvent body. One can see that for increasing potentials the process is more and more differentiated from the original free rotational diffusion (FRD), that is, the rotational diffusive motion of a spherical body in the absence of any coupling. The slow mode becomes more and more effective when the potential strength is increased (i.e., the weight goes from 0.096 for $v_1 = 1$ to 0.560 for $v_1 = 4$, for $D_2 = 0.01$). This is the cause of the dramatic increase of the autocorrelation time, since the solute rotation is heavily damped by the large cage.

The composition of the eigenvectors corresponding to the dominant modes is analyzed in two cases ($D_2 = 0.01$ and $v_1 = 2, 4$) in terms of the basis sets used in the representation of the time evolution operator (see Table II). The square moduli of the coefficients c_Λ^i, each of them representing the contribution of the basis set function labeled by the collective index Λ to the ith eigenvector, are shown together with the index Λ itself. In the present case, only the quantum numbers J_1 and J_2 are nondiagonal, while the total angular momentum quantum number J is a constant, and it is equal to 1 (2) for first (second) rank correlation functions. From the entry of Table II to the eigenvalue labeled 2a ($D_2 = 0.01$ and $v_1 = 2$), one can see that the slow mode is largely a FRD of the solvent body ($|c_\Lambda^i|^2$ equal to 0.76 for $J_1 = 0$, $J_2 = 1$) with a small component of "dynamic interaction" between the two bodies ($|c_\Lambda^i|^2$ equal to 0.14 for $J_1 = 1$, $J_2 = 2$). From entry 2b it is seen that the fast mode is mostly due to FRD of the solute body ($|c_\Lambda^i|^2$ equal to 0.70 for $J_1 = 1$, $J_2 = 0$), again with a dynamic interaction contribution ($|c_\Lambda^i|^2$ equal to 0.23 for $J_1 = 1$, $J_2 = 2$). The dynamic interaction becomes more important for the case of a tighter interaction ($v_1 = 4$); cf. entries 2c and 2d.

Table III contains correlation times and dominant eigenvalues for a second rank observable in a first rank potential. There are still roughly two ranges of decay rates when the solvent body is slow ($D_2 \leq 1$). The slower range is mostly due to the FRD of the solvent body, while the faster one is described by motions of the solute body and/or dynamic interactions. This faster decay is hardly described by a single frequency, unlike the case of a first rank correlation function. Rather, it is controlled by a few eigenvalues of the same order of magnitude. Thus for $D_2 = 0.01$ and $v_1 = 2$ the slow mode, entry 3a in Table III, is largely described by a $J_1 = 0$, $J_2 = 2$ term; the fast mode 3b is mostly due to dynamic interactions ($J_1 = 1$, $J_2 = 1$ and $J_1 = 1$, $J_2 = 3$ are the important terms); and the fast mode 3c is mainly due to FRD of the solute ($J_1 = 1$, $J_2 = 0$ is dominant). Dynamic interaction terms are more important in the eigenvectors when the potential is stronger ($v_1 = 4$): see entries 3d (similar to 3a), 3e (similar to 3b) and 3f (similar to 3c), and note that mode 3e, which is dominated by dynamic interactions, is now heavily weighted in the correlation function.

When a second rank potential is considered, the previous description must be modified, particularly when odd rank autocorrelation functions are involved, as we have pointed out above. We present results here that confirm our previous interpretation [40]. In Table IV we show correlation times and eigenmodes for a first rank observable. As in Table II, two dominant modes are present for $D_2 \lesssim 1$; the fast one is again a FRD of the solute body, whereas the slow one is a thoroughly "mixed" nature (i.e., dynamic interactions), and may be loosely related, for very slow cages, to the jump motion of body 1 from one metastable orientation to another (cf. the cases in Table IV for $D_2 = 0.01$, $v_2 = 2$ and $v_2 = 4$).

Finally, results on second rank correlation functions for a second rank potential are collected in Table V. The situation is now very similar to the corresponding set of data for a first rank potential (Table III), since even rank correlation functions are not sensitive, for symmetry reasons, to jump motions. The slow mode is then again mostly due to the FRD of the larger solvent body while the fast modes are mainly dominated by motions of the first body (cf. the entries for $D_2 = 0.01$, $v_2 = 2$ and 4 in Table V). Note that other faster eigenvalues are present, with smaller weights, whose nature is mostly mixed, but are not listed in the table. Their individual contribution to the overall decay of the correlation function is small, but their cumulative weights may be around 0.1–0.3 or even more.

In Figs. 2a–d, we show the time decay of the first rank correlation function $G_1(t)$ for a first rank potential. In Fig. 2a results for different values of v_1 for $D_2 = 10$ are shown (they correspond to the first column in Table II). Observe that even for large potentials the effect of the light solvent body is negligible. For intermediate values of D_2 (cf. Figs. 2b and 2c) the contribution of the slow decay mode is more effective. A complete separation of time scales is evident in Fig. 2d ($D_2 = 0.01$). Similar behavior is obtained in the case of a second rank potential (Figs. 4a–d). Finally, the same features are observed in the case of second rank correlation functions $G_2(t)$ both for a first rank potential (Figs. 3a–d) and a second rank one (Figs. 5a–d), although the sensitivity of second rank correlation functions to the size of the solvent body seems to be less pronounced than for first rank correlation functions (compare for example Fig. 2c with Fig. 3c).

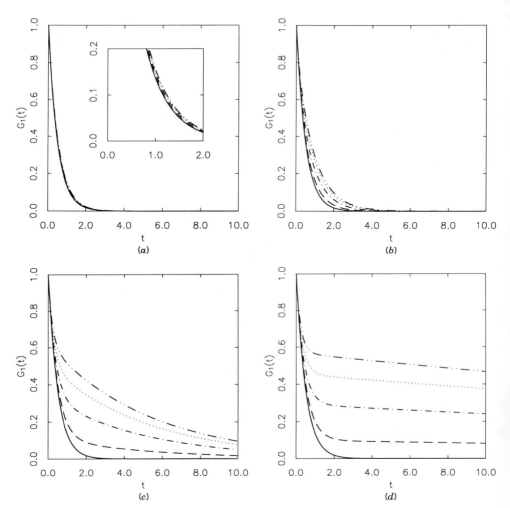

Figure 2. 2BSM First rank correlation functions for a first rank potential coupling:
———, $v_1 = 0$; ----, $v_1 = 1$; —·—·—, $v_1 = 2$; ·····, $v_1 = 3$; —··—··—, $v_1 = 4$. (a) $D_2 = 10$; (b)
$D_2 = 1$; (c) $D_2 = 0.01$. The unit of time in Figs. 2–7 has been taken by setting $D_1^{-1} = 1$.

From the strictly computational point of view, we may note that all the
computations were made with truncation parameters $J_{1_{max}}$, $J_{2_{max}}$ ranging
from 4 to 8; the number of Lanczos steps necessary to achieve convergence (with respect to the correlation times *and* the dominant eigenvalues) was usually less than 50.

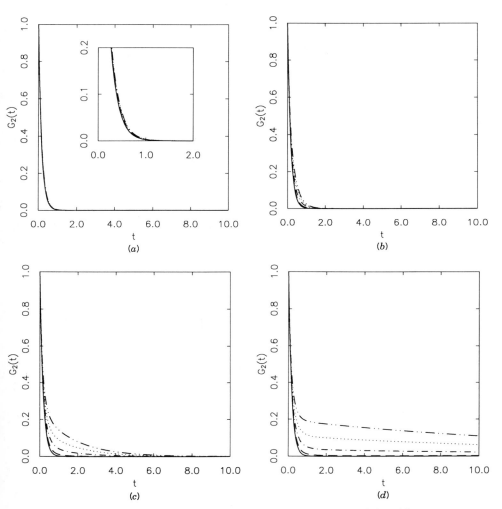

Figure 3. Second rank correlation functions for a first rank potential coupling: ———, $v_1 = 0$; — — — —, $v_1 = 1$; — $\cdot$ — $\cdot$ —, $v_1 = 2$; $\cdots\cdots$, $v_1 = 3$; — $\cdot\cdot$ — $\cdot\cdot$ —, $v_1 = 4$. (a) $D_2 = 10$; (b) $D_2 = 1$; (c) $D_2 = 0.1$; (d) $D_2 = 0.01$.

3. Three-Body Smoluchowski Model

The next model that we have treated in order of complexity is a three-body Smoluchowski model (3BSM). A field **X** has been included, coupled exclusively through first rank (dipole–field) interactions to the two spherical rotators. No direct coupling has been taken to exist

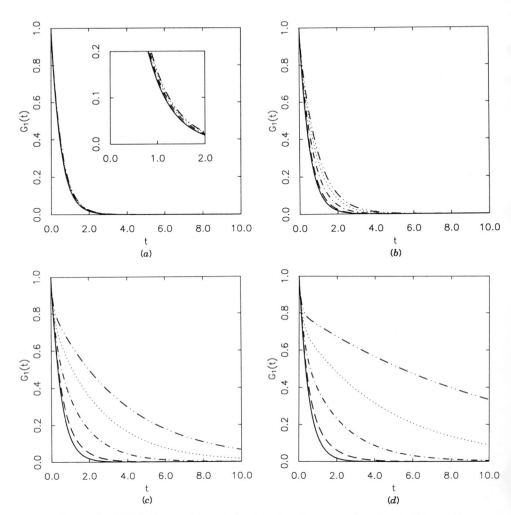

Figure 4. 2BSM. First rank correlation functions for a second rank potential coupling: ———, $v_2 = 0$; — — — —, $v_2 = 1$; — · — · —, $v_2 = 2$; · · · · ·, $v_2 = 3$; — · · — · · —, $v_2 = 4$. (a) $D_2 = 10$; (b) $D_2 = 1$; (c) $D_2 = 0.1$; (d) $D_2 = 0.01$.

between the probe and the solvent body in order to show the effect of the field on the motion of the two bodies, and to examine its role in providing an indirect coupling between them. According to Eq. (2.34) the only parameters that now define the system energies are μ_1 and μ_2. We have kept $\mu_2 = 10\mu_1$ in all the computations. Then μ_1 has been varied from 0.0

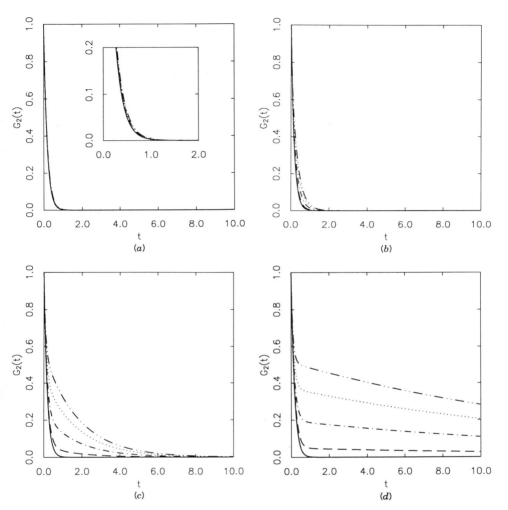

Figure 5. 2BSM. Second rank correlation functions for a second rank potential coupling: ———, $v_2 = 0$; – – – –, $v_2 = 1$; –·–·–, $v_2 = 2$; ·····, $v_2 = 3$; –··–··–, $v_2 = 4$. (a) $D_2 = 10$; (b) $D_2 = 1$; (c) $D_2 = 0.1$; (d) $D_2 = 0.01$.

to 0.5 in 0.1 steps. The diffusion coefficient D_1 for the first body has been taken as the unit of frequency, while D_2, the diffusion coefficient of the solvent body, has been set at 0.1. That is, we are simulating the effect of increasing coupling between two rotating spherical dipoles in a polar medium, with the second dipole ten times slower than the first one. Three sets of results have been obtained: (1) a fast interacting field ($D_X = 10$); (2) a field with a correlation time comparable to that of the lighter body ($D_X = 1$); (3) a slow field ($D_X = 0.1$).

From our results, one can see that the departure from simple single-exponential decay is even more evident than for the 2BSM case. The correlation functions of both ranks are greatly affected by the motion of the field, so that a third decay constant is almost always necessary to fit the decay. Notice that the effect is most pronounced, as expected, for the case $D_X = 0.1$. In Table VI the correlation times and the most important decay frequencies (eigenvalues) are collected for each set of values of μ_1 and D_X, for a first rank rotational observable. When the field is relaxing rapidly ($D_X = 10$, first column), the system is always biexponential for a significant coupling ($\mu_1 \geq 0.2$): that is, the fast third body just provides an effective coupling between the two bodies. For slower fields ($D_X = 1$ and 0.1) the decay is roughly triexponential, since now the timescale of the field is interfering with the motional timescales of the rotators. The dominant modes are described largely as pure motions of the first and the second body, without any appreciable component of the field. This may be seen from the composition of the corresponding eigenvectors in Table VI. The eigenvector corresponding to the slow mode labeled 6a (for $\mu_1 = 0.2$ and $D_X = 1$) is almost completely described as a FRD of body 2, whereas the fastest one (6b) is a FRD of body 1. An increase in coupling leads to dynamic interaction terms; for example, for $\mu_1 = 0.5$ the slowest mode (6c) is more than half composed of a FRD of body 2, and the fastest one (6d) of a FRD of body 1, but there are significant contributions to both from mixed terms. Note that in Table VI the field related quantum numbers (i.e., n and j) are always less than 2. Terms with n equal to 1 contribute almost negligibly to the dominant eigenmodes; i.e., relaxation of first rank observables seems to be largely independent of fluctuations in the magnitude of the field, and more affected by fluctuations in its orientation.

In Table VII, numerical results are shown for second rank correlation functions. For low values of the potential coupling, the motions are largely FRD. Some new features arise for large couplings. Let us look more closely, for example, at the eigenvectors associated with eigenvalues 7b, 7c, 7d (the dominant modes for $D_X = 1$ and $\mu_1 = 0.5$). One can see that the slowest mode (7b) is mainly the FRD of the slow, large second body (the largest coefficient being for the case of $n = 0$, $j = 0$, $J_1 = 0$ and $J_2 = 2$; and $J = 2$). The second mode has a dominant term with $n = 0$ and $j = 0$, $J_1 = 1$ and $J_2 = 1$, (i.e., a "mixed" motion involving only the two rotators). Finally, the third and fastest one has as its most important basis function (but with a weighting coefficient of only 0.380): $n = 1$, $j = 0$,

TABLE VI

3BSM: First Rank Correlation Times (Left Column), Dominant Eigenvalues (Right Column)[a] and Some of the Corresponding Eigenvectors[b]

μ_1	10.0		1.0		0.1	
0.0	0.500	2.000 (1.000)	0.500	2.000 (1.000)	0.500	2.000 (1.000)
0.1	0.506	1.999 (0.947)	0.512	0.180 (0.002) 1.068 (0.012) 2.014 (0.984)	0.565	0.065 (0.005) 2.006 (0.993)
0.2	0.597	0.191 (0.022) 2.023 (0.933)	0.669	0.140 (0.024)[6a] 1.163 (0.032) 1.603 (0.025) 2.077 (0.905)[9b]	1.341	0.035 (0.031) 2.041 (0.921) 2.193 (0.039)
0.3	0.972	0.180 (0.099) 2.141 (0.886)	1.457	0.101 (0.105) 1.131 (0.025) 1.922 (0.190) 2.255 (0.656)	5.933	0.019 (0.105) 2.130 (0.660) 2.327 (0.208)
0.4	1.814	0.165 (0.150) 2.452 (0.713)	3.766	0.072 (0.249) 1.081 (0.010) 2.397 (0.445) 2.749 (0.168)	21.97	0.011 (0.244) 2.318 (0.367) 2.649 (0.296)
0.5	3.037	0.149 (0.437) 3.056 (0.522)	8.224	0.052 (0.409)[6e] 2.921 (0.403)[6d]	57.33	0.007 (0.409) 2.695 (0.178) 3.228 (0.266)

Eigenvalue 6a

$|c_i|^2$	n	j	J_1	J_2	J
0.942	0	0	0	1	1
0.009	0	0	1	2	1
0.006	0	1	0	0	0
0.035	0	0	1	0	2
0.006	0	1	0	2	2

Eigenvalue 6b

$|c_i|^2$	n	j	J_1	J_2	J
0.002	0	0	0	1	1
0.001	1	0	0	1	1
0.918	0	1	0	0	1
0.002	0	0	2	1	1
0.070	0	1	0	0	0

Eigenvalue 6c

$|c_i|^2$	n	j	J_1	J_2	J
0.644	0	0	0	1	1
0.125	0	0	1	0	1
0.163	0	0	2	2	1
0.016	0	0	2	1	2
0.012	0	0	2	3	1

Eigenvalue 6d

$|c_i|^2$	n	j	J_1	J_2	J
0.044	0	0	0	1	1
0.001	1	0	0	1	1
0.669	0	0	1	0	1
0.002	1	0	1	0	1
0.137	0	0	1	2	1

[a] These are calculated for $D_1 = 1$, $D_2 = 0.1$ and $\mu_2 = 10\mu_1$. For increasing μ_1, for each dominant eigenvalue the relative weight is given (in parentheses).

[b] J_T is constant and equal to 1.

TABLE VII

3BSM: Second Rank Correlation Times (Left Column), Dominant Eigenvalues (Right Column)[a] and Some of the Corresponding Eigenvectors[b]

μ_1	10.0		1.0		0.1	
0.0	0.167	6.000 (1.000)	0.167	6.000 (1.000)	0.167	6.000 (1.000)
0.1	0.167	5.994 (0.996)	0.167	6.010 (0.994)	0.167	6.006 (0.994)
0.2	0.170	2.218 (0.013); 6.003 (0.978)	0.171	2.225 (0.012); 6.060 (0.965)[7a]	0.176	0.086 (0.001); 2.089 (0.018); 6.041 (0.930)
0.3	0.185	0.541 (0.004); 2.335 (0.058); 6.097 (0.903)	0.197	0.302 (0.005); 2.398 (0.033); 6.205 (0.875)	0.283	0.054 (0.005); 2.222 (0.060); 6.108 (0.891)
0.4	0.240	0.497 (0.022); 2.654 (0.137); 6.382 (0.698)	0.326	0.214 (0.021); 2.589 (0.127); 6.521 (0.669)	1.068	0.033 (0.030); 2.554 (0.129); 6.515 (0.746)
0.5	0.383	0.450 (0.094); 3.276 (0.241); 6.969 (0.533)	0.776	0.156 (0.095)[7b]; 3.039 (0.165)[7c]; 7.088 (0.344)[7d]	4.220	0.022 (0.096); 3.173 (0.232); 7.112 (0.456)

Eigenvalue 7a

$\|c_i\|^2$	n	j	J_1	J_2	J
0.001	0	0	1	1	2
0.978	0	0	2	0	2
0.016	0	1	1	0	1
0.002	0	1	3	0	3

Eigenvalue 7b

$\|c_i\|^2$	n	j	J_1	J_2	J
0.586	0	0	0	2	2
0.001	1	0	0	2	2
0.162	0	0	1	1	2
0.119	0	0	1	3	2
0.011	0	0	2	0	2
0.011	0	0	2	2	2
0.008	0	0	2	4	2
0.054	0	1	0	1	1

Eigenvalue 7c

$\|c_i\|^2$	n	j	J_1	J_2	J
0.090	0	0	0	2	2
0.002	1	0	0	2	2
0.611	0	0	1	1	2
0.005	1	0	1	1	2
0.124	0	0	1	3	2
0.041	0	1	2	0	2
0.008	0	0	2	2	2
0.010	0	0	2	4	2

Eigenvalue 7d

$\|c_i\|^2$	n	j	J_1	J_2	J
0.006	0	0	0	2	2
0.360	1	0	0	2	2
0.008	0	0	1	1	2
0.106	1	0	1	1	2
0.003	0	0	1	3	2
0.029	0	1	1	3	2
0.020	0	0	2	0	2
0.010	0	1	2	0	2

[a] These are calculated for $D_1 = 1$, $D_2 = 0.1$ and $\mu_2 = 10\mu_1$, for increasing μ_1. For each dominant eigenvalue the relative weight is given (in parentheses).

[b] J_T is constant and equal to 2.

$J_1 = 0$ and $J_2 = 2$, it is a mode in which the fluctuation of the field magnitude, and not only its orientation in space, is important.

Figures 6 and Fig. 7 contain respectively first rank autocorrelation functions $G_1(t)$ and second rank autocorrelation functions $G_2(t)$ versus time for the three values of $D_{\mathbf{x}}$ considered. Note that for the time range

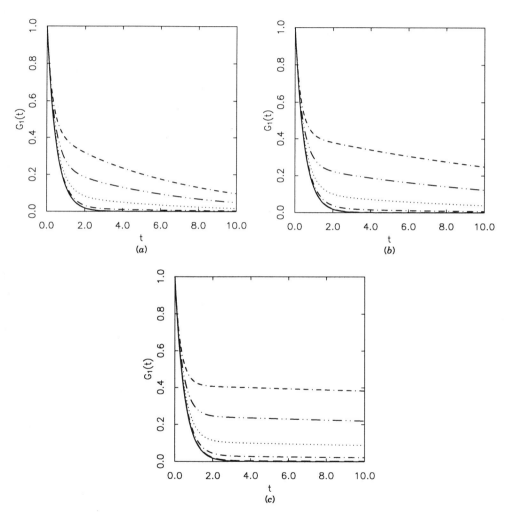

Figure 6. 3BSM. First rank correlation functions: ——, $\mu_1 = 0$; ————, $\mu_1 = 0.1$; —·—·—, $\mu_1 = 0.2$; ·····, $\mu_1 = 0.3$; —···—···—, $\mu_1 = 0.4$; —··——··—, $\mu_1 = 0.5$. (a) $D_{\mathbf{x}} = 10$; (b) $D_{\mathbf{x}} = 1$; (c) $D_{\mathbf{x}} = 0.1$.

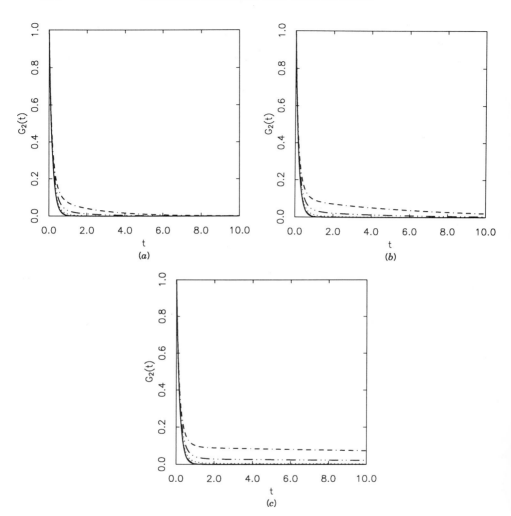

Figure 7. 3BSM. Second rank correlation functions: ———, $\mu_1 = 0$; – – – –, $\mu_1 = 0.1$; – · – · –, $\mu_1 = 0.2$; · · · · · ·, $\mu_1 = 0.3$; – · · – · · –, $\mu_1 = 0.4$; – – · – – · – –, $\mu_1 = 0.5$. (a) $D_x = 10$; (b) $D_x = 1$; (c) $D_x = 0.1$

considered (ten times the inverse of D_1), first rank correlation functions are much more affected by a large coupling via the fluctuating field than are second rank functions. This is primarily due to the slow eigenvalues which are strongly dependent on μ_1. This drastically changes the long-time behavior of G_1, such that τ_1 for the first rank processes is much larger for large μ_1.

The additional two parameters due to the presence of the field, n_{max} and j_{max} were both equal to 1 in the fast field case, 2 in the intermediate case, and 5 in the slow field case, while $J_{1_{max}}$ and $J_{2_{max}}$ were fixed at 6; J_{max} has been set equal to the maximum value given by the triangle rule, that is from 14 to 17. Note that a careful analysis of the eigenvector tables suggests that if one is interested only in evaluating the dominant modes of the system (with a relative error, say, less than 20%), much smaller matrices could be used. The number of Lanczos steps was always between 50 and 100.

4. *Two-Body Fokker–Planck–Kramers Model: SRLS Case*

In the previous two subsections the coupling between two bodies in a completely diffusional regime was investigated. It was seen that, for the two-body model with direct coupling at least two characteristic decay times are always present (and their order of magnitude and physical interpretation depend strongly on the rank of the interaction potential). When a third, translational degree of freedom was added as a source of indirect coupling, a third characteristic time was often observed.

In this subsection we include the conjugate momentum degrees of freedom in the two-body model. Thus, we obtain a multidimensional rotational Fokker–Planck–Kramers equation for the stochastic motion of the two bodies. According to Section II.F, we have now to deal with a phase space of dimension equal to 12, specified by the orientations of the two bodies Ω_1 and Ω_2 and by their angular momentum vectors L_1 and L_2. The one-body Fokker–Planck–Kramers model for rotational motions has been studied (in the absence of potentials) by many authors including Fixman and Rider [4] and McClung [6]. Physically, inertial effects (i.e., the effects due to the explicit inclusion of momenta) will be negligible when the collision frequency of the rotational body is much greater than its streaming frequency. In this case the relaxation of the momentum vector is much faster than the reorientation of the body. But inertial effects are important for smaller collision frequencies. In a two-body model one must also consider the collision frequency of the second body, which can be in an inertial regime. Also, strong potential couplings will yield inertial effects, especially for short times.

In all our 2BKM calculations, we had a physical picture in mind in which the first body is in a diffusive or inertial regime, while the surrounding, massive, solvent cage is always in a heavily damped regime. Thus, by varying the frictional parameter (collision frequency) of the solute body, we have studied its motion in the cage provided by the second body, from the Smoluchowski regime to an almost inertial regime in which librational modes become important. The only source of coupling is assumed to be due to the interaction potential. No "third body" effects are included for simplicity, and the model can be regarded as a generalization of the "slowly relaxing local structure" (SRLS) models of Freed and co-workers [33, 35].

Throughout this set of simulations, the unit of frequency has been chosen as the streaming frequency of body 1, that is, $\omega_1^s = 1$; the ratio between the moments of inertia has been set equal to 10, that is, $I_2/I_1 = 10$, so that the streaming frequency of the second body is given by $\omega_2^s = 1/\sqrt{10}$. Finally, the collision frequency ω_2^c of body 2 has been maintained at 100. The only parameters varied were the collision frequency of body 1, ω_1^c (for values of 50 (damped case), 5 (intermediate case), 0.5 (inertial case)). The computations were performed both for a first rank potential ($v_1 = 0, 1, \ldots, 3$) and for a second rank potential ($v_2 = 0, 1, \ldots, 3$). Orientational correlation functions of rank 1 and 2 for body 1 have been computed; also, correlation functions for the reorientation of the conjugate momentum $\mathbf{L}_1$ have been evaluated.

Table VIII contains the autocorrelation times and the dominant modes for first rank correlation observables. Note that in this table, and in the following tables for Kramers models, the eigenvalues and their weights are complex numbers (but the real part of any eigenvalue is nonnegative). In this and succeeding tables we write the real and imaginary parts for each and we use the convention of placing a bar over the first figure of a negative number. Since the correlation function must be real, each complex eigenvalue is accompanied by its conjugate, which is not shown in the table.

As was the case in the 2BSM, a slow eigenmode (equal to twice the diffusion coefficient of the solvent body) is always present. It represents the FRD of the large cage in the diffusive regime. The only exception is for zero coupling ($v_1 = 0$) where the model reduces to a one-body case (that is completely equivalent to the spherical rotational Kramers case treated by McClung). The motion of the solute body is responsible for the other fast modes whether in the diffusive regime ($\omega_1^c = 50$), the intermediate regime ($\omega_1^c = 5$) or the inertial regime ($\omega_1^c = 0.5$). In the last case there are eigenmodes with nonzero imaginary parts having a significant weight. These motions are of a librational kind. But there are also fast modes whose eigenvalues are purely real, and they correspond to solute modes that are largely diffusional (i.e., the coupling to angular momentum is not very significant). Thus the model seems to provide *three* different types of decay process: namely, a pure rotation of the solvent body (slow mode), a librational motion of the solute body (complex mode), and a fast reorientation of larger amplitude, more or less related to the FRD of the solute body.

Table IX gives the equivalent results for a second rank observable.

TABLE VIII

2BKM-SRLS: First Rank Correlation Times (Left Column) and Dominant Eigenvalues (Right Column)[a]

υ_1	ω_1^c = 50.0		ω_1^c = 5.0		ω_1^c = 0.5	
0.0	25.00	0.040, 0.000 (1.000, 0.000)	2.646	0.377, 0.000 (1.000, 0.000)	1.036	0.702, 0.000 (0.438, 0.000) 0.145, 1.822 (0.276, 0.349)
1.0	74.59	0.002, 0.000 (0.107, 0.000) 0.043, 0.000 (0.879, 0.000)	51.76	0.002, 0.000 (0.098, 0.000) 0.445, 0.000 (0.910, 0.000)	49.94	0.002, 0.000 (0.097, 0.000) 0.577, 0.000 (0.143, 0.000) 0.686, 0.431 (0.167, 0.133)
2.0	171.5	0.002, 0.000 (0.311, 0.000) 0.053, 0.000 (0.638, 0.000)	147.4	0.002, 0.000 (0.290, 0.000) 0.551, 0.000 (0.634, 0.000) 0.573, 0.000 (0.116, 0.000)	145.4	1.313, 1.861 (0.110, 0.308) 0.002, 0.000 (0.287, 0.000) 0.489, 0.000 (0.097, 0.000)
3.0	253.9	0.002, 0.000 (0.477, 0.000) 0.067, 0.000 (0.456, 0.000)	228.8	0.002, 0.000 (0.453, 0.000) 0.726, 0.000 (0.514, 0.000)	226.9	1.431, 1.360 (0.283, 0.171) 0.002, 0.000 (0.451, 0.000) + faster modes

[a] These are calculated for $\omega_1^s = 1$ (which defines the frequency scale), $\omega_2^s = 1/\sqrt{10}$ and $\omega_2^c = 100$ for increasing first rank potential coupling. For each dominant eigenvalue the relative weight is given (in parentheses).

TABLE IX

2BKM-SRLS: Second Rank Correlation Times (Left Column) and Dominant Eigenvalues (Right Column)[a]

v_1	ω_1^c = 50.0		ω_1^c = 5.0		ω_1^c = 0.5	
0.0	8.340	0.120, 0.000 (1.000, 0.000)	1.005	1.213, 0.000 (1.276, 0.000) 6.542, 0.000 (0.401, 0.000)	0.537	1.431, 0.000 (0.200, 0.000) 2.361, 0.000 ($\bar{0}$.131, 0.000) 2.881, 2.889 (0.175, 0.021) 3.250, 1.868 (0.117, $\bar{0}$.093)
1.0	9.718	0.006, 0.000 (0.005, 0.000) 0.045, 0.000 (0.067, 0.000) 0.123, 0.000 (0.907, 0.000)	1.689	0.006, 0.000 (0.004, 0.000) 0.448, 0.000 (0.065, 0.000) 1.248, 0.000 (1.165, 0.000) 8.380, 0.000 (0.143, 0.000)	1.158	0.006, 0.000 (0.003, 0.000) 0.582, 0.000 (0.021, 0.000) 1.362, 0.315 (0.074, 0.017) 1.979, 2.588 (0.532, 0.155)
2.0	16.73	0.006, 0.000 (0.046, 0.000) 0.056, 0.000 (0.188, 0.000) 0.134, 0.000 (0.678, 0.000) 0.140, 0.000 (0.032, 0.000)	7.469	0.006, 0.000 (0.038, 0.000) 0.558, 0.000 (0.188, 0.000) 1.356, 0.000 (0.911, 0.000) 8.374, 0.000 (0.214, 0.000)	6.805	0.006, 0.000 (0.037, 0.000) 0.493, 0.000 (0.060, 0.000) 1.265, 0.542 (0.027, 0.013) 1.833, 1.297 (0.244, $\bar{0}$.036)
3.0	29.39	0.006, 0.000 (0.126, 0.000) 0.071, 0.000 (0.273, 0.000) 0.151, 0.000 (0.437, 0.000) 0.159, 0.000 (0.073, 0.000)	19.25	0.006, 0.000 (0.109, 0.000) 0.733, 0.000 (0.284, 0.000) 1.517, 0.000 (0.696, 0.000)	18.46	0.006, 0.000 (0.109, 0.000) 0.482, 0.000 (0.078, 0.000) 1.320, 0.997 (0.015, 0.040) 1.721, 1.384 (0.225, $\bar{0}$.085)

[a]These are calculated for $\omega_1^s = 1$, $\omega_2^s = 1/\sqrt{10}$ and $\omega_2^c = 100$ for increasing first rank potential coupling. For each dominant eigenvalue the relative weight is given (in parentheses).

Here the slow eigenvalue is equal to six times the diffusion coefficient of the solvent body (since we are looking at a second rank property). For nonzero values of the coupling parameter we find a larger number of fast eigenmodes than in the first rank correlation case; but it is usually possible to put them together as "clusters" of similar magnitude. We can again identify at least three processes.

In Fig. 8 corresponding to Table VIII, the first rank correlation functions $G_1(t)$ have been plotted for the various values of v_1 and ω_1^c. The overdamped and intermediate cases (Figs. 8a and 8b) are close to the 2BSM. As was expected, the situation is rather different for the inertial case. Here the librational motion of the light first body, that is only slightly damped by an effective friction, becomes important at least for short times. The presence of librations is indicated by the damped oscillations in the graph, which are more pronounced for an increased potential (dotted line in Fig. 8c). The effect of the first rank coupling potential on the second rank correlation functions shown in Figs. 9a ($\omega_1^c = 50$), 9b ($\omega_1^c = 5$) and 9c ($\omega_1^c = 0.5$), is somewhat weaker, as was the case for the two-body Smoluchowski model. The librational peaks in Fig. 9c are still present, but they are less pronounced.

Table X contains numerical data concerning the temporal decay of momentum correlation functions (for body 1, i.e., L_1). One realizes immediately that in this case the influence of the cage body is much weaker than it was for orientational observables. For $\omega_1^c = 50$ the relaxation of the momentum of body 1 is almost totally decoupled from reorientation of body 2, even for large potentials. For $\omega_1^c = 5$, a cluster of eigenvalues close in value to the collision frequency is present. This is also the case for $\omega_1^c = 0.5$, but librational modes are beginning to play a nonnegligible role.

These features are confirmed by an analysis of the correlation function plots for the momentum, $G_J(t)$, in Fig. 10. The coupling to a second body is almost ineffective both in the Smoluchowski regime (Fig. 10a) and in the intermediate regime (Fig. 10b). The departure from monoexponential decay, which is rigorously observed for the uncoupled case, is quite small. On the other hand, a strong effect on the angular momentum relaxation is observed in the inertial regime (Fig. 10c). Note that the potential coupling makes the decay of the momentum vector faster, and the librational motion is more prominent.

When a second rank potential ($v_2 \neq 0$) is considered, there are significant differences in behavior of both the reorientational correlation functions (as in the 2BSM) and in the momentum correlation functions. Tables XI and XII give the results for first and second rank correlation functions, respectively. In both cases we have at least three decay modes.

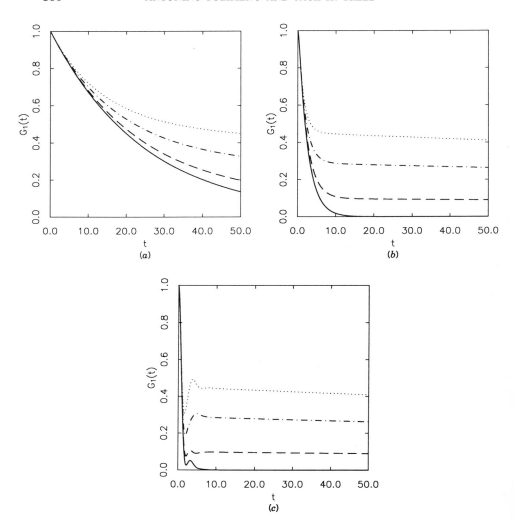

Figure 8. 2BKM-SRLS. First rank correlation functions for a first rank potential coupling: ———, $v_1 = 0$; – – – –, $v_1 = 1$; – · – · –, $v_1 = 2$; · · · · ·, $v_1 = 3$. (*a*) $\omega_1^c = 50$; (*b*) $\omega_1^c = 5$; (*c*) $\omega_1^c = 0.5$.

One of them is much slower than the others, and the fastest one becomes librational (i.e., it acquires a detectable imaginary part) in the inertial regime ($\omega_1^c = 0.5$). Note, however, that since the second rank potential coupling provides two potential minima in which the solute can reorient (with the possibility of "jump" motions), the nature of the slow mode in

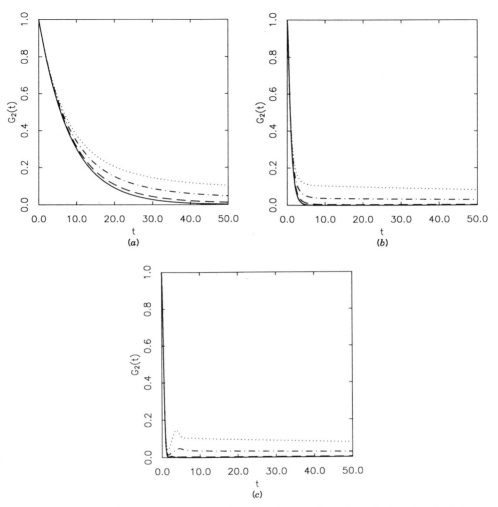

Figure 9. 2BKM-SRLS. Second rank correlation functions for a first rank potential coupling: ——, $v_1 = 0$; – – – –, $v_1 = 1$; – · – · –, $v_1 = 2$; · · · · ·, $v_1 = 3$. (a) $\omega_1^c = 50$; (b) $\omega_1^c = 5$; (c) $\omega_1^c = 0.5$.

the case of first rank correlation functions is no longer simply the overall relaxation of the solvent body. The situation is very close to the 2BSM for a second rank potential for $\omega_1^c = 50$. The relaxation of the momentum vector $\mathbf{L}_1$ is so fast that we are virtually in a completely diffusive regime.

In the intermediate regime ($\omega_1^c = 5$) inertial effects become more

TABLE X

2BKM-SRLS: Momentum Correlation Times (Left Column) and Dominant Eigenvalues (Right Column)[a]

v_1	ω_1^c = 50.0		ω_1^c = 5.0		ω_1^c = 0.5	
0.0	0.020	50.00, 0.000 (1.000, 0.000)	0.200	5.00, 0.000 (1.000, 0.000)	2.000	0.500, 0.000 (1.000, 0.000)
1.0	0.020	50.01, 0.000 (1.000, 0.000)	0.197	4.882, 0.000 (0.413, 0.000)	1.425	0.499, 0.000 (0.267, 0.000)
				5.200, 0.000 (0.355, 0.000)		0.684, 0.431 (0.240, 0.162)
				5.661, 0.000 (0.263, 0.000)		1.231, 0.000 (0.304, 0.000)
				7.096, 0.000 (0.023, 0.000)		2.430, 0.000 (0.090, 0.000)
2.0	0.020	50.02, 0.000 (0.999, 0.000)	0.189	4.656, 0.000 (0.094, 0.000)	0.904	0.500, 0.000 (0.142, 0.000)
				5.151, 0.000 (0.226, 0.000)		0.733, 0.918 (0.099, 0.085)
				5.794, 0.000 (0.697, 0.000)		1.033, 0.000 (0.090, 0.000)
				7.563, 0.000 (0.402, 0.000)		1.730, 0.025 (0.920, 0.153)
3.0	0.020	50.02, 0.000 (0.998, 0.000)	0.181	4.423, 0.000 (0.024, 0.000)	0.654	0.500, 0.000 (0.054, 0.000)
				5.915, 0.000 (0.817, 0.000)		0.741, 1.324 (0.026, 0.041)
				8.160, 0.627 (0.092, 0.942)		0.978, 0.000 (0.072, 0.000)
						1.415, 0.000 (0.305, 0.000)

[a] These are calculated for $\omega_1^s = 1$, $\omega_2^s = 1/\sqrt{10}$ and $\omega_2^c = 100$ for increasing first rank potential coupling. For each dominant eigenvalue the relative weight is given (in parentheses).

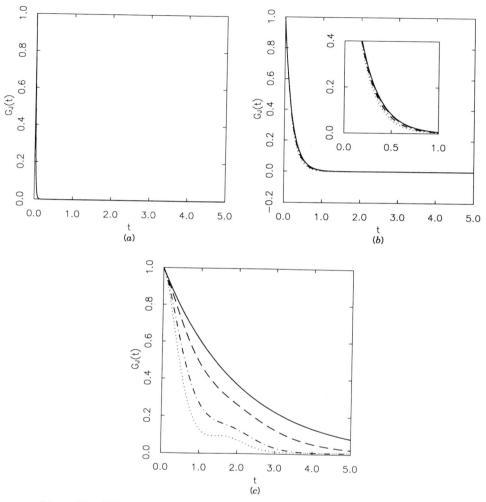

Figure 10. 2BKM-SRLS. Momentum correlation functions for a first rank potential coupling: ———, $v_1 = 0$; ————, $v_1 = 1$; —·—·—, $v_1 = 2$; ·····, $v_1 = 3$. (a) $\omega_1^c = 50$; (b) $\omega_1^c = 5$; (c) $\omega_1^c = 0.5$.

important. The relaxation of the momentum $\mathbf{L}_1$ is coupled to the slow mode corresponding to the jump motion. The net result is a decreased effective friction acting on this mode, so that the dominant frequency of first rank correlation functions is increased. This effect is also present when the collision frequency is further reduced ($\omega_1^c = 0.5$). Note however

TABLE XI

2BKM-SRLS: First Rank Correlation Times (Left Column), Dominant Eigenvalues (Right Column)[a] and Some Corresponding Eigenvectors[b]

v_2	ω_1^c = 50.0		ω_1^c = 5.0		ω_1^c = 0.5	
0.0	25.00	0.040, 0.000 (1.000, 0.000)	2.646	0.377, 0.000 (1.000, 0.000)	1.036	0.702, 0.000 (0.438, 0.000) 1.145, 1.822 (0.276, 0.349)
1.0	31.27	0.023, 0.000 (0.559, 0.000) 0.058, 0.000 (0.435, 0.000)	3.485	0.198, 0.000 (0.502, 0.000) 0.597, 0.000 (0.576, 0.000) 2.314, 0.000 (0.017, 0.000)	1.848	0.270, 0.000 (0.417, 0.000) 1.064, 1.720 (0.388, 0.215) 1.200, 2.270 (0.027, 0.049)
2.0	59.73	0.012, 0.000 (0.659, 0.000) 0.084, 0.000 (0.319, 0.000) 0.245, 0.000 (0.015, 0.000)	7.843	0.085, 0.000 (0.630, 0.000) 0.895, 0.000 (0.405, 0.000) 2.724, 0.000 (0.220, 0.000)	6.431	0.094, 0.000 (0.596, 0.000) 0.956, 1.944 (0.463, 0.012) 1.349, 1.795, (0.118, 0.139)
3.0	126.8	0.006, 0.000 (0.753, 0.000) 0.121, 0.000 (0.208, 0.000) 0.267, 0.000 (0.035, 0.000)	21.55	0.034, 0.00 (0.733, 0.000) 1.283, 0.000 (0.208, 0.000) 2.961, 1.181 (0.026, 0.176)	21.02	0.034, 0.000 (0.722, 0.000)[11a] + faster modes

Eigenvalue 11a

$|c_i|^1$	n_1	n_2	j_1	j_2	j	J_1	J_2	J
0.267	0	0	0	0	0	1	0	1
0.003	1	0	0	0	0	1	0	1
0.526	0	0	0	0	0	1	2	1
0.007	1	0	0	0	0	1	2	1
0.069	0	0	0	0	0	3	2	1
0.102	0	0	0	0	0	3	4	1
0.003	0	0	1	0	0	5	4	1
0.002	0	0	1	1	1	1	0	1

[a] These are calculated $\omega_1^s = 1$, $\omega_2^s = 1/\sqrt{10}$ and $\omega_2^c = 100$ for increasing second rank potential coupling. For each dominant eigenvalue the relative weight is given (in parentheses).

[b] J_T is constant and equal to 1.

TABLE XII

2BKM-SRLS: Second Rank Correlation Times (Left Column) and Dominant Eigenvalues (Right Column)[a]

v_2	ω_1^c					
	50.0		5.0		0.5	
0.0	8.340	0.120, 0.000 (1.000, 0.000)	1.005	1.213, 0.000 (1.276, 0.000) 6.542, 0.000 (0.401, 0.000)	0.537	1.431, 0.000 (0.200, 0.000) 2.361, 0.000 (0.131, 0.000) 2.881, 2.889 (0.175, 0.021) 3.250, 1.868 (0.117, 0.093)
1.0	16.69	0.006, 0.000 (0.053, 0.000) 0.116, 0.000 (0.655, 0.000) 0.132, 0.000 (0.123, 0.000) 0.147, 0.000 (0.160, 0.000)	9.138	0.006, 0.000 (0.049, 0.000) 1.104, 0.000 (0.300, 0.000) 1.251, 0.000 (0.557, 0.000) 1.438, 0.000 (0.353, 0.000)	8.618	0.006, 0.000 (0.050, 0.000) 0.693, 0.000 (0.094, 0.000) 1.707, 1.047, (0.177, 0.051) 1.976, 1.887, (0.264, 0.036)
2.0	41.38	0.006, 0.000 (0.211, 0.000) 0.126, 0.000 (0.399, 0.000) 0.140, 0.000 (0.237, 0.000) 0.177, 0.000 (0.120, 0.000)	32.92	0.006, 0.000 (0.195, 0.000) 1.123, 0.000 (0.207, 0.000) 1.582, 0.000 (0.753, 0.000) 1.778, 0.000 (0.082, 0.000)	32.67	0.006, 0.000 (0.192, 0.000) 0.489, 0.000 (0.111, 0.000) 1.269, 2.377 (0.064, 0.211)
3.0	70.07	0.006, 0.000 (0.394, 0.000) 0.148, 0.000 (0.100, 0.000) 0.163, 0.000 (0.387, 0.000) 0.217, 0.000 (0.109, 0.000)	56.52	0.006, 0.000 (0.370, 0.000) 1.260, 0.000 (0.108, 0.000) 1.993, 0.000 (0.125, 0.000) 2.576, 0.000 (0.732, 0.000)	61.34	0.006, 0.000 (0.366, 0.000) 0.428, 0.000 (0.080, 0.000) 1.147, 2.772 (0.019, 0.448)

[a]These are calculated for $\omega_1^s = 1$, $\omega_2^s = 1/\sqrt{10}$ and $\omega_2^c = 100$ for increasing second rank potential coupling. For each dominant eigenvalue the relative weight is given (in parentheses).

that the rate of change of the eigenvalue with decreasing ω_1^c is slowed down, and it is negligible when the potential is high; that is, for $v_2 = 1$ the eigenvalue goes from 0.023 ($\omega_1^c = 50$), to 0.198 ($\omega_1^c = 5$) and 0.270 ($\omega_1^c = 0.5$); for $v_2 = 2$ it goes from 0.012 to 0.085 and 0.094; finally for $v_2 = 3$ it goes from 0.006 to 0.034 and 0.034, that is, it remains unchanged when the collision frequency is reduced by a factor of ten. This may be due to an incipient Kramers turnover effect. It is possible that for a larger potential, the jump eigenmode would invert its dependence versus ω_1^c by starting to increase when the collision frequency is decreased.

Nothing of this sort is observed for second rank correlation functions, since the dominant slow mode is simply a FRD of the solvent body. In both first and second rank correlation functions one notes that librational modes are slightly more important when the potential coupling is second rank than they were for a first rank potential coupling. This may be due to the increased curvature of the potential near the minima.

The complex nature of the slow mode responsible for the long-time behavior of first rank correlation functions for a first rank interaction potential is illustrated by the composition of the eigenvector corresponding to the slow mode 11a in Table XI, for $v_2 = 3$ and $\omega_1^c = 0.5$. Note that n_1, n_2, j_1, j_2 describe the magnitudes and the orientations of the momentum vectors $\mathbf{L}_1$ and $\mathbf{L}_2$; j is referred to the orientation of $\mathbf{L}_1 + \mathbf{L}_2$, J_1 and J_2 are related to the orientations of the two bodies, and the total orientational angular operator defines the quantum number J; finally J_T, which is not included in this table, is the total angular momentum quantum number, and it is always equal to 1 for first rank orientational and momentum correlation functions, and to 2 for second rank correlation functions. In Fig. 11 we show the first rank correlation functions for different collision frequencies of body 1. The second rank correlation function decays are plotted in Fig. 12. The librational motions in the wells are more important than they were in the first rank potential case (since there is now a more accentuated curvature of the potential wells).

In Table XIII we show momentum correlation functions. One finds that there are increased librational effects from the second rank potential. Compare for instance the case of $v_2 = 3$ and $\omega_1^c = 5$ with the corresponding entry in Table X ($v_1 = 3$ and $\omega_1^c = 5$). In the present case the librational mode is dominant and the simple decay mode has a weight only half that of the case in Table X, for which most of the decay is by a nonlibrational mode. The interpretation of the dominant modes is complicated when the potential is large and the regime of motion of the solute body is inertial. In Table XIII some of the eigenvectors corresponding to the dominant eigenvalues for $v_2 = 3$ and $\omega_1^c = 0.5$ are shown. It is not possible to isolate a single component having a coefficient larger than 0.5

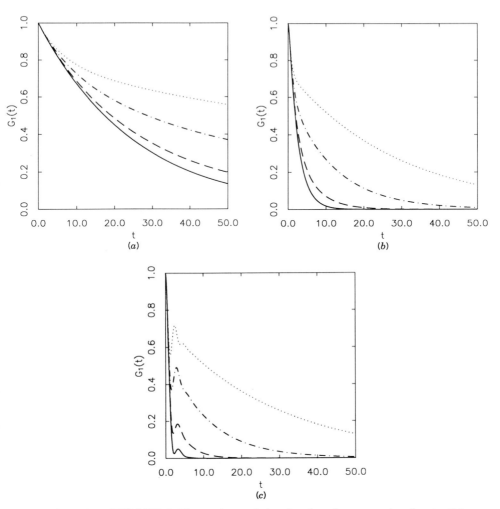

Figure 11. 2BKM-SRLS. First rank correlation functions for a second rank potential coupling: ———, $v_2 = 0$; ————, $v_2 = 1$; —·—·—, $v_2 = 2$; ·····, $v_2 = 3$. (a) $\omega_1^c = 50$; (b) $\omega_1^c = 5$; (c) $\omega_1^c = 0.5$.

in the three eigenvectors given. However, the angular momentum quantum numbers for the solvent body, n_2 and j_2, are always zero, given the large viscosity imposed on it.

Figure 13 shows $G_J(t)$ for a second rank potential coupling. The effect of the second body is still negligible in the overdamped case (Fig. 13a),

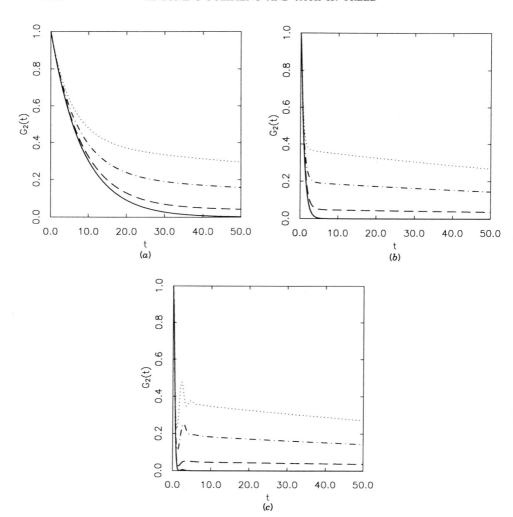

Figure 12. 2BKM-SRLS. Second rank correlation functions for a second rank potential coupling: ———, $v_2 = 0$; – – – –, $v_2 = 1$; – · – · –, $v_2 = 2$; · · · · ·, $v_2 = 3$. (a) $\omega_1^c = 50$; (b) $\omega_1^c = 5$; (c) $\omega_1^c = 0.5$.

since the momentum relaxation is so fast that it is not affected by the details of the solvent. But even for the intermediate case shown in Fig. 13b, the librational motions in the cage have a large enough amplitude to make the momentum reorient in the opposite direction with respect to the starting orientation. This is reflected in the negative part of $G_J(t)$.

TABLE XIII

2BKM-SRLS: Momentum Correlation Times (Left Column), Dominant Eigenvalues (Right Column)[a] and Some Corresponding Eigenvectors[b]

	ω_1^c					
v_2	50.0		5.0		0.5	
0.0	0.020	50.00, 0.000 (1.000, 0.000)	0.200	5.000, 0.000 (1.000, 0.000)	2.000	0.500, 0.000 (1.000, 0.000)
1.0	0.018	49.94, 0.000 (0.978, 0.000)	0.182	4.642, 0.000 (0.672, 0.000) 5.003, 0.000 (0.326, 0.000)	1.449	0.500, 0.000 (0.333, 0.000) 0.808, 0.000 (0.506, 0.000) 1.275, 0.823 (0.136, 0.020)
2.0	0.015	49.86, 0.000 (0.972, 0.000)	0.141	3.678, 0.537 (0.404, 0.233) 5.003, 0.000 (0.330, 0.000)	0.975	0.500, 0.000 (0.334, 0.000) 1.076, 0.000 (0.196, 0.000) 1.044, 1.725 (0.304, $\bar{0}$.146) 1.309, 2.373 (0.584, 0.143)
3.0	0.011	49.83, 0.000 (0.892, 0.000)	0.106	3.172, 1.122 (0.656, 0.223) 5.003, 0.000 (0.327, 0.000)	0.775	0.500, 0.000 (0.33, 0.000)[13a] 1.235, 0.000 (0.069, 0.000)[13b] 1.145, 2.791, ($\bar{0}$.132, 0.366)[13c] 1.644, 2.323 (0.027, 0.070)

Eigenvalue 13a

$|c_i|^2$	n_1	n_2	j_1	j_2	j	J_1	J_2	J
0.212	0	0	1	0	1	0	0	0
0.409	0	0	1	0	1	0	2	2
0.043	0	0	1	0	1	2	0	2
0.110	0	0	1	0	1	2	2	0
0.060	0	0	1	0	1	2	2	2
0.141	0	0	1	0	1	2	4	2
0.005	0	0	1	0	1	2	4	4
0.010	0	0	1	0	1	4	0	4
0.005	0	0	1	0	1	4	4	2

Eigenvalue 13b

$|c_i|^2$	n_1	n_2	j_1	j_2	j	J_1	J_2	J
0.018	0	0	0	0	0	0	0	0
0.004	0	0	0	0	0	4	4	1
0.005	1	0	0	0	0	4	4	1
0.171	0	0	1	0	1	0	0	0
0.096	1	0	1	0	1	0	0	0
0.005	2	0	1	0	1	0	0	0
0.002	3	0	1	0	1	0	0	0
0.083	0	0	1	0	1	2	2	2
0.056	0	0	1	0	1	2	4	2

Eigenvalue 13c

$|c_i|^2$	n_1	n_2	j_1	j_2	j	J_1	J_2	J
0.050	0	0	0	0	0	2	2	1
0.042	1	0	0	0	0	2	2	1
0.008	2	0	0	0	0	2	2	1
0.043	0	0	0	0	0	4	4	1
0.037	1	0	0	0	0	4	4	1
0.006	2	0	0	0	0	4	4	1
0.011	0	1	0	1	1	0	0	0
0.011	1	1	0	1	1	0	0	0
0.003	0	1	0	1	1	0	2	2

[a] These are calculated for $\omega_1^s = 1$, $\omega_2^s = 1/\sqrt{10}$, $\omega_2^c = 100$ for increasing second rank potential coupling. For each dominant eigenvalue the relative weight is given (in parentheses).

[b] J_T is constant and equal to 1.

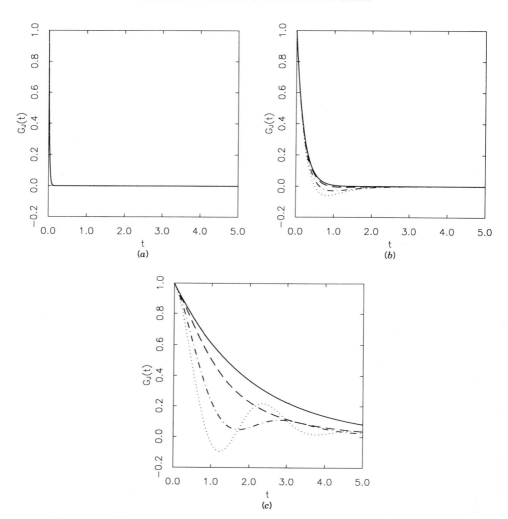

Figure 13. 2BKM-SRLS. Momentum correlation functions for a second rank potential coupling: ———, $v_2 = 0$; – – – –, $v_2 = 1$; – · – · –, $v_2 = 2$; · · · · · , $v_2 = 3$. (a) $\omega_1^c = 50$; (b) $\omega_1^c = 5$; (c) $\omega_1^c = 0.5$.

When the first body is in an underdamped regime of motion (Fig. 13c) and the potential is high (dotted line), the momentum vector actually fluctuates back and forth for a while before decaying toward zero.

For all the computations, $J_{1_{max}} = J_{2_{max}} = 5$ and $n_{2_{max}} = j_{2_{max}} = 1$ (since body 2 is always in an overdamped regime); $n_{1_{max}}$ and $j_{1_{max}}$ have been both set equal to 1 for $\omega_1^c = 50$, to 2 for $\omega_1^c = 5$ and to 5 for $\omega_1^c = 0.5$ and the number of Lanczos steps was between 100 and 500. Note that the largest matrices treated (for $\omega_1^c = 0.5$ and $v_2 = 3$) had dimensions of order 10^5!

5. Two-Body Fokker–Planck–Kramers Model: FT Case

The last model considered in this work is a variation on the previous inertial two-body approach. Instead of allowing a direct source of coupling between the two bodies via a simple interaction potential, we now introduce a *frictional* coupling between them. This is the residual effect after the elimination as fast variables, of a stochastic field vector and its conjugate linear momentum (see Section II.G). The model is an inertial counterpart of the 3BSM described above, provided the "third body" is relaxing fast enough that only its averaged effect on the torques acting on the two principal bodies is left. This case is equivalent to similar models with "fluctuating torques" (FT) features (cf. Stillman and Freed [33]).

Both the SRLS and the FT inertial models were discussed in the context of the Hubbard–Einstein relation, that is, the relation between the momentum correlation time τ_J and the rotational correlation time (second rank) τ_R for a stochastic Brownian rotator [39]. It was shown that both models can cause a substantial departure from the simple expression predicted by a one-body Fokker–Planck–Kramers equation:

$$\tau_J \tau_2 = \frac{I}{6k_B T} \tag{2.93}$$

In the FT case, it was found that the additional friction due to the fast field has a different effect on the rotational versus momentum relaxation, such that, whereas τ_2 still behaves in a "normal" fashion (i.e., it is roughly proportional to the *total* friction, from both the solvent terms and the field terms), τ_J is not much influenced by the friction generated by the fast field. These comments apply to the case in which the sources of friction are large, so that the system is always in a diffusional regime.

These matters are described in more detail in the last set of calculations included in the present work. We have considered a fixed "core" friction (from the unspecified fast solvent modes) and fixed dimensions for the second body: $\omega_2^s = 1/\sqrt{10}$, $\omega_2^c = 100$, with $\omega_1^s = 1$ for the first body (so it is ten times smaller than body 2). We have investigated two cases: $\omega_1^c = 50$ and $\omega_1^c = 5$. The additional source of coupling, according to Section II.F, is specified by the frictional parameters ω_1, ω_2. To further simplify the analysis we have kept $\omega_2 = 10\omega_1$, and we have varied ω_1 from 0 to 400.

In Tables XIV and XV we show the dominant eigenvalues and correlation times for a first rank and for a second rank orientational observable, respectively. Only the real parts of the eigenvalues have been written, since we have just explored a range of parameters for which all imaginary parts are negligible. (The same is also true for the relative weights.) The existence of slow modes is due to the large values of the frictional parameters, both for the solvent and the solute body. In all cases at least four important decay frequencies are reported. Note the great difference in magnitude between the first and second rank autocorrelation times, due to the presence of a slow mode in the first rank case that is absent in the second rank case. The effect of the core frictional parameter ω_1^c is less relevant than in the SRLS model, since for the range of parameters used, most of the friction comes from the fast relaxing stochastic field. Let us look at the case of $\omega_1 = 200$ and $\omega_1^c = 50$. In Tables XIV and XV the eigenvectors corresponding to the most important eigenvalues for each case are shown. The very slow mode (entry 14a) in the first rank decay is dominated by a FRD of the solvent cage. The next eigenvalue corresponds to a dynamic interaction mode (entry 14b), with an important component of FRD of body 1. Finally the eigenvalue labelled 14c, which is the one with the highest weight, is mostly described as the relaxation frequency of body 1, with a component of mixed dynamics. For second rank correlation functions, the decay process for the same set of parameters is governed by a set of frequencies which are difficult to relate to simple motions of the two isolated bodies. That is, for all entries in Table XV one sees that the eigenvectors always have a mixed character. Not surprisingly, the momentum quantum numbers do not appear to influence the rotational properties (i.e., there are no eigenvectors with a significant projection on basis set functions with nonzero values of n_1, n_2, j_1, j_2 or j).

Table XVI contains numerical data for the momentum correlation functions. As previously shown, one finds that by increasing the coupling parameter ω_1 the correlation time tends to reach a constant value that appears to be only a function of the core frictions ω_1^c and ω_2^c. Analysis of the eigenvectors suggests a strong dynamic interaction between the two bodies. In Table XVI we show the eigenvectors for the same set of parameters given above. In all cases, components depending on basis functions with quantum numbers j_1 and/or j_2 equal to 1 are present, while n_1 and n_2 are almost always equal to 0. That is, the motions corresponding to the eigenvalues of Table XVI are coupled modes of the vectors $\mathbf{L}_1$ and $\mathbf{L}_2$ involving their (mutual) orientations, but unaffected by fluctuations in their magnitudes.

In Fig. 14 we show first rank correlation functions, $G_1(t)$, for $\omega_1^c = 50$

TABLE XIV

2BKM-FT: First Rank Correlation Times (Left Column) and Dominant Eigenvalues (Right Column)[a] and Some Corresponding Eigenvectors[b]

ω_1	ω_1^c = 50.0		ω_1^c = 5.0	
0.0	25.00	0.400(−1) (1.000)	2.646	0.377 (1.000)
100.0	64.30	0.181(−3) (0.004)	40.69	0.183(−3) (0.003)
		0.153(−1) (0.338)		0.230(−1) (0.328)
		0.300(−1) (0.644)		0.104 (0.628)
				0.285 (0.018)
200.0	101.6	0.952(−4) (0.004)[14a]	76.99	0.953(−4) (0.004)
		0.951(−2) (0.331)[14b]		0.119(−1) (0.326)
		0.272(−1) (0.642)[14c]		0.916(−1) (0.593)
		0.945(−1) (0.015)		0.213 (0.020)
300.0	129.7	0.718(−4) (0.004)	112.8	0.646(−4) (0.004)
		0.691(−2) (0.264)		0.809(−2) (0.326)
		0.256(−1) (0.583)		0.861(−1) (0.573)
		0.780(−1) (0.019)		0.243 (0.041)
400.0	168.9	0.524(−4) (0.004)	152.5	0.466(−4) (0.004)
		0.539(−2) (0.327)		0.611(−2) (0.325)
		0.244(−1) (0.623)		0.826(−1) (0.516)
		0.663(−1) (0.019)		0.218 (0.051)

Eigenvalue 14a

$|c_i|^2$	n_1	n_2	j_1	j_2	j	J_1	J_2	J
0.993	0	0	0	0	0	0	1	1
0.004	0	0	0	0	0	1	0	1
0.002	0	0	0	0	1	1	2	1

Eigenvalue 14b

$|c_i|^2$	n_1	n_2	j_1	j_2	j	J_1	J_2	J
0.331	0	0	0	0	0	1	0	1
0.655	0	0	0	0	0	1	2	1
0.006	0	0	0	0	0	2	3	1
0.003	0	0	0	0	0	3	2	1
0.003	0	0	0	0	0	3	4	1

Eigenvalue 14c

$|c_i|^2$	n_1	n_2	j_1	j_2	j	J_1	J_2	J
0.006	0	0	0	0	0	0	1	1
0.642	0	0	0	0	0	1	0	1
0.315	0	0	0	0	0	1	2	1
0.002	0	0	0	0	0	2	1	1
0.008	0	0	0	0	0	2	3	1
0.015	0	0	0	0	0	3	2	1
0.010	0	0	0	0	0	3	4	1

[a] These are calculated $\omega_1^s = 1$, $\omega_2^s = 1/\sqrt{10}$, $\omega_1^c = 100$ and $\omega_2 = 10\omega_1$, for increasing ω_1. For each dominant eigenvalue the relative weight is given.

[b] J_T is constant and equal to 1.

TABLE XV

2BKM-FT: Second Rank Correlation Times (Left Column) and Dominant Eigenvalues (Right Column)[a] and Some of the Corresponding Eigenvectors[b]

ω_1	ω_1^c	
	50.0	5.0
0.0	8.340 0.120 (1.000)	1.005 1.213 (1.276) 6.542 (0.401)
100.0	13.90 0.539(−1) (0.209) 0.684(−1) (0.380) 0.920(−1) (0.387) 0.200 (0.010)	5.377 0.234(−1) (0.002) 0.883(−1) (0.192) 0.173 (0.077) 0.195 (0.259) 0.346 (0.364)
200.0	17.30 0.970(−2) (0.002)[15a] 0.350(−1) (0.196)[15b] 0.552(−1) (0.360)[15c] 0.860(−1) (0.378)[15d] 0.144 (0.018)	7.850 0.120(−1) (0.002) 0.469(−1) (0.190) 0.147 (0.250) 0.317 (0.250) 0.334 (0.078)
300.0	20.21 0.702(−2) (0.002) 0.260(−1) (0.195) 0.476(−1) (0.281) 0.827(−1) (0.367) 0.114 (0.025)	10.09 0.817(−2) (0.003) 0.320(−1) (0.189) 0.127 (0.168) 0.196 (0.118) 0.315 (0.245)
400.0	22.88 0.554(−2) (0.002) 0.206(−1) (0.193) 0.440(−1) (0.333) 0.801(−1) (0.359) 0.979(−1) (0.034)	12.24 0.630(−2) (0.003) 0.243(−1) (0.189) 0.163 (0.214) 0.229 (0.078) 0.308 (0.244)

Eigenvalue 15a

$|c_i|^2$	n_1	n_2	j_1	j_2	j	J_1	J_2	J
0.395	0	0	0	0	0	1	1	2
0.583	0	0	0	0	0	1	3	2
0.002	0	0	0	0	0	2	0	2
0.011	0	0	0	0	0	2	4	2
0.002	0	0	0	0	0	3	1	2
0.002	0	0	0	0	0	3	3	2
0.002	0	0	0	0	0	3	5	2

Eigenvalue 15b

$|c_i|^2$	n_1	n_2	j_1	j_2	j	J_1	J_2	J
0.003	0	0	0	0	0	1	3	2
0.197	0	0	0	0	0	2	0	2
0.277	0	0	0	0	0	2	2	2
0.464	0	0	0	0	0	2	4	2
0.001	0	0	0	0	0	3	3	2
0.036	0	0	0	0	0	3	5	2
0.011	0	0	0	0	0	4	2	2
0.010	0	0	0	0	0	4	4	2

Eigenvalue 15c

$|c_i|^2$	n_1	n_2	j_1	j_2	j	J_1	J_2	J
0.009	0	0	0	0	0	1	3	2
0.360	0	0	0	0	0	2	0	2
0.127	0	0	0	0	0	2	2	2
0.358	0	0	0	0	0	2	4	2
0.029	0	0	0	0	0	3	5	2
0.002	0	0	0	0	0	4	4	2

Eigenvalue 15d

$|c_i|^2$	n_1	n_2	j_1	j_2	j	J_1	J_2	J
0.002	0	0	0	0	0	1	1	2
0.002	0	0	0	0	0	1	3	2
0.377	0	0	0	0	0	2	0	2
0.522	0	0	0	0	0	2	2	2
0.022	0	0	0	0	0	2	4	2
0.001	0	0	0	0	0	3	1	2
0.029	0	0	0	0	0	3	3	2
0.006	0	0	0	0	0	3	5	2

[a] These are calculated for $\omega_1^s = 1$, $\omega_1^s = 1/\sqrt{10}$, $\omega_2^c = 100$, $\omega_2 = 10\omega_1$ for increasing ω_1. For each dominant eigenvalue the relative weight is given (in parentheses).

[b] J_T is constant and equal to 2.

TABLE XVI

2BKM-FT: Momentum Correlation Times (Left Column) and Dominant Eigenvalues (Right Column)[a] and Some of the Corresponding Eigenvectors[b]

ω_1	ω_1^c	
	50.0	5.0
0.0	0.020 — 50.00 (1.000) 50.00 (0.333) 54.36 (0.201) 59.70 (0.071) 143.9 (0.158)	0.200 — 5.000 (1.000) 5.000 (0.333) 13.00 (0.203) 60.20 (0.105) 98.82 (0.114)
100.0	0.014 — 50.00 (0.333)[16a] 54.45 (0.202)[16b] 153.8 (0.126)[16c] 236.8 (0.160)[16d]	0.090 — 5.000 (0.333) 13.30 (0.202) 111.7 (0.126) 192.2 (0.160)
200.0	0.013 — 50.00 (0.331) 54.47 (0.270) 205.3 (0.118) 330.2 (0.159)	0.086 — 5.000 (0.333) 13.41 (0.201) 162.8 (0.127) 285.6 (0.159)
300.0	0.012 — 50.00 (0.333) 54.590 (0.271) 256.4 (0.118) 423.6 (0.159)	0.084 — 5.000 (0.332) 13.46 (0.201) 213.9 (0.127) 378.9 (0.159)
400.0	0.012	0.083

Eigenvalue 16a

$|c_i|^2$	n_1	n_2	j_1	j_2	j	J_1	J_2	J
0.333	0	0	1	0	1	0	0	0
0.667	0	0	1	0	1	2	0	2

Eigenvalue 16b

$|c_i|^2$	n_1	n_2	j_1	j_2	j	J_1	J_2	J
0.043	0	0	0	1	1	1	1	1
0.026	0	0	0	1	1	1	1	2
0.017	0	0	0	1	1	1	3	2

Eigenvalue 16b (Continued)

$|c_i|^2$	n_1	n_2	j_1	j_2	j	J_1	J_2	J
0.202	0	0	1	0	1	0	0	0
0.400	0	0	1	0	1	0	2	2
0.001	0	0	1	0	1	1	1	0
0.001	0	0	1	0	1	1	1	1
0.002	0	0	1	0	1	1	1	2
0.003	0	0	1	0	1	1	3	2
0.101	0	0	1	1	1	2	0	2
0.038	0	0	1	1	1	2	2	0
0.091	0	0	1	1	1	2	2	1
0.070	0	0	1	1	1	2	2	2

Eigenvalue 16c

$|c_i|^2$	n_1	n_2	j_1	j_2	j	J_1	J_2	J
0.003	1	0	0	0	0	1	1	1
0.034	1	0	0	0	0	2	2	1
0.011	1	0	0	0	0	3	3	1
0.025	1	0	0	0	0	4	4	1
0.007	0	0	0	1	1	1	1	0
0.001	0	0	0	1	1	1	1	2
0.002	0	0	0	1	1	1	3	2
0.002	0	0	0	1	1	3	1	2
0.002	0	0	1	1	1	3	1	2
0.106	0	0	1	0	1	0	0	0
0.053	0	0	1	0	1	0	2	2
0.168	0	0	1	0	1	4	4	0
0.388	0	0	1	0	1	4	4	2

Eigenvalue 16d

$|c_i|^2$	n_1	n_2	j_1	j_2	j	J_1	J_2	J
0.002	1	0	0	0	0	2	2	1
0.001	1	0	0	0	0	4	4	1
0.001	0	0	0	1	1	3	3	1
0.001	0	0	0	1	1	5	5	2
0.003	0	1	0	1	1	5	5	0
0.159	0	0	1	1	1	5	5	2
0.080	0	0	1	0	1	0	0	0
0.080	0	0	1	0	1	2	2	2
0.133	0	0	1	0	1	2	2	2
0.376	0	0	1	0	1	2	2	0
0.006	0	0	1	0	1	2	4	2
0.038	0	0	1	0	1	4	4	0
0.110	0	0	1	0	1	4	4	2

[a] These are calculated for $\omega_1^s = 1$, $\omega_1^s = 1/\sqrt{10}$, $\omega_2^c = 100$, $\omega_2 = 10\omega_1$ for increasing ω_1. For each dominant eigenvalue the relative weight is given (in parentheses).

[b] J_T is constant and equal to 1.

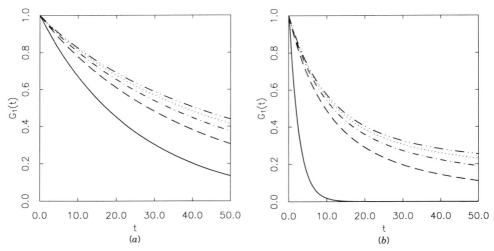

Figure 14. 2BKM-FT. First rank correlation functions: ———, $\omega_1 = 0$; — — — —, $\omega_1 = 100$; — · — · —, $\omega_1 = 200$; · · · · ·, $\omega_1 = 300$; — · · — · · —, $\omega_1 = 400$. (a) $\omega_1^c = 50$; (b) $\omega_1^c = 5$.

and $\omega_1^c = 5$, in Fig. 15 second rank correlation functions, $G_2(t)$, and in Fig. 16 momentum correlation functions, $G_J(t)$. Slower modes appear to be more important than in the SRLS model (but this may be due to the range of frictional parameters utilized). Note the significant difference between the zero coupling (one-body) case and the other ones, especially when the core friction is small. Neither negative tails are present in the momentum correlation functions, nor librational oscillations in the orientational ones. Since the potential coupling is set equal to zero, no "cages" are present in which the light probe can librate.

All the computational parameters were chosen in this set of calculations exactly as they were in the SRLS case; and $n_{1_{max}}$, $n_{2_{max}}$, $j_{1_{max}}$ and $j_{2_{max}}$ were always equal to 2.

H. Discussion and Summary

In the final section of this paper we discuss some of our results in comparison with the studies of other authors. We also consider available experimental data and MD results.

1. Asymptotic Forms for Spectral Densities

We start by considering the works of Freed and co-workers [10, 59]. ESR relaxation studies of small deuterated nitroxide probes have been performed in their laboratory, showing the sensitivity of this spectroscopic

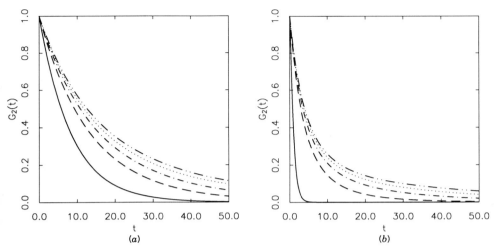

Figure 15. 2BKM-FT. Second rank correlation functions: ———, $\omega_1 = 0$; – – – –, $\omega_1 = 100$; –·–·–, $\omega_1 = 200$; · · · · ·, $\omega_1 = 300$; –··–··–, $\omega_1 = 400$. (a) $\omega_1^c = 50$; (b) $\omega_1^c = 5$.

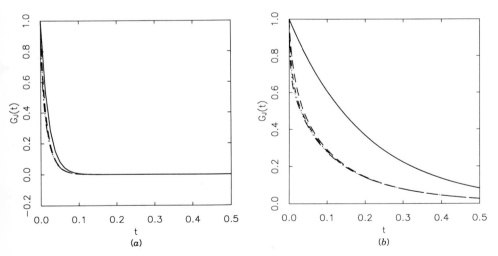

Figure 16. 2BKM-FT. Momentum correlation functions: ———, $\omega_1 = 0$; – – – –, $\omega_1 = 100$; –·–·–, $\omega_1 = 200$; · · · · ·, $\omega_1 = 300$; –··–··–, $\omega_1 = 400$. (a) $\omega_1^c = 50$; (b) $\omega_1^c = 5$.

technique to molecular reorientational dynamics in liquids. Hwang, Mason, Hwang and Freed conducted an analysis of line shapes for the nitroxide radical PD–Tempone in deuterated solvents, and they discussed simple asymptotic formulas for fitting the observed reorientational spectral densities, based on the theoretical analysis of Hwang and Freed [35]. Zager and Freed [59] have conducted ESR relaxation studies to rationalize (i) the solvent and pressure dependence of non-Debye spectral densities and (ii) the relation between rotational and momentum correlation times (compared with the existing simple one-body prediction, i.e., the Hubbard–Einstein relation; see below).

They have shown that a simple SRLS model predicts, in the limit of very slow relaxation of the solvent body, the following form for spectral densities of rank L [59]:

$$J_L(\omega) \sim \frac{\tau_L(1 - S_L^2)}{1 + \omega^2\tau_L^2} + \frac{\tau_x S_L^2}{1 + \omega^2\tau_x^2} \tag{2.94}$$

where τ_L is the correlation time for the isolated solute, while τ_x is the correlation time (of the same rank) for the isolated solvent body; S_L is the order parameter (i.e., the equilibrium average of the Lth Legendre polynomial in Ω_1 assuming the second body is fixed). That is, in the limit of a very large solvent cage, the motion is expected to be a linear combination of the fast FRD of the isolated solute and the slow FRD of the isolated cage. One may expect this limiting expression to be adequate when compared to actual computations based on our 2BS and 2BK-SRLS models when D_1 is much larger than D_2.

In Fig. 17 we show how computed spectral densities compare, in a few cases, with Eq. (2.94) for second rank correlation functions. Figure 17a corresponds to the 2BSM case for $D_2 = 0.1$ and for a first rank potential coupling $v_1 = 4$ (cf. Table III), whereas Fig. 17b refers to the equivalent 2BSM case with a second rank potential coupling $v_2 = 4$. One observes some deviation both for $J(0)$ and the frequency dependence of $J(\omega)$. Note that the asymptotic formula underestimates the spectral densities in the low frequency region, while it overestimates it in the high frequency region.

Spectral densities are less sensitive to inertial effects than correlation functions. We show the spectral density for the 2BK-SRLS case $\omega_1^c = 0.5$ and $v_1 = 2.0$ (cf. Table IX) in Fig. 17c, at $v_2 = 2$ in Fig. 17d. The asymptotic formula (2.94) provides a good fit, especially for a second rank coupling potential. This is due to the large difference between the correlation times for the isolated FRD of the two bodies (i.e., 0.08 for body 1 and close to 167 for body 2).

Note that Eq. (2.94) fails completely when one attempts to reproduce first rank spectral densities calculated in a second rank potential coupling (see Fig. 17e), since it is based on a model in which the solute is a FRD when the solvent body is frozen. One could probably use Eq. (2.94) for $L = 1$ when the potential contains different minima by redefining τ_1 as the inverse of the jump rate in the fixed potential provided by an infinitely damped cage.

Zager and Freed have also compared their experimental data against line shapes predicted by perturbational treatments of simple FT models [35]. To lowest order, such models predict (cf. also Hwang et al. [10]) that the original Lorentzian shape of a pure FRD for the isolated first body should be replaced by a modified function

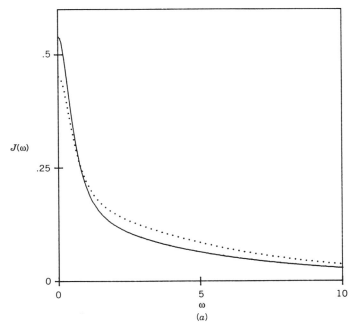

Figure 17. Comparison between exact spectral densities (———) and asymptotic spectral densities given by Eq. (94) ($\cdots\cdots$). (a) Second rank, 2BSM, $D_2 = 0.1$ and $v_1 = 4$; (b) second rank, 2BSM, $D_2 = 0.1$ and $v_2 = 4$; (c) second rank, 2BKM – SRLS, $\omega_1^c = 0.5$ and $v_1 = 2$; (d) second rank, 2BKM-SRLS, $\omega_1^c = 0.5$ and $v_2 = 2$; (e) first rank, 2BSM, $D_2 = 0.01$ and $v_2 = 3$ (note that $J(0) = 3.5$ whereas $J_{\text{asymp}}(0) = 18.6$). For (a) and (b) unit of frequency is relative to $D_1 = 1$; for (c)–(e) it is relative to $\omega_1^s = 1$.

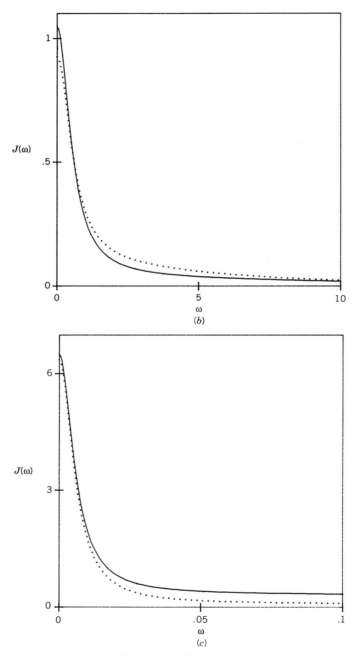

Figure 17.   (*Continued*).

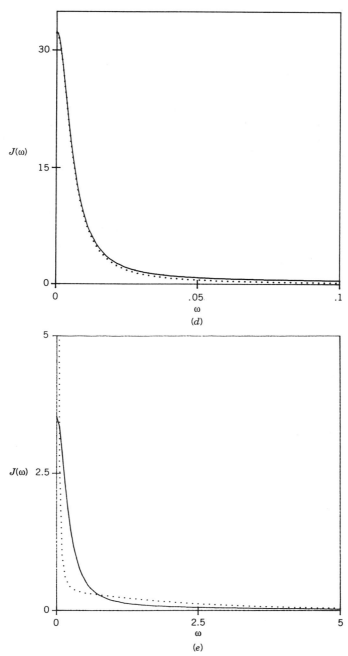

Figure 17. (*Continued*).

$$J(\omega) = \frac{\tau'_L}{1 + \epsilon \tau'^2_L \omega^2} \qquad (2.95)$$

where $\epsilon \geq 1$ and it should be a constant.

Note that Eq. (2.95) is valid only for smaller values of ω and relatively rapidly fluctuating torques. Also, τ'_L need not be the correlation time for the Lth rank FRD of the isolated solute, but depends on the relaxation time of the process providing the FT effect. We except Eq. (2.95) to be acceptable in reproducing the low frequency region of the 3BS spectral densities when the diffusion coefficient of the field is large; 2BK-FT spectral densities are likely to obey Eq. (2.95) if ω is not too large. Note however that in both cases we can expect to apply Eq. (2.95) only for small values of the coupling between the solute and the solvent cage. Equation (2.95) is an adequate approximation for spectral densities (in a limited range of frequencies) only when SRLS effects are absent or negligible; and when the sources of the fluctuating torques are fast relaxing and weakly coupled to the solute.

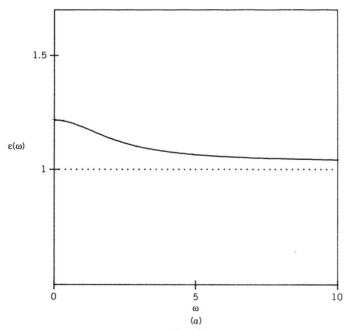

Figure 18. Plots of $[J(\omega)/J(0) - 1]/[\omega^2 J(0)^2]$ versus ω. (a) Second rank, 3BSM, $\mu_1 = 0.2$ and $D_{\mathbf{x}} = 1$; (b) second rank, 2BKM-FT, $\omega_1 = 100$, $\omega^c_1 = 50$; (c) second rank, 2BKM-FT, $\omega_1 = 300$, $\omega^c_1 = 50$. For (a) unit of frequency is relative to $D_1 = 1$; for (b) and (c) it is relative to $\omega^s_1 = 1$.

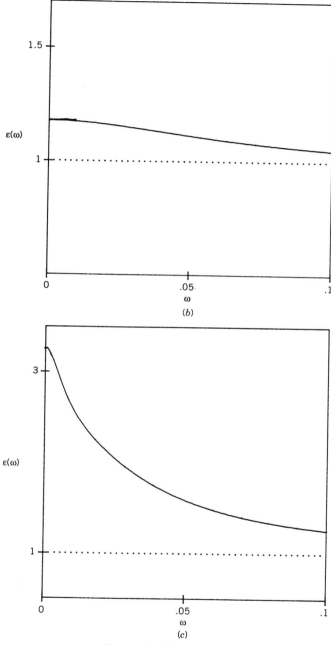

Figure 18. (*Continued*).

We show in Fig. 18 plots of the function $\epsilon(\omega) = [J(\omega)/J(0) - 1/[\omega^2 J(0)^2]$ versus ω. If Eq. (2.95) were valid we should have a horizontal line corresponding to the value of ϵ. In practice one obtains a slowly decreasing plot at low frequencies, which eventually goes asymptotically to one. In fact the actual spectral densities are sums of a finite number of Lorentzian functions, each of them corresponding to a dominant eigenvalue, and for larger frequencies only the Lorentzian corresponding to the largest eigenvalue is nonnegligible. In Fig. 18a we show what we get for a second rank spectral rank spectral density obtained by a 3BS calculation ($\mu_1 = 0.2$ and $D_X = 1$). One may note that the validity of Eq. (2.95) is limited to short frequencies; a rough evaluation of ϵ is in the range 1.1–1.2. In Figs. 18b and 18c we show similar plots for the 2BK-FT model; Fig. 18b is for $\omega_1 = 100$ and Fig. 18c is for $\omega_1 = 300$ (ω_1^c has been taken equal to 50). In the weak coupling case (Fig. 17b the ϵ-fitting is much better than in the strong coupling one (Fig. 17c); in this last case one can approximately use a value of ϵ close to 3. The departure from Eq. (2.95) is then much more evident at lower frequencies when the coupling is increased.

2. The Hubbard–Einstein Relation

The next application we discuss is the interpretation of the anomalous behavior of the product $\tau_J \tau_2$ observed by Freed and co-workers [10] in isotropic and ordered liquid phases. In the absence of mean field effects, a simple one-body Fokker–Planck treatment predicts that the product of the second rank correlation time and the momentum correlation time for a spherical rotator obeys the Hubbard–Einstein relation Eq. (2.93). It is correct in a diffusive (high friction) regime only, where τ_2 is linearly dependent on the viscosity η. Since τ_J is proportional to $1/\eta$, then in order to satisfy Eq. (2.93) for large η τ_J must be short, that is $\tau_J \ll \tau_2$. According to Hwang et al., ESR studies give $\tau_2 > 10^{-11}$ s for PD–Tempone in several solvents, corresponding to $\tau_J < 5 \times 10^{-14}$, that is, of the same order as molecular vibrational periods. A careful analysis of the experimental data suggests that for decreasing temperatures (i.e., increasing viscosities) the left hand side of Eq. (2.93) tends to be larger than the right hand side. One may expect that this is due to a τ_J which has a weaker than linear dependence on $1/\eta$.

Since one-body models fail to reproduce such behavior, even if large mean field potentials are included, one must turn to a many-body description. One would expect that the solute body should be described as coupled to a collective solvent body in such a way that the potential energy of the system is not affected, in order to maintain the normal diffusive behavior of τ_2 (i.e., proportionality to η). We may then introduce a friction tensor affecting the motion of the molecule and the first solvation sphere, such that the variation of τ_J is "damped" when the friction is large.

We choose to describe our coupled system by using the 2BK-FT model, without any torque contribution (zero potential). The collisional matrix is provided by Eq. (2.72). The diagonal terms ω_1^c and ω_2^c are kept constant, whereas the coupling terms ω_1 and ω_2 are changed, for a fixed ratio of the moments of inertia of the two bodies. In this way one expects to model the effect of a fast fluctuating torque (directly related, in our approximation, to the fast relaxing reaction field) which provides the largest friction, and which rapidly varies with temperature and/or pressure. The rest of the solvent provides merely a constant damping that is supposed to be less affected by a change in temperature and pressure, at least in the range of parameters considered in the few experiments that are available.

An analysis of this kind has been made in our recent paper [39], for a solvent cage ten times larger than the solute, a streaming frequency for the solute equal to 10^{12} and an overall friction, parametrized by ω_1 ranging from 10^{12} to $10^{15} \, \text{s}^{-1}$, using the same numerical techniques described in this chapter. In that study, it has been confirmed that for such pure FT models, the correlation time τ_2 behaves approximately in a "diffusive" way, that is, it increases with the increase of the total friction acting on the solute (proportional to $\omega_1 + \omega_1^c$).

An entirely different behavior is observed for the angular momentum correlation time. The coupling terms in the collisional matrix, causing the mutual friction between body 1 and body 2 are much more important. The momentum correlation function is largely dominated by the eigenvalues of the collisional matrix. This means that for large coupling ($\omega_1 \gg \omega_1^c$) the dominant eigenmode for the momentum tends to be proportional to the smallest eigenvalue of the collisional matrix, which is practically equal to ω_1^c, the "core" friction. Thus the particular structure of the friction tensor of a 2BKM-FT provides a way of interpreting the slow change with temperature of τ_J. Our present, more extensive study confirms this analysis (cf. Table XVI).

The 2BKM-SRLS model can also cause a substantial departure from the Hubbard–Einstein relation [39]. This is because $\tau_R \approx \tau_2(1 - S_2^2) + \tau_x S_2^2$ [cf. Eq. (2.94)], so τ_R increases with increased potential coupling and with increase in size of the solvent cage. However, momentum relaxation is dominated by eigenmodes that are primarily the FRD of the isolated solute (cf. Tables X and XIII).

3. Molecular Dynamics Simulations

Molecular dynamics (MD) simulations are an important way of providing insight into motions in liquid phases. In recent years, such simulations have been extensively employed to study the properties of model fluids consisting of interacting molecules and to obtain reorientational and angular velocity correlation functions. Stochastic models can be thought of as complementary theoretical tools to MD, since they may provide (i) general models to interpret results from MD observations, which may be regarded as ideal experiments, and (ii) information at long-times, where for computational reasons MD simulations are not feasible.

The complex rotational behavior of interacting molecules in the liquid state has been studied by a number of authors using MD methods. In particular we consider here the work of Lynden–Bell and co-workers [60–62] on the reorientational relaxation of tetrahedral molecules [60] and cylindrical top molecules [61]. In [60], both rotational and angular velocity correlation functions were computed for a system of 32 molecules of CX_4 (i.e., tetrahedral objects resembling substituted methanes, like CBr_4 or $C(CH_3)_4$) subjected to periodic boundary conditions and interacting via a simple Lennard–Jones potential, at different temperatures. They observe substantial departures of both $G_{1,2}(t)$ and $G_J(t)$ from predictions based on simple theoretical models, such as small-step diffusion or J-diffusion [58]. Although we have not attempted to quantitatively reproduce their results with our mesoscopic models, we have found a close resemblance to our 2BK-SRLS calculations. Compare for instance our Fig. 13 with their Fig. 1 in [60].

In particular they consider a set of simulations for a system of CX_4 molecules at three different temperatures ("hot", "intermediate" and "cool") which bears a close resemblance to our computations made in the presence of a second rank interaction potential. Their "hot" case corresponds to our low potential coupling cases, whereas their "cool" simulation is related to our high potential results: that is, a decrease in temperature corresponds in our rescaled coordinates to an increase in the potential coupling. One may note that the presence of a negative tail, assigned by Lynden-Bell to librational motion of the observed molecule in an instantaneous cage, causes the momentum correlation functions to behave differently in the "cool" state with respect to the purely diffusive decay observed for the "hot" state. This behavior is very similar to our 2BKM-SRLS case for $\omega_1^c = 5$ and $v_2 = 3$ (cf. Fig. 13b).

4. *Impulsive Stimulated Scattering Experiments*

In the last few years Nelson and co-workers [63–65] have presented a new approach to light scattering spectroscopy, named impulsive stimulated light scattering (ISS), which seems to be able to detect one particle rotational correlation functions. In ISS, one induces coherent vibrational motion by irradiating the sample with two femtosecond laser pulses, and then observes a light scattering intensity signal decaying in time. The ISS spectrum is resolved in the time domain and can be directly related to second rank rotational correlation function $G_2(t)$ [65]. Thus ISS is one of the few spectroscopic techniques which appears to give, at least in some cases, direct information on single-molecule rotational dynamics, together with nuclear magnetic resonance (NMR), electron spin resonance (ESR) and neutron scattering.

In particular, Nelson and co-workers have collected a set of experimental data concerning the reorientational dynamic of CS_2 both in temperature-dependent [64] and pressure-dependent [65] ISS experiments. In both cases they observed "weakly oscillatory responses" in the signal either for low temperature regimes or for high pressure regimes. These have been identified as librational motions of the probe molecule in the transient local potential minima inside the instantaneous cages formed by its neighbors.

Comparable behavior has been observed by Fayer et al. in a series of subpicosecond transient grating optical Kerr effect measurements on the reorientation of byphenyl molecules in neat biphenyl and n-heptane solutions [66, 67]. They have shown that on the ultrafast timescale ($t < 2$ ps) the dynamics of the probe is controlled by librational motions having an inertial character, although diffusive reorientational relaxation of the whole molecule and internal torsional motions can also have a role.

The analysis of local librations in terms of the few existing tractable theoretical models (e.g., IOM) have shown that although a qualitative agreement can be reached with experiments, the interpretation of the short time dynamic behavior remains an open problem. We think that our methodology could help to clarify some aspects of the experimental observations.

5. Summary

A careful analysis has been performed on several stochastic models for rotational relaxation of rigid molecules in complex liquids. These include two-body rotational diffusion in the overdamped (Smoluchowski) regime (2BSM), as well as a related three-body model (3BSM). Inertial effects have been considered in two other models which are two-body Fokker–Planck–Kramers models in the full phase space of rotational coordinates and momenta (2BKM). In one, the two bodies interact via an orientation-dependent interaction potential, and this leads to a "slowly relaxing local structure" (SRLS) description. In the other there is an orientation-dependent frictional coupling, derivable from other faster solvent modes, which leads to a "fluctuating torque" (FT) description. The computational challenge of solving multidimensional Fokker–Planck equations has been dealt with by (i) constructing efficient sets of basis functions utilizing angular momentum coupling techniques; (ii) utilizing the complex symmetric Lanczos algorithm to obtain the orientational and angular momentum correlation functions. These correlation functions have been analyzed in terms of the dominant "normal modes" with their associated decay constants.

For the 2BSM, the effect of a large solvent cage yields biexponential behavior with significantly different decay rates. While this behavior may be approximated by modes related to the original free rotational diffusion (FRD) of each body in the absence of coupling, these modes become more influence by "dynamic interactions" for increased interaction potential and/or more nearly equal rotational diffusion coefficients of the two bodies. It has been shown that first rank versus second rank potentials lead to significantly different behaviors, especially for first rank correlation functions (i.e., $G_1(t)$). In this case, a second rank potential leads to an apparent "strong collision effect", that is, a two-body small-step diffusion which exhibits features typical of a one-body strong collision model. Previous simpler SRLS models are inconsistent with this effect. Also, one finds that second rank correlation functions [$G_2(t)$] have somewhat complex behavior with several decay modes, and with increased importance of dynamic interactions. The 3BSM leads to more pronounced departure from single exponential decay. When inertial effects are included via the 2BKM-SRLS case, there are still fast modes for orientational relaxation with purely real decay constants (corresponding to solute modes that are largely diffusional), but now there are solute modes with complex decay constants corresponding to librational motion. For $G_1(t)$ with second rank potentials, the coupling of the angular momentum to the jump motion leads to unusual behavior that may be an incipient Kramers turnover effect. Angular momentum correlation functions [$G_L(t)$] have been found to be much less influenced by the solvent cage than are orientational observables, except for the importance of librational motion in nearly inertial regimes with such motion being enhanced by second rank potentials. These librational modes have been found to have a complex character. In the 2BKM-FT case there are no librational motions. Instead one observes that the FT has little effect on the solute angular momentum correlation time despite the fact that it leads to strong dynamic coupling of the two angular momenta. However, the FT makes an important frictional contribution to the orientational relaxation, such that there is a significant breakdown of the Hubbard–Einstein relation.

These results have been compared with previous studies to show: (i) a simple SRLS model used in ESR is reasonable in the asymptotic limit of a very slow solvent cage except when a "strong collision effect" (cf. above) is important; (ii) the ϵ correction to a Debye spectral density, used in ESR to account for FT, only has a limited validity for low frequencies and relatively rapid but weak torques; (iii) the $G_L(t)$ with a second rank potential resembles molecular dynamics simulations on spherical tops in showing librational motion in an instantaneous cage; (iv) new light scattering results for $G_2(t)$ appear to have features accountable with the present models.

APPENDIX A: CUMULANT PROJECTION PROCEDURE

In this appendix we review briefly the TTOC (total time ordered cumulant) procedure applied to a general linear time evolution operator. The same technique was used by Stillman and Freed [33]; for other details see Yoon et al. [28] and Hwang and Freed [35], and references quoted therein. Also we show how to apply the TTOC procedure for projecting out a subset of fast momenta, from a phase space of coordinates and momenta.

1. General Algorithm

We start by considering a system described by the set of generalized coordinates (and momenta) $(\mathbf{q}_s, \mathbf{q}_f)$

$$\frac{\partial P(\mathbf{q}_s, \mathbf{q}_f, t)}{\partial t} = -\hat{\Gamma}(\mathbf{q}_s, \mathbf{q}_f) P(\mathbf{q}_s, \mathbf{q}_f, t) \tag{A.1}$$

The time evolution operator is supposed to be given by

$$\hat{\Gamma} = \hat{\Gamma}_s(\mathbf{q}_s) + \hat{\Gamma}_f(\mathbf{q}_f) + \hat{\Gamma}_{\text{int}}(\mathbf{q}_s, \mathbf{q}_f) \tag{A.2}$$

We now introduce a biorthonormal complete set of functions defined in the $\mathbf{q}_f$ subspace

$$\langle \mathbf{n} | \mathbf{n}' \rangle = \delta(\mathbf{n} - \mathbf{n}') \tag{A.3}$$

where $\mathbf{n}$ is a collective index for the set of quantum numbers labeling these functions. Note that (1) $|\mathbf{n}\rangle$ and $\langle \mathbf{n}|$ could be the set of eigenfunctions of $\hat{\Gamma}_f$ and its adjoint, respectively, or at this stage, of any other operator acting on the phase space spanned by $\mathbf{q}_f$; (2) in general we do not suppose here that $\mathbf{n}$ is a collection of integers, that is, we can consider a continuum of quantum numbers. The function $|\mathbf{0}\rangle$ is supposed to be unique and to fulfill the following properties:

$$\hat{\Gamma}_f|\mathbf{0}\rangle = 0 \tag{A.4}$$

$$\langle\mathbf{0}|\hat{\Gamma}_f^\dagger = 0 \tag{A.5}$$

that is, $|\mathbf{0}\rangle$ is the unique eigenfunction of zero eigenvalue of $\hat{\Gamma}_f$ (while this may be not necessarily true for $\mathbf{n} \neq \mathbf{0}$, according to the previous remark). Note that here we are always dealings with symmetrized operators: for example, if the subsystem defined by $\mathbf{q}_f$ tends to the equilibrium distribution $P_{eq}(\mathbf{q}_f)$ for $t \to +\infty$ then $|\mathbf{0}\rangle = P_{eq}^{1/2} = \langle\mathbf{0}|$. Although not necessary from a mathematical point of view, in all the physical applications we have considered, the following equation holds:

$$\langle\mathbf{0}|\hat{\Gamma}_{int}|\mathbf{0}\rangle = 0 \tag{A.6}$$

Following [33] closely, we now take the time evolution equation for the reduced probability density in just $\mathbf{q}_s$ as the average over $\mathbf{q}_f$ obtained by computing the "expectation value" with respect to $|\mathbf{0}\rangle$; that is

$$\frac{\partial P(\mathbf{q}_s, t)}{\partial t} = -\langle\mathbf{0}|\hat{\Gamma} P(\mathbf{q}_s, \mathbf{q}_f, t)|\mathbf{0}\rangle . \tag{A.7}$$

After Laplace transformation we easily recover the following exact multidimensional equivalent of the result shown in [33]:

$$\tilde{P}(\mathbf{q}_s, s) = \langle\mathbf{0}|(s + \hat{\Gamma})^{-1}|\mathbf{0}\rangle P(\mathbf{q}_s, 0) \tag{A.8}$$

where the resolvent $\langle\mathbf{0}|(s + \hat{\Gamma})^{-1}|\mathbf{0}\rangle$ can be evaluated according to [28, 35] as

$$\langle\mathbf{0}|(s + \hat{\Gamma})^{-1}|\mathbf{0}\rangle = (s + \hat{\Gamma}_s - \hat{G})^{-1} \tag{A.9}$$

and $\hat{G}$ is defined as

$$\hat{G} = \sum_{k=0}^{+\infty} (-)^{k+1}\langle\mathbf{0}|\hat{\Gamma}_{int}[(s + \hat{\Gamma}_s + \hat{\Gamma}_f)^{-1}(1 - |\mathbf{0}\rangle\langle\mathbf{0}|)\hat{\Gamma}_{int}]^k|\mathbf{0}\rangle . \tag{A.10}$$

If we may assume that the $|\mathbf{n}\rangle$ are the eigenfunctions of $\hat{\Gamma}_f$, then

$$\hat{\Gamma}_f|\mathbf{n}\rangle = E_\mathbf{n}|\mathbf{n}\rangle \tag{A.11}$$

$$\langle\mathbf{n}|\hat{\Gamma}_f^\dagger = \langle\mathbf{n}|E_\mathbf{n}^* \tag{A.12}$$

Then $\hat{G}$ can be further expanded in

$$\hat{G} = \sum_{j=1}^{+\infty} \mathfrak{Y} \, \mathbf{n}_1 \ldots \mathfrak{Y} \, \mathbf{n}_j \langle 0|\hat{\Gamma}_{\mathrm{int}}|\mathbf{n}_1 \rangle \langle \mathbf{n}_1|(s + \hat{\Gamma}_s + E_{\mathbf{n}_1})^{-1}\hat{\Gamma}_{\mathrm{int}}|\mathbf{n}_2 \rangle \ldots$$

$$\langle \mathbf{n}_j|(s + \hat{\Gamma}_s + E_{\mathbf{n}_j})^{-1}\hat{\Gamma}_{\mathrm{int}}|0 \rangle \tag{A.13}$$

where $\mathfrak{Y} \, \mathbf{n}_i$ is a restricted sum (or integral) over all possible $\mathbf{n}_i \neq \mathbf{0}$. If we consider the first order correction only (in the approximation $|\hat{\Gamma}_f| \gg |\hat{\Gamma}_s|$), and we restrict our analysis to low frequencies $(s \sim 0)$ we obtain

$$\hat{G} \sim \mathfrak{Y} \, \mathbf{n} \, \frac{1}{E_{\mathbf{n}}} \langle 0|\hat{\Gamma}_{\mathrm{int}}|\mathbf{n} \rangle \langle \mathbf{n}|\hat{\Gamma}_{\mathrm{int}}|0 \rangle \tag{A.14}$$

as the first perturbation correction to $\hat{\Gamma}_s$.

2. Elimination of Some Momenta from a MFPKE

We apply the technique reviewed in the previous section to a MFPKE defined for a set of general coordinates $(\mathbf{x}_1, \mathbf{x}_2)$ and their conjugate momenta $(\mathbf{p}_1, \mathbf{p}_2)$. The system is divided into two subsystems interacting via a general potential function V and a friction matrix $\boldsymbol{\omega}^c$

$$\boldsymbol{\omega}^c = \begin{pmatrix} \boldsymbol{\omega}_1^c & \boldsymbol{\omega} \\ \boldsymbol{\omega}^{\mathrm{tr}} & \boldsymbol{\omega}_2^c \end{pmatrix} \tag{A.15}$$

$\boldsymbol{\omega}^c$ is a symmetric definite positive $(N_1 + N_2) \times (N_1 + N_2)$ dimensional matrix, and it depends on $\mathbf{x}_1$, $\mathbf{x}_2$. We want to obtain a reduced equation after eliminating all momenta $\mathbf{p}_2$. That is, according to the previous section we are considering $\mathbf{q}_s = (\mathbf{x}_1, \mathbf{p}_1, \mathbf{x}_2)$ and $\mathbf{q}_f = \mathbf{p}_2$. The initial MFPK operator is written as the sum of

$$\hat{\Gamma}_s = \hat{\mathbf{S}}_1^+ \boldsymbol{\omega}_1^s \hat{\mathbf{R}}_1^- - \hat{\mathbf{R}}_1^+ \boldsymbol{\omega}_1^s \hat{\mathbf{S}}_1^- + \hat{\mathbf{S}}_1^+ \boldsymbol{\omega}_1^c \hat{\mathbf{S}}_1^- \tag{A.16}$$

$$\hat{\Gamma}_f = \hat{\mathbf{S}}_2^+ \boldsymbol{\omega}_2^c \hat{\mathbf{S}}_2^- \tag{A.17}$$

$$\hat{\Gamma}_{\mathrm{int}} = \hat{\mathbf{S}}_2^+ \boldsymbol{\omega}_2^s \hat{\mathbf{R}}_2^- - \hat{\mathbf{R}}_2^+ \boldsymbol{\omega}_2^s \hat{\mathbf{S}}_2^- + \hat{\mathbf{S}}_1^+ \boldsymbol{\omega}^{\mathrm{tr}} \hat{\mathbf{S}}_2^- + \hat{\mathbf{S}}_2^+ \boldsymbol{\omega} \hat{\mathbf{S}}_1^- \tag{A.18}$$

where the vector operators $\hat{\mathbf{R}}_m$ and $\hat{\mathbf{S}}_m$ are defined as

$$(\hat{\mathbf{S}})_{m_i}^{\pm} = \frac{1}{2} \, p_{m_i} \mp \frac{\partial}{\partial p_{m_i}} \tag{A.19}$$

$$(\hat{\mathbf{R}})_{m_i}^{\pm} = \frac{1}{2} \left(\frac{\partial V}{\partial \mathbf{x}_m} \right)_i \mp \left(\frac{\partial}{\partial \mathbf{x}_m} \right)_i \tag{A.20}$$

For instance, this is the compact form for the symmetrized time evolution MFPK operator for two Brownian particles (or rotators; see below) coupled via a potential V and a frictional (collisional) matrix $\boldsymbol{\omega}^c$. Although both the terms acting on the momentum space and the positional space are written, for the sake of simplicity, as formal raising and lowering operators, actually only the properties of the $\hat{\mathbf{S}}_m^{\pm}$ operators will be used in the following. Note that we have not specified the nature of the gradient operators in $\mathbf{x}_m$, so they could be a set of rotational coordinates (in this case we should include a precession-like term in $\hat{\Gamma}_{\text{int}}$; but we shall see in the next section that the presence of the precession operator is irrelevant). We define $|\mathbf{n}\rangle$ as the direct product of the eigenfunctions of $\hat{S}_{2_i}^{+}\hat{S}_{2_i}^{-}$

$$|\mathbf{n}\rangle = |n_1\rangle |n_2\rangle \ldots |n_{N_2}\rangle \tag{A.21}$$

$$\hat{S}_{2_i}^{+}\hat{S}_{2_i}^{-}|n_i\rangle = n_i |n_i\rangle \tag{A.22}$$

Then $\mathbf{n}$ is a collection of integers and the set of functions is orthonormal [52] (i.e., we can neglect the integral symbol in Eq. (A.9)); $\hat{S}_{2_i}^{\pm}$ are the raising and lowering operators with respect to the ith momentum in $\mathbf{p}_2$; $|0\rangle$ is the Boltzmann distribution on the momenta $\mathbf{p}_2$. However, we cannot apply Eq. (A.13) directly, because $\boldsymbol{\omega}_2^c$ is not diagonal. We then utilize Eq. (A.9) under the assumption that $\hat{\Gamma}_f \sim |\boldsymbol{\omega}_2^c|$ is the dominant term (i.e., $\mathbf{p}_2$ relaxes very fast relative to the remaining coordinates). Then for low frequencies

$$\hat{G} \sim \langle 0|\hat{\Gamma}_{\text{int}}\hat{\Gamma}_f^{-1}\hat{\Gamma}_{\text{int}}|0\rangle \tag{A.23}$$

where we have used Eq. (A.6) twice. Given that $|\mathbf{n}\rangle$ is a complete set of basis functions in the subspace $\mathbf{p}_2$, we then rewrite Eq. (A.23) in the form

$$\hat{G} = \sum_{\mathbf{n},\mathbf{n}'} \langle 0|\hat{\Gamma}_{\text{int}}|\mathbf{n}\rangle \langle \mathbf{n}|\hat{\Gamma}_f^{-1}|\mathbf{n}'\rangle \langle \mathbf{n}'|\hat{\Gamma}_{\text{int}}|0\rangle . \tag{A.24}$$

When $\hat{\Gamma}_{\text{int}}$ acts on $|0\rangle$, it generates only single excited states, for example, $|0\ldots1\ldots0\rangle$. If we call $|1_j\rangle$ the singly excited function in the jth position, it is easy to rewrite the previous expression for $\hat{G}$ in the form

$$\hat{G} = \sum_{j,j'} \langle 0|\hat{\Gamma}_{\text{int}}|1_j\rangle \langle 1_j|\hat{\Gamma}_f^{-1}|1_{j'}\rangle \langle 1_{j'}|\hat{\Gamma}_{\text{int}}|0\rangle \tag{A.25}$$

the summation indexes run from 1 to N_2. From the equations

$$\langle 0|\hat{\Gamma}_{\text{int}}|1_j\rangle = -\sum_{i=1}^{N_2} \hat{R}_{2_i}^+ \omega_{2_{ji}}^s + \sum_{i=1}^{N_1} \hat{S}_{1_i}^+ \omega_{ji} \qquad (A.26)$$

$$\langle 1_{j'}|\hat{\Gamma}_{\text{int}}|0\rangle = \sum_{i=1}^{N_2} \hat{R}_{2_i}^- \omega_{2_{ij'}}^s + \sum_{i=1}^{N_1} \hat{S}_{1_i}^- \omega_{ij'} \qquad (A.27)$$

it follows that the final reduced operator is given by

$$\hat{\Gamma} = \hat{S}_1^+ \omega_1^s \hat{R}_1^- - \hat{R}_1^+ \omega_1^s \hat{S}_1^- + \hat{S}_1^+ \omega_1' \hat{S}_1^- + \hat{S}_1^+ f \hat{R}_2^- - \hat{R}_2^+ f'' \hat{S}_1^- + \hat{R}_2^+ D_2^0 \hat{R}_2^- \qquad (A.28)$$

where

$$\omega_1' = \omega_1^c - \omega \tilde{\omega} \omega^{\prime r} \qquad (A.29)$$

$$f = \omega \tilde{\omega} \omega_2^s \qquad (A.30)$$

$$D_2^0 = \omega_2^s \tilde{\omega} \omega_2^s \qquad (A.31)$$

The new matrix $\tilde{\omega}$ *is defined as*

$$(\tilde{\omega})_{jj'} = \langle 1_j|\tilde{\Gamma}_f^{-1}|1_{j'}\rangle \qquad (A.32)$$

It is now relatively simple to see that $\tilde{\omega}$ *is exactly equal to* ω_2^{-1}. *Let us consider the matrix representation of* $\hat{\Gamma}_f$ *on* $|\mathbf{n}\rangle$: by inspection, one soon realizes that $\hat{\Gamma}_f$ mixes $|\mathbf{n}\rangle$ and $|\mathbf{n}'\rangle$ if and only if $\Sigma n_i = \Sigma n_i'$; that is, only states equally excited are mixed. The matrix is then partitioned in diagonal blocks; the first block is 1×1 (fundamental state); the second one is $N_2 \times N_2$, mixes only the states $\Sigma n_i = 1$, that is the $|1_j\rangle$ functions, and it is given by ω_2^c.

3. Precessional Operator

For a rotational system one has to include the precessional operator in the rotational FPK operator, in case a nonspherical top is considered. In terms of the raising and lowering operators $\hat{S}^\pm$ defined in the last section (systematically suppressing the subscript 2 since it is understood here that we are dealing entirely only with the subspace $\mathbf{p}_2$), we can write the precessional operator as

$$\hat{P}\omega^s\nabla = \Delta_1(\hat{S}_1^- - \hat{S}_1^+)(\hat{S}_2^- + \hat{S}_2^+)(\hat{S}_3^- + \hat{S}_3^+)$$
$$+ \Delta_2(\hat{S}_1^- + \hat{S}_1^+)(\hat{S}_2^- - \hat{S}_2^+)(\hat{S}_3^- + \hat{S}_3^+)$$
$$+ \Delta_3(\hat{S}_1^- + \hat{S}_1^+)(\hat{S}_2^- + \hat{S}_2^+)(\hat{S}_3^- - \hat{S}_3^+) \qquad (A.33)$$

where Δ_1, Δ_2, Δ_3 are functions of the streaming frequency matrix elements. Note that

$$\Delta_1 + \Delta_2 + \Delta_3 = 0 \tag{A.34}$$

The crucial point in the TTOC expansion delineated in the last section is that the interaction operator $\hat{\Gamma}_{int}$ acting on $|0\rangle$ generates only single excited states. In this rotational case, we may include the precessional term in $\hat{\Gamma}_{int}$, and it is easy to see that

$$\hat{P}\omega^s \nabla |0,0,0\rangle = -(\Delta_1 + \Delta_2 + \Delta_3)|0,0,0\rangle \tag{A.35}$$

In fact all the factors containing a lowering operator go to zero; and one obtains zero because of Eq. (A.34). This means that it is not necessary to consider the precessional effects in projecting out to lowest order the role of angular momentum.

APPENDIX B: ELIMINATION OF HARMONIC DEGREES OF FREEDOM

Here we show how to implement the TTOC procedure for eliminating in a single step a set of harmonic degrees of freedom together with their conjugate momenta from an initial MFPKE. This technique is applied in Section I.C to project out the fast field $\mathbf{X}$ and its momentum $\mathbf{P}$ from the initial three body Fokker–Planck–Kramers equation.

1.   Elimination of One Harmonic Degree of Freedom

We start by considering a one-dimensional example given by the rescaled symmetrized MK evolution operator in the coordinates (x_1, x_2) and conjugate momenta (p_1, p_2),

$$\begin{aligned}
\hat{\Gamma} = \omega_1^s \left(p_1 \frac{\partial}{\partial x_1} - \frac{\partial V}{\partial x_1} \frac{\partial}{\partial p_1} \right) - \omega_1^c \exp(p_1^2/4) \frac{\partial}{\partial p_1} \exp(-p_1^2/2) \frac{\partial}{\partial p_1} \\
\times \exp(p_1^2/4) \\
+ \omega_2^s \left(p_2 \frac{\partial}{\partial x_2} - \frac{\partial V}{\partial x_2} \frac{\partial}{\partial p_2} \right) - \omega_2^c \exp(p_2^2/4) \frac{\partial}{\partial p_2} \exp(-p_2^2/2) \frac{\partial}{\partial p_2} \\
\times \exp(p_2^2/4)
\end{aligned} \tag{B.1}$$

where the potential V is defined as

$$V = V_0(x_1) - \mu(x_1)x_2 + \frac{1}{2}x_2^2 \tag{B.2}$$

We introduce the shifted coordinates $\tilde{x}_2 = x_2 - \mu$. This canonical transformation enables us to obtain a more suitable form for the operator, in which V is decoupled. Neglecting the tilde symbol in the following, we identify $\mathbf{q}_s$ with (x_1, p_1) and $\mathbf{q}_f$ with (x_2, p_2)

$$\hat{\Gamma}_s = \omega_1^s \left(p_1 \frac{\partial}{\partial x_1} - \frac{\partial V}{\partial x_1} \frac{\partial}{\partial p_1} \right) - \omega_1^c \exp(p_1^2/4) \frac{\partial}{\partial p_1} \exp(-p_1^2/2) \frac{\partial}{\partial p_1}$$
$$\times \exp(p_1^2/4) \tag{B.3}$$

$$\hat{\Gamma}_f = \omega_2^s \left(p_2 \frac{\partial}{\partial x_2} - x_2 \frac{\partial}{\partial p_2} \right) - \omega_2^c \exp(p_2^2/4) \frac{\partial}{\partial p_2} \exp(-p_2^2/2) \frac{\partial}{\partial p_2}$$
$$\times \exp(p_2^2/4) \tag{B.4}$$

$$\hat{\Gamma}_{int} = -\omega_1^s \frac{\partial \mu}{\partial x_1} \left(p_1 \frac{\partial}{\partial x_2} - x_2 \frac{\partial}{\partial p_1} \right) \tag{B.5}$$

and the potential V is now

$$V = V_0 - \frac{1}{2} \mu^2 \tag{B.6}$$

It is useful now to recall the general properties of the harmonic Kramers operator. We utilize the summary provided by Risken [43]. The raising and lowering operators for the momentum p_2 and the position x_2 are given by

$$\hat{S}^{\pm} = \frac{1}{2} p \mp \frac{\partial}{\partial p} \tag{B.7}$$

$$\hat{R}^{\pm} = \frac{1}{2} x \mp \frac{\partial}{\partial x} \tag{B.8}$$

where in Eqs. (B7) and (B8) and below in this subsection we suppress the subscript 2 for convenience. The quantities $\lambda_{1,2}$, solutions of the secular equation $\lambda^2 - \omega^c \lambda + \omega^{s^2} = 0$, are calculated. We also define a parameter $\bar{\omega}$

$$\lambda_{1,2} = \tfrac{1}{2}(\omega \pm \bar{\omega}) \tag{B.9}$$

$$\bar{\omega} = (\omega^{c^2} - \omega^{s^2})^{1/2} = \lambda_1 - \lambda_2 \tag{B.10}$$

The following operators are then defined:

$$\hat{c}_{1_+} = \frac{1}{\bar{\omega}^{1/2}} (\lambda_1^{1/2} \hat{S}^+ - \lambda_2^{1/2} \hat{R}^-) \tag{B.11}$$

$$\hat{c}_{1_-} = \frac{1}{\bar{\omega}^{1/2}} (\lambda_1^{1/2} \hat{S}^- - \lambda_2^{1/2} \hat{R}^+) \tag{B.12}$$

$$\hat{c}_{2_+} = \frac{1}{\bar{\omega}^{1/2}} (-\lambda_1^{1/2} \hat{S}^+ + \lambda_2^{1/2} \hat{R}^+) \tag{B.13}$$

$$\hat{c}_{2_-} = \frac{1}{\bar{\omega}^{1/2}} (\lambda_1^{1/2} \hat{S}^- + \lambda_2^{1/2} \hat{R}^-) \tag{B.14}$$

The following identity is deduced:

$$\hat{\Gamma} = \lambda_1 \hat{c}_{1_+} \hat{c}_{1_-} + \lambda_2 \hat{c}_{2_+} \hat{c}_{2_-} \tag{B.15}$$

and eigenfunctions and eigenvalues are easily obtained as

$$|n_1, n_2\rangle = (n_1! n_2!)^{-1/2} (\hat{c}_{1_+})^{n_1} (\hat{c}_{2_+})^{n_2} |0, 0\rangle \tag{B.16}$$

$$E_{n_1, n_2} = \lambda_1 n_1 + \lambda_2 n_2 \tag{B.17}$$

where

$$|0, 0\rangle = P_{eq}^{1/2}(x, p) = \frac{1}{(2\pi)^{1/2}} \exp(-p^2/4 - x^2/4) \tag{B.18}$$

and

$$\hat{c}_{1_+} |n_1, n_2\rangle = (n_1 + 1)^{1/2} |n_1 + 1, n_2\rangle \tag{B.19}$$

$$\hat{c}_{1_-} |n_1, n_2\rangle = (n_1)^{1/2} |n_1 - 1, n_2\rangle \tag{B.20}$$

$$\hat{c}_{2_+} |n_1, n_2\rangle = (n_2 + 1)^{1/2} |n_1, n_2 + 1\rangle \tag{B.21}$$

$$\hat{c}_{2_-} |n_1, n_2\rangle = (n^2)^{1/2} |n_1, n_2 - 1\rangle \tag{B.22}$$

For the adjoint operator similar equations hold:

$$\hat{\Gamma}^\dagger = \lambda_1 \hat{c}_{1_-}^\dagger \hat{c}_{1_+}^\dagger + \lambda_2 \hat{c}_{2_-}^\dagger \hat{c}_{2_+}^\dagger. \tag{B.23}$$

$$\langle n_1, n_2| = \langle 0, 0| (n_1! n_2!)^{-1/2} (\hat{c}_{1_-}^\dagger)^{n_1} (\hat{c}_{2_-}^\dagger)^{n_2} \tag{B.24}$$

where

$$|0, 0\rangle = \langle 0, 0| \tag{B.25}$$

and

$$\langle n_1, n_2| \hat{c}_{1_+}^\dagger = (n_1)^{1/2} \langle n_1, n_2| \tag{B.26}$$

$$\langle n_1, n_2 | \hat{c}_{1_-}^\dagger = (n_1 + 1)^{1/2} \langle n_1, n_2 | \qquad (B.27)$$

$$\langle n_1, n_2 | \hat{c}_{2_+}^\dagger = (n_2)^{1/2} \langle n_1, n_2 | \qquad (B.28)$$

$$\langle n_1, n_2 | \hat{c}_{2_-}^\dagger = (n_2 + 1)^{1/2} \langle n_1, n_2 | \qquad (B.29)$$

We have so defined a biorthonormal set of functions

$$\langle n_1, n_2 | n_1', n_2' \rangle = \delta_{n_1, n_1'} \delta_{n_2, n_2'} \qquad (B.30)$$

We may now use the method of Appendix A for the case of a biorthonormal discrete set of *eigen*functions. We then have that

$$\hat{G} = \sum_{(n_1, n_2) \neq (0\ 0)} \frac{1}{E_{n_1, n_2}} \langle 0, 0 | \hat{\Gamma}_{int} | n_1, n_2 \rangle \langle n_1, n_2 | \hat{\Gamma}_{int} | 0, 0 \rangle \qquad (B.31)$$

From the identities

$$\hat{\Gamma}_{int} | 0, 0 \rangle = -\frac{\omega_1^s}{\bar{\omega}^{1/2}} \frac{\partial \mu}{\partial x_1} \left[-\lambda_2^{1/2} \left(\frac{1}{2} p_1 + \frac{\partial}{\partial p_1} \right) | 1, 0 \rangle \right.$$

$$\times - \lambda_1^{1/2} \left(\frac{1}{2} p_1 + \frac{\partial}{\partial p_1} \right) | 0, 1 \rangle \Big] \qquad (B.32)$$

$$\hat{\Gamma}_{int} | 0, 1 \rangle = -\frac{\omega_1^s}{\bar{\omega}^{1/2}} \frac{\partial \mu}{\partial x_1} \left[-\lambda_2^{1/2} \left(\frac{1}{2} p_1 - \frac{\partial}{\partial p_1} \right) | 0, 0 \rangle + \cdots \right]$$

$$\qquad (B.33)$$

$$\hat{\Gamma}_{int} | 1, 0 \rangle = -\frac{\omega_1^s}{\bar{\omega}^{1/2}} \frac{\partial \mu}{\partial x_1} \left[\lambda_1^{1/2} \left(\frac{1}{2} p_1 - \frac{\partial}{\partial p_1} \right) | 0, 0 \rangle + \cdots \right] \qquad (B.34)$$

one obtains easily the reduced operator

$$\hat{\Gamma} = \omega_1^s \left(p_1 \frac{\partial}{\partial x_1} - \frac{\partial V}{\partial x_1} \frac{\partial}{\partial p_1} \right) - \omega_1^{c'} \exp(p_1^2/4) \frac{\partial}{\partial p_1}$$

$$\times \exp(-p_1^2/2) \frac{\partial}{\partial p_1} \exp(p_1^2/4) \qquad (B.35)$$

where the effective collisional frequency is defined as

$$\omega_1^{c'} = \omega_1^c + \frac{\omega_2^c \omega_1^{s2}}{\omega_2^2} \left(\frac{\partial \mu}{\partial x_1} \right)^2 \qquad (B.36)$$

Note that both reversible effects (correction to the potential function) and irreversible ones (correction to the initial friction) are obtained.

2.   Elimination of N_2 Harmonic Degrees of Freedom

We now generalize this result to a multidimensional case. The initial rescaled and symmetrized operator is split into three parts:

$$\hat{\Gamma}_s = \hat{S}_1^+ \, \omega_1^s \hat{R}_1^- - \hat{R}_1^+ \, \omega_1^s \hat{S}_1^- + \hat{S}_1^+ \, \omega_1^s \hat{S}_1^- \tag{B.37}$$

$$\hat{\Gamma}_f = \hat{S}_2^+ \, \omega_2^s \hat{R}_2^- - \hat{R}_2^+ \, \omega_2^s \hat{S}_2^- + \hat{S}_2^+ \, \omega_2^s \hat{S}_2^- \tag{B.38}$$

$$\hat{\Gamma}_{int} = -\mathbf{p}_1 \omega_1^s \left(\frac{\partial \boldsymbol{\mu}}{\partial \mathbf{x}_1} \right) \left(\frac{\partial}{\partial \mathbf{x}_2} \right) + \left(\frac{\partial}{\partial \mathbf{p}_1} \right) \omega_1^s \left(\frac{\partial \boldsymbol{\mu}}{\partial \mathbf{x}_1} \right) \mathbf{x}_2 \tag{B.39}$$

where the averaged potential, on which $\hat{R}_2^+$ and $\hat{R}_2^-$ are defined, is a quadratic function of the vector dipole $\boldsymbol{\mu}$,

$$V = V_0 - \frac{1}{2} \, \boldsymbol{\mu}^2 \tag{B.40}$$

and $\hat{R}_2^\pm$, $\hat{S}_2^\pm$ are the vector equivalents of the previous similar one-dimensional operators. For the sake of simplicity we choose $\omega_{1,2}^s$ and $\omega_{1,2}^s$ as diagonal and constant. We generalize the previous definitions introducing the matrices $\bar{\omega}$, $\boldsymbol{\lambda}_{1,2}$ *and the vector operator* $\hat{c}_{1_\pm}$, $\hat{c}_{2_\pm}$ *and their adjoints. The eigenfunctions of* $\hat{\Gamma}_f$ *are the direct product of the eigenfunc-*tions of the one-dimensional harmonic Kramers operators. We label each member of the set with the obvious symbol $|\mathbf{n}_1, \mathbf{n}_2\rangle$. The zero eigenvalue function is

$$|\mathbf{0}, \mathbf{0}\rangle = \langle \mathbf{0}, \mathbf{0}| = \frac{1}{(2\pi)^{N/2}} \exp(-\mathbf{p}_2^2/4 - \mathbf{x}_2^2/4) \tag{B.41}$$

and we call $|1_j, \mathbf{0}\rangle$ the first excited state with respect to n_{1_j}, etcetera; it is easy to show that

$$\hat{\Gamma}_{int}|\mathbf{0}, \mathbf{0}\rangle = \sum_{i=1}^{N_1} \sum_{j=1}^{N_2} \hat{S}_{1_i}^- \left(\frac{\partial \boldsymbol{\mu}}{\partial \mathbf{x}_1} \right)_{ij} \frac{\omega_{1_i}^s \lambda_{2_j}^{1/2}}{\bar{\omega}_j^{1/2}} |1_j, \mathbf{0}\rangle$$

$$+ \hat{S}_{1_i}^- \left(\frac{\partial \boldsymbol{\mu}}{\partial \mathbf{x}_1} \right)_{ij} \frac{\omega_{1_i}^s \lambda_{1_j}^{1/2}}{\bar{\omega}_j^{1/2}} |\mathbf{0}, 1_j\rangle \tag{B.42}$$

$$\hat{\Gamma}_{int}|\mathbf{0}, 1_j\rangle = \hat{S}_{1_i}^+ \left(\frac{\partial \boldsymbol{\mu}}{\partial \mathbf{x}_1} \right)_{ij} \frac{\omega_{1_i}^s \lambda_{2_j}^{1/2}}{\bar{\omega}_k^{1/2}} |\mathbf{0}, \mathbf{0}\rangle \tag{B.43}$$

$$\hat{\Gamma}_{int}|1_j, \mathbf{0}\rangle = -\hat{S}_{1_i}^+ \left(\frac{\partial \boldsymbol{\mu}}{\partial \mathbf{x}_1} \right)_{ij} \frac{\omega_{1_i}^s \lambda_{1_j}^{1/2}}{\bar{\omega}_j^{1/2}} |\mathbf{0}, \mathbf{0}\rangle \tag{B.44}$$

$\hat{G}$ is now given by

$$\hat{G} = \sum_{k=1}^{N_1} \frac{1}{\lambda_{1_k}} \langle 0, 0^{\dagger} | \hat{\Gamma}_{int} | 1_k, 0 \rangle \langle 1_k, 0^{\dagger} | \hat{\Gamma}_{int} | 0, 0 \rangle$$

$$+ \frac{1}{\lambda_{2_k}} \langle 0, 0^{\dagger} | \hat{\Gamma}_{int} | 0, 1_k \rangle \langle 0, 1_k^{\dagger} | \hat{\Gamma}_{int} | 0, 0 \rangle \qquad (B.45)$$

and the final reduced operator has the form

$$\hat{\Gamma} = \hat{S}_1^+ \omega_1^s \hat{R}_1^- - \hat{R}_1^+ \omega_1^s \hat{S}_1^- + \hat{S}_1^+ \omega_1^{c'} \hat{S}_1^- \qquad (B.46)$$

where the frictional (collisional) matrix is

$$\omega_1^{c'} = \omega_1^c + \omega_1^s \left(\frac{\partial \mu}{\partial x_1} \right) \omega_2^{s^{-1}} \omega_2^c \omega_2^{s^{-1}} \left(\frac{\partial \mu}{\partial x_1} \right) \omega_1^s \qquad (B.47)$$

APPENDIX C: THE REDUCED MATRIX ELEMENTS

We evaluate in this appendix the reduced matrix elements employed in the WE calculations throughout the main text.

1. Reduced Matrix Element of the Torque

To evaluate the reduced matrix element of the torque T, we first rewrite Eq. (2.17) as

$$\mathbf{T} = -i[(\hat{\mathbf{J}}_1 V)_{op} - (V \hat{\mathbf{J}}_1)_{op}] \qquad (C.1)$$

where for $(\ )_{op}$ what is contained within acts as an operator. From the WE theorem (weak form for noncommuting operators)

$$J_1 J_2 J \| (\hat{\mathbf{J}}_1 V)_{op} \| J_1' J_2' J') = [1]^{1/2} (-)^{J+J'+1} \sum_{J_1'' J_2'' J''} \begin{Bmatrix} 0 & 1 & 1 \\ J & J'' & J \end{Bmatrix}$$

$$\times (J_1 J_2 J \| \hat{\mathbf{J}}_1 \| J_1'' J_2'' J'')(J_1'' J_2'' J'' \| V \| J_1' J_2' J') \qquad (C.2)$$

The $6j$ symbol is readily reduced

$$\begin{Bmatrix} 0 & 1 & 1 \\ J & J'' & J' \end{Bmatrix} = (-)^{J+J'+1} [1J]^{-1/2} \delta_{J'J''} \qquad (C.3)$$

The reduced matrix element of $\hat{\mathbf{J}}_1$ is given by Eq. (2.31), while the reduced matrix element of the potential V is

$$(J_1''J_2''J''\|V\|J_1'J_2'J') = \sum_R V_R(-)^{J_1''+J_1'+J'}[J']^{1/2}\begin{Bmatrix} J_1'' & J_2'' & J' \\ J_2' & J_1' & R \end{Bmatrix}\frac{[J_1''RJ_1']^{1/2}}{(8\pi^2)^{1/2}}$$

$$\times \begin{pmatrix} J_1'' & R & J_1' \\ 0 & 0 & 0 \end{pmatrix}\frac{[J_2''RJ_2']^{1/2}}{(8\pi^2)^{1/2}}\begin{pmatrix} J_2'' & R & J_2' \\ 0 & 0 & 0 \end{pmatrix} \quad (C.4)$$

where the reduced matrix element in the Ω_m subspace only was used ($m = 1, 2$)

$$(J_m\|J_m'\|J_m'') = (-)^{J_m}\frac{[J_mJ_m'J_m'']^{1/2}}{(8\pi^2)^{1/2}}\begin{pmatrix} J_m & J_m' & J_m'' \\ 0 & 0 & 0 \end{pmatrix} \quad (C.5)$$

Finally, one obtains

$$J_1J_2J\|(\hat{\mathbf{J}}_1V)_{\text{op}}\|J_1'J_2'J') = -v_R(-)^{J_1+J_2}[JJ'J_1J_1'J_2J_2']^{1/2}$$

$$\times [J_1(J_1+1)(2J_1+1)]^{1/2}$$

$$\times \begin{pmatrix} J_1 & R & J_1' \\ 0 & 0 & 0 \end{pmatrix}\begin{pmatrix} J_2 & R & J_2' \\ 0 & 0 & 0 \end{pmatrix}$$

$$\times \begin{Bmatrix} J_1 & J & J_2 \\ J' & J_1 & 1 \end{Bmatrix}\begin{Bmatrix} J_1 & J_2 & J' \\ J_2' & J_1' & R \end{Bmatrix} \quad (C.6)$$

where the definition $v_R = [R]V_R/8\pi^2$ was used. An analogous formula holds for the reduced matrix element of $(V\hat{\mathbf{J}}_1)_{\text{op}}$

$$(J_1J_2J\|(v\hat{\mathbf{J}}_1)_{\text{op}}\|J_1'J_2'J') = -v_R(-)^{J_1+J_2+J+J'}[JJ'J_1J_2J_2']^{1/2}$$

$$\times [J_1'(J_1'+1)(2J_1'+1)]^{1/2}$$

$$\times \begin{pmatrix} J_1 & R & J_1' \\ 0 & 0 & 0 \end{pmatrix}\begin{pmatrix} J_2 & R & J_2' \\ 0 & 0 & 0 \end{pmatrix}$$

$$\times \begin{Bmatrix} J_1' & J & J_2' \\ J' & J_1' & 1 \end{Bmatrix}\begin{Bmatrix} J_1 & J_2 & J' \\ J_2' & J_1' & R \end{Bmatrix} \quad (C.7)$$

so that, finally

$$(J_1J_2J\|\mathbf{T}\|J_1'J_2'J') = iv_R[JJ'J_1J_1'J_2J_2']^{1/2}\begin{pmatrix} J_1 & R & J_1' \\ 0 & 0 & 0 \end{pmatrix}\begin{pmatrix} J_2 & R & J_2' \\ 0 & 0 & 0 \end{pmatrix}$$

$$\times \left[(-)^{J_1+J_2}[J_1(J_1+1)(2J_1+1)]^{1/2}\begin{Bmatrix} J_1 & J & J_2 \\ J' & J_1 & 1 \end{Bmatrix}\right.$$

$$\times \begin{Bmatrix} J_1 & J_2 & J' \\ J_2' & J_1' & R \end{Bmatrix}$$

$$-(-)^{J_1+J_2+J+J'}[J_1'(J_1'+1)(2J_1'+1)]^{1/2}$$

$$\times \left.\begin{Bmatrix} J_1' & J & J_2' \\ J' & J_1' & 1 \end{Bmatrix}\begin{Bmatrix} J_1 & J_2 & J \\ J_2' & J_1' & R \end{Bmatrix}\right] \quad (C.8)$$

2. Vector representation of $P_{eq}^{1/2}$

Since $P_{eq}^{1/2}$ is a zero rank tensor, we can simply write Eq. (2.26) as

$$(\mathbf{v}_0)_\Lambda = \langle \Lambda | P_{eq}^{1/2} \rangle \propto \langle J_1 J_2 00 | P_{eq}^{1/2} \rangle \delta_{J0} \delta_{M0} \qquad (C.9)$$

since we have already found $J = 0$ and $M = 0$. By inspection, one can see that $|J_1 J_2 00\rangle$ (coupled basis set function) is proportional to $\mathscr{D}_{00}^{J_1^*}(\Omega_2 - \Omega_1)\delta_{J_1 J_2}$ (just write explicitly the coupled basis set function in terms of the uncoupled basis set functions). Then, by making the (canonical) change of variables $(\Omega_1, \Omega_2) \rightarrow (\Omega_2 - \Omega_1, \Omega_2 + \Omega_1)$, and integrating over $\Omega_2 + \Omega_1$, after a few algebraic manipulations the following expression is found:

$$\langle \Lambda | P_{eq}^{1/2} \rangle = \mathscr{F} \delta_{J_1 J_2} \left[\int_{-1}^{+1} dx \exp\left(-\sum_R P_R(x)\right) \right]^{-1/2} \frac{[J_1]^{1/2}}{(2)^{1/2}}$$

$$\times \int_{-1}^{+1} dx P_{J_1}(x) \exp\left(-\sum_R P_r(x)/2\right) \qquad (C.10)$$

where the factor $\mathscr{F}$ is simply $\delta_{J0}\delta_{M0}$. Note that the original 4-variable integral is thereby reduced to a simple integral in the dummy variable x.

3. Reduced Matrix Elements in the $|nj\rangle$ Subspace

The reduced matrix elements of $\hat{\mathbf{S}}^\pm$ are suitably evaluated as linear combinations of the reduced matrix elements of $\mathbf{X}$ and $\nabla_\mathbf{X}$

$$(nj\|\hat{\mathbf{S}}^\pm\|n'j') = \frac{1}{2}(nj\|\mathbf{X}\|n'j') \mp (nj\|\nabla_\mathbf{X}\|n'j') \qquad (C.11)$$

The explicit evaluation of these reduced matrix elements is simple, taking into account the properties of Laguerre polynomials (cf. [52]); the only nonzero cases are

$$(nj\|\mathbf{X}\|n'j-1) = (-)^j (2)^{1/2} [jj-1]^{1/2} \begin{pmatrix} j & 1 & j-1 \\ 0 & 0 & 0 \end{pmatrix}$$

$$\times \left[\left(j+n+\frac{1}{2}\right)^{1/2} \delta_{nn'} - (n+1)^{1/2} \delta_{nn'-1} \right] \qquad (C.12)$$

$$(nj\|\mathbf{X}\|n'j+1) = (-)^j (2)^{1/2} [jj+1]^{1/2} \begin{pmatrix} j & 1 & j+1 \\ 0 & 0 & 0 \end{pmatrix} (2)^{1/2}$$

$$\times \left[\left(j+n+\frac{3}{2}\right)^{1/2} \delta_{nn'} - (n)^{1/2} \delta_{nn'-1} \right] \qquad (C.13)$$

$$(nj\|\nabla_{\mathbf{X}}\|n'j-1) = \frac{(-)^j j}{(2)^{1/2} \begin{pmatrix} j & 1 & j-1 \\ 0 & 0 & 0 \end{pmatrix} [(2j-1)(2j+1)]^{1/2}}$$

$$\times \left[\left(j + n + \frac{1}{2} \right)^{1/2} \delta_{nn'} + (n+1)^{1/2} \delta_{nn'-1} \right] \tag{C.14}$$

$$(nj\|\nabla_{\mathbf{X}}\|n'j+1) = \frac{(-)^j (j+1)}{(2)^{1/2} \begin{pmatrix} j & 1 & j+1 \\ 0 & 0 & 0 \end{pmatrix} [(2j+1)(2j+3)]^{1/2}}$$

$$\times \left[\left(j + n + \frac{3}{2} \right)^{1/2} \delta_{nn'} + (n)^{1/2} \delta_{nn'-1} \right] \tag{C.15}$$

Finally, the reduced matrix elements of $\hat{\mathbf{S}}$ are evaluated using the weak form of the WE theorem

$$(nj\|\hat{\mathbf{S}}\|n'j') = (nj\|[\hat{\mathbf{S}}^+ \otimes \hat{\mathbf{S}}^-]^{(2)}\|n'j') = [2]^{1/2} (-)^{j+j'}$$

$$\times \sum_{n''j''} \begin{Bmatrix} 1 & 2 & 1 \\ j & j'' & j' \end{Bmatrix} (nj\|\hat{\mathbf{S}}^+\|n''j'')(n''j''\|\hat{\mathbf{S}}^-\|n'j') \tag{C.16}$$

Acknowledgments

A. P. acknowledges the Italian Ministry of the University and the Scientific and Technological Research, and in part the National Research Council through its Centro Studi sugli Stati Molecolari supporting his stay at Cornell University. The computations reported here were performed at the Cornell National Supercomputer Facility.

References

1. P. Debye, *Polar Molecules* (Dover, 1929).
2. F. Perrin, *J. Phys. Radium* **5**, 497 (1934).
3. R. A. Sack, *Proc. Phys. Soc.* **708**, 402 (1957).
4. M. Fixman and K. Rider, *J. Chem. Phys.* **51**, 2425 (1969).
5. P. S. Hubbard, *Phys. Rev. A* **6**, 2421 (1972).
6. R. E. D. McClung, *J. Chem. Phys.* **75**, 5503 (1981).
7. A. Morita, *J. Chem. Phys.* **76**, 3198 (1981).
8. L. D. Favro, in *Fluctuation Phenomena in Solids*, R. E. Burgess (ed.) (Academic Press, New York, 1965) p. 79.
9. T. Dorfmuller and R. Pecora (eds.), *Rotational Dynamics of Small and Macro-molecules*, Lecture Notes in Physics (Springer, Berlin, 1987).
10. J. S. Hwang, R. P. Mason, L. P. Hwang, and J. H. Freed, *J. Phys. Chem.* **79**, 489 (1975).

11. J. H. Freed, in *Rotational Dynamics of Small and Macromolecules*, T. Dorfmuller and R. Pecora (eds.), Lecture Notes in Physics (Springer, Berlin, 1987) p. 89.

12. H. W. Spiess, in *Rotational Dynamics of Small and Macromolecules*, T. Dorfmuller and R.Pecora (eds.), Lecture Notes in Physics (Springer, Berlin, 1987) p. 89.

13. R. A. McPhail and D. Kivelson, *J. Chem. Phys.* **90**, 6549 (1989).

14. R. A. McPhail and D. Kivelson, *J. Chem. Phys.* **90**, 6555 (1989).

15. C. J. Reid and M. Evans, *J. Chem. Soc., Faraday Trans. 2* **74**, 1218 (1978).

16. C. J. Reid and M. Evans, *Adv. Mol. Rel. Int. Proc.* **15**, 281 (1979).

17. G. P. Johari, *J. Chem. Phys.* **58**, 1766 (1973).

18. G. P. Johari, *Ann. N. Y. Acad. Sci.* **279**, 117 (1976).

19. G. Williams, in *Dielectric and Related Molecular Processes*, vol. 2, Spec. Periodical Report (Chemical Society, London, 1977).

20. G. Williams, *Molecular Liquids*, A. Barnes, W. Orville-Thomas and J. Yarwood (eds.), NATO ASI Series C (Reidel, Boston, 1984).

21. M. A. Floriano and C. A. Angell, *J. Chem. Phys.* **91**, 2537 (1989).

22. V. Nagarajan, A. M. Brearly, T. J. Kang, and P. F. Barbara, *J. Chem. Phys.* **86**, 8183 (1987).

23. W. T. Coffey, P. M. Corcoran, and M. W. Evans, *Mol. Phys.* **61**, 15 (1987).

24. W. T. Coffey, M. W. Evans, and G. J. Evans, *Mol. Phys.* **38**, 477 (1979).

25. P. M. Corcoran, W. T. Coffey, and M. W. Evans, *Mol. Phys.* **61**, 1 (1987).

26. M. W. Evans, P. Grigolini, and F. Marchesoni, *Chem. Phys. Lett.* **95**, 548 (1983).

27. F. Marchesoni and P. Grigolini, *J. Chem. Phys.* **78**, 6287 (1983).

28. B. Yoon, J. M. Deutch, and J. H. Freed, *J. Chem. Phys.* **62**, 4687 (1975).

29. P. Grigolini, *J. Chem. Phys.* **89**, 4300 (1988).

30. G. Van der Zwan and J. T. Hynes, *J. Chem. Phys.* **78**, 4174 (1982).

31. B. J. Gertner, J. P. Bergsma, K. R. Wilson, S. Lee, and J. T. Hynes, *J. Chem. Phys.* **86**, 1377 (1987).

32. B. J. Gertner, K. R. Wilson, and J. T. Hynes, *J. Chem. Phys.* **90**, 3537 (1989).

33. A. E. Stillman and J. H. Freed, *J. Chem. Phys.* **72**, 550 (1980).

34. J. H. Freed, *J. Chem. Phys.* **66**, 4183 (1977).

35. L. P. Hwang and J. H. Freed, *J. Chem. Phys.* **63**, 118 (1975).

36. D. Kivelson and R. Miles, *J. Chem. Phys.* **88**, 1925 (1988).

37. D. Kivelson and A. Kivelson, *J. Chem. Phys.* **90**, 4464 (1989).

38. D. Kivelson and T. Keyes, *J. Chem. Phys.* **57**, 4599 (1972).

39. A. Polimeno and J. H. Freed, *Chem. Phys. Lett.* **174**, 338 (1990).

40. A. Polimeno and J. H. Freed, *Chem. Phys. Lett.* **174**, 481 (1990).

41. G. Moro, P. L. Nordio and A. Polimeno, *Mol. Phys.* **68**, 1131 (1989).

42. D. J. Schneider and J. H. Freed, *Adv. Chem. Phys.* **73**, 387 (1989).

43. H. Risken, *The Fokker Planck Equation*, Springer Series in Synergetics 18 (Springer, New York, 1984).

44. R. Kubo, *Adv. Chem. Phys.* **15**, 101 (1969).

45. J. H. Freed, G. V. Bruno, and C. F. Polnaszek, *J. Phys. Chem.* **75**, 3385 (1971).

46. H. Haken, *Rev. Mod. Phys.* **47**, 67 (1975).

47. B. Bagchi, A. Chandra, and S. A. Rice, *J. Chem. Phys.* **93**, 8991 (1990).
48. D. Wei and G. N. Patey, *J. Chem. Phys.* **91**, 7113 (1989) refs. therein.
49. D. Wei and G. N. Patey, *J. Chem. Phys.* **93**, 1399 (1990) and refs. therein.
50. (a) G. Moro and J. H. Freed, *J. Chem. Phys.* **74**, 3757 (1981); (b) G. Moro and J. H. Freed, in *Large Scale Eigenvalue Problems*, J. Cullum and R. A. Willoughby (eds.) (North-Holland, Amsterdam, 1986).
51. K. V. Vasavada, D. S. Schneider, and J. H. Freed, *J. Chem. Phys.* **86**, 647 (1987).
52. M. Abramovitz and I. A. Stegun, *Handbook of Mathematical Functions* (Dover, New York, 1980).
53. R. N. Zare, *Angular Momentum* (Wiley-Interscience, New York, 1988).
54. L. C. Biedenharn and J. D. Louck, *Angular Momentum in Quantum Physics*, G. C. Rota (ed.) (Addison-Wesley, Reading, MA, 1981).
55. R. Pavelle, M. Rothstein, and J. Fitch, *Sci. Am.* **245** (6), 136 (1981).
56. S. Wolfram, *Mathematica: a System for Doing Mathematics by Computer* (Addison-Wesley, Reading, MA, 1988).
57. J. K. Cullum and R. A. Willoughby, *Lanczos Algorithms for Large Sparse Eigenvalue Computations* (Birkhäuser, Boston, 1985).
58. R. G. Gordon, *J. Chem. Phys.* **44**, 1830 (1966).
59. S. A. Zager and J. H. Freed, *J. Chem. Phys.* **77**, 3344 (1982); 3360 (1982).
60. R. M. Lynden-Bell and W. A. Steele, *J. Phys. Chem.* **88**, 6514 (1984).
61. R. M. Lynden-Bell and I. R. McDonald, *Mol. Phys.* **43**, 1429 (1981).
62. R. M. Lynden-Bell, I. R. McDonald, D. T. Stott, and R. J. Tough, *Mol. Phys.* **58**, 193 (1986).
63. S. Ruhman and K. Nelson, *J. Chem. Phys.* **94**, 859 (1991) and refs. therein.
64. S. Ruhman, B. Kohler, A. G. Joly, and K. A. Nelson, *Chem. Phys. Lett.* **141**, 16 (1987).
65. B. Kohler and K. A. Nelson, *J. Phys.* **2**, 109 (1990).
66. F. W. Deeg, J. S. Stankus, S. R. Greenfield, V. J. Newell, and M. D. Fayer, *J. Chem. Phys.* **90**, 6893 (1989).
67. F. W. Deeg, S. R. Greenfield, J. S. Stankus, V. J. Newell, and M. D. Fayer, *J. Chem. Phys.* **93**, 3503 (1990).

SOME STRUCTURAL-ELECTRONIC ASPECTS OF HIGH TEMPERATURE SUPERCONDUCTORS

JEREMY K. BURDETT

Department of Chemistry, The James Franck Institute, and The NSF Center for Superconductivity, The University of Chicago, Chicago, Illinois

CONTENTS

Advances in Chemical Physics, Volume LXXXIII, Edited by I. Prigogine and Stuart A. Rice.
ISBN 0-471-54018-8 © 1993 John Wiley & Sons, Inc.

I. INTRODUCTION

"Never before in human history has a chemical substance been the object of collaborative research by so many scientists of different disciplines as in the case of the first high temperature (HT_c) oxide superconductor $La_{2-x}Sr_xCuO_4$. If the importance of a class of materials is measured by the number of papers published in a given period, the new superconductors hold the all-time record; moreover this flood of publications also reflects the high expectations that are held worldwide for this class of substances. Every day seems to bring further surprises. From the two phases of HT_c oxide superconductors that were recognized at the outset, there have been developed within a period of scarcely two years a series of new materials which all show superconductivity at temperatures above the boiling point of liquid nitrogen; this property is undoubtedly related to the crystal chemistry of the individual phases and compounds" [1].

The importance of structure in chemistry is an all-pervading one. Structure controls function in a dominant way in the biochemical world; mutant enzyme strains, with apparently small geometrical differences often leading to dramatic physiological effects. Similar small differences in structure in solids often have significant effects on physical properties, but in the area of high temperature superconductivity, the details of the structural chemistry of the materials frequently dominate the electronic picture. A single example will suffice at this point. Consider the two systems $TlSrRECuO_5$ and $TlBaRECuO_5$ where RE = La, Nd. The former is a superconductor, but its barium analog is not. The two systems are isostructural, and appear virtually identical in almost every way. However, there is a small difference between the two. The Cu–O bond lengths in the two materials differ by about 0.05 Å. (The crystallographic a parameters (roughly twice the Cu–O distance) are 3.849 Å (Ba) and 3.761 Å (Sr). As we will see later, the striking differences in physical properties of these two materials may be traced to this difference [2].

Irrespective of whether we understand why these systems are super-conductors or not (at present we appear to be a long way off this goal), the large number of detailed structural and physical studies that have been made in recent years of these oxides have provided us with an unprecedented opportunity to explore in detail aspects of chemistry and physics which just would not have been possible with the large, but woefully inadequate set of observations available to us a decade ago. This article will therefore not be a review of high temperature superconductivity, per se, neither will it be comprehensive, but really some explorations into how nature controls the structure and metallic properties of oxide materials, a vital first step in generating a superconductor. It will begin with the premise that the history and basic chemistry and physics of the area are known to the reader. References [1] and [3–9] provide useful background material.

II. ELECTRONIC STRUCTURE

A. The General Problem

The global electronic description of these materials has proven to be a significant challenge. $La_{2-x}Sr_xCuO_4$, for example, in its undoped state is an antiferromagnetic insulator where the unpaired copper 3d electron is clearly localized on the metal. On doping however, it becomes a super-conductor and then eventually a normal metal. Although on doping its chemical composition changes (in strontium, of course but probably in oxygen as well for high x), the electronic description of the copper electrons has changed during the process to one where a localized model is now inappropriate, and the delocalized band model invariably used. What therefore are the energetic factors which favor one over the other? One can in principle lead to metallic behavior (and thus the *possibility* of superconductivity), the other cannot. A brief description of the two ends of this spectrum and the energetic features of each will be useful for our discussion below. It will be simplest to use the hydrogen molecule as an example.

The molecular orbital approach (Mulliken–Hund = MH) to the chemical bonding problem in the H_2 molecule leads to a delocalized picture. Ignoring normalization and antisymmetrization, and using ϕ_1 and ϕ_2 as the two hydrogen 1s orbitals, we can write a singlet wave function as in Eq. (2.1) for the two electrons located in the H–H bonding orbital $(\phi_1 + \phi_2)$.

$$
\begin{aligned}
{}^1\Psi_{MH} &= (\phi_1 + \phi_2)(1)(\phi_1 + \phi_2)(2) \\
&= \phi_1(1)\phi_1(2) + \phi_2(1)\phi_2(2) + \phi_1(1)\phi_2(2) + \phi_2(1)\phi_1(2)
\end{aligned}
\tag{2.1}
$$

The one-electron energy can be readily evaluated by setting $\langle \phi_1|\mathcal{H}^{eff}|\phi_1 \rangle = \langle \phi_2|\mathcal{H}^{eff}|\phi_2 \rangle = \alpha$ and $\langle \phi_1|\mathcal{H}^{eff}|\phi_2 \rangle = \beta$, the interaction or hopping integral, using the Hückel model. It is just $2(\alpha + \beta)$ where α, $\beta < 0$. However, one result of this electronic description is that two electrons may reside simultaneously on one atom since terms such as $\phi_1(1)\phi_1(2)$ and $\phi_2(1)\phi_2(2)$ occur in the wave function. If the Coulombic repulsion of two electrons located on the same atom, $\langle \phi_a|1/r_{ij}|\phi_a \rangle = U$, then the total energy is readily evaluated as $e_{MH} = 2(\alpha + \beta) + \frac{1}{2}U$. However, as the H–H distance in the molecule increases, the chance of two electrons residing on the same atom becomes less likely and the molecular orbital model less appropriate. At finite separation each neutral H atom holds just one electron and never two in the lowest energy arrangement, and is described by a localized (Heitler–London = HL) wave function of the form in Eq. (2).

$$^1\Psi_{HL} = \phi_1(1)\phi_2(2) + \phi_2(1)\phi_1(2) \qquad (2.2)$$

The total energy of this arrangement is then $e_{HL} = 2\alpha$. Of course there is a higher energy state where the two electrons are forced to be localized on the same atom with an energy of $2\alpha + U$. So when asking which arrangement will be more stable, the lower energy localized HL state with an energy of 2α, or the delocalized MH state with an energy of $2(\alpha + \beta) + \frac{1}{2}U$, it is clear that the critical ratio is that of the two-electron to one-electron energy parameters U/β. When $|U/\beta|$ is large, then the localized arrangement is appropriate, when it is small, then the delocalized (molecular orbital) picture is a better description. In a solid, a system described by the localized picture cannot be a metal since there is a Coulombic repulsion which prevents two electrons being on the same atom at the same time. Such a material is described as an antiferromagnetic insulator. The delocalized model in solids gives rise to the generation of energy bands constructed via the interaction of Bloch sums to give the solid state analog, crystal orbitals, of the molecular orbitals. Both pictures may coexist in the same material. Consider a transition metal oxide, for example. The interactions between the metal (n + 1)s and (n + 1)p orbitals and the oxygen 2s and 2p orbitals may well be strong enough (i.e., large β) to lead to a MH picture, but the corresponding interactions involving the nd orbitals may be much weaker and result in a HL description. This is probably the case for undoped La_2CuO_4. The picture may change too with internuclear separation. In general, the shorter the internuclear distance, the larger the overlap which in turn leads to a larger β. In fact we will see some examples later where, by changing the relevant internuclear separation in a series of related compounds the electronic description changes.

Such a balance between the one- and two-electron terms in the energy play important roles in several places in chemistry. For example the relative stability of high spin (large $|P/\Delta|$) and low spin (small $|P/\Delta|$) octahedral transition metal complexes is set by a similar ratio. Here P is the "pairing energy" associated with a pair of electrons (a combination of exchange and coulomb terms) and Δ the e_g/t_{2g} splitting. In fact there are observations from molecular coordination chemistry which are of use to us in providing insights into the electronic behavior of solids. We know from electronic spectra that as the size of an atom increases the magnitude of the electron–electron interactions, U, P (measured by the Racah parameters) decreases. In terms of the one-electron part of the energy we know too that the e_g/t_{2g} splitting in octahedral complexes of the transition metals is larger for second and third row metals than for first. Thus $|U/\beta|$ is expected to be larger for first row transition metal systems than their heavier analogs. Indeed complexes of Ni(II) are sometimes paramagnetic and octahedral, and sometimes diamagnetic and square planar, but those of Pt(II) are always diamagnetic and square planar. As we will see, analogous arguments allow us to understand why the nickel-containing system of Fig. 1 is an antiferromagnetic insulator,

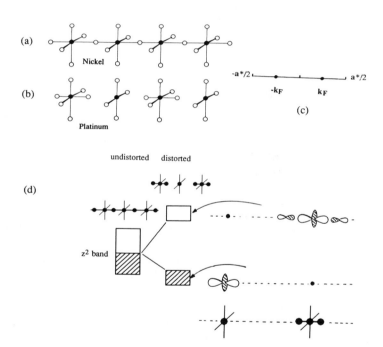

Figure 1. (a) Schematic structure of the infinite chain compound [Ni(cyclohexanediamine)$_2$Br]Br$_2$ where all of the bridging Ni–Br distances are equal and the compound is an antiferromagnetic insulator. (b) The structure of platinum compounds of this type which experience a Peierls distortion leading to a set of bridging Pt–Halogen distances which alternaite ..ssllssllssll... (s = short, l = long). (c) The very simple Fermi surface of the undistorted platinum compound. (d) The change in electronic structure accompanying the Peierls distortion. Notice that after distortion, the upper (empty) band is located on octahedral Pt(IV) and the lower (full) band on square planar Pt(II).

but the structure of the comparable platinum system accessible by a delocalized or band model.

A major theoretical problem in providing a coherent picture for the class of superconducting metal oxides lies in the generation of a single theoretical model which is able to move from the delocalized to localized regime smoothly. Hubbard Hamiltonians, which in their most basic form treat the one-electron part of the energy (β) as a perturbation of the wave function of Eq. (2.2), readily incorporate correlation effects and appear to be reasonable for explaining photoelectron spectra and other high energy excitations [10]. The band model predicts [11] undoped La_2CuO_4 to be a metal which it is not. One way which is useful to visualize the transition is to start from the MH end of the spectrum and correct for electron correlation using configuration interaction, writing a wave function of the form of Eq. (2.3) which mixes higher energy

$$\Psi_{MH}(0) = \sum_j c_i \Psi_{MH}(i) \qquad (2.3)$$

configurations ($\Psi_{MH}(i)$) into the ground configuration ($\Psi_{MH}(0)$). This implies in a crude way that when the system is described by the delocalized model, $\Psi_{MH}(0)$ dominates, but even when correlation effects start to become important, there is still a "memory" of the delocalized wave function in the intermediate region. In our discussions in this article we will rely heavily on this viewpoint and will use the band model to study these metallic cuprates, even though correlation may be quite important. Similar ideas come from Fermi liquid theory [12]. (Parenthetically we note that even though LDA calculations [11] predict, incorrectly, that undoped La_2CuO_4 is a metal, the agreement between the calculated and observed phonon spectra is a good one.)

The description of the transition from one regime to the other was studied extensively by Hubbard [13] and his picture is shown in Fig. 2. At the left hand side are the two Hubbard bands (described by the HL wave functions with energies of 2α, $2\alpha + U$) separated by a correlation gap of U. For the solid β is the hopping or transfer integral which connects one orbital of the infinite solid to the next. The bandwidth of the energy band which results is proportional to β. As $|\beta/U|$ increases, a critical point (indicated by an arrow) is reached when the two Hubbard bands touch and an energy band is formed. Thus the important parameters are the Coulomb repulsion term U, and β. A partially filled band, in principle will lead to a metal of course. This can occur in principle for any partially filled band to the right of this point, or addition or subtraction of electrons from the half-filled arrangement to the left of this point.

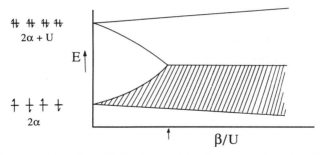

Figure 2. The Hubbard model for the metal insulator transition using the parameter $|U/\beta|$. At the left hand side are the two Hubbard bands separated by a correlation gap. At the right hand side a band model is appropriate.

However, within the band model there are geometrical instabilities which have to be taken into consideration before a metal is formed.

A partially filled band with a Fermi wave vector $\mathbf{k}_F$, is in general unstable with respect to a distortion which stabilizes the energy of the occupied energy levels. This is usually described in terms of the nesting of the Fermi surface. If large parts of the Fermi surface may be translated by a unique vector $\mathbf{q}$ so as to be superimposable on other parts of the surface then such nesting often leads to structural instabilities. For an extremely readable discussion of this problem, see [14]. The classic case is the Peierls distortion of linear chain systems such as polyacetylene where the half-filled energy band leads to $\mathbf{q} = \mathbf{k}_F - (-\mathbf{k}_F) = 2\mathbf{k}_F$. In general for a one-dimensional system described by a unit cell of length a, if $\mathbf{k}_F = (1/q)\mathbf{a}^*$, then the new unit cell will be $2q$ times the old. Polyacetylene, where $1/q = 1/4$, is stabilized by a distortion which leads to alternating short and long C–C bonds and thus a doubling of the cell length. The distortion does not necessarily need to be commensurate with the lattice and incommensurate charge density waves (CDW) are found in a variety of materials. The platinum-containing system of Fig. 1 is described by a distortion of this type. Simple electron counting indicates that the platinum is present as Pt(III) with a d^7 configuration, leading to a situation where the z^2 band, the highest occupied band of the solid, is half full. Such an arrangement has the Fermi surface shown in Fig. 1c, given by the two k-points $\pm 0.25\mathbf{a}^*$, and is unstable to the Peierls distortion shown where the structure distorts to give a square planar center (usefully identified as Pt(II)) and an octahedral center (identified with Pt(IV)). This description leads us immediately to the identification of some of the energetic ingredients which make up this problem. The Peierls distortion here is associated with the disproportionation reaction,

$2Pt(III) \rightarrow Pt(II) + Pt(IV)$. This involves the coulombic penalty of placing two electrons on one of the atoms, namely, $Pt(II)$. However there is an energetic stabilization from the reduction in energy of the occupied orbital levels. So, once again, the distortion energetics are controlled by the ratio of one-electron to two-electron energy terms.

There are therefore two effects (at least) which compete with the delocalized metallic state for these systems with partially filled bands, namely localization of electrons and geometric distortions. They are both controlled by the ratio of terms of the type U/β, sensitive to the nature of the atoms concerned, the band filling and the chemical environment they find themselves in. It is particularly interesting, recalling the connection made above to coordination chemistry that the Peierls distorted solid is found for the heavier transition metal in Fig. 1, where $|U/\beta|$ is expected to be smaller than for the first row analog. Beyond such rather broad generalizations however, it is very difficult to predict, or mimic numerically the behavior actually found for a given system. Although it is often relatively simple, working within a band model, to study aspects of CDWs, showing when this state of affairs is favored over the localized system has not in general been possible. The problem lies in being able to accurately estimate the ratio $|U/\beta|$. One thing we do know however, is that moving away from the half-filled band tends to relieve the driving force for a CDW distortion by removing electron density from that part of the band (the top) which is most strongly stabilized. The metallic state (and perhaps a superconductor) remains as the electronic ground state. We leave this section by noting that the cuprate superconductors, containing a first row transition metal, like in an interesting regime where the ratio of U/β, just as for the case of molecular complexes of $Ni(II)$, allows transitions from one type of behavior to another. It is probable that $|U/\beta|$ in fact lies close to unity for these superconducting cuprates, the very worst value for a simple theoretical description.

It is interesting at this point see how some of these ideas apply to transition metal oxides in general, using the series of electronic pictures shown in Fig. 3 for metal containing systems. It has been suggested [15] that three parameters, Δ, W and U' are important in describing the gross features of these materials. U' is some two-electron energy term, related to the U of our discussion above and shown in Fig. 2, W is the bandwidth (proportional to β of the above discussion) and Δ is a new parameter which measures the separation between the oxygen and metal states. Two types of electrons are included, electrons lying in metal d orbitals which may or may not be localized, and a set of electrons filling the oxygen 2p bands. Five possibilities are immediately apparent, of which two give rise to metallic behavior. Δ and U' may be quantified using a simple ionic

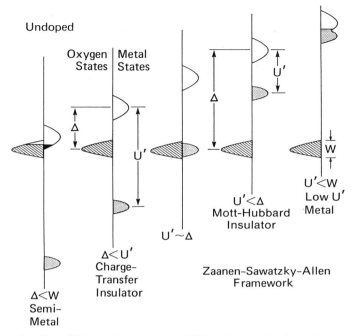

Figure 3. Five different electronic possibilities for metal oxides depending on the relative sizes of U', W and Δ (from [16] with permission).

model [16]. U' may be approximated (call it U_0') by noting that it measures the energy of removing an electron from one metal site (the ionization potential, I_{v+1}), attaching it to an adjacent one (the electron affinity, $-I_v$) plus the corresponding electrostatic energy of the pair created ($-e^2/d_{MM}$). A similar expression may be written for Δ_0 as an approximation to Δ if we represent it as the energy difference involved in moving an electron this time from O to M. This energy term is then just sum of the difference in Madelung site energy that the electron experiences when it moves from O to M (ΔV_M), the electron affinity of M ($-I_v$), the ionization potential of O^{2-} (I_O) and the corresponding electrostatic energy of the pair created ($-e^2/d_{MO}$). All of these terms are either readily available or may be simply calculated, except for I_O. It however, remains constant for all examples. Figure 4 shows the results [16] for some 76 oxides. A broad separation into metals and insulators is found, with the systems that show metal-insulator transitions lying close to the two boundaries set by the pair of lines at $\Delta_0 \sim 10$ eV and $U_0' \sim 11$ eV. The plot suggests that the observation of metallic behavior in $LaCuO_3$ (the

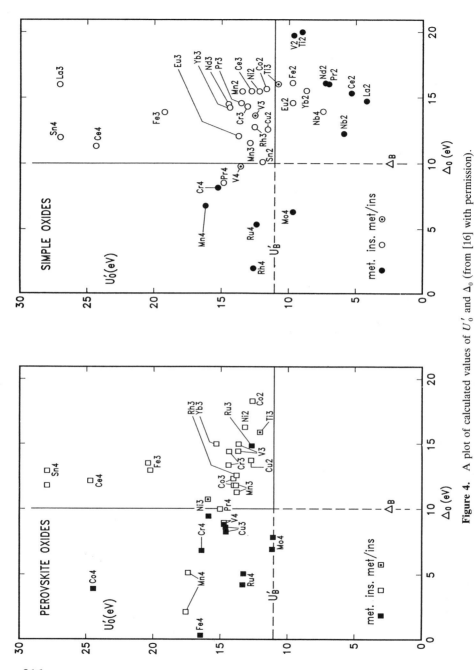

Figure 4. A plot of calculated values of U'_0 and Δ_0 (from [16] with permission).

point labeled Cu3) is associated with metal d/oxygen 2p overlap such that holes appear in the oxygen 2p levels, but that LaO (La2) is a metal because it has a low U. Note in this plot no attention has been paid to the size of the bandwidth or as to how the parameters mentioned will vary with band filling or other things. However, it shows quite clearly how gross *atomic* properties control in a broad way the properties of these materials.

B. The Square Planar Cuprates

We will describe the geometrical structures of the cuprates in more detail later. However, one essential ingredient of all known high temperature cuprate superconductors is the structural unit shown in Fig. 5. In this CuO_2 sheet the copper atoms are four coordinate in a square planar environment, the Cu–O distances varying from 1.86 to 1.94 Å in the hole-doped superconductors to around 1.98 Å in the electron-doped materials. Usually, but not always, there are one or two additional oxygen atoms coordinated above and below this plane, leading to a square pyramidal or octahedral geometry at copper. In all of these compounds copper is close to being formally Cu^{2+}, and this d^9 ion is well known for the interesting structural chemistry associated with the Jahn–Teller instability of a regular octahedron with this electron count. Accordingly the fifth and sixth oxygen atoms which are coordinated to this plane have long Cu–O distances, ~2.3 Å in the $YBa_2Cu_3O_{7-\delta}$ system, and ~2.45 Å in $La_{2-x}Sr_xCuO_4$, with variations which depend upon the stoichiometry. Usually the coordination geometries are distorted from the ideal ones. The apical-basal angle for the square pyramids in

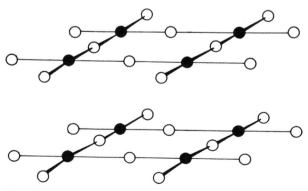

Figure 5. The CuO_2 sheets common to all high temperature superconductors. Frequently these sheets are not planar, and may be puckered, rumpled or folded.

$YBa_2Cu_3O_{7-\delta}$ is around 97°, the exact value dependent on δ, and in the orthorhombic form of $La_{2-x}Sr_xCuO_4$ the octahedra are tilted with respect to each other. In the electron-doped systems the planes are unusual in that there are no axial oxygen atoms at all. There does not seem to be any correlation of T_c with the Cu–O–Cu angle and thus superconductivity is found in both flat and distorted sheets.

The band structure of these square planar units is in principle very simple. Figure 6 shows the d orbital level splitting pattern which we would expect for an isolated planar CuO_4 fragment. The $x^2 - y^2$ orbital of the metal is involved in σ interactions with the ligands and it lies at the highest energy of the set. With a d^9 configuration this is the highest occupied orbital of the fragment. Since the two axial ligands of the octahedron, when present, are attached at long Cu–O distances, the z^2 orbital of the unit lies at low energy. (In the regular octahedron z^2 and $x^2 - y^2$ are degenerate.) Of considerable importance in these compounds

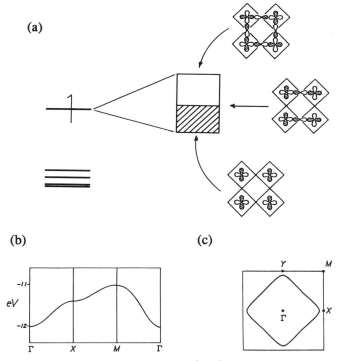

Figure 6. (a) Schematic deviation of the $x^2 - y^2$ band for the CuO_2 sheet. (b) Dispersion properties of the $x^2 - y^2$ band. (c) Schematic Fermi surface for the CuO_2 sheet.

is the juxtaposition of the copper 3d and oxygen 2p orbitals, set by their relative atomic energies. This leads to the possibility of strong hybridization between them (in the delocalized regime) leading to a description of the "metal $x^2 - y^2$" orbital as being one where there is a heavy mixture of the two. The level occupied by the single unpaired electron of Cu(II), is antibonding between the two atoms. As we will see, this shows up experimentally as a strong dependence of Cu–O distance in a given system as a function of doping, in the sense that decreasing electron density in this orbital leads to shorter bonds. Figure 6 also shows the generation of an energy band from the $x^2 - y^2$ orbital of this CuO$_4$ fragment, and how the orbital wave function changes as the energy increases. The energies at both the top and bottom of the band increase as the Cu–O distance decreases, a direct result of the general increase in overlap as distances shorten. There is also a small calculated change in its width. A half-filled configuration is shown for the energy band, but this is a situation in fact which probably never exists because of localization, as we have described above. In the diagram we have focussed attention on the $x^2 - y^2$ band of the system, and believe it is the key player in the electronic structure of these materials. A model has been proposed [7] however, in which there are two metal orbitals which are important, $x^2 - y^2$ involved in σ interactions, and the set of π orbitals (specifically xz and yz of the metal). In this article we shall pursue a model in which only one metal-based band, $x^2 - y^2$, has a direct role to play.

The dispersive properties of the $x^2 - y^2$ band and the Fermi surface which one calculates [17] from such a picture is shown in Fig. 6c. At the half-filled point (no doping) strong nesting is seen as shown, which suggests the possibility of a Peierls distortion, the two-dimensional analog of that shown in Fig. 1b. However, here the effective U is high. As electrons are removed from the $x^2 - y^2$ band, it changes its shape significantly. A curvature develops on doping and leads to a loss of such nesting and a reduced energetic stabilization for this distortion. Such distorted CuO$_2$ planes are in fact never seen as static geometries in these compounds. All of these materials are either antiferromagnetic insulators at the half-filled band situation (high U) and remain so or become metallic on either adding or removing electrons. The widespread occurrence of disorder, defects and incommensurate modulations in the non-cuprate part of the structure of these materials has received much comment (see, for example, [18]), and the suggestion has been made that it is the presence of such disorder which suppresses CDW distortions of the CuO$_2$ sheets. However there are now very good structural determinations of materials such as the YBa$_2$Cu$_3$O$_7$ superconductor which show no disorder at all; neither do they show any indication of distortion of the planes appropriate for a CDW. Thus we believe that the structural stability of these CuO_2 sheets towards an in-plane distortion is an intrinsic feature of these units.

III. CHARGE TRANSFER IN THE CUPRATES

A. The Basic Structural Features

Metals are systems that contain energy bands which are partially filled. We have noted above that such a state of affairs is best met in the cuprates by taking a system with a half-filled band and removing (usually) electron density in order to move away from the half-filled point, especially susceptible to localization or to a Peierls distortion, both routes leading to the creation of an insulator. In fact all of the high T_c materials that are presently known are of the general form shown in Fig. 7 and consist of sheets of stoichiometry CuO_2, containing approximately Cu^{2+}, interleaved with other oxide sheets or slabs which act as charge reservoirs. The less than half full copper $x^2 - y^2$ band is achieved (Fig. 8) by either (a) cation substitution, cation vacancies, the presence of extra oxygen (effects easily understandable by consideration of the chemical formula) or (b) by overlap with an empty band (for which the structure is important).

Such charge transfer is intimately connected with the crystal structure and in principle occurs in a straightforward way. The structures of all of the known high T_c oxides are based on variations of the structure of the high symmetry perovskite parent, ABO_3. Its structure (Fig. 9) may be described in several different ways. Thus it consists of BO_6 octahedra joined by vertex fusion in all three cartesian directions with a large A atom located in the interstice created in this way. Alternatively it may be

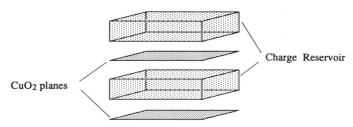

Figure 7. Schematic structure of all high T_c systems. CuO_2 sheets are separated by reservoirs of other material. The latter may be rocksalt layers, fluorite layers, layers of metal atoms alone, or combinations of these.

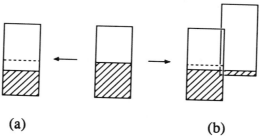

(a) **(b)**

Figure 8. Two mechanisms for generating a metal: (a) chemical stoichiometric depletion of the half filled $x^2 - y^2$ band; (b) overlap of the $x^2 - y^2$ band with an empty band.

described as a defect rocksalt structure where 1/4 of the oxygen atoms and 3/4 of the metal atoms are missing (this is the ReO_3 structure) followed by insertion of a large A ion into one of the oxygen vacancies. Another way of visualizing the structure (Fig. 9b) is to note that it is made up of sheets of BO_2 stoichiometry, containing vertex-fused CuO_4 squares, which alternate with rocksalt-like AO layers. This latter description is often a good one since in the high T_c materials the B(=Cu)–O distances perpendicular to these planes are always long. The A and B atoms occupy distinctly different sites, twelve and six coordinate, respectively. An interesting feature of the perovskite structure is that it can accommodate a large number of defects, often ordered as in $CaMnO_{2.5}$ and the brownmillerite structure of $CaFeO_{2.5}$. Many of the oxides we will be concerned with here are simple intergrowths of these CuO_2 sheets with additional rocksalt layers. Thus the K_2NiF_4 structure adopted by La_2CuO_4 is simply written $(LaO)_2CuO_2$, and $Bi_2Sr_2Ca_{n-1}Cu_nO_{2n+4}$ as $(BiO)_2(SrO)_2Ca_{n-1}(CuO_2)_n$. The second example contains extra layers of Ca atoms located between the CuO_2 layers. The system where the rocksalt layers are absent such that the structure just consists of simple alternation of the CuO_2 layers with layers of metal atoms, has also been made. $Ca_{0.86}Sr_{0.14}CuO_2$ [19] and $Nd_ySr_{1-y}CuO_2$ are known for $0 < y < 0.15$. The second material is a 40 K superconductor. Early suggestions that T_c increases with the number of adjacent CuO_2 planes in these oxides have not stood the test of time; the electronic properties seem to be controlled more by the geometric details of the sheets rather than how many are arranged consecutively in the structure.

An interesting feature of these structures is the degree of mismatch between the set of layers of different types. Geometrical considerations (see Fig. 9c) suggest that the size match will be perfect if the so-called Goldschmidt tolerance factor $t = (r_{AO})/\sqrt{2}(r_{BO}) = 1$. Since $\sqrt{2}(r_{CuO}) \sim$

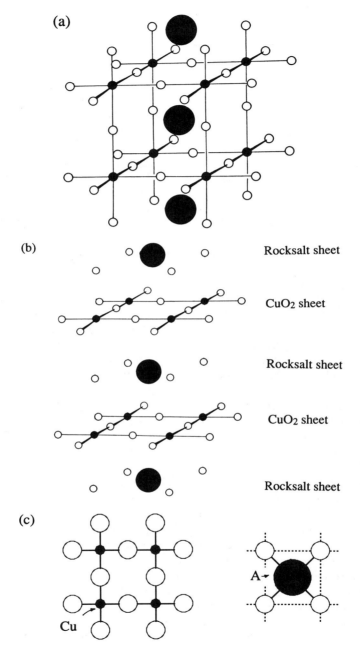

Figure 9. (*a*) The perovskite structure. (*b*) The alternating CuO$_2$ sheets and rocksalt sheets. (*c*) Size requirements for (*b*).

2.7 Å and with typical values of $r_{BiO} = 2.43$ Å and $r_{TlO} = 2.29$ Å, the bismuth and thallium systems are under considerable tension. As we will see these planes relax to take care of this problem. The distances r_{BaO}, r_{SrO} and r_{LaO} of 2.87 Å, 2.71 Å, and 2.62 Å lead to less of a mismatch, but the CuO_2 sheet is rumpled in many oxides perhaps as a result of the difference. The K_2NiF_4 structure is in fact only found for the La member of the RE_2CuO_4 series, a direct result of the size of the RE atom. For the smaller rare earths the T' structure is found where the interplanar material has the fluorite structure. Analogous distortions occur too of the perovskite structure itself as a function of the A atom size as we will see later for $RENiO_3$.

$La_{2-x}Sr_xCuO_4$, perhaps the simplest (in principle) system of all, consists of such CuO_2 sheets interleaved with rocksalt $(AO)_2$ layers where the La and Sr atoms are randomly arranged on the metal sites (Fig. 10). Here the holes in the sheets are created by cation substitution, namely replacement of some of the three-valent La ion by some two-valent Sr ions. Thus the $x^2 - y^2$ band of the copper, half full when $x = 0$, becomes $\frac{1}{2}x$ full when partially substituted. In the orthorhombic $Bi_2Sr_{3-x}Y_xCu_2O_{8+y}$ structure (Fig. 11) isostructural to that of the $Bi_2Sr_{3-x}Ca_xCu_2O_{8+y}$ superconductor [20], there are CuO_2 sheets, SrO rocksalt layers and a complex modulated BiO_{1+y} region. One mechanism for conductivity here is thus by nonstoichiometry. In the $YBa_2Cu_3O_{7-\delta}$ system (Fig. 12), really

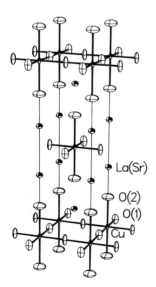

La(Sr)

O(2)

O(1)

Cu

Figure 10. The structure of $La_{2-x}Sr_xCuO_4$.

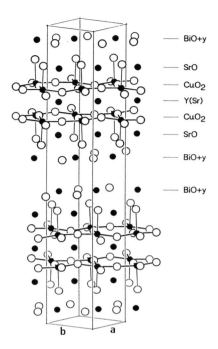

BiO+y
SrO
CuO₂
Y(Sr)
CuO₂
SrO
BiO+y

BiO+y

Figure 11. The structure of $Bi_2S-r_{3-x}Y_xCu_2O_{8+y}$ (from [20] with permission).

b   a

a defect perovskite, in addition to YO layers there are slabs of a different type of copper oxide unit. For $\delta = 0$, as shown in the figure, this structural unit is a set of infinite CuO_3 chains which run along one direction for the orthorhombic cell. For $\delta = 1$ the sites labeled $O(1)$ are empty and there are now linear two coordinate units arranged on the same sites in a tetragonal arrangement. The mechanism of hole generation here is by band overlap [21] as shown in Fig. 8. In general terms, an empty band of the reservoir material overlaps the $x^2 - y^2$ band of the CuO_2 sheet and charge transfer takes place as shown. The overlap between the two bands will depend on their relative energy and on the density of states of the two structural units. Both of these will depend crucially on the details of the crystal structure. In the broadest terms the overlap may be adjusted by either changing the geometry of the CuO_2 sheet, by changing the nature and structure of the intergrowth material or both. The details of this charge transfer process will depend critically on the geometry of the system, and calculations designed to reproduce it will be worthless unless the correct geometry is used. Thus for example, calculations [22] on $Bi_2Sr_2Ca_{n-1}Cu_nO_{2n+4}$ in which the ideal rocksalt structure was used for the BiO layers suggested that the bottom of the Bi

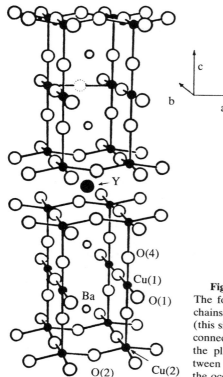

Figure 12. The structure of $YBa_2Cu_3O_{7-\delta}$. The following atom labels are used: Cu(1) in the chains, Cu(2) in the planes, O(1) forms the chains (this site is full at $\delta = 0$ and empty for $\delta = 1$), O(4) connects Cu(1) and Cu(2), O(2) and O(3) are in the planes, O(5) is the site, usually vacant, between the chains. It is indicated on our diagram by the occupation of a single example of this site with a dotted atom.

6p block bands lay below the Fermi level. Use of the correct BiO geometry [23] pushes these levels up in energy so that band overlap is not a doping mechanism. The holes in this material come in part from extra oxygen atoms in the Bi–O layers. Evidence that the band structure calculations using the wrong geometry are incorrect, also comes from chemical ideas. It would imply that Cu^{2+} is oxidized by Bi^{3+}. Bi^{3+} is not a strong enough oxidant. In the Tl and Bi containing systems there are intergrowths of different sizes. Thus $Tl_2Ba_2Ca_{n-1}Cu_nO_{2n+4}$ contains distorted Tl_2O_2 double rocksalt layers between the CuO_2 sheets, whereas $TlBa_2Ca_{n-1}Cu_nO_{2n+3}$ contains single such TlO layers. In the former there is band overlap [24], but in the latter there is not. In the oxygen rich superconductor $La_2CuO_{4+\delta}$, the doping mechanism comes about through the presence of excess oxygen. In the $Bi_2Sr_2Ca_{n-1}Cu_nO_{2n+4}$ series, the holes appear to be created both by the off-stoichiometry of strontium and by the presence of extra oxygen atoms as we have just noted.

The crystal structures of these materials invariably provide challenges for the experimentalist. These are not the "one-day wonders" of the organometallic chemist accustomed to collecting a data set overnight which often leads to a definitive geometry in a straightforward manner. In many of these structures there are problems associated with the presence of partial site occupancy sometimes for the cations and often for the oxygen atoms, problems generated by the presence of incommensurate superstructures of various types and the effects of stacking faults and of chemical disorder. Thus for example, in the thallium containing materials, refinement [25] of the TlO region of the structure leads to a rocksalt arrangement with Tl–O bond lengths of around 2.7 Å. As we noted above, these are much too long for such Tl–O contacts; 2.2–2.3 Å are the expected values. In the tetragonal unit cells found for these compounds, large thermal parameters are found for these atoms, indicative of disorder. Anisotropic refinement shows that the disorder lies within the TlO sheets and thus the diffraction experiment is actually seeing a structure averaged over many unit cells. Studies of the radial distribution function show much shorter Tl–O distances. Such "disorder" is quite likely an experimental artifact; the movements of the Tl atoms are almost certainly highly correlated. Similar observations are made in the bismuth containing systems. As we will see below there are several different ways in which the bismuth or thallium atoms may move in these structures to satisfy these observations. None of them may be definitively identified from the diffraction studies. Each of them lead to a very different electronic state of affairs than that for the rocksalt parent.

In some of these oxide materials, the rocksalt layers may be incommensurate with the CuO_2 planes and noted above. Diffraction studies of the $(BiO)_2Sr_2Ca_{n-1}Cu_nO_{2n+2}$ phases show such behavior is commonplace [25]. This may be one way to relieve the size mismatch problem, and is reminiscent of some recent results in metal sulfide chemistry [26]. There, systems of stoichiometry $(MS)_{1+x}(NbS_2)$ have been identified where $x \sim 0.14$ for the cases of M = RE or Pb. In these cases the size mismatch in the MS and NbS_2 sheets is achieved by stacking different numbers of unit cells together, the value of x being determined by the size mismatch.

Another problem of vital importance in any deliberations concerning these materials is their oxygen content. For example, the addition of less than 2% oxygen to the La_2CuO_4 species leads to $La_2CuO_{4.07}$, a superconductor [27] with a T_c close to the maximum observed for the $La_{2-x}Sr_xCuO_4$ system. Yet the determination of the location of this excess oxygen has been a difficult structural problem. By analogy with the corresponding nickel compound [28] (Fig. 13) the extra oxygen probably

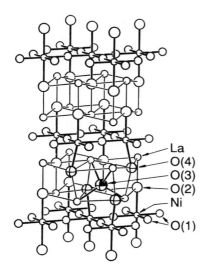

La
O(4)
O(3)
O(2)
Ni
O(1)

Figure 13. The structure of $La_2NiO_{4+\delta}$ (from [28] with permission).

lies in a site some way away from the copper atom and thus acts as a simple recipient of two electrons. The deleterious effects of small oxygen deficiency on superconductivity has been noted in the $La_{2-x}Na_xCuO_4$ compound [29]. In some recent experiments [30] T_c could be changed at will by changing the oxygen stoichiometry by tiny amounts.

The question of whether, for a given gross chemical stoichiometry, there will be oxygen vacancies of interstitials, or indeed the presence of defects on the metal sites, is one which crystal chemists have only begun to address. Our theoretical understanding of the energetics of such processes is nonexistent. The examples which follow in this section will therefore be associated with that part of the electronic structure problem where we have made progress, namely the influence of a given geometrical arrangement on the band structure and electronic properties of the whole.

B. Variation in the Cu–O Distance

The $x^2 - y^2$ band of the CuO_2 sheet is metal–oxygen antibonding. Addition of electrons (as in the n-doped systems) should therefore lengthen these Cu–O distances but removal of electrons as in the p-doped systems) should result in bond length shortening. Figure 14 shows [31] that this is true in general, but also shows that structural changes are found in the reservoir material too, and thus that these CuO_2 sheets may not be taken in isolation. In Fig. 14a for the $La_{2-x}Sr_xCuO_4$ series, the effect of substitution of the smaller La^{3+} by the larger Sr^{2+} is an increase

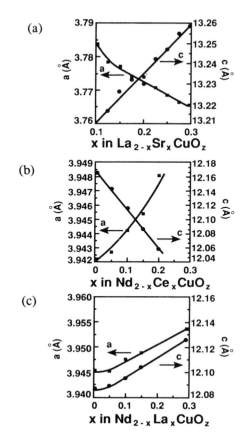

Figure 14. The variation in crystallographic cell parameters for the series (a) La_{2-x}

in the crystallographic c parameter (perpendicular to the sheets) as the "size" of the rocksalt layer expands. However there is a comparable decrease in the a-axis parameter which measures the Cu–O distance in the sheets. This strengthening of the CuO bonds is the result expected by removal of electron density from the $x^2 - y^2$ band. Figure 14b shows that the converse is true in the electron doped series $Nd_{2-x}Ce_xCuO_4$. Here, as the smaller Ce^{4+} ion replaces the larger Nd^{3+} ion the c-axis contracts. However the a-axis increases, a result in accord with the addition of electrons to the antibonding $x^2 - y^2$ band and a weakening of the Cu–O bonds. (This result does in fact need to be regarded with some care since in this particular system we are not exactly certain of the geometrical structure of this oxide [32] as we will see below.) Figure 14c shows the control experiment, that of substitution of Nd^{3+} by the larger La^{3+}. Here there is no change in the electron density (in first order at least) in the $x^2 - y^2$ band of the sheets. As a result *both* a and c increase. The picture that results is therefore one where the CuO_2 sheets are stiffer than the interlayer material, and where the electronic demands of the sheets is the dominant effect. It means that the reservoir material is much more easily deformed, but comparison of the details of the last two plots shows too that the actual Cu–O distance is also influenced by steric effects in additional to the electronic ones. The interplay of the two is seen in a broad study of the Cu–O distances in these superconductors.

There are now quite a sizeable number of high T_c superconductors, and in the plot [33] shown in Fig. 15 it is apparent that the Cu–O distance is a useful parameter to describe their superconducting properties. A connection can be made between these Cu–O distances and the number of holes in the CuO_2 layer by consideration of the bond valence sum, v, at the copper atoms. This concept dates from Pauling's idea of the electrostatic bond sum rule and has been refined in the intervening period [34]. Modern usage defines it as the sum of the bond strengths over all bonds coordinated to copper in the following way. The bond valence sum, $v = \Sigma_i \exp(r_i - r_0)/0.37$, where the parameter r_0 depends on the

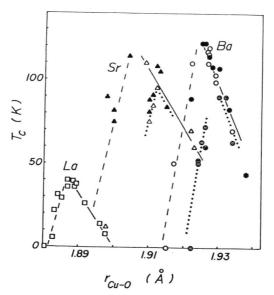

Figure 15. Variation of T_c with Cu–O distance for three major series of superconductors, differentiated by the identity of the large cation they contain. The presence of some subseries is also indicated (from [33] with permission).

chemical identity of the atoms A, B which make up the AB bonds in question, and also in more detailed applications of the approach on their oxidation state. Importantly the quantity v has been claimed for many years to be equal to the "valency" of the atom concerned. The theoretical origin of this sum rule is unclear but it often seems to work quite well. Obviously the steric effects just described will complicate matters, but if systems which are chemically similar are compared, then the bond sum trends ought to follow the changes in planar copper valence and hence the number of holes in the $x^2 - y^2$ band of the sheets. The inverted parabolic dependence of Fig. 16 is very striking for the various series [33]. The trends of this figure await a theoretical explanation, but show quite clearly the dependence of T_c on the number of holes in this band, the latter being detected by the variation in internuclear separation. Figure 17 shows schematically how the electronic properties of the $La_{2-x}Sr_xCuO_4$ superconductor vary with x. The system moves from an antiferromagnetic insulator, to a superconductor, to a metal. Thus the electronic description varies in the same way as on moving from left to right in Fig. 2. As x increases the Cu–O distances shorten, the result of removing electron density from the antibonding band. The parameter β is thus expected to correspondingly increase. Assuming that U remains unchanged on moving away from the half-filled band, the system then moves across the diagram. (Parenthetically we note that we will have to use a different model for the electron-doped systems where the Cu–O distance apparently increases with doping.) It puts the superconductor close to the critical point where the electronic description is complex.

C. The TIMRECuO$_5$ and RENiO$_3$ Systems

We mentioned in the Introduction how the TIMRECuO$_5$ system led to superconductors for RE = La, Nd and M = Sr, but not for its M = Ba

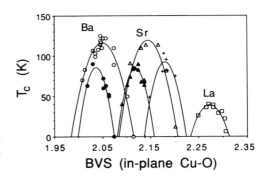

Figure 16. The data of Fig. 15 replotted but having converted the Cu–O distances to bond valence sums (from [33] with permission).

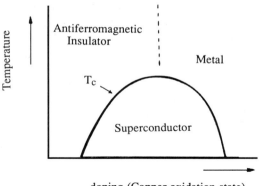

Figure 17. Typical behavior of a cuprate superconductor on doping. The break between metal and insulator above T_c is variable and indicated by a dashed line.

analogues. The solid solutions, $TlSr_{1-x}Ba_xRECuO_5$ lose their superconductivity when $x > 0.6$. In qualitative terms, such a result is readily understood in terms of band overlap [2]. As we noted the Cu–O bond lengths in the two materials are different. The crystallographic a parameters (roughly twice the Cu–O distance) are 3.849 Å (Ba) and 3.761 Å (Sr). This difference of about 0.05 Å in the Cu–O distance is sufficient to raise the energy of the $x^2 - y^2$ band of the CuO_2 sheet so that band overlap with the TlO layer may occur, as shown qualitatively in Fig. 18. It is important to point out that the increase in Cu–O distance on moving from Sr to Ba is just caused by the introduction of the larger cation. At this stage we point out that our model here says absolutely nothing about the superconductivity in the Sr system. The mechanism we have described here is one that simply generates a *metal* by band overlap.

Similar control of T_c is found in the $Tl_2Ba_2CuO_6$ system [35]. Simple electron counting tells us that either there is some nonstoichiometry not

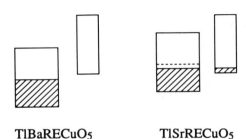

TlBaRECuO5   **TlSrRECuO5**

Figure 18. Schematic showing how the $x^2 - y^2$ band in the TlMRECuO$_5$ system (RE = La, Nd) is sensitive to whether M = Sr or Ba.

shown in the chemical formula or there must be some band overlap. The latter is true as shown below. The compound is an interesting one because it has the highest T_c (90 K) for a single sheet material. However substitution of Sr to give $Tl_2Ba_{2-x}Sr_xCuO_6$ leads to a shortening of the crystallographic a axis, presumably an increase in the extent of band overlap but now a *decrease* in T_c. This may be indicative of the overdoping effect seen in Fig. 16, where there is an optimal hole concentration for superconductivity.

The idea of controlling the properties of materials of this type by substitution of ions of different sizes is also shown in a related series [36], that of $RENiO_3$ where RE = La, Pr, Nd and Sm. These compounds are not superconductors, but the last three compounds of the series show metal–insulator transitions whose temperature may be controlled by the structural parameters of the solid. These materials crystallize in distorted form of the perovskite structure, the $GdFeO_3$ structure, where as shown in Fig. 19, adjacent pairs of NiO_3 octahedra are bent at the common oxygen atom to improve the oxygen coordination to the rare earth atom.

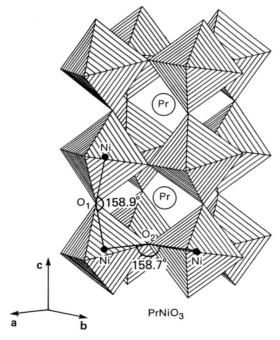

Figure 19. The $GdFeO_3$ structure showing its distortion from the perovskite structure (from [36] with permission).

The degree of bending is determined by the size of the rare earth, the Ni–O–Ni angle being close to 180° for La (here though the structure is distorted rhombohedrally), 158.1° for Pr, 156.8° for Nd and 151.8° for Sm. The Ni–O distances change slightly 1.940(8) Å, 1.943(9) Å and 1.955(6) Å for Pr–Sm. The tighter angle and longer Ni–O distance both contribute to a smaller transfer or hopping integral for the Ni–O bands. Assuming that the Coulomb U remains constant in this series, we expect to see values of $|U/\beta|$ increasing in the series La $\ll$ Pr $<$ Nd $<$ Sm, a result in qualitative accord with the measured conductivity data of Fig. 20. The size of the rare earth ion and the structural change it induces in the NiO_3 framework is of crucial importance in determining the electronic parameters of the system. Via control of the bandwidth, effective control of the metallic properties of the material is obtained using the ideas of Fig. 2.

D. The $Tl_2Ba_2Ca_{n-1}Cu_nO_{2n+4}$ and $TlBa_2Ca_{n-1}Cu_nO_{2n+3}$ Systems

The gross structures of these systems are relatively straightforward, but a detailed   geometrical   picture   has   yet   to   be   achieved.

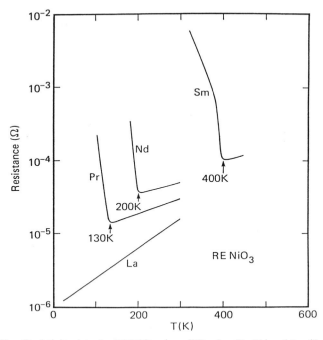

Figure 20.   Resistivity data for $RENiO_3$ where RE = La, Pr, Nd and Sm (from [36] with permission).

$Tl_2Ba_2Ca_{n-1}Cu_nO_{2n+4}$ contains Tl_2O_2 double layers which lie between the CuO_2 sheets, whereas $TlBa_2Ca_{n-1}Cu_nO_{2n+3}$ contains single TlO layers. The question which is usefully asked where, is whether the Tl/O sheets are involved in charge transfer, especially since the Tl 6s orbital is quite deep lying. If we assume that the oxidation states of Tl^{3+}, Ba^{2+}, Ca^{2+}, and O^{2-} are as indicated, then the average oxidation state of the stoichiometric "12" series is simply $(2n+1)/n$ ($=300$, 2.50 and 2.33 for $n = 1, 2, 3$) but 2.0 for the stoichiometric "22" series. Thus the latter, to be metals, need to introduce holes, either by some type of nonstoichiometry or by the overlap of the Tl 6s bands with the Fermi Level of the CuO_2 planes. The results of band structure calculations on these materials [24] are shown in pictorial form in Fig. 21. For the double layers there is indeed band overlap so that this mechanism of hole generation is indeed open to these materials. For the single layers the bottom of the Tl band lies at too high an energy for this to occur. The explanation for this is of exactly the same type as that for the generation of the energy levels of square planar chains of various lengths as shown in the next section in Figs. 29, 30 for the $YBa_2Cu_3O_{7-\delta}$ system. As the number of building blocks in a given direction increases so the spread of energy levels increases until it reaches that expected for the infinite system. Thus the bandwidth for the one layer slab is smaller than that for the two layer. Of course, although the electronic features we have described are on firm ground, it is a rather different matter to claim that this is the only mechanism for the generation of holes in the CuO_2 planes in these systems. As noted above the quality of the crystallographic refinement which is possible for the Tl and Bi containing materials often leaves a lot to be desired, especially in terms of the oxygen content.

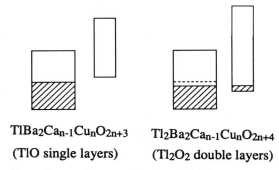

$TlBa_2Ca_{n-1}Cu_nO_{2n+3}$

(TlO single layers)

$Tl_2Ba_2Ca_{n-1}Cu_nO_{2n+4}$

(Tl_2O_2 double layers)

Figure 21. The effect of reservoir layer thickness on band overlap in $Tl_2Ba_2Ca_{n-1}Cu_nO_{2n+4}$ (Tl_2O_2 double layers) and $TlBa_2Ca_{n-1}Cu_nO_{2n+3}$ (single TlO layers).

E.   The $Bi_2Sr_2CaCu_2O_8$ System

This system was mentioned earlier as one example of the importance of local geometry on the location of energy bands. Experimentally it is found that the BiO layers in this material are quite distorted. We described some of the experimental evidence above. The local geometry around bismuth, rather than being octahedral with six long Bi–O distances, has in fact distorted to one where the distances are much shorter. Figure 22 shows three possibilities. The first still has some long Bi–O distances in the plane but the island and ladder structures do not. The figure also shows the calculated [23] energies of the bands of these units relative to that of a CuO_2 sheet. Clearly, as soon as the distortion takes place charge transfer via band overlap is switched off.

F.   The $YBa_2Cu_3O_{7-\delta}$ System

This compound was the second high temperature superconductor to be discovered and the first with a T_c above liquid nitrogen temperature. It has a wealth [37, 38] of interesting structural features which control its properties, and these are intimately connected with its wide range of stoichiometry, $0 < \delta < 1$. Even though there are many structural problems which are still unsolved in this system, our overall level of knowledge concerning the geometrical structure of these systems, especially their variation with δ, is much higher than for the bismuth and thallium containing systems. We shall spend more time on this system than on the others as a result.

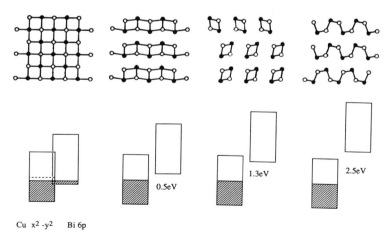

Cu x^2-y^2 Bi 6p

Figure 22.   Effect of reservoir distortion on band overlap in the $Bi_2Sr_2CaCu_2O_8$ system.

When $\delta = 0$ the compound is a superconductor, but when $\delta = 1$ it is a semiconductor, and in between its properties vary as shown in Fig. 23. At around $\delta = 0$ a prominent feature is the "90 K plateau" and at around $\delta = 0.5$ there is a similar "60 K plateau". Close to $\delta = 0.65$ there is a precipitous drop of T_c to zero. At around this point there is also a structural change, the material becomes tetragonal. The structure of this fascinating material is shown in Fig. 12. Unlike most of the other high T_c materials which are dubbed "perovskite-like", this compound is in fact a defect perovskite. Its formula, $YBa_2Cu_3O_{7-\delta}$, when rewritten as $(YBa_2)(Cu_3)O_{7-\delta}$, may be readily compared with the $(A_3)(B_3)O_9$ appropriate for a triple formula perovskite. The size difference between Ba^{2+} and Y^{3+} means that these ions order strongly on the A sites and (unlike the $La_{2-x}Sr_xCuO_4$ system) solid solutions of Ba and Y are not known. Clearly $(2 + \delta)/9$ of the oxygen atoms of the parent perovskite structure are missing. When $\delta = 0$ this deficit of oxygen atoms has led to the generation of square pyramidal and square planar copper coordination in the ratio of $2:1$. In fact one of the striking structural features when $\delta = 0$ is the presence of CuO_3 chains of such square planar units running along the b-axis of an orthorhombic unit cell. These chains are linked to the CuO_2 planes by a long bond $Cu(2)-O(4)$ which is around 2.30 Å. The variation in some of the relevant distances with δ are shown [38] in Fig.

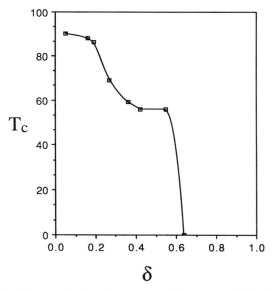

Figure 23. Variation in T_c with oxygen stoichiometry in $YBa_2Cu_3O_{7-\delta}$.

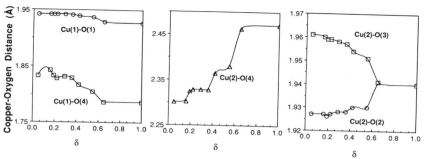

Figure 24. Variation in some of the Cu–O distances with oxygen stoichiometry in $YBa_2Cu_3O_{7-\delta}$ (the figures are taken from [38]). The following atom labels are used (see Fig. 12); Cu(1) in the chains, Cu(2) in the planes, O(1) forms the chains (this site is full at $\delta = 0$ and empty for $\delta = 1$), O(4) connects Cu(1) and Cu(2), O(2) and O(3) are in the planes, O(5) is the site, usually vacant, between the chains.

24. As δ increases from 0, oxygen atoms are lost from the chains so that at $\delta = 1$ the interplanar units are CuO_2 dumbells containing two-coordinate copper. The presence of the chains in the structure sets up an orthorhombic strain in the material and it is therefore not surprising that at some critical value of δ, corresponding to a critical chain concentration, there is a transition to a tetragonal structure where the oxygen atoms are apparently disordered amongst the two sites (0, 1/2, 0) and (1/2, 0, 0). These are the O(1) and O(5) sites of Fig. 12. The O(5) site is the one located between the copper atoms of the chains present at the $\delta = 1$ stoichiometry. A second view of the structure is shown in Fig. 25. We use the word apparently here because this structural information comes from diffraction studies, sensitive only to an averaged picture. It is quite possible from the data to have short chain lengths along both a and b directions within the same layer, or indeed longer chains which alternate along a and b directions on moving up c. The actual microscopic description of the tetragonal phase will difficult to really establish, but the degree of local ordering is probably of great importance in controlling structural coherence vital for superconductivity. At around this stoichiometry in some samples there is a sharp change in some of the Cu–O distances along the c direction, an event termed [38, 39] the c-axis anomaly (Fig. 24). It is not clear yet whether these two structural changes are related. In samples where $\delta \neq 0$ it appears that the chains do not always run in the same direction. The site between the chains at (1/2, 0, 0) (labeled O(5) and shown dotted in Fig. 12) is often occupied, the degree of occupancy depending of the method of synthesis. Where the synthesis is a high temperature one [37], this site is invariably occupied.

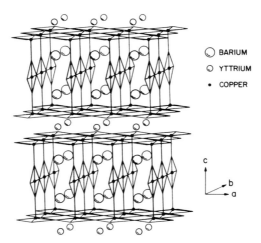

Figure 25.   A second picture of $YBa_2Cu_3O_{7-\delta}$ (compare with Fig. 12). (From [6] with permission.)

In a lower temperature synthetic route using a zirconium getter to remove oxygen [38], this site is empty. Thus we conclude that the thermo-dynamically more stable arrangement is the one where the chains run in the same direction and do not turn corners.

The assignment of valency and oxidation state in the freshman chemis-try sense to these systems is often a problem, since as we have noted, electron transfer between units is sometimes a vital feature of the electronic picture. But it will be useful to make comparisons with the mode of copper coordination seen in other solids with that found in this system. Square planar coordination is found in copper crystal chemistry for both Cu(II) and Cu(III) but square pyramidal geometries, always with long apical bonds, only for Cu(II). Using these observations, the formula $YBa_2(Cu(II)O_2)_2(Cu(III)O_3)$ is a chemically satisfying way to describe the $\delta = 0$ compound, reflecting the 2:1 ratio of copper atoms in square pyramidal geometries to those in square planar geometries in the crystal. In a similar way, the formula $YBa_2(Cu(II)O_2)_2(Cu(I)O_2)$ is a chemically satisfying way to describe the $\delta = 1$ compound, in turn reflect-ing the 2:1 ratio of copper atoms in square pyramidal geometries to those in linear, two coordinate dumbell geometries in the crystal. As a very crude approximation between these two extremes we could imagine a mixture of four and two coordinate geometries of this type, their numbers set by the oxygen stoichiometry. Domains of ordered material can be observed at many stoichiometries by electron microscopy and by electron diffraction, but no long range order is found by neutron diffraction. At

$\delta = 0.5$ is there clear evidence from electron diffraction experiments [38] of long range ordering in all three directions associated with the structure (labeled Ortho II to distinguish it from the Ortho I structure for the $\delta = 0$ stoichiometry) where complete rows of chains alternate along the a direction with complete rows of dumbells. Our electronic description however, for both the $\delta = 0$ and $\delta = 1$ compounds, would predict that both are insulators since a half-filled planar Cu(II) situation is found for both. As noted above such an arrangement is expected to localize the unpaired electrons at copper. Although this is the correct description indeed for the $\delta = 1$ system it is the incorrect one for the $\delta = 0$ material. It is a superconductor, becoming a metal by band overlap.

In order to understand the electronic structure of this material, especially the location of the relevant levels of the square plane and dumbells present at $\delta = 0$, 1, respectively, we need at least a qualitative feel for the d levels of the range of geometries we are likely to encounter when oxygen atoms are removed from the chains. Some calculated values are shown in Fig. 26. We will make use of the relative energetic placement of the energy levels rather than any absolute values. In this light, notice that isolated square planar, square pyramidal, and octahedral

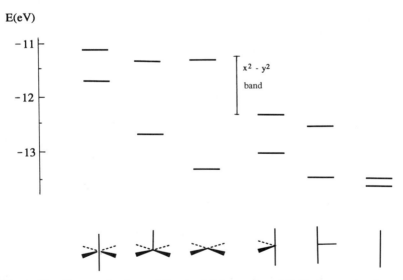

Figure 26. Computed values of the d orbital energies of CuO_x fragments of varying shapes and stoichiometry, compared with the location of the $x^2 - y^2$ band of the planar CuO_2 system. The distances used for the fragments are those appropriate for the interplanar region.

units give rise to energy levels about the calculated for that for the square pyramids appropriate for the planes, but the two (dumbell), three (T-shape) and four-coordinate (butterfly) structures lead to deep-lying levels. The last two are ideal geometries in the sense that the butterfly structure is expected to relax towards a tetrahedron and the T-shape towards a trigonal plane. Importantly for our discussion, the levels of both of these units and those of the two-coordinate dumbell lie deeper than those for the $x^2 - y^2$ levels of the planar copper atoms. These energy levels will move of course as the relevant internuclear distances change but the picture remains a useful guide. The results imply that all three geometries, butterfly (tetrahedral), T-shape (trigonal planar) and the dumbell structure should be identified with Cu(I) in the systems studied here since the plane $x^2 - y^2$ levels are approximately half full and these atoms are regarded as Cu(II). For the dumbell geometry there is substantial crystal chemical evidence for this view. Linear two-coordination is prevalent in Cu(I) chemistry, is found in cuprite, Cu_2O, $YBa_2(Cu(II)O_2)_2(Cu(I)O_2)$, and the copper delafossites, $CuMO_2$. There is an excellent review in [40]. There are several distorted trigonal planar geometries known for Cu(I), some of which are close to the T-shape (see, for example [41]) but the T-shaped geometry has been described [42] as an "abnormal" one for Cu/O. Tetrahedral copper coordination is found in both Cu(II) and Cu(I) chemistry, but the ideal Cu(II) geometry is Jahn–Teller unstable. Square planar geometries are prevalent for Cu(II) and Cu(III) but never Cu(I) as noted above.

Using these ideas, the band structure of the material is developed in Fig. 27. The $x^2 - y^2$ band of the square planes arises in the same fashion as shown in Fig. 6. The long axial distance means that the z^2 orbital lies deep in energy and is of no concern of us (in a first approximation at least). For the square planar copper atoms of the chains then an analogous $x^2 - y^2$ band may be constructed. However since these chains run in the yz plane the relevant label for this orbital is now $z^2 - y^2$. In terms of the energy bands of the solid description of the $\delta = 1$ system as $YBa_2(Cu(II)O_2)_2(Cu(III)O_3)$ forces the relative locations of the $x^2 - y^2$ and $z^2 - y^2$ bands of the square pyramids and the planes to be as shown in Fig. 27a. (We shall call these the valence bands of the system.) The $z^2 - y^2$ band corresponding to the square planes is empty (Cu(III) and the $x^2 - y^2$ band is half full (Cu(II)). However, from calculation, the two bands overlap (Fig. 27b) to such an extent that plane-to-chain transfer occurs. Thus the role of the interplanar CuO_3 unit is to remove electron density from the planes to give a result similar to the substitution of Sr for La in the La_2CuO_4 system. The details of the local geometry at the two copper atoms are crucial in controlling this charge transfer. Imagine what

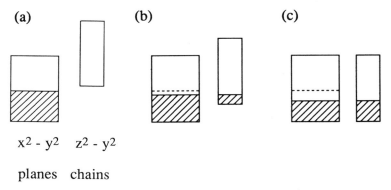

(a)   (b)   (c)

$x^2 - y^2$ $z^2 - y^2$

planes   chains

Figure 27. Some possibilities for the relative location of the $x^2 - y^2$ and $z^2 - y^2$ bands of $YBa_2(Cu_3O_7)$. (a) Ensuring that the $z^2 - y^2$ band is empty, that is, forcing the description $YBa_2(Cu(II)O_2)_2(Cu(III)O_3)$. (b) The actual state of affairs. (c) The approximate arrangement if the Cu–O distances in the planes and chains were equal.

would happen if the square planar bond lengths and the in-plane square pyramidal bond lengths were the same. The $x^2 - y^2$ and $z^2 - y^2$ bands, while being of different widths would be very similarly located energetically as in Fig. 27c and all three copper atoms would carry the same number of electrons. This would lead on the crudest model to 0.33 holes on each planar copper atom. However the bond lengths in the planes (1.927 Å (twice) and 1.961 Å (twice) with an average of 1.944 Å) are considerably longer than those in the chains 1.833 Å (twice) and 1.942 Å (twice) with an average of 1.888 Å). The result is a strong destabilization of the $z^2 - y^2$ band relative to $x^2 - y^2$ but still with some overlap (Fig. 27b). The electron transfer which results is vital in generating a less than half full plane band, an essential ingredient for the generation of a metal.

How then might we expect [43] the extent of plane chain transfer to vary with δ? At some critical doping level the system will become unstable with respect to electron localization at copper and a semiconductor will result. It is not clear at present whether this occurs at the point where the orthorhombic to tetragonal transition takes place or whether the c-axis anomaly is a result of such a localization process. But how does the broad picture of plane-to-chain charge transfer change with stoichiometry? On the simplest model, allowing square planes and dumbells only, the band structure will be as shown in Fig. 28a. Importantly the Cu(I) levels for the dumbells lie deeper in energy than either the $z^2 - y^2$ and $x^2 - y^2$ bands of the chains and planes and are always filled. With reference to Figs. 28b, d, c, where the dumbell levels are left off the diagram for clarity there are obviously twice as many levels in the $z^2 - y^2$

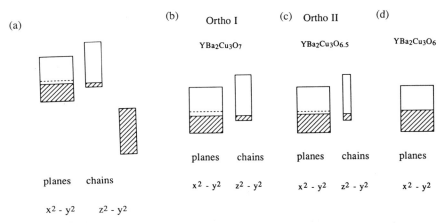

Figure 28. The relative location of the energy bands. (*a*) For an arbitrary value of δ showing the location of the $x^2 - y^2$, $z^2 - y^2$, and z^2 bands of the square planes, chains, and dumbells respectively. (*b*)–(*d*) The arrangement for $\delta = 0$ (Ortho I), $\delta = 0.5$ (Ortho II) and $\delta = 1$ stoichiometries. The number of levels in each band is indicated qualitatively by the width of the box.

band in the so-called Ortho I structure (of stoichiometry $YBa_2Cu_3O_7$) than in the idealized Ortho II structure (of stoichiometry $YBa_2Cu_3O_{6.5}$) where every other chain contains two-coordinate rather than four-coordinate copper atoms. At $\delta = 1.0$ (Fig. 28*d*) there are no chain levels to be involved. On this simple model, the total number of electrons in the bands shown in (*b*) − (*d*) are the same. Thus, removal of oxygen is expected to lead to a dependence, decreasing linearly with δ, of the extent of plane-to-chain transfer associated with the monotonic decrease in the number of $z^2 - y^2$ levels, and a corresponding decrease in the number of holes in the planes. This however, does not take into account the way the oxygen atoms are ordered.

We know that there is a strong tendency for the oxygen atoms in the interplanar region to order into chains. Experimentally the data are particularly strong at the $\delta = 0$ and $\delta = 0.5$ stoichiometries corresponding to the Ortho I and II structures, respectively. However, what are the charge transfer possibilities for a system where the interplanar region is composed of a completely disordered arrangement of square planar copper atoms? (We note that this is geometrically impossible but will not dwell on the matter.) Figure 29 shows the result of calculations [43] on chain fragments of various lengths. The $z^2 - y^2$ orbital of each isolated square planar copper atom is strongly antibonding between copper and oxygen 2p and thus its energy lies in the middle of the $z^2 - y^2$ *band* of the

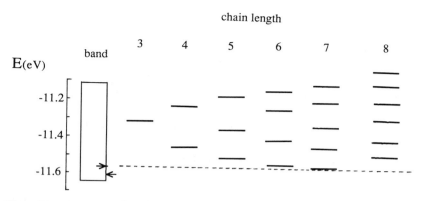

Figure 29. Calculated $z^2 - y^2$ levels for chains of various lengths compared to the width of the $z^2 - y^2$ energy band for the infinite system. The bond lengths for all geometries except for the chain of length 8 are those appropriate for the $\delta = 0$ stoichiometry. To highlight the dependence of the energy levels on geometry, the values for the chain of length 8 were calculated using the bond lengths appropriate for $\delta = 0.5$. The dashed line shows the Fermi level expected for the $\delta = 0$ stoichiometry assuming a planar copper valence of 2.15; the lower arrow shows the level expected for $\delta = 0.5$ where the magnitude of the electron transfer has been halved.

ordered system. Thus the levels of the collection of isolated square planes are well separated energetically from that part of the $x^2 - y^2$ plane band involved in electron transfer, and thus plane-to-chain transfer will not occur. As a result, a half-filled $x^2 - y^2$ band is found for the planar copper atoms and an antiferromagnetic insulator results. This is quite similar to the electronic situation in $REBa_2(Cu(II)O_2)_2(Ga(III)O_3)$ where the levels of Ga(III) lie too high in energy to participate in such electron transfer [44]. An increase in chain length is then associated with the gradual generation of a set of energy levels, and finally an energy band capable of charge transfer. The quantitative details will vary with the geometric details but for transfer of 0.30 electrons to the $z^2 - y^2$ band, the Fermi level lies as shown and a chain length of 8 or 9 copper atoms should be a critical length for transfer to begin. Such a mechanism may be a part of the explanation behind some recent experiments on samples which had been rapidly quenched from high temperature, and then studied at room temperature with time [45]. Very interestingly it was found that T_c for these samples increased with time. Studies by neutron diffraction showed no change in the average occupation of the sites O(1) and O(5) and was interpreted in terms of the generation of short range order, namely an increase in the average chain length. The generation with time of chains long enough to allow plane-to-chain electron transfer allows holes to develop in the planar copper $x^2 - y^2$ band. Support for the idea that T_c reflects the plane-chain transfer and planar copper valence comes from calculation of the bond valence sum at these atoms. The plot [38], shown in Fig. 30, has a strong resemblance to that of Fig. 23. However the rapid drop at around $\delta = 0.6$ is largely associated with the "c-axis anomaly" which shows up in Fig. 24 with the jump in the Cu(2)–O(4) distance. Since this is a bond which lies perpendicular to the CuO$_2$ planes, and is not therefore associated with orbital interactions with the $x^2 - y^2$ orbital, it is difficult to suggest an explanation for such a change in copper valence. This is a point which has been noted by others [33].

There is also some support for such an idea of a critical chain length from the results [46] shown in Fig. 31. These authors used calculated interatomic potentials corresponding to those shown in Fig. 32 to model the local geometry in the YBa$_2$Cu$_3$O$_{7-\delta}$ system by Monte Carlo methods. They then found that the variation of T_c with δ followed the simple expression $n_1 93 + n_{11} 58$ where n_1 and n_{11} are the fractions of oxygen atoms in clusters of the solid which have the local geometry of the Ortho I and Ortho II arrangement, but importantly are larger than the critical size shown in the figure (93 K and 58 K are the values appropriate for the two plateau regions where the dominant structural features are the Ortho I and Ortho II arrangements, respectively). The direct correlation between their approach and the one used here is difficult to make but it is interesting that the largest chain length in the smallest cluster of the

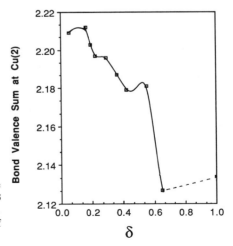

Figure 30. Variation in bond valence sum for the planar copper atoms with δ using the bond lengths reported in [38]. Notice the strong resemblance to the plot of Fig. 23.

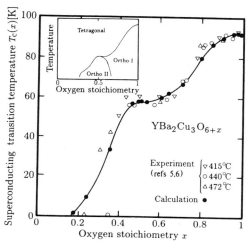

Figure 31. Calculated variation in T_c with δ using the expression $n_I\,93 + n_{II}\,58$ where n_I and n_{II} are the fractions of oxygen atoms in clusters of the solid which have the local geometry of the Ortho I and Ortho II arrangement, but are larger than the critical size shown in Fig. 32, found from the results of Monte Carlo simulations (from [46] with permission).

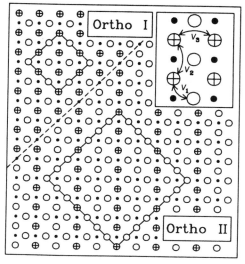

Figure 32. The three potential parameters which go into the Monte Carlo simulations of [46], and the minimum size clusters which need to be chosen (arbitrarily) to provide the fit to the experimental data shown in Fig. 31 (from [46] with permission).

Ortho II arrangement contains eight oxygen atoms and therefore nine copper atoms. Of course the experimental data are fit to the model. The critical size clusters are those which need to be chosen to reproduce the observed plateau widths.

However let us move to a different model and investigate in more detail the electronic consequences of removing a single oxygen atom from a copper oxide chain as in Fig. 33a. The result is the generation of two T-shaped copper atoms (labeled A, B). Now, since the results of calculations on long chain fragments set the d levels for these end copper atoms close to what is reported in Fig. 26, they should be viewed as Cu(I). Simple electron counting [47] allows an estimate of the electronic demands on the planar structure. On removal of oxygen two levels from the $z^2 - y^2$ band of the chains drop in energy below the bottom of the $x^2 - y^2$ band of the planes and are filled with two pairs of electrons. Since one oxygen atom is lost from the solid, two extra electrons are needed to fill these two levels. Electrons are thus removed from the valence bands shown in Fig. 28 and placed in deeper-lying ones appropriate for the T-shape. Overall, the presence of isolated oxygen vacancies in a CuO_3 chain then leads to the generation of extra holes in the planes. If random loss of oxygen atoms continues from the chains, T-shaped copper atoms continue to be generated as in Fig. 33a, each time with the removal of two more electrons from the valence band and more holes created in the planes. If however a second oxygen atom is lost in a site adjacent to the first (Fig. 33b) then the electronic demands are quite different. Now there are three Cu(I) atoms (two T-shaped units (A, C) and one dumbell (B)). With two oxygen atoms lost, the overall demand on the planes is the same as in Fig. 33a, so that the second oxygen atom is lost without any change in the number of electrons in the valence bands. Energetically the presence of these T-shape geometries is unfavorable, from both

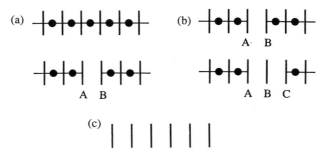

Figure 33.   Vacancy patterns for the chains.

crystal chemical considerations and also from calculations. So the process in which oxygen atoms are lost from existing "nucleation" sites in the crystal is more likely than one where oxygen atoms are lost randomly. This result may be extended to the "empty" chain (Fig. 33c), where there are an equal number of two-coordinate Cu(I) dumbells and oxygen vacancies. Here the charge balance is perfect and there is no extra demand for electrons from the plane levels (beyond that from the effect shown in Figs. 28b, c, d). However what happens when the chain lengths become short? As seen in Figs. 29, 30, the energy levels associated with these fragments may not be in the energy region required for plane-to-chain electron transfer to occur. From these calculations the chain has to contain at least eight or nine copper atoms before it becomes effective in plane-to-chain transfer via band overlap. This figure is obviously open to modification since the location of the energy bands and levels is sensitive to the Cu–O distance which will change gradually with δ. Assuming that half as many electrons are transferred from the plane-to-chain when $\delta = 0.5$, the size of the critical unit is even larger.

As should now be clear, the details of the plane-to-chain transfer will depend quite strongly on the local geometry at copper and the short and long range ordering in the chains [43]. On the simplest model, however, the number of holes in the $x^2 - y^2$ band will just respond to the loss of levels in the $z^2 - y^2$ band as shown in Fig. 28 in a linear way. Since the Cu(1)–O(4) distance decreases quite sharply with δ this process will be somewhat accelerated as a result of the raising in energy of the $z^2 - y^2$ band. At some critical value of δ, either electron transfer will be switched off entirely as a result of this change in geometry (such that there are no holes in the plane band), or the hole concentration in the planes reaches a critical value which leads to the metallic state being untenable, as we suggested above. In each case a nonmetallic state is produced. This may or may not be associated with the orthorhombic to tetragonal transition but the "c-axis anomaly" (Fig. 24), the apparently sharp change in Cu(1)–O(4) distance in some samples, may well be the result of localization of this type. Assuming that T_c depends directly on the number of holes in the planes, the variation of T_c we expect to see with δ on this, the simplest model is shown in Fig. 34a. Modifications to this picture depend crucially on how the oxygen atoms are removed.

At $\delta = 0$ all the chains are full. As δ increases we expect to find initially that the oxygen removed comes from a variety of isolated sites in the chains. This initial loss of oxygen, as in Fig. 33a, leads to the introduction of extra holes in the planes beyond that expected via the gradual loss of $z^2 - y^2$ levels shown in Fig. 28, but the generation of a completely empty chain (Fig. 33c) does not. Such a process introduces

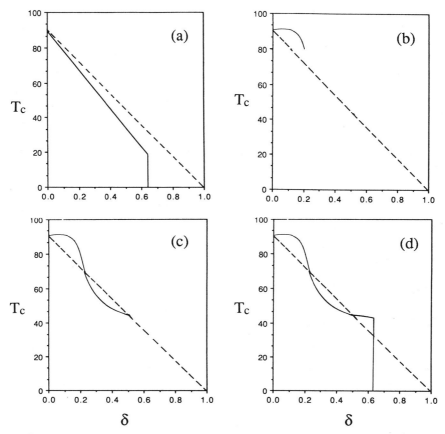

Figure 34. Generation of the plot showing variation in T_c with δ using a microscopic model. (*a*) Behavior expected from Figs. 28*b*, *c*, *d* and localization at $\delta \sim 0.6$. (*b*) Generation of excess holes at low δ as a result of the presence of T-shapes (Fig. 33*a*). (*c*) As δ increases a smaller number of holes than expected are generated because the chain lengths are too short to take part in plane-chain transfer (Fig. 29). (*d*) As δ increases past 0.5 the same behavior as that found at low δ starts to occur before localization destroys superconductivity.

extra holes into the planes (Fig. 34*b*), and is responsible for the 90 K plateau of Fig. 23. As more oxygen is removed then these nucleation sites start to grow and the chain length decreases. Energetically this process is driven by the instability associated with the T-shape geometry. We noted above the strong evidence from electron diffraction of the medium range ordering of the Ortho II structure close to $\delta = 0.5$ (not seen in neutron diffraction) for the zirconium gettered samples. Before this stage is reached however, the influence of chain length on the charge transfer process (Fig. 30) has become apparent. From Fig. 29 since there is a critical size before such units may act as reservoirs of charge, the number of holes in the planes will be smaller than expected. The solid line in Fig. 34c has now dropped below that of Fig. 34a (shown dashed in this figure), and the 60 K plateau has started to occur. We presume that the whole process which started at $\delta = 0$ and continued to $\delta = 0.5$ will repeat itself, and thus for a short while the plateau continues for $\delta > 0.5$ (Fig. 34d). Thus there are two factors determining the 60 K plateau depending on whether $\delta < 0.5$ or $\delta > 0.5$.

IV. PHASE SEPARATION

A. The $La_2CuO_{4+\delta}$ System

Oxygen rich La_2CuO_4 (i.e., $La_2CuO_{4+\delta}$ with $\delta > 0$) is a superconductor with a T_c around 35 K, only the stoichiometric material with $\delta = 0$ being an insulator. However, at room temperature and below, this material is not phase pure [27]. As shown in Fig. 35, at around this temperature a phase separation occurs [28] of the form $La_2CuO_{4+\delta} \rightarrow La_2CuO_4 + La_2CuO_{4.07}$, the proportions of the two phases being set by the value of δ. Importantly it is the $O_{4.07}$ material which is the superconductor, with a hole concentration and T_c very similar indeed to the optimum values for the $La_{2-x}Sr_xCuO_4$ series. This is chemically a fascinating process, associated with oxygen atom movement at room temperature, found also of course in the experiments described above for the $YBa_2Cu_3O_{7-\delta}$ system [50]. The structure of this oxygen rich material has been difficult to establish but is probably analogous to that of the corresponding nickel compound in Fig. 13.

Several important questions come to mind stimulated by these observations. None of them have found really satisfactory answers. What is the energetic driving force for the separation? At first sight, bringing together a collection of oxygen ions should be energetically unfavorable on a simple ionic model. Also, if the usual quadratic dependence of copper d ionization potential on charge is assumed, then all disproportionation reactions are disfavored, as described earlier for the Pt(III) system in Fig. 1. One asks too therefore whether there is some special stability associated with the $O_{4.07}$ structure conferred by the electronic configuration at copper. (In Fig. 36 we show the phase separation schematically but remind the reader that for La_2CuO_4 itself the band model is inapplicable.) This is an appropriate question to ask since T_c is very

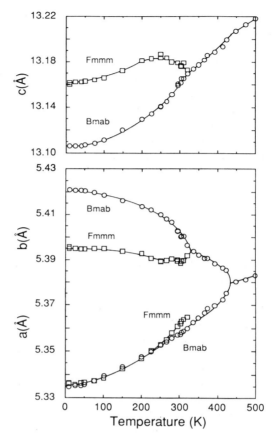

Figure 35. Results of neutron diffraction studies on the phase separation $La_2CuO_{4+\delta} \rightarrow La_2CuO_4 + La_2CuO_{4.07}$. The temperature of the orthorhombic/tetragonal transition at around 420 K is sensitive to the oxygen concentration. The phase separation into two orthorhombic phases which occurs close to room temperature is not (from [27] with permission).

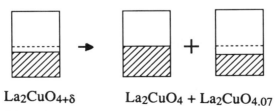

$La_2CuO_{4+\delta}$ $La_2CuO_4 + La_2CuO_{4.07}$

Figure 36. Schematic showing the phase separation $La_2CuO_{4+\delta} \rightarrow La_2CuO_4 + La_2CuO_{4.07}$.

similar indeed to the maximum value in the $La_{2-x}Sr_xCuO_4$ series which occurs close to $x = 0.15$. It also brings up an important question concerning the connection between superconductivity and structural stability. Many existing low T_c superconductors appear to lie close to a structural instability. Indeed the competition between superconductivity and CDW instabilities is a well established one. By contrast some of the high temperature superconductors appear to be generated by some sort of rearrangement leading to a more stable structure. This is true also in the $YBa_2Cu_3O_{7-\delta}$ system where T_c increases with time for the rapidly quenched samples when allowed to stand at room temperature. In other words the more stable structure is associated with the higher T_c. We shall see another system of this type below.

Of some interest is the mobility of oxygen in the structure which triggers the phase separation at room temperature. One can ask whether a similar effect should occur in principle with other systems. Specifically for the $La_{2-x}Sr_xCuO_4$ series for example, are the series of solid solutions with different x metastable, the diffusion of metal ions at a low enough temperature foiled by the rigidity of the lattice? Is the thermodynamically stable structure, one with a single composition? This is a viewpoint which has attracted considerable attention (see, for example, [48]). These are important chemical questions and ones to which we have very few answers at the present time.

B.   The Electron-Doped $RE_{2-x}M(IV)_xCuO_4$ System

The majority of the high T_c materials are hole-doped but two systems, $RE_{2-x}M(IV)_xCuO_4$ [49] and $RE_2CuO_{4-x}F_x$ [50], are known where formally there is electron transfer to rather than from the CuO_2 planes. M may be Ce or Th. The fact that similar behavior seems to occur for both $RE_{2-x}M(IV)_xCuO_4$ and $RE_2CuO_{4-x}F_x$ suggests that the structural and electronic properties are dominated by the demands of the electron-doped CuO_2 sheets. The dimensions of these sheets increase on addition of electrons for the oxide material as shown in Fig. 14. There is an optimal Cu–O distance for superconductivity [31], similar to the behavior shown in Fig. 16, but the details of the structure are certainly much more complex than appear at first sight. Figure 37 shows [32] the dependence of T_c and the superconducting fraction on x for $Nd_{2-x}Ce_xCuO_4$. Notice the rather narrow range of stoichiometry for which the material is a superconductor. Cerium is insoluble for $x > 0.2$. Figure 38 shows that in fact the material is much more complex [32]. These powder neutron diffraction results indicate the presence of two phases with slightly different axial ratios. The presence of more than one phase is also indicated by electron diffraction, electron microscopy, and NMR experi-

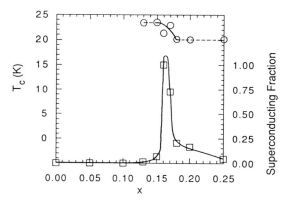

Figure 37. Variation of T_c and the superconducting fraction with x in $Nd_{2-x}Ce_xCuO_4$ (from [32] with permission).

ments [52]. An extremely interesting observation is that the c/a ratio of one of the phases varies with x and that at the superconducting composition, the two phases are indistinguishable. A plot of the volume fraction of the second phase is shown in Fig. 39, a picture with immediate similarities to Fig. 37. The conclusion is a very striking one. There are two phases, one of which is a superconductor. Thus $Nd_{2-x}Ce_xCuO_4 \rightarrow a + b$, the proportions of the two varying with x. In contrast to the oxygen-rich $La_2CuO_{4+\delta}$ system above, we have very little detailed information concerning the difference between the two phases,

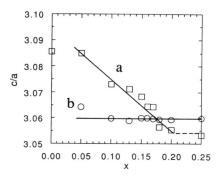

Figure 38. Variation with x in the c/a ratios of the two phases determined from neutron diffraction in $Nd_{2-x}Ce_xCuO_4$ (from [32] with permission).

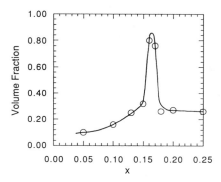

Figure 39. Variation with x in the fraction of phase b in $Nd_{2-x}Ce_xCuO_4$ determined from neutron diffraction (from [32] with permission).

including their stoichiometry. For this system, NMR gives us very useful information concerning the local environment at copper [52]. Figure 40 shows how the spectrum changes with x. Notice that there are three resonances. A, identified with the square planar copper environment for $x = 0$, changes its frequency and loses intensity as x increases. Concurrently resonance B, of fixed frequency as x changes, increases in intensity. On the simplest model by comparison with Fig. 38 we would associate resonance A with phase a, which changes structure on doping and resonance B with phase b, which does not. We have interpreted [53] the behavior of phase a with a geometry change at copper, moving from a square planar towards a tetrahedral structure as x increases. Addition of electrons to square planar Cu(II) is expected to eventually lead to a tetrahedral (or perhaps dumbell) geometry typical for Cu(I). Thus, whereas the oxygen ordering process formed the basis for the phase separation in $La_2CuO_{4+\delta}$, here the structural change may be the bending of the CuO_2 sheets towards a tetrahedral geometry at copper. How this occurs in detail is not clear; presently available diffraction data do not allow generation of a detailed structural picture. The results do suggest however, that there is a specific geometry (that corresponding to $x = 0.15$) which is stable, identified with the NMR peak B, in a similar way to the $O_{4.07}$ structure (Fig. 36) in $La_2CuO_{4+\delta}$. (We do not have an assignment for the third NMR peak C which grows in at high x). We have suggested [53] that one possibility for this locally stable structure is the minimum in the Madelung energy as the oxygen atoms are moved.

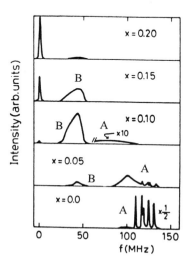

Figure 40. Variation with x of the copper NMR spectrum in $Nd_{2-x}Ce_xCuO_4$. We suggest that the NMR resonances A and B are associated with the phases a and b of Fig. 38 (from [52] with permission).

C. The $YBa_2Cu_3O_{7-\delta}$ System

This system can also be regarded as being generated via a phase separation, but in two different ways associated with two different length scales. As we have described above, the most stable arrangement for δ smaller than about 0.6 is the one where the oxygen atoms have segregated into full and empty chains. This first separation process occurs at the microscopic level, and is driven clearly by the energetic stabilization associated with minimization of the number of T-shaped units. The stability of the Ortho II structure itself may be understood in terms of interchain oxygen repulsions. In this structure where full and empty chains alternate, the oxygen atoms along the chain direction are furthest apart from their partners in neighboring chains. The second phase separation in this material is associated with a larger length scale and is concerned with the ubiquitous orthorhombic to tetragonal transition noted earlier. We noted that since the presence of the chains in the structure set up an orthorhombic strain in the material, it is not surprising that at some value of δ, corresponding to a critical chain concentration, there is a transition to a tetragonal structure. Studies [54] on $ErBa_2Cu_3O_{7-\delta}$, find two-phase behavior over the composition range $0.75 < \delta < 0.55$, a tetragonal phase and an orthorhombic phase, presumed to be Ortho II. One may wonder whether a part of the driving force for this process is localization within the tetragonal phase. In this light it is useful to note several recent ideas (see [55] and refs. therein) concerning such a phase separation by study of the energetics demanded by an extended Hubbard Hamiltonian. It has been shown simply by consideration of the functional behavior of the U parameters for both oxygen p and copper d, this d-p energy separation, the corresponding hopping integral, β, and V the nearest neighbor copper oxygen repulsion (an intersite parameter of the same form as U) that there is a region of the phase diagram, close to the metal–insulator transition, and close to where pairing mechanisms can be established, where the system is unstable towards phase separation (i.e., the process of Fig. 36 for $La_2CuO_{4+\delta}$). Once again the energetics of the structural problem are determined by a subtle balance between one- and the various types of two-electron energy terms. This leads then to an answer to one of the questions asked earlier. A phase separation can in principle occur, close to the metal–insulator transition and *generate a superconductor* in the process. Importantly the variation in "β" with electron count will be a crucial ingredient in the process. With reference to Fig. 36 the systems with two different electron counts will have two different geometries with different bandwidths. The structural change which is easiest to see is of course the one associated with changes in Cu–O distance as the electron concentration in the $x^2 - y^2$ band changes, but in general more dramatic effects are found. We may summarize these as follows:

1. $La_2CuO_{4+\delta}$: oxygen ordering (clustering);
2. $REBa_2Cu_3O_{7-\delta}$: chain formation by minimization of the number of T-shapes;
3. $La_{2-x}Sr_xCuO_4$: rumpling of CuO_2 sheets [56, 57];
4. $Nd_{2-x}Ce_xCuO_4$: bending of the CuO_2 plane towards a tetrahedron [53];
5. Bi, $Tl/Sr/Cu_nO_m$: distortions within rocksalt sheets.

We have [56] described the electronic origin of the orthorhombic to tetragonal phase transition in $La_{2-x}Sr_xCuO_4$ and structural transformations in $La_{2-x}Ba_xCuO_4$ have been shown [57] to be driven electronically too. The suggestion for the large series of bismuth and thallium superconductors is a speculative one, but we do note that this reservoir region is almost invariably poorly determined crystallographically.

D. A Theoretical Model

We have noted above how phase separation is possible using a three band Hubbard model with judiciously chosen parameters [55]. Here we present a more chemical picture that not only gives some insight into the nature of the superconducting state but also shows why it is so extremely sensitive to changes in doping levels. In our discussions above we have referred in general to the "$x^2 - y^2$" band in a general way without much consideration as to its orbital composition and how that might change on doping. We know of course that the band is in fact a hybrid of copper $3d$ and oxygen $2p$ orbitals. In the insulating state (undoped La_2CuO_4 for example) the wavefunction is localized in the Heitler-London sense on the metal atom. Thus the electrons lie in an orbital that is largely copper $3d$. The electronic description of the heavily doped metallic state though is quite different. This delocalized wavefunction has a large oxygen character such that the holes now lie in oxygen $2p$ orbitals. So on doping there are two electronic changes that take place simultaneously along with a shortening of the Cu-O distance and hence an increase in the overlap between copper and oxygen orbitals.

Such a change in electronic description leads us to make comparisons with a similar situation in the alkali halide molecules. There is an avoided crossing at large internuclear separation between ionic and covalent curves (Fig. 41) with a rapid change in electronic character on moving through the crossing region. We can envisage an analogous process in the high temperature superconducting cuprates. Figure 42 shows four possible situations that arise via the avoided crossing of the "oxygen" and

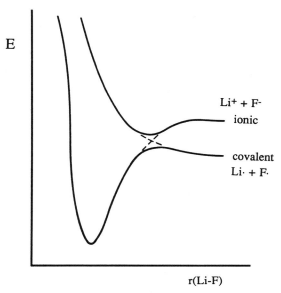

Figure 41. The avoided crossing found for the "ionic" and "covalent" curves of alkali halide molecules as a function of internuclear separation.

"copper" curves that describe the states where the holes are located on oxygen $2p$ and copper $3d$ orbitals respectively. At long distances of course such states are localized as we have described above, but at shorter distances they are delocalized. The "copper" curve lies lowest in energy at long Cu-O distances. The interaction between the two curves depends on their energy separation and the overlap between them. Since we know that the addition of electrons to a state moves it up in energy, the "oxygen" state drops and the "copper" state rises in energy as a result of doping. Figure 42 shows four plots that differ in the energy separation of the two curves at large $r(Cu-O)$. (This becomes smaller as the doping level increases.) Figure 42a shows the undoped case. The avoided crossing takes place here at large $r(Cu-O)$; the system lies in a minimum set by the copper curve with a localized electronic description. Figure 42b shows the other extreme where with heavy doping the separation between the two curves at large $r(Cu-O)$ is reversed. The electronic description at the energy minimum shows a delocalised, largely oxygen located band. Two possible intermediate cases are shown where the crossing is very important. In Fig. 42c there are two states, close in energy; the state with the shorter internuclear separation is probably metallic and the state with the longer distance is perhaps a semi-

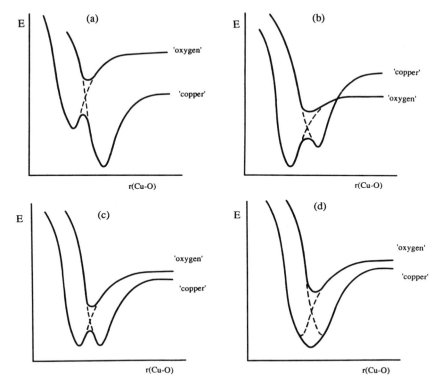

Figure 42. Four possibilities for the intersection of "copper" and "oxygen" curves in the cuprates, by analogy with the behavior in Fig. 41. (*a*) The undoped case where the energy separation at large *r*(Cu-O) is large. (*b*) The heavily doped case where the energies of the curves are reversed. (*c*) and (*d*) Two intermediate cases, which differ in the magnitude of the interaction between the two curves. In (*c*) it is small but large enough in (*d*) to generate a single minimum.

conductor. Figure 41*d* shows the most interesting case where the interaction between the two curves is so large that a single minimum results.

The properties of such a "magic" state are unique and worth exploring. There is a strong stabilization of the special electron configuration that leads to the degenerate crossing of Fig. 42*d*, which is absent at other electron counts. If this "magic" electron count is 0.15 holes, the model leads directly to a mechanism for the phase separation noted above for $La_2CuO_{4+\delta}$ and for $La_{2-x}Sr_xCuO_4$. We may readily see that the generation of the single minimum is very sensitive indeed to the location of the two diabatic curves. Small changes in their location will lead to a single well with no such stabilization. Thus small changes in the doping level or geometrical structure that change the location of the two intersecting curves can destroy it entirely. We mentioned several results above that fit into this category. One example is the case of the TlMRECuO$_5$ compounds where superconductors are found [2] for RE = La, Nd and M = Sr, but not when M = Ba, even though the systems are isostructural. Here the difference in r(Cu-O) between the two is 0.05 Å.

A final question that needs to be asked at this point is whether there are any clues as to the nature of the superconducting state from our model. The answer is an interesting one. The wavefunction at the bottom of the minimum of Fig. 41d is a rather special one. It is made up of equal contributions from the two starting states. However, small changes in the internuclear separation around this point will lead to large changes in the electronic description. In fact we could expect a substantial electron–phonon interaction from such a state of affairs. Such an electronic state is thus indeed a very interesting one. Its properties are investigated further elsewhere.

V. EPILOGUE

We have shown two mechanisms by which it has proven possible to generate a metal in these oxides—the first step towards a superconductor. However, the crystal chemistry often sets a very complex phase diagram for these systems, which means that a rational superconductor design is still a long way off. For example, doping of these materials may often be achieved by changing the chemical stoichiometry, but although this occurs in a controlled fashion in La$_{2-x}$Sr$_x$CuO$_4$, in many of the Bi and Tl containing compounds, the stoichiometry is still not well established, and it is often the presence of defects of various types that are responsible for the generation of holes in the planar copper $x^2 - y^2$ band. We can ask whether there are any rules which might give us information concerning the ease of doping these oxides. There have been two suggestions which have been made. The first [25] recognizes that high oxidation states (e.g., Cu^{3+}) are generally found when the cations in the lattice are highly electropositive ones. A requirement for hole doping is therefore the presence of such ions in the solid, a feature of all of the high temperature superconductors made to date. In fact there is a correlation [25] between T_c and the sum of the electronegativities of the cations (excluding copper). The second [58] suggestion builds in some structure dependence via a study of the Madelung potentials for a whole series of structures. There is a correlation between the ease of doping found experimentally and the Madelung potential at various sites in the crystal.

Even if a doped material is synthesized, it is not always a metal, let alone a superconductor. In this category consider the two systems [59]

$La_{1-x}Sr_{1+x}ECuO_5$ where E = aluminum or gallium. Both systems contain CuO_2 planes, but folded in a slightly different way to those in $La_{2-x}Sr_xCuO_4$. For the system where E = Al, the compound has a structure related to that of brownmillerite, can be made stoichiometric, and is an antiferromagnetic insulator with a half-filled band. For the analogous gallium compound, with the brownmillerite structure, a stoichiometric material cannot be made up as a single phase, but a compound with $x \sim 0.1$ can be. The number of holes in this compound is similar to that for superconducting $La_{2-x}Sr_xCuO_4$, yet the material is a semiconductor. A good structural determination of these materials has yet to be achieved, but we can speculate that the bond lengths are sufficiently different between $La_{1-x}Sr_{1+x}GaCuO_5$ and $La_{2-x}Sr_xCuO_4$ that the band-width is quite different in the two cases. This would lead to a $|U/\beta|$ which would be significantly different for the two systems. The presence of the AlO_3 chains will certainly lead to stiffness in one of the chain directions (the steric effect of [33]). This system brings up an interesting comparison with some of the ideas of the previous section. Synthesis of a solid of nominal composition $LaSrGaCuO_5$ leads to the generation of more than one phase, $La_{1-x}Sr_{1+x}GaCuO_5$ ($x \sim 0.1$) being one of them. In contrast though to the situation for $La_2CuO_{4+\delta}$, for example, such phase separation does not lead to the generation of a superconductor.

These comments should be sufficient to emphasize that the factors behind nature's design of a high temperature superconductor involve many principles of crystal chemistry and physics which are as yet not established. It is our hope that the intense study of this relatively small area of chemistry will lead to the discovery and understanding of a larger whole.

Acknowledgments

This research was supported by the National Science Foundation under NSF DMR-8809854. I would like to thank C. C. Torardi, D. G. Hinks, J. D. Jorgensen, K. Levin, K. Poeppelmeier, B. W. Veal, and M.-H. Whangbo for many useful conversations, and the RER of Paris where this article was written.

References

1. H. Müller-Buschbaum, *Angew. Chem., Int. Ed.* **28**, 1472 (1989).

2. M.-H. Whangbo, M. A. Subramanian, *J. Solid State Chem.* **91**, 403 (1991).

3. A. W. Sleight, *Science*, **242**, 1519 (1988); R. J. Cava, *Science* **247**, 656 (1990).

4. J. M. Williams et al., *Acc. Chem. Res.* **21**, 1 (1988).

5. C. N. R. Rao, B. Raveau, *Acc. Chem. Res.* **22**, 106 (1989).

6. D. L. Nelson, M. S. Whittinghan, and T. F. George (eds.), *Chemistry of High-Temperature Superconductors*, ACS Symposium No. 351 (ACS, 1987); D. L. Nelson and T. F. George (ed.), No. 377 (ACS, 1988).

7. J. B. Goodenough, *Phase Transitions* **22**, 79 (1990).

8. T. A. Vanderah (ed.) *Chemistry of High-Temperature Superconductors* (Noyes, 1991).

9. C. N. R. Rao, *Annu. Rev. Phys. Chem.* **40**, 291 (1989).

10. S. Hüfner, *Solid State Comm.* **74**, 969 (1990).

11. D. Vaknin et al., *Phys. Rev. Lett.* **58**, 2802 (1987).

12. K. Levin, J. H. Kim, J. P. Lu and Q. Si, *Physica C* **175**, 449 (1991).

13. J. Hubbard, *Proc. R. Soc., Ser. A* **276**, 238 (1963); **277**, 237 (1964); **281**, 401 (1964); **285**, 542 (1965); **296**, 82 (1966); **296**, 100 (1966).

14. E. Canadell and M.-H. Whangbo, *Chem. Rev.* **91**, 965 (1991).

15. J. Zaanen, G. A. Sawatzky and J. W. Allen, *Phys. Rev. Lett.* **55**, 418 (1985).

16. J. B. Torrance, P. Lacorro, C. Asavaroengchai and R. M. Metzger, *J. Solid State Chem.* **90**, 168 (1991).

17. M.-H. Whangbo, M. Evain, M. A. Beno, and J. M. Willians, *Inorg. Chem.* **26**, 1829 (1987).

18. C. C. Torardi, M. A. Subramanian, J. Gopalkrishnan, and A. W. Sleight, *Physica C* **158**, 465 (1989).

19. T. Siegrist, S. M. Zahurak, D. W. Murphy, and R. S. Roth, *Nature* **334**, 231 (1988); M. G. Smith, A. Manthiran, J. Zhou, J. B. Goodenough, and J. T. Markert, *Nature* **351**, 549 (1991); R. J. Cava, *Nature* **351**, 518 (1991).

20. C. C. Torardi, J. B. Parise, M. A. Subramanian, J. Gopalkrishnan, and A. W. Sleight, *Physica C* **157** 115 (1989).

21. J. K. Burdett and G. V. Kulkarni, *Phys. Rev. B* **40**, 8908 (1989).

22. M. S. Hybertson and L. F. Mattheiss, *Phys. Rev. Lett.* **60**, 1661 (1998); F. Herman, R. V. Kasowski, and W. Y. Hsu, *Phys. Rev. B* **38**, 204 (1988).

23. J. Ren, D. Jung, M.-H. Whangbo, J.-M. Tarascon, Y. Le Page, W. R. McKinnon, and C. C. Torardi, *Physica C* **158**, 501 (1989), **159**, 151 (1989); C. C. Torardi, D. Jung, D. B. Kang, J. Ren, and M.-H. Whangbo, *Mar. Res. Soc. Symp. Proc.* **156**, 295 (1989).

24. D. Jung, M.-H. Whangbo, N. Herron and C. C. Torardi, *Physica C* **160**, 381 (1989).

25. M. A. Subramanian, C. C. Torardi, J. Gopalkrishnan, P. L. Gai, and A. W. Sleight, in *Physics and Materials Science of High Temperature Superconductors*, R. Kossowsky et al. (eds.) (Kluwer, 1990) p 261; A. W. Sleight, M. A. Subramanian, and C. C. Torardi, *MRS Bull.* 45 (1989).

26. A. Meerschaut, C. Auriel, A. Lafond, C. Deudon, P. Gressier, and J. Rouxel, *Eur. J. Solid State Inorg. Chem.* **28**, 581 (1991).

27. J. D. Jorgensen, B. Dabrowski, S. Pei, D. G. Hinks, L. Soderholm, B. Morosin, J. E. Schirber, E. L. Venturini, and D. S. Ginley, *Phys. Rev. B* **38**, 11337 (1988).

28. J. D. Jorgensen, B. Dabrowski, D. R. Richards, and D. G. Hinks, *Phys. Rev. B* **40**, 2187 (1989).

29. M. A. Subramanian, J. Gopalkrishnan, C. C. Torardi, T. R. Askew, R. B. Flippen, A. W. Sleight, J. J. Lin, and S. J. Poon, *Science* **240**, 495 (1988).

30. C. Michel, C. Martin, M. Hervieu, A. Maignan, J. Provost, M. Huve, and B. Raveau, *J. Solid State Chem.* **96**, 271 (1992).

31. E. Wang, J.-M. Tarrascon, L. H. Greene, G. W. Hull, and W. R. McKinnon, *Phys. Rev. B* **41**, 6582 (1990).

32. P. Lightfoot, D. R. Richards, B. Dabrowski, D. G. Hinks, S. Pei, D. T. Marx, A. W. Mitchell, Y. Zheng, and J. D. Jorgensen, *Physica C* **168**, 627 (1990).

33. M.-H. Whangbo and C. C. Torardi, *Science* **249**, 1143 (1990); *Acc. Chem. Res.* **24**, 127 (1991).

34. I. D. Brown, *J. Solid State Chem.* **90**, 155 (1991); J. K. Burdett, *Chem. Rev.* **88**, 3 (1988).

35. A. K. Ganguli and M. A. Subramanian, *J. Solid State Chem.* **90**, 382 (1991).

36. P. Lacorre, J. B. Torrance, J. Pannetier, A. Nazzal, P. W. Wang, and T. C. Huang, *J. Solid State Chem.* **91**, 225 (1991).

37. J. D. Jorgensen, M. A. Beno, D. B. Hinks, L. Soderholm, K. J. Volin, R. L. Hitterman, J. D. Grace, I. K. Schuller, C. U. Segre, K. Zhang, and M. S. Kleefisch, *Phys. Rev. B* **36**, 3608 (1987).

38. R. J. Cava, A. W. Hewat, E. A. Hewat, B. Batlogg, M. Marezio, K. M. Rabe, J. J. Krajewski, W. F. Peck, and L. W. Rupp, *Physica C* **165**, 419 (1990).

39. A. Renault, J. K. Burdett, and J.-P. Pouget, *J. Solid State Chem.* **71**, 587 (1987).

40. H. Müller-Buschbaum, *Angew. Chem., Int. Ed.* **30**, 723 (1991).

41. J. F. Riehl, I. El-Idrissi Rachidi, Y. Jean, and M. Pelissier, *New J. Chem.* **15**, 239 (1991).

42. B. Raveau, C. Michel, M. Hervieu, and J. Provost, *Rev. Solid State Sci.* **2**, 115 (1988).

43. J. K. Burdett, *Physica C* **191**, 282 (1992).

44. J. T. Vaughey, J. P. Thiel, E. R. Hasty, D. A. Groenke, C. L. Stern, and K. Poeppelmeier, *Chem. Materials* **3**, 935 (1991).

45. B. W. Veal, H. You, A. P. Paulikas, H. Shi, Y. Fang, and J. W. Downey, *Phys. Rev. B* **42**, 4770 (1990); H. Claus, S. Yang, A. P. Paulikas, J. W. Downey, and B. W. Veal, *Physica C* **171**, 205 (1990); J. D. Jorgensen, S. Pei, P. Lightfoot, H. Shi, A. P. Paulikas, and B. W. Veal, *Physica C* **167**, 571 (1990).

46. H. F. Poulsen, N. A. Andersen, J. V. Andersen, H. Bohr, and O. G. Mouritsen, *Nature* **349**, 594 (1991); J. D. Jorgensen, *Nature* **349**, 565 (1991).

47. J. K. Burdett, G. V. Kulkarni, K. Levin, *Inorg. Chem.* **26**, 3650 (1987).

48. J. B. Torrance, A. Bezinge, A. I. Nazzal, and S. S. P. Parkin, *Physica C* **162**, 291 (1989).

49. Y. Tokura, H. Takagi, S. Uchida, *Nature* **337**, 345 (1989).

50. A. C. W. P. James, S. M. Zahurak, and D. W. Murphy, *Nature* **338**, 240 (1989).

51. C. H. Chen, D. J. Werder, A. C. W. P. James, D. W. Murphy, S. M. Zahurak, R. M. Fleming, B. Batlogg, and L. F. Schneemyer, *Physica C* **160**, 375 (1989); F. Izumi, Y. Matsui, H. Takagi, S. Uchida, Y. Tokura, and H. Asano, *Physica C* **168**, 627 (1990).

52. M. Abe, K. Kumagai, S. Awagi, and T. Fujita, *Physica C* **160**, 8 (1989).

53. N. C. Baird and J. K. Burdett, *Physica C* **168**, 637 (1990).

54. J. D. Jorgensen, D. G. Hinks, P. G. Radaelli, S. Pei, P. Lightfoot, B. Dabrowski, C. U. Segre, and B. A. Hunter, *Physica C* **185–189**, 184 (1991).

55. Y. Bang, G. Kotliar, C. Castellani, M. Grilli, and R. Raimondi, *Phys. Rev. B* **43**, 13724 (1991).

56. J. K. Burdett and G. V. Kulkarni, *Chem. Phys. Lett.* **160**, 350 (1989).

57. W. E. Pickett et al., *Phys. Rev. Lett.* **67**, 228 (1991).

58. J. B. Torrance and R. M. Metzger, *Phys. Rev. Lett.* **63**, 1515 (1989).

59. J. T. Vaughey, J. D. Wiley, K. Poeppelmeir, *Z. Anorg. Allge. Chem.* **589–599**, 327 (1991).

ON THE THEORY OF DEBYE AND NÉEL RELAXATION OF SINGLE DOMAIN FERROMAGNETIC PARTICLES

W. T. COFFEY
P. J. CREGG

School of Engineering, Department of Microelectronics and Electrical Engineering, Trinity College, Dublin, Ireland

and

Yu. P. KALMYKOV

The Institute of Radioengineering and Electronics, Russian Academy of Sciences, Fryazino, Moscow Region, Russia

CONTENTS

Advances in Chemical Physics, Volume LXXXIII, Edited by I. Prigogine and Stuart A. Rice.
ISBN 0-471-54018-8   © 1993 John Wiley & Sons, Inc.

LIST OF MAJOR SYMBOLS FOR PHYSICAL QUANTITIES IN THE ORDER IN WHICH THEY APPEAR IN THE TEXT

Section I.C

γ	gyromagnetic ratio
$\boldsymbol{\mu}_e$	Bohr magneton

$\boldsymbol{\omega}_e$	angular momentum of an electron
g	Landé splitting factor
$\mathbf{L}_e$	torque on an electron
$\mathbf{H}$	magnetic field
$\mathbf{M}$	magnetization of a single domain ferromagnetic particle
M_s	magnitude of $\mathbf{M}$ (saturation magnetization)
λ	the Landau-Lifshitz dissipation constant
η	the Gilbert dissipation constant
α	$\eta \gamma M_s$ the dimensionless damping factor

Section I.D

g'	$\dfrac{\gamma}{(1+\alpha^2)M_s}$
h'	$\dfrac{\alpha\gamma}{(1+\alpha^2)M_s} = \alpha g'$
ϑ	polar angle
ϕ	azimuth
V	free energy per unit volume of a magnetic particle
υ	volume of a single domain magnetic particle

Section I.E

τ_{eff}	effective relaxation time of a single domain ferromagnetic particle suspended in a fluid
τ_N	Néel relaxation time
τ_D	Debye (Brownian) relaxation time
$\boldsymbol{\mu}$	magnetic moment of a ferromagnetic particle of volume $\upsilon (\mathbf{M}\upsilon = \boldsymbol{\mu})$
$\boldsymbol{\mu}$	electric dipole moment of a rigid dipolar molecule
ζ	drag coefficient of a rigid dipolar molecule

Section I.F

V_a	free anisotropy energy of unit volume
K	anisotropy constant
σ	ratio of height of anisotropy energy barrier to the thermal energy i.e. $K\upsilon/kT$
ξ	ratio of maximum external field energy to the thermal energy
W_0	equilibrium probability density function of orientations of the magnetization vector on the surface of the unit sphere in the absence of a potential

n	unit vector in direction of the internal anisotropy axis of a ferromagnetic particle
r	unit vector in the direction of the magnetization
h	direction of external magnetic field
v_s	volume of a magnetic sample containing N domains
N	number of domains each of volume v
n	number of domains per unit sample volume that is the domain concentration
$\langle M_\parallel \rangle$	mean equilibrium magnetic moment of sample of volume v_s in the direction of a longitudinal DC field
$\langle M_\perp \rangle$	mean equilibrium magnetic moment of sample of volume v_s in the direction of a transverse DC field
χ_0	static linear magnetic susceptibility
$\chi_\parallel^s$	static longitudinal magnetic susceptibility
$\chi_\perp^s$	static transverse magnetic susceptibility
χ	static susceptibility averaged over random particle anisotropy axes
d	particle magnetic diameter
$\phi(d)$	particle size distribution

Section I.G

$L(\xi)$	the Langevin function

Section II.A

m	mass of a Brownian particle
ζ	$6\pi\eta a$; drag on a *translating* spherical particle of radius a in a fluid of viscosity η
$\lambda(t)$	white noise driving force
$2D$	spectral density of the white noise force
$x(t)$	random variable describing displacement of a Brownian particle

Section II.B

$u(t)$	time derivative of mean square displacement of a Brownian particle
x_0	initial position of a Brownian particle

Section II.C

$\theta(t)$	polar angle in plane polar coordinates describing angular position of a rigid rotator
ζ	$8\pi\eta a^3$; drag on a *rotating* spherical particle in a fluid of viscosity η
$\lambda(t)$	white noise driving torque
I	moment of inertia of a two dimensional rotator
p	random variable describing instantaneous dipole moment of a rigid dipole molecule
F_0	magnitude of an applied DC electric field

Section II.D

$\mathbf{h}(t)$	the random magnetic field term in Gilbert's equation arising from thermal agitation
μ	spectral density of the random field $\mathbf{h}(t)$

Section II.E

$W(x, t)$	number density of translating Brownian particles
$\mathbf{J}(x, t)$	current density of translating Brownian particles

Section II.F

$W(\mathbf{r}, t)$	density of representative points (representing the magnetic moments) on the surface of the unit sphere
$k' = 1/2\tau_N$	diffusion coefficient in Brown's equation
$\mathbf{J}(\mathbf{r}, t)$	surface current density of representative points

Section III.B

$\tau_\parallel$	longitudinal magnetic relaxation time

Section III.C

$\tau_\perp$	transverse magnetic relaxation time
ω_0	frequency of the gyromagnetic precession

Section III.G

β	volume of magnetic particle divided by thermal energy

$\tau_{\parallel}(\sigma)$ longitudinal relaxation time in the presence of uniaxial anisotropy

$\tau_{\parallel}(\sigma, \xi)$ longitudinal relaxation time in the presence of uniaxial anisotropy and an applied field

Section IV.A

$\mathbf{H}_1 U(t)$ a small perturbing DC field

ζ_1 ratio of maximum perturbing field energy to the thermal energy

Section IV.B

$\chi(\omega)$ complex susceptibility

$\chi'(\omega)$ phase component of complex susceptibility

$\chi''(\omega)$ quadrature component of complex susceptibility

$\chi_{\parallel}(\omega)$ longitudinal component of complex susceptibility

Section IV.C

$\chi_{\perp}(\omega)$ transverse component of complex susceptibility

$\chi_{\parallel}^{s}$ static longitudinal susceptibility

$\chi_{\perp}^{s}$ static transverse susceptibility

Section IV.E

δ loss angle

$\tan \delta$ loss tangent

$\mu(\omega)$ complex permeability

$\mu'(\omega)$ phase component of complex permeability

$\mu''(\omega)$ quadrature component of complex permeability

μ_0 permeability of free space

μ_s static relative permeability at zero frequency and zero field strength

$(\tan \delta)_{\parallel}$ loss tangent for a longitudinal field

$(\tan \delta)_{\perp}$ loss tangent for a transverse field

$(\tan \delta)_m$ maximum value of loss tangent

x normalised frequency

ω_N angular frequency of maximum Néel absorption

Section V.A

$\mathbf{H}_T$	total magnetic field acting on a ferroparticle
μ	spectral density of the random white noise field in Gilbert's equation
$\mathbf{\Omega}$	angular velocity of a ferrofluid particle
$\mathbf{H}_0$	magnetic DC bias field
I	moment of inertia of a spherically symmetric ferrofluid particle
$\mathbf{N}$	magnetic torque acting on a ferrofluid particle
α	dynamic viscosity of a ferrofluid
ν	volume of a ferrofluid particle
$\mathbf{e}$	unit vector in direction of the magnetic field acting on a ferrofluid particle
$(\omega_1, \omega_2, \omega_3)$	components of the angular velocity vector $\mathbf{\Omega}$
(ϑ, ϕ, ψ)	the Eulerian angles
$C_{\parallel}(t)$	autocorrelation function of the parallel component of the magnetization of a single domain particle
$C_{\perp}(t)$	autocorrelation function of the perpendicular component of the magnetization of a single domain particle
$\mathbf{E}_0$	electric DC bias field

Section VI.A

$\mathbf{H}_{ef}(\mathbf{M})$	axially symmetric magnetic field
$\{\xi\}$	a set of random variables
$\Gamma_k(t)$	k^{th} component of a white noise driving term
$W(\{\mathbf{M}\}, t)$	probability density function of the orientations of $\mathbf{M}$ at an initial time t

Section VI.B

W_0	equilibrium distribution of orientations in the presence of a longitudinal field

Section VI.C

$g_{\parallel}$	longitudinal retardation factor
$g_{\perp}$	transverse retardation factor
λ_{eff}	effective eigenvalue
T_γ	relaxation time of the magnetization calculated from the effective eigenvalue

Section VIII

v_c	critical value of magnetic particle volume
$\tau_N(\sigma)$	Néel relaxation time in the presence of an anisotropy potential
V'	hydrodynamic value of the ferrofluid particle volume
$\tau_N(0)$	Néel relaxation time for very low barrier heights
v	magnetic volume of particle
μ	Néel or internal relaxation parameter i.e. the analogue of the fluid viscosity in Debye relaxation
η	fluid viscosity
$W(\mathbf{r}, \mathbf{n}, t)$	joint probability density function of orientations of magnetic moment and easy axes orientations
ω_L	Larmor frequency
ω^*	cross-over frequency
$F((\mathbf{r.n}), t)$	probability density function of the projection of the magnetic vector on the easy axis
$\Phi(\mathbf{n}, t)$	probability density of the directions of the easy axes.

Appendix A.1

$x(t)$	position of a Brownian particle
$v(t)$	velocity of a Brownian particle
$\mathbf{X}(t)$	state vector
β	friction coefficient per unit mass
v_0	initial velocity of a Brownian particle
x_0	initial position of a Brownian particle
Δx	displacement of a Brownian particle
D	$m\beta kT$

Appendix A.2

$\xi(t)$	arbitrary random variable
$\Phi_\xi(\omega)$	spectral density of $\xi(t)$
$\rho(\tau)$	autocorrelation function of $\xi(t)$
$\rho_v(\tau)$	velocity autocorrelation function
$\chi(\omega)$	transfer function of the system
$\Phi_v(\omega)$	spectral density of the velocity of a Brownian particle
$\Phi_\lambda(\omega)$	spectral density of a white noise

Appendix B.1

$f(y_1, t_1; y_2, t_2)$ joint probability density function of y_1 and y_2

$P(y_1, t_1 \mid y_2, t_2)$ transition probability

$D^{(1)}(z, t)$ drift coefficient in the Fokker-Planck equation

$D^{(2)}(z, t)$ diffusion coefficient in the Fokker-Planck equation

$y(t)$ initial value of the random variable ξ in a Langevin equation

Appendix B.4

D_1 diffusion coefficient

$\boldsymbol{\kappa}$ wave vector

$\Gamma_s(\boldsymbol{\kappa}, \omega)$ dynamic structure factor

P_v transition probability of the velocity distribution

$\mathbf{y} = (y_1 \ldots y_n)$ vector representing the currents y_i in an electric circuit having n meshes

Appendix C

$f(\theta, t)$ probability density distribution of orientation of dipoles on the unit circle

$a(t)$ electric dipole moment induced in a dielectric body by a unit electric field

$\mathbf{E}(t)$ inducing electric field

$\mathbf{m}(t)$ instantaneous electric dipole moment of the body

$\mathbf{E}_0$ a DC electric field

$b(t)$ after-effect function

$\alpha(\omega)$ complex polarizability

$\alpha'(\omega)$ real part of complex polarizability

$\alpha''(\omega)$ imaginary part of polarizability

$R(t)$ normalised response function

$\mathbf{M}(t)$ instantaneous dipole moment of a dielectric body in the absence of an external field

$\tilde{\mathbf{M}}(\omega)$ Fourier transform of $\mathbf{M}(t)$

$\mathfrak{M}(\omega)$ spectral density of $\mathbf{M}(t)$

$C_M(t)$ autocorrelation function of $\mathbf{M}(t)$

μ_i dipole moment in itinerant oscillator model

ϕ_1 angular position of dipole in itinerant oscillator model

ϕ_2 angular position of cage

$\lambda_1(t)$ noise torque on dipole

$\lambda_2(t)$	noise torque on cage
χ	sum angle variable in an itinerant oscillator model
η	difference angle variable
ζ_1	drag coefficient of dipole in itinerant oscillator model
ζ_2	drag coefficient of cage
$V'(\phi_1 - \phi_2)$	mutual torque in itinerant oscillator model
ζ_1	drag coefficient of yolk in egg model
ζ_2	drag coefficient of egg-shell
$\mathbf{T}_1$	sum of magnetic and random torques acting on yolk
$\mathbf{T}_2$	sum of magnetic and random torques on egg-shell
$\boldsymbol{\omega}_1$	angular velocity of yolk
$\boldsymbol{\omega}_2$	angular velocity of egg-shell
$\boldsymbol{\Omega}$	local angular velocity of fluid surrounding egg

Appendix E

W	joint probability density function of the angular variables in configuration – angular velocity space in the itinerant oscillator model
$\boldsymbol{\omega}(t)$	angular velocity of a rigid electric dipole
$\boldsymbol{\lambda}(t)$	noise torque on a dipole
$\mathbf{F}(t)$	electric field acting on a dipole
$\mathbf{F}_0$	DC bias field
(μ_x, μ_y, μ_z)	components of dielectric dipole moment
$W(\{\boldsymbol{\mu}\}, t)$	probability density of the orientations of $\boldsymbol{\mu}$ at an initial time t

I.   INTRODUCTORY CONCEPTS

A.   Ferromagnetic Materials

A large ferromagnetic particle consists of many magnetic domains, each of which possesses a magnetic moment. The domains, however, are randomly oriented so that the particle's overall magnetic moment is zero.

The existence of these distinct magnetic domains separated by domain boundaries can be explained by the presence of both surface and volume energies in the total free energy of the system. The minimization of the total free energy of the system results in the energetic favorability, and hence, creation of domains of a critical size [1, 2]. Within each magnetic domain the individual magnetic moments of the electrons all tend to align so that the domain is magnetized even in the absence of an external magnetic field. This alignment tends to be disrupted by thermal agitation,

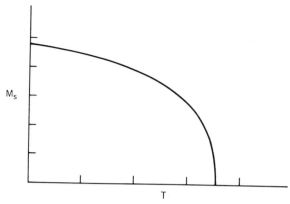

Figure 1. Variation of the magnetization M_s of a single domain ferromagnetic particle with temperature.

and indeed, beyond the Curie temperature the disruption is total so that the domain possesses no magnetization above this temperature. Far below the Curie temperature, however, the domain is magnetized to saturation and the magnitude of the magnetization of the domain is essentially constant. This temperature dependence is illustrated in Fig. 1.

Frenkel and Dorfman [3] predicted that by reducing the dimensions of a ferromagnetic specimen, a point in this reduction would be reached at which domain boundaries would no longer be energetically favorable so that the whole specimen would become a single domain. The prediction was found to be correct and particles of this nature are known as single domain ferromagnetic particles. Each individual single domain ferromagnetic particle behaves as a permanent magnet possessing a magnetic moment. The magnetization, which is the magnetic moment per unit volume, can be described by the vector $\mathbf{M} = M_s\mathbf{r}$, where $\mathbf{r}$ is a unit vector in Euclidean space, and M_s is the magnitude of the magnetization. The magnetic moment of each domain is therefore $\mathbf{M}v$ where v is the volume of the particle. The critical radius of a spherical single domain ferromagnetic particle was estimated by Kittel [1], by a minimization of the total free energy, as 150 Å, and this estimation has been verified by experiment where single domain ferromagnetic particles exist with radii of the order of 100 Å [4].

B. The Purpose of this Review

The behavior of the magnetization of a single domain ferromagnetic particle has been the subject of much study. Landau and Lifshitz in their 1935 paper [5] gave an equation of motion describing the average behavior of the magnetization. Gilbert in his 1955 papers [6, 7] presented a similar equation. Brown in 1963 [8] extended these equations of motion to describe not the average but the specific behavior of the magnetization of an individual single domain ferromagnetic particle. He based his work on Langevin's approach [9] to the theory of Brownian motion, which starts from the equation of motion of a Brownian particle in a fluid. Brown took as the "Langevin equation", the equation of motion describing the dynamic behavior of an individual particle's magnetization. He was able to construct from it, the underlying probability density diffusion equation, which is the Fokker–Planck equation. In order to accomplish this, Brown wrote down the equation of motion in spherical polar coordinates. He then used the methods of Wang and Uhlenbeck [10, 11] coupled with the Stratonovich [12] definition of the derivative of a stochastic variable to form the Fokker–Planck equation in these coordinates. This method is very lengthy and in order to circumvent it, he also gave an alternative approach to writing down the Fokker–Planck equation using a continuity equation argument similar to that used by Einstein in his 1905 [13] treatment of translational Brownian movement.

The form of the Fokker–Planck equation obtained by Brown closely resembles that obtained by Debye [14] in his study of the dielectric relaxation of assemblies of noninteracting polar molecules. It will be convenient to call this Fokker–Planck equation for the probability density of orientations of the magnetization of a single domain ferromagnetic particle, Brown's equation. Having obtained this equation, Brown applied the method of separation of the variables to convert the solution to a Sturm–Liouville problem. He restricted himself, for simplicity, to axially symmetric problems, where there is no azimuthal (ϕ) dependence of the distribution function, where the particle magnetization possesses a uniaxial anisotropy, and where a longitudinal field is applied. This results in a Sturm–Liouville equation which is related to Legendre's equation. The quantity in this equation of greatest interest is the (lowest eigenvalue)$^{-1}$ which corresponds to the longest relaxation time of the process. Brown finds explicit solutions for the cases of low and high (anisotropy) energy barrier height. The low barrier height solution he obtains by perturbation theory while the high barrier height solution is found by adapting Kramers's transition state method [15]. The solid state relaxation mechanism described by Brown is the rotation of the magnetic moment with respect to the body of the particle and is called Néel relaxation.

Further developments of Brown's approach were made by the group of Shliomis [16–19], who showed how Brown's work could be adapted to describe the dynamic behavior of suspensions of single domain ferromagnetic particles in fluids, that is, ferrofluids (see Section I.E.).

The Fokker–Planck equation method is extremely lengthy to use, in practice, since it involves many mathematical manipulations, especially for rotation in three dimensions [16–19]. Thus, an alternative approach is desirable. The main thrust of this review is to show how the results of the previous investigators may be obtained in a simple manner directly from the "Langevin equation" of the process, using the methods of the present authors [20]. The work of the previous authors is also expanded upon and treated in some detail, in order to provide an elementary introduction to the subject for the reader.

We may summarize the contents of this chapter in more detail as follows. In Section I we demonstrate how the explicit form of Gilbert's equation describing Néel relaxation may be written down from the gyromagnetic equation and how, in the limit of low damping, this becomes the Landau–Lifshitz equation. Next the application of this equation to ferrofluid relaxation is discussed together with the analogy to dielectric relaxation.

In Section II it is shown how the effects of thermal agitation may be included in Gilbert's equation and how the Fokker–Planck equation for the density of orientations of the magnetic moments on the unit sphere may be written down in an intuitive manner from Gilbert's equation. (The rigorous derivation of the Fokker–Planck equation from Gilbert's equation is given in Appendix D). We coin the term Brown's equation for this particular form of the Fokker–Planck equation.

In Section III we expand the density of orientations in Brown's equation in spherical harmonics to write that equation as a set of differential-difference equations. The following problems are reduced to the solution of a set of differential-difference equations.

1. A single domain ferromagnetic particle in a longitudinal field excluding internal anisotropy
2. A single domain ferromagnetic particle in a transverse field excluding internal anisotropy
3. A single domain ferromagnetic particle with uniaxial anisotropy but excluding the effects of an external field

The results of (1) and (2) apply to both Néel and Debye relaxation (see Section I.E). In each of the three cases it is shown how the longest relaxation time of the system may be obtained numerically by writing the set of differential-difference equations (called in the literature the Brinkman equations) as a matrix differential equation and successively increasing the size of the matrix until convergence is attained. It is also demonstrated how expressions for the relaxation times may be obtained by perturbation theory in the limit of low potential barriers. In Section

III.G., the WKBJ approximation is applied to retrieve the analytic expression derived by Brown using the Kramers transition state theory for the longest relaxation time for uniaxial anisotropy in the limit of high potential barriers. This is also extended to yield the formula for the relaxation time for high uniaxial anisotropy in the presence of a longitudinal field.

Section IV is concerned with the determination of analytic expressions for the relaxation times for a weak AC field superimposed on a DC bias field of arbitrary strength for a single domain ferromagnetic particle excluding internal anisotropy. It is shown how the transverse and longitudinal relaxation times may be written as functions of the Langevin function and how closed form expressions for the dynamic susceptibilities and loss tangents may be obtained. The results are also applicable to ferrofluids (Debye relaxation). The solution is effected using linear response theory, that is, the response to a small step change in the bias field is first calculated; this may then be related to the AC response using linear response theory as described in Appendix C.

In Section V we are concerned with Gilbert's equation as applied to the Debye relaxation of a ferrofluid particle with the inertia of the particle included. It is shown, by averaging Gilbert's equation for Debye relaxation corrected for inertia and proceeding to the noninertial limit, how analytic expressions for the transverse and longitudinal relaxation times for Debye relaxation may be obtained directly from that equation thus bypassing the Fokker–Planck equation entirely. These expressions coincide with the previous results of the group of Shliomis [16].

In Section VI we consider how the relaxation times of the magnetization of a single domain ferromagnetic particle may be obtained directly from Gilbert's equation. (Problem (3) of Section III as outlined above). In the Néel relaxation of the magnetization within a single domain ferromagnetic particle, inertia plays no role [8]. Consequently for Néel relaxation, Gilbert's equation must be averaged using the Stratonovich interpretation rule. This procedure leads to expressions for the parallel and perpendicular relaxation times for uniaxial anisotropy in terms of Dawson's integral. The expression for the parallel relaxation time coincides with that of Raĭkher and Shliomis, obtained by separating the variables in the Fokker–Planck equation and employing Leontovich's closure procedure [17]. Expressions for the relaxation times that are valid for any axially symmetric potential of the crystalline anisotropy are also given in terms of the equilibrium averages of the first and second moments of the distribution function.

Section VII is concerned with the theory of dispersion of the magnetic susceptibility of fine ferromagnetic particles. Brown's equation, including the gyromagnetic terms, is reduced to a set of differential-difference equations by expanding the distribution function in spherical harmonics. An approximate expression for the longitudinal relaxation time $\tau_\parallel$ in the presence of a weak external field, which coincides with the result of Section VI, is obtained in terms of Dawson's integral. The corresponding susceptibility is obtained in closed form using the closure method of Section IV. The results are compared with the approximate formula of Brown. The transverse relaxation time $\tau_\perp$ and susceptibility are also investigated using the closure methods of Section IV. This allows us to investigate the transition from relaxational to ferromagnetic resonance behavior as a function of the anisotropy parameter.

The appendices contain an account of those parts of the theory of Brownian motion and linear response theory which are essential for the reader in order to achieve an understanding of relaxational phenomena in magnetic domains and in ferrofluid particles. The analogy with dielectric relaxation is emphasized throughout these appendices. Appendix D contains the rigorous derivation of Brown's equation.

C. The Gyromagnetic Equation

Let us consider the behavior of an individual uncompensated electron in a ferromagnetic material. It possesses both a magnetic moment and an angular momentum. The magnetic moment and the angular momentum of an electron are related phenomena. The former is due to the motion of the charge of an electron and the latter to the motion of the mass. We can therefore relate the two vector quantities by a scalar quantity γ as follows,

$$\boldsymbol{\mu}_e = \gamma \boldsymbol{\omega}_e \tag{1.1}$$

where $\boldsymbol{\mu}_e$ and $\boldsymbol{\omega}_e$ are the magnetic moment and angular momentum of an electron, respectively (the magnetic moment of a spin electron is also known as a Bohr magneton), and γ is a scalar quantity known as the gyromagnetic ratio, given by [2, 5, 21]

$$\gamma = \frac{-ge}{2mc} \tag{1.2}$$

where e is the electron charge, m the electron mass, c the velocity of light and g the Landé splitting factor. This factor equals one when the electrons in question are orbital electrons and two for spin electrons. The electrons largely responsible for the magnetic moments in ferromagnetic bodies are spin electrons with $g = 2$, and so

$$\gamma = \frac{-e}{mc} \qquad (1.3)$$

It is worth pointing out that since the charge on an electron is negative, γ is a positive constant. Landau and Lifshitz [5] used a negative γ and consequently so did many researchers [16–19] who based their work on that of Landau and Lifshitz. It is, however, merely a matter of convention as to which direction of precessional rotation is taken as positive, and as such does not affect our results. When placed in a magnetic field, the magnetic moment of each electron in a single domain ferromagnetic particle tries to align itself with the direction of that field. This alignment, however, is opposed by the angular momentum of the electron. In other words the mass wishes to continue spinning on its own axis, while the charge wishes to spin along the axis of the field. The result is not alignment with the field, but rather precession about the axis of the field. The field causes a torque $\mathbf{L}_e$ to act on each electron, given by

$$\mathbf{L}_e = \mathbf{\mu}_e \times \mathbf{H} \qquad (1.4)$$

where $\mathbf{H}$ is the resultant magnetic field. This torque must also equal the rate of change of angular momentum $d\mathbf{\omega}_e/dt$ of the electron. Returning to Eq. (1.1) and differentiating with respect to time, we have

$$\dot{\mathbf{\mu}}_e = \gamma \dot{\mathbf{\omega}}_e \qquad (1.5)$$

and so from the above we obtain the simple gyromagnetic equation for each electron

$$\dot{\mathbf{\mu}}_e = \gamma \mathbf{\mu}_e \times \mathbf{H} \qquad (1.6)$$

Since in a single domain ferromagnetic particle, all the uncompensated electrons are aligned, it is possible to use the above equation of motion, which is for an individual electron, to describe the dynamics of the total magnetization. In so doing we have,

$$\dot{\mathbf{M}} = \gamma \mathbf{M} \times \mathbf{H} \qquad (1.7)$$

This simple gyromagnetic equation (Eq. (1.7)) [22] is the most basic equation describing the dynamic behavior of the magnetization of a single domain ferromagnetic particle. It represents uniform undamped precession of the vector $\mathbf{M}$ about the axis of the field $\mathbf{H}$. The observable behavior, however, of the magnetization of a single domain ferromagnetic particle, is that of alignment of $\mathbf{M}$ with $\mathbf{H}$. This alignment is due to collisions between precessing electrons which take place within the particle. It is apparent, therefore, that the field does not directly cause alignment; rather it causes a precession of $\mathbf{M}$ about the axis of the field which along with collisions will produce alignment. With the aim of including this fact in the mathematical analysis, Landau and Lifshitz introduced a second term into the gyromagnetic equation (Eq. (1.7)), the tendency of which is to align $\mathbf{M}$ with $\mathbf{H}$. In their 1935 paper they proposed an equation for the dynamic behavior of $\mathbf{M}$ which included a term proportional to

$$(\mathbf{M} \times \mathbf{H}) \times \mathbf{M} = \mathbf{H}(\mathbf{M} \cdot \mathbf{M}) - (\mathbf{H} \cdot \mathbf{M})\mathbf{M} \qquad (1.8)$$

When examined on the vector diagram of $\mathbf{M}$ and $\mathbf{H}$, this is seen to be directed along the plane of $\mathbf{M}$ and $\mathbf{H}$. The Landau–Lifshitz equation in the form in which it originally appeared is

$$\dot{\mathbf{M}} = -\gamma\left[\mathbf{H} \times \mathbf{M} + \lambda\left(\mathbf{H} - \frac{(\mathbf{H} \cdot \mathbf{M})\mathbf{M})}{M_s^2}\right)\right] \qquad (1.9)$$

where λ is a constant of the same dimensions as M_s and limited by the condition that $\lambda \ll M_s$. Noting Eq. (1.8), the Landau–Lifshitz equation is more commonly written in the following form

$$\dot{\mathbf{M}} = \gamma\mathbf{M} \times \mathbf{H} - \frac{\gamma\lambda}{M_s^2}(\mathbf{M} \times \mathbf{H}) \times \mathbf{M} \qquad (1.10)$$

The equation now represents damped precessional motion, wherein alignment ultimately takes place between the two vectors, $\mathbf{M}$ and $\mathbf{H}$.

More recently, Gilbert (1955) [6, 7] proposed an equation describing the dynamic behavior of $\mathbf{M}$ which incorporated the collision damping incurred by the precessional motion, in an effective damping field term. He assumed that the damping field is

$$\mathbf{H}_{damping} = -\eta\dot{\mathbf{M}} \qquad (1.11)$$

where η is a damping constant with units such that $\eta\gamma M_s = \alpha$ is dimensionless. Including this damping field in the simple gyromagnetic equation, we have

$$\dot{\mathbf{M}} = \gamma\mathbf{M} \times (\mathbf{H} - \eta\dot{\mathbf{M}}) \qquad (1.12)$$

This is Gilbert's equation. It is implicit in $d\mathbf{M}/dt$. It is possible, for small values of the dimensionless damping factor, α, to express it explicitly in $\dot{\mathbf{M}}$ by iterating, as follows:

$$\dot{\mathbf{M}} = \gamma\mathbf{M} \times \mathbf{H} - \gamma\mathbf{M} \times \eta\dot{\mathbf{M}} \qquad (1.13)$$

$$= \gamma\mathbf{M} \times \mathbf{H} - \gamma^2\eta\mathbf{M} \times [\mathbf{M} \times (\mathbf{H} - \eta\dot{\mathbf{M}})] \qquad (1.14)$$

Neglecting terms $O(\eta^2)$ and higher and noting that $\mathbf{M} \times \mathbf{H} = -\mathbf{H} \times \mathbf{M}$, we have

$$\dot{\mathbf{M}} = \gamma\mathbf{M} \times \mathbf{H} + \gamma^2\eta(\mathbf{M} \times \mathbf{H}) \times \mathbf{M} \qquad (1.15)$$

This explicit form of Gilbert's equation for the case of low damping is of the same form as the previous Landau–Lifshitz equation. The neglect of the terms $O(\eta^2)$ and higher corresponds to the assumption of Landau and Lifshitz that $\lambda \ll M_s$, that is, of small damping. This correspondence becomes more apparent if we equate the prefixes present before the aligning terms in both equations. Equating these gives us

$$-\lambda/M_s = \eta\gamma M_s = \alpha \qquad (1.16)$$

The negative sign here arises from the (opposing) direction of precession conventions of Gilbert and Landau–Lifshitz.

It is, however, possible to obtain an explicit form of Gilbert's equation which involves no assumptions about the level of the internal magnetic damping acting on the precessing magnetization within the single domain ferromagnetic particle. This explicit form was used, though neither derived nor stated in Brown's 1963 paper [8]. It was stated though not derived in the 1956 paper of Kikuchi [23]. Its derivation, therefore, is given here.

D. Derivation of the Explicit Form of Gilbert's Equation

Gilbert's equation is

$$\dot{\mathbf{M}} = \gamma\mathbf{M} \times (\mathbf{H} - \eta\dot{\mathbf{M}}) \qquad (1.17)$$

Transposing the η term, we have

$$\dot{\mathbf{M}} + \eta\gamma\mathbf{M} \times \dot{\mathbf{M}} = \gamma\mathbf{M} \times \mathbf{H} \qquad (1.18)$$

The explicit solution to this can be found as follows. We cross-multiply vectorially by **M** on both sides. Thus we have

$$\dot{\mathbf{M}} \times \mathbf{M} + \eta\gamma(\mathbf{M} \times \dot{\mathbf{M}}) \times \mathbf{M} = \gamma(\mathbf{M} \times \mathbf{H}) \times \mathbf{M} \tag{1.19}$$

The triple vector product $(\mathbf{M} \times \dot{\mathbf{M}}) \times \mathbf{M}$ is

$$(\mathbf{M} \times \dot{\mathbf{M}}) \times \mathbf{M} = -\mathbf{M}(\dot{\mathbf{M}} \cdot \mathbf{M}) + \dot{\mathbf{M}}M_s^2 \tag{1.20}$$

The first term on the right hand side contains $\dot{\mathbf{M}} \cdot \mathbf{M}$. This can be evaluated by multiplying Eq. (1.18) by $\cdot \mathbf{M}$ to give

$$\dot{\mathbf{M}} \cdot \mathbf{M} + \eta\gamma(\mathbf{M} \times \dot{\mathbf{M}}) \cdot \mathbf{M} = \gamma(\mathbf{M} \times \mathbf{H}) \cdot \mathbf{M} \tag{1.21}$$

From the properties of vectors it is known that $(\mathbf{P} \times \mathbf{Q}) \cdot \mathbf{R} = 0$ if any two of the vectors $\mathbf{P}, \mathbf{Q}, \mathbf{R}$ are parallel. Hence

$$(\mathbf{M} \times \dot{\mathbf{M}}) \cdot \mathbf{M} = 0 \tag{1.22}$$

and

$$(\mathbf{M} \times \mathbf{H}) \cdot \mathbf{M} = 0 \tag{1.23}$$

and consequently from Eq. (1.21),

$$\dot{\mathbf{M}} \cdot \mathbf{M} = 0 \tag{1.24}$$

also. Equation (1.20) therefore reads

$$(\mathbf{M} \times \dot{\mathbf{M}}) \times \mathbf{M} = \dot{\mathbf{M}}M_s^2 \tag{1.25}$$

Substitution of this into Eq. (1.19) gives

$$\dot{\mathbf{M}} \times \mathbf{M} + \eta\gamma\,\dot{\mathbf{M}}M_s^2 = \gamma(\mathbf{M} \times \mathbf{H}) \times \mathbf{M} \tag{1.26}$$

Equation (1.18) can be rearranged to read

$$\dot{\mathbf{M}} \times \mathbf{M} = (\dot{\mathbf{M}} - \gamma\mathbf{M} \times \mathbf{H})(\eta\gamma)^{-1} \tag{1.27}$$

Use of this in Eq. (1.26) leads to

$$(\dot{\mathbf{M}} - \gamma\mathbf{M} \times \mathbf{H})(\eta\gamma)^{-1} + \eta\gamma\,\dot{\mathbf{M}}M_s^2 = \gamma(\mathbf{M} \times \mathbf{H}) \times \mathbf{M} \tag{1.28}$$

and multiplying across by $\eta\gamma$, we have

$$\dot{\mathbf{M}} + (\eta\gamma)^2\,\dot{\mathbf{M}}M_s^2 = \gamma\mathbf{M}\times\mathbf{H} + \eta\gamma^2(\mathbf{M}\times\mathbf{H})\times\mathbf{M} \qquad (1.29)$$

or

$$\dot{\mathbf{M}}[1 + (\eta\gamma M_s)^2] = \gamma\mathbf{M}\times\mathbf{H} + \eta\gamma^2(\mathbf{M}\times\mathbf{H})\times\mathbf{M} \qquad (1.30)$$

and finally we have the explicit solution for $\dot{\mathbf{M}}$

$$\dot{\mathbf{M}} = \frac{\gamma\mathbf{M}\times\mathbf{H}}{1 + (\eta\gamma M_s)^2} + \eta\,\frac{\gamma^2(\mathbf{M}\times\mathbf{H})\times\mathbf{M}}{1 + (\eta\gamma M_s)^2} \qquad (1.31)$$

By writing

$$g' = \frac{\gamma}{(1 + \eta^2\gamma^2 M_s^2)M_s} = \frac{\gamma}{(1 + \alpha^2)M_s} \qquad (1.32)$$

and

$$h' = \frac{\eta\gamma^2}{1 + \eta^2\gamma^2 M_s^2} = \frac{\alpha\gamma}{(1 + \alpha^2)M_s} = \alpha g' \qquad (1.33)$$

we can write

$$\dot{\mathbf{M}} = M_s g'\mathbf{M}\times\mathbf{H} + h'(\mathbf{M}\times\mathbf{H})\times\mathbf{M} \qquad (1.34)$$

This is the form of Gilbert's equation used, though not stated, by Brown in [8]. It is essentially of the same form as the Landau–Lifshitz equation except that both prefixes g' and h' depend on the damping level.

In his analysis of the dynamic behavior of $\mathbf{M}$, Brown made use of the spherical polar coordinate system which relates to the Cartesian coordinate system as follows:

$$\mathbf{M} = M_x\mathbf{i} + M_y\mathbf{j} + M_z\mathbf{k} \qquad (1.35)$$

where $\mathbf{i}, \mathbf{j}, \mathbf{k}$, are unit vectors in Cartesian (x, y, z) space and

$$M_x = M_s\sin\vartheta\cos\phi \qquad (1.36)$$

$$M_y = M_s\sin\vartheta\sin\phi \qquad (1.37)$$

$$M_z = M_s\cos\vartheta \qquad (1.38)$$

It is assumed throughout this analysis that only the orientation and not the magnitude of the magnetization is subject to variation in the presence of a magnetic field. Although the magnitude does in fact undergo some variation, this variation is, however, negligible and can be ignored. This reduces Gilbert's equation in the spherical polar coordinates, from three component equations in (r, ϑ, ϕ) to two equations, one in each of the angular variables ϑ and ϕ. These two component differential equations are

$$\dot{\vartheta} = M_s[-g'H_\phi + h'H_\vartheta] \tag{1.39}$$

$$\dot{\phi} \sin \vartheta = M_s[g'H_\vartheta + h'H_\phi] \tag{1.40}$$

The resultant magnetic field $\mathbf{H}$ is given by

$$\mathbf{H} = -\partial V/\partial \mathbf{M} \tag{1.41}$$

where V is the free energy per unit volume. In terms of V, therefore, the component equations are

$$\dot{\vartheta} = g' \frac{1}{\sin \vartheta} \frac{\partial V}{\partial \phi} - h' \frac{\partial V}{\partial \vartheta} \tag{1.42}$$

$$\dot{\phi} \sin \vartheta = -g' \frac{\partial V}{\partial \vartheta} - h' \frac{1}{\sin \vartheta} \frac{\partial V}{\partial \phi} \tag{1.43}$$

E. Ferrofluids

We will also consider in this review the dynamic behavior of the magnetization of a colloidal suspension of single domain ferromagnetic particles in a fluid, that is, of a ferrofluid. (A distinction should be made here between a ferromagnetic fluid and a ferrofluid. We say that a ferromagnetic fluid is such that its viscosity is significantly altered by the presence of a magnetic field whereas in a ferrofluid its viscosity is not [24].) We assume that the particle to volume ratio of the ferrofluid is small enough so that the particles are *noninteracting*. In so doing we reduce the study of the magnetization of an *assembly* of single domain ferromagnetic particles to the study of the magnetization of an *individual* single domain ferromagnetic particle.

A single domain ferromagnetic particle suspended in a fluid is subject to two orientational variational mechanisms. The first is that discussed already and takes place inside the particle, it is the solid state mechanism known as Néel [25] relaxation. The second is due to the rotation of the particle within the fluid, since changes in the orientation of the particle effect changes in the orientation of its magnetization. This mechanism is similar to the Debye relaxation of a polar molecule and consequently is known as Debye relaxation.

In our study we treat the two mechanisms separately. For each we assume the dominance of that mechanism. This means that for Debye relaxation we assume that the magnetization vector is fixed to the particle, that is the Néel relaxation is blocked or "frozen" due to an insurmountable energy barrier preventing its operation. On the other hand for the Néel relaxation, we assume that the particle is fixed in space.

For a single domain ferromagnetic particle in suspension in a fluid and undergoing orientational variations due to both mechanisms, the effective relaxation time τ_{eff} is [26]

$$1/\tau_{\text{eff}} = 1/\tau_N + 1/\tau_D \qquad (1.44)$$

where the subscripts N and D signify Néel and Debye relaxation times, respectively. It is apparent here that the mechanism with the shortest relaxation time is dominant. It should also be noted that at a critical particle radius we have $\tau_N = \tau_D$, thus allowing us to determine the particle size distribution.

In a typical ferrofluid the particles will have median radii 2–10 nm so that thermal energy overcomes magnetostatic interaction between the particles and prevents aggregation. The single domain particles are considered to be in a state of *uniform* magnetization with magnetic moment

$$\mu = v\mathbf{M}$$

where v is the volume of the particle.

In Section II, Gilbert's equation describing the Néel relaxation is augmented by a random field term, representing thermal fluctuations. The underlying Fokker–Planck equation is then constructed from this augmented equation. The time constant in this equation is the Néel relaxation time

$$\tau_N = \frac{v}{kT} \frac{M_s}{\gamma} \frac{1+\alpha^2}{2\alpha} \qquad (1.45)$$

Noting this, Gilbert's equation can be rewritten as

$$2\tau_N \dot{\mathbf{M}} = \frac{v}{kT} [\alpha^{-1} M_s \mathbf{M} \times \mathbf{H} + (\mathbf{M} \times \mathbf{H}) \times \mathbf{M}] \qquad (1.46)$$

The last term in Gilbert's equation above is the aligning term and consequently the term of interest in relaxation of the magnetization with respect to a magnetic field. For low damping ($\alpha \ll 1$), we see that precessional motion is the dominant motion and that the relaxation time is approximately given by

$$\tau_N = \frac{v}{kT} \frac{M_s}{\gamma} \frac{1}{2\alpha} \tag{1.47}$$

whereas for high damping ($\alpha \gg 1$), alignment is dominant and

$$\tau_N = \frac{v}{kT} \frac{M_s}{\gamma} \frac{\alpha}{2} \tag{1.48}$$

We see that in both the limits $\alpha \to 0$ and $\alpha \to \infty$, the relaxation time approaches infinity and alignment does not take place. This is due in the former case to infinitely persisting undamped precessional motion and in the latter to the absence of all motion as a result of the high damping. The minimization of the Néel relaxation time was a goal of Kikuchi's 1956 paper [23] and clearly occurs at $\alpha = 1$ where neither motion is dominant and

$$\tau_N = \frac{vM_s}{kT\gamma} \tag{1.49}$$

In a ferrofluid particle, the Néel relaxation of which is blocked, orientational changes in **M** are due to the rotational motion of the particle not to changes in the axis of the spinning electrons within the particle. Consequently the magnetization behaves like that of a bar magnet when placed in a magnetic field. No gyromagnetic or precessional terms are present in its equation of motion and the averaged equation resembles Gilbert's equation with τ_N replaced by τ_D and the precessional terms dropped. The equation of motion is therefore

$$2\tau_D \dot{\mathbf{M}} = \frac{v}{kT} [(\mathbf{M} \times \mathbf{H}) \times \mathbf{M}] \tag{1.50}$$

This behavior is analogous to that of a polar molecule in a fluid under the influence of an electric field. This was studied by Debye [14], who obtained the relaxation time

$$\tau_D = \zeta/2kT \tag{1.51}$$

where ζ is the drag coefficient proportional to the viscosity of the surroundings. Substitution of this into Eq. (1.50) therefore gives

$$\zeta \dot{\mathbf{M}} = v[(\mathbf{M} \times \mathbf{H}) \times \mathbf{M}] \qquad (1.52)$$

The presence of the prefix v on the right hand side of Eq. (1.52) is explained by the fact that the quantity that directly corresponds here to the dipole moment in the Debye theory is the magnetic moment of an individual ferrofluid particle, not the magnetization which is the magnetic moment of the particle per unit volume. Writing this equation for the magnetic moment $\boldsymbol{\mu}$ of the particle, where

$$\boldsymbol{\mu} = \mathbf{M}v \qquad (1.53)$$

we have

$$\zeta \dot{\boldsymbol{\mu}} = (\boldsymbol{\mu} \times \mathbf{H}) \times \boldsymbol{\mu} \qquad (1.54)$$

which directly corresponds to the equation used by Debye for a polar molecule.

It should be emphasized that both Gilbert's equation and the earlier Landau–Lifshitz equation are merely phenomenological equations which are used to explain the time decay of the average magnetization. Brown [8] suggested that the Gilbert equation should be augmented by a white-noise driving term, in order to explain the effect of thermal fluctuations of the surroundings on the magnetization.

F.   Relevant Magnetic Potential Parameters

Easy axes of alignment of the magnetization, or an anisotropy, often exist within a single domain ferromagnetic particle. The simplest form of anisotropy is uniaxial anisotropy in which the magnetization favors either of two opposite directions. This may be represented in the framework of our analysis by means of an anisotropy potential V_a of the form

$$V_a = K \sin^2 \vartheta \qquad (1.55)$$

where K is an anisotropy constant. The parameter which is of importance here is

$$\sigma = Kv/kT \qquad (1.56)$$

which is the ratio of anisotropy energy to thermal energy. (The full significance of the thermal energy is discussed in Section II). When σ is small, thermal agitation and external fields can easily overcome the potential barrier that exists between the two favored directions. But when

σ is large, this becomes difficult. More significantly with high values of σ, the distribution of the orientations of **M** tends to be concentrated close to the favored directions because of the high energetic favorability of those directions. This fact suggests the use of a discrete orientation model such as that used by Kramers for calculating the escape rate of particles from potential wells. Such a method was employed by Brown and a similar calculation is performed by us [27] in Section III using the Wentzel–Kramers–Brillouin–Jeffreys (WKBJ) [28] method.

For the case of externally applied magnetic fields the parameter of importance is

$$\xi = vHM_s/kT \tag{1.57}$$

where **H** is the magnetic field. We consider two directions for the external field in this chapter:

1. An applied field parallel to the easy axis of magnetization taken as the z axis yielding the free energy density

$$V = -HM_s \cos \vartheta \tag{1.58}$$

2. An applied field perpendicular to the z axis so that

$$V = -HM_s \sin \vartheta \cos \phi \tag{1.59}$$

This is the transverse field.

The analysis which we have just given ignores the internal anisotropy of the ferrofluid particle. It is instructive to review a calculation of Shliomis [16–18] which considers the static susceptibility in a monodispersed colloid taking account of the internal anisotropy of the particles. The direction of the magnetization vector will be represented by $\mathbf{r} = \mathbf{M}/M_s$ and the direction of the internal anisotropy axis by **n**. The equilibrium distribution function of the orientations taking account of the internal anisotropy and the applied DC field is

$$W = C \exp[\sigma(\mathbf{r} \cdot \mathbf{n})^2 + \xi(\mathbf{r} \cdot \mathbf{h})]$$
$$= C \exp[\sigma(\mathbf{r} \cdot \mathbf{n})^2]\{1 + \xi(\mathbf{r} \cdot \mathbf{h})\}$$

in the linear approximation of ξ. **h** denotes the direction of the applied DC field. We wish to calculate $\langle \mathbf{h} \cdot \mathbf{r} \rangle$. The magnetization clearly depends on the angle between **h** and **n**. We consider two directions for **h**, **h** $\parallel$ **n** and **h** $\perp$ **n**. For **h** $\parallel$ **n** we have with

$$\mathbf{r} \cdot \mathbf{h} = \cos \vartheta = x = \mathbf{r} \cdot \mathbf{n}$$

so that for N domains each of volume v the mean magnetic dipole moment is

$$\langle M_\parallel \rangle = N v M_s \langle \mathbf{h} \cdot \mathbf{r} \rangle = N v M_s \frac{\int_{-1}^{+1} x \exp(\sigma x^2)(1 + \xi x) \, dx}{\int_{-1}^{+1} \exp(\sigma x^2) \, dx}$$

$$= \frac{N M_s^2 v^2 H}{kT} \frac{\int_0^1 x^2 e^{\sigma x^2} \, dx}{\int_0^1 e^{\sigma x^2} \, dx}$$

so that the susceptibility $\chi_\parallel^s$ for $\mathbf{h}_\parallel \mathbf{n}$ is

$$\chi_\parallel^s = \frac{n M_s^2 v^2}{kT} \frac{F'}{F}, \qquad F(\sigma) = \int_0^1 e^{\sigma x^2} \, dx \qquad (1.60)$$

where n is the number of particles per unit volume v_s of the sample. In the same way the susceptibility for $\mathbf{h} \perp \mathbf{n}$ is from

$$\langle M_\perp \rangle = N v M_s \langle \sin \vartheta \cos \phi \rangle = N v M_s \frac{\int_0^{2\pi} \xi \cos^2 \phi \, d\phi \int_0^1 (1 - x^2) e^{\sigma x^2} \, dx}{2\pi \int_0^1 e^{\sigma x^2} \, dx},$$

$$\chi_\perp^s = \frac{n M_s^2 v^2}{kT} \frac{F - F'}{2F} \qquad (1.61)$$

Following Shliomis [18], Eqs. (1.60) and (1.61) may be conveniently written as

$$\chi_\parallel^s = \chi_0 (1 + 2\langle P_2 \rangle_0), \qquad \chi_\perp^s = \chi_0 (1 - \langle P_2 \rangle_0)$$

$$\chi_0 = \frac{n M_s^2 v^2}{3kT}, \qquad \langle P_2 \rangle_0 = \frac{3}{2} \left(\frac{F'}{F} - \frac{1}{3} \right) \qquad (1.62)$$

$\langle P_2 \rangle_0$ is the average of the Legendre polynomial of order 2. We note that [17]

$$F = 1 + \frac{\sigma}{3} + \frac{\sigma^2}{10} + \frac{\sigma^3}{42} + \cdots, \qquad \sigma \ll 1$$

$$F = \frac{e^\sigma}{2\sigma} \left(1 + \frac{1}{2\sigma} + \frac{3}{4\sigma^2} + \cdots \right), \qquad \sigma \gg 1$$

so that for small σ

$$\chi_\parallel^s = \chi_0$$

while for large σ

$$\chi_\parallel^s = 3\chi_0$$

We further note that according to Shliomis [16–18] the particle anisotropy axes are oriented in a *random* fashion therefore the susceptibility averaged over particle orientations is given by the formula

$$\chi = \frac{1}{3}\,(\chi_\parallel^s + 2\chi_\perp^s) \tag{1.63}$$

so that with Eq. (1.62)

$$\chi = \chi_0$$

Thus the susceptibility in the linear approximation does *not depend on the internal anisotropy*. We note that the calculation of $\langle \mathbf{h} \cdot \mathbf{r} \rangle$ described above, interpreted in terms of the itinerant oscillator model described in Appendix C, amounts to the calculation of $\langle \cos(\phi_1 - \phi_2) \rangle = \langle \cos 2\eta \rangle$ (field applied to the anisotropy axis) and $\langle \sin 2\eta \rangle$ (field applied $\perp$ to the anisotropy axis).

The analysis given above assumes a *monodispersed* colloid. Actual colloids are *polydispersed*, that is, they contain a range of particle sizes (with corresponding effects on τ_N and τ_D, see VIII). The particle size distribution is according to [81] accurately described by the Γ distribution

$$\phi(d) = \frac{1}{d_0} \left(\frac{d}{d_0} \right)^\beta \frac{\exp\!\left(\dfrac{d}{d_0} \right)}{\Gamma(\beta + 1)}$$

d is the particle magnetic diameter (see VIII), $\Gamma(x)$ is the gamma function, β and d_0 are the parameters of the distribution. The average diameter is then

$$\bar{d} = \beta + d_0$$

G. Calculation of the Static Magnetization

The magnetization at equilibrium of a ferrofluid particle, in the presence of a DC external magnetic field $\mathbf{H}$ is given by

$$\langle M_z \rangle = M_s \langle \cos \vartheta \rangle$$

The quantity $\cos \vartheta$ can be calculated at equilibrium as follows. Writing $x = \cos \vartheta$ and noting that $\xi = vHM_s/kT$, we have

$$\langle \cos \vartheta \rangle = \langle x \rangle = \frac{\int_{-1}^{1} xe^{\xi x}\, dx}{\int_{-1}^{1} e^{\xi x}\, dx}$$

$$= \frac{\int_{-1}^{1} xe^{\xi x}\, dx}{\dfrac{e^{\xi} - e^{-\xi}}{\xi}}$$

Note that

$$\frac{d}{d\xi} \int_{-1}^{+1} e^{\xi x}\, dx = \int_{-1}^{+1} xe^{\xi x}\, dx$$

and that

$$\int_{-1}^{+1} e^{\xi x}\, dx = \frac{e^{\xi} - e^{-\xi}}{\xi}$$

Using the quotient rule we have

$$\frac{d}{d\xi}\left[\frac{e^{\xi} - e^{-\xi}}{\xi} \right] = \frac{\xi[e^{\xi} + e^{-\xi}] - [e^{\xi} - e^{-\xi}]}{\xi^2}$$

Hence

$$\langle x \rangle = \frac{\dfrac{1}{\xi}\left[\xi(e^{\xi} + e^{-\xi}) - (e^{\xi} - e^{-\xi}) \right]}{e^{\xi} - e^{-\xi}}$$

$$= \frac{\left[(e^{\xi} + e^{-\xi}) - (e^{\xi} - e^{-\xi})/\xi \right]}{e^{\xi} - e^{-\xi}}$$

$$= \left\{ \frac{e^{\xi} + e^{-\xi}}{e^{\xi} - e^{-\xi}} - \frac{1}{\xi} \right\} \tag{1.64}$$

Using

$$\coth \xi = \frac{\cosh \xi}{\sinh \xi} = \frac{e^{\xi} + e^{-\xi}}{e^{\xi} - e^{-\xi}}$$

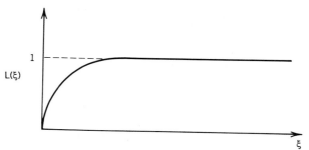

Figure 2. The Langevin function $L(\xi)$. The behavior shown can be observed at room temperature even for moderate fields as vMs is of the order 10^4 to 10^5 Bohr magnetons.

we obtain

$$\langle x \rangle = \coth \xi - \frac{1}{\xi} = L(\xi)$$

The function $L(\xi)$ here is the Langevin function which is plotted in Fig. 2. The static magnetization in the direction of an external magnetic field is therefore

$$\langle M_z \rangle = M_s L(\xi)$$

The behavior of ferrofluid particles subject to a *constant* magnetizing field is adequately described by this Langevin theory of paramagnetism suitably modified to take account of a distribution of particle sizes and particle interactions. Thus ferrofluids have a magnetization curve which *does not exhibit hysteresis.*

II. INCLUSION OF THERMAL AGITATION

A single domain ferromagnetic particle and also a ferrofluid particle experience both a systematic damping of their magnetization and a random fluctuation of the magnetization caused by the presence of thermal energy. In this section we consider two methods of incorporating thermal agitation into the analysis:

1. By using Langevin's treatment of Brownian motion
2. By Brown's intuitive method (which is an adaptation of the 1905 arguments of Einstein)

A. Langevin's Treatment of Brownian Motion: The Langevin Equation

In his treatment of Brownian motion, Langevin began by writing down the equation of motion of a Brownian particle in a suspension. He assumed the forces acting on it could be divided into two parts:

1. A systematic part, $\zeta \dot{x}(t)$ representing a dynamical friction experienced by the particle
2. A fluctuating part $\lambda(t)$ that is characteristic of Brownian motion ($\lambda(t)$ here should not be confused with the constant λ of Landau and Lifshitz)

His equation is thus

$$m\ddot{x}(t) + \zeta \dot{x}(t) = \lambda(t) \tag{2.1}$$

The frictional term $\zeta \dot{x}(t)$ is assumed to be governed by Stokes's Law, which states that the frictional force decelerating a spherical particle of radius a and mass m is

$$\zeta \dot{x} = 6\pi \eta a \dot{x} \tag{2.2}$$

where η is the viscosity of the surrounding fluid (η here should not be confused with Gilbert's damping constant). For the fluctuating part $\lambda(t)$, the following assumptions are made:

1. that it is statistically independent of x;
2. that it varies extremely rapidly compared with the variations of $x(t)$. Such a force is now called a white noise driving force. It has an autocorrelation function

$$\langle \lambda(t_1)\lambda(t_2) \rangle = 2D\delta(t_1 - t_2) \tag{2.3}$$

where D is a constant, $\delta(\tau)$ is the Dirac-delta function and the angular brackets $\langle\ \rangle$ denote statistical averages [10, 29].

The Langevin equation is the very first example of a stochastic differential equation. In that equation, $x(t)$ is a random variable; in other words it is a variable which can only take on particular values, its realizations, with a certain probability. A very thorough account of the meaning of the Langevin equation has been given by Doob (1942) [30]; he showed that the Langevin equation should be interpreted properly not as a differential equation, but as an integral equation. These considerations, however, do not affect the elementary use of the Langevin equation in deriving a formula for the mean-square displacement which we give here.

B.   The Mean-Square Displacement of a Brownian Particle

In the mathematical literature, the random variables are often denoted by $\xi(t)$ and the realizations of those variables by $x(t)$ [31, 32]. An alternative notation is to denote the random variables by capital letters, and the realizations by lower case ones. These distinctions are often ignored by physicists, for economy of notation. We shall, in our discussion of the Langevin equation, adapt the physicists's notation. The distinction becomes more important in the calculation of drift and diffusion coefficients and is consequently adhered to in the appendices.

Multiplying Eq. (2.1) by $x(t)$ we have

$$m\ddot{x}x = -\zeta\dot{x}x + \lambda(t)x \tag{2.4}$$

The quantity

$$\dot{x}x = \frac{1}{2}\frac{d}{dt}(x^2) \tag{2.5}$$

and further

$$\ddot{x}x = \frac{1}{2}\frac{d}{dt}\frac{d(x^2)}{dt} - \dot{x}^2 \tag{2.6}$$

Thus,

$$\frac{m}{2}\frac{d}{dt}\frac{d(x^2)}{dt} - m\dot{x}^2 = \frac{-\zeta}{2}\frac{d(x^2)}{dt} + \lambda x \tag{2.7}$$

The foregoing equations refer to one Brownian particle. If we average this equation over a large number of particles which all start with the same velocity, we obtain

$$\frac{m}{2}\frac{d}{dt}\frac{d\langle x^2\rangle}{dt} - m\langle\dot{x}^2\rangle = -\frac{\zeta}{2}\frac{d\langle x^2\rangle}{dt} + \langle\lambda x\rangle \tag{2.8}$$

We now assume that $\langle\lambda x\rangle$ vanishes, because the force λ varies in a completely irregular manner. In other words, the random force λ and the displacement x are completely uncorrelated. We also assume that the equipartition theorem holds (i.e., the velocity process has reached its equilibrium value, given by the Maxwellian distribution) so that,

$$\frac{1}{2} m \langle \dot{x}^2 \rangle = \frac{1}{2} kT \tag{2.9}$$

Thus Eq. (2.8) becomes

$$\frac{m}{2} \frac{d}{dt} \frac{d\langle x^2 \rangle}{dt} + \frac{\zeta d\langle x^2 \rangle}{2 \, dt} = kT \tag{2.10}$$

Let us now write

$$\frac{d\langle x^2 \rangle}{dt} = u \tag{2.11}$$

and assume that the order of differentiation and averaging may be interchanged. Thus

$$\frac{m}{2} \frac{du}{dt} + \frac{\zeta}{2} u = kT \tag{2.12}$$

The solution to this is

$$u = \frac{2kT}{\zeta} + C \exp\left(\frac{-\zeta t}{m}\right), \quad t > 0 \tag{2.13}$$

where C is a constant of integration. Here, if t is large compared with m/ζ, that is, if a small mass is combined with a large friction constant, the exponential term has no influence after the first extremely small time interval, and

$$u = \frac{d\langle x^2 \rangle}{dt} = \frac{2kT}{\zeta} \tag{2.14}$$

(our neglect of the term $\exp(-\zeta t/m)$ amounts to ignoring the effect of inertia of the Brownian particle entirely; hence we say inertial effects are excluded).

Integrating Eq. (2.14) from $t = 0$ to $t = \tau$, we obtain

$$\langle x^2 \rangle - \langle x_0^2 \rangle = 2kT\tau/\zeta \tag{2.15}$$

If we now set $x_0 = 0$ when $t = 0$, and because of its small value write $\langle (\Delta x)^2 \rangle$ instead of $\langle x^2 \rangle$, then

$$\langle (\Delta x)^2 \rangle = \frac{2kT\tau}{\zeta} \tag{2.16}$$

which is Einstein's formula for the mean-square displacement of a Brownian particle as derived by Langevin. The Langevin equation is treated in detail in Appendix A using the concept of white noise.

C.   Langevin's Method Applied to Rotational Relaxation

The arguments given in Sections II.A and II.B for translational Brownian motion may also be applied to the rotational Brownian motion of a rigid body, such as the magnetization vector **M** of a single domain ferromagnetic particle. We illustrate this by referring to the method used by one of the authors [33] to obtain the Debye theory of dielectric relaxation of an assembly of non-interacting rotators in two dimensions.

The Langevin equation for a body free to rotate about an axis normal to itself is

$$I\ddot{\theta}(t) + \zeta\dot{\theta}(t) + \mu F(t) \sin \theta(t) = \lambda(t) \tag{2.17}$$

In Eq. (2.17), I is the moment of inertia of the rotator about the axis of rotation, θ is the angle the rotator makes with the direction of the driving field $F(t)$, and $\zeta\dot{\theta}$ and $\lambda(t)$ are the frictional and white noise torques due to the Brownian motion of the surroundings. $\lambda(t)$ has the property of Eq. (2.3) where

$$D = kT\zeta \tag{2.18}$$

In order to specialize Eq. (2.17) to the step-on field we write

$$F(t) = F_0 U(t) \tag{2.19}$$

where $U(t)$ is the unit step function and F_0 the amplitude. We require to calculate, for this model, the statistical average $\langle \mu \cos \theta \rangle$ when the inertial effects are ignored.

The problem which presents itself when treating the model using the Langevin equation in the form of Eq. (2.17) is that it is not apparent how that equation may be linearized to yield the solution for small $\mu F_0/kT$. Frood and Lal [34] have suggested that this difficulty may be circumvented by rewriting Eq. (2.17) as an equation of motion for the instantaneous dipole moment

$$p = \mu \cos \theta \tag{2.20}$$

so that

$$\dot{\theta} = -\dot{p}(\mu^2 - p^2)^{-1/2} = -\dot{p}(\mu \sin \theta)^{-1} \tag{2.21}$$

$$\ddot{\theta} = -\ddot{p}(\mu^2 - p^2)^{-1/2} - \dot{p}^2 p(\mu^2 - p^2)^{-3/2} \tag{2.22}$$

$$= -\ddot{p}(\mu \sin \theta)^{-1} - \dot{\theta}^2 p(\mu \sin \theta)^{-1} \tag{2.23}$$

The Langevin equation (2.17) with this change of variable becomes

$$I\ddot{p} + \zeta\dot{p} + (I\dot{\theta}^2 + pF_0)p = \mu^2 F_0 - \mu \sin \theta(t)\lambda(t) \tag{2.24}$$

which is the exact Langevin equation for the motion of the instantaneous dipole moment. Equation (2.24) is nonlinear in p due to the last term on the left and right hand sides. In order to linearize it we first form its statistical average over a large number of rotators which all start with the same value of θ. It then becomes

$$I\langle \ddot{p}\rangle + \zeta\langle \dot{p}\rangle + \langle (I\dot{\theta}^2 + pF_0)p\rangle = \mu^2 F_0 - \langle \mu \sin \theta(t)\lambda(t)\rangle \tag{2.25}$$

The quantity $\langle \mu \sin \theta\lambda(t)\rangle$ can be evaluated as follows. We have from the binomial theorem,

$$\langle \lambda(t) \sin \theta \rangle = \left\langle \lambda(t)\left[1 - \frac{1}{2}\cos 2\theta \ldots \right]\right\rangle \tag{2.26}$$

We also note that Eq. (2.25) may be written as

$$I\langle \ddot{p}\rangle + \zeta\langle \dot{p}\rangle + \langle I\dot{\theta}^2 p\rangle + \frac{1}{2}\mu^2 F_0[\langle \cos 2\theta\rangle - 1] = -\langle \mu\sin \theta(t)\lambda(t)\rangle \tag{2.27}$$

We note further that

$$q = \mu \sin \theta(t) \tag{2.28}$$

satisfies

$$I\langle \ddot{q}\rangle + \zeta\langle \dot{q}\rangle + \langle I\dot{\theta}^2 q\rangle + \frac{1}{2}\mu^2 F_0\langle \sin 2\theta\rangle = \langle \mu \cos \theta(t)\lambda(t)\rangle \tag{2.29}$$

Since we are only concerned with the dipole moment in linear approximation we can discard $\langle \cos 2\theta(t)\lambda(t)\rangle$ because this is of lowest order $(\mu F_0/kT)^2$. Thus

$$\langle \lambda(t) \sin \theta(t)\rangle = \langle \lambda(t)\rangle = 0 \tag{2.30}$$

The remaining terms in Eq. (2.25) which cause difficulty are $\langle p^2\rangle F_0$ and

$\langle I\dot{\theta}^2 p \rangle$. The term $\langle p^2 \rangle F_0$ when written in terms of $\cos \theta$ is

$$\langle p^2 \rangle F_0 = \left(\frac{1}{2} \mu^2 F_0 \right) \langle (1 + \cos 2\theta) \rangle \tag{2.31}$$

Here the term $\langle \cos 2\theta \rangle$ which is the average pertaining to the Kerr-effect relaxation is at least of the order $(\mu F_0/kT)^2$ [35] so that for the linear response we are fully justified in setting

$$\langle p^2 \rangle F_0 = \frac{1}{2} \mu^2 F_0 \tag{2.32}$$

Since the Debye theory pertains to the situation where

$$t \gg I/\zeta \tag{2.33}$$

which implicitly means [36] that a Maxwellian distribution of angular velocities has been achieved, we may now write

$$\langle I\dot{\theta}^2 p \rangle = kT \langle p \rangle \tag{2.34}$$

since the orientation and the angular velocity variables, when equilibrium of the angular velocities has been reached, are decoupled from each other, as far as the time behavior of the orientations is concerned [36]. The assumption of a Maxwellian distribution of velocities also means that

$$I\langle \ddot{p} \rangle = 0 \tag{2.35}$$

in Eq. (2.25), so that finally the linearized form of that equation in the limit of long times is

$$\zeta \langle \dot{p} \rangle + kT \langle p \rangle = \frac{1}{2} \mu^2 F_0 U(t) \tag{2.36}$$

If we take the Laplace transform of $\langle p \rangle$, we find that

$$\mathscr{L}\{\langle p \rangle\} = \frac{\mu^2 F_0}{2\zeta s \left[s + \dfrac{kT}{\zeta} \right]} \tag{2.37}$$

which immediately yields

$$\langle p \rangle = \frac{\mu^2 F_0}{2kT} U(t) \left[1 - \exp\left(-\frac{kT}{\zeta} t \right) \right] \tag{2.38}$$

which is the time response of the average of the moment $\langle \mu \cos \theta \rangle$.

D.   Application of Langevin's Method to Rotational Brownian Motion

We have shown, in Sections II.A–II.C, how the translational and rotational motion of a Brownian particle in a viscous fluid can be treated using Langevin's method. The damping effects in the rotational motion of the magnetization of a single domain ferromagnetic particle can be interpreted as due to Brownian torques. Viewing the damping torques in this way, Gilbert's equation, Eq. (1.12), represents a statistical average of Brownian motion fluctuations since only a systematic term is present (the damping field term). If we wish to use Gilbert's equation to describe not the average but the specific behavior of **M** for an individual particle it is necessary to include the random fluctuations outlined by Langevin. This can be achieved by augmenting the damping field with a random field term, possessing statistical properties similar to those of the white noise driving field of Eq. (2.3).

Gilbert's equation so augmented is

$$\frac{d\mathbf{M}}{dt} = \gamma \mathbf{M} \times \left(\mathbf{H} - \eta \frac{d\mathbf{M}}{dt} + \mathbf{h}(t) \right) \tag{2.39}$$

where the random field is represented by a vector $\mathbf{h}(t)$ such that

$$\mathbf{h}(t) = h_1(t)\mathbf{i} + h_2(t)\mathbf{j} + h_3(t)\mathbf{k} \tag{2.40}$$

where $h_i(t)$ has the properties

$$\langle h_i(t) \rangle = 0 \tag{2.41}$$

that is, a statistical average of zero, and for $j = 1, 2, 3$

$$\langle h_i(t) h_j(t) \rangle = \begin{cases} \mu, & i = j \\ 0, & \text{otherwise}. \end{cases} \tag{2.42} \\ \tag{2.43}$$

that is, noise torques about different Cartesian axes are uncorrelated, and also

$$\langle h_i(t) h_i(t + \Delta t) \rangle = 0, \quad \Delta t \neq 0 \tag{2.44}$$

E.   The Fokker–Planck Equation Method (Intuitive Treatment)

We have very briefly outlined Langevin's approach to the theory of a random process in which attention is focussed on the random variation in time of the system variable of interest [10].

The second method is the diffusion equation or Fokker–Planck method. To quote Wang and Uhlenbeck [10], "macroscopically, for an ensemble of particles or systems, the variations which occur are like that of a diffusion process. The distribution function of the random variables of the system will, therefore, satisfy a partial differential equation of diffusion type". A rigorous derivation of such an equation requires one to precisely define its stochastic process, which we shall do later. Initially, however, we will give an intuitive derivation of the diffusion equation for a translating Brownian particle.

This procedure is based on Einstein's argument that in equilibrium, the rate of diffusion under a concentration gradient must be exactly balanced by the effect of the applied force. Thus if $W(x, t)$ is the number density of Brownian particles, then we have

$$D \, \partial W(x, t)/\partial x = W(x, t)F/6\pi\eta a \qquad (2.45)$$

where D is the diffusion coefficient, F the applied force, η the drag coefficient and a the radius of the particle. The velocity $\dot{x}$ of the particle from Stokes's Law, (see Eq. (2.2)), is

$$\dot{x} = F/6\pi\eta a \qquad (2.46)$$

Thus

$$0 = D \, \partial W/\partial x - W\dot{x} \qquad (2.47)$$

in equilibrium. Under nonequilibrium conditions, however, a nonzero current density $\mathbf{J}$ exists, such that,

$$\mathbf{J} = -(D \, \partial W/\partial x - W\dot{x})\mathbf{i} \qquad (2.48)$$

By Gauss's divergence theorem, we must have, for a volume v bounded by a closed surface S,

$$\int_s \mathbf{J} \cdot \mathbf{n} \, dS = \int_v \text{div} \, \mathbf{J} \, dv \qquad (2.49)$$

which by noting $\text{div} \, \mathbf{J} = -\partial W/\partial t$ where

$$\dot{x} = -\frac{1}{\zeta} \frac{\partial V}{\partial x}$$

gives

$$\frac{\partial W}{\partial t} = \frac{\partial}{\partial x} \left(D \frac{\partial W}{\partial x} + W \frac{1}{\zeta} \frac{\partial V}{\partial x} \right) \qquad (2.50)$$

This is the Fokker–Planck equation for diffusion in one dimension under the influence of a force F.

A precise derivation of the Fokker–Planck equation requires the introduction of the notion of a stochastic process. This is given in Appendix B.

F. Brown's Intuitive Derivation of the Fokker–Planck Equation

The manner in which Brown expressed his intuitive method of deriving the Fokker–Planck equation is to consider the effect of thermal fluctuations on $W(\mathbf{r}, t)$, the probability density of orientations of $\mathbf{M}$. Brown suggests that thermal agitation causes $W(\mathbf{r}, t)$ to become more uniform, so that in an equation describing the time evolution of W thermal agitation gives rise to a diffusion term in W. The equation which governs the time evolution of W in the absence of thermal agitation is based on the principle of continuity of the representative points comprising W. It is

$$\partial W / \partial t = -\text{div }\mathbf{J} \tag{2.51}$$

where $\mathbf{J}$ is the current density which in the absence of thermal agitation is

$$\mathbf{J} = W\dot{\mathbf{r}} \tag{2.52}$$

where $\dot{\mathbf{r}} = \dot{\mathbf{M}}/M_s$ is obtained from Gilbert's equation written in spherical polar coordinates. Here the presence of thermal agitation will result in a diffusion current density proportional to ∇W, so that

$$\mathbf{J} = W\dot{\mathbf{r}} - k'\nabla W \tag{2.53}$$

where k' is an arbitrary constant which determines the extent of thermal disruption and which is found by imposing the Maxwell–Boltzmann distribution at equilibrium, so that the continuity equation, including thermal fluctuations on writing div, grad, and $\dot{\mathbf{r}}$ in spherical polar coordinates having discarded the radial dependence, is (for details see Appendix D)

$$\frac{\partial W}{\partial t} = \frac{1}{\sin\vartheta}\frac{\partial}{\partial\vartheta}\left\{\sin\vartheta\left[\left(h'\frac{\partial V}{\partial\vartheta} - \frac{g'}{\sin\vartheta}\frac{\partial V}{\partial\phi}\right)W + k'\frac{\partial W}{\partial\vartheta}\right]\right\}$$
$$+ \frac{1}{\sin\vartheta}\frac{\partial}{\partial\phi}\left\{\left(g'\frac{\partial V}{\partial\vartheta} + \frac{h'}{\sin\vartheta}\frac{\partial V}{\partial\phi}\right)W + \frac{k'}{\sin\vartheta}\frac{\partial W}{\partial\phi}\right\} \tag{2.54}$$

which is Brown's equation. The arbitrary constant k' can now be found by imposition of the Maxwell–Boltzmann equilibrium distribution which is

$$W(\vartheta, \phi) = W_0 \exp(-vV(\vartheta, \phi)/kT) \tag{2.55}$$

where W_0 is the zero potential equilibrium value of W. In equilibrium $\partial W/\partial t = 0$ and so Brown's equation is

$$0 = \frac{1}{\sin \vartheta} \frac{\partial}{\partial \vartheta} \left\{ \sin \vartheta \left[\left(h' \frac{\partial V}{\partial \vartheta} - \frac{g'}{\sin \vartheta} \frac{\partial V}{\partial \phi} \right) W - k' \frac{v}{kT} \frac{\partial V}{\partial \vartheta} W \right] \right\}$$
$$+ \frac{1}{\sin \vartheta} \frac{\partial}{\partial \vartheta} \left\{ \left(g' \frac{\partial V}{\partial \vartheta} + \frac{h'}{\sin \vartheta} \frac{\partial V}{\partial \phi} \right) W - \frac{k'}{\sin \vartheta} \frac{v}{kT} \frac{\partial V}{\partial \phi} W \right\} \tag{2.56}$$

Considering each partial differential component separately, we find that

$$h' = vk'/kT \tag{2.57}$$

which by noting Eq. (1.33), defining h', gives

$$k' = \frac{kT}{v} \frac{\gamma}{M_s} \frac{\alpha}{1 + \alpha^2} \tag{2.58}$$

and so Brown's equation can be rewritten as

$$\frac{1}{k'} \frac{\partial W}{\partial t} = \frac{1}{\sin \vartheta} \frac{\partial}{\partial \vartheta} \left\{ \sin \vartheta \left[\frac{v}{kT} \left(\frac{\partial V}{\partial \vartheta} - \frac{1}{\sin \vartheta} \frac{1}{\alpha} \frac{\partial V}{\partial \phi} \right) W + \frac{\partial W}{\partial \vartheta} \right] \right\}$$
$$+ \frac{1}{\sin \vartheta} \frac{\partial}{\partial \phi} \left\{ \frac{v}{kT} \left(\frac{1}{\alpha} \frac{\partial V}{\partial \vartheta} + \frac{1}{\sin \vartheta} \frac{\partial V}{\partial \phi} \right) W + \frac{1}{\sin \vartheta} \frac{\partial W}{\partial \phi} \right\} \tag{2.59}$$

III. SOLUTION OF BROWN'S EQUATION USING SPHERICAL HARMONICS

A. Introduction

In this section we summarize the approach used by previous authors [8, 16–19] to find expressions for the relaxation times of single domain ferromagnetic and ferrofluid particles. We begin with the Fokker–Planck equation obtained from Gilbert's equation, in spherical polar coordinates, augmented by a random field term, that is, with Brown's equation. We then expand the probability density of orientations of **M**, that is,

$W(\vartheta, \phi, t)$, in spherical harmonics, by assuming W to be of the form

$$W(\vartheta, \phi, t) = \sum_{n=0}^{\infty} \sum_{m=-n}^{n} a_{nm}(t) P_n^{|m|}(\cos \vartheta) \exp im\phi$$

$$|m| \le n, \qquad a_{n,-m} = a_{nm}^* \tag{3.1}$$

where $P_n^m(\cos \vartheta)$ are the associated Legendre functions and $a_{nm}(t)$ are the time-dependent functions describing the time evolution (the average value) of each spherical harmonic. The condition $a_{n,-m} = a_{nm}^*$ means that the solution need only be determined for $m > 0$. (The asterisk $*$ here denotes the complex conjugate).

This leads to an infinite set of differential-difference equations, on substitution into Brown's equation. The calculation of the relaxation times is then achieved by selecting the relevant indices n and m, from which we obtain a differential-difference equation describing the time evolution of the average value of the desired spherical harmonic. Those spherical harmonics which are of interest to us here are $P_1(\cos \vartheta)$, with $n = 1$ and $m = 0$, which describes the evolution of the alignment with the z axis and $P_1^1(\cos \vartheta) \cos \phi$, with $n = 1$ and $m = 1$ which describes alignment perpendicular to the z axis.

The main difficulty which arises in solving these differential-difference equations is that the equation describing the time average behaviour of a given spherical harmonic is linked to all the others so that the solution of one equation involves the solution of all the others. One method of coping with this is to write the differential-difference equation in the form of a continued fraction and iterate this fraction until convergence within certain limits occurs. This method is equivalent to the construction of a matrix equation out of the infinite set of differential-difference equations, and then to increasing the matrix size until convergence occurs. Another method is to assume that the largest order spherical harmonic in the equation has reached equilibrium so that it is no longer time dependent. This method will result in closed form expressions for the relaxation times. The last method is to convert the solution into a Sturm–Liouville problem. They are all considered below.

In the following subsections we will study the differential-difference equations for the cases of an external uniform field $(\cos \vartheta)$ potential and anisotropy $(\sin^2 \vartheta)$ potential.

B. Use of a Spherical Harmonic Expansion for a Longitudinal Field

The first differential-difference equation derived is that pertaining to a single domain ferromagnetic particle in an external magnetic field directed along the z axis so that the potential is

$$V(\vartheta) = -HM_s \cos \vartheta \qquad (3.2)$$

The spherical harmonic of interest to us here is that describing the relaxation in the direction of the applied field, namely $P_1(\cos \vartheta)$. Since the field has no azimuthal (or ϕ) dependence it is possible to use an expansion, in only one space coordinate of the form

$$W(\vartheta, t) = \sum_{n=0}^{\infty} a_n(t) P_n(\cos \vartheta) \qquad (3.3)$$

where $P_n(\cos \vartheta)$ are the Legendre polynomials and $a_n(t)$ are the corresponding time-dependent separation coefficients. Furthermore since V and W are independent of ϕ, Brown's equation reduces to

$$\frac{1}{k'} \sin \vartheta \, \frac{\partial W}{\partial t} = \frac{\partial}{\partial \vartheta} \left[\sin \vartheta \left(\frac{v}{kT} \frac{\partial V}{\partial \vartheta} W + \frac{\partial W}{\partial \vartheta} \right) \right] \qquad (3.4)$$

For convenience we now make the substitution $x = \cos \vartheta$ and also use the potential given in Eq. (3.2) to give

$$\frac{1}{k'} \frac{\partial W}{\partial t} = \frac{\partial}{\partial x} \left[(1 - x^2) \left(\frac{\partial W}{\partial x} - \xi W \right) \right] \qquad (3.5)$$

where the parameter $\xi = vHM_s/kT$ is the ratio of external magnetic field energy to thermal energy. Now substituting the Legendre polynomial expansion for $W(\vartheta, t)$ we have

$$\frac{1}{k'} \sum_{n=0}^{\infty} P_n \frac{da_n}{dt} = \sum_{n=0}^{\infty} a_n \left\{ \frac{d}{dx} \left[(1 - x^2) \left(\frac{dP_n}{dx} - \xi P_n \right) \right] \right\} \qquad (3.6)$$

We make use of the following recurrence relations for the Legendre polynomials, in order to simplify the left hand side; these are [37]

$$\frac{d}{dx} \left[(1 - x^2) \frac{dP_n}{dx} \right] = -n(n + 1) P_n \qquad (3.7)$$

and

$$(1 - x^2) dP_n/dx = nP_{n-1} - nxP_n \qquad (3.8)$$

so that

$$\frac{1}{k'} \sum_{n=0}^{\infty} P_n \frac{da_n}{dt} = \sum_{n=0}^{\infty} a_n \{ P_n[-n(n + 1) + \xi(2 + n)x] - \xi nP_{n-1} \} \qquad (3.9)$$

Here we use another recurrence relation namely

$$(2n + 1)xP_n = (n + 1)P_{n+1} + nP_{n-1} \tag{3.10}$$

to simplify the xP_n term thus

$$\frac{1}{k'} \sum_{n=0}^{\infty} P_n \frac{da_n}{dt} = \sum_{n=0}^{\infty} a_n \left[- n(n + 1)P_n + \frac{\xi(n + 1)(n + 2)}{2n + 1} P_{n+1} \right. $$
$$\left. + \xi \left(\frac{n(n + 2)}{2n + 1} - n \right) P_{n-1} \right] \tag{3.11}$$

which after simplifying the last term is

$$\frac{1}{k'} \sum_{n=0}^{\infty} P_n \frac{da_n}{dt} = \sum_{n=0}^{\infty} a_n \left[-n(n + 1)P_n + \frac{\xi(n + 1)(n + 2)}{2n + 1} P_{n+1} \right. $$
$$\left. - \xi \left(\frac{n(n - 1)}{2n + 1} \right) P_{n-1} \right] \tag{3.12}$$

The orthogonality properties of the Legendre polynomials namely

$$\int_{-1}^{+1} P_n(x)P_m(x) \, dx = \frac{2}{2n + 1} \delta_{m,n} \tag{3.13}$$

(where $\delta_{m,n}$ is the Kronecker delta) allow us to reduce the infinite summation to an infinite set of differential-difference equations, by writing

$$\frac{1}{k'} \sum_{n=0}^{\infty} \int_{-1}^{1} \frac{da_n}{dt} P_n P_m \, dx = \sum_{n=0}^{\infty} \int_{-1}^{1} -n(n + 1)a_n P_n P_m \, dx$$
$$+ \xi \sum_{n=0}^{\infty} \int_{-1}^{1} a_n P_{n+1} P_m \frac{(n + 1)(n + 2)}{2n + 1} \, dx$$
$$- \xi \sum_{n=0}^{\infty} \int_{-1}^{1} a_n P_{n-1} P_m \frac{n(n - 1)}{2n + 1} \, dx \tag{3.14}$$

which results in

$$\frac{1}{k'n(n + 1)} \dot{a}_n + a_n = \xi \left[\frac{a_{n-1}}{2n - 1} - \frac{a_{n+1}}{2n + 3} \right] \tag{3.15}$$

This is the infinite set of differential-difference equations which describes the time evolution of all the coefficients corresponding to all the Legendre polynomials. The polynomial for the relaxation in the direction of the field is $P_1(\cos \vartheta) = \cos \vartheta$, the corresponding time-dependent function is $a_1(t)$ and so letting $n = 1$ for this description, we have

$$\frac{1}{2k'} \dot{a}_1 + a_1 = \xi \left[a_0 - \frac{a_2}{5} \right] \tag{3.16}$$

We see from the zero field time behavior of a_1 which has the exponential form

$$a_1(t) = a_1(0) \exp(-t/\tau_N) \tag{3.17}$$

that the Néel relaxation time τ_N is related to the diffusion constant k' by

$$\tau_N = (2k')^{-1} \tag{3.18}$$

and from Eq. (2.58), we have

$$\tau_N = \frac{v}{kT} \frac{M_s(1 + \alpha^2)}{\gamma 2\alpha} \tag{3.19}$$

The difficulty in solving these equations is now apparent, since we see that the evolution of $a_1(t)$ depends on $a_2(t)$ which would in turn depend upon $a_3(t)$ and so on. The exact solution therefore involves *all coefficients* $a_n(t)$. The calculation of the exact solution is consequently impossible. It is possible nevertheless to formulate the equation as a continued fraction which can be iterated until convergence, within certain limits, of expressions obtained for n and $n + 1$, is obtained.

The formulation of the continued fraction is best achieved using Laplace transforms. The general equation (Eq. (3.15)) so transformed is

$$\frac{2\tau_N s}{n(n + 1)} A_n(s) + A_n(s) = \frac{2\tau_N}{n(n + 1)} a_n(0) + \xi \left[\frac{A_{n-1}(s)}{2n - 1} - \frac{A_{n+1}(s)}{2n + 3} \right] \tag{3.20}$$

which rearranged becomes

$$\left(\frac{2\tau_N s}{n(n + 1)} + 1 \right) A_n(s) = \frac{2\tau_N}{n(n + 1)} a_n(0) + \xi \left[\frac{A_{n-1}(s)}{2n - 1} - \frac{A_{n+1}(s)}{2n + 3} \right] \tag{3.21}$$

We then define

$$R_n(s) = A_n(s) / A_{n-1}(s) \tag{3.22}$$

so that

$$\left(\frac{2\tau_N s}{n(n+1)} + 1\right)R_n(s) = \frac{2\tau_N}{n(n+1)}\frac{a_n(0)}{A_{n-1}(s)} + \xi\left[\frac{1}{2n-1} - \frac{R_n R_{n+1}}{2n+3}\right] \tag{3.23}$$

Transposing the last term we have

$$\left(\frac{2\tau_N s}{n(n+1)} + 1\right)R_n(s) + \frac{\xi R_n R_{n+1}}{2n+3} = \frac{\xi}{2n-1} + \frac{2\tau_N}{n(n+1)}\frac{a_n(0)}{A_{n-1}(s)} \tag{3.24}$$

and

$$R_n(s) = \frac{\dfrac{2\tau_N}{n(n+1)}\dfrac{a_n(0)}{A_{n-1}(s)} + \dfrac{\xi}{2n-1}}{\dfrac{2\tau_N s}{n(n+1)} + 1 + \dfrac{\xi R_{n+1}(s)}{2n+3}} \tag{3.25}$$

which means that for $n = 1$,

$$A_1 = R_1 A_0 = \frac{\tau_N a_1(0) + \xi A_0}{\tau_N s + 1 + \dfrac{\xi R_2(s)}{5}} \tag{3.26}$$

where $R_2(s)$ is to be found from Eq. (3.25). The field-dependent relaxation time is now found by equating the denominator of the left hand side with zero and solving the polynomial for s, which is related to $\tau_\parallel$ by

$$s = -1/\tau_\parallel \tag{3.27}$$

The exact solution for the largest relaxation time in this context is found by an iteration of $R_n(s)$. The convergent iteration has been performed by our colleague K.P. Quinn and is graphed in Fig. 3.

C.   Use of a Spherical Harmonic Expansion for a Transverse Field

To enable us to calculate the relaxation time of the magnetization in a direction perpendicular to the applied field it is necessary to expand W in spherical harmonics involving the two space coordinates ϑ and ϕ. For this we use the form given in the introduction to this section which is

$$W(\vartheta, \phi, t) = \sum_{n=0}^{\infty} \sum_{m=-n}^{n} a_{nm}(t) P_n^{|m|}(\cos\vartheta) \exp im\phi \tag{3.28}$$

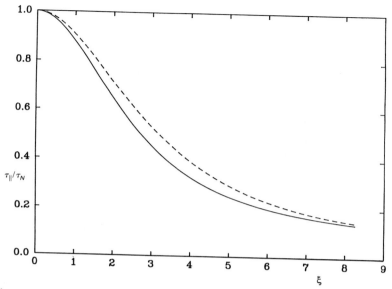

Figure 3. The parallel relaxation time $\tau_\parallel / \tau_N$ as a function of the field strength parameter, $\xi = vHMs/kT$. The dotted line is the result obtained from the iteration of the continued fraction, Eq. (3.26). The solid line is the closed form expression of Eq. (4.46).

We begin our use of this expansion with the calculation of the field dependence of the spherical harmonic $P_1^1(\cos \vartheta)e^{i\phi} = \sin \vartheta(\cos \phi + i \sin \phi)$ in a field again directed along the z axis. (Since it is only the eigenvalue that is of interest it is not necessary to impose a transverse field). Brown's equation for $V(\vartheta, t)$ and $W(\vartheta, \phi, t)$ (noting Eq. (3.18)) is

$$2\tau_N \sin \vartheta \, \frac{\partial W}{\partial t} = \frac{\partial}{\partial \vartheta} \left\{ \sin \vartheta \left[\frac{v}{kT} \frac{\partial V}{\partial \vartheta} W + \frac{\partial W}{\partial \vartheta} \right] \right\}$$
$$+ \frac{\partial}{\partial \phi} \left[\frac{v}{kT\alpha} \frac{\partial V}{\partial \vartheta} W + \frac{1}{\sin \vartheta} \frac{\partial W}{\partial \phi} \right] \quad (3.29)$$

Here we note that the complex quantity resulting from $\partial W/\partial \phi$ describes the oscillation of the x and y components of the magnetization on the x–y plane; that is, the gyromagnetic precession. The characteristic frequency of this is apparent from the explicit form of Gilbert's equation and is obtainable from the above by writing

$$2\tau_N \omega_0 = \frac{v}{kT} \frac{H}{\alpha} M_s \quad (3.30)$$

which with Eq. (3.19) for τ_N yields

$$\omega_0 = \gamma H/(1 + \alpha^2) \qquad (3.31)$$

However this will not affect the calculation of the real quantities in the relaxation time and can be ignored in this analysis (see also Section VI). The equation which we consider is therefore

$$2\tau_N \sin\vartheta \, \frac{\partial W}{\partial t} = \frac{\partial}{\partial\vartheta}\left[\sin\vartheta\left(\frac{v}{kT}\frac{\partial V}{\partial\vartheta}W + \frac{\partial W}{\partial\vartheta}\right)\right] + \frac{1}{\sin\vartheta}\frac{\partial^2 W}{\partial\phi^2} \qquad (3.32)$$

We have once again, making the substitution $x = \cos\vartheta$ and using the potential $V(\vartheta) = -HM_s \cos\vartheta$

$$2\tau_N \frac{\partial W}{\partial t} = \frac{\partial}{\partial x}\left[(1 - x^2)\left(\frac{\partial W}{\partial x} - \xi W\right)\right] + \frac{1}{(1 - x^2)}\frac{\partial^2 W}{\partial\phi^2} \qquad (3.33)$$

We substitute the expansion $W = W(\vartheta, \phi, t)$ to obtain

$$2\tau_N \frac{\partial W}{\partial t} = \sum_{n=0}^{\infty}\sum_{m=-n}^{n} a_{nm}(t)e^{im\phi}\left\{\frac{d}{dx}\left[(1 - x^2)\frac{dP_n^{|m|}}{dx}\right] - \xi P_n^{|m|}\right.$$
$$\left. - \frac{m^2}{(1 - x^2)}P_n^{|m|}\right\} \qquad (3.34)$$

The associated Legendre functions satisfy the equation [37]

$$\frac{d}{dx}\left[(1 - x^2)\frac{dP_n^m(x)}{dx}\right] + \left[n(n + 1) - \frac{m^2}{(1 - x^2)}\right]P_n^m(x) = 0 \qquad (3.35)$$

and so we can write

$$2\tau_N \frac{\partial W}{\partial t} = \sum_{n=0}^{\infty}\sum_{m=-n}^{n} a_{nm}e^{im\phi}\left\{-n(n + 1)P_n^{|m|} - \xi\frac{d}{dx}[(1 - x^2)P_n^{|m|}]\right\} \qquad (3.36)$$

The last term in this equation contains

$$-\frac{d}{dx}[(1 - x^2)P_n^{|m|}] = 2xP_n^{|m|} - (1 - x^2)\frac{dP_n^{|m|}}{dx} \qquad (3.37)$$

From the recurrence relations [37]

$$(2n + 1)xP_n^m(x) = (n - m + 1)P_{n+1}^m(x) + (n + m)P_{n-1}^m(x) \quad (3.38)$$

and

$$(1 - x^2)\frac{dP_n^m}{dx} = (n + m)P_{n-1}^m - nxP_n^m \quad (3.39)$$

we have

$$-\frac{d}{dx}[(1 - x^2)P_n^{|m|}] = \frac{2}{2n + 1}[(n - |m| + 1)P_{n+1}^{|m|}(x) + (n + |m|)P_{n-1}^{|m|}(x)]$$

$$+ [nxP_n^{|m|} - (n + |m|)P_{n-1}^{|m|}] \quad (3.40)$$

and using Eq. (3.38) to eliminate the $xP_n^{|m|}$ term, this is

$$\frac{n + 2}{2n + 1}[(n - |m| + 1)P_{n+1}^{|m|}(x) + (n + |m|)P_{n-1}^{|m|}(x)] - (n + |m|)P_{n-1}^{|m|}$$
$$(3.41)$$

which on separating the coefficients and substituting into Eq. (3.36) is

$$2\tau_N \frac{\partial W}{\partial t} = \sum_{n=0}^{\infty} \sum_{m=-n}^{n} a_{nm}e^{im\phi}\left[-n(n + 1)P_n^{|m|}\right.$$

$$+ \xi \frac{(n + 2)(n + |m| + 1)}{2n + 1} P_{n+1}^{|m|} - \xi \frac{(n - 1)(n + |m|)}{2n + 1} P_{n-1}^{|m|}\right]$$
$$(3.42)$$

Use of the orthogonality properties of the associated Legendre functions, which are

$$\int_{-1}^{+1} P_n^m(x)P_{n'}^{m'}(x)\, dx = \frac{(n + m)!}{(n - m)!}\frac{2}{2n + 1}\delta_{n,n'}\delta_{m,m'} \quad (3.43)$$

finally reduces the infinite series summation to the following infinite set of differential-difference equations:

$$2\tau_N \dot{a}_{nm} = -n(n + 1)a_{nm}$$

$$+ \xi \frac{(n + 1)(n - |m|)}{2n - 1} a_{n-1m} - \xi \frac{n(n + |m| + 1)}{2n + 3} a_{n+1m} \quad (3.44)$$

As would be expected this yields the same result for $n = 1$, $m = 0$ as does the equation gained from the Legendre polynomial expansion. Here for $P_1^1(\cos \vartheta)e^{i\phi}$ we use $n = m = 1$, which gives

$$2\tau_N \dot{a}_{11}(t) = -2a_{11}(t) - \frac{3}{5}\xi a_{21}(t) \tag{3.45}$$

D. The Calculation of the Parallel and Perpendicular Relaxation Times for Small ξ

In this section we consider an equivalent procedure to the construction of a continued fraction, that is, the construction of an infinite matrix equation

$$2\tau_N \dot{\mathbf{a}}_m = \boldsymbol{\lambda}_m \mathbf{a}_m \tag{3.46}$$

where

$$\mathbf{a}_m = \begin{pmatrix} a_{1m}(t) \\ a_{2m}(t) \\ \vdots \\ a_{nm}(t) \\ \vdots \end{pmatrix} \tag{3.47}$$

where the coefficients $a_{nm}(t)$ are the same as the coefficients in the previous section, and $\boldsymbol{\lambda}_m$ is the three-diagonal matrix determined by Eq. (3.44). Thus the full equation is

$$2\tau_N \begin{pmatrix} \dot{a}_{1m} \\ \dot{a}_{2m} \\ \dot{a}_{3m} \\ \vdots \\ \dot{a}_{nm} \\ \vdots \end{pmatrix}$$

$$= \begin{pmatrix} -2 & -\dfrac{\xi(2+m)}{5} & 0 & \cdots & 0 & \cdots \\ 3\xi(2-m) & -6 & -\dfrac{2\xi(3+m)}{7} & \cdots & 0 & \cdots \\ 3 & \dfrac{4\xi(3-m)}{5} & -12 & \cdots & \dfrac{n(n+m+1)}{2n+3} & \cdots \\ 0 & & \dfrac{\xi(n+1)(n-m)}{2n-1} & -n(n+1) & \cdots \\ \cdots & & & & \\ 0 & 0 & & & \end{pmatrix} \begin{pmatrix} a_{1m} \\ a_{2m} \\ a_{3m} \\ \vdots \\ a_{nm} \\ \vdots \end{pmatrix} \tag{3.48}$$

Taking the Laplace transform of this, we have

$$2\tau_N s\mathbf{A}(s) = \boldsymbol{\lambda}_m \mathbf{A}(s) + \mathbf{a}_m(0)2\tau_N \tag{3.49}$$

The system matrix is therefore

$$(2\tau_N s\mathbf{I} - \boldsymbol{\lambda}_m)$$

Here the field-dependent relaxation time $\tau = 1/s$ is determined by the lowest eigenvalue of the system matrix $(2\tau_N s\mathbf{I} - \boldsymbol{\lambda}_m)$. The eigenvalues are obtained by Gaussian elimination of all the nondiagonal terms in the system matrix. Once again, since the matrix is infinite the exact solution will involve an infinite number of eliminations. An analogous procedure to the iteration of the continued fraction until convergence occurs, is the expansion of the system matrix, until convergence, within certain limits, of the eigenvalue expressions for $n \times n$ and $(n + 1) \times (n + 1)$ is achieved.

A result that may be obtained by this method is that for low field strengths, $\xi \ll 1$; it is possible to limit the matrix size to 2×2 so that terms $O(\xi^2)$ may be neglected.

This calculation is performed here for the relaxation times $\tau_\parallel(\xi)$ and $\tau_\perp(\xi)$ (the time constants of the time-dependent functions corresponding to $\langle P_1(\cos\vartheta)\rangle$ and $\langle P_1^1(\cos\vartheta)e^{i\phi}\rangle$). The differential-difference equations for these are

$$2\tau_N\dot{a}_{10} = -2a_{10} + 2\xi a_{00} - \frac{2}{5}\xi a_{20} \tag{3.50}$$

and

$$2\tau_N\dot{a}_{11} = -2a_{11} - \frac{3}{5}\xi a_{21} \tag{3.51}$$

respectively. To construct the 2×2 matrix for these, it is necessary to use the differential-difference equations for $n = 2$ in each case. These are

$$2\tau_N\dot{a}_{20} = -6a_{20} + 2\xi a_{10} - \frac{6\xi}{7}a_{30} \tag{3.52}$$

and

$$2\tau_N\dot{a}_{21} = -6a_{21} + \xi a_{11} - \frac{8\xi}{7}a_{31} \tag{3.53}$$

The two 2×2 system matrices are therefore

$$\begin{pmatrix} 2\tau_N s + 2 & \dfrac{2}{5}\xi \\ -2\xi & 2\tau_N s + 6 \end{pmatrix} \tag{3.54}$$

and

$$\begin{pmatrix} 2\tau_N s + 2 & \dfrac{3}{5}\xi \\ -\xi & 2\tau_N s + 6 \end{pmatrix} \tag{3.55}$$

We consider the first of these. Equating the determinant to zero produces the quadratic equation

$$(2\tau_N s)^2 + 8(2\tau_N s) + 12 + 4\xi^2/5 = 0 \tag{3.56}$$

the roots of which are

$$2\tau_N s = -4 \pm 2\sqrt{1 - \frac{\xi^2}{5}} \tag{3.57}$$

with the root corresponding to a_1 being

$$2\tau_N s = -4 + 2\sqrt{1 - \frac{\xi^2}{5}} \tag{3.58}$$

The use of the binomial theorem for small ξ gives

$$2\tau_N s = -2(1 + \xi^2/10) \tag{3.59}$$

The relation between s and the relaxation time τ is

$$\tau = -1/s \tag{3.60}$$

giving a low field strength ($\xi \ll 1$) relaxation time $\tau_\parallel(\xi)$ of

$$\tau_\parallel(\xi) = \tau_N/(1 + \xi^2/10) \tag{3.61}$$

which by further use of the binomial expansion is

$$\tau_\parallel(\xi) = \tau_N(1 - \xi^2/10) \tag{3.62}$$

A similar procedure for the other system matrix gives the characteristic equation

$$(2\tau_N s)^2 + 8(2\tau_N s) + 12 + 3\xi^2/5 = 0 \tag{3.63}$$

which in turn leads to the low field strength ($\xi \ll 1$) relaxation time $\tau_\perp(\xi)$ of

$$\tau_\perp(\xi) = \tau_N(1 - 3\xi^2/40) \tag{3.64}$$

These expressions coincide with those obtained by Martsenyuk et al. [16] which were for the Debye relaxation of a ferrofluid particle.

E.   Use of the Legendre Polynomial Expansion for Uniaxial Anisotropy

The field-dependent expressions obtained in the previous sections for relaxation in an external magnetic field are applicable to both Néel and Debye relaxation and their specific application merely requires the use of the relevant time constant τ_N or τ_D.

In this section we consider the dependence of the relaxation time on the internal anisotropy of a single domain ferromagnetic particle. Thus it is applicable only to Néel relaxation. We consider the simplest form of anisotropy, namely uniaxial anisotropy described by the potential

$$V_a = K \sin^2\vartheta \tag{3.65}$$

where K is an anisotropy constant.

We approach the derivation in a similar way to that of Section III.B. We expand $W = W(\vartheta, t)$ as

$$W(\vartheta, t) = \sum_{n=0}^{\infty} a_n(t) P_n(\cos \vartheta) \tag{3.66}$$

where P_n and a_n are again the Legendre polynomials and the corresponding time-dependent separation coefficients. Again there is no azimuthal dependence and Brown's equation reduces to

$$2\tau_N \sin \vartheta \, \frac{\partial W}{\partial t} = \frac{\partial}{\partial \vartheta} \left[\sin \vartheta \left(\frac{v}{kT} \frac{\partial V}{\partial \vartheta} W + \frac{\partial W}{\partial \vartheta} \right) \right] \tag{3.67}$$

Substitution of $V(\vartheta, t)$ from above and $x = \cos \vartheta$ gives

$$2\tau_N \frac{\partial W}{\partial t} = \frac{\partial}{\partial x} \left[(1 - x^2)\left(\frac{\partial W}{\partial x} - 2\sigma x W \right) \right] \tag{3.68}$$

where $\sigma = Kv/kT$. Substitution of the Legendre polynomial expansion for W leads to

$$2\tau_N \frac{\partial W}{\partial t} = \sum_{n=0}^{\infty} a_n \left\{ \frac{d}{dx} \left[(1 - x^2) \left(\frac{dP_n}{dx} - 2\sigma x P_n \right) \right] \right\} \qquad (3.69)$$

From Eq. (3.7), this is

$$2\tau_N \frac{\partial W}{\partial t} = \sum_{n=0}^{\infty} a_n \left\{ -n(n+1)P_n - 2\sigma \frac{d}{dx} [x(1 - x^2)P_n] \right\} \qquad (3.70)$$

The last term in this equation is

$$-2\sigma \left[x(1 - x^2) \frac{dP_n}{dx} + (1 - 3x^2)P_n \right] \qquad (3.71)$$

Noting Eq. (3.8), this is

$$-2\sigma[x(nP_{n-1} - nxP_n) + P_n - 3x^2 P_n] \qquad (3.72)$$

The quantity $x^2 P_n$ can be evaluated from Eq. (3.10) and is

$$x^2 P_n = \frac{1}{2n+1} \left\{ \frac{n+1}{2n+3} [(n+2)P_{n+2} + (n+1)P_n] \right.$$

$$\left. + \frac{n}{2n+1} [nP_n + (n-1)P_{n-2}] \right\} \qquad (3.73)$$

Using this in conjunction with Eq. (3.10) for xP_{n-1}, Eq. (3.72) is

$$-2\sigma[x(nP_{n-1} - nxP_n) + P_n - 3x^2 P_n] =$$

$$-2\sigma \left\{ \frac{n^2 P_n + n(n-1)P_{n-2}}{2n-1} \right.$$

$$+ P_n - \frac{n+3}{2n+1} \left[\frac{n+1}{2n+3} ((n+2)P_{n+2} + (n+1)P_n) \right.$$

$$\left. \left. + \frac{n}{2n-1} (nP_n + (n-1)P_{n-2}) \right] \right\} \qquad (3.74)$$

Separating the coefficients for each polynomial we have

$$-2\sigma \left\{ P_n \left[\frac{n^2}{2n-1} + 1 - \frac{n+3}{2n+1} \left(\frac{(n+1)^2}{2n+3} + \frac{n^2}{2n-1} \right) \right] \right.$$

$$- P_{n+2} \left[\frac{-(n+1)(n+2)(n+3)}{(2n+1)(2n+3)} \right]$$

$$\left. + P_{n-2} \left[\frac{n(n-1)}{2n-1} - \frac{n(n-1)(n+3)}{(2n+1)(2n-1)} \right] \right\} \qquad (3.75)$$

The P_n coefficient can be rewritten as

$$\frac{(n^2 + 2n - 1)(2n + 1)(2n + 3) - (n + 3)(n + 1)^2(2n - 1) - n^2(n + 3)(2n + 3)}{(2n - 1)(2n + 1)(2n + 3)}$$

(3.76)

Evaluation of the numerator in this equation leads to

$$\frac{(2n^3 + 3n^2 + n)}{(2n - 1)(2n + 1)(2n + 3)}$$

$$= \frac{-n(2n + 1)(n + 1)}{(2n - 1)(2n + 1)(2n + 3)}$$

$$= \frac{-n(n + 1)}{(2n - 1)(2n + 3)}$$

(3.77)

Similarly the polynomial coefficient before P_{n-2} simplifies to

$$\frac{n(n - 1)(n - 2)}{(2n + 1)(2n - 1)}$$

(3.78)

Equation (3.71) can therefore be rewritten as

$$2\sigma\left\{\frac{n(n + 1)}{(2n - 1)(2n + 3)} P_n \right.$$

$$+ \frac{(n + 1)(n + 2)(n + 3)}{(2n + 1)(2n + 3)} P_{n+2}$$

$$\left. - \frac{n(n - 1)(n - 2)}{(2n + 1)(2n - 1)} P_{n-2}\right\}$$

(3.79)

Substitution of this into Eq. (3.70) and using the orthogonality properties of Legendre polynomials given in Eq. (3.13), we obtain the infinite set of differential-difference equations

$$\frac{2\tau_N}{n(n + 1)} \dot{a}_n + \left[1 - \frac{2\sigma}{(2n - 1)(2n + 3)}\right]a_n$$

$$= \frac{2\sigma(n - 1)}{(2n - 3)(2n - 1)} a_{n-2} - \frac{2\sigma(n + 2)}{(2n + 5)(2n + 3)} a_{n+2}$$

(3.80)

This set can be written in the form of a continued fraction. Taking the Laplace transform and using the relation

$$R_n(s) = A_n(s)/A_{n-2}(s)$$

(3.81)

we obtain

$$\left[\frac{2\tau_N s}{n(n+1)} + 1 - \frac{2\sigma}{(2n-1)(2n+3)}\right]R_n(s)$$

$$= \frac{2\tau_N}{n(n+1)}\frac{a_n(0)}{A_{n-2}(s)} + \frac{2\sigma(n-1)}{(2n-3)(2n-1)} - \frac{2\sigma(n+2)}{(2n+5)(2n+3)}R_n(s)R_{n+2}(s)$$

(3.82)

and transposing the $R_n(s)$ term on the right hand side, we can write

$$R_n(s) = \frac{\dfrac{2\tau_N}{n(n+1)}\dfrac{a_n(0)}{A_{n-2}} + \dfrac{2\sigma}{(2n-3)(2n-1)}}{\dfrac{2\tau_N s}{n(n+1)} + 1 - \dfrac{2\sigma}{(2n-1)(2n+3)} + \dfrac{2\sigma(n+2)R_{n+2}(s)}{(2n+3)(2n+5)}}$$

(3.83)

while Eq. (3.80) yields

$$A_1(s) = \frac{a_1(0)\tau_N}{n(n+1)\tau_N s + 1 - \dfrac{2\sigma}{5} + \dfrac{6\sigma}{35}R_3(s)}$$

(3.84)

The relaxation time $\tau_{\parallel}(\sigma)$ is now obtained by finding the roots of the equation formed by equating the denominator with zero and where $R_3(s)$ is given by Eq. (3.83). Iteration of this fraction and subsequent plotting of the results leads to the solid line in Fig. 4.

F. Calculation of the Relaxation Time for Small σ

The relaxation time for small σ (the ratio of uniaxial anisotropy energy to thermal energy defined as $\sigma = Kv/kT$) can be calculated in a similar way to the calculation for small ξ, performed in Section III.D, that is, by limiting the size of the system matrix. From the differential-difference equations, Eqs. (3.80), obtained in the previous section it is possible to construct the matrix equation

$$2\tau_N \dot{\mathbf{a}} = \boldsymbol{\lambda}\mathbf{a} + \mathbf{c}$$

(3.85)

where

$$\mathbf{a}(t) = \begin{pmatrix} a_1(t) \\ a_2(t) \\ \vdots \\ a_n(t) \end{pmatrix}$$

(3.86)

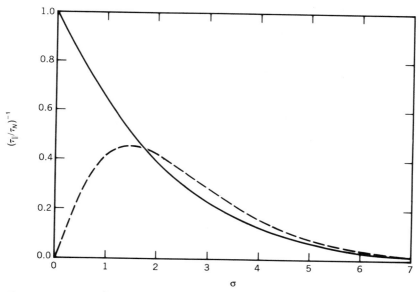

Figure 4. $(\tau_\parallel/\tau_N)^{-1}$ where $\tau_\parallel$ is the parallel relaxation time as a function of the anisotropy parameter, $\sigma = Kv/kT$. Here the solid line is the result obtained from iteration of the continued fraction, Eq. (3.84). The dotted line is the closed form expression obtained by the WKBJ method, Eq. (3.131).

$$\mathbf{c} = \begin{pmatrix} 0 \\ 4\sigma \\ 0 \\ \vdots \\ 0 \\ \vdots \end{pmatrix} \tag{3.87}$$

and

$$\boldsymbol{\lambda} =$$

$$\begin{pmatrix} -\left(2 - \dfrac{4\sigma}{5}\right) & 0 & -\dfrac{12\sigma}{5} & 0 \\ 0 & -\left(6 - \dfrac{4\sigma}{7}\right) & 0 & \cdots \\ \dfrac{16\sigma}{5} & 0 & \cdots & -\dfrac{2n(n+1)(n+2)\sigma}{(2n+3)(2n+5)} \\ 0 & \cdots & \dfrac{2n(n+1)(n-1)\sigma}{(2n-3)(2n-1)} & -n(n+1)\left(1 - \dfrac{2\sigma}{(2n-1)(2n+3)}\right) \end{pmatrix}$$

$$\tag{3.88}$$

Taking the Laplace transform of this matrix equation we obtain the system matrix $(2\tau_N s \mathbf{I} - \boldsymbol{\lambda})$ which on ignoring terms $O(\sigma^3)$ and higher we can truncate at 3×3. Equating the determinant to zero, we then obtain the relaxation time $\tau_{\parallel}(\sigma)$ from the roots of the characteristic equation. The 3×3 truncation of the system matrix is

$$
\begin{pmatrix}
2\tau_N s + 2 - \dfrac{4}{5}\,\sigma & 0 & \dfrac{12\sigma}{35} \\[2ex]
0 & 2\tau_N s + 6 - \dfrac{4}{7}\,\sigma & 0 \\[2ex]
-\dfrac{16\sigma}{5} & 0 & 2\tau_N s + 12 - \dfrac{8\sigma}{15}
\end{pmatrix}
\tag{3.89}
$$

Elimination of the upper right hand entry requires

$$
\left(2\tau_N s + 2 - \frac{4}{5}\,\sigma\right)\left(2\tau_N s + 12 - \frac{8}{15}\,\sigma\right) + \frac{12}{35}\frac{16}{5}\,\sigma^2 = 0 \tag{3.90}
$$

Setting

$$
p = 2\tau_N s \tag{3.91}
$$

and multiplying, we have

$$
p^2 + p\left(14 - \frac{4}{3}\,\sigma\right) + \left(2 - \frac{4}{5}\,\sigma\right)\left(12 - \frac{8}{15}\,\sigma\right) + \frac{192}{175}\,\sigma^2 = 0 \tag{3.92}
$$

The solution of the quadratic is

$$
p = -\left(7 - \frac{2}{3}\,\sigma\right) \pm \sqrt{\frac{\left(14 - \dfrac{4}{3}\,\sigma\right)^2}{4} - \left(24 - \frac{32}{3}\,\sigma + \frac{32}{75}\,\sigma^2\right) - \frac{192}{175}\,\sigma^2}
\tag{3.93}
$$

or

$$
p = -\left(7 - \frac{2}{3}\,\sigma\right) \pm \sqrt{25 + \frac{4}{3}\,\sigma + \frac{4}{225}\,\sigma^2 - \frac{192}{175}\,\sigma^2} \tag{3.94}
$$

that is

$$
p = -\left(7 - \frac{2}{3}\,\sigma\right) \pm \sqrt{\left(5 + \frac{2}{15}\,\sigma\right)^2 - \frac{192}{175}\,\sigma^2} \tag{3.95}
$$

or

$$p = -\left(7 - \frac{2}{3}\,\sigma\right) \pm 5\sqrt{\left(1 + \frac{2}{75}\,\sigma\right)^2 - \frac{192}{4375}\,\sigma^2} \qquad (3.96)$$

Using the binomial expansion we have

$$p = -\left(7 - \frac{2}{3}\,\sigma\right) \pm \left[5\left(1 + \frac{2}{75}\,\sigma\right) - \frac{96}{875}\,\sigma^2\right] \qquad (3.97)$$

and taking the higher of the roots

$$p = -\left(2 - \frac{4}{5}\,\sigma + \frac{96}{875}\,\sigma^2\right) \qquad (3.98)$$

in agreement with the result of Brown [8]. The relaxation time

$$\tau(\sigma) = -\frac{2\tau_N}{p} \qquad (3.99)$$

is therefore

$$\tau(\sigma) = \tau_N\left(1 - \frac{2}{5}\,\sigma + \frac{48}{875}\,\sigma^2\right)^{-1} \qquad (3.100)$$

which using the binomial expansion, again leads to (see also Eq. (6.78))

$$\tau(\sigma) = \tau_N\left(1 + \frac{2}{5}\,\sigma\right) + O(\sigma^2) \qquad (3.101)$$

G. Calculation of the Longest Relaxation Time for Large σ

When the uniaxial anisotropy energy is large in comparison with thermal energy, that is when σ is large, the distribution of orientations of **M** will be concentrated at the two opposing minima situated along the easy axis of magnetization. Consequently the relaxation time can be calculated using a discrete orientation model. Brown [8] performed such a calculation using Kramers's transition state method which describes the escape rate of particles from a potential well. The method employed here is that of Wentzel–Kramers–Brillouin–Jeffreys (WKBJ) [27, 28].

Using this method it is possible to calculate the longest relaxation time of **M** for a single domain ferromagnetic particle with large uniaxial anisotropy. It is also possible to calculate the longest relaxation time when the particle is in the presence of an external magnetic field applied along the easy axis. The potential for this is a summation of Eqs. (1.55) and (1.58) given by

$$V(\vartheta) = K \sin^2\vartheta - HM_s \cos \vartheta \qquad (3.102)$$

which for $x = \cos \vartheta$ is

$$V(x) = -Kx^2 - HM_s x + K \qquad (3.103)$$

The two minima structure will remain in the presence of an external magnetic field provided

$$HM_s < 2K \qquad (3.104)$$

For

$$HM_s > 2K \qquad (3.105)$$

the external field provides sufficient energy so that the discrete orientation model is no longer applicable. It is useful therefore to write

$$h = HM_s/2K \qquad (3.106)$$

which will serve as an indicator of the model's applicability.

We begin with Brown's equation in x for V and W independent of ϕ, which is

$$2\tau_N \frac{\partial W}{\partial t} = \frac{\partial}{\partial x}\left[(1-x^2)\left(\frac{\partial W}{\partial x} + \frac{v}{kT}\frac{\partial V}{\partial x}W\right)\right] \qquad (3.107)$$

Following Brown we use an expansion of the form

$$W(x, t) = \sum_{n=0}^{\infty} a_n(t)F_n(x) \qquad (3.108)$$

with

$$a_n(t) = a_n(0) \exp(-p_n t) \qquad (3.109)$$

where here the p_n are the eigenvalues, and the F_n the corresponding eigenfunctions, of the Fokker–Planck operator, determined by the boundary conditions that F must be finite at $x = \pm 1$. Equation (3.107) then becomes

$$\frac{d}{dx}\left\{(1-x^2)e^{-\beta V}\frac{d}{dx}(e^{\beta V}F_n(x))\right\} + \lambda_n F_n(x) = 0 \qquad (3.110)$$

where

$$\lambda_n = 2\tau_N p_n \qquad (3.111)$$

and

$$\beta = \frac{v}{kT} \qquad (3.112)$$

By writing

$$f_n(x) = e^{\beta V}F_n(x) \qquad (3.113)$$

Eq. (3.110) becomes

$$\frac{d}{dx}\left\{(1-x^2)e^{-\beta V}\frac{df}{dx}\right\} + \lambda e^{-\beta V}f = 0 \qquad (3.114)$$

subject to $f(\pm 1)$ being finite. Substituting Eqs. (3.106) and (1.56), this is

$$\frac{d}{dx}\left\{(1-x^2)\exp[\sigma(x^2+2hx)]\frac{df}{dx}\right\} + \lambda \exp[\sigma(x^2+2hx)]f = 0 \qquad (3.115)$$

where $0 \le h < 1$. The value $h = 0$ corresponds to zero applied field and at $h = 1$ the two-minima of the potential on which the method is based disappears. Multiplying across by $\exp(\sigma h^2)$ to complete the square and integrating once, we have

$$(1-x^2)\exp[\sigma(x+h)^2]\frac{df}{dx} = -\lambda\int_{\pm 1}^{x}\exp[\sigma(t+h)^2]f(t)\,dt \qquad (3.116)$$

from which it follows immediately that

$$\int_{-1}^{1}\exp\sigma[(t+h)^2]f(t)\,dt = 0 \qquad (3.117)$$

which is Brown's equation (4.15). Integrating Eq. (3.116) we have the singular integral equations

$$f(x) - f(d) = -\lambda \int_d^x \frac{dy}{(1-y^2)} \exp[-\sigma(y+h)^2] \int_c^y \exp \sigma[(t+h)^2] f(t)\, dt$$

(3.118)

where c and d are both $+1$ or -1. For $c = d = -1$, assuming real transition points ϑ_1 and ϑ_2, where following Brown, $0 < \vartheta_1 < \vartheta_2 < \pi$, we may assume that

$$f(x) = f(-1) \exp[-\sigma(x+1)^2].$$

(3.119)

The t integral in Eq. (3.118) has no point of stationary phase and may be integrated to give the semi-classical result

$$\frac{f(-1)}{2\sigma(h-1)} \{\exp\{\sigma[(y+h)^2 - (y+1)^2]\} - \exp[\sigma(h-1)^2]\}$$

(3.120)

Substitution of this into Eq. (3.118) gives convergence at $y = -1$. Application of the method of steepest descents at the saddle point $y = -h$ gives the semi-classical result

$$f(x) = f(-1) + \frac{\lambda f(-1)}{2\sigma(h-1)} \frac{\sqrt{\pi}}{\sqrt{\sigma}} \frac{\exp[\sigma(h-1)^2]}{(1-h^2)} \quad \text{for } x > -h$$

(3.121)

$$= f(-1) + \frac{\lambda f(-1)}{2\sigma(h-1)} \frac{\sqrt{\pi}}{2\sqrt{\sigma}} \frac{\exp[\sigma(h-1)^2]}{(1-h^2)} \quad \text{for } x = -h$$

(3.122)

where in Eq. (3.122), the extra factor of two arises since the saddle point is then an end point in the domain of integration.

We may repeat the process taking $c = d = +1$. We assume that

$$f(x) = f(1) \exp[-\sigma(x-1)^2]$$

(3.123)

and we obtain

$$f(x) = f(1) - \frac{\lambda \phi(1)}{2\sigma(h+1)} \frac{\sqrt{\pi}}{\sqrt{\sigma}} \frac{\exp[\sigma(1+h)^2]}{(1-h^2)} \quad \text{for } x < -h$$

(3.124)

$$= f(1) - \frac{\lambda \phi(1)}{2\sigma(h+1)} \frac{\sqrt{\pi}}{2\sqrt{\sigma}} \frac{\exp[\sigma(1+h)^2]}{(1-h^2)} \quad \text{for } x = -h$$

(3.125)

Matching $f(-h)$ via Eqs. (3.122) and (3.125) gives

$$f(1)\left[1 - \frac{\lambda \exp[\sigma(1+h)^2]\sqrt{\pi}}{4\sigma^{3/2}(h+1)(1-h^2)}\right] = f(-1)\left[1 - \frac{\lambda \exp[\sigma(1-h)^2]\sqrt{\pi}}{4\sigma^{3/2}(1-h)(1-h^2)}\right]$$

(3.126)

Using Eqs. (3.117), (3.119) and (3.122), we may show that

$$0 = \int_{\cos\theta_1}^{1} dx \, f(1) \exp[\sigma(x+h)^2 - \sigma(x-1)^2]$$

$$+ \int_{\cos\theta_2}^{\cos\theta_1} dx \, f(-h)\left\{1 + \frac{\lambda}{2\sigma(1-h^2)} \left\{\exp[\sigma(x+h)^2] - 1\right\}\right\}$$

$$+ \int_{-1}^{\cos\theta_2} dx \, f(-1) \exp[\sigma(x+h)^2 - \sigma(x+1)^2]$$

(3.127)

$$\approx \frac{f(1) \exp[\sigma(h+1)^2]}{2\sigma(h+1)} - \frac{f(-1) \exp[\sigma(h-1)^2]}{2\sigma(h-1)}, \quad \sigma \gtrsim\gtrsim 1$$

(3.128)

In Eq. (3.127) we have integrated the first and third integrals by parts to order $1/\sigma$ and have neglected the middle integral as being of higher order, anticipating that λ is exponentially decreasing with σ and where we obtained its integrand by taking $c = d = -h$ in Eq. (3.118) and correctly assuming $f(-h) = 0$. Equations (3.127) and (3.128) give the result

$$\lambda \approx 2\sigma^{3/2}(1-h^2)\pi^{-1/2}\{(1+h) \exp[-\sigma(1+h)^2]$$

$$+ (1-h) \exp(-\sigma(1-h)^2)\}, \quad \sigma \gtrsim\gtrsim 1$$

(3.129)

which for $h = 0$ (no applied field) is

$$\lambda \approx 4\sigma^{3/2}\pi^{-1/2}e^{-\sigma}$$

(3.130)

which for the relaxation time $\tau_\| = 1/p = 2\tau_N/\lambda$ gives

$$\tau_\|(\sigma) \approx \frac{\tau_N}{2} \sqrt{\pi}\sigma^{-3/2}e^{\sigma}$$

(3.131)

These formulae agree with the formulae given by Aharoni [38] and Brown [8]. The notation given by $\gtrsim\gtrsim$ indicates "reasonably greater than" and indeed Eq. (3.131) holds well down to $\sigma = 1.5$, as can be seen from

Fig. 4. For values of σ less than 1.5, Eq. (29) of Raĭkher and Shliomis [17] that is Eq. (6.75), which is based on the effective eigenvalue method, (see Sections VI and VII) is an extremely good approximation to the exact numerical result. Equation (3.131) is plotted against the numerical result obtained from iteration of the continued fraction in Fig. 4.

The relaxation time $\tau_\|(\sigma, \xi)$ for the magnetization of a single domain ferromagnetic particle with uniaxial anisotropy in the presence of an external field from $\tau_\| = 1/p = 2\tau_N/\lambda$ and Eq. (3.129) is

$$\tau_\|(\sigma, \xi)$$

$$= \frac{\tau_N}{\sigma^{3/2}(1 - h^2)\pi^{-1/2}\{(1 + h)\exp[-\sigma(1 + h)^2] + (1 - h)\exp[-\sigma(1 - h)^2]\}}$$

(3.132)

which by noting that $\xi = 2\sigma h$, can be rewritten as

$$\tau_\|(\sigma, \xi) = \frac{1}{2}\tau_N\pi^{1/2}\sigma^{-3/2}e^\sigma\left[\frac{\exp(\xi^2/4\sigma)}{\left(1 - \frac{\xi^2}{4\sigma^2}\right)\left(\cosh \xi - \frac{\xi}{2\sigma}\sinh \xi\right)}\right]$$

(3.133)

which is in agreement with the formula obtainable from Brown's equations (4.43) and (4.44) [8].

IV. CLOSED FORM EXPRESSIONS FOR THE LONGITUDINAL AND TRANSVERSE RELAXATION TIMES EXCLUDING INTERNAL ANISOTROPY

A. Introduction

The object of this section is to show how one may obtain analytic expressions for the rise transient of the magnetization following the application of a small DC field, a large DC field having been applied at $t = -\infty$. Since the increment due to the DC field is very small relative to the DC bias field, the response to a weak AC field superimposed on the bias field can be found from the response to the increment in the DC field by linear response theory. It will be shown that using a certain approximation, closed form expressions may be obtained in terms of the Langevin function for the longitudinal and transverse relaxation times and the corresponding susceptibilities. Furthermore it is shown how the loss tangent is affected by the presence of the bias field.

We first outline the approach of the following sections. Starting with Brown's equation we expand the distribution $W(\mathbf{r}, t)$ of orientations of $\mathbf{M}$ in spherical harmonics as in the previous chapter whence we obtain an infinite set of differential-difference equations. We then select the spherical harmonic of interest; P_1 for the longitudinal relaxation and $P_1^1 e^{i\phi}$ for the transverse relaxation. The relaxation times can then be expressed as in the previous section in terms of a continued fraction. Since we consider only the response with respect to a small applied field $\mathbf{H}_1$ such that $\xi_1 \ll 1$, it is only necessary to evaluate quantities linear in ξ_1. Furthermore we assume that equilibrium has been attained for the ratio $R_{nm}(s)$ in the continued fraction and that this ratio is expressible in terms of the equilibrium values of the relevant spherical harmonics. This expression is achieved as follows. The average value of any spherical harmonic is

$$\langle P_{n'}^{|m'|} e^{im'\phi} \rangle = \frac{\int_0^{2\pi} \int_0^\pi P_{n'}^{|m'|} e^{im'\phi} W \sin \vartheta \, d\vartheta \, d\phi}{\int_0^{2\pi} \int_0^\pi W \sin \vartheta \, d\vartheta \, d\phi} \tag{4.1}$$

$$= \frac{\sum_{n=0}^\infty \sum_{m=-n}^n \int_0^{2\pi} \int_0^\pi P_n^{|m|} e^{im\phi} P_{n'}^{|m'|} e^{im'\phi} a_{nm} \sin \vartheta \, d\vartheta \, d\phi}{\sum_{n=0}^\infty \sum_{m=-n}^{m=n} \int_0^{2\pi} \int_0^\pi P_n^{|m|} e^{im\phi} a_{nm} \sin \vartheta \, d\vartheta \, d\phi} \tag{4.2}$$

From the orthogonality properties of the associated Legendre functions, which are

$$\int_{-1}^{+1} P_n^{|m|}(x) P_{n'}^{|m'|}(x) \, dx = \frac{(n+|m|)!}{(n-|m|)!} \frac{2}{2n+1} \delta_{n,n'} \delta_{m,m'} \tag{4.3}$$

we obtain a relation between a_{nm} and the average value of $P_n^{|m|} e^{im\phi}$ which is

$$\langle P_n^{|m|} e^{im\phi} \rangle = \frac{(n+|m|)!}{(n-|m|)!} \frac{1}{2n+1} \frac{a_{nm}}{a_{00}} \tag{4.4}$$

Noting this the ratio $a_{nm}(t)/a_{n-1,m}(t)$ is easily expressible in terms of $\langle P_n^{|m|} e^{im\phi} \rangle$ and $\langle P_{n-1}^{|m|} e^{im\phi} \rangle$.

Our starting point in this section is again Brown's equation,

$$\frac{\partial W}{\partial t} = \frac{1}{\sin \vartheta} \frac{\partial}{\partial \vartheta} \left\{ \sin \vartheta \left[\left(h' \frac{\partial V}{\partial \vartheta} - \frac{g'}{\sin \vartheta} \frac{\partial V}{\partial \phi} \right) W + k' \frac{\partial W}{\partial \vartheta} \right] \right\}$$
$$+ \frac{1}{\sin \vartheta} \frac{\partial}{\partial \phi} \left\{ \left(g' \frac{\partial V}{\partial \vartheta} + \frac{h'}{\sin \vartheta} \frac{\partial V}{\partial \phi} \right) W + \frac{k'}{\sin \vartheta} \frac{\partial W}{\partial \phi} \right\} \tag{4.5}$$

B. A Closed Form Expression for the Relaxation Time in a Longitudinal Field

The first analytic expression we will obtain is for a small longitudinal DC field $\mathbf{H}_1$ which is suddenly applied to a system which is in equilibrium in a larger DC field $\mathbf{H}$. We begin with Brown's equation, which as we have seen when there is no azimuthal or ϕ dependence, reduces to

$$2\tau_N \sin\vartheta \, \frac{\partial W}{\partial t} = \frac{\partial}{\partial\vartheta}\left[\sin\vartheta\left(\frac{v}{kT}\frac{\partial V}{\partial\vartheta}W + \frac{\partial W}{\partial\vartheta}\right)\right]$$

which noting that $x = \cos\vartheta$ is

$$2\tau_N \frac{\partial W}{\partial t} = \frac{\partial}{\partial x}\left[(1 - x^2)\left(\frac{\partial W}{\partial x} + \frac{v}{kT}\frac{\partial V}{\partial x}W\right)\right] \tag{4.6}$$

If we now substitute in the potential

$$V(\vartheta) = -[HM_s + H_1 M_s U(t)]\cos\vartheta \tag{4.7}$$

where $U(t)$ is the unit step function, we can write Brown's equation in the form

$$\frac{\partial W}{\partial t} = LW + L_{\text{ext}}(t)W \tag{4.8}$$

where

$$L = \frac{1}{2\tau_N}\left\{\frac{1}{\sin\vartheta}\frac{\partial}{\partial\vartheta}\left[\sin\vartheta\left(\xi\sin\vartheta + \frac{\partial}{\partial\vartheta}\right)\right]\right\} \tag{4.9}$$

and

$$L_{\text{ext}} = \frac{1}{2\tau_N}\frac{1}{\sin\vartheta}\frac{\partial}{\partial\vartheta}[\xi_1\sin^2\vartheta]U(t) \tag{4.10}$$

where $\xi_1 = vH_1 M_s/kT$. As given in Appendix C, we can expand W as

$$W = W^{(0)} + W^{(1)} + W^{(2)} \tag{4.11}$$

and write Brown's equation as

$$\frac{\partial W^{(0)}}{\partial t} + \frac{\partial W^{(1)}}{\partial t} = LW^{(0)} + LW^{(1)} + L_{\text{ext}}W^{(1)} + O(\xi_1^2) \tag{4.12}$$

Equilibrium for $W^{(0)}$ has been achieved and so

$$\frac{\partial W^{(0)}}{\partial t} = LW^{(0)} = 0 \tag{4.13}$$

The linear response to ξ_1 is therefore

$$\frac{\partial W^{(1)}}{\partial t} = LW^{(1)} + L_{\text{ext}} W^{(0)} \tag{4.14}$$

We next expand $W^{(0)}$ and $W^{(1)}$ in Legendre polynomials with

$$W^{(0)} = \sum_{n=0}^{\infty} a_n^{(0)}(0) P_n(x) \tag{4.15}$$

and

$$W^{(1)} = \sum_{n=0}^{\infty} a_n^{(1)}(t) P_n(x) \tag{4.16}$$

Substitution of these expansions into Eq. (4.14) yields the set of forced differential-difference equations

$$\frac{da_n^{(1)}}{dt} = \frac{1}{2\tau_N} \left\{ -n(n+1) a_n^{(1)} + n(n+1)\xi \left[\frac{a_{n-1}^{(1)}}{2n-1} - \frac{a_{n+1}^{(1)}}{2n+3} \right] \right\}$$
$$+ \frac{\xi_1}{2\tau_N} n(n+1) \left[\frac{a_{n-1}^{(0)}}{2n-1} - \frac{a_{n+1}^{(0)}}{2n+3} \right] U(t) \tag{4.17}$$

The differential-difference equation for $n = 1$ corresponding to $\langle \cos \vartheta \rangle$ is

$$\frac{da_1^{(1)}}{dt} = \frac{1}{2\tau_N} \left\{ -2a_1^{(1)} + 2\xi \left[a_0^{(1)} - \frac{a_2^{(1)}}{5} \right] \right\} + \frac{\xi_1}{\tau_N} \left[a_0^{(0)} - \frac{a_2^{(0)}}{5} \right] U(t) \tag{4.18}$$

which can be written as

$$\frac{da_1^{(1)}}{dt} + a_1^{(1)} \lambda_{\parallel} = \frac{\xi_1}{\tau_N} \left[a_0^{(0)} - \frac{a_2^{(0)}}{5} \right] U(t) \tag{4.19}$$

where

$$\lambda_{\parallel} = \frac{1}{\tau_{\parallel}} = \frac{1}{\tau_N} \left[1 - \xi \left(a_0^{(1)} - \frac{a_2^{(1)}}{5} \right) \right] \tag{4.20}$$

We first consider the right hand side of Eq. (4.19). For this it is necessary to evaluate $a_2^{(0)}$. From Eq. (4.4) which relates the average values of spherical harmonics to the corresponding time coefficients, we have

$$a_2^{(0)} = 5\langle P_2 \rangle_0 a_0^{(0)} \tag{4.21}$$

The right hand side of Eq. (4.19) is therefore

$$\xi_1 \frac{a_0^{(0)}}{\tau_N} [1 - \langle P_2 \rangle_0] \tag{4.22}$$

where $\langle P_2 \rangle_0$ is the *equilibrium* value of $\langle P_2(x) \rangle = \langle (3x^2 - 1)/2 \rangle$ in the field **H** which is given by

$$\langle P_2 \rangle_0 = \frac{3}{2} \frac{\int_{-1}^{1} x^2 e^{\xi x}\, dx}{\int_{-1}^{1} e^{\xi x}\, dx} - \frac{1}{2} \tag{4.23}$$

Integrating by parts, this is

$$\langle P_2 \rangle_0 = \frac{3}{2} \left[1 - \frac{2\langle P_1 \rangle_0}{\xi} \right] - \frac{1}{2} \tag{4.24}$$

Noting that $\langle P_1 \rangle_0 = L(\xi)$, we have

$$\langle P_2 \rangle_0 = 1 - \frac{3L(\xi)}{\xi} \tag{4.25}$$

Equation (4.19) is therefore

$$\dot{a}_1^{(1)} + \lambda_\parallel a_1^{(1)} = \frac{a_0^{(0)} \xi_1}{\tau_N} \frac{3L(\xi)U(t)}{\xi} \tag{4.26}$$

Noting Eq. (4.4), again we have

$$a_1^{(0)} = 3\langle P_1 \rangle_0 a_0^{(0)} \tag{4.27}$$

Dividing across by $a_0^{(0)}$ and writing

$$\langle P_n(x) \rangle_{\xi_1} = \langle P_n(x) \rangle_1 \tag{4.28}$$

we have

$$\frac{d}{dt} \langle P_1 \rangle_1 + \lambda_\parallel \langle P_1 \rangle_1 = \frac{\xi_1}{\tau_N} \left[\frac{L(\xi)}{\xi} \right] U(t) \tag{4.29}$$

Multiplying across by M_s we then have the equation describing the dynamics of the magnetization in the direction of the field $\mathbf{H}_1$ (to terms linear in ξ_1) which is

$$\frac{d}{dt} \langle M_z \rangle_1 + \lambda_{\parallel} \langle M_z \rangle_1 = M_s \xi_1 \frac{L(\xi)}{\tau_N \xi} U(t) \qquad (4.30)$$

By writing

$$H_1 \chi_0 = \frac{n v M_s}{3} \xi_1 = \frac{n v^2 M_s^2}{3kT} H_1 \qquad (4.31)$$

this can be written to give the magnetic moment of unit sample volume

$$\frac{d}{dt} \langle M_{\parallel} \rangle_1 + \lambda_{\parallel} \langle M_{\parallel} \rangle_1 = \chi_0 \frac{3L(\xi)}{\tau_N \xi} U(t) H_1 \qquad (4.32)$$

The eigenvalue $\lambda_{\parallel}$ can be evaluated from the continued fraction already obtained in Section III which is for $\lambda_{\parallel} = -s$

$$A_1^{(1)}(s)(0) = R_1 A_0^{(1)}(s) = \frac{\tau_N a_1^{(1)}(0) + \xi A_0^{(1)}(s)}{\tau_N s + 1 + \dfrac{\xi R_2(s)}{5}} \qquad (4.33)$$

Here $R_2(s)$ is the ratio $R_2(s) = A_2^{(1)}(s)/A_1^{(1)}(s)$. This can be evaluated to first order in ξ_1 as follows. The orthogonality properties of the Legendre polynomials yield

$$\frac{5}{3} \frac{\langle P_2 \rangle_1}{\langle P_1 \rangle_1} = \frac{a_2^{(1)}(t)}{a_1^{(1)}(t)} \qquad (4.34)$$

and so taking Laplace transforms we have

$$\frac{5}{3} \frac{\mathscr{L}\langle P_2 \rangle_1}{\mathscr{L}\langle P_1 \rangle_1} = \frac{A_2^{(1)}(s)}{A_1^{(1)}(s)} \qquad (4.35)$$

Here we use the final value theorem whence we can replace the right hand side above by

$$\lim_{s \to 0} \frac{s\mathscr{L}\langle P_2 \rangle_1}{s\mathscr{L}\langle P_1 \rangle_1} = \lim_{t \to \infty} \frac{\langle P_2 \rangle_1}{\langle P_1 \rangle_1} \qquad (4.36)$$

The equilibrium averages $\langle P_2 \rangle_1$ and $\langle P_1 \rangle_1$ are now evaluated. This is done as follows. The equilibrium value of the first Legendre polynomial corresponding to the relaxation in the direction of the field **H**, prior to the switching on of $\mathbf{H}_1$ is

$$\langle P_1 \rangle_0 = \frac{\int_{-1}^{1} x e^{\xi x}\, dx}{\int_{-1}^{1} e^{\xi}\, dx} \tag{4.37}$$

(where $\xi = vHM_s/kT$). From Section I.G, this is the Langevin function, and so

$$\langle P_1(\cos \vartheta) \rangle_0 = L(\xi) \tag{4.38}$$

Let us now consider the equilibrium solution long after $\mathbf{H}_1$ has been applied. This will be

$$\langle P_1(\cos \vartheta) \rangle = L(\xi + \xi_1) \tag{4.39}$$

This can be expanded as

$$\langle P_1(\cos \vartheta) \rangle = L(\xi) + \xi_1 L'(\xi) + O(\xi_1)^2 \tag{4.40}$$

The response to the first order in ξ_1 can therefore be expressed as

$$\langle P_1(\cos \vartheta) \rangle_1 = \langle P_1(\cos \vartheta) \rangle - \langle P_1(\cos \vartheta) \rangle_0 = \xi_1 L'(\xi) \tag{4.41}$$

Similarly the equilibrium value of the second Legendre polynomial in the field **H** prior to the application of $\mathbf{H}_1$ is

$$\langle P_2 \rangle_0 = \frac{\int_{-1}^{1} \dfrac{3x^2 - 1}{2} e^{\xi x}\, dx}{\int_{-1}^{1} e^{\xi x}\, dx} = 1 - \frac{3L(\xi)}{\xi} \tag{4.42}$$

the integrals being evaluated using integration by parts in both cases. As $t \to \infty$, the Langevin argument ξ is replaced by $\xi + \xi_1$ and so expanding in a Taylor series about ξ, we again obtain

$$L(\xi + \xi_1) = L(\xi) + \xi_1 L'(\xi) + \cdots \tag{4.43}$$

We retain only terms linear in ξ_1 to yield

$$\langle P_2 \rangle - \langle P_2 \rangle_0 = \frac{3}{\xi} \left[\xi_1 L'(\xi) - \frac{\xi_1}{\xi} L(\xi) \right] = \langle P_2 \rangle_1 \tag{4.44}$$

The eigenvalue $\lambda_\|$ with this procedure is then

$$\lambda_\| = \frac{1}{\tau_N} + \frac{\xi}{3\tau_N} \frac{\langle P_2 \rangle - \langle P_2 \rangle_0}{\langle P_1 \rangle - \langle P_1 \rangle_0}$$

$$= \frac{1}{\tau_N} \frac{L(\xi)}{\xi L'(\xi)} \tag{4.45}$$

and thus the relaxation time $\tau_\| = 1/\lambda_\|$ is

$$\tau_\| = \frac{\tau_N \xi L'(\xi)}{L(\xi)} \tag{4.46}$$

This may be further simplified by differentiating the Langevin function (it is easiest to perform this differentiation using the integral form of the Langevin function given in Eq. (4.37)) to yield

$$\tau_\| = \frac{\tau_N \xi \left[1 - \dfrac{2}{\xi} L(\xi) - L^2(\xi) \right]}{L(\xi)} \tag{4.47}$$

For small values of ξ, the Langevin function can be approximated by

$$L(\xi) = \xi/3 \tag{4.48}$$

and so for small ξ we find that $\tau_\| = \tau_N$ as expected. If terms $O(\xi^3)$ are retained in the Langevin function, we have

$$L(\xi) = \xi/3 - \xi^3/45 \tag{4.49}$$

which results in

$$\tau_\| = \tau_N(1 - 2\xi^2/15) \tag{4.50}$$

For the other limit of large ξ, the predicted relaxation is virtually instantaneous, the asymptotic behavior being

$$\tau_\| = \tau_N/\xi \tag{4.51}$$

Having obtained an expression for the eigenvalue in the equation describing $\langle M_\| \rangle_1$, we can substitute this into that equation and so

$$\frac{d}{dt} \langle M_\| \rangle_1 + \frac{L(\xi)\langle M_\| \rangle_1}{\tau_N \xi L'(\xi)} = \chi_0 \frac{3L(\xi)}{\tau_N \xi} U(t)H_1 \tag{4.52}$$

Here the equilibrium value with

$$\frac{d\langle M_\parallel \rangle_1}{dt} = 0$$

is

$$\frac{L(\xi)\langle M_\parallel \rangle_1}{\tau_N \xi L'(\xi)} = \chi_0 \frac{3L(\xi)}{\tau_N \xi} U(t)H_1 \tag{4.53}$$

which gives

$$\langle M_\parallel \rangle_1 = M_s \xi_1 L'(\xi)H_1 \tag{4.54}$$

as expected. The frequency domain response can be obtained by taking the Laplace transform of Eq. (4.52)

$$\mathcal{L}\langle M_\parallel \rangle_1 \left[s + \frac{L(\xi)}{\xi \tau_N L'(\xi)} \right] = \chi_0 \frac{3L(\xi)H_1}{\tau_N \xi s} \tag{4.55}$$

which can be rearranged as

$$\mathcal{L}\langle M_\parallel \rangle_1 = \chi_0 \frac{3L(\xi)}{\tau_N \xi} \frac{H_1}{s\left(s + \dfrac{L(\xi)}{\xi \tau_N L'(\xi)}\right)}$$

$$= \chi_0 \frac{3L(\xi)}{\tau_N \xi} \frac{\dfrac{\xi \tau_N L'(\xi)}{L(\xi)} H_1}{s\left(s \dfrac{\xi \tau_N L'(\xi)}{L(\xi)} + 1\right)} \tag{4.56}$$

and so

$$\mathcal{L}\langle M_\parallel \rangle_1 = \frac{3\chi_0 L'(\xi)H_1}{s(1 + s\tau_\parallel)} \tag{4.57}$$

whence the rise transient is

$$\langle M_\parallel(t) \rangle = 3\chi_0 L'(\xi)U(t)\left[1 - \exp\left(-\frac{\xi L'(\xi)t}{L(\xi)\tau_N}\right)\right] H_1 \tag{4.58}$$

Furthermore since the response is *linear*, the decay transient that will follow the removal of $\mathbf{H}_1$ must be the *mirror image* of the rise transient so

$$\langle M_{\parallel}(t) \rangle = 3\chi_0 L'(\xi) U(t) \exp\left[-\frac{\xi L'(\xi) t}{L(\xi)\tau_N} \right] H_1 \qquad (4.59)$$

The after-effect function is thus

$$3\chi_0 L'(\xi) \exp\left(\frac{-t}{\tau_{\parallel}(\xi)} \right) U(t) \qquad (4.60a)$$

The frequency dependent susceptibility arising from the imposition of an AC field $H_1 e^{i\omega t}$ may then be written down (since we have limited the solution to terms linear in ξ_1) from linear response theory as

$$\chi_{\parallel}(\omega) = \chi'_{\parallel}(\omega) - i\chi''_{\parallel}(\omega) \qquad (4.60b)$$

$$= \chi_0 3 L'(\xi) \left[1 - i\omega \int_0^{\infty} \exp\left(\frac{-t}{\tau_{\parallel}(\xi)} \right) \exp(-i\omega t)\, dt \right] \qquad (4.60c)$$

Thus

$$\chi_{\parallel}(\omega) = \frac{\chi_0 3 L'(\xi)}{1 + i\omega\tau_{\parallel}} = \frac{\chi_{\parallel}^s}{1 + i\omega\tau_{\parallel}} \qquad (4.60d)$$

C. A Closed Form Expression for the Relaxation Time in a Transverse Field

In this section we consider the time behavior of the magnetization following the imposition at $t = 0$ of a small DC field in a direction perpendicular to a larger DC bias field which was applied at $t = -\infty$.

Since we first require eigenvalues only, we begin with the unforced set of differential-difference equations for $m = 1$ since the spherical harmonic of interest will be $P_1^1(\cos \vartheta)e^{i\phi}$. From Eq. (4.14), this set is

$$2\tau_N \frac{da_{n,1}^{(1)}(t)}{dt} = -n(n+1)a_{n,1}^{(1)} + \xi\left[\frac{n^2 - 1}{2n - 1} a_{n-1,1}^{(1)} - \frac{n(n+2)}{2n+3} a_{n+1,1}^{(1)} \right] \qquad (4.61)$$

The Laplace transform of this is

$$A_{n,1}^{(1)}(s) = \frac{2\tau_N a_{n,1}^{(1)}(0) + \xi \dfrac{n^2 - 1}{2n - 1} A_{n-1,1}^{(1)}(s)}{2s\tau_N + n(n+1) + \xi \dfrac{n(n+2)}{2n+3} R_{n+1}(s)} \qquad (4.62)$$

with $R_{n+1}(s)$ obtainable from

$$R_n(s) = \frac{2\tau_N \dfrac{a_{n,1}^{(1)}(0)}{A_{n-1,1}(s)} + \xi \dfrac{n^2 - 1}{2n - 1}}{2\tau_N s + n(n+1) + \xi n \dfrac{(n+2)}{2n+3} R_{n+1}(s)} \tag{4.63}$$

We note that since there is initially no field in the transverse direction, $a_{n,1}^{(1)}(0) = 0$. Thus Eq. (4.62) may be rearranged to read, in terms of spherical harmonics $X_{nm} = P_n^{|m|} e^{im\phi}$

$$\frac{\mathscr{L}\langle X_{n,1}\rangle_1}{\mathscr{L}\langle X_{n-1,1}\rangle_1} = \frac{\dfrac{\xi(n+1)^2}{2\tau_N(2n+1)}}{s + \dfrac{n(n+1)}{2\tau_N} + \dfrac{\xi}{2\tau_N}\dfrac{n^2}{2n+1}\dfrac{\mathscr{L}\langle X_{n+1,1}\rangle_1}{\mathscr{L}\langle X_{n,1}\rangle_1}} \tag{4.64}$$

The perpendicular relaxation time is obtained by setting $n = 1$ in this equation and setting the denominator of the resulting equation to zero. Thus the lowest eigenvalue is the solution of

$$s + \frac{1}{\tau_N}\left(1 + \frac{\xi}{6}\frac{\mathscr{L}\langle X_{21}\rangle_1}{\mathscr{L}\langle X_{11}\rangle_1}\right) = 0 \tag{4.65}$$

The equilibrium averages are

$$\langle X_{11}\rangle = \langle \sin\vartheta\,\cos\phi\rangle$$

$$= \frac{\int_0^{2\pi}\int_0^\pi \sin\vartheta\,\cos\phi\,\exp(\xi\cos\vartheta + \xi_1\sin\vartheta\,\cos\phi)\sin\vartheta\,d\vartheta\,d\phi}{\int_0^{2\pi}\int_0^\pi \exp(\xi\cos\vartheta + \xi_1\sin\vartheta\,\cos\phi)\sin\vartheta\,d\vartheta\,d\phi} \tag{4.66a}$$

which in the linear approximation of ξ_1 is

$$\langle X_{11}\rangle_1 = \frac{\xi_1 \int_{-1}^{+1}(1 - x^2)e^{\xi x}\,dx}{2\int_{-1}^{+1}e^{\xi x}\,dx}$$

$$= \xi_1 L(\xi)/\xi \tag{4.66b}$$

In the same way

$$\langle X_{21}\rangle_1 = \frac{\xi_1 \int_{-1}^{+1}3x(1 - x^2)e^{\xi x}\,dx}{2\int_{-1}^{+1}e^{\xi x}\,dx} \tag{4.67}$$

and thus

$$\frac{\langle X_{21}\rangle_1}{\langle X_{11}\rangle_1} = \frac{\int_{-1}^{+1} 3x(1-x^2)e^{\xi x}\,dx}{\int_{-1}^{+1}(1-x^2)e^{\xi x}\,dx} \tag{4.68}$$

whence

$$\lambda_\perp = \frac{1}{\tau_N}\left[1 + \frac{\xi}{6}\frac{\langle X_{21}\rangle_1}{\langle X_{11}\rangle_1}\right]$$

$$= \frac{1}{\tau_N}\left[1 + \xi\frac{L(\xi) - (\int_{-1}^{+1} x^3 e^{\xi x}\,dx/\int_{-1}^{+1} e^{\xi x}\,dx)}{4L(\xi)/\xi}\right] \tag{4.69}$$

which on integration by parts reduces to

$$\lambda_\perp = \frac{1}{\tau_N}\left[1 + \xi^2\frac{-\dfrac{1}{\xi} + \dfrac{3}{\xi} - \dfrac{6L(\xi)}{\xi^2}}{4L(\xi)}\right] \tag{4.70}$$

or

$$\lambda_\perp = \frac{1}{\tau_N}\frac{\xi - L(\xi)}{2L(\xi)} \tag{4.71}$$

whence the relaxation time $\tau_\perp$ is

$$\tau_\perp(\xi) = \tau_N\frac{2L(\xi)}{\xi - L(\xi)} \tag{4.72}$$

which is in agreement with Martsenyuk et al. [16]. In the limit $\xi \ll 1$, we find again as expected, that the relaxation time is simply the Néel relaxation time. For small values of ξ, retaining terms $O(\xi^3)$ in the Langevin function, we obtain the low field approximation

$$\tau_\perp = \tau_N(1 - \xi^2/10) \tag{4.73}$$

For large values of ξ, the Langevin function approaches unity (see Fig. 2) and so we obtain the asymptotic solution

$$\tau_\perp = 2\tau_N/\xi \tag{4.74}$$

The equation of motion of the magnetization in the transverse direction, on applying the same method involving the differential operators L and L_{ext} as before, is

$$\langle \dot{X}_{11}\rangle_1 + \frac{1}{\tau_N} \langle X_{11}\rangle_1 = -\frac{\xi}{6\tau_N} \langle X_{21}\rangle_1 + \frac{\xi_1 U(t)}{6\tau_N} [2 - \tfrac{1}{2}\langle X_{22}\rangle_0 + \langle P_2\rangle_0] \tag{4.75}$$

which noting our eigenvalue calculation is

$$\langle \dot{X}_{11}\rangle_1 + \lambda_\perp \langle X_{11}\rangle_1$$
$$= \frac{\xi_1 U(t)}{6\tau_N} [2 - \tfrac{1}{2}\langle X_{22}\rangle_0 + \langle P_2\rangle_0] \tag{4.76}$$

We note that

$$\langle X_{22}\rangle_0 = 0 \tag{4.77}$$

since

$$\int_0^{2\pi} \cos 2\phi \, d\phi = 0 \tag{4.78}$$

Equation (4.76) then simplifies with the aid of Eq. (4.42) to

$$\langle \dot{X}_{11}\rangle_1 + \frac{1}{\tau_\perp} \langle X_{11}\rangle_1 = \frac{\xi_1 U(t)}{2\tau_N} \left(\frac{\xi - L(\xi)}{\xi} \right) \tag{4.79}$$

We can then find just as for the parallel case, that the complex suscep-tibility for a small transverse AC field $\mathbf{H}_1 e^{i\omega t}$ is

$$\chi_\perp(\omega) = \frac{\chi_0 \dfrac{3L(\xi)}{\xi}}{1 + i\omega\tau_\perp} = \frac{\chi_\perp^s}{1 + i\omega\tau_\perp} \tag{4.80}$$

These formulae for the parallel and perpendicular susceptibilities, Eqs. (4.60d) and (4.80), were written down without derivation by Shliomis and Raĭkher for the Debye relaxation of ferrofluid particles [19]. Note that in that paper there appears to be a transcription error where $\chi_\parallel$ is written as $\chi_\perp$ and vice versa.

D.   Comparison of the Closed Form Expressions for the Relaxation Times with Numerical Solutions

We have derived closed form expressions for the relaxation times $\tau_\parallel$ and $\tau_\perp$ by assuming that spherical harmonics (in the differential-difference equations) of order two and higher reach their equilibrium values considerably faster than those of order one. In this section, we compare these expressions with numerical solutions obtained by iteration of the continued fraction expression for the relaxation times until convergence occurs. As we have pointed out, this procedure is equivalent to finding the eigenvalues of the system matrix for progressively larger matrix sizes until a similar convergence occurs.

In Fig. 3 we compare the closed form expression for the parallel relaxation time $\tau_\parallel$ given by Eq. (4.46) with the numerical solution obtained by 20 convergents of the continued fraction given by Eq. (4.33) with $a_n^{(1)}(0) = 0$ and written in terms of $\langle P_n \rangle_1$, namely

$$\frac{\mathscr{L}\langle P_n \rangle_1}{\mathscr{L}\langle P_{n-1} \rangle_1} = \frac{\dfrac{\xi}{2\tau_N} \dfrac{n(n+1)}{2n+1}}{s + \dfrac{n(n+1)}{2\tau_N} + \dfrac{\xi}{2\tau_N} \dfrac{n(n+1)}{2n+1} \dfrac{\mathscr{L}\langle P_{n+1} \rangle_1}{\mathscr{L}\langle P_n \rangle_1}} \qquad (4.81)$$

It is thus obvious that $\mathscr{L}\langle P_1 \rangle_1 / \mathscr{L}\langle P_0 \rangle_1$ may be written from the above equation to any desired degree of accuracy as a continued fraction. The characteristic equation determining the eigenvalues is

$$s + \frac{1}{\tau_N} + \frac{\xi}{3\tau_N} \frac{\mathscr{L}\langle P_2 \rangle_1}{\mathscr{L}\langle P_1 \rangle_1} = 0 \qquad (4.82)$$

In Fig. 5 we compare the closed form expression for the perpendicular relaxation time $\tau_\perp$ given by Eq. (4.72) with the numerical solution obtained by 20 convergents of Eq. (4.64) in

$$s + \frac{1}{\tau_N} \left(1 + \frac{\xi}{6} \frac{\mathscr{L}\langle X_{21} \rangle_1}{\mathscr{L}\langle X_{11} \rangle_1} \right) = 0 \qquad (4.83)$$

E. Calculation of the Field-Dependent Loss Tangent

One application of the above formulae for $\tau_\parallel$ and $\tau_\perp$ is to the calculation of the loss tangent defined as [39]

$$\tan \delta = \mu''/\mu' \qquad (4.84)$$

where μ' and μ'' are defined by the equation

$$\mu = \mu'(\omega) - i\mu''(\omega) \qquad (4.85)$$

where μ is the complex permeability. The problem of the loss tangent for a weak AC field superimposed upon a strong DC field for a polar

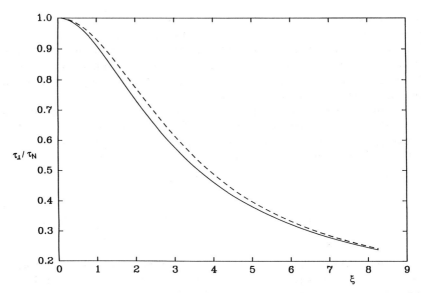

Figure 5. The perpendicular relaxation time $\tau_\perp/\tau_N$ as a function of the field strength parameter, $\xi = vHM_s/kT$. The dotted line is the result obtained from iteration of the continued fraction, Eq. (4.65). The solid line is the closed form expression, Eq. (4.72).

dielectric was considered by Coffey and Paranjape [40] in a discussion of the solution of the Smoluchowski equation using perturbation theory. They gave results correct to second order in the strong DC field only, so that their solution was limited to the small perturbation approximation. The analytic formulae derived for $\tau_\parallel$ and $\tau_\perp$ allow us to give a simple analytic expression for $\tan \delta$ correct to any order in the strong DC field. We first note that [40]

$$\tan \delta = \chi''/(\mu_0 + \chi') \qquad (4.86)$$

and

$$\chi_0 = \mu_0(\mu_s - 1) = \frac{nv^2M_s^2}{3kT} \qquad (4.87)$$

so that for the parallel case

$$(\tan \delta)_\parallel = \frac{\dfrac{\chi_0 3L'(\xi)\omega\tau_\parallel}{1 + \omega^2\tau_\parallel^2}}{\mu_0 + \dfrac{3L'(\xi)\chi_0}{1 + \omega^2\tau_\parallel^2}} \qquad (4.88)$$

which simplifies to (on use of Eq. (4.87))

$$(\tan \delta)_{\parallel} = \frac{\xi L'(\xi)}{L(\xi)} \frac{\omega \tau_N (\mu_s - 1)}{\xi^2 \omega^2 \tau_N^2 \dfrac{L'(\xi)}{3L^2(\xi)} + \mu_s + \dfrac{1}{3L'(\xi)} - 1} \qquad (4.89)$$

where $\mu_s = \mu'(0)$ is the static relative permeability at $\xi = 0$. This should be compared with $\tan \delta$ for negligibly small fields as follows:

$$(\tan \delta)_{\parallel} = \frac{\omega \tau_N (\mu_s - 1)}{\omega^2 \tau_N^2 + \mu_s} \qquad (4.90)$$

In the perpendicular case we find that

$$(\tan \delta)_{\perp} = \frac{2L(\xi)}{\xi - L(\xi)} \frac{\omega \tau_N (\mu_s - 1)}{\mu_s - 1 + \dfrac{\xi}{3L(\xi)} + \dfrac{4\omega^2 \tau_N^2 \xi L(\xi)}{3\{\xi - L(\xi)\}^2}} \qquad (4.91)$$

with the same expression for negligibly small fields as Eq. (4.90).

The analytical formulae for $\tan \delta$ also allow us to predict the field dependence of the maximum absorption itself. Consider first the behavior of $\tan \delta$ when $\xi = 0$. It is obvious from Eq. (4.90) that $\tan \delta$ is a maximum when, for $x = \omega \tau_N$,

$$x = x_m = \sqrt{\mu_s} \qquad (4.92)$$

giving a maximum value of

$$(\tan \delta)_m = \frac{\mu_s - 1}{2\sqrt{\mu_s}} \qquad (4.93)$$

Here it has been possible to drop the subscripts $\parallel$ and $\perp$ since the formula is the same for both when $\xi = 0$. Also since the formulae for $\xi \neq 0$ have the same mathematical form as the formula for negligible fields, we can write down expressions for the frequency of maximum absorption when $\xi \neq 0$. For this we note that

$$(\tan \delta)_{\parallel}$$

$$= \frac{\xi L'(\xi)}{L(\xi)} \frac{x(\mu_s - 1)}{\mu_s + \left(\dfrac{1}{3L'(\xi)} - 1\right)\left[1 + \dfrac{\xi^2 L'(\xi)}{3L^2(\xi)} \dfrac{x^2}{\mu_s + \left(\dfrac{1}{3L'(\xi)} - 1\right)}\right]}$$

$$(4.94)$$

which yields

$$x_{\|_m}^2(\xi) = \frac{3L^2(\xi)}{\xi^2 L'(\xi)} \left(x_m^2(0) + \frac{1}{3L'(\xi)} - 1 \right) \tag{4.95}$$

so that

$$\omega_{\|_m} = \tau_\|^{-1} \sqrt{1 + 3(\mu_s - 1)L'(\xi)} \tag{4.96}$$

The corresponding expression for $\tan \delta$ is

$$(\tan \delta)_{\|_m} = \frac{\sqrt{3L'(\xi)}(\mu_s - 1)}{2\sqrt{\mu_s + \dfrac{1}{3L'(\xi)} - 1}} \tag{4.97}$$

We note that, for small ξ,

$$3L'(\xi) \to 1 \tag{4.98a}$$

and

$$L(\xi) \to \xi/3 \tag{4.98b}$$

so that these formulae reduce to Eqs. (4.92) and (4.93). For large ξ, however,

$$\tau_\| \to \tau_N/\xi \tag{4.98c}$$

which indicates that $\omega_m \propto \xi$ for large ξ. (This behavior when the formulae are applied for polar dielectrics with appropriate changes of quantities is in qualitative agreement with the observations of Block and Hayes [41] and Ullman [42].) Thus the effect of the DC field is to shift the absorption maximum to higher frequencies. Such behavior is also evident in the perturbation treatment of Coffey and Paranjape [40]. For large ξ,

$$(\tan \delta)_{\|_m} = \frac{3\sqrt{\mu_s} (\tan \delta)_m}{\xi^2} \tag{4.99}$$

again in qualitative agreement with observations [43, 44, 45] on ferrofluids where the amplitude of the maximum absorption decreases as the strength of the bias field is increased. In the application to ferrofluids τ_N must be replaced by the Debye time τ_D. The maximum amplitude in the perpendicular case is

$$x^2_{\perp_m} = \frac{3[\xi - L(\xi)]^2\left[\mu_s - 1 + \frac{\xi}{3L(\xi)}\right]}{4\xi L(\xi)} \qquad (4.100)$$

with

$$(\tan \delta)_{\perp_m} = \frac{(\mu_s - 1)\sqrt{\frac{3L(\xi)}{\xi}}}{2\sqrt{\mu_s - 1 + \frac{\xi}{3L(\xi)}}} \qquad (4.101)$$

For small ξ

$$(\tan \delta)_{\perp_m} = (\tan \delta)_m \qquad (4.102)$$

and for large ξ

$$(\tan \delta)_{\perp_m} = \frac{3\sqrt{\mu_s}(\tan \delta)_m}{\xi} \qquad (4.103)$$

We have shown how analytic expressions may be obtained for the Debye relaxation (or zero anisotropy Néel relaxation) of a single domain ferromagnetic particle subject to a weak DC field $\mathbf{H}_1$ superimposed on a strong DC field $\mathbf{H}$ so as to cause only linear behavior in response to the weak field $\mathbf{H}_1$. This allows us to treat in a simple manner the relaxation effects caused by the coupling of the weak and strong DC fields. There is reasonable agreement between the analytical expressions found for the relaxation times and numerical solutions calculated from the continued fraction as shown in Figs. 3 and 5. Use of this method also has the advantage over the perturbation method employed in Section III.D for small ξ in that closed form expressions are also available for the susceptibility and the loss tangent valid for all values of ξ.

We reiterate that the availability of the closed form expressions for the relaxation behavior depends on the assumption that the ratio of the Laplace transform of the first and second order spherical harmonic averages has reached its final value appreciably faster than the ratio of the first and zero order. This means that for step-on and step-off solutions, we are considering processes that take place in times of order of magnitude

$$t > \frac{2\tau_N}{n(n+1)} \qquad (4.104)$$

where n is the lowest order spherical harmonic ratio that we assume to reach equilibrium faster than for $n = 1$. In this case therefore for $n = 2$, we consider processes taking place in times of order of magnitude

$$t > \tau_N/3 \qquad (4.105)$$

The implication of this when viewed in the context of the frequency domain is that the angular frequencies under consideration are such that

$$\omega < 3\omega_N \qquad (4.106)$$

V. GILBERT'S EQUATION AS MODIFIED TO INCLUDE THE INERTIA OF A FERROFLUID PARTICLE

A. Introduction

The primary purpose of this section is to show how the equation of motion of the average magnetic dipole moment of a ferrofluid may be written down directly from Gilbert's equation as modified to include the inertia of a ferrofluid particle. A knowledge of the dynamic behavior of the average magnetic dipole moment will allow us to determine the complex magnetic susceptibility of the ferrofluid in the noninertial or Debye limit as a particular case. This approach is similar to the one used previously by us [33, 48, 49] in the theory of dielectric and Kerr effect relaxation. It allows one to obtain in simple fashion (with the aid of linear response theory) closed form expressions for the longitudinal and transverse relaxation times. The result coincides precisely with that obtained from the Fokker–Planck equation by Martsenyuk et al. [16].

In order to discuss the Langevin equation for a single domain ferrofluid particle we first consider Gilbert's equation for the dynamic behavior of the particle's magnetization vector $\mathbf{M}$ in the presence of thermal agitation, which is Eq. (1.12),

$$\frac{d\mathbf{M}}{dt} = \gamma \mathbf{M} \times \left(\mathbf{H}_T - \eta\,\frac{d\mathbf{M}}{dt} \right) \qquad (5.1)$$

where γ is the "gyromagnetic" ratio and η is a phenomenological damping parameter, $\mathbf{H}_T$ is the magnetic field acting on the particle which consists of the applied fields, the demagnetizing field, and a random white noise field $\mathbf{h}(t)$, which has the properties defined by Eqs. (2.40)–(2.44), where $\mu = 2kT\eta/\nu$, k is the Boltzmann constant, T is the absolute temperature, ν is the volume of the particle.

Gilbert's equation (5.1) may be rearranged explicitly as shown in

Section I.D using the properties of the triple product formula to yield that equation in the Landau–Lifshitz form:

$$\frac{d\mathbf{M}}{dt} = g'M_s\mathbf{M} \times \mathbf{H}_T + h'(\mathbf{M} \times \mathbf{H}_T) \times \mathbf{M} \tag{5.2}$$

where the constants g' and h' are

$$g' = \frac{\gamma}{(1 + \eta^2\gamma^2M_s^2)M_s}, \quad h' = \frac{\eta\gamma^2}{1 + \eta^2\gamma^2M_s^2} \tag{5.3}$$

M_s is the (constant) magnitude of the magnetization; only its direction varies with time.

As we have already discussed, a single domain ferromagnetic particle suspended in a fluid is subject to two separate mechanisms which cause variations in its magnetization. The first is due to the thermal fluctuations inside the particle (spontaneous remagnetization). This is the solid state mechanism known as Néel relaxation [25]. The second is due to the rotation of the particle within the fluid, since changes in the orientation of the particle will effect changes in the orientation of its magnetization. This mechanism is similar to that governing the Debye relaxation of a polar molecule and is consequently known as Debye relaxation [14].

In a ferrofluid particle where the Néel relaxation mechanism is blocked, orientational changes in $\mathbf{M}$ will be due to the rotational motion of the particle only. Consequently the magnetization behaves like that of a bar magnet when placed in a magnetic field. The gyromagnetic or precessional term is not present in its equation of motion. Consequently that equation is Eq. (5.2) with the precessional terms set equal to zero corresponding to $\eta\gamma M_s \gg 1$ so that now (see Section 1 of [16])

$$\frac{d\mathbf{M}}{dt} = h'(\mathbf{M} \times \mathbf{H}_T) \times \mathbf{M} \tag{5.4}$$

with

$$h' = 1/\eta M_s^2$$

Equation (5.4) is the Langevin equation of the problem. This equation is now analogous to the equation of motion of a polar molecule under the influence of an electric field [52]. The quantity that directly corresponds to the dipole moment of the polar molecule $\boldsymbol{\mu}$ in the Debye theory is the magnetic moment $\nu\mathbf{M}$ of an *individual* ferrofluid particle, not the magnetization $\mathbf{M}$ which is the magnetic moment *per unit volume*. Equation

(5.4) also means that we are now treating each particle as a rigid magnetic dipole.

The form of Eq. (5.4) suggests that one should introduce an angular velocity vector

$$\mathbf{\Omega} = h'\mathbf{M} \times \mathbf{H}_T \tag{5.5}$$

Thus Eq. (5.4) has the form of the kinematic relation

$$\frac{d\mathbf{M}}{dt} = \mathbf{\Omega} \times \mathbf{M} \tag{5.6}$$

whence

$$\ddot{\mathbf{M}} = \dot{\mathbf{\Omega}} \times \mathbf{M} + \mathbf{\Omega} \times \dot{\mathbf{M}} \tag{5.7}$$

or using Eq. (5.6),

$$\ddot{\mathbf{M}} = \dot{\mathbf{\Omega}} \times \mathbf{M} + \mathbf{\Omega} \times (\mathbf{\Omega} \times \mathbf{M}) \tag{5.8}$$

The quantity $\dot{\mathbf{\Omega}}$ is determined by including an inertial term in the Langevin equation for $\mathbf{\Omega}$ so that [16]

$$I\dot{\mathbf{\Omega}} + \zeta\mathbf{\Omega} = \mathbf{N} \tag{5.9}$$

where

$$\mathbf{N} = \nu\mathbf{M} \times \mathbf{H}_T \tag{5.10}$$

is the torque acting on the particle. The drag coefficient ζ which is proportional to the viscosity of the surroundings may be found by setting $\dot{\mathbf{\Omega}} = 0$ in Eq. (5.9) so that

$$\mathbf{\Omega} = \nu\mathbf{M} \times \mathbf{H}_T/\zeta$$

whence on comparison with Eq. (5.5),

$$\zeta = \nu/h' = \nu\eta M_s^2 \tag{5.11}$$

(ζ can also be expressed in terms of the dynamic viscosity α of the fluid as follows, $\zeta = 6\nu\alpha$ [16]). We have supposed in writing down Eq. (5.9) that the particle is a sphere with moment of inertia I.

Thus

$$\dot{\boldsymbol{\Omega}} = -\frac{\zeta}{I}\,\boldsymbol{\Omega} + \frac{\nu}{I}\,\mathbf{M} \times \mathbf{H}_T \tag{5.12}$$

Equation (5.8) can therefore be written as

$$\ddot{\mathbf{M}} = \left(-\frac{\zeta}{I}\,\boldsymbol{\Omega} + \frac{\nu}{I}\,\mathbf{M} \times \mathbf{H}_T\right) \times \mathbf{M} + \boldsymbol{\Omega} \times (\boldsymbol{\Omega} \times \mathbf{M}) \tag{5.13}$$

Multiplying across by I and again noting Eq. (5.6), this becomes

$$I\ddot{\mathbf{M}} + \zeta\dot{\mathbf{M}} + I(\boldsymbol{\Omega} \times \mathbf{M}) \times \boldsymbol{\Omega} = \nu(\mathbf{M} \times \mathbf{H}_T) \times \mathbf{M} \tag{5.14}$$

The total magnetic field $\mathbf{H}_T$ acting on the ferrofluid particle (the Néel mechanism being been blocked) is made up of a large constant field term $\mathbf{H}_0$ representing the DC bias field, a small time-dependent external field $\mathbf{H}(t)$ and a random field term $\mathbf{h}(t)$ so that

$$\mathbf{H}_T(t) = \mathbf{H}_0 + \mathbf{H}(t) + \mathbf{h}(t) \tag{5.15}$$

Substituting the above expression for the total field $\mathbf{H}_T$, Eq. (5.14) may be written

$$I\ddot{\mathbf{M}} + \zeta\dot{\mathbf{M}} + I(\boldsymbol{\Omega} \times \mathbf{M}) \times \boldsymbol{\Omega} = \nu(\mathbf{M} \times \mathbf{H}_0) \times \mathbf{M}$$
$$+ \nu(\mathbf{M} \times \mathbf{H}(t)) \times \mathbf{M} + \nu(\mathbf{M} \times \mathbf{h}(t)) \times \mathbf{M} \tag{5.16}$$

Rearranging the triple vector products in Eq. (5.16), we can write this equation as

$$I\ddot{\mathbf{M}} + \zeta\dot{\mathbf{M}} + I(\Omega^2\mathbf{M} - \boldsymbol{\Omega}(\boldsymbol{\Omega} \cdot \mathbf{M})) = \nu(M_s^2\mathbf{H}_0 - \mathbf{M}(\mathbf{M} \cdot \mathbf{H}_0)) + \mathbf{F}H(t) + \mathbf{L}(t) \tag{5.17}$$

where

$$\mathbf{F} = \nu M_s^2\mathbf{e} - \nu\mathbf{M}(\mathbf{M} \cdot \mathbf{e})\,, \quad \mathbf{e} = \frac{\mathbf{H}(t)}{H(t)} \tag{5.18}$$

$$\mathbf{L}(t) = \nu M_s^2\mathbf{h}(t) - \nu\mathbf{M}(\mathbf{M} \cdot \mathbf{h}(t)) \tag{5.19}$$

B. Components of the Inertial Equation

We can now write Eq. (5.17) in terms of the Euler angles $(\vartheta, \varphi, \psi)$ and the corresponding angular velocity components [47]. It is evident from Eq. (5.17) that the only problem is to evaluate the term $I(\Omega^2\mathbf{M} - \mathbf{\Omega}(\mathbf{\Omega}\cdot\mathbf{M}))$. This procedure which is rather lengthy is described at the end of this section.

In the laboratory XYZ coordinate system, Eq. (5.17) becomes (we use Eqs. (5.70)–(5.72))

$$I\ddot{M}_z + \zeta\dot{M}_z + IM_z(\omega_1^2 + \omega_2^2) - IM_s\omega_2\omega_3\sin\vartheta$$
$$= \nu H_0[M_s^2 - M_z^2] + F_zH(t) + L_z(t) \tag{5.20}$$

$$I\ddot{M}_x + \zeta\dot{M}_x + IM_x(\omega_1^2 + \omega_2^2) - IM_s\omega_1\omega_3\cos\varphi + IM_s\omega_2\omega_3\cos\vartheta\sin\varphi$$
$$= -\nu H_0 M_x M_z + F_xH(t) + L_x(t) \tag{5.21}$$

$$I\ddot{M}_y + \zeta\dot{M}_y + IM_y(\omega_1^2 + \omega_2^2) - IM_s\omega_1\omega_3\sin\varphi - IM_s\omega_2\omega_3\cos\vartheta\cos\varphi$$
$$= -\nu H_0 M_y M_z + F_yH(t) + L_y(t) \tag{5.22}$$

where F_i and L_i are the projections of the vectors $\mathbf{F}$ and $\mathbf{L}$ onto the x, y and z axes, $\omega_1 = \dot{\vartheta}$, $\omega_2 = \dot{\varphi}\sin\vartheta$ and $\omega_3 = \dot{\varphi}\cos\vartheta + \dot{\psi}$ are the components of the angular velocity $\mathbf{\Omega}$, $M_z = M_s\cos\vartheta$, $M_x = M_s\sin\vartheta\sin\varphi$ and $M_y = -M_s\sin\vartheta\cos\varphi$ are the projections of $\mathbf{M}$ onto the X, Y and Z axes.

We form the average of Eqs. (5.20)–(5.22) noting that $\langle\mathbf{L}(t)\rangle$ will vanish throughout because, in the inertia corrected Langevin equation, $\mathbf{M}$ is statistically independent of the white noise field $\mathbf{h}(t)$. This is not, however, true of the noninertial Langevin equation where the multiplicative noise term $\mathbf{L}(t)$ contributes a noise induced drift term to the average (see Section VI). The averages so formed are

$$I\langle\ddot{M}_z\rangle + \zeta\langle\dot{M}_z\rangle + I\langle M_z(\omega_1^2 + \omega_2^2)\rangle - M_sI\langle\omega_1\omega_2\sin\vartheta\rangle$$
$$= \nu H_0[M_s^2 - \langle M_z^2\rangle] + H(t)\langle F_z\rangle \tag{5.23}$$

$$I\langle\ddot{M}_x\rangle + \zeta\langle\dot{M}_x\rangle + I\langle M_x(\omega_1^2 + \omega_2^2)\rangle - M_sI\langle\omega_1\omega_3\cos\varphi - \omega_2\omega_3\cos\vartheta\sin\varphi\rangle$$
$$= -\nu H_0\langle M_x M_z\rangle + \langle F_x\rangle H(t) \tag{5.24}$$

$$I\langle\ddot{M}_y\rangle + \zeta\langle\dot{M}_y\rangle + I\langle M_y(\omega_1^2 + \omega_2^2)\rangle - M_sI\langle\omega_1\omega_3\sin\varphi + \omega_2\omega_3\cos\vartheta\cos\varphi\rangle$$
$$= -\nu H_0\langle M_y M_z\rangle + \langle F_y\rangle H(t) \tag{5.25}$$

This system of equations can be solved by the method suggested by us [20, 48, 49] in the theory of dielectric and Kerr effect relaxation.

We will consider Eqs. (5.23)–(5.25) in the noninertial limit where

$I \to 0$ or equivalently the drag coefficient ζ becomes very large. (The noninertial limit typically corresponds to $I/\zeta \le 10^{-10}$ s.) In this limit the Maxwell–Boltzmann distribution for the angular velocities has set in so that orientation and angular velocity variables are decoupled from each other as far as the time behavior of the particle orientations is concerned. Thus on setting $I = 0$ in Eqs. (5.23)–(5.25), we have

$$\zeta \langle \dot{M}_z \rangle + 2kT \langle M_z \rangle = \nu H_0 [M_s^2 - \langle M_z^2 \rangle] + \langle F_z \rangle H(t) \qquad (5.26)$$

$$\zeta \langle \dot{M}_x \rangle + 2kT \langle M_x \rangle = -\nu H_0 \langle M_x M_z \rangle + \langle F_x \rangle H(t) \qquad (5.27)$$

$$\zeta \langle \dot{M}_y \rangle + 2kT \langle M_y \rangle = -\nu H_0 \langle M_y M_z \rangle + \langle F_y \rangle H(t) \qquad (5.28)$$

In writing these we have used the facts that angular velocity components about different axes are statistically independent and that on account of the decoupling between functions of angular orientations and angular velocities in the diffusion limit,

$$\frac{I}{2} \langle \omega_i \omega_j f(\mathbf{M}) \rangle = \delta_{ij} \frac{1}{2} kT \langle f(\mathbf{M}) \rangle \qquad (5.29)$$

C. Calculation of the Relaxation Times $\tau_\parallel$ and $\tau_\perp$

The magnetization decay of a ferrofluid which is under the influence of a constant field $\mathbf{H}_0$, a small constant external field $\mathbf{H}$ such that $\nu(\mathbf{M} \cdot \mathbf{H})/kT \ll 1$, having been switched off at time $t = 0$ is, from linear response theory (Appendix C)

$$\langle M_\parallel(t) - M_\parallel^0 \rangle = \chi_\parallel^s H C_\parallel(t) \qquad (5.30)$$

(for the case of $\mathbf{H} \| \mathbf{H}_0$) and

$$\langle M_\perp(t) - M_\perp^0 \rangle = \chi_\perp^s H C_\perp(t) \qquad (5.31)$$

(for the case of $\mathbf{H} \perp \mathbf{H}_0$), where

$$\chi_\parallel^s = \frac{\nu^2 n}{kT} [\langle M_z^2(0) \rangle_0 - \langle M_z(0) \rangle_0^2] \qquad (5.32)$$

and

$$\chi_\perp^s = \frac{\nu^2 n}{kT} [\langle M_x^2(0) \rangle_0 - \langle M_x(0) \rangle_0^2] \qquad (5.33)$$

are the components of the static magnetic susceptibility,

$$C_{\|}(t) = \frac{\langle M_z(0)M_z(t)\rangle_0 - \langle M_z(0)\rangle_0^2}{\langle M_z(0)^2\rangle_0 - \langle M_z(0)\rangle_0^2} \qquad (5.34)$$

and

$$C_{\perp}(t) = \frac{\langle M_x(0)M_x(t)\rangle_0 - \langle M_x(0)\rangle_0^2}{\langle M_x(0)^2\rangle_0 - \langle M_x(0)\rangle_0^2} \qquad (5.35)$$

are the autocorrelation functions of the components of the magnetization

$$M_{\|}^0 = \nu n \langle M_z(0)\rangle_0 \qquad (5.36)$$

and

$$M_{\perp}^0 = \nu n \langle M_x(0)\rangle_0 \qquad (5.37)$$

are the components of the equilibrium magnetization, the brackets $\langle\ \rangle_0$ designate the equilibrium ensemble average, n is the number of particles per unit volume.

The corresponding complex magnetic susceptibilities $\chi_{\|}(\omega)$ and $\chi_{\perp}(\omega)$ are

$$\chi_\gamma(\omega) = \chi_\gamma^s \left[1 - i\omega \int_0^\infty e^{-i\omega t} C_\gamma(t)\, dt \right] \quad (\gamma = \|, \perp) \qquad (5.38)$$

In the limit of low frequencies, Eq. (5.38) may be written as

$$\chi_\gamma(\omega) \cong \chi_\gamma^s [1 - i\omega\tau_\gamma] \quad (\gamma = \|, \perp) \qquad (5.39)$$

where

$$\tau_\gamma = \int_0^\infty C_\gamma(t)\, dt \quad (\gamma = \|, \perp) \qquad (5.40)$$

are the relaxation times. Equation (5.39) can be written down to the same order of accuracy in the form of the Debye equation,

$$\chi_\gamma(\omega) \cong \frac{\chi_\gamma^s}{1 + i\omega\tau_\gamma} \quad (\gamma = \|, \perp) \qquad (5.41)$$

It should be noted that Eq. (5.41) is valid only in the limit of low frequencies, namely $\omega\tau_\gamma \ll 1$.

This is the rotational diffusion limit where the behavior of both $C_\gamma(t)$ and $\langle M_\gamma(t) - M_\gamma^0 \rangle$ ($\gamma = \parallel, \perp$) may be approximated by the pure exponential

$$C_\gamma(t) = \exp(-t/\tau_\gamma) \tag{5.42}$$

In this limit both the nonequilibrium part of the magnetization $\langle M_\perp(t) - M_\perp^0 \rangle$ and autocorrelation function $C_\gamma(t)$ obey the equations

$$\frac{d}{dt} \langle M_\gamma(t) - M_\gamma^0 \rangle + \frac{1}{\tau_\gamma} \langle M_\gamma(t) - M_\gamma^0 \rangle = 0 \tag{5.43}$$

$$\frac{d}{dt} C_\gamma(t) + \frac{1}{\tau_\gamma} C_\gamma(t) = 0 \tag{5.44}$$

Thus in order to calculate the relaxation time τ_γ, we may use the formula

$$\tau_\gamma = -\frac{\langle M_\gamma(0) - M_\gamma^0 \rangle}{\langle \dot{M}_\gamma(0) \rangle} = -\frac{C_\gamma(0)}{\dfrac{d}{dt} C_\gamma(0)} = -\frac{\langle M_\gamma^2(0)\rangle_0 - \langle M_\gamma(0)\rangle_0^2}{\langle M_\gamma(0) \dfrac{d}{dt} M_\gamma(0)\rangle_0} \tag{5.45}$$

In order to calculate the relaxation time $\tau_\parallel$, we have to set $H(t) = 0$ (the switch off case) in Eq. (5.26). Using Eq. (5.45) we obtain

$$\tau_\parallel = -\frac{\langle M_z(0) - M_z^0 \rangle}{\langle \dot{M}_z(0) \rangle} = -\zeta \frac{\langle M_z(0) - M_z^0 \rangle}{\langle -2kTM_z(0) + \nu M_s^2 H_0 - \nu M_z^2(0)H_0 \rangle}$$

$$= \tau_D \frac{\langle \cos^2\vartheta \rangle_0 - \langle \cos\vartheta \rangle_0^2}{\langle \cos^2\vartheta \rangle_0 - \frac{1}{2}\xi \langle \cos\vartheta - \cos^3\vartheta \rangle_0}, \tag{5.46}$$

where

$$\tau_D = \nu(h'2kT)^{-1} = \zeta/2kT \tag{5.47}$$

is the relaxation time at $\mathbf{H}_0 = 0$, that is, the Debye relaxation time and

$$\xi = \frac{\nu M_s H_0}{kT} \tag{5.48}$$

The second term in the denominator of Eq. (5.44) may be reduced to a simpler form as follows:

$$\frac{1}{2} \xi \langle \cos \vartheta - \cos^3 \vartheta \rangle_0 = \frac{1}{2} \xi \frac{\int_0^\pi e^{\xi \cos \vartheta} (\cos \vartheta - \cos^3 \vartheta) \sin \vartheta \, d\vartheta}{\int_0^\pi e^{\xi \cos \vartheta} \sin \vartheta \, d\vartheta}$$

$$= \frac{1}{2} \frac{[-(\cos \vartheta - \cos^3 \vartheta) e^{\xi \cos \vartheta} |_0^\pi - \int_0^\pi (1 - 3 \cos^2 \vartheta) e^{\xi \cos \vartheta} \sin \vartheta \, d\vartheta]}{\int_0^\pi e^{\xi \cos \vartheta} \sin \vartheta \, d\vartheta}$$

$$= \frac{3}{2} \langle \cos^2 \vartheta \rangle_0 - \frac{1}{2} \tag{5.49}$$

We now have on substituting Eq. (5.49) into (5.46),

$$\tau_\parallel = 2\tau_D \frac{\langle \cos^2 \vartheta \rangle_0 - \langle \cos \vartheta \rangle_0^2}{1 - \langle \cos^2 \vartheta \rangle_0} \tag{5.50}$$

The same procedure can be applied to the transverse relaxation. We have

$$\tau_\perp = \tau_D \frac{\langle \sin^2 \vartheta \sin^2 \varphi \rangle_0 - \langle \sin \vartheta \sin \varphi \rangle_0^2}{\langle \sin^2 \vartheta \sin^2 \varphi + \frac{1}{2} \xi \cos \vartheta \sin^2 \vartheta \sin^2 \varphi \rangle_0}$$

$$= 2\tau_D \frac{1 - \langle \cos^2 \vartheta \rangle_0}{1 + \langle \cos^2 \vartheta \rangle_0} \tag{5.51}$$

Now

$$\langle \cos \vartheta \rangle_0 = \frac{\int_0^\pi \cos \vartheta e^{\xi \cos \vartheta} \sin \vartheta \, d\vartheta}{\int_0^\pi e^{\xi \cos \vartheta} \sin \vartheta \, d\vartheta} = L(\xi) \tag{5.52}$$

and

$$\langle \cos^2 \vartheta \rangle_0 = \frac{\int_0^\pi \cos^2 \vartheta e^{\xi \cos \vartheta} \sin \vartheta \, d\vartheta}{\int_0^\pi e^{\xi \cos \vartheta} \sin \vartheta \, d\vartheta} = 1 - \frac{2}{\xi} L(\xi) \tag{5.53}$$

where

$$L(\xi) = \coth \xi - \frac{1}{\xi} \tag{5.54}$$

is the Langevin function. Hence

$$\tau_\parallel = \tau_D \frac{\xi - 2L(\xi) - \xi L^2(\xi)}{L(\xi)} \tag{5.55}$$

$$\tau_\perp = \tau_D \frac{2L(\xi)}{\xi - L(\xi)} \tag{5.56}$$

Equations (5.55) and (5.56) coincide precisely with the formulae obtained by Martsenyuk et al. [16] (Eq. 29 of their paper) which is apparent if we note that

$$\frac{\xi - 2L(\xi) - \xi L^2(\xi)}{L(\xi)} = \frac{\xi L'(\xi)}{L(\xi)} \qquad (5.57)$$

Substituting Eqs. (5.55) and (5.57) into (5.41) we obtain the longitudinal and transverse components of the complex magnetic susceptibility, namely

$$\chi_\parallel(\omega) \cong \frac{\chi_\parallel^s}{1 + i\omega\tau_\parallel} \qquad (5.58)$$

and

$$\chi_\perp(\omega) \cong \frac{\chi_\perp^s}{1 + i\omega\tau_\perp} \qquad (5.59)$$

where

$$\chi_\parallel^s = \frac{v^2 n M_s^2}{kT} [\langle \cos^2 \vartheta \rangle_0 - \langle \cos \vartheta \rangle_0^2] = \frac{v^2 n M_s^2}{kT} \left(1 - \frac{2}{\xi} L(\xi) - L^2(\xi)\right) \qquad (5.60a)$$

and

$$\chi_\perp^s = \frac{v^2 n M_s^2}{kT} [\langle \sin^2\vartheta \cos^2\phi \rangle_0 - \langle \sin \vartheta \cos \phi \rangle_0^2] = \frac{v^2 n M_s^2}{kT} \frac{1}{\xi} L(\xi) \qquad (5.60b)$$

We have shown how analytic expressions may be obtained for the relaxation behavior of an assembly of noninteracting ferrofluid particles subject to a strong constant magnetic field $\mathbf{H}_0$ superimposed on which is an alternating field $\mathbf{H}(t)$ which is so weak as to cause only linear behavior in the response to that field. We have treated the relaxation behavior by computing the desired averages directly from the Langevin equation. This allows us to treat in a simple manner the relaxation effects caused by the coupling between the constant field and the weak alternating field. The method also has the advantage, over the perturbation calculation, that closed form expressions are available for the susceptibility for all values of ξ.

It is of interest to compare these results with those for the field dependencies of the relaxation times $\tau_\perp$ and for $\tau_\|$ for the longitudinal and for the transverse polarization components of a polar fluid in a constant electric field $\mathbf{E}_0$. As shown in [52, 55] the relaxation times $\tau_\perp$ and $\tau_\|$ are also given by Eqs. (5.55) and (5.56), where $\xi = \mu E_0 / kT$, $\boldsymbol{\mu}$ is the dipole moment of a polar molecule and τ_D is the Debye rotational diffusion time with $\mathbf{E}_0 = 0$. Thus, Eqs. (5.55) and (5.56) predict the same field dependencies of the relaxation times $\tau_\perp$ and $\tau_\|$ for both a ferrofluid and a polar fluid. This is not unexpected because from a physical point of view the behavior of a suspension of fine ferromagnetic particles in a constant magnetic field $\mathbf{H}_0$ is similar to that of a system of electric dipoles (polar molecules) in a constant electric field $\mathbf{E}_0$.

We remark that a similar approach to the one used here for the calculation of $\tau_\|$ and $\tau_\perp$ has been used by San Miguel et al. [78] in the calculation of the intensity correlations of single and two mode lasers.

D.   Evaluation of $\Omega^2 \mathbf{M} - \Omega(\Omega \cdot \mathbf{M})$

The coordinate systems are those used by Landau and Lifshitz [50, p.111]. In terms of the Euler angles (ϑ, ϕ, ψ) and the angular velocity components

$$\omega_1 = \dot{\vartheta}, \quad \omega_2 = \dot{\varphi} \sin \vartheta, \quad \omega_3 = \dot{\varphi} \cos \vartheta + \dot{\psi} \tag{5.61}$$

we have

$$\Omega^2 = \omega_1^2 + \omega_2^2 + \omega_3^2 \tag{5.62}$$

$$\Omega_x = \dot{\vartheta} \cos \varphi + \dot{\psi} \sin \vartheta \sin \varphi = \omega_1 \cos \varphi - (\omega_2 \cos \vartheta - \omega_3 \sin \vartheta) \sin \varphi \tag{5.63}$$

$$\Omega_y = \dot{\vartheta} \sin \varphi - \dot{\psi} \sin \vartheta \cos \varphi = \omega_1 \sin \varphi + (\omega_2 \cos \vartheta - \omega_3 \sin \vartheta) \cos \varphi \tag{5.64}$$

$$\Omega_z = \dot{\phi} + \dot{\psi} \cos \vartheta = \dot{\varphi} \sin^2 \vartheta + (\dot{\varphi} \cos \vartheta + \dot{\psi}) \cos \vartheta = \omega_2 \sin \vartheta + \omega_3 \cos \vartheta \tag{5.65}$$

Further

$$(\Omega \cdot \mathbf{M}) = M_s \omega_3 \tag{5.66}$$

and

$$M_z = M_s \cos \vartheta \tag{5.67}$$

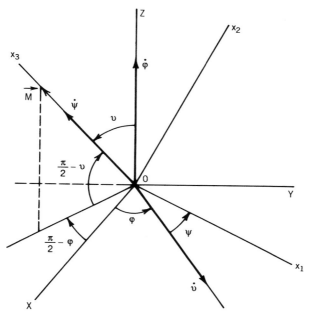

Figure 6. The coordinate system used for the resolution of $\Omega^2 \mathbf{M} - \mathbf{\Omega}(\mathbf{\Omega} \cdot \mathbf{M})$.

$$M_x = M_s \sin \vartheta \sin \varphi \qquad (5.68)$$

$$M_y = - M_s \sin \vartheta \cos \varphi \qquad (5.69)$$

Now from (5.61)–(5.69),

$$
\begin{aligned}
{[\Omega^2 \mathbf{M} - \mathbf{\Omega}(\mathbf{M} \cdot \mathbf{\Omega})]}\big|_z &= M_s(\omega_1^2 + \omega_2^2 + \omega_3^2) \cos \vartheta - M_s \omega_2 \omega_3 \sin \vartheta \\
&\quad - M_s \omega_3^2 \cos \vartheta \\
&= M_s[(\omega_1^2 + \omega_2^2) \cos \vartheta - \omega_2 \omega_3 \sin \vartheta] \\
&= (\omega_1^2 + \omega_2^2)M_z - \omega_2 \omega_3 \sin \vartheta M_s \qquad (5.70)
\end{aligned}
$$

$$
\begin{aligned}
{[\Omega^2 \mathbf{M} - \mathbf{\Omega}(\mathbf{M} \cdot \mathbf{\Omega})]}\big|_x &= M_s(\omega_1^2 + \omega_2^2 + \omega_3^2) \sin \vartheta \sin \varphi \\
&\quad - M_s \omega_3[\omega_1 \cos \varphi - \omega_2 \cos \vartheta \sin \varphi \\
&\quad + \omega_3 \sin \vartheta \sin \varphi]
\end{aligned}
$$

$$= M_s\{(\omega_1^2 + \omega_2^2) \sin \vartheta \sin \varphi$$

$$- \omega_3 \omega_1 \cos \varphi \quad + \omega_2 \omega_3 \cos \vartheta \sin \varphi\}$$

$$= (\omega_1^2 + \omega_2^2) M_x - \omega_3 \omega_1 M_s \cos \varphi$$

$$+ \omega_2 \omega_3 M_s \cos \vartheta \sin \varphi \tag{5.71}$$

$$[\Omega^2 \mathbf{M} - \mathbf{\Omega}(\mathbf{M} \cdot \mathbf{\Omega})]|_y = M_s\{-(\omega_1^2 + \omega_2^2 + \omega_3^2) \sin \vartheta \cos \varphi$$

$$- \omega_3[\omega_1 \sin \varphi + \omega_2 \cos \vartheta \cos \varphi - \omega_3 \sin \vartheta \cos \varphi]\}$$

$$= M_s\{-(\omega_1^2 + \omega_2^2) \sin \vartheta \cos \varphi - \omega_3 \omega_1 \sin \varphi$$

$$- \omega_3 \omega_2 \cos \vartheta \cos \varphi\}$$

$$= (\omega_1^2 + \omega_2^2) M_y - M_s \omega_3 \omega_1 \sin \varphi$$

$$- M_s \omega_3 \omega_2 \cos \vartheta \cos \varphi \tag{5.72}$$

which are the desired components of Eq. (5.17) as written in Eqs. (5.20), (5.21).

VI.   THE CALCULATION OF RELAXATION TIMES FOR A SINGLE-DOMAIN FERROMAGNETIC PARTICLE FROM GILBERT'S EQUATION

A.   Introduction

In Section V we have shown how the longitudinal $\tau_\parallel$ and transverse $\tau_\perp$ relaxation times of the magnetization of an assembly of ferrofluid particles may be found by averaging the inertial Langevin equation and proceeding to the noninertial limit. In the present section we calculate the relaxation times $\tau_\parallel$ and $\tau_\perp$ of the magnetization of single domain ferromagnetic particles. Ordinary inertia plays no role [8] in Néel relaxation unlike relaxation of ferrofluid particles. Thus it will be necessary to have an interpretation rule (Itô [51] or Stratonovich [12]) for the nonlinear Langevin equations. This is also necessary when using the Fokker–Planck equation. Indeed the interpretation rule used by Brown [8] is the Stratonovich one.

The method we describe is analogous to the one used by us [52] in order to study dielectric relaxation of polar fluids in the presence of a DC electric field using the Langevin equation. Our calculations are carried out by interpreting the Cartesian components of Gilbert's equation as a set of stochastic nonlinear differential equations of Stratonovich type [12].

The desired averages may then be computed in accordance with the Stratonovich rule.

We then obtain from the averaged Gilbert equation, general expressions for $\tau_\parallel$ and $\tau_\perp$ in terms of equilibrium averages of the magnetization components for the relaxation times for an arbitrary uniaxial potential of the crystalline anisotropy. These may be applied to several particular cases. We first consider a $\cos \vartheta$ potential. The general expressions then yield the same field dependence for $\tau_\parallel$ and $\tau_\perp$ (in terms of the Langevin function) as that given by Martsenyuk et al. [16] for Debye relaxation of suspensions of fine ferromagnetic particles in a constant magnetic field $\mathbf{H}_0$. Next we obtain an expression in terms of Dawson's integral for $\tau_\parallel$ for Néel relaxation in the uniaxial potential of the crystalline anisotropy $V = K \sin^2\vartheta$. This result is in agreement with that obtained by Raĭkher and Shliomis in [17]. It is further extended to calculate $\tau_\perp$. We also derive equations for $\tau_\parallel$ and $\tau_\perp$ for the uniaxial potential $V(\vartheta) = -MH_0 \cos \vartheta + K \sin^2\vartheta$. This potential includes the effects of both an applied constant magnetic field $\mathbf{H}_0$ and the crystalline anisotropy.

We begin with Gilbert's equation

$$\frac{d\mathbf{M}}{dt} = \gamma \mathbf{M} \times \left(\mathbf{H}_T - \eta \frac{d\mathbf{M}}{dt} \right) \tag{6.1}$$

where

$$\mathbf{H}_T = \mathbf{H}_{ef} + \mathbf{H}(t) + \mathbf{h}(t) \tag{6.2}$$

is the total magnetic field which consists of an axially symmetrical field $\mathbf{H}_{ef}$ and a small external applied field $\mathbf{H}(t)$ ($\nu(\mathbf{M} \cdot \mathbf{H}) \ll kT$, k is the Boltzmann constant, T is the temperature, ν is the volume of the particle) and $\mathbf{h}(t)$ is a random field term. The field $\mathbf{h}(t)$ has the properties (as discussed in Section II.D)

$$\overline{h_i(t)} = 0 \tag{6.3}$$

that is, it has a statistical average of zero, and for $i, j = x, y, z$

$$\overline{h_j(t)h_i(t')} = \frac{2kT\eta}{\nu} \delta(t - t')\delta_{ij} \tag{6.4}$$

where the symbol ——— means the statistical average [8], $\delta(t)$ is the Dirac delta function and δ_{ij} is Kronecker's delta.

The effective field $\mathbf{H}_{ef}$ includes the applied constant magnetic field $\mathbf{H}_0$ and the effect of the crystalline anisotropy [8]. If $V(\mathbf{M})$ is the free energy per unit volume expressed as a function of **M**, then [8]

$$\mathbf{H}_{ef}(\mathbf{M}) = -\frac{\partial V}{\partial \mathbf{M}} \tag{6.5}$$

where

$$\frac{\partial}{\partial \mathbf{M}} = \mathbf{i}\,\frac{\partial}{\partial M_x} + \mathbf{j}\,\frac{\partial}{\partial M_y} + \mathbf{k}\,\frac{\partial}{\partial M_z}$$

Because $|\mathbf{M}| = \text{constant} = M_s$, V is indeterminate by an arbitrary function of $\mathbf{M}^2$ and $\mathbf{H}_{ef}$ by an arbitrary vector along **M** which contributes nothing to $\mathbf{M} \times \mathbf{H}_{ef}$ [4]. We confine ourselves to a uniaxial potential, where $\mathbf{H}_{ef}(\mathbf{M})$ has only a **k**-component, namely

$$\mathbf{H}_{ef}(\mathbf{M}) = -\mathbf{k}\,\frac{\partial}{\partial M_z}\,V(\mathbf{M}) \tag{6.6}$$

As we already know Gilbert's equation, (6.1) can be transformed to the equivalent Landau–Lifshitz form

$$\frac{d\mathbf{M}}{dt} = g' M_s \mathbf{M} \times \mathbf{H}_T + h'(\mathbf{M} \times \mathbf{H}_T) \times \mathbf{M} \tag{6.7}$$

where

$$g' = \frac{\gamma}{(1 + \eta^2 \gamma^2 M_s^2)M_s}, \quad h' = \frac{\eta \gamma^2}{1 + \eta^2 \gamma^2 M_s^2} \tag{6.8}$$

On using Eq. (6.2), we can rewrite Eq. (6.7) as

$$\frac{d\mathbf{M}}{dt} = g' M_s \mathbf{M} \times \mathbf{H}_{ef} - h'\mathbf{M} \times (\mathbf{M} \times \mathbf{H}_{ef}) + \mathbf{F}H(t) + \mathbf{L} \tag{6.9}$$

where

$$\mathbf{F} = g' M_s \mathbf{M} \times \mathbf{e} - h'\mathbf{M} \times (\mathbf{M} \times \mathbf{e}), \quad \mathbf{e} = \mathbf{H}(t)/H(t) \tag{6.10}$$

$$\mathbf{L} = g' M_s \mathbf{M} \times \mathbf{h} - h'\mathbf{M} \times (\mathbf{M} \times \mathbf{h}) \tag{6.11}$$

We can rewrite Eq. (6.9) using the triple product formula as

$$\frac{d\mathbf{M}}{dt} = g' M_s \mathbf{M} \times \mathbf{H}_{ef} + h'(M_s^2 \mathbf{H}_{ef} - \mathbf{M}(\mathbf{M} \cdot \mathbf{H}_{ef})) + \mathbf{F}H(t) + \mathbf{L} \tag{6.12}$$

or

$$\frac{d}{dt} M_x = g'M_s H_{ef} M_y - h'M_x M_z H_{ef} + F_x H(t) + L_x \tag{6.13}$$

$$\frac{d}{dt} M_y = -g'M_s H_{ef} M_x - h'M_y M_z H_{ef} + F_y H(t) + L_y \tag{6.14}$$

$$\frac{d}{dt} M_z = h'[M_s^2 - M_z^2]H_{ef} + F_z H(t) + L_z \tag{6.15}$$

where F_i and L_i are the projections of the vectors $\mathbf{F}$ and $\mathbf{L}$ onto the x, y and z axes.

We may further rewrite Eqs. (6.13)–(6.15) as

$$\frac{d}{dt} M_x = g'H_{ef}M_s M_y - h'H_{ef}M_x M_z + F_x H(t) + h'(M_s^2 - M_x^2)h_x$$
$$- (g'M_s M_z + h'M_x M_y)h_y + (g'M_y M_s - h'M_z M_x)h_z \tag{6.16}$$

$$\frac{d}{dt} M_y = - g'H_{ef}M_s M_x - h'H_{ef}M_z M_y + F_y H(t) + h'(M_s^2 - M_y^2)h_y$$
$$- (g'M_s M_x + h'M_z M_y)h_z + (g'M_s M_z - h'M_x M_y)h_x \tag{6.17}$$

$$\frac{d}{dt} M_z = h'H_{ef}(M_s^2 - M_z^2) + F_z H(t) + h'(M_s^2 - M_z^2)h_z$$
$$- (g'M_s M_y + h'M_z M_x)h_x + (g'M_s M_x - h'M_z M_y)h_y \tag{6.18}$$

It is evident that Eqs. (6.16)–(6.18) have the form of the set of nonlinear Langevin equations (Appendix D)

$$\frac{d}{dt} M_i = \Phi_i(\mathbf{M}) + G_{ik}(\mathbf{M})h_k(t), \quad (i = x, y, z) \tag{6.19}$$

$$\overline{h_k(t)} = 0, \quad \overline{h_k(t)h_m(t')} = \delta_{km}\mu\delta(t - t') \tag{6.20}$$

where according to Eq. (6.4),

$$\mu = \frac{2kT\eta}{\nu}$$

and summation over k is understood.

Equation (6.19) contains multiplicative noise terms $G_{ik}(\mathbf{M})h_k(t)$. This poses an interpretation problem as discussed by Risken [31]. Risken has shown, taking the Langevin equation for N stochastic variables $\{\xi\} = \{\xi_1, \xi_2, \ldots, \xi_N\}$ as

$$\dot{\xi}_i = h_i(\{\xi\}, t) + g_{ik}(\{\xi\}, t)\Gamma_k(t) \qquad (6.21)$$

with

$$\overline{\Gamma_k(t)} = 0, \quad \overline{\Gamma_k(t)\Gamma_m(t')} = 2\delta_{km}\delta(t - t') \qquad (6.22)$$

and interpreting it as a Stratonovich [12] equation, that the drift coefficient is

$$D_i(\{x\}, t) = \dot{x}_i = \lim_{\tau \to 0} \left\{ \frac{1}{\tau} \overline{(\xi_i(t + \tau) - x_i)} \right\} \Big|_{\xi_k(t) = x_k}$$

$$= h_i(\{x\}, t) + g_{kj}(\{x\}, t) \frac{\partial}{\partial x_k} g_{ij}(\{x\}, t), \quad k, j = 1, 2, \ldots, N \qquad (6.23)$$

In Eq. (6.23), $\xi_i(t + \tau)$ ($\tau > 0$) is a solution of (6.21) which at time t has the sharp value $\xi_k(t) = x_k$ for $k = 1, 2, \ldots, N$. The last term in Eq. (6.18) is called the noise induced or spurious drift [31]. It should be noted that the quantities x_k in Eq. (6.23) are themselves random variables with the probability density function $W(\{x\}, t)$ defined such that $W \, dx_k$ is the probability of finding x_k in the interval $(x_k, x_k + dx_k)$.

We now use this theorem to evaluate the average of the multiplicative noise terms in Eqs. (6.16)–(6.20). In our case $\xi_1(t) = M_x$, $\xi_2(t) = M_y$, $\xi_3(t) = M_{\|}$. Thus on taking into account Eqs. (6.16)–(6.23) and averaging Eq. (6.23) with the density distribution function $W(\{\mathbf{M}\}, t)$ of magnetization orientations in the configuration space at time t, we obtain the equations of motion for the averaged components $\langle M_i \rangle$ ($i = x, y, z$) of magnetization

$$\frac{d}{dt} \langle M_x \rangle = g' \langle H_{ef} M_s M_y \rangle - h' \langle H_{ef} M_x M_z \rangle + \langle F_x \rangle H(t)$$
$$+ \frac{\mu}{2} \left\langle G_{jk} \frac{\partial}{\partial M_j} G_{1k} \right\rangle \qquad (6.24)$$

$$\frac{d}{dt} \langle M_y \rangle = -g' \langle H_{ef} M_s M_x \rangle - h' \langle H_{ef} M_y M_z \rangle + \langle F_y \rangle H(t)$$
$$+ \frac{\mu}{2} \left\langle G_{jk} \frac{\partial}{\partial M_j} G_{2k} \right\rangle \qquad (6.25)$$

$$\frac{d}{dt} \langle M_z \rangle = h' \langle H_{ef}(M_s^2 - M_z^2) \rangle + \langle F_z \rangle H(t) + \frac{\mu}{2} \left\langle G_{jk} \frac{\partial}{\partial M_j} G_{3k} \right\rangle \qquad (6.26)$$

where the symbol $\langle f \rangle$ designates averaging f over the density distribution function $W(\{\mathbf{M}\}, t)$, namely

$$\langle f(\{\mathbf{M}\}) \rangle = \int f(\{\mathbf{M}\}) W(\{\mathbf{M}\}, t)\, d\mathbf{M}$$

The evaluation of the terms $\frac{1}{2}\mu \langle G_{jk}\, \partial/\partial M_j)G_{ik} \rangle$ is given in Section VI.D. The results are

$$\frac{\mu}{2}\left\langle G_{jk}\frac{\partial}{\partial M_j} G_{1k}\right\rangle = -\mu(g'^2 + h'^2)M_s^2\langle M_x \rangle \qquad (6.27)$$

$$\frac{\mu}{2}\left\langle G_{jk}\frac{\partial}{\partial M_j} G_{2k}\right\rangle = -\mu(g'^2 + h'^2)M_s^2\langle M_y \rangle \qquad (6.28)$$

$$\frac{\mu}{2}\left\langle G_{jk}\frac{\partial}{\partial M_j} G_{3k}\right\rangle = -\mu(g'^2 + h'^2)M_s^2\langle M_z \rangle \qquad (6.29)$$

According to Brown [8, Eqs. (3.19)–(3.22)],

$$\mu M_s^2(h'^2 + g'^2) = \frac{2kTh'}{\nu} = \tau_N^{-1} \qquad (6.30)$$

where τ_N is the relaxation time for $H_{\mathrm{ef}} = 0$. We find on taking account of Eqs. (6.24)–(6.30) that

$$\frac{d}{dt}\langle M_x \rangle + \frac{1}{\tau_N}\langle M_x \rangle = g'M_s\langle H_{\mathrm{ef}}M_y \rangle - h'\langle H_{\mathrm{ef}}M_xM_z \rangle + \langle F_x \rangle H(t)$$
$$(6.31)$$

$$\frac{d}{dt}\langle M_y \rangle + \frac{1}{\tau_N}\langle M_y \rangle = -g'M_s\langle H_{\mathrm{ef}}M_x \rangle - h'\langle H_{\mathrm{ef}}M_yM_z \rangle + \langle F_y \rangle H(t)$$
$$(6.32)$$

$$\frac{d}{dt}\langle M_z \rangle + \frac{1}{\tau_N}\langle M_z \rangle = h'\langle H_{\mathrm{ef}}(M_s^2 - M_z^2) \rangle + \langle F_z \rangle H(t)$$
$$(6.33)$$

These equations govern the Néel relaxation of a single domain ferromagnetic particle. They bear a resemblance to equations (5.26)–(5.28) for the Debye relaxation of ferrofluid particles (with the Néel mechanism blocked) subjected to a weak AC field superimposed on a strong DC magnetic field $\mathbf{H}_0$. They differ from the ferrofluid equations, however, insofar as they contain precessional terms $g'M_s\langle H_{\mathrm{ef}}M_x \rangle$ and $g'M_s\langle H_{\mathrm{ef}}M_y \rangle$ which are absent from those equations. Nevertheless, these terms do not affect the calculation of relaxation times. (The precessional terms are responsible for ferromagnetic resonance.)

B. Derivation of Equations for the Relaxation Times $\tau_\parallel$ and $\tau_\perp$

Equations (6.31)–(6.33) are the first terms in an infinite hierarchy of differential equations. It is obvious that in order to solve these equations, we must also obtain equations for $\langle H_{ef}M_x \rangle$, $\langle H_{ef}M_yM_z \rangle$, $\langle H_{ef}M_y \rangle$, $\langle H_{ef}M_xM_z \rangle$, $\langle H_{ef} \rangle$, $\langle H_{ef}M_z^2 \rangle$ and so on. (We have shown in [52] how the next member of the hierarchy of equations can also be obtained from the Langevin equation by means of transformation of variables). However, if we merely wish to calculate the relaxation time, this difficulty may be circumvented by means of a procedure which has been suggested by Morita [56] in connection with the calculation of relaxation times from the Fokker–Planck equation. His method was also used by Coffey et al. [52] in conjunction with the Langevin equation and linear response theory in the study of dielectric relaxation of a polar fluid under the influence of a constant electric field superimposed on which is a weak AC field. It may also be used for the present problem in conjunction with the Langevin equation for a single domain ferromagnetic particle. The procedure also corresponds to that used by San Miguel et al. [78] in the calculation of intensity correlations of single and two mode lasers. The technique is commonly known as the effective eigenvalue method.

We first consider the equation of motion of $\langle M_z \rangle$, namely Eq. (6.33). We suppose that a small field $\mathbf{H}(t) = HU(t)\mathbf{k}$, where $U(t)$ is the unit step function, is applied along the z-axis at time $t = 0$. We require the linear response to $\mathbf{H}(t)$. We therefore assume that in Eq. (6.33), $\langle M_z \rangle$ and $\langle H_{ef}(M_s^2 - M_z^2) \rangle$ can be represented as

$$\langle M_z \rangle = \langle M_z \rangle_0 + \langle M_z \rangle_1$$
$$\langle H_{ef}(M_s^2 - M_z^2) \rangle = \langle H_{ef}(M_s^2 - M_z^2) \rangle_0 + \langle H_{ef}(M_s^2 - M_z^2) \rangle_1$$

(6.34)

where the subscript 0 denotes the equilibrium ensemble average in the absence of the field $\mathbf{H}(t)$ and the subscript 1 denotes the portion of the ensemble average which is linear in $\mathbf{H}(t)$. Thus we have from Eqs. (6.33) and (6.34)

$$\frac{d}{dt} (\langle M_z \rangle_0 + \langle M_z \rangle_1) + \frac{1}{\tau_N} (\langle M_z \rangle_0 + \langle M_z \rangle_1) = h'(\langle H_{ef}(M_s^2 - M_z^2) \rangle_0$$
$$+ \langle H_{ef}(M_s^2 - M_z^2) \rangle_1) + (\langle F_z \rangle_0 + \langle F_z \rangle_1)H(t)$$

(6.35)

Now $\langle M_z \rangle_0$ has reached equilibrium, so that

$$\langle M_z \rangle_0 = \int_0^{2\pi} \int_0^\pi M_s \cos \vartheta \, W_0(\vartheta) \sin \vartheta \, d\vartheta \, d\phi \qquad (6.36)$$

where

$$W_0(\vartheta) = C \exp\left(-\frac{\nu V(\vartheta)}{kT}\right) \qquad (6.37)$$

is the equilibrium distribution function, ϑ and ϕ are the polar and azimuthal angles, respectively, and C is the normalizing constant. In turn, the ensemble average $\langle M_z \rangle_1$ satisfies

$$\frac{d}{dt} \langle M_z \rangle_1 + \frac{1}{\tau_N} \langle M_z \rangle_1 = h' \langle H_{\text{ef}}(M_s^2 - M_z^2) \rangle_1 + h' \langle M_s^2 - M_z^2 \rangle_0 HU(t) \qquad (6.38)$$

Equation (6.38) represents a three term recurrence relation driven by a forcing function, namely the $H(t)$ term. In order to determine $\tau_\parallel$, we consider the unforced equation, namely

$$\frac{d}{dt} \langle M_z \rangle_1 + \frac{1}{\tau_N} \langle M_z \rangle_1 = h' \langle H_{\text{ef}}(M_s^2 - M_z^2) \rangle_1 \qquad (6.39)$$

The Laplace transform of Eq. (6.39) with the initial condition

$$\langle M_z(0) \rangle_1 = 0 \qquad (6.40)$$

is

$$(s + \tau_N^{-1})\mathscr{L}\langle M_z \rangle_1 = h' \mathscr{L}\langle H_{\text{ef}}(M_s^2 - M_z^2) \rangle_1 \qquad (6.41)$$

which may be rearranged to give

$$\mathscr{L}\langle M_z \rangle_1 \left(s + \tau_N^{-1} - h' \frac{\mathscr{L}\langle H_{\text{ef}}(M_s^2 - M_z^2) \rangle_1}{\mathscr{L}\langle M_z \rangle_1}\right) = 0 \qquad (6.42)$$

where the symbol $\mathscr{L}f$ denotes the Laplace transform of a function $f(t)$, namely

$$\mathscr{L}f = \int_0^\infty \exp(-st) f(t) \, dt$$

The characteristic equation of the system is

$$s + \tau_N^{-1} - h' \frac{\mathcal{L}\langle H_{ef}(M_s^2 - M_z^2)\rangle_1}{\mathcal{L}\langle M_z\rangle_1} = 0 \tag{6.43}$$

If we suppose, following Morita [56], that

$$\frac{\mathcal{L}\langle H_{ef}(M_s^2 - M_z^2)\rangle_1}{\mathcal{L}\langle M_z\rangle_1} = \frac{\mathcal{L}\langle H_{ef}(M_s^2 - M_z^2)\rangle - \mathcal{L}\langle H_{ef}(M_s^2 - M_z^2)\rangle_0}{\mathcal{L}\langle M_z\rangle - \mathcal{L}\langle M_z\rangle_0} \tag{6.44}$$

may be replaced by its final (equilibrium) value (i.e., its value as t tends to infinity), namely

$$\lim_{t \to \infty} \frac{\langle H_{ef}(M_s^2 - M_z^2)\rangle_1}{\langle M_z\rangle_1} = \lim_{s \to 0} \frac{s\mathcal{L}\langle H_{ef}(M_s^2 - M_z^2)\rangle_1}{s\mathcal{L}\langle M_z\rangle_1} \tag{6.45}$$

Equation (6.43) may then be evaluated as follows. At equilibrium $(t \to \infty)$

$$\langle M_z\rangle = M_s \frac{\int_0^\pi \cos\vartheta \, \exp\left(-\frac{\nu V}{kT} + \frac{\nu H M_s}{kT} \cos\vartheta\right) \sin\vartheta \, d\vartheta}{\int_0^\pi \exp\left(-\frac{\nu V}{kT} + \frac{\nu H M_s}{kT} \cos\vartheta\right) \sin\vartheta \, d\vartheta}$$

which becomes in the linear approximation of H

$$\langle M_z\rangle \cong M_s \frac{\int_0^\pi \cos\vartheta\left(1 + \frac{\nu H M_s}{kT} \cos\vartheta\right) \exp\left(-\frac{\nu V}{kT}\right) \sin\vartheta \, d\vartheta}{\int_0^\pi \left(1 + \frac{\nu H M_s}{kT} \cos\vartheta\right) \exp\left(-\frac{\nu V}{kT}\right) \sin\vartheta \, d\vartheta}$$

$$= M_s\langle \cos\vartheta\rangle_0 + \frac{\nu H M_s^2}{kT} (\langle \cos^2\vartheta\rangle_0 - \langle \cos\vartheta\rangle_0^2)$$

$$= \langle M_z\rangle_0 + \frac{\nu H}{kT} (\langle M_z^2\rangle_0 - \langle M_z\rangle_0^2) \tag{6.46}$$

Further

$$\langle H_{ef}(M_s^2 - M_z^2)\rangle$$

$$= M_s^2 \frac{\int_0^\pi (1 - \cos^2\vartheta) H_{ef} \exp\left(-\frac{\nu V}{kT} + \frac{\nu H M_s}{kT} \cos\vartheta\right) \sin\vartheta \, d\vartheta}{\int_0^\pi \exp\left(-\frac{\nu V}{kT} + \frac{\nu H M_s}{kT} \cos\vartheta\right) \sin\vartheta \, d\vartheta}$$

$$\cong M_s^2 \frac{\int_0^\pi (1 - \cos^2\vartheta) H_{\mathrm{ef}} \left(1 + \frac{\nu H M_s}{kT} \cos \vartheta\right) \exp\left(-\frac{\nu V}{kT}\right) \sin \vartheta \, d\vartheta}{\int_0^\pi \exp\left(-\frac{\nu V}{kT}\right)\left(1 + \frac{\nu H M_s}{kT} \cos \vartheta\right) \sin \vartheta \, d\vartheta}$$

$$= \langle H_{\mathrm{ef}}(M_s^2 - M_z^2)\rangle_0 + \frac{\nu H M_s^3}{kT} \langle H_{\mathrm{ef}}(\cos \vartheta - \langle \cos \vartheta \rangle_0)(1 - \cos^2\vartheta)\rangle_0$$

$$\tag{6.47}$$

The second term on the right hand side of Eq. (6.47) can be evaluated on noting that

$$\langle H_{\mathrm{ef}}(\cos \vartheta - \langle \cos \vartheta \rangle_0)(1 - \cos^2\vartheta)\rangle_0$$

$$= \frac{\int_0^\pi -\frac{\partial V}{M_s \, \partial \cos \vartheta} [\cos \vartheta - \langle \cos \vartheta \rangle_0][1 - \cos^2\vartheta] \exp\left(-\frac{\nu V}{kT}\right) \sin \vartheta \, d\vartheta}{\int_0^\pi \exp\left(-\frac{\nu V}{kT}\right) \sin \vartheta \, d\vartheta}$$

$$= \frac{kT}{M_s \nu} \left\{ [\cos \vartheta - \langle \cos \vartheta \rangle_0][1 - \cos^2\vartheta] \exp\left(-\frac{\nu V}{kT}\right) \Big|_0^\pi \Big/ \right.$$

$$\times \int_0^\pi \exp\left(-\frac{\nu V}{kT}\right) \sin \vartheta \, d\vartheta$$

$$\left. - \frac{\int_0^\pi (1 - 3\cos^2\vartheta + 2\cos \vartheta \langle \cos \vartheta \rangle_0)\exp\left(-\frac{\nu V}{kT}\right) \sin \vartheta \, d\vartheta}{\int_0^\pi \exp\left(-\frac{\nu V}{kT}\right) \sin \vartheta \, d\vartheta} \right\}$$

$$= -\frac{kT}{M_s \nu} [1 - 3\langle \cos^2\vartheta \rangle_0 + 2\langle \cos \vartheta \rangle_0^2]$$

$$\tag{6.48}$$

Substituting Eq. (6.48) into (6.47), we obtain

$$\langle H_{\mathrm{ef}}(M_s^2 - M_z^2)\rangle = \langle H_{\mathrm{ef}}(M_s^2 - M_z^2)\rangle_0 - H(M_s^2 - 3\langle M_z^2 \rangle_0 + 2\langle M_z \rangle_0^2)$$

$$\tag{6.49}$$

The effective eigenvalue $\lambda_\parallel$ with this procedure is then

$$\lambda_\parallel = \tau_N^{-1} - h' \frac{\langle H_{\mathrm{ef}}(M_s^2 - M_z^2)\rangle - \langle H_{\mathrm{ef}}(M_s^2 - M_z^2)\rangle_0}{\langle M_z \rangle - \langle M_z \rangle_0}$$

$$= \tau_N^{-1} - h' \frac{M_s^2 H(1 - 3\langle \cos^2\vartheta \rangle_0 + 2\langle \cos \vartheta \rangle_0^2)}{\frac{\nu H M_s^2}{kT} (\langle \cos^2\vartheta \rangle_0 - \langle \cos \vartheta \rangle_0^2)}$$

$$= (2\tau_N)^{-1} \frac{1 - \langle \cos^2 \vartheta \rangle_0}{\langle \cos^2 \vartheta \rangle_0 - \langle \cos \vartheta \rangle_0^2} \tag{6.50}$$

We have used Eqs. (6.30), (6.46) and (6.49) here. Thus the longitudinal relaxation time $\tau_\parallel = \lambda_\parallel^{-1}$ may be expressed in terms of the equilibrium averages $\langle \cos \vartheta \rangle_0$ and $\langle \cos^2 \vartheta \rangle_0$ only.

We have

$$\tau_\parallel = 2\tau_N \frac{\langle \cos^2 \vartheta \rangle_0 - \langle \cos \vartheta \rangle_0^2}{1 - \langle \cos^2 \vartheta \rangle_0} \tag{6.51}$$

We now calculate the transverse relaxation time $\tau_\perp$. We consider the same problem as above but this time the step change in the field $\mathbf{H}(t) = \mathbf{i} U(t) H(t)$ is applied parallel to the x axis so that we need to determine the behavior of $\langle M_x \rangle$ from Eq. (6.31). We assume as before that

$$\langle M_x \rangle = \langle M_x \rangle_0 + \langle M_x \rangle_1$$

$$\langle H_{ef} M_x M_z \rangle = \langle H_{ef} M_x M_z \rangle_0 + \langle H_{ef} M_x M_z \rangle_1 \tag{6.52}$$

$$\langle H_{ef} M_y \rangle = \langle H_{ef} M_y \rangle_0 + \langle H_{ef} M_y \rangle_1$$

which with Eq. (6.31) leads us to the linearized equation for $\langle M_x \rangle_1$

$$\frac{d}{dt} \langle M_x \rangle_1 + \frac{1}{\tau_N} \langle M_x \rangle_1 = g' M_s \langle H_{ef} M_y \rangle_1 - h' \langle H_{ef} M_x M_z \rangle_1$$

$$+ h' H U(t) \langle M_s^2 - M_x^2 \rangle_0 \tag{6.53}$$

Note that

$$\langle M_x \rangle_0 = \langle H_{ef} M_x M_z \rangle_0 = \langle H_{ef} M_y \rangle_0 = 0 \tag{6.54}$$

in this case. We find just as before that the eigenvalue equation is

$$s + \tau_N^{-1} - \frac{g' M_s \mathcal{L} \langle H_{ef} M_y \rangle_1 - h' \mathcal{L} \langle H_{ef} M_x M_z \rangle_1}{\mathcal{L} \langle M_x \rangle_1} = 0 \tag{6.55}$$

or

$$s + \tau_N^{-1} - \frac{g' M_s \langle H_{ef} M_y \rangle - h' \langle H_{ef} M_x M_z \rangle}{\langle M_x \rangle} = 0 \quad (t \to \infty) \tag{6.56}$$

Now

$$\langle M_x \rangle = M_s \frac{\int_0^\pi \int_0^{2\pi} \sin\vartheta \cos\phi \exp\left(-\frac{\nu V}{kT} + \frac{\nu HM_s}{kT} \sin\vartheta \cos\phi\right) \sin\vartheta \, d\vartheta \, d\phi}{\int_0^\pi \int_0^{2\pi} \exp\left(-\frac{\nu V}{kT} + \frac{\nu HM_s}{kT} \sin\vartheta \cos\phi\right) \sin\vartheta \, d\vartheta \, d\phi}$$

(6.57)

which in the linear approximation of H is

$$\langle M_x \rangle \cong \frac{\nu HM_s^2}{kT} \langle \sin^2\vartheta \cos^2\phi \rangle_0 = \frac{\nu HM_s^2}{2kT} (1 - \langle \cos^2\vartheta \rangle_0) \quad (6.58)$$

In the same way

$$\langle H_{\text{ef}} M_x M_z \rangle \cong \frac{\nu HM_s^2}{kT} \langle H_{\text{ef}} \cos\vartheta \sin^2\vartheta \cos^2\phi \rangle_0$$

$$= -M_s^2 H(1 - 3\langle \cos^2\vartheta \rangle_0)/2 \quad (6.59)$$

$$\langle H_{\text{ef}} M_y \rangle = 0 \quad (6.60)$$

We have used the same procedure as in Eq. (6.48) in order to obtain Eq. (6.59).

Thus the effective eigenvalue $\lambda_\perp$ is given by

$$\lambda_\perp = \tau_N^{-1} - \frac{g'M_s\langle H_{\text{ef}} M_y \rangle - h'\langle H_{\text{ef}} M_x M_z \rangle}{\langle M_x \rangle}$$

$$= \tau_N^{-1} - \frac{h'M_s^2 H(1 - 3\langle \cos^2\vartheta \rangle_0)}{\frac{\nu M_s^2 H}{kT}(1 - \langle \cos^2\vartheta \rangle_0)}$$

$$= (2\tau_N)^{-1} \frac{1 + \langle \cos^2\vartheta \rangle_0}{1 - \langle \cos^2\vartheta \rangle_0}$$

whence the transverse relaxation time $\tau_\perp = \lambda_\perp^{-1}$ may be expressed in terms of the equilibrium average $\langle \cos^2\vartheta \rangle_0$ as

$$\tau_\perp = 2\tau_N \frac{1 - \langle \cos^2\vartheta \rangle_0}{1 + \langle \cos^2\vartheta \rangle_0} \quad (6.61)$$

It should be noted that Eqs. (6.51) and (6.61) for the relaxation times $\tau_\parallel$ and $\tau_\perp$ are valid for any axially symmetric potential of the crystalline anisotropy.

C.   Calculation of $\tau_{\parallel}$ and $\tau_{\perp}$ for Different Cases

In the simplest uniaxial case, the crystalline anisotropy energy density is $K \sin^2 \vartheta$ $(K > 0)$, where ϑ is the angle between $\mathbf{M}$ and the positive z axis [8]. Hence the total free energy density for a constant field $\mathbf{H}_0$ applied along the z axis is [8]

$$V(\vartheta) = -H_0 M_s \cos \vartheta - K \cos^2 \vartheta + K \qquad (6.62)$$

We may now calculate the relaxation times for particular cases.

1. Setting $K = 0$ when

$$V(\vartheta) = -M_s H_0 \cos \vartheta \qquad (6.63)$$

so that we ignore the crystalline anisotropy, we have

$$\langle \cos \vartheta \rangle_0 = \frac{\int_0^\pi \cos \vartheta e^{\xi \cos \vartheta} \sin \vartheta \, d\vartheta}{\int_0^\pi e^{\xi \cos \vartheta} \sin \vartheta \, d\vartheta} = L(\xi) \qquad (6.64)$$

$$\langle \cos^2 \vartheta \rangle_0 = \frac{\int_0^\pi \cos^2 \vartheta e^{\xi \cos \vartheta} \sin \vartheta \, d\vartheta}{\int_0^\pi e^{\xi \cos \vartheta} \sin \vartheta \, d\vartheta} = 1 - \frac{2}{\xi} L(\xi) \qquad (6.65)$$

where

$$\xi = \frac{\nu M_s H_0}{kT} \qquad (6.66)$$

and

$$L(\xi) = \coth \xi - 1/\xi \qquad (6.67)$$

is the Langevin function. We obtain on substituting Eqs. (6.64) and (6.65) into (6.51) and (6.61),

$$\tau_{\parallel} = \tau_N \frac{\xi - 2L(\xi) - \xi L^2(\xi)}{L(\xi)} \qquad (6.68)$$

$$\tau_{\perp} = \tau_N \frac{2L(\xi)}{\xi - L(\xi)} \qquad (6.69)$$

Equations (6.68) and (6.69) coincide precisely as far as their field dependencies are concerned with the formulae obtained in Section V.C. Note that in our formulae the relaxation time τ_N is the Néel time rather than the Debye time τ_D used in Section V.C as that discussion pertained to a ferrofluid where the Néel relaxation mechanism was blocked.

2. In the absence of the external field $\mathbf{H}_0 = 0$, so that for the relaxation in the purely crystalline potential,

$$V = -K \cos^2 \vartheta + K \tag{6.70}$$

we have

$$\langle \cos \vartheta \rangle_0 = 0 \tag{6.71}$$

and

$$\langle \cos^2 \vartheta \rangle_0 = \frac{\int_0^\pi e^{\sigma \cos^2 \vartheta} \cos^2 \vartheta \sin \vartheta \, d\vartheta}{\int_0^\pi e^{\sigma \cos^2 \vartheta} \sin \vartheta \, d\vartheta} = \frac{1}{2\sqrt{\sigma} D(\sqrt{\sigma})} - \frac{1}{2\sigma} \tag{6.72}$$

where

$$\sigma = \frac{\nu K}{kT} \tag{6.73}$$

and

$$D(x) = \exp(-x^2) \int_0^x \exp(t^2) \, dt \tag{6.74}$$

is the Dawson integral [53]. According to Eqs. (6.51) and (6.61), then

$$\tau_{\parallel} = 2\tau_N \frac{\sqrt{\sigma} - D(\sqrt{\sigma})}{D(\sqrt{\sigma})(1 + 2\sigma) - \sqrt{\sigma}} = \tau_N \frac{(1 + 2\langle P_2 \rangle_0)}{1 - \langle P_2 \rangle_0} = g_{\parallel} \tau_N \tag{6.75}$$

$$\tau_{\perp} = 2\tau_N \frac{D(\sqrt{\sigma})(2\sigma + 1) - \sqrt{\sigma}}{D(\sqrt{\sigma})(2\sigma - 1) + \sqrt{\sigma}} = 2\tau_N \frac{1 - \langle P_2 \rangle_0}{2 + \langle P_2 \rangle_0} = g_{\perp} \tau_N = \frac{2\tau_N}{g_{\parallel} + 1} \tag{6.76}$$

where the retardation factor is

$$g_{\parallel} = \frac{\chi_{\parallel}^s}{\chi_{\perp}^s}$$

with $\chi_{\parallel}^s$ and $\chi_{\perp}^s$ given by Eqs. (1.60) and (1.61).

Equation (6.75) for the relaxation time $\tau_\parallel$ coincides with that obtained by Raĭkher and Shliomis [17], that is, Eq. (29) of their paper [17], which becomes apparent if we note that

$$D(\sqrt{\sigma}) = \sqrt{\sigma} e^{-\sigma} F(\sigma)$$

where the function

$$F(\sigma) = \int_0^1 e^{\sigma x^2} \, dx$$

was used in [17].

Equations (6.75) and (6.76) are simplified (on using the asymptotic expansion of the Dawson integral, namely

$$D(y) = \frac{1}{2y} \left(1 + \frac{1}{2y^2} + \cdots \right)$$

for large y) to

$$\tau_\parallel \cong 2\tau_N \sigma , \quad \tau_\perp \cong \tau_N / \sigma \tag{6.77}$$

in the case of high potential barriers, that is, $\sigma \gg 1$, while in the opposite case ($\sigma \ll 1$), using the Taylor series expansion of $D(y)$ for small y,

$$D(y) = y - \frac{2}{3} y^3 + \frac{4}{15} y^5 + \cdots$$

so that (see also Eq. (3.101))

$$\tau_\parallel \cong \tau_N(1 + 2\sigma/5) , \quad \tau_\perp \cong \tau_N(1 - \sigma/5) \tag{6.78}$$

It is apparent from Eqs. (6.77)–(6.78) that with an increase in the barrier height parameter σ, the relaxation time $\tau_\parallel$ is enhanced while the relaxation time $\tau_\perp$ is decreased in contrast to the $\cos \vartheta$ potential where both $\tau_\parallel$ and $\tau_\perp$ decrease as ξ is increased.

It is interesting to note that in the theory of dielectrics, the relaxation times of polarization found for the Meier–Saupe potential which has the form of Eq. (6.70) and for a DC electric field E_0 [52, 54, 55] are similar to those found here for the pure crystalline anisotropy and for a constant magnetic field H_0, respectively.

3. In the general case the potential $V(\vartheta)$ is

$$\frac{vV}{kT} = -\xi \cos \vartheta - \sigma \cos^2 \vartheta + \sigma \tag{6.79}$$

We may also evaluate the quantities $\langle \cos \vartheta \rangle_0$ and $\langle \cos^2 \vartheta \rangle_0$ for this potential. We have

$$
\langle \cos \vartheta \rangle_0 = \frac{\int_0^\pi e^{\xi \cos \vartheta + \sigma \cos^2 \vartheta} \cos \vartheta \sin \vartheta \, d\vartheta}{\int_0^\pi e^{\xi \cos \vartheta + \sigma \cos^2 \vartheta} \sin \vartheta \, d\vartheta}
$$

$$
= \left\{ \frac{\xi}{\sqrt{\sigma} \left[(\xi L(\xi) + 1 + \xi) D\left(\sqrt{\sigma} + \frac{\xi}{2\sqrt{\sigma}}\right) + (\xi L(\xi) + 1 - \xi) D\left(\sqrt{\sigma} - \frac{\xi}{2\sqrt{\sigma}}\right) \right]} - \frac{\xi}{2\sigma} \right\}
\tag{6.80}
$$

and

$$
\langle \cos^2 \vartheta \rangle_0 = \frac{\int_0^\pi e^{\xi \cos \vartheta + \sigma \cos^2 \vartheta} \cos^2 \vartheta \sin \vartheta \, d\vartheta}{\int_0^\pi e^{\xi \cos \vartheta + \sigma \cos^2 \vartheta} \sin \vartheta \, d\vartheta}
$$

$$
= \left\{ \frac{\xi L(\xi) + 1 - \dfrac{\xi^2}{2\sigma^2}}{\sqrt{\sigma} \left[(\xi L(\xi) + 1 + \xi) D\left(\sqrt{\sigma} + \frac{\xi}{2\sqrt{\sigma}}\right) + (\xi L(\xi) + 1 - \xi) D\left(\sqrt{\sigma} - \frac{\xi}{2\sqrt{\sigma}}\right) \right]} + \frac{\dfrac{\xi^2}{2\sigma} - 1}{2\sigma} \right\}
\tag{6.81}
$$

On substituting Eqs. (6.80) and (6.81) into Eqs. (6.51) and (6.62) and on writing

$$
T(\xi, \sigma) = \sqrt{\sigma} \left[(\xi L(\xi) + 1 + \xi) D\left(\sqrt{\sigma} + \frac{\xi}{2\sqrt{\sigma}}\right) \right.
$$
$$
\left. + (\xi L(\xi) + 1 - \xi) D\left(\sqrt{\sigma} - \frac{\xi}{2\sqrt{\sigma}}\right) \right]
\tag{6.82}
$$

we obtain equations for the parallel and perpendicular relaxation times:

$$
\tau_\parallel = \tau_N \frac{\dfrac{2\sigma}{T(\xi, \sigma)} \left(\xi L(\xi) + 1 + \dfrac{\xi^2}{2\sigma} \right) - 1 - \dfrac{2\xi^2}{T^2(\xi, \sigma)}}{\sigma - \dfrac{\sigma}{T(\xi, \sigma)} \left(\xi L(\xi) + 1 - \dfrac{\xi^2}{2\sigma} \right) - \dfrac{\xi^2}{4\sigma} + \dfrac{1}{2}}
\tag{6.83}
$$

$$
\tau_\perp = 2\tau_N \frac{\sigma - \dfrac{\sigma}{T(\xi, \sigma)} \left(\xi L(\xi) + 1 - \dfrac{\xi^2}{2\sigma} \right) - \dfrac{\xi^2}{4\sigma} + \dfrac{1}{2}}{\sigma + \dfrac{\sigma}{T(\xi, \sigma)} \left(\xi L(\xi) + 1 - \dfrac{\xi^2}{2\sigma} \right) + \dfrac{\xi^2}{4\sigma} - \dfrac{1}{2}}
\tag{6.84}
$$

Equations (6.83) and (6.84) reduce to the previous formulae (6.68), (6.69), and (6.75), (6.76) when $\xi = 0$ and $\sigma = 0$, respectively.

Thus we have shown how general formulae for $\tau_{\parallel}$ and $\tau_{\perp}$ (valid for an arbitrary uniaxial potential of the crystalline anisotropy) for single domain particles may be calculated directly in terms of equilibrium averages $\langle \cos \vartheta \rangle_0$ and $\langle \cos^2 \vartheta \rangle_0$ from Gilbert's equation. That equation is regarded as a stochastic nonlinear equation of the Stratonovich type. This eliminates the complicated mathematical analysis which arises from the Fokker–Planck equation. Our approach is based on a well-defined method (that of Morita [56] or the effective eigenvalue [78]) of closing the hierarchy at the first equation and of reducing the nth order characteristic equation of the system to one of the first order. The relaxation of the magnetization components is thus characterized by a single effective eigenvalue or weighted decay rate so allowing a precise definition of the term "relaxation time of magnetization" when an external potential is present. The effective eigenvalues $\lambda_{\parallel}$ and $\lambda_{\perp}$ give precise information on the initial decay of the magnetization components.

The effective eigenvalue may be defined in the context of the Sturm–Liouville equation as [78]

$$\lambda_{\text{eff}} = \frac{\Sigma_k \lambda_k C_k}{\Sigma_k C_k} \tag{6.85}$$

where λ_k and C_k are the eigenvalues and their corresponding weight coefficients (amplitudes). It is very difficult to evaluate λ_{eff} from this formula using the Sturm–Liouville equation as a knowledge of the law of formation of the eigenvalues and their corresponding amplitudes is required. Such information is rarely available. The approach we have used in this section does not attempt to calculate λ_{eff} by explicitly calculating the eigenvalue spectrum as required by (6.85), rather it gives λ_{eff} in terms of the equilibrium averages $\langle \cos \vartheta \rangle_0$ and $\langle \cos^2 \vartheta \rangle_0$. It should be noted that a global characterization of the magnetization decay is given by the relaxation times $T_{\parallel}$ and $T_{\perp}$, defined as

$$T_{\gamma} = \frac{\int_0^{\infty} \langle M_{\gamma}(t) \rangle \, dt}{\langle M_{\gamma}(0) \rangle} \quad (\gamma = \parallel, \perp)$$

As noted in [78], this correlation time also includes contributions from all the eigenvalues, but it gives no information on possible different time regimes of relaxation. The behavior of T_{γ} and τ_{γ} is sometimes similar. In fact, if a single eigenvalue dominates the relaxation of magnetization, $T_{\gamma} = \tau_{\gamma}$. However if different timescales are involved, the behavior of T_{γ}

and τ_y can be different and in this case τ_y gives precise information on the initial relaxation of the magnetization.

D. Evaluation of the Spurious Drift Terms $G_{jk}(\partial G_{ik}/\partial M_j)$

We have in the case under consideration

$$G_{jk} \frac{\partial}{\partial M_j} G_{1k} = G_{11} \frac{\partial}{\partial M_x} G_{11} + G_{21} \frac{\partial}{\partial M_y} G_{11} + G_{31} \frac{\partial}{\partial M_z} G_{11}$$

$$+ G_{12} \frac{\partial}{\partial M_x} G_{12} + G_{22} \frac{\partial}{\partial M_y} G_{12} + G_{32} \frac{\partial}{\partial M_z} G_{12}$$

$$+ G_{13} \frac{\partial}{\partial M_x} G_{13} + G_{23} \frac{\partial}{\partial M_y} G_{13} + G_{33} \frac{\partial}{\partial M_z} G_{13}$$

$$(6.86)$$

$$G_{jk} \frac{\partial}{\partial M_j} G_{2k} = G_{11} \frac{\partial}{\partial M_x} G_{21} + G_{21} \frac{\partial}{\partial M_y} G_{21} + G_{31} \frac{\partial}{\partial M_z} G_{21}$$

$$+ G_{12} \frac{\partial}{\partial M_x} G_{22} + G_{22} \frac{\partial}{\partial M_y} G_{22} + G_{32} \frac{\partial}{\partial M_z} G_{22}$$

$$+ G_{13} \frac{\partial}{\partial M_x} G_{23} + G_{23} \frac{\partial}{\partial M_y} G_{23} + G_{33} \frac{\partial}{\partial M_z} G_{23}$$

$$(6.87)$$

$$G_{jk} \frac{\partial}{\partial M_j} G_{3k} = G_{11} \frac{\partial}{\partial M_x} G_{31} + G_{21} \frac{\partial}{\partial M_y} G_{31} + G_{31} \frac{\partial}{\partial M_z} G_{31}$$

$$+ G_{12} \frac{\partial}{\partial M_x} G_{32} + G_{22} \frac{\partial}{\partial M_y} G_{32} + G_{32} \frac{\partial}{\partial M_z} G_{32}$$

$$+ G_{13} \frac{\partial}{\partial M_x} G_{33} + G_{23} \frac{\partial}{\partial M_y} G_{33} + G_{33} \frac{\partial}{\partial M_z} G_{33}$$

$$(6.88)$$

Taking into account the explicit form of G_{ik} and that

$$\frac{\partial}{\partial M_y} G_{11} = \frac{\partial}{\partial M_z} G_{11} = \frac{\partial}{\partial M_x} G_{22} = \frac{\partial}{\partial M_z} G_{22} = \frac{\partial}{\partial M_x} G_{33} = \frac{\partial}{\partial M_y} G_{33} = 0$$

$$(6.89)$$

$$\frac{\partial}{\partial M_x} G_{12} = \frac{\partial}{\partial M_x} G_{21} = \frac{\partial}{\partial M_z} G_{32} = \frac{\partial}{\partial M_z} G_{23} = \frac{1}{2} \frac{\partial}{\partial M_x} G_{22} = -h' M_y$$

$$(6.90)$$

$$\frac{\partial}{\partial M_y} G_{12} = \frac{\partial}{\partial M_y} G_{21} = \frac{\partial}{\partial M_z} G_{13} = \frac{\partial}{\partial M_z} G_{31} = \frac{1}{2} \frac{\partial}{\partial M_x} G_{11} = -h'M_x$$
(6.91)

$$\frac{\partial}{\partial M_x} G_{13} = \frac{\partial}{\partial M_x} G_{31} = \frac{\partial}{\partial M_y} G_{32} = \frac{\partial}{\partial M_y} G_{23} = \frac{1}{2} \frac{\partial}{\partial M_z} G_{33} = -h'M_z$$
(6.92)

$$\frac{\partial}{\partial M_z} G_{12} = -\frac{\partial}{\partial M_z} G_{21} = -\frac{\partial}{\partial M_y} G_{13} = \frac{\partial}{\partial M_y} G_{31} = \frac{\partial}{\partial M_x} G_{23} = -\frac{\partial}{\partial M_x} G_{32}$$

$$= -g'M_s$$
(6.93)

we obtain

$$G_{jk} \frac{\partial}{\partial M_j} G_{1k} = -2(h'^2 + g'^2)M_s^2 M_x$$
(6.94)

$$G_{jk} \frac{\partial}{\partial M_j} G_{2k} = -2(h'^2 + g'^2)M_s^2 M_y$$
(6.95)

$$G_{jk} \frac{\partial}{\partial M_j} G_{3k} = -2(h'^2 + g'^2)M_s^2 M_z$$
(6.96)

which lead to Eqs. (6.27)–(6.29).

VII. NONAXIALLY SYMMETRIC PROBLEMS FOR THE UNIAXIALLY ANISOTROPIC POTENTIAL

A. Introduction

The spherical harmonic analysis so far presented for uniaxial anisotropy is mainly concerned with the relaxation in a direction parallel to the easy axis of the uniaxial anisotropy. We have not considered in detail the behavior resulting from the transverse application of an external field and the relaxation in that direction for uniaxial anisotropy. Thus we have only considered potentials of the form $V(\mathbf{r}, t) = V(\vartheta, t)$ where the azimuthal or ϕ dependence in Brown's equation is irrelevant to the calculation of the relaxation times. This has simplified the reduction of that equation to a set of differential-difference equations. In this section we consider the reduction when the azimuthal dependence is included. This is of importance in the transition of the system from magnetic relaxation to ferromagnetic resonance. The original study [17] was made using the method of separation of variables on Brown's equation which reduced the solution to an eigenvalue problem. We reconsider the solution by casting the problem in the form of a set of differential-difference equations and we rederive many of the results of Raĭkher and Shliomis by this method.

The problem considered first is $V(\vartheta) = K \sin^2\vartheta$ and $W = W(\vartheta, \phi, t)$ not $W = W(\vartheta, t)$ as before. We assume a solution of Brown's equation of the form given in Section III, namely

$$W(\vartheta, \phi, t) = \sum_{n=0}^{\infty} \sum_{m=-n}^{n} a_{nm}(t) P_n^{|m|}(\cos \vartheta) e^{im\phi}, \quad |m| \leq n \quad (7.1)$$

$W(\vartheta, \phi, t)$ must be entirely real so that the $a_{nm}(t)$ must satisfy

$$a_{n,-m} = a_{nm}^* \quad (7.2)$$

Thus, the ensemble averages with $x = \cos \vartheta$ may now be calculated from

$$\langle P_{n'}^{|m'|}(x)e^{im'\phi} \rangle$$
$$= \frac{\sum_{n=0}^{\infty} \sum_{m=-n}^{n} a_{nm}(t) \int_{-1}^{1} \int_{0}^{2\pi} P_n^{|m|}(x) P_{n'}^{|m'|}(x) e^{i(m+m')\phi} \, d\phi \, dx}{\sum_{n=0}^{\infty} \sum_{m=-n}^{n} a_{nm}(t) \int_{-1}^{1} \int_{0}^{2\pi} P_n^{|m|}(x) e^{im\phi} \, d\phi \, dx} \quad (7.3)$$

The orthogonality properties of the associated Legendre functions are

$$\int_{-1}^{+1} P_n^m(x) P_{n'}^m(x) \, dx = \frac{(n+|m|)!}{(n-|m|)!} \frac{2}{2n+1} \delta_{n,n'} \delta_{m,m'} \quad (7.4)$$

Noting these in conjunction with Eq. (7.3), we obtain

$$\langle P_n^{|m|}(x)e^{-im\phi} \rangle = \frac{1}{2n+1} \frac{(n+|m|)!}{(n-|m|)!} \frac{a_{n,-m}}{a_{00}}$$

$$\langle P_n^{|m|}(x)e^{im\phi} \rangle = \frac{1}{2n+1} \frac{(n+|m|)!}{(n-|m|)!} \frac{a_{nm}}{a_{00}} \quad (7.5)$$

We can resolve (7.5) into parallel and perpendicular components using the above property, so that

$$\langle P_n^{|m|}(x) \cos m\phi \rangle = \frac{1}{2n+1} \frac{(n+|m|)!}{(n-|m|)!} \frac{a_{nm} + a_{n,-m}}{2a_{00}} \quad (7.6)$$

and

$$\langle P_n^{|m|}(x) \sin m\phi \rangle = \frac{1}{2n+1} \frac{(n+|m|)!}{(n-|m|)!} \frac{a_{nm} + a_{n,-m}}{2ia_{00}} \quad (7.7)$$

B. Separation of the Variables in Brown's Equation

Brown's equation for $V = V(\vartheta)$ and $W = W(\vartheta, \phi, t)$ becomes

$$2\tau_N \frac{\partial W}{\partial t} = \frac{1}{\sin \vartheta} \frac{\partial}{\partial \vartheta} \left\{ \sin \vartheta \left[\frac{v}{kT} \left(\frac{\partial V}{\partial \vartheta} \right) W + \frac{\partial W}{\partial \vartheta} \right] \right\}$$
$$+ \frac{1}{\sin \vartheta} \frac{\partial}{\partial \phi} \left\{ \frac{v}{kT} \left(\frac{1}{\alpha} \frac{\partial V}{\partial \vartheta} \right) W + \frac{1}{\sin \vartheta} \frac{\partial W}{\partial \phi} \right\} \qquad (7.8)$$

The potential is

$$V = K \sin^2 \vartheta \qquad (7.9)$$

so that

$$\partial V / \partial \vartheta = 2K \sin \vartheta \cos \vartheta \qquad (7.10)$$

Substituting this potential into (7.8), we obtain

$$2\tau_N \frac{\partial W}{\partial t} = \frac{1}{\sin \vartheta} \frac{\partial}{\partial \vartheta} \left\{ \sin \vartheta \left[2\sigma \sin \vartheta \cos \vartheta W + \frac{\partial W}{\partial \vartheta} \right] \right\}$$
$$+ \frac{1}{\sin \vartheta} \frac{\partial}{\partial \phi} \left\{ \frac{2\sigma}{\alpha} \sin \vartheta \cos \vartheta W + \frac{1}{\sin \vartheta} \frac{\partial W}{\partial \phi} \right\} . \qquad (7.11)$$

The term

$$\frac{1}{\sin \vartheta} \frac{\partial}{\partial \vartheta} \left(\sin \vartheta \frac{\partial W}{\partial \vartheta} \right) + \frac{1}{\sin^2 \vartheta} \frac{\partial^2 W}{\partial \phi^2} , \qquad (7.12)$$

is the $\nabla^2 W$ term and the term

$$\frac{1}{\sin \vartheta} \frac{\partial}{\partial \vartheta} \sin \vartheta (2\sigma \sin \vartheta \cos \vartheta W) \qquad (7.13)$$

can be rewritten as

$$2\sigma \sin \vartheta \cos \vartheta \frac{\partial W}{\partial \vartheta} + 2\sigma(3 \cos^2 \vartheta - 1)W \qquad (7.14)$$

and the gyromagnetic term is

$$\frac{2\sigma}{\alpha} \cos \vartheta \frac{\partial W}{\partial \phi} \qquad (7.15)$$

Combining these we can rewrite (7.11) as

$$2\tau_N \frac{\partial W}{\partial t} = \frac{1}{\sin \vartheta} \frac{\partial}{\partial \vartheta} \left(\sin \vartheta \frac{\partial W}{\partial \vartheta} \right) + \frac{1}{\sin^2 \vartheta} \frac{\partial^2 W}{\partial \phi^2}$$

$$+ 2\sigma \sin \vartheta \cos \vartheta \frac{\partial W}{\partial \vartheta} + 2\sigma(3\cos^2\vartheta - 1)W + \frac{2\sigma}{\alpha} \cos \vartheta \frac{\partial W}{\partial \phi}$$

$$(7.16)$$

which is the same as Eq. (11) of Raĭkher and Shliomis [17] (recalling that α in the Landau–Lifshitz equation has an opposite sign to that of Gilbert's equation). We now proceed to solve each of the three components identified above separately.

The first component is the $\nabla^2 W$ term. Any spherical harmonic $X_{nm}(\vartheta, \phi) = P_n^{|m|}(\cos \vartheta)e^{im\phi}$ satisfies

$$\nabla^2 X_{nm}(\vartheta, \phi) = -n(n + 1)X_{nm}(\vartheta, \phi) \qquad (7.17)$$

therefore, the $\nabla^2 W$ term can be written as

$$\nabla^2 W = -\sum_{n=0}^{\infty} \sum_{m=-n}^{n} a_{nm}(t)n(n + 1)P_n^{|m|}(\cos \vartheta)e^{im\phi} \qquad (7.18)$$

Expanding the term

$$\frac{2\sigma}{\alpha} \cos \vartheta \frac{\partial W}{\partial \phi}$$

we obtain for $x = \cos \vartheta$

$$\frac{2\sigma}{\alpha} \sum_{n=0}^{\infty} \sum_{m=-n}^{n} a_{nm}(t)n(n + 1)xP_n^{|m|}(x)ime^{im\phi} \qquad (7.19)$$

Now the associated Legendre functions satisfy [37]

$$(2n + 1)xP_n^m(x) = (n - m + 1)P_{n+1}^m(x) + (n + m)P_{n-1}^m(x) \qquad (7.20)$$

so that Eq. (7.19) is

$$\frac{2\sigma}{\alpha} \sum_{n=0}^{\infty} \sum_{m=-n}^{m} \frac{a_{nm}e^{im\phi}}{2n + 1} [(n - |m| + 1)P_{n+1}^{|m|} + (n + |m|)P_{n-1}^{|m|}]$$

$$(7.21)$$

Finally we consider the term

$$2\sigma \sin \vartheta \cos \vartheta \, \frac{\partial W}{\partial \vartheta} + 2\sigma(3 \cos^2\vartheta - 1)W \tag{7.22}$$

This can be rewritten using the substitution $x = \cos \vartheta$ as

$$2\sigma \left[\sum_{n=0}^{\infty} \sum_{m=-n}^{n} a_{nm} e^{im\phi} \left[(x^2 - 1)x \, \frac{dP_n^{|m|}}{dx} + (3x^2 - 1)P_n^{|m|} \right] \right] \tag{7.23}$$

which on noting that [37]

$$(1 - x^2) \, \frac{dP_n^m}{dx} = (n + m)P_{n-1}^m - nxP_n^m \tag{7.24}$$

yields

$$2\sigma \left[\sum_{n=0}^{\infty} \sum_{m=-n}^{n} a_{nm} e^{im\phi} [-x(n + |m|)P_{n-1}^{|m|} + (n + 3)x^2 P_n^{|m|} - P_n^{|m|}] \right] \tag{7.25}$$

Again on using the property

$$(2n + 1)xP_n^m(x) = (n - m + 1)P_{n+1}^m(x) + (n + m)P_{n-1}^m \tag{7.26}$$

we may obtain a recurrence relation for $x^2 P_n^m$ which reduces Eq. (7.25) to

$$2\sigma \sum_{n=0}^{\infty} \sum_{m=-n}^{n} a_{nm} e^{im\phi} \Bigg\{ \frac{-(n + |m|)}{2n - 1} \, [(n - |m|)P_n^{|m|} + (n + |m| - 1)P_{n-2}^{|m|}]$$

$$+ \frac{(n + 3)}{2n + 1} \left\{ \frac{(n - |m| + 1)}{2n + 3} \, [(n - |m| + 2)P_{n+2}^{|m|} \right.$$

$$+ (n + |m| - 1)P_n^{|m|}]$$

$$+ \frac{(n + |m|)}{2n - 1} \, [(n - |m|)P_n^{|m|}$$

$$+ (n + |m| - 1)P_{n-2}^{|m|}] \Bigg\}$$

$$- P_n^{|m|} \Bigg\} \tag{7.27}$$

On combining the three solutions of Eqs. (7.18), (7.21), and (7.27), we now have

$$
2\tau_N \frac{\partial W}{\partial t} = \sum_{n=0}^{\infty} \sum_{m=-n}^{n} a_{nm} e^{im\phi} [-n(n+1)P_n^{|m|}]
$$

$$
+ \sum_{n=0}^{\infty} \sum_{m=-n}^{n} a_{nm} e^{im\phi} 2\sigma \left\{ P_n^{|m|} \left\{ -\frac{(n+|m|)(n-|m|)}{2n-1} - 1 \right. \right.
$$

$$
+ \frac{(n+3)(n-|m|+1)(n+|m|+1)}{(2n+1)(2n+3)}
$$

$$
\left. + \frac{(n+3)(n+|m|)(n-|m|)}{(2n+1)(2n-1)} \right\}
$$

$$
+ P_{n+2}^{|m|} \left\{ \frac{(n+3)(n-|m|+1)(n-|m|+2)}{(2n+1)(2n+3)} \right\}
$$

$$
+ P_{n-2}^{|m|} \left\{ \frac{(n+3)(n+|m|)(n+|m|-1)}{(2n+1)(2n-1)} \right.
$$

$$
\left. - \frac{(n+|m|)(n+|m|-1)}{(2n-1)} \right\}
$$

$$
+ \frac{i}{\alpha} \left[\frac{|m|(n+|m|)}{2n+1} P_{n-1}^{|m|} \right.
$$

$$
\left. \left. + \frac{|m|(n-|m|+1)}{2n+1} P_{n+1}^{|m|} \right] \right\} \tag{7.28}
$$

The $P_n^{|m|}$ coefficient can be simplified as follows. On factoring out the quantity $(n^2 - m^2)$, it is

$$
\frac{(n^2-m^2)[(n+3)(2n+3)-(2n+1)(2n+3)+(n+3)(2n-3)]}{(2n+1)(2n+3)(2n-1)}
$$

$$
+ \frac{(n+3)}{(2n+3)} - 1
$$

$$
= \frac{3(n^2-m^2) - n(2n-1)}{(2n+3)(2n-1)}
$$

$$
= \frac{n(n+1) - 3m^2}{(2n+3)(2n-1)} \tag{7.29}
$$

Note that the coefficients of the $P_{n-2}^{|m|}(x)$ term can be rewritten as

$$\frac{(n - |m|)(n + |m| - 1)[(n + 3) - (2n + 1)]}{(2n + 1)(2n - 1)}$$

$$= -\frac{(n - |m|)(n + |m| - 1)(n - 2)}{(2n + 1)(2n - 1)} \tag{7.30}$$

Thus using the orthogonality properties of the associated Legendre functions, we obtain the differential-difference equation

$$2\tau_N \frac{da_{nm}(t)}{dt} = \left[-n(n + 1) + 2\sigma \frac{n(n + 1) - 3m^2}{(2n + 3)(2n - 1)}\right] a_{nm}(t)$$

$$+ \frac{2\sigma}{\alpha} \left[\frac{im(n + |m| + 1)}{2n + 3} a_{n+1,m} + \frac{im(n - |m|)}{2n - 1} a_{n-1,m}\right]$$

$$+ 2\sigma \left[\frac{(n - |m|)(n + 1)(n - |m| - 1)}{(2n - 1)(2n - 3)} a_{n-2,m}\right.$$

$$\left. - \frac{n(n + |m| + 1)(n + |m| + 2)}{(2n + 3)(2n + 5)} a_{n+2,m}\right] \tag{7.31}$$

This is the same result as that obtained by Stepanov and Shliomis in [79] which is apparent if we note that the equation presented there is for $\langle P_n^{|m|}\rangle$ where these are related to a_{nm} by Eq. (7.5). The imaginary part implies that in the transverse direction the dynamics of the magnetization (with $m = 1$) are oscillatory. On setting $m = 0$ we obtain the same differential-difference equation as in Section III, Eq. (3.80), namely

$$2\tau_N \frac{da_n}{dt} + \left[n(n + 1) - \frac{2\sigma n(n + 1)}{(2n - 1)(2n + 3)}\right] a_n$$

$$= \frac{2\sigma n(n + 1)(n - 1)}{(2n - 3)(2n - 1)} a_{n-2}$$

$$- \frac{2\sigma n(n + 1)(n + 2)}{(2n + 5)(2n + 3)} a_{n+2} \tag{7.32}$$

We shall now show how this set may be closed as in Section IV to yield an expression for the longitudinal susceptibility with the relaxation time as in Section VI.

C. The Longitudinal Susceptibility $\chi_{\parallel}(\omega)$

Brown's equation for a weak field superimposed on a large field applied along the z or easy anisotropy axis is

$$\partial W^{(1)}/\partial t = LW^{(1)} + L_{\text{ext}}W^{(0)} \tag{7.33}$$

where

$$W^{(1)} = \sum_{n=0}^{\infty} a_n^{(1)}(t)P_n(\cos\vartheta) \tag{7.34}$$

and

$$W^{(0)} = \sum_{n=0}^{\infty} a_n^{(0)}(t)P_n(\cos\vartheta) \tag{7.35}$$

The differential-difference equation for uniaxial anisotropy in conjunction with the application of a small field along the z or easy axis of magnetization from Eqs. (3.80) and (4.17) with $n = 1$, $m = 0$, is therefore (to terms $O(\xi)$) see Eq. (3.50),

$$\tau_N \dot{a}_1^{(1)} = -\left(1 - \frac{2\sigma}{5}\right)a_1^{(1)} - \frac{6\sigma}{35}a_3^{(1)} + \xi\left(a_0^{(0)} - \frac{1}{5}a_2^{(0)}\right) \tag{7.36}$$

where $\xi = vHM_s/kT$.

The parallel relaxation time may be found by setting $L_{\text{ext}} = 0$, whence using the equation derived in Section III, Eq. (3.84),

$$A_1(s) = \frac{\tau_N a_1(0)}{\tau_N s + 1 - \dfrac{2\sigma}{5} + \dfrac{6\sigma}{35}R_3(s)} \tag{7.37}$$

The parallel relaxation time $\tau_{\parallel}$ is found by setting the denominator of this equation to zero and solving for s. The relaxation time is therefore

$$\tau_{\parallel} = \tau_N \frac{1}{1 - \dfrac{2\sigma}{5} + \dfrac{6\sigma}{35}R_3(s)} \tag{7.38}$$

As has been shown previously throughout Section IV, it is possible to replace the ratio $R_3(s)$ by

$$R_3(s) = \frac{7\langle P_3 \rangle}{3\langle P_1 \rangle} \tag{7.39}$$

which can be evaluated at equilibrium in the external field $\mathbf{H}$ (where $\xi = vHM_s/kT$) by again assuming that this ratio reaches equilibrium considerably faster than $A_1(s)$. The relaxation time is therefore

$$\tau_\parallel = \tau_N \frac{1}{1 - \dfrac{2\sigma}{5} + \sigma \dfrac{2\langle P_3 \rangle_\xi}{5\langle P_1 \rangle_\xi}} \tag{7.40}$$

Here the quantities $\langle P_1 \rangle_\xi$ and $\langle P_3 \rangle_\xi$ can be calculated (subscripts meaning to terms $O(\xi)$) as follows

$$\frac{\langle P_3 \rangle_\xi}{\langle P_1 \rangle_\xi} = \frac{\displaystyle\int_{-1}^{1} \frac{1}{2} (5x^3 - 3x) \exp(\sigma x^2 + \xi x)\, dx}{\displaystyle\int_{-1}^{1} x \exp(\sigma x^2 + \xi x)\, dx} \tag{7.41}$$

where $x = \cos\vartheta$. Expanding the exponential in powers of ξ and truncating after terms $O(\xi)$ both above and below we have

$$\frac{\langle P_3 \rangle_\xi}{\langle P_1 \rangle_\xi} = \frac{\xi \displaystyle\int_{-1}^{1} \frac{1}{2} (5x^4 - 3x^2) e^{\sigma x^2}\, dx}{\xi \displaystyle\int_{-1}^{1} x^2 e^{\sigma x^2}\, dx} \tag{7.42}$$

Using

$$F = \int_0^1 e^{\sigma x^2}\, dx$$

so that

$$F' = \frac{dF}{d\sigma} = \int_0^1 x^2 e^{\sigma x^2}\, dx$$

$$F'' = \frac{d^2 F}{d\sigma^2} = \int_0^1 x^4 e^{\sigma x^2}\, dx \tag{7.43}$$

given in Section VI, means that Eq. (7.42) can be written as

$$\frac{\langle P_3 \rangle_\xi}{\langle P_1 \rangle_\xi} = \frac{5F'' - 3F'}{2F'} = \frac{5F''}{2F'} - \frac{3}{2} \tag{7.44}$$

thus

$$\tau_\parallel = \tau_N \frac{1}{1 - \frac{2\sigma}{5} + \sigma \frac{F''}{F'} - \frac{3\sigma}{5}} = \tau_N \frac{1}{1 - \sigma + \sigma \frac{F''}{F'}} \qquad (7.45)$$

$$= 2\tau_N \left[\frac{\exp(\sigma) - F}{F(1 + 2\sigma) - \exp \sigma} \right]$$

$$= 2\tau_N \left[\frac{\sqrt{\sigma} - D(\sqrt{\sigma})}{D(\sqrt{\sigma})(1 + 2\sigma) - \sqrt{\sigma}} \right]$$

which is the same formula as that obtained in the previous section, and given by Raĭkher and Shliomis in [17] (see Eq. (6.75)).

Having obtained $\tau_\parallel$ we may proceed as in Section IV to derive an expression for the susceptibility. The differential-difference equation may be written

$$\tau_N \dot{a}_1^{(1)} = -\frac{\tau_N}{\tau_\parallel} a_1^{(1)} + \xi \left(a_0^{(0)} - \frac{1}{5} a_2^{(0)} \right) \qquad (7.46)$$

We can again replace the coefficient $a_2^{(0)}$ by the average value of the corresponding Legendre polynomial $\langle P_2 \rangle_0$, see Eq. (4.4), and so

$$a_2^{(0)} = 5\langle P_2 \rangle_0 a_0^{(0)} \qquad (7.47)$$

We must evaluate $1 - \langle P_2 \rangle_0$ according to Eq. (7.46). This is

$$1 - \langle P_2 \rangle_0 = 1 - \frac{\frac{1}{2} \int_{-1}^{1} (3x^2 - 1)e^{\sigma x^2} \, dx}{\int_{-1}^{1} e^{\sigma x^2} \, dx}$$

$$= \frac{3}{2} \left(1 - \frac{\int_{-1}^{1} x^2 e^{\sigma x^2} \, dx}{\int_{-1}^{1} e^{\sigma x^2} \, dx} \right) \qquad (7.48)$$

whence

$$1 - \langle P_2 \rangle_0 = \frac{3}{2} \left(1 - \frac{F'}{F} \right) \qquad (7.49)$$

The differential-difference equation is now

$$\tau_N \dot{a}_1^{(1)} = -\frac{\tau_N}{\tau_\parallel} a_1^{(1)} + \xi a_0^{(0)} \frac{3}{2} \left(1 - \frac{F'}{F} \right) \qquad (7.50)$$

If we now replace the coefficients by their respective Legendre polynomial averages recalling that

$$a_1 = 3a_0\langle P_1 \rangle \tag{7.51}$$

we have

$$\tau_N \langle \dot{P}_1 \rangle_\xi = -\frac{\tau_N}{\tau_\|} \langle P_1 \rangle_\xi + \frac{\xi}{2}\left(1 - \frac{F'}{F}\right) \tag{7.52}$$

which on dividing across by the Néel relaxation time given by Eq. (1.45) is

$$\langle \dot{P}_1 \rangle_\xi = -\frac{1}{\tau_\|} \langle P_1 \rangle_\xi + \frac{\alpha \gamma H}{1 + \alpha^2}\left(1 - \frac{F'}{F}\right) \tag{7.53}$$

where we note that

$$\lim_{t \to \infty} \langle P_1 \rangle_\xi = \frac{vHM_s}{2kT} \frac{\left(1 - \dfrac{F'}{F}\right)}{1 - \sigma + \sigma \dfrac{F''}{F}} = \frac{vHM_s}{kT}\frac{F'}{F}$$

This is the equation describing the dynamics of the longitudinal component of the magnetization obtained by the same method of truncation of the continued fraction as that employed in Section IV. This method, also used by Morita for dielectric relaxation [56], is a consequence of the final value theorem for Laplace transforms, which is

$$\lim_{t \to \infty} g(t) = \lim_{s \to 0} sG(s) \tag{7.54}$$

Equation (7.53) for the dynamics of the longitudinal magnetization component is the same as Raĭkher and Shliomis's equation (29) [17], with the quantity α from the Landau–Lifshitz equation replaced by $\alpha/(1 + \alpha^2)$ (the corresponding quantity in Brown's explicit form of Gilbert's equation Eqs. ((1.31)–(1.34))). We note that since the quantity α comes directly from the Néel relaxation time, there is no sign change necessary. We also note that the expression obtained here for $\tau_\|(\sigma)$ by the truncation of the continued fraction is the same as the closed form expression for $\tau_\|(\sigma)$ which we derived in the previous chapter using linear response theory and direct averaging of the Gilbert–Langevin equation. A comparison between this closed form expression, the low field expression obtained by truncation of the system matrix at 3×3, Brown's high anisotropy expression (both derived in Section III) and that obtained by numerical methods has already been made in [17].

In a periodic field $\mathbf{H}(t) = \mathbf{H}_0 e^{i\omega t}$ parallel to the anisotropy field, we find from Eq. (7.53) the magnetic susceptibility $\chi_\parallel(\omega)$:

$$\chi_\parallel(\omega) = \frac{nM_s v}{Hv} \langle P_1 \rangle = \frac{\chi_\parallel^s}{1 + i\omega\tau_\parallel} \tag{7.55}$$

where the static susceptibility $\chi_\parallel^s$ is (see Eq. (1.60))

$$\chi_\parallel^s = \frac{nM_s^2 v^2}{kT} \frac{F'}{F} \tag{7.56}$$

Raĭkher and Shliomis (Eq. (31) [17]) write this result as

$$\chi_0 = \frac{M_s^2}{2Kv} \frac{\sigma\left(1 - \dfrac{F'}{F}\right)}{1 - \sigma + \sigma\dfrac{F''}{F'}}$$

D. The Transverse Susceptibility $\chi_\perp(\omega)$

We now evaluate the magnetic susceptibility in a transverse external field directed along the x axis (the z axis as above, is oriented along the anisotropy field). We confine ourselves to the case of small σ, that is $\sigma \ll 1$ and we evaluate the critical value of σ at which the characteristic frequency of precession of the magnetic moment vanishes.

The differential-difference equation Eq. (7.31) for $m = 1$ is, following the switching off of the field,

$$2\tau_N \frac{da_{n,1}(t)}{dt} = \left\{ -n(n+1) + 2\sigma\left[\frac{n(n+1) - 3}{(2n-1)(2n+3)} \right] \right\} a_{n,1}(t)$$

$$+ \frac{2\sigma}{\alpha} \left[\frac{i(n+2)}{2n+3} a_{n+1,1} + \frac{i(n-1)}{2n-1} a_{n-1,1} \right]$$

$$+ 2\sigma\left[\frac{(n-1)(n+1)(n-2)}{(2n-1)(2n-3)} a_{n-2,1} - \frac{n(n+2)(n+3)}{(2n+3)(2n+5)} a_{n+2,1} \right] \tag{7.57}$$

The corresponding formula for $m = -1$ differs only in the sign preceding the $a_{n-1,-1}$ term since this is the only place where the absence of the modulus bars $||$ is significant. From this we obtain

$$\tau_N \dot{a}_{11} = -\left(1 + \frac{\sigma}{5}\right) a_{11} + i\frac{3\sigma}{5\alpha} a_{21}$$

$$\tau_N \dot{a}_{1,-1} = -\left(1 + \frac{\sigma}{5}\right) a_{1,-1} - i\,\frac{3\sigma}{5\alpha}\, a_{2,-1}$$

$$\tau_N \dot{a}_{21} = -\left(3 - \frac{\sigma}{7}\right) a_{21} + i\,\frac{\sigma}{3\alpha}\, a_{11}$$

$$\tau_N \dot{a}_{2,-1} = -\left(3 - \frac{\sigma}{7}\right) a_{2,-1} - i\,\frac{\sigma}{3\alpha}\, a_{1,-1} \tag{7.58}$$

In the above we have neglected the terms $a_{3,\pm 1}$ and $a_{4,\pm 1}$ because they are $O(\sigma^2)$ and higher. We combine these equations to obtain

$$\tau_N \frac{d}{dt}(a_{11} + a_{1,-1}) = -\left(1 + \frac{\sigma}{5}\right)(a_{11} + a_{1,-1}) + \frac{3\sigma i}{5\alpha}(a_{21} - a_{2,-1})$$

$$\tau_N \frac{d}{dt}\, i(a_{21} - a_{2,-1}) = -\left(3 - \frac{\sigma}{7}\right) i(a_{21} - a_{2,-1}) - \frac{\sigma}{3\alpha}(a_{11} + a_{1,-1}) \tag{7.59}$$

On making the substitution

$$a = \frac{1}{3}(a_{11} + a_{1,-1}), \qquad b = \frac{i}{2}(a_{21} - a_{2,-1}) \tag{7.60}$$

we obtain

$$\dot{a}(t) = -\frac{1}{\tau_N}\left(1 + \frac{\sigma}{5}\right) a(t) + \frac{2\sigma}{5\alpha\tau_N}\, b(t)$$

$$\dot{b}(t) = -\frac{1}{\tau_N}\left(3 - \frac{\sigma}{7}\right) b(t) - \frac{\sigma}{2\alpha\tau_N}\, a(t) \tag{7.61}$$

We have based our analysis throughout on Gilbert's equation. Raĭkher and Shliomis [17] based their analysis of this problem on the Landau–Lifshitz equation which has a different direction of precession convention. This results in a dimensionless damping factor α of opposite sign. If we bear this in mind and make use of the low σ approximations of Eq. (6.78) we see that the above equations coincide precisely with those of Raĭkher and Shliomis's equation (33) [17] for small σ, when we neglect all terms of order σ^2 and higher.

According to linear response theory (Appendix C) and Eq. (1.61)

$$\chi_\perp(\omega) = \frac{nv^2 M_s^2}{kT}\langle \sin^2\vartheta\,\cos^2\phi\rangle_0[1 - i\omega A(\omega)] \tag{7.62}$$

where

$$A(\omega) = \int_0^\infty a(t)e^{-i\omega t}\,dt \tag{7.63}$$

and $a(0) = 1$. This procedure temporarily normalises the $a(t)$ for the purposes of the calculation. Taking the Fourier transform ($s = i\omega$) of Eq. (7.61) with initial conditions $a(0) = 1$ and $b(0) = 0$, we have

$$\left[s + \frac{1}{\tau_N}\left(1 + \frac{\sigma}{5}\right)\right]A(s) - \frac{2\sigma}{5\alpha\tau_N}B(s) = 1$$

$$\left(s + \frac{1}{\tau_N}\left(3 - \frac{\sigma}{7}\right)\right)B(s) + \frac{\sigma}{2\alpha\tau_N}A(s) = 0 \tag{7.64}$$

The characteristic equation of the system is

$$\left[s + \frac{1}{\tau_N}\left(1 + \frac{\sigma}{5}\right)\right]\left[s + \frac{1}{\tau_N}\left(3 - \frac{\sigma}{7}\right)\right] + \frac{\sigma^2}{5\alpha^2\tau_N^2} = 0$$

that is

$$s^2 + \frac{2}{\tau_N}\left(2 + \frac{\sigma}{35}\right)s + \frac{1}{\tau_N^2}\left(3 + \frac{16}{35}\sigma + \frac{\sigma^2}{5\alpha^2} - \frac{\sigma^2}{35}\right) = 0$$

with roots

$$s = \frac{1}{2}\left\{-\frac{2}{\tau_N}\left(2 + \frac{\sigma}{35}\right) \pm \frac{1}{\tau_N}\sqrt{4\left(2 + \frac{\sigma}{35}\right)^2 - 4\left(3 + \frac{16\sigma}{35} + \frac{\sigma^2}{5\alpha^2} - \frac{\sigma^2}{35}\right)}\right\}$$

For small σ the behavior of the discriminant is dominated by the term (since for all ferromagnets $\alpha \ll 1$)

$$\left(1 - \frac{\sigma^2}{5\alpha^2}\right)$$

so that oscillatory behavior occurs if

$$\frac{\sigma^2}{5\alpha^2} > 1$$

that is the critical value of σ at which ferromagnetic resonance disappears is

$$\sigma^* = \alpha\sqrt{5}$$

in agreement with the result of [17]. This equation [17] also determines the critical volume of the ferromagnetic particle, namely

$$v^* = \frac{\sigma^* kT}{K}$$

at which the characteristic frequency of precession of the magnetic moment vanishes. For $\alpha = 0.1$ we find that $\sigma^* \cong 0.22$. In [17] it is shown that d^* the critical diameter usually varies from 30Å to 120Å. We may now obtain a formula for the transverse susceptibility $\chi_\perp(\omega)$ in the limit of small σ and non vanishing σ^2/α^2.

Eliminating $B(s)$ in Eq. (7.64) and putting $s = i\omega$ one obtains

$$A(\omega) = \cfrac{1}{i\omega + \cfrac{1}{\tau_N}\left(1 + \cfrac{\sigma}{5}\right) + \cfrac{\sigma^2/5\alpha^2\tau_N^2}{i\omega + \cfrac{1}{\tau_N}\left(3 - \cfrac{\sigma}{7}\right)}}$$

$$= \cfrac{i\omega + \cfrac{1}{\tau_N}\left(3 - \cfrac{\sigma}{7}\right)}{\left[i\omega + \cfrac{1}{\tau_N}\left(1 + \cfrac{\sigma}{5}\right)\right]\left[i\omega + \cfrac{1}{\tau_N}\left(3 - \cfrac{\sigma}{7}\right)\right] + \cfrac{\sigma^2}{5\alpha^2\tau_N^2}}$$

$$(7.65)$$

In the above equation we retain the terms $\sigma^2/5\alpha^2\tau_N^2$ since for single domain ferromagnetic particles $\alpha \ll 1$ [17]. Therefore for nonvanishing but small values of σ, it is necessary to keep this term. Now, substituting $A(\omega)$ from above into Eq. (7.61) and taking into account that for small σ,

$$\langle \sin^2\vartheta \, \cos^2\phi \rangle_0 = \frac{1}{3}\left(1 - \frac{2\sigma}{15}\right) \tag{7.66}$$

we have for the response to a small transverse AC field characterised by the parameter $\xi \ll 1$,

$$\frac{\chi_\perp(\omega)}{\chi_\perp^s} = \frac{\left(1 + \dfrac{\sigma}{5}\right)i\omega + \dfrac{1}{\tau_N}\left(3 + \dfrac{16\sigma}{35} + \dfrac{\sigma^2}{5\alpha^2}\right)}{\dfrac{1}{\tau_N}\left(3 + \dfrac{16\sigma}{35} + \dfrac{\sigma^2}{5\alpha^2}\right) - \omega^2\tau_N + 2i\omega\left(2 + \dfrac{\sigma}{35}\right)} \tag{7.67}$$

where

$$\chi_\perp^s = \frac{nv^2 M_s^2}{3kT} \left(1 - \frac{2\sigma}{15}\right) \tag{7.68}$$

Eq. (7.67) exhibits resonance for $\sigma > \sigma^*$ note also, that if we neglect the term $\sigma^2/5\alpha^2\tau_N^2$ then we obtain from Eqs. (7.62) and (7.67), the pure relaxational spectrum

$$\frac{\chi_\perp(\omega)}{\chi_\perp^s} = \frac{1}{1 + i\omega\tau_\perp} \tag{7.69}$$

Here we note that this result differs from that of Raĭkher and Shliomis [17] (Eq. (42) of their paper) where the relaxation time used in the denominator is simply the Néel relaxation time $\tau_N(\sigma)$. The relaxation time used here is $\tau_\perp(\sigma)$, the (perpendicular field-dependent) relaxation time in the presence of uniaxial anisotropy which from Eq. (7.58) to $O(\sigma)$ from the binomial expansion is (see Eq. 6.78)

$$\tau_\perp = \tau_N\left(1 - \frac{\sigma}{5}\right) \tag{7.70}$$

It should also be noted that the ferromagnetic resonance arises because spherical harmonics of order 2 are retained in the truncation of Eq. (7.57). This behavior is termed "entanglement' of the dipole and quadrupole branches of the response by Raĭkher and Shliomis [17].

VIII. CONCLUSIONS

The purpose of this review is to provide both an elementary introduction to and detailed notes on aspects of the theory of magnetic relaxation with particular emphasis on the pioneering work of William Fuller Brown Jr., Mark Shliomis and Yuri Raĭkher. In order to achieve this goal we have given a lengthy account of how a number of problems associated with magnetic relaxation may be treated both by the Fokker–Planck equation method and by averaging of the Langevin equation. The Fokker–Planck equation is particularly difficult to use for nonaxially symmetric problems involving anisotropy, because of the complicated manipulations of spherical harmonics which are required. It is evident that the Langevin equation provides a much easier and more physical way of treating such problems and in forthcoming work we shall show how this method may be applied to study the change from relaxation to ferromagnetic resonance behavior outlined in Section VII, in more detail. We have also emphasized throughout, the analogy with dielectric relaxation as many of the results for that theory may also be applied to magnetic relaxation. This is particularly important from an experimental point of view in relation to ferrofluids as it is much easier to observe both nonlinear effects such as magnetic saturation and frequency dependent effects such as magnetic relaxation for a ferrofluid than for a polar dielectric fluid because of the far lower fields and probe frequencies required.

We remark that the analysis is based throughout on the assumption that the Brownian and Néel mechanisms may be treated *separately*. For example in the work of Martsenyuk et al. [16] the magnetic moment is assumed to be "frozen" into the body of the particle thus only the Debye (Brownian) mechanism of relaxation is operative. The other limiting case is that considered by Raĭkher and Shliomis [17] where the liquid matrix is regarded as becoming solid so that the particles are deprived of their mechanical mobility ($\tau_D = \infty$) so that only the Néel relaxation mechanism is brought into play.

In actual polydisperse magnetic fluids [80], there are always particles with $\tau_D > \tau_N$ and $\tau_D < \tau_N$ therefore both diffusional processes will make comparable contributions to the magnetic relaxation. Shliomis [18] addressed this problem of distribution of particle sizes in a ferrofluid taking into account both mechanisms of relaxation by defining an effective relaxation time (not to be confused with the effective eigenvalue described earlier)

$$\tau_{\text{eff}} = \frac{\tau_N(\sigma)\tau_D}{\tau_N(\sigma) + \tau_D} \tag{8.1}$$

Shliomis [18] has also indicated that at critical value for the particle volume $\tau_D = \tau_N$ so that from (8.1)

$$\tau_{\text{eff}} \cong \tau_D, \quad v > v_c, \tau_N \gg \tau_D \tag{8.2}$$

and

$$\tau_{\text{eff}} \cong \tau_N, \quad v < v_c, \tau_N \ll \tau_D \tag{8.3}$$

The first case, Eq. (8.2) corresponds to the magnetic moment being *frozen* or *blocked* as considered in [16]. Since **M** will maintain its direction *relative to axes fixed* in the particle for a long time compared with the Debye time τ_D. The second case corresponds to the calculation in [17] where the effect of the rotational Brownian motion of the fluid on the magnetic susceptibility is ignored since the directional fluctuations of

M are fast compared with the random changes in particle orientation induced by the thermal agitation of the surrounding fluid molecules. In Eq. (8.1),

$$\tau_D = \frac{\zeta}{2kT} = \frac{8\pi\eta a^3}{2kT} = \frac{3V'\eta}{kT} \tag{8.4}$$

where V' is the *hydrodynamic* value of the particle volume which includes the surfactant layer and in this instance η is the dynamic viscosity of the liquid carrier. It is further assumed that τ_N is adequately represented by Brown's expressions for the longest relaxation time for high and low barrier heights written in the form [86]

$$\tau_N(\sigma) = \left\{ \begin{array}{ll} \tau_N(0) \dfrac{\sqrt{\pi}}{2} \sigma^{-3/2} e^{\sigma}, & \sigma > 2 \\[2mm] \tau_N(0), & \sigma \ll 1 \end{array} \right\} \tag{8.5}$$

$$\tau_N(0) = \frac{v}{kT} \frac{M_s}{\gamma} \frac{1+\alpha^2}{2\alpha} \cong \frac{v}{kT} \frac{M_s}{2\alpha\gamma} \tag{8.6}$$

which for convenience in comparison with experiment is often written [26]

$$\tau_N = \left\{ \begin{array}{ll} \tau_1 v^{-1/2} \exp\left(\dfrac{Kv}{kT}\right), & \dfrac{Kv}{kT} > 2 \\[3mm] \tau_2 v, & \dfrac{Kv}{kT} \ll 1 \end{array} \right\} \tag{8.7}$$

K is the anisotropy constant τ_1 and τ_2 are constants independent of v. Kv is, as in Section VI and [17], the barrier to rotation of the magnetic vector. The value V' always differs from v, the magnetic volume of the particle, because [80] of the surfactant layer coating the particle and the thin non-magnetic layer which is often created on the particle surface owing to the interaction with the surfactant molecules.

In order to further facilitate comparison with experiment, Shliomis [18, 80] writes $\tau_N(0)$ in the same form as (8.4), namely

$$\tau_N(0) = \frac{3\mu v}{kT}, \qquad \mu = \frac{M_s(1+\alpha^2)}{6\alpha\gamma} \cong \frac{M_s}{6\alpha\gamma} \tag{8.8}$$

μ, the internal relaxation or Néel parameter, plays the same role as the fluid viscosity η in Debye relaxation. Usually $\mu/\eta \leq 10^{-2}$ [80] so that $\tau_N(0) \gg \tau_D$. The critical particle volume v_c which determines the predominant relaxation mechanism is then computed by setting

$$\tau_N(\sigma) = \tau_D \qquad (8.9)$$

so that

$$\frac{v_c}{V'} = \frac{\mu}{\eta} \frac{\sqrt{\pi}}{2} \sigma^{-3/2} e^{\sigma} \qquad (8.10)$$

According to Eq. (8.5) virtually exponential growth of σ occurs with increase of particle volume so that for large particles $\tau_N \gg \tau_D$ and the relaxation is purely of the Debye type. For small particles so that $v < v_c$ on the other hand, the relaxation is purely of the Néel type.

We reiterate that the analysis given above (quite apart from the question of identification of the largest relaxation time of Brown's equation with that of the magnetization, in the Néel limit, a justification for which is provided by the numerical calculations presented in [17]), makes the assumption that the Debye and Néel processes may be treated independently.

Stepanov and Shliomis [80] have proposed an "egg model" akin to a three-dimensional form of the itinerant oscillator model [72]. (We recall [72] that in the itinerant oscillator, one relaxation mechanism occurs due to rotation of the oscillator as a whole, the other due to jumping of the inner rotator over a potential barrier) whereby the interdependence of the two relaxation mechanisms is taken into account. In the egg model [80], the ferrofluid particle is represented as an egg of volume V' (the hydrodynamic volume) embedded in the liquid with viscosity η. The magnetic moment of the ferroparticle is modeled by the yolk of volume $v < V'$; the ferroparticle, however, is surrounded by the white of viscosity μ, the magnetic torque acts directly on the yolk but due to the viscosity of the white it is also transmitted to the eggshell. Such a model allows one to take into account the coupling between the two mechanisms in a quantitative way so that its influence on τ_N and τ_D may be estimated.

The model is analyzed by writing down the form of Brown's equation appropriate to the distribution function $W(\mathbf{r}, \mathbf{n}, t)$ of orientations $\mathbf{r}$ of the magnetic moment (yolk) and $\mathbf{n}$ easy magnetization (egg) axes. In the absence of an external field, the magnetic energy is simply the anisotropy energy $(\mathbf{r} \cdot \mathbf{n})^2$. In this case, $W(\mathbf{r}, \mathbf{n}, t)$ may be written just as in the itinerant oscillator model (treated in Appendix C) as the product of the distribution function $F((\mathbf{r} \cdot \mathbf{n}), t)$ of the projection of the magnetic moment vector $\mathbf{r}$ and the distribution function $\Phi(\mathbf{n}, t)$ of the easy axes $\mathbf{n}$. The Φ equation is Brown's equation in the absence of any potential and has a spectrum of eigenvalues $l(l+1)/2\tau_D$. The F equation is Eq. (7.16) which generates the set of differential-difference equations (7.31). The eigen-

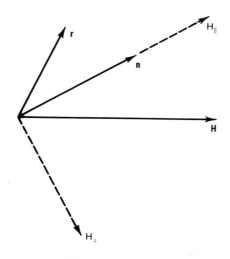

Figure 7. The **r** (magnetic moment vector) and **n** (easy axis vector). **H** may be resolved into vectors $\mathbf{H}_\parallel$ and $\mathbf{H}_\perp$ parallel and perpendicular to the **n** axis.

values of W are the sum of the eigenvalues of F and Φ so that the $\parallel$ and $\perp$ relaxation times obey the relations

$$\frac{1}{\tau_\parallel} = \frac{1}{\tau_{10}} + \frac{1}{\tau_D} \tag{8.11}$$

$$\frac{1}{\tau_\perp} = \frac{1}{\tau_{11}} + \frac{1}{\tau_D} \tag{8.12}$$

where τ_{10} is given by Eq. (6.75) and τ_{11} by Eq. (6.76) so confirming the Shliomis formula (8.1). Ferromagnetic resonance occurs in the F equation just as in Section VII.D at frequencies of the order ω_L where ω_L is the Larmor frequency

$$\omega_L = \gamma \left| \frac{\partial V}{\partial \mathbf{M}} \right| \tag{8.13}$$

Stepanov and Shliomis assert [80] that in the frequency range $0 < \omega < \omega_L$, the susceptibility is well approximated by

$$\chi(\omega) = \frac{1}{3} \left(\frac{\chi_\parallel^s}{1 + i\omega\tau_\parallel} + \frac{2\chi_\perp^s}{1 + i\omega\tau_\perp} \right) \tag{8.14}$$

with $\chi_\parallel^s$ and $\chi_\perp^s$ given by Eqs. (1.60) and (1.61). For ultra fine particles where $\sigma \ll 1$, this formula goes over into

$$\chi = \frac{\chi_0}{1 + i\omega\tau_N(0)} \tag{8.15}$$

with χ_0 given by Eq. (1.62). As σ (effectively the particle volume) increases $\tau_\perp$ reduces as $\tau_N(0)/\sigma$ (see Eq. (6.77)) and ultimately for large σ

$$\chi = \frac{1}{3}\left(2\chi_\perp + \frac{\chi_\parallel}{1 + i\omega\tau_\parallel}\right) \tag{8.16}$$

Equation (8.16) [80] predicts two plateaux in the frequency dependence of χ', a transition from the upper one to the lower one occurring near

$$\omega^* = \tau_\parallel^{-1} = \frac{(\tau_N(\sigma) + \tau_D)}{\tau_N(\sigma)\tau_D} \tag{8.17}$$

with an accompanying pronounced maximum in the spectrum of χ'' at $\omega = \omega^*$.

If an external field characterized by the parameter ξ is now applied, $W(\mathbf{r}, \mathbf{n}, t)$ can no longer be written as $F\Phi$. However if $\xi \ll 1$, we may expand $W(\mathbf{r}, \mathbf{n}, t)$ in spherical harmonics to obtain [80] a set of differential-difference equations analogous to (7.31) which may be solved by a numerical procedure similar to that described in Section III. According to Stepanov and Shliomis [80], the results of this calculation confirm the simple analytical formulae (8.11)–(8.17).

Having regard to the above discussion of the effective relaxation time τ_{eff}, it is useful to refer the reader to recent measurements of τ_N and τ_D. Fannin and Charles [83] have measured the complex susceptibility of ferrofluids consisting of colloidal suspensions of small magnetite particles in a hydrocarbon with anisotropy constant K in the region 10^4 to $4 \times 10^4\,\mathrm{J\,m^3}$ and saturation magnetization $M_s \cong 0.44\,\mathrm{T}$, the particles of each fluid are of median radii 2.6, 3.05 and 3.6 nm, respectively, with standard deviation 0.2.

The measurements are carried out in the frequency range 1 kHz to 160 MHz and show loss peaks at 34, 31 and 29 MHz, respectively. They conclude that in each case the loss peaks are due to the Néel mechanism only. This work is of significance since up to the present, susceptibility measurements have been in general limited [84, 85] to the frequency region below 13 MHz where the Néel mechanism ceases to be the dominant one. Referring to Debye relaxation, Fannin and Charles [84] report measurements of the complex susceptibility of water-based ferrofluids with saturation magnetization 0.0175 T. They report loss peaks at 250 Hz and 2.2 MHz, respectively, corresponding to average hydrodynamic radii of 59.4 nm and 2.9 nm. They conclude that the low frequency peak is due solely to Debye relaxation and the high frequency one due to a combination of Néel and Debye relaxation.

It is our opinion that these results provide a powerful impetus towards the further development of the egg model (already developed in its main features by Stepanov and Shliomis [80]) for the purpose of gaining a simple analytic formula for the effective relaxation time τ_{eff} (analogous to Eqs. (8.1) and (8.5)). This approach, as it is essentially based on a two particle model, allows one to take account of the interdependence of the Néel and Debye relaxation mechanisms and simultaneously eliminates the undesirable features of the present theory which is based on the two extremes of a frozen liquid matrix [17] and a frozen magnetic moment [16].

Acknowledgments

We thank Professor B. K. P. Scaife for introducing us to the subject of ferrofluids and for many helpful discussions. We further thank Professor D. S. F. Crothers, Dr. P. C. Fannin, Professor V. I. Gaiduk, Dr. Yu. L. Raĭkher, Professor M. I. Shliomis, and Mr. S. Sparrow for helpful conversations and Dr. K. P. Quinn for numerical computations. We thank Mr. E. S. Massawe for his careful preparation of the manuscript. We would also like to thank The British Council, Trinity College Dublin, The Institute of Radioengineering and Electronics of the Russian Academy of Sciences, Professor O. V. Betsky, EOLAS and the Wicklow County Council for financial support for this work.

APPENDIX   A

A.1.   The Langevin Equation Method

The kernel of Langevin's argument is contained in the statements (1) and (2) which are given in Section II concerning the nature of the random force $\lambda(t)$. The modern name for such a force is a white noise force (so called because it has a flat spectrum). It has an autocorrelation function (which measures the degree to which the present value of the force depends on its past value),

$$\langle \lambda(t_1) \, \lambda(t_2) \rangle = 2D\delta(t_1 - t_2) \tag{A.1}$$

where D is a constant to be determined. It is useful to recast Langevin's argument using the concept of white noise. To do this, we write the Langevin equation as

$$\dot{x} = v$$

$$\dot{v} = -\frac{\zeta}{m} v + \frac{\lambda(t)}{m} \tag{A.2}$$

or

$$\dot{\mathbf{X}} = \mathbf{A}\mathbf{X} + \mathbf{B}u(t) \tag{A.3}$$

that is,

$$\begin{pmatrix} \dot{x} \\ \dot{v} \end{pmatrix} = \begin{pmatrix} 0 & 1 \\ 0 & -\beta \end{pmatrix} \begin{pmatrix} x \\ v \end{pmatrix} + \begin{pmatrix} 0 \\ 1 \end{pmatrix} \frac{\lambda(t)}{m} \tag{A.4}$$

where $\beta = \zeta/m$.

The general solution of this set is

$$\mathbf{X}(t) = (\exp \mathbf{A}t)\mathbf{X}_0 + \int_0^t \exp[\mathbf{A}(t - t')]\mathbf{B} \frac{\lambda(t')}{m} dt' \tag{A.5}$$

Next by calculating

$$\mathscr{L}^{-1}\{(s\mathbf{I} - \mathbf{A})^{-1}\} = \exp(\mathbf{A}t) \tag{A.6}$$

we have

$$\begin{pmatrix} x \\ v \end{pmatrix} = \begin{pmatrix} 1 & \dfrac{1 - e^{-\beta t}}{\beta} \\ 0 & e^{-\beta t} \end{pmatrix} \begin{pmatrix} x_0 \\ v_0 \end{pmatrix} + \int_0^t \begin{pmatrix} 1 & \dfrac{1 - e^{-\beta(t-t')}}{\beta} \\ 0 & e^{-\beta(t-t')} \end{pmatrix} \begin{pmatrix} 0 \\ 1 \end{pmatrix} \frac{\lambda(t')}{m} dt' \tag{A.7}$$

It is assumed that the particle starts off at a *definite* phase point (x_0, v_0) so that the state vector has components

$$x(t) = x_0 + \frac{v_0}{\beta} (1 - e^{-\beta t}) + \frac{1}{\beta} \int_0^t (1 - e^{-\beta(t-t')}) \frac{\lambda(t')}{m} dt'$$

$$\dot{x}(t) = v_0 e^{-\beta t} + \int_0^t e^{-\beta(t-t')} \frac{\lambda(t')}{m} dt' \tag{A.8}$$

Note, $\langle \lambda(t) \rangle = 0$, so that taking the average over the time variable

$$\langle v(t) \rangle = v_0 e^{-\beta t} \tag{A.9}$$

$$\langle v^2(t) \rangle = v_0^2 e^{-2\beta t} + \left\langle 2v_0 e^{-\beta t} \int_0^t e^{-\beta(t-t')} \frac{\lambda(t')}{m} dt' \right\rangle$$

$$+ \left\langle \int_0^t e^{-\beta(t-t')} \frac{\lambda(t')}{m} dt' \int_0^t e^{-\beta(t-t'')} \frac{\lambda(t'')}{m} dt'' \right\rangle \tag{A.10}$$

$$= v_0^2 e^{-2\beta t} + e^{-2\beta t} \frac{2D}{m^2} \int_0^t \int_0^t dt' \, dt'' \, e^{\beta(t'+t'')} \delta(t'-t'') \tag{A.11}$$

since v_0 and $\lambda(t)$ are statistically independent. Now

$$\int_0^t e^{\beta(t'+t'')} \delta(t'-t'') \, dt'' = e^{2\beta t'} \tag{A.12}$$

whence

$$\langle v^2(t) \rangle = v_0^2 e^{-2\beta t} + e^{-2\beta t} \frac{2D}{m^2} \int_0^t e^{2\beta t'} dt' \tag{A.13}$$

whence

$$\langle v^2(t) \rangle = v_0^2 e^{-2\beta t} + \frac{D}{\beta m^2} (1 - e^{-2\beta t}) \tag{A.14}$$

$$\langle v^2(t) \rangle - \langle v(t) \rangle^2 = \frac{D}{\beta m^2} (1 - e^{-2\beta t}) \tag{A.15}$$

The value of the constant D is found as follows. We have

$$\lim_{t \to \infty} \langle v^2(t) \rangle = \frac{D}{\beta m^2} \tag{A.16}$$

But it is also assumed that the Maxwell–Boltzmann [11] distribution sets in so that

$$\lim_{t \to \infty} \langle v^2(t) \rangle = \frac{kT}{m} \tag{A.17}$$

whence

$$D = m\beta kT \tag{A.18}$$

and

$$\langle \lambda(t_1) \, \lambda(t_2) \rangle = 2kT\zeta\delta(t_1 - t_2) \tag{A.19}$$

Having determined the mean square velocities and the constant D we may write the mean squared displacement as follows. We have

$$x(t) = x_0 + \frac{v_0}{\beta} (1 - e^{-\beta t}) + \frac{1}{m\beta} \int_0^t [1 - e^{-\beta(t-t')}] \lambda(t') \, dt' \tag{A.20}$$

Hence

$$\Delta x = x(t) - x_0 = \frac{v_0}{\beta}(1 - e^{-\beta t}) + \frac{1}{m\beta} \int_0^t [1 - e^{-\beta(t-t')}] \lambda(t') \, dt'$$

(A.21)

Now

$$\langle \Delta x \rangle = \frac{v_0}{\beta}(1 - e^{-\beta t})$$

$$\langle (\Delta x)^2 \rangle = \frac{v_0^2}{\beta^2}(1 - e^{-\beta t})^2$$

$$+ \frac{1}{m^2\beta^2} \int_0^t dt' \int_0^t dt''[1 - e^{-\beta(t-t')}][1 - e^{-\beta(t-t'')}] \langle \lambda(t')\lambda(t'') \rangle$$

(A.22)

$$= \frac{v_0^2}{\beta^2}(1 - e^{-\beta t})^2$$

$$+ \frac{2D}{m^2\beta^2} \int_0^t dt' \int_0^t dt''[1 - e^{-\beta(t-t')}][1 - e^{-\beta(t-t'')}]\delta(t'' - t')$$

(A.23)

Since

$$\int_0^t dt''\delta(t'' - t')[1 - e^{-\beta(t-t'')}] = 1 - e^{-\beta(t-t')}$$

(A.24)

we have from Eq. (A.23)

$$\langle (\Delta x)^2 \rangle = \frac{v_0^2}{\beta^2}(1 - e^{-\beta t})^2 + \frac{2D}{m^2\beta^2} \int_0^t dt'[1 - 2e^{-\beta(t-t')} + e^{-2\beta(t-t')}]$$

$$= \frac{v_0^2}{\beta^2}(1 - e^{-\beta t})^2 + \frac{2Dt}{m^2\beta^2} + \frac{D}{m^2\beta^3}[-3 + 4e^{-\beta t} - e^{-2\beta t}]$$

(A.25)

This is the solution given that the collection of particles started off with the *definite* velocity v_0.

For long times, the term in t is all that is significant so that

$$\langle (\Delta x)^2 \rangle = \frac{2Dt}{m^2\beta^2} = \frac{2kTt}{m\beta}$$

(A.26)

which is the result of Einstein [13] and Langevin [9]. If we suppose that instead of an ensemble of particles which all start with the same velocity v_0 we have a Maxwell–Boltzmann distribution of initial velocities, we find that

$$\langle (\Delta x)^2 \rangle = \frac{2kT}{m\beta^2} (\beta t - 1 + e^{-\beta t}), \quad t > 0 \tag{A.27}$$

so that $t = |t|$, which is the Ornstein–Fürth formula [11]. It should be noted that, for short times, Eq. (A.26) is *nondifferentiable*, so that in the noninertial approximation, the *velocity* does not exist. If inertia is included, however, $\langle (\Delta x)^2 \rangle$ is differentiable and the velocity exists. This question has been discussed at length by Doob (1942) [30] who showed that the Langevin equation should be regarded not as a differential equation but as an integral equation.

A.2. The Stationary Solution of the Langevin Equation

We have illustrated the calculation of the averages from the Langevin equation for sharp initial conditions. The solution of the Langevin equation subject to a Maxwell–Boltzmann distribution of velocities is called the stationary solution. Clearly for the stationary solution

$$\langle v^2 \rangle = kT/m \tag{A.28}$$

The quantity of interest now is the velocity correlation function. The stationary solution may be found by extending the lower limit of integration to $-\infty$ and discarding the term in v_0 in Eq. (A.7). Thus,

$$v(t) = \int_{-\infty}^{t} e^{-\beta(t-t')} \frac{\lambda(t')}{m} \, dt' \tag{A.29}$$

and for distinct times t_1 and t_2,

$$v(t_1) = \int_{-\infty}^{t_1} e^{-\beta(t_1-t')} \frac{\lambda(t')}{m} \, dt' \tag{A.30}$$

$$v(t_2) = \int_{-\infty}^{t_2} e^{-\beta(t_2-t'')} \frac{\lambda(t'')}{m} \, dt'' \tag{A.31}$$

$$\langle v(t_1)v(t_2) \rangle = \int_{-\infty}^{t_1} \int_{-\infty}^{t_2} e^{-\beta(t_1+t_2)} e^{\beta(t'+t'')} \frac{\langle \lambda(t')\lambda(t'') \rangle}{m^2} \, dt' \, dt''$$

$$= \int_{-\infty}^{t_1} \int_{-\infty}^{t_2} e^{-\beta(t_1+t_2)} e^{\beta(t'+t'')} \frac{2D}{m^2} \, \delta(t'' - t') \, dt' \, dt''$$

$$= e^{-\beta(t_1+t_2)} \int_{-\infty}^{t_1} \frac{2D}{m^2} e^{2\beta t'} \, dt' = \frac{kT}{m} e^{-\beta|t_1-t_2|} \tag{A.32}$$

The modulus bars must be inserted in order to ensure a decaying covariance. We write $|t_2 - t_1| = \tau$. This is the velocity correlation function of a free Brownian particle. The mean square displacement may be found using the formula

$$\langle (\Delta x)^2 \rangle = 2 \int_0^t (t - u) \langle v(t)v(u) \rangle \, du \qquad \text{(A.33)}$$

yielding the same result (Eq. (A.27)) as before. The velocity correlation function may also be computed using the Wiener–Khinchin theorem. We give a short exposition of this. We follow closely the paper of Wang and Uhlenbeck [10]. Consider, for a very long time T, a *stationary* random process; that is, a process where the mechanism which causes the fluctuations does not change with the course of time. In order to achieve a general treatment, we specify the process by the random variable $\xi(t)$. We now form

$$\overline{\xi(t)} = \lim_{T \to \infty} \frac{1}{T} \int_{-T/2}^{T/2} y(t) \, dt \qquad \text{(A.34)}$$

This is the time average of $\xi(t)$. We may also write

$$\overline{\xi(t)\xi(t + \tau)} = \lim_{T \to \infty} \frac{1}{T} \int_{-T/2}^{T/2} \xi(t)\xi(t + \tau) \, dt \qquad \text{(A.35)}$$

We also note the equality of time and ensemble averages which is a fundamental tenet of statistical mechanics (ergodic theorem)

$$\langle \xi(t)\xi(t + \tau) \rangle = \overline{\xi(t)\xi(t + \tau)} \qquad \text{(A.36)}$$

Now

$$\xi(t) = \frac{1}{2\pi} \int_{-\infty}^{\infty} d\omega \, \tilde{\xi}(\omega)e^{i\omega t} \qquad \text{(A.37)}$$

where

$$\tilde{\xi}(\omega) = \int_{-\infty}^{\infty} \xi(t)e^{-i\omega t} \, dt \qquad \text{(A.38)}$$

Hence using the shift theorem for Fourier transforms

$$\langle \xi(t)\xi(t + \tau) \rangle = \lim_{T \to \infty} \frac{1}{T} \int_{-T/2}^{T/2} dt \int_{-\infty}^{\infty} \frac{d\omega}{2\pi} \tilde{\xi}(\omega)e^{i\omega t}$$

$$\times \int_{-\infty}^{\infty} \frac{d\omega_1}{2\pi} e^{i\omega_1(t+\tau)}\tilde{\xi}(\omega_1) \qquad \text{(A.39)}$$

Now

$$\frac{1}{2\pi} \int_{-\infty}^{\infty} e^{\pm ixy} \, dy = \delta(x) \tag{A.40}$$

whence assuming that one may perform the integration over the t variables first

$$\langle \xi(t)\xi(t + \tau) \rangle = \lim_{T \to \infty} \frac{1}{T} \int_{-\infty}^{\infty} \int_{-\infty}^{\infty} \frac{\tilde{\xi}(\omega)\tilde{\xi}(\omega_1)}{2\pi} \, e^{i\omega_1 \tau}\delta(\omega_1 + \omega) \, d\omega \, d\omega_1 \tag{A.41}$$

and so with

$$\int_{-\infty}^{\infty} f(x)\delta(x + a) \, dx = f(-a) \tag{A.42}$$

$$\langle \xi(t)\xi(t + \tau) \rangle = \lim_{T \to \infty} \frac{1}{T} \int_{-\infty}^{\infty} \tilde{\xi}(\omega)\tilde{\xi}(-\omega)e^{-i\omega\tau} \, d\omega \tag{A.43}$$

and since $\langle \xi(t)\xi(t + \tau) \rangle$ is real,

$$\tilde{\xi}(-\omega) = \tilde{\xi}^*(\omega)$$

so that

$$\rho(\tau) = \langle \xi(t)\xi(t + \tau) \rangle = \lim_{T \to \infty} \int_{-\infty}^{\infty} \frac{|\tilde{\xi}(\omega)|^2}{2\pi T} e^{-i\omega\tau} \, d\omega \tag{A.44}$$

$$= \int_{-\infty}^{\infty} \lim_{T \to \infty} \frac{|\tilde{\xi}(\omega)|^2}{2\pi T} e^{-i\omega\tau} \, d\omega \tag{A.45}$$

or

$$\langle \xi(t)\xi(t + \tau) \rangle = \frac{1}{2\pi} \int_{-\infty}^{\infty} \Phi_{\xi}(\omega)e^{-i\omega\tau} \, d\omega \tag{A.46}$$

$$\Phi_{\xi}(\omega) = \lim_{T \to \infty} \left(\frac{|\tilde{\xi}(\omega)|^2}{T} \right) \tag{A.47}$$

$\Phi_{\xi}(\omega)$ is the spectral density of the random function $\xi(t)$. By Fourier's integral theorem, we have

$$\Phi_{\xi}(\omega) = \int_{-\infty}^{\infty} \rho(\tau)e^{-i\omega\tau} \, d\tau \tag{A.48}$$

Since $\rho(\tau)$ is an even function of τ and Φ_ξ is an even function of ω, we also have

$$\Phi_\xi(\omega) = 2 \int_0^\infty \rho(\tau) \cos \omega\tau \, d\tau \tag{A.49}$$

or

$$\frac{\Phi_\xi(\omega)}{2} = \int_0^\infty \rho(\tau) \cos \omega\tau \, d\tau \tag{A.50}$$

This is the Wiener–Khinchin theorem where the spectral density is the Fourier transform of the autocorrelation function.

We illustrate its use by evaluating the velocity correlation function of a Brownian particle. We have

$$\dot{x} = v \tag{A.51}$$

$$\dot{v} = -\beta v + \lambda(t)/m \tag{A.52}$$

$$\langle \lambda(t_1)\lambda(t_2) \rangle = 2D\delta(t_1 - t_2) \tag{A.53}$$

The Fourier transform of $v(t)$ is

$$\tilde{v}(\omega) = \int_{-\infty}^\infty v(t)e^{i\omega t} \, dt \tag{A.54}$$

Hence

$$\tilde{v}(\omega)[\beta + i\omega] = \tilde{\lambda}(\omega)/m \tag{A.55}$$

where $\tilde{\lambda}(\omega)$ is the Fourier transform of $\lambda(t)$. Thus writing

$$\chi(\omega) = (\beta + i\omega)^{-1} \tag{A.56}$$

$$\tilde{v}(\omega) = \tilde{\chi}(\omega)\tilde{\lambda}(\omega)/m \tag{A.57}$$

$\chi(\omega)$ is the transfer function of the system. Now

$$\lim_{T\to\infty} \frac{\tilde{v}(\omega)\tilde{v}(\omega)^*}{T} = \Phi_v(\omega) \tag{A.58}$$

and with Eq. (A.56)

$$\Phi_v(\omega) = \chi(\omega)\chi^*(\omega) \frac{\Phi_\lambda(\omega)}{m^2} \tag{A.59}$$

$$= \frac{\Phi_\lambda(\omega)}{m^2(\beta^2 + \omega^2)} \tag{A.60}$$

Now

$$\Phi_\lambda(\omega) = \int_{-\infty}^{\infty} 2D\delta(\tau)e^{-i\omega\tau}\, d\tau \tag{A.61}$$

Thus the spectral density of the noise is

$$\Phi_\lambda(\omega) = 2D \tag{A.62}$$

whence

$$\Phi_v(\omega) = \frac{2D}{m^2} \frac{1}{\beta^2 + \omega^2} \tag{A.63}$$

Thus

$$\rho_v(\tau) = \frac{1}{2\pi} \int_{-\infty}^{\infty} \frac{2D}{m^2} \frac{e^{i\omega\tau}\, d\omega}{\beta^2 + \omega^2} \tag{A.64}$$

Also

$$\int_0^{\infty} \frac{\cos mx}{1 + x^2}\, dx = \frac{\pi}{2} e^{-|m|} \tag{A.65}$$

hence

$$\rho_v(\tau) = \frac{2D}{\pi m^2} \int_0^{\infty} \frac{\cos \omega\tau}{\beta^2 + \omega^2}\, d\omega$$

$$= \frac{kT}{m} e^{-\beta|\tau|} \tag{A.66}$$

since $D = \beta kTm$.

This is the velocity correlation function as obtained by the Wiener–Khinchin theorem. The above arguments may be extended to multi-dimensional systems, as discussed in some detail by Wang and Uhlenbeck [10] and by McConnell [57] in the context of rotational Brownian motion.

We shall allude to the Wiener–Khinchin theorem later in this appendix in our discussion of linear response theory.

The Langevin approach has been used by many authors in order to treat nonlinear systems. This is of importance to us since the equations of rotational motion are intrinsically nonlinear. The concept of a nonlinear Langevin equation is also subject to a number of criticisms. These have been discussed extensively by van Kampen [58] (Chapters 8 and 14). In our calculations, we shall encounter stochastic differential equations of the form

$$\dot{\xi} = h(\xi, t) + g(\xi, t)\lambda(t) \qquad (A.67)$$

This is a nonlinear Langevin equation of the first order. It contains a multiplicative noise term. The noise $\lambda(t)$ may be represented, according to van Kampen [58], by a random sequence of delta functions. Thus each delta function jump in $\lambda(t)$ causes a jump in $\xi(t)$. Hence, the value of ξ at the time the delta function arrives is indeterminate and consequently so is g at this time also. A problem arises, as the equation does not indicate which value of ξ one should substitute in g; whether the value of ξ before the jump, the value after or a mean of both.

Itô [51] interpreted this Langevin equation by prescribing that in g the value of ξ before the jump should be taken so that

$$\xi(t + \Delta t) - x(t) = h(x, t)\Delta t + g(x, t) \int_t^{t+\Delta t} \lambda(t') \, dt' \qquad (A.68)$$

This is countered by Stratonovich [12] who takes the mean values before and after the jump and writes

$$\xi(t + \Delta t) - x(t) = h(x, t)\Delta t + g\left(\frac{[x(t) + \xi(t + \Delta t)]}{2}\right) \int_t^{t+\Delta t} \lambda(t') \, dt' \qquad (A.69)$$

It follows from this that in the Itô definition (the angle brackets denoting an average over the ξ variables),

$$\langle g(\xi, t)\lambda(t) \rangle = 0 \qquad (A.70)$$

and in the Stratonovich one, that

$$\langle g(\xi, t)\lambda(t) \rangle = g(x, t) \, dg(x, t)/dx \qquad (A.71)$$

In the Stratonovich definition, we may use the ordinary rules of calculus, whereas in the Itô definition this is not so. One has to use new rules for differentiation and integration (the Itô calculus).

The Stratonovich definition has the disadvantage that

$$\langle g(\xi, t)\lambda(t)\rangle \neq 0$$

which leads to an alteration of the drift coefficient in the Fokker–Planck equation while the Itô definition allows the drift coefficient to remain the same. The Stratonovich definition leads to the concept of noise induced or spurious drift [12]. In Eqs. (A.68) and (A.69), $\xi(t + \Delta t)$ is the solution of Eq. (A.67) which at time t has the *sharp* value $\xi(t) = x$. It should be noted that x itself is a random variable with probability density function $W(x, t)$ defined such that $W\,dx$ is the probability of finding x in the interval x, $x + dx$. It will be necessary when applying the Langevin equation method to take a second average over the probability distribution of x at time t. The dual average is often denoted by a single pair of angular braces for economy of notation. In Eqs. (A.70) and (A.71), the angle brackets mean an average over the probability distribution of ξ only. In other parts of the text, e.g. Section V the average over the distribution of ξ is denoted by an overbar and the dual average by single angle brackets.

The Itô–Stratonovich dilemma is essentially a consequence of ignoring the inertia of the Brownian particle so that $\dot{\xi}$ does not exist as observed by Doob [11] in his criticism of the *linear* Langevin equation. This dilemma may be circumvented by including inertia so that in this case ξ becomes differentiable. In the problems which we shall treat, the white noise is the limiting case of a physical noise with finite noise power. In this case it appears that the Stratonovich definition is the correct one to use [58]. For example, the Debye theory of dielectric relaxation for rotation in two and three dimensions with inertial effects [33, 49] may be obtained by direct averaging of the Langevin equation. If the limit $I \to 0$ of this result is taken, then the hierarchy of differential-difference equations obtained, corresponds to the one obtained by starting from the noninertial Langevin equation taking the Stratonovich definition.

We have illustrated the problem of multiplicative noise by referring to motion in one dimension, however the dilemma is present, for practical purposes only for motion in several dimensions since a *one-dimensional* equation with *multiplicative noise* can always be transformed into an equation with *additive noise*.

The multiplicative noise problem in several dimensions will presently arise in the rigorous derivation of Brown's equation. It has already arisen in the main text in connection with the calculation of the relaxation times.

APPENDIX B: THE FOKKER–PLANCK EQUATION FOR TRANSLATIONAL MOTION

B.1.   Stochastic Processes

A stochastic process is a family of random variables $\{\xi(t), t \in T\}$ where t is some parameter, generally the time, defined on a set T. It is convenient to decompose this set T into instants $t_1 < t_2 < t_3 < \cdots < t_n < T$ and then to approximate the family of random variables $\{\xi(t)\}$ by $\xi(t_1)$, $\xi(t_2)$, and so on. We may describe the process by the following family of *joint* probability distributions: $f_1(y_1, t_1) \, dy_1$ is the probability of finding $\xi(t_1)$ in $(y_1, y_1 + dy_1)$; $f_2(y_1, t_1; y_2, t_2) \, dy_1 \, dy_2$ that of finding $\xi(t_1)$ in $(y_1, y_1 + dy_1)$ *and* $\xi(t_2)$ in $(y_2, y_2 + dy_2)$; $f(y_1, t_1; y_2, t_2; y_3, t_3) \, dy_1 \, dy_2 \, dy_3$ that of finding $\xi(t_1)$ in $(y_1, y_1 + dy_1)$, $\xi(t_2)$ in $(y_2, y_2 + dy_2)$ *and* $\xi(t_3)$ in $(y_3, y_3 + dy_3)$; and so on to $f_n(y_1, t_1; y_2, t_2; y, t, \ldots; y_n, t_n)$.

The process is *stationary* when the probability distribution underlying the process during a given interval of time depends only on the length of that interval and not on when the interval began. Another way of saying this is that the underlying mechanism that causes random variables $[\xi(t)]$ to fluctuate does not change with the course of time. This means that a shift of the time axis does not influence the functions f_n and as a result our set becomes $f_1(y)dy$, which is the probability of finding ξ in $(y, y + dy)$, $f_2(y_1, y_2, t) \, dy_1 \, dy_2$ which is the joint probability of finding a pair of values of ξ in ranges $(y_1, y + dy_1)$ and $(y_2, y_2 + dy_2)$, which are a time interval apart from each other (where $t = |t_2 - t_1|$) and so on.

The functions f_n may now be determined [10] by experiment from a single record $\xi(t)$ taken over a sufficiently long time. One may then cut the record into pieces of length T with T long in comparison to all periods contained in the process. The different pieces may then be considered as the different records of an *ensemble* or *collection* of observations. One has *in general* to distinguish between an ensemble average and a time average. The two methods of averaging for a *stationary process* will nevertheless always give the same result [10].

We may classify random processes as follows.

A *purely random process* is

$$f_1(y_1, t_1)$$

$$f_2(y_1, t_1; y_2, t_2) = f_1(y_1, t_1)f_1(y_2, t_2)$$

$$f_3(y_1, t_1; y_2, t_2; y_3, t_3) = f_1(y_1, t_1)f_1(y_2, t_2)f_1(y_3, t_3) \tag{B.1}$$

All the information is contained in the first probability density function, f_1.

A *Markov process* is such that *all* the information about the process is contained in f_2. It is convenient at this point to introduce the conditional probability density function.

If the probability that $\xi(t_n)$ is in $(y_n, y_n + dy_n)$ at time t_n depends only on $\xi(t_{n-1})$, y_n and t_n then (we use the symbol P for conditional probabilities)

$$P_n(y_n, t_n \mid y_{n-1}, t_{n-1}; \ldots; y_1, t_1) = P_2(y_n, t_n \mid y_{n-1}, t_{n-1}) \quad \text{(B.2)}$$

This defines the Markov concept.

For two events A and B the conditional probability is [32]

$$P(A \mid B) = \frac{P(A \cap B)}{P(B)} \quad \text{(B.3)}$$

$P(A \cap B)$ means the probability measure of both A and B occurring and $P(A \mid B)$ means the probability measure of event A occurring if event B occurs.

Thus, setting $t = |t_2 - t_1|$ (the elapsed time),

$$f_2(y_1, y_2, t) = f_1(y_1) P_2(y_2, t \mid y_1) \quad \text{(B.4)}$$

where $P_2 dy_2$ is the probability that given $\xi(t_1) = y_1$ one finds $\xi(t)$ in the range $(y_2, y_2 + dy_2)$ at time t later and P_2 satisfies

$$P_2(y_2, t \mid y_1) \geq 0 \quad \text{(B.5)}$$

$$\int dy_2 P_2(y_2, t \mid y_1) = 1 \quad \text{(B.6)}$$

$$f_1(y_2) = \int f_1(y_1) P_2(y_2, t \mid y_1) \, dy_1 \quad \text{(B.7)}$$

$$\lim_{t \to \infty} P_2(y_2, t \mid y_1) = f_1(y_2) \quad \text{(B.8)}$$

The latter condition means that the conditional probability ceases to be conditional as t approaches infinity. P_2 is the *transition probability*. It is evident that if we know P_2 we can calculate all the desired properties of the system. We may now derive the Fokker–Planck equation as follows.

Consider a set of instants $t_1 < t_2 < t_3$, where we suppose for the present that t_1 and y_1 are fixed. Define $P_2(y_2, t_2 \mid y_1, t_1) \, dy_2$ as the probability that $\xi(t_2)$ is in $(y_2, y_2 + dy_2)$ given that $\xi(t_1)$ had the value y_1 at time t_1, and $P_3(y_3, t_3 \mid y_2, t_2; y_1, t_1) \, dy_3$ as the probability that $\xi(t_3)$ is in

$(y_3, y_3 + dy_3)$ given that $\xi(t_2)$ had the value y_2 at time t_2 and that $\xi(t_1)$ had the value y_1 at time t_1.

If we multiply P_2 and P_3 and integrate with respect to y_2, all dependence on y_2 vanishes. The new probability density function depends only on y_1 at t_1:

$$P_3(y_3, t_3 \mid y_1, t_1)\, dy_3 = \int_{-\infty}^{\infty} P_2(y_2, t_2 \mid y_1, t_1)$$
$$\times P_3(y_3, t_3 \mid y_2, t_2; y_1, t_1)\, dy_2\, dy_3 \quad \text{(B.9)}$$

or

$$P_3(y_3, t_3 \mid y_1, t_1) = \int_{-\infty}^{\infty} P_2(y_2, t_2 \mid y_1, t_1) P_3(y_3, t_3 \mid y_2, t_2; y_1, t_1)\, dy_2$$
$$\text{(B.10)}$$

If we confine ourselves to *Markov* processes,

$$P_3(y_3, t_3 \mid y_2, t_2; y_1, t_1) = P_2(y_3, t_3 \mid y_2, t_2)$$

and

$$P_2(y_3, t_3 \mid y_1, t_1) = \int_{-\infty}^{\infty} P_2(y_2, t_2 \mid y_1, t_1) P_2(y_3, t_3 \mid y_2, t_2)\, dy_2$$
$$\text{(B.11)}$$

Equation (B.10) is termed the *Chapman–Kolmogorov equation* and Eq. (B.11) the *Smoluchowski equation* (Smoluchowski's integral equation).

In Eq. (B.11) let $P_2 = P$, $y_3 = y$, $y_2 = z$, $y_1 = x$, $t = t_2$, $t_3 = t + \Delta t$, and let us further suppress the t_1 dependence, so that

$$P(y, t + \Delta t \mid x) = \int_{-\infty}^{\infty} P(y, t + \Delta t \mid z) P(z, t \mid x)\, dz \quad \text{(B.12)}$$

We write

$$P(y, t + \Delta t \mid x) = P(y, \Delta t \mid z) \quad \text{(B.13)}$$

so that

$$P(y, t + \Delta t \mid x) = \int_{-\infty}^{\infty} P(y, \Delta t \mid z) P(z, t \mid x)\, dz \quad \text{(B.14)}$$

We wish to derive a partial differential equation for $P(y, t \mid x)$ the *transition probability*. Consider

$$\int_{-\infty}^{\infty} dy \, R(y) \, \frac{\partial P(y, t \mid x)}{\partial t} \tag{B.15}$$

where

$$\lim_{y \to \pm \infty} R(y) = 0$$

and $R^{(n)}(y)$ exists at $y = \pm \infty$. $R^{(n)}$ denotes the nth derivative of $R(y)$ with respect to y. Now

$$\int_{-\infty}^{\infty} dy \, R(y) \dot{P} = \int_{-\infty}^{\infty} dy \, R(y) \lim_{\Delta t \to 0} \left[\frac{P(y, t + \Delta t \mid x) - P(y, t \mid x))}{\Delta t} \right] \tag{B.16}$$

by the definition of the partial derivative of P with respect to t. In Eq. (B.16) we assume that

$$\int_{-\infty}^{\infty} \lim_{\Delta t \to 0} = \lim_{\Delta t \to 0} \int_{-\infty}^{\infty}$$

so that Eq. (B.16) becomes

$$\int_{-\infty}^{\infty} dy \, R(y) \dot{P} = \lim_{\Delta t \to 0} \left\{ \int_{-\infty}^{\infty} dy \, R(y) \left[\frac{P(y, t + \Delta t \mid x) - P(y, t \mid x)}{\Delta t} \right] \right\} \tag{B.17}$$

since

$$\int_{-\infty}^{\infty} dy \, R(y) P(y, t \mid x) = \int_{-\infty}^{\infty} dz \, R(z) P(z, t \mid x) \tag{B.18}$$

we now substitute for $P(y, t + \Delta t \mid x)$ in Eq. (B.17) using the Smoluchowski equation (B.11) to obtain

$$\int_{-\infty}^{\infty} dy \, R(y) \dot{P} = \lim_{\Delta t \to 0} \frac{1}{\Delta t} \left[\int_{-\infty}^{\infty} dy \, R(y) \int_{-\infty}^{\infty} P(y, \Delta t \mid z) P(z, t \mid x) \, dz \right.$$
$$\left. - \int_{-\infty}^{\infty} dz \, R(z) P(z, t \mid x) \right]. \tag{B.19}$$

We now interchange the order of integration in the leading term on the right hand side of the above equation to obtain

$$
\int_{-\infty}^{\infty} dy \, R(y)\dot{P} = \lim_{\Delta t \to 0} \frac{1}{\Delta t} \left[\int_{-\infty}^{\infty} P(z, t \,|\, x) \, dz \int_{-\infty}^{\infty} dy \, R(y) P(y, \Delta t \,|\, z) \right.
$$
$$
\left. - \int_{-\infty}^{\infty} dz \, R(z) P(z, t \,|\, x) \right] \tag{B.20}
$$

We now expand $R(y)$ in a Taylor series about the point $y = z$ to obtain

$$
R(y) = R(z) + (y - z)R'(z) + \frac{(y - z)^2}{2!} R''(z) + \cdots \tag{B.21}
$$

Therefore

$$
\int_{-\infty}^{\infty} dy \, R(y)\dot{P} = \lim_{\Delta t \to 0} \frac{1}{\Delta t} \left\{ \int_{-\infty}^{\infty} P(z, t \,|\, x) \, dz \int_{-\infty}^{\infty} dy \left[R(z) + (y - z)R'(z) \right. \right.
$$
$$
\left. + \frac{(y - z)^2}{2!} R''(z) + \cdots \right] P(y, \Delta t \,|\, z)
$$
$$
\left. - \int_{-\infty}^{\infty} dz \, R(z) P(z, t \,|\, x) \right\} \tag{B.22}
$$

Now $P(y, \Delta t \,|\, z)$ is a probability density function, hence

$$
\int_{-\infty}^{\infty} P(y, \Delta t \,|\, z) \, dy = 1
$$

thus

$$
\int_{-\infty}^{\infty} P(z, t \,|\, x) \, dz \int_{-\infty}^{\infty} dy \, R(z) P(y, \Delta t \,|\, z) = \int_{-\infty}^{\infty} P(z, t \,|\, x) R(z) \, dz \tag{B.23}
$$

The leading and end terms of the right hand side of Eq. (B.22) now cancel and we are left with

$$
\int_{-\infty}^{\infty} dy \, R(y)\dot{P} = \lim_{\Delta t \to 0} \frac{1}{\Delta t} \left\{ \int_{-\infty}^{\infty} P(z, t \,|\, x) \, dz \int_{-\infty}^{\infty} dy \left[(y - z)R'(z) \right. \right.
$$
$$
\left. \left. + \frac{(y - z)^2}{2!} R''(z) + \cdots \right] P(y, \Delta t \,|\, z) \right\} \tag{B.24}
$$

Let us define

$$a_n(z, \Delta t) = \int_{-\infty}^{\infty} (y - z)^n P(y, \Delta t \,|\, z) \, dy$$

$$= \langle (\Delta z)^n \rangle, \quad \Delta z = z - y$$

then

$$\int_{-\infty}^{\infty} dy \, R(y) \dot{P} = \lim_{\Delta t \to 0} \frac{1}{\Delta t} \left\{ \int_{-\infty}^{\infty} dz \, P(z, t \,|\, x) \right.$$

$$\left. \times \left[a_1(z, \Delta t) R'(z) + \frac{a_2(z, \Delta t)}{2!} R''(z) + \cdots \right] \right\}$$

$$(B.25)$$

and interchanging $\lim\limits_{\Delta t \to 0}$ and $\int_{-\infty}^{\infty}$ again

$$\int_{-\infty}^{\infty} dy \, R(y) \dot{P} = \int_{-\infty}^{\infty} dz \, P(z, t \,|\, x) \left\{ \lim_{\Delta t \to 0} \frac{a_1(z, \Delta t)}{\Delta t} R'(z) \right.$$

$$\left. + \lim_{\Delta t \to 0} \frac{a_2(z, \Delta t)}{2!} R''(z) + \cdots \right\}$$

$$(B.26)$$

We now suppose that

$$\lim_{\Delta t \to 0} \frac{a_n(z, \Delta t)}{\Delta t} = 0 \quad \text{for } n > 2$$

Thus

$$\int_{-\infty}^{\infty} dy \, R(y) \dot{P} = \left\{ \int_{-\infty}^{\infty} dz \, P(z, t \,|\, x) \left[D^{(1)}(z, t) R'(z, t) + \frac{D^{(2)}(z,t)}{2} R''(z) \right] \right\}$$

$$(B.27)$$

where we have written

$$D^{(1)}(z, t) = \lim_{\Delta t \to 0} \frac{a_1(z, \Delta t)}{\Delta t} = \lim_{\Delta t \to 0} \frac{\langle \Delta z \rangle}{\Delta t} \qquad (B.28)$$

$$D^{(2)}(z, t) = \lim_{\Delta t \to 0} \frac{a_2(z, t)}{\Delta t} = \lim_{\Delta t \to 0} \frac{\langle (\Delta z)^2 \rangle}{\Delta t} \qquad (B.29)$$

We need to factor $R(z)$ out of the right hand side, so consider

$$\int_{-\infty}^{\infty} dz\, P(z, t\,|\,x) D^{(1)}(z, t) R'(z) = \int u\, dv = uv - \int v\, du$$

$$dv = R'(z)\, dz\,, \qquad u = D^{(1)}(z, t) P$$

$$v = R(z)\,, \qquad du = \frac{\partial}{\partial z}\, [D^{(1)}(z, t) P]\, dz \qquad \text{(B.30)}$$

whence

$$\int_{-\infty}^{\infty} R'(z) D^{(1)}(z, t) P\, dz = D^{(1)}(z, t) R(z, t) P\,\Big|_{-\infty}^{\infty}$$

$$- \int_{-\infty}^{\infty} R(z)\, \frac{\partial}{\partial z}\, [D^{(1)}(z, t) P]\, dz \qquad \text{(B.31)}$$

and so

$$\int_{-\infty}^{\infty} R'(z) P D^{(1)}(z, t)\, dz = -\int_{-\infty}^{\infty} R(z)\, \frac{\partial}{\partial z}\, [D^{(1)}(z, t) P]\, dz$$
$$\text{(B.32)}$$

Similarly, because $R(z)$ vanishes as $z \to \pm\infty$, by repeated integration by parts

$$\int_{-\infty}^{\infty} R''(z) D^{(2)}(z, t) P\, dz = \int_{-\infty}^{\infty} R(z)\, \frac{\partial^2}{\partial z^2}\, [D^{(2)}(z, t) P]\, dz \qquad \text{(B.33)}$$

$R(z)$ can now be factored out of the right hand side of Eq. (B.27) to yield

$$\int_{-\infty}^{\infty} dy\, R(y) \dot{P} = \int_{-\infty}^{\infty} dz\, R(z) \left\{ -\frac{\partial}{\partial z}\, [D^{(1)}(z, t) P] + \frac{1}{2}\, \frac{\partial^2}{\partial z^2}\, [D^{(2)}(z, t) P] \right\}$$

$$= \int_{-\infty}^{\infty} dy\, R(y) \left\{ -\frac{\partial}{\partial y}\, [D^{(1)}(y, t) P] + \frac{1}{2}\, \frac{\partial^2}{\partial y}\, [D^{(2)}(y, t) P] \right\}$$
$$\text{(B.34)}$$

since z is a dummy variable and further since $R(y)$ is an arbitrary function

$$\frac{\partial P(y, t\,|\,x)}{\partial t} = -\frac{\partial}{\partial y}\, [D^{(1)}(y, t) P] + \frac{1}{2}\, \frac{\partial^2}{\partial y^2}\, [D^{(2)}(y, t) P] \qquad \text{(B.35)}$$

which is the Fokker–Planck equation for the one-dimensional Markov process characterized by the random variable ξ. The condition that the Taylor series may be truncated at $n = 2$ will be satisfied, if (in the underlying stochastic differential equation) the *driving stimulus is white noise*. If this is not so, higher order terms *must* be included in the Kramers–Moyal [31] expansion (B.22).

In general [10] the equation for the transition probability is always an integral equation of the same type as the Boltzmann equation in the kinetic theory of gases, and only in certain limits (such as considered here) may it be written as a partial differential equation.

We shall return to the question of the truncation at $n = 2$ later. Since we will be dealing with the multivariable form of the Fokker–Planck equation it is necessary to quote the form of that equation for many dimensions. The multivariable form of the Fokker–Planck equation [31] is

$$\frac{\partial P}{\partial t} = -\sum_i \frac{\partial}{\partial y_i} [D_i^{(1)}(\mathbf{y}, t)P] + \frac{1}{2} \sum_{k,l} \frac{\partial^2}{\partial y_k \, \partial y_l} [D_{k,l}^{(2)}(\mathbf{y}, t)P] \quad \text{(B.36)}$$

Let us suppose that the process is characterized by a state vector $\mathbf{Y}$ having components (y_1, y_2) (these, for example, could be the position and velocity of a Brownian particle), and so for this two variable case the Fokker–Planck equation written in full is

$$\frac{\partial P}{\partial t} = -\sum_{i=1}^{2} \frac{\partial}{\partial y_i} [D_i^{(1)}(y_1, y_2, t)P] + \frac{1}{2} \sum_{k=1}^{2} \sum_{l=1}^{2} \frac{\partial^2}{\partial y_k \, \partial y_l} [D_{k,l}^{(2)}(y_1, y_2, t)P]$$

$$\text{(B.37)}$$

or

$$\frac{\partial P}{\partial t} = -\frac{\partial}{\partial y_1} [D_1^{(1)}(y_1, y_2, t)P] - \frac{\partial}{\partial y_2} [D_2^{(1)}(y_1, y_2, t)P]$$

$$+ \frac{1}{2} \sum_{k=1}^{2} \left\{ \frac{\partial^2}{\partial y_k \, \partial y_1} [D_{k,1}^{(2)}(y_1, y_2, t)P] + \frac{\partial^2}{\partial y_k \, \partial y_2} [D_{k,2}^{(2)}(y_1, y_2, t)P] \right\}$$

$$\text{(B.38)}$$

$$= -\frac{\partial}{\partial y_1} [D_1^{(1)}(y_1, y_2, t)P] - \frac{\partial}{\partial y_2} [D_2^{(1)}(y_1, y_2, t)P]$$

$$+ \frac{1}{2} \left\{ \frac{\partial^2}{\partial y_1^2} [D_{1,1}^{(2)}(y_1, y_2, t)P] + \frac{\partial^2}{\partial y_1 \, \partial y_2} [D_{1,2}^{(2)}(y_1, y_2, t)P] \right.$$

$$\left. + \frac{\partial^2}{\partial y_2 \, \partial y_1} [D_{2,1}^{(2)}(y_1, y_2, t)P] + \frac{\partial^2}{\partial y_2^2} [D_{2,2}^{(2)}(y_1, y_2, t)P] \right\} \quad \text{(B.39)}$$

In general we find that $D_{1,2}^{(2)} = D_{2,1}^{(2)}$ and so

$$\frac{\partial P}{\partial t} = -\frac{\partial}{\partial y_1} [D_1^{(1)}(y_1, y_2, t)P] - \frac{\partial}{\partial y_2} [D_2^{(2)}(y_1, y_2, t)P]$$

$$+ \frac{1}{2} \left\{ \frac{\partial^2}{\partial y_1^2} [D_{1,1}^{(2)}P] + \frac{\partial^2}{\partial y_2^2} [D_{2,2}^{(2)}P] + 2 \frac{\partial^2}{\partial y_1 \partial y_2} [D_{1,2}^{(2)}P] \right\}$$

(B.40)

where

$$D_1^{(1)} = \lim_{\Delta t \to 0} \frac{\langle \Delta y_1 \rangle}{\Delta t} \qquad D_{1,2}^{(2)} = \lim_{\Delta t \to 0} \frac{\langle \Delta y_1 \Delta y_2 \rangle}{\Delta t}$$

$$D_2^{(1)} = \lim_{\Delta t \to 0} \frac{\langle \Delta y_2 \rangle}{\Delta t} \qquad D_{2,2}^{(2)} = \lim_{\Delta t \to 0} \frac{\langle (\Delta y_2)^2 \rangle}{\Delta t}$$

$$D_{1,1}^{(2)} = \lim_{\Delta t \to 0} \frac{\langle (\Delta y_1)^2 \rangle}{\Delta t}$$

(B.41)

We have assumed in writing down our Fokker–Planck equation that

$$D^{(n)}(z, t) = \lim_{\Delta t \to 0} \frac{a_n(z, \Delta t)}{\Delta t} = \lim_{\Delta t \to 0} \frac{\int dy(y - z)^n P(y, \Delta t \mid z)\, dy}{\Delta t}$$

$$= 0$$

(B.42)

for $n > 2$. This allows us to truncate the Kramers–Moyal expansion (B.26). We write

$$D^{(1)}(z, t) = \lim_{\Delta t \to 0} \frac{a_1(z, \Delta t)}{\Delta t}$$

(B.43)

and

$$D^{(2)}(z, t) = \lim_{\Delta t \to 0} \frac{a_2(z, \Delta t)}{\Delta t}$$

(B.44)

In the Fokker–Planck equation these quantities (which express the fact that in small times in the process under consideration the space coordinate can only change by a small amount, which is the central idea underlying the theory of the Brownian motion) are to be calculated from the Langevin equation. Thus that equation is the basic equation of the theory of the Brownian movement. We have assumed in writing down the

Langevin equation that the noise $\lambda(t)$ has the following statistical properties:

$$\langle \lambda(t_1) \rangle = 0$$

$$\langle \lambda(t_1)\lambda(t_2) \rangle = 2D\delta(t_1 - t_2) \tag{B.45}$$

for an ensemble of particles which all start at the same point in velocity space. This is however not sufficient for the truncation of the Fokker–Planck equation at the second derivative. In order for this to happen the $\lambda(t)$'s must also satisfy Isserlis's Theorem, [39] that in $2n + 1$ observations,

$$\langle \lambda(t_1)\lambda(t_2), \ldots, \lambda(t_{2n+1}) \rangle = 0$$

$$\langle \lambda(t_1)\lambda(t_2), \ldots, \lambda(t_{2n}) \rangle = \sum_{\text{all pairs}} \langle \lambda(t_i)\lambda(t_j) \rangle \cdot \langle \lambda(t_k)\lambda(t_l) \rangle \ldots .$$

$$\tag{B.46}$$

where the sum is to be taken over all the different ways in which one can divide the $2n$ time points $t_1, \ldots, t_{2n}$ into n pairs [39].

To apply the theorem, consider for example $n = 2$. Then [58]

$$\langle \lambda(t_1)\lambda(t_1)\lambda(t_3)\lambda(t_4) \rangle = \langle \lambda(t_1)\lambda(t_2) \rangle \cdot \langle \lambda(t_3)\lambda(t_4) \rangle + \cdots$$

$$= 4D^2 \{ \delta(t_1 - t_2)\delta(t_3 - t_4) + \delta(t_1 - t_3)\delta(t_2 - t_4)$$

$$+ \delta(t_1 - t_4)\delta(t_2 - t_3) \} \tag{B.47}$$

We can now calculate $D^{(1)}(y, t)$ and $D^{(2)}(y, t)$ the drift and diffusion coefficients in the Fokker–Planck equation.

The Langevin equation for the process characterized by $\xi(t)$ is

$$d\xi/dt + \beta\xi = \lambda(t) \tag{B.48}$$

If we integrate this over a short time Δt, we have (with $\xi(t + \Delta t)$ being the solution of (B.48) which at time t has the sharp value $y(t)$, so that $\langle y \rangle = y$),

$$\xi(t + \Delta t) - y(t) = -\int_t^{t+\Delta t} \beta\xi(t') \, dt' + \int_t^{t+\Delta t} \lambda(t') \, dt' \tag{B.49}$$

thus (for a more detailed exposition, see our detailed calculation, further on),

$$\langle \xi(t + \Delta t) - y(t) \rangle = -\beta y(t) \, \Delta t + O(\Delta t)^2$$

so that

$$\lim_{\Delta t \to 0} \frac{\langle \xi(t + \Delta t) - y(t) \rangle}{\Delta t} = D^{(1)}(y, t)$$

$$= -\beta y \qquad (B.50)$$

This is the drift coefficient $D^{(1)}(y, t)$. In order to calculate the diffusion coefficient $D^{(2)}(y, t)$, we square $[\xi(t + \Delta t) - y]$ to obtain (over a small time Δt)

$$[\xi(t + \Delta t) - y]^2 = \beta^2 y^2 (\Delta t)^2 - 2\Delta t \beta y \int_t^{t+\Delta t} \lambda(t') \, dt'$$

$$+ \int_t^{t+\Delta t} \int_t^{t+\Delta t} \lambda(t') \lambda(t'') \, dt' \, dt'' \qquad (B.51)$$

The first term on the right hand side is of the order $(\Delta t)^2$. The middle term vanishes because λ and y are statistically independent. The last term is treated as follows. It is, on averaging,

$$2D \int_t^{t+\Delta t} \int_t^{t+\Delta t} \delta(t' - t'') \, dt' \, dt''$$

We note here that

$$\int_t^{t+\Delta t} \delta(t' - t'') \, dt' = 1, \quad t < t'' < t' < t + \Delta t \qquad (B.52)$$

because the integral is (in effect) from $-\infty$ to ∞. Thus the value of the double integral is $2D\Delta t$. Hence

$$D^{(2)}(y, t) = \lim_{\Delta t \to 0} \frac{\langle [\xi(t + \Delta t) - y]^2 \rangle}{\Delta t} = 2D \qquad (B.53)$$

$D^{(3)}(y, t)$ is calculated as follows. We form

$$\langle [\xi(t + \Delta t) - y]^3 \rangle = -\beta y^3 (\Delta t)^3 + \int_t^{t+\Delta t} \int_t^{t+\Delta t} \int_t^{t+\Delta t} \cdots$$

$$- 2y^2 \beta^2 (\Delta t)^2 \int_t^{t+\Delta t} \cdots + 2y\beta \Delta t \left(\int_t^{t+\Delta t} \cdots \right)^2 \cdots$$

$$(B.54)$$

The only term which will contribute to the average in this equation is the one involving the triple integral. This will vanish for a white noise driving force because by Isserlis's Theorem, all odd values are zero. The same applies for all odd powers of $\xi - y$. For even powers of $\xi - y$, all terms will be $O(\Delta t)^2$ or higher.

The integral involving the $\lambda(t)$'s for example for $(\Delta y)^4$ will be

$$\int_t^{t+\Delta t} \int_t^{t+\Delta t} \int_t^{t+\Delta t} \int_t^{t+\Delta t} \lambda(t_1) \cdots \lambda(t_4) \, dt_1 \, dt_2 \, dt_3 \, dt_4$$

which will be of the order $(\Delta t)^2$ and will consequently vanish in the limit $\Delta t \to 0$. Thus the transition probability $P(y, t \mid x)$ satisfies the Fokker–Planck equation

$$\frac{\partial P}{\partial t} = \beta \frac{\partial}{\partial y} (yP) + D \frac{\partial^2 P}{\partial y^2}$$

corresponding to the Langevin equation

$$\dot{\xi}(t) + \beta \xi(t) = \lambda(t) , \qquad \langle \lambda(t)\lambda(t + \tau) \rangle = 2D\delta(\tau)$$

P must also satisfy, since it is a transition probability (f denotes the stationary solution)

$$\lim_{t \to 0} P(y, t \mid x) = \delta(y - x) , \qquad \lim_{t \to \infty} P(y, t \mid x) = f(y)$$

B.2. Drift and Diffusion Coefficients for Multiplicative Noise

We have shown how to obtain the drift and diffusion coefficients for simple additive noise. A difference arises in the case of multiplicative noise, which we encounter in Gilbert's equation as augmented by a random noise field. In that case the system is governed by a Langevin equation of the form (we take the one-dimensional case for simplicity)

$$\dot{\xi} = h(\xi, t) + g(\xi, t)\lambda(t) \tag{B.55}$$

and in general we have to evaluate

$$
\begin{aligned}
D^{(n)}(y, t) &= \frac{1}{n!} \lim_{\Delta t \to 0} \frac{\langle [\xi(t + \Delta t) - y]^n \rangle}{\Delta t} \bigg|_{\xi(t) = y} \\
&= \lim_{\Delta t \to 0} \frac{\langle (\Delta \xi)^n \rangle}{\Delta t}
\end{aligned}
$$

where again $\xi(t + \Delta t)$ is the solution of the Langevin equation above which at the time t has the sharp value $y(t)$ so that

$$\langle y(t) \rangle = y(t)$$

The treatment which follows is similar to that of Risken [31] as that is extremely clear. Let us first write our nonlinear Langevin equation as an integral equation

$$\xi(t + \Delta t) - y(t) = \int_t^{t+\Delta t} [h(\xi(t'), t') + g(\xi(t'), t')\lambda(t')] \, dt' \tag{B.56}$$

and expand h and g as a Taylor series about the sharp point $\xi = y$ so that recalling that the increment during the interval (t, t') is

$$\xi(t') - y(t)$$

$$h(\xi(t'), t') = h(y, t') + [\xi(t') - y] \frac{\partial h(y, t')}{\partial y} + \cdots$$

$$g(\xi(t'), t') = g(y, t') + [\xi(t') - y] \frac{\partial g(y, t')}{\partial y} + \cdots$$

where

$$\left\{ \frac{\partial}{\partial \xi(t')} [h(\xi(t'), t')] \right\} \Big|_{\xi(t')=y} \equiv \frac{\partial h(y, t')}{\partial y}$$

and thus with the integral equation, Eq. (B.56), to first order in the increment,

$$\xi(t + \Delta t) - y(t) = \int_t^{t+\Delta t} h(y, t') \, dt' + \int_t^{t+\Delta t} [\xi(t') - y] \frac{\partial h(y, t')}{\partial y} \, dt' + \cdots$$

$$+ \int_t^{t+\Delta t} g(\xi, t')\lambda(t') \, dt' + \int_t^{t+\Delta t} [\xi(t') - y] \frac{\partial g(y, t')}{\partial y}$$

$$\times \lambda(t') \, dt' + \cdots$$

$$t < t' < t + \Delta t. \tag{B.57}$$

We now iterate for $\xi(t') - y$ in the integrand using the Langevin equation to obtain

$$\xi(t + \Delta t) - y(t) = \int_t^{t+\Delta t} h(y, t')\, dt' + \int_t^{t+\Delta t} \frac{\partial h(y, t')}{\partial y} \int_t^{t'} h(y, t'')\, dt'\, dt''$$

$$+ \int_t^{t+\Delta t} \frac{\partial h(y, t')}{\partial y} \int_t^{t'} g(y, t'')\lambda(t'')\, dt'\, dt'' + \cdots$$

$$+ \int_t^{t+\Delta t} g(y, t')\lambda(t')\, dt'$$

$$+ \int_t^{t+\Delta t} \frac{\partial g(y, t')}{\partial y} \int_t^{t'} h(y, t'')\lambda(t')\, dt'\, dt''$$

$$+ \int_t^{t+\Delta t} \frac{\partial g(y, t')}{\partial y} \int_t^{t'} g(y, t'')\lambda(t'')\lambda(t')\, dt'\, dt'' + \cdots \tag{B.58}$$

Now

$$\langle \lambda(t) \rangle = 0$$

$$\langle \lambda(t')\lambda(t'') \rangle = 2D\delta(t' - t'')$$

Thus

$$\langle \xi(t + \Delta t) - y(t) \rangle = \int_t^{t+\Delta t} h(y, t')\, dt'$$

$$+ \int_t^{t+\Delta t} \int_t^{t'} \frac{\partial h(y, t')}{\partial y} h(y, t'')\, dt'\, dt'' + \cdots$$

$$+ \int_t^{t+\Delta t} \frac{\partial g(y, t')}{\partial y} \int_t^{t'} g(y, t'')2D\delta(t' - t'')\, dt'\, dt'' + \cdots \tag{B.59}$$

Now

$$\int_t^{t'} \delta(t' - t'')\, dt'' = \int_0^\infty \delta(x)\, dx = \frac{1}{2} \tag{B.60}$$

with $t < t'' < t' < t + \Delta t$ and so with

$$\int_0^\infty f(x)\delta(x)\, dx = \frac{1}{2} f(0)$$

$$\int_t^{t'} g(y, t'')2D\delta(t' - t'')\, dt'' = Dg(y, t') \tag{B.61}$$

and thus

$$\lim_{\Delta t \to 0} \frac{\langle \xi(t+\Delta t) - y(t) \rangle}{\Delta t} \bigg|_{\xi(t)=y} = D^{(1)}(y, t) = \lim_{\Delta t \to 0} \frac{\langle \Delta \xi \rangle}{\Delta t}$$

$$= h(y, t) + D\, g(y, t)\, \frac{\partial g(y, t)}{\partial y} = h(y, t) + Dgg' \qquad \text{(B.62)}$$

Similarly we may prove that

$$D^{(2)}(y, t) = D^2 g^2(y, t) \qquad \text{(B.63)}$$

and on use of Isserlis's Theorem

$$D^{(n)}(y, t) = 0 \quad \text{for } n > 2 \qquad \text{(B.64)}$$

and we note that

$$D^{(1)}(y, t) = h(y, t) + \frac{1}{2D}\, \frac{\partial}{\partial y}\, D^{(2)}(y, t) \qquad \text{(B.65)}$$

The term

$$\frac{1}{2D}\, \frac{\partial}{\partial y}\, D^{(2)}(y, t) \qquad \text{(B.66)}$$

is the *noise induced* drift or *spurious* drift referred to earlier in our discussion of the Langevin equation with multiplicative noise, in Appendix A. This term arises because during a change in $\lambda(t)$ a change in $\xi(t)$ also occurs, so that $\langle g(\xi, t)\lambda(t) \rangle$ is no longer zero. In general we may say since g is arbitrary, that in the Langevin equation

$$\dot{\xi} = h(\xi(t), t) + g(\xi(t), t)\lambda(t) \qquad \text{(B.67)}$$

$$\langle g(\xi(t), t)\lambda(t) \rangle = D\, \frac{\partial g}{\partial y}\, g = Dgg' \qquad \text{(B.68)}$$

and

$$\dot{y} = h(y, t) + Dg(y, t)\, \frac{\partial g(y, t)}{\partial y}$$

The results obtained here pertain to the one-dimensional problem. The multidimensional case is discussed in Appendix E.

B.3.   Examples of the Calculation of Drift and Diffusion Coefficients

Having illustrated the problem associated with multiplicative noise we will now illustrate how the procedure is applied to obtain the drift and diffusion coefficients for the two-dimensional Fokker–Planck equation in phase space for a free Brownian particle and for the Brownian motion in a one-dimensional potential. This equation is often called the Kramers equation or Klein–Kramers equation [31].

The Langevin equation for a free Brownian particle may be represented as the system

$$\dot{x} = v$$

$$\dot{v} = -\beta v + \frac{\lambda(t)}{m} \tag{B.69}$$

The corresponding two-dimensional Fokker–Planck equation for the transition probability density $P(x, v, t \mid x_0, v_0, t_0)$ in phase space with $x = y_1$, $v = y_2$, in Eq. (B.40) is

$$\frac{\partial P}{\partial t} = -\frac{\partial}{\partial x}\left[D_1^{(1)}(x, v)P\right] - \frac{\partial}{\partial v}\left[D_2^{(1)}(x, v)P\right]$$

$$+ \frac{1}{2}\left\{\frac{\partial^2}{\partial x^2}\left[D_{1,1}^{(2)}(x, v)P\right] + \frac{\partial^2}{\partial v^2}\left[D_{2,2}^{(2)}(x, v)P\right] + 2\frac{\partial^2}{\partial x\,\partial v}\left[D_{1,2}^{(2)}(x, v)P\right]\right\} \tag{B.70}$$

Since $x = y_1$, $\Delta y_1 = \Delta x$ and proceeding as in (B.48) and (B.49)

$$\lim_{\Delta t \to 0} \frac{\langle \Delta y_1 \rangle}{\Delta t} = \lim_{\Delta t \to 0} \frac{\langle \Delta x \rangle}{\Delta t} = v$$

so that

$$D_1^{(1)} = v \tag{B.71}$$

Now

$$\Delta v = -\beta v \Delta t + \int_t^{t + \Delta t} dt' \frac{\lambda(t')}{m}$$

Thus

$$\lim_{\Delta t \to 0} \frac{\langle \Delta v \rangle}{\Delta t} = -\beta v, \qquad D_2^{(1)} = -\beta v \tag{B.72}$$

$$D_{1,1}^{(2)}(x, v) = \lim_{\Delta t \to 0} \frac{\langle (\Delta x)^2 \rangle}{\Delta t} = \lim_{\Delta t \to 0} \frac{\langle v^2 (\Delta t)^2 \rangle}{\Delta t} = 0 \qquad (B.73)$$

$$D_{1,2}^{(2)}(x, v) = \lim_{\Delta t \to 0} \frac{\langle \Delta x \Delta v \rangle}{\Delta t} = \lim_{\Delta t \to 0} \frac{\langle v \Delta t \Delta v \rangle}{\Delta t}$$

$$= \lim_{\Delta t \to 0} \left\langle v \left[-\beta v \Delta t + \int_t^{t+\Delta t} \frac{\lambda(t')}{m} \, dt' \right] \right\rangle = 0 \qquad (B.74)$$

because all the terms are of the order $(\Delta t)^2$. In order to evaluate

$$D_{2,2}^{(2)}(x, v) = \lim_{\Delta t \to 0} \frac{\langle (\Delta v)^2 \rangle}{\Delta t} \qquad (B.75)$$

consider

$$(\Delta v)^2 = \beta^2 v^2 (\Delta t)^2 - 2 \beta v \Delta t \int_t^{t+\Delta t} \frac{\lambda(t')}{m} \, dt'$$

$$+ \int_t^{t+\Delta t} \int_t^{t+\Delta t} \frac{\lambda(t') \lambda(t'')}{m^2} \, dt' \, dt'' \qquad (B.76)$$

The two leading terms on the left hand side vanish on averaging and

$$\int_t^{t+\Delta t} \int_t^{t+\Delta t} \frac{\langle \lambda(t') \lambda(t'') \rangle}{m^2} \, dt' \, dt'' = \int_t^{t+\Delta t} \int_t^{t+\Delta t} \frac{2kT\zeta}{m^2} \delta(t' - t'')$$

$$= \int_t^{t+\Delta t} \frac{2kT\zeta}{m^2} \, dt' = \frac{2kT\zeta}{m^2} \Delta t$$

whence

$$D_{2,2}^{(2)}(x, v) = \frac{2kT\zeta}{m^2} = \frac{2kT\beta}{m} \qquad (B.77)$$

B.4. Solutions of the Fokker–Planck Equation

The simplest form of the Fokker–Planck equation is that for diffusion in configuration space, namely the Smoluchowski equation, which was originally derived by Einstein [13] in 1905 in the context of the theory of the Brownian movement of a particle in one dimension under no external forces. We have seen earlier that this equation is (Eq. 2.50)

$$\frac{\partial P(x, t \mid 0)}{\partial t} = \frac{kT}{m\beta} \frac{\partial^2 P(x, t \mid 0)}{\partial x^2} \qquad (-\infty < x < \infty)$$

where for convenience we place the particle at the origin so that $x_0 = 0$.

By writing $D_1 = kT/m\beta$, for the diffusion coefficient this is

$$\partial P/\partial t = D_1\, \partial^2 P/\partial x^2 \quad (-\infty < x < \infty) \tag{B.78}$$

Mathematically our problem is to solve the above equation subject to the initial condition

$$P(x, 0\,|\,0) = \delta(x) \tag{B.79}$$

where $\delta(x)$ denotes the Dirac delta function. This is a way of stating that all the particles of the ensemble were definitely at $x = 0$ at $t = 0$. The solution of this equation subject to the delta function initial condition is called the *fundamental* solution of the equation, and mathematically speaking this solution is the *Green's function* of Eq. (B.78).

The solution is best effected using Fourier transforms. On taking the Fourier transform over the variable x, we have

$$\tilde{P}(u, t) = \int_{-\infty}^{\infty} P(x, t\,|\,0)e^{iux}\, dx \tag{B.80}$$

whence

$$\frac{\partial \tilde{P}(u, t)}{\partial t} = -D_1 u^2 \tilde{P}(u, t) \tag{B.81}$$

and thus

$$\tilde{P}(u, t) = A(u)e^{-D_1 u^2 t} \quad (t > 0) \tag{B.82}$$

the initial condition is

$$\tilde{P}(u, 0) = \int_{-\infty}^{\infty} \delta(x)e^{iux}\, dx = 1 \tag{B.83}$$

whence $A(u) = 1$ and we have

$$\tilde{P}(u, t) = e^{-D_1 u^2 t} \quad (t > 0) \tag{B.84}$$

In order to invert this integral, we make use of the formula

$$\int_0^{\infty} e^{-a^2 y^2} \cos 2py\, dy = \frac{\sqrt{\pi}}{2a}\, e^{-p^2/a^2} \quad (a^2 > 0) \tag{B.85}$$

Thus

$$P(x, t \mid 0) = \frac{1}{2\pi} \int_{-\infty}^{\infty} e^{-D_1 u^2 t} e^{-iux} \, dx = \frac{1}{(4\pi D_1 t)^{1/2}} e^{-x^2/4D_1 t}$$

(B.86)

The above equation is a one-dimensional Gaussian distribution with mean zero and variance (note that P tends to zero as x tends to infinity and to the delta function as t tends to zero),

$$\sigma^2 = 2D_1 t \quad (t > 0)$$

(B.87)

or

$$\sigma^2 = 2D_1 |t|$$

(B.88)

Thus

$$\langle x^2 \rangle = 2D_1 |t|$$

(B.89)

or

$$\langle x^2 \rangle = \frac{2kT|t|}{m\beta} = \frac{kT|t|}{3\pi\eta a}$$

(B.90)

where η is the viscosity of the suspension and a is the radius of the Brownian grain. This result is the same as that obtained in 1908 by Langevin [9] by simply writing down the equation of motion of the Brownian particle and averaging it directly.

$P(\mathbf{r}, t \mid \mathbf{r}_0, 0)$ in three dimensions is a quantity of interest to spectroscopists as it is the dynamic structure factor, defined by Van Hove [59] as

$$\Gamma_s(\mathbf{\kappa}, \omega) = \iint G_s(\mathbf{r}, t) \exp[i(\mathbf{\kappa} \cdot \mathbf{r} - \omega t)] \, d\mathbf{r} \, dt$$

(B.91)

where

$$G_s(\mathbf{r}, t) = P(\mathbf{r}, t \mid \mathbf{r}_0, 0)$$

(B.92)

The form of the Smoluchowski equation appropriate to the Brownian movement of a particle moving in one dimension under the influence of an external potential $V(x)$ is, as stated earlier (see Eq. (2.50)),

$$\frac{\partial P}{\partial t} = \frac{kT}{m\beta} \frac{\partial^2 P}{\partial x^2} + \frac{1}{m\beta} \frac{\partial}{\partial x} \left(\frac{\partial V(x)}{\partial x} P \right) \tag{B.93}$$

and as before we seek the fundamental solution $P(x, t \mid x_0, 0)$ of this equation. So we have to solve

$$\frac{\partial P}{\partial t} = \frac{kT}{m\beta} \frac{\partial^2 P}{\partial x^2} + \frac{1}{m\beta} \frac{\partial}{\partial x} \left(\frac{\partial V(x)}{\partial x} P \right) \tag{B.94}$$

subject to the initial condition $P(x, 0) = \delta(x - x_0)$.

We assume that the complete set of eigensolutions, ψ_n, of the stationary equation may be expressed as (this procedure is the one used by Brown)

$$\Psi_n = e^{-V(x)/2kT} \varphi_n(x) \tag{B.95}$$

Thus one sees that our Smoluchowski equation degenerates into a Schrödinger equation with a temperature-dependent potential [60], namely

$$\left[-\frac{\partial^2}{\partial x^2} + U(x) \right] \varphi_n(x) = \frac{\lambda_n}{D} \varphi_n \tag{B.96}$$

where the potential $U(x)$ is given by

$$U(x) = -\frac{V''(x)}{2kT} + \left(\frac{V'(x)}{2kT} \right)^2 \tag{B.97}$$

and where the primes denote differentiation with respect to x. The eigenfunctions are thus labeled by a wave vector κ (restricted to the first Brillouin zone) and a band index $\nu : n = (\kappa, \nu)$. The ground state has the eigenvalue $\lambda_0 = 0$. Its wave function is

$$\varphi_0 = C \exp\left(-\frac{V(x)}{2kT} \right) \tag{B.98}$$

We now restrict ourselves to *periodic* functions $V(x)$ of x, so that we may impose *periodic* boundary conditions. Thus φ_0 is now normalized within the length of the system and the corresponding eigenfunction is directly proportional to the equilibrium distribution. The probability density P may now be written in terms of the so-called biorthogonal expansion given by Morse and Feschbach [61]

$$P(x, t) = \sum_{\kappa, \nu} \varphi_{\kappa, \nu}(x) \varphi_0(x) \varphi_{\kappa, \nu}^*(x) \varphi_0(x_0)^{-1} e^{-\lambda_\nu(\kappa)t} \tag{B.99}$$

Substituting this equation into our expression for $\Gamma_s(\kappa, \omega)$, one then obtains the dynamic structure factor as a sum of Lorentzians:

$$\Gamma_s(\kappa, \omega) = \sum_{\nu=0}^{\infty} \frac{|M_\nu(\kappa^2)|^2 \lambda_\nu(\tilde{\kappa})}{\omega^2 + [\lambda_\nu(\tilde{\kappa})]^2}, \tag{B.100}$$

where

$$M_\nu(\kappa) = \int dx \, e^{i\kappa x} \varphi_{\kappa,\nu}(x) \varphi_0(x) \tag{B.101}$$

and $\tilde{\kappa}$ is the wave vector reduced to the first Brillouin zone. The most striking consequence of introducing the periodic or indeed any type of potential $V(x)$, then, is the appearance of a *discrete set* of Lorentzian spectra rather than just a single one as in the case of the free Brownian particle. The discrete set of Lorentzians will always be present whatever the form of V (excluding the trivial case of $V = 0$) because the Sturm–Liouville equation [61] will always possess a discrete set of eigenvalues $\{\lambda_n\}$ whenever $V \neq 0$. We thus have seen that the problem of solving the Smoluchowski equation can be reduced to the solution of a known Sturm–Liouville problem in quantum mechanics. Notwithstanding this, the eigenvalues λ_n and the corresponding eigenfunctions of the Sturm–Liouville equation can usually be found only by numerical techniques. We remark on two cases where it is possible to find exact solutions: (a) the case where $V(x) \propto x^2$ so that $V'(x) \propto x$ (the parabolic potential); and (b) the case where $V'(x) \propto -x + x^{-1}$. In case (a) the governing Sturm–Liouville equation reduces to Hermite's equation whereas in case (b) it has been shown by Stratonovich [62] that the Sturm–Liouville equation reduces to a form of Laguerre's equation. Finally Mörsch et al. [63] have considered a number of problems in which the potential has the form of a square wave. Again an exact solution is possible in these cases as in quantum mechanics.

We now consider the Brownian motion of a free particle in velocity space. The Fokker–Planck equation for this problem, is from Eqs. (B.35) et seq.,

$$\frac{\partial P_v}{\partial t} = \frac{\partial}{\partial v}(\beta v P_v) + \frac{D}{m^2}\frac{\partial^2 P_v}{\partial v^2} \tag{B.102}$$

where $D = m\beta kT$. This equation may be solved by expanding the function $P_v(v, t \,|\, v_0, 0)$ as follows

$$P_v(v, t \,|\, v_0, 0) = \sum_{n=0}^{\infty} \varphi_n(t) D_n\left(\sqrt{\frac{m}{kT}}\, v\right) \exp\left(-\frac{mv^2}{4kT}\right) \tag{B.103}$$

where the D_n are the harmonic oscillator or Weber functions [64] and we proceed as in the original work of Uhlenbeck and Ornstein [11]. A more transparent method of arriving at the solution, however, is to calculate the characteristic function $\tilde{P}(u, t)$ rather than P_v directly; that is, we proceed in the same manner as we did for the Smoluchowski equation in the previous section. Thus, on writing

$$\tilde{P}(u, t) = \int_{-\infty}^{\infty} P_v(v, t \mid v_0, 0) e^{-iuv} \, dv \qquad (B.104)$$

we have

$$\int_{-\infty}^{\infty} \frac{\partial}{\partial v} (\beta v P_v) e^{-iuv} \, dv = -\beta u \frac{\partial \tilde{P}}{\partial u} \qquad (B.105)$$

and

$$\int_{-\infty}^{\infty} \frac{\partial^2 P_v}{\partial v^2} e^{-iuv} \, dv = -u^2 \tilde{P} \qquad (B.106)$$

So our original Fokker–Planck equation is transformed into the *first order* linear partial differential equation

$$\frac{\partial \tilde{P}}{\partial t} = -\frac{D}{m^2} u^2 \tilde{P} - \beta u \frac{\partial \tilde{P}}{\partial u} \qquad (B.107)$$

or

$$\frac{\partial \tilde{P}}{\partial t} + \beta u \frac{\partial \tilde{P}}{\partial u} = -\frac{D}{m^2} u^2 \tilde{P} \qquad (B.108)$$

We make a small aside here concerning the solution of first order linear partial differential equations such as the above. An equation of the form

$$P(x, y, z) \frac{\partial z}{\partial x} + Q(x, y, z) \frac{\partial z}{\partial y} = R(x, y, z) \qquad (B.109)$$

is satisfied by the function defined by the equation

$$\varphi(x, y, z) = 0 \qquad (B.110)$$

if [65]

$$P \frac{\partial \varphi}{\partial x} + Q \frac{\partial \varphi}{\partial y} + R \frac{\partial \varphi}{\partial z} = 0 \qquad (B.111)$$

To solve the equation, we form the subsidiary system [65]

$$\frac{d\varphi}{0} = \frac{dx}{P(x, y, z)} = \frac{dy}{Q(x, y, z)} = \frac{dz}{R(x, y, z)} \qquad \text{(B.112)}$$

In the problem at hand, $P = 1$, $Q = \beta u$, $R = -(Du^2/m^2)P$, hence our subsidiary system is

$$\frac{dt}{1} = \frac{du}{\beta u} = -\frac{dP}{(Du^2/m^2)P} \qquad \text{(B.113)}$$

and the general solution of this system is

$$\varphi\left\{ ue^{-\beta t}, \exp\left(\frac{Du^2}{2\beta m^2}\right)P\right\} = 0 \qquad \text{(B.114)}$$

which may be written as

$$\tilde{P}(u, t) = \Psi(ue^{-\beta t}) \exp\left(\frac{Du^2}{2\beta m^2}\right) \qquad \text{(B.115)}$$

where Ψ is an arbitrary function, the value of which for the present problem is found as follows. Since the initial distribution of velocities has the form

$$P_v(v, 0 \mid v_0, 0) = \delta(v - v_0) \qquad \text{(B.116)}$$

then

$$\tilde{P}(u, 0) = e^{-iuv_0} \qquad \text{(B.117)}$$

and on setting $t = 0$ in Eq. (B.115)

$$e^{-iuv_0} = \Psi(u) \exp\left(\frac{Du^2}{2\beta m^2}\right) \qquad \text{(B.118)}$$

and therefore

$$\Psi(u) = e^{-iuv_0} \exp\left(-\frac{Du^2}{2\beta m^2}\right) \qquad \text{(B.119)}$$

Hence it follows that

$$\Psi(ue^{-\beta t}) = \exp(-iv_0 ue^{-\beta t}) \exp\left[-\frac{Du^2}{2\beta m^2} e^{-2\beta t}\right] \qquad \text{(B.120)}$$

whence

$$\tilde{P}(u, t) = \exp\left[-iv_0 u e^{-\beta t} + \frac{Du^2}{2\beta m^2}(1 - e^{-2\beta t})\right] \qquad (B.121)$$

The above equation is the characteristic function of a one-dimensional Gaussian random variable having mean

$$\langle v(t) \rangle = v_0 e^{-\beta t} \qquad (B.122)$$

and variance σ^2, where

$$\sigma^2 = \langle [v(t) - \langle v(t) \rangle]^2 \rangle = \frac{D}{\beta m^2}(1 - e^{-2\beta t}) \qquad (B.123)$$

(We have shown in Appendix A how these results can be obtained directly from the Langevin equation). Thus the conditional probability distribution of the velocities in the Ornstein–Uhlenbeck [11] process is

$$P(v, t \mid v_0, 0) = \frac{1}{\sigma\sqrt{2\pi}} \exp\left[-\frac{(v(t) - \langle v(t) \rangle)^2}{2\sigma^2}\right] \qquad (B.124)$$

with σ^2 and μ given by the two preceding equations.

The stationary solution is found by taking $\lim\limits_{t \to \infty} P$ which by inspection is

$$\lim_{t \to \infty} P(v, t \mid v_0, 0) = \sqrt{\frac{\beta m^2}{2\pi D}} \exp\left(-\frac{v^2 \beta m^2}{2D}\right) \qquad (B.125)$$

which is independent of v_0. Let us now suppose that v_0 also has this distribution. Then

$$P(v_0) = \sqrt{\frac{\beta m^2}{2\pi D}} \exp\left(-\frac{m^2 v_0^2 \beta}{2D}\right) \qquad (B.126)$$

which is a one-dimensional Gaussian distribution. Thus from

$$P(A \mid B)P(B) = P(A \cap B) \qquad (B.127)$$

we have the *joint* distribution

$$f(v, t, v_0, 0) = \frac{\beta m^2}{2\pi D\sqrt{1 - e^{-2\beta t}}} \exp\left[-\frac{m^2 \beta(v^2 + v_0^2 - 2v_0 v e^{-\beta t})}{2D(1 - e^{-2\beta t})}\right].$$

$$\qquad (B.128)$$

Since the process is *stationary* (that is the underlying mechanism does not depend on when the process began), we can write

$$v(t_1) = v_0 , \qquad v(t_2) = v(t)$$

and $\tau = |t_2 - t_1|$ thus f has the *two*-dimensional Gaussian distribution

$$f(v(t_2), v(t_1), \tau)$$

$$= \frac{\beta m^2}{2\pi D\sqrt{1 - \rho^2}} \exp\left\{ -\frac{m^2\beta}{2D} \frac{[v^2(t_1) - 2v(t_1)v(t_2)\rho + v^2(t_2)]}{(1 - \rho^2)} \right\} \qquad \text{(B.129)}$$

where ρ is the autocorrelation function (or correlation coefficient)

$$\rho(\tau) = e^{-\beta|\tau|} \qquad \text{(B.130)}$$

$(D = m\beta kT)$ which we obtained earlier using the Wiener–Khinchin theorem and the Langevin equation. This procedure may be also applied for many dimensions as in the following example.

Wang and Uhlenbeck in 1945 [10] showed how the solution procedure just outlined could be applied to the more general Fokker–Planck equation pertaining to an electric circuit having n meshes, each mesh being driven by a white noise voltage. Their form of the Fokker–Planck equation is (dropping the conditional probability bar for economy of notation)

$$\frac{\partial P(\mathbf{y}, t)}{\partial t} = -\sum_{i=1}^{n} \lambda_i \frac{\partial}{\partial y_i} [y_i P(\mathbf{y}, t)] + \frac{1}{2} \sum_{i,j=1}^{n} \sigma_{ij} \frac{\partial^2 P(\mathbf{y}, t)}{\partial y_i \, \partial y_j} \qquad \text{(B.131)}$$

for which it is required to find that solution (λ_1 and σ_{ij} are parameters) which for $t = 0$ becomes

$$P(\mathbf{y}, 0) = \delta(y_1 - y_{10})\delta(y_2 - y_{20}) \cdots \delta(y_n - y_{n0}) \qquad \text{(B.132)}$$

$\mathbf{y}$ is the vector, $\mathbf{y} = (y, \ldots, y_n)$ in $\mathbb{R}^n$. Again we introduce instead of P its Fourier transform

$$\tilde{P}(\mathbf{u}, t) = \int_{\mathbb{R}^n} P(\mathbf{y}, t) \exp(-i\mathbf{u} \cdot \mathbf{y}) \, d\mathbf{y} \qquad \text{(B.133)}$$

from which it follows (in exactly the same manner as we outlined for the one-dimensional case above) that $\tilde{P}(\mathbf{u}, t)$ must satisfy the linear, first order partial differential equation

$$\frac{\partial \tilde{P}(\mathbf{u}, t)}{\partial t} - \sum_{i=1}^{n} \lambda_i u_i \frac{\partial \tilde{P}(\mathbf{u}, t)}{\partial u_i} = -\frac{1}{2} \sum_{i,j} \sigma_{ij} u_i u_j \tilde{P}(\mathbf{u}, t) \qquad (B.134)$$

The subsidiary equations are

$$\frac{dt}{1} = -\frac{du_1}{\lambda_1 u_1} = -\frac{du_2}{\lambda_2 u_2} = \cdots = -\frac{du_n}{\lambda_n u_n} = -\frac{d\tilde{P}}{\frac{1}{2} \sum_{i,j} \sigma_{ij} u_i u_j \tilde{P}}$$

$$(B.135)$$

These may be integrated in exactly the same manner as before, leading to the general solution

$$\tilde{P}(\mathbf{u}, t) = \Psi(u_1 e^{\lambda_1 t}, u_2 e^{\lambda_2 t}, \ldots, u_n e^{\lambda_n t}) \exp\left(\frac{1}{2} \sum_{i,j} \frac{\sigma_{ij} u_i u_j}{\lambda_i + \lambda_j}\right) \qquad (B.136)$$

where Ψ is again an arbitrary function. Again for $t = 0$ with the initial conditions given, we have

$$\tilde{P}(\mathbf{u}, 0) = \exp(-i\mathbf{u} \cdot \mathbf{y}_0) \qquad (B.137)$$

and we obtain for $\tilde{P}(\mathbf{u}, t)$

$$\tilde{P}(\mathbf{u}, t) = \exp\left(-i \sum_j u_j y_{j0} e^{\lambda_j t} + \frac{1}{2} \sum_{i,j} \frac{\sigma_{ij} u_i u_j}{\lambda_i + \lambda_j} [1 - e^{(\lambda_i + \lambda_j)t}]\right) \qquad (B.138)$$

This is the Fourier transform of an n-dimensional Gaussian distribution with the mean values

$$\langle y_i \rangle = y_{i0} e^{\lambda_i t} \qquad (B.139)$$

and the variances

$$\langle (y_i - \langle y_i \rangle)(y_j - \langle y_j \rangle) \rangle = -\frac{\sigma_{ij}}{\lambda_i + \lambda_j} [1 - e^{(\lambda_i + \lambda_j)t}] \qquad (B.140)$$

APPENDIX C: LINEAR RESPONSE THEORY

The basic concepts of linear response theory are best illustrated by considering the rotational diffusion model of an assembly of electric dipoles constrained to rotate in two dimensions due to Debye [14] which is governed by the Smoluchowski equation

$$\zeta \frac{\partial f(\theta, t)}{\partial t} = \frac{\partial}{\partial \theta} \left[kT \frac{\partial f}{\partial \theta} + \mu E \sin \theta f \right] \tag{C.1}$$

where $f(\theta, t) = P(\theta, t \mid \theta_0, 0) \lim_{t \to \infty} P$. (The concepts described here are also applicable to magnetic fluid relaxation). θ is the angle between the dipole and the field.

Debye obtained two solutions for the Smoluchowski equation given above. The first type of solution is the after effect solution where it is supposed that the dielectric consisting of an assembly of noninteracting dipolar molecules has been influenced for a long time by a steady external field; thus on average the axes of the dipoles are oriented mainly in the direction of the field. Let us now suddenly switch off the field; then the axes of the dipoles revert to their original random orientations. This phenomenon of reverting from the polarized state to the chaotic state is called dielectric relaxation. If only terms linear in the field strength are considered (linear response), the response following the imposition of a step field is the mirror image of the after-effect response. The second problem considered by Debye is the behavior when an alternating (AC) electric field is applied. We suppose that an AC field has been applied to the dielectric for a long time so that the transient effect associated with the switching of the field can be ignored. The after-effect and alternating field solutions may be related by means of a method which has been described by Scaife [66, 67]. Let us suppose that at a time $t = 0$, say, a unit electric field is applied to a dielectric body; then an electric dipole moment $a(t)$ is induced in that body. The quantity $a(t)$ is called the response function of the body. Let us now suppose that the inducing electric field at time t is $E(t)$ and $0 \le t' \le t$; then the increase in the field intensity in an infinitesimal time $\delta t'$ is $(dE/dt')\delta t$ and this in its turn contributes $\dot{E}(t')\delta t' a(t - t')$ to the induced moment at time t. Thus at time t, the instantaneous dipole moment of the body, $\mathbf{m}(t)$, is

$$\mathbf{m}(t) = \int_0^t \frac{d\mathbf{E}(t')}{dt'} \, dt' a(t - t') \tag{C.2}$$

Now on integrating the above equation by parts,

$$\mathbf{m}(t) = \mathbf{E}(t')a(t - t')\Big|_{t'=0}^{t} - \int_0^t \mathbf{E}(t') \frac{d}{dt'} [a(t - t')] \, dt' \qquad \text{(C.3)}$$

We now suppose that $\mathbf{E}(t')$ vanishes for $t' \le 0$ and so $\mathbf{E}(0) = 0$. Also we suppose that when a field is switched on, there is no instantaneous response so that $a(0) = 0$; thus the first term on the right hand side of the above vanishes and we may write with $t - t' = x$

$$\mathbf{m}(t) = \int_0^t \mathbf{E}(t - x) \frac{da(x)}{dx} \, dx \qquad \text{(C.4)}$$

Our discussion so far refers to the case where the field is being switched on. Let us now consider the opposite case where a constant field $\mathbf{E}_0$ had been operative for a long time so that the induced moment is $a(\infty)$ and let us suppose that $\mathbf{E}_0$ is switched off at time $t = 0$. For $t > 0$ the induced moment $\mathbf{m}(t)$ is $\mathbf{E}_0(a(\infty) - a(t))$. This leads us to define the after-effect function $b(t)$ by the relation

$$\begin{aligned} b(t) &= a(\infty) - a(t) \quad t > 0) \\ &= 0 \quad\quad\quad\quad (t < 0) \end{aligned} \qquad \text{(C.5)}$$

whence

$$\mathbf{m}(t) = \mathbf{E}_0 b(t) \qquad \text{(C.6)}$$

Let us now put

$$\begin{aligned} \mathbf{E}(t) &= \mathbf{E}_m \cos \omega t \quad\quad (t > 0) \\ &= 0 \quad\quad\quad\quad\quad (t < 0) \\ &= \mathbf{E}_m U(t) \cos \omega t \end{aligned} \qquad \text{(C.7)}$$

whence

$$\mathbf{m} = \int_0^t \mathbf{E}_m \cos \omega(t - x) \frac{da(x)}{dx} \, dx \qquad \text{(C.8)}$$

$$= \mathbf{E}_m \cos \omega t \int_0^t \cos \omega x \, \frac{da(x)}{dx} \, dx + \mathbf{E}_m \sin \omega t \int_0^t \sin \omega x \, \frac{da(x)}{dx} \, dx \qquad \text{(C.9)}$$

Now if t is very large $da(x)/dx$, $t \le x \le \infty$, is negligibly small and thus the integrals

$$\int_t^\infty [da(x)/dx]\cos \omega x \, dx$$

and

$$\int_t^\infty [da(x)/dx]\sin \omega x \, dx$$

are also negligible. Thus we may write

$$\mathbf{m}(t) = \mathbf{E}_m \alpha'(\omega)\cos \omega t + \mathbf{E}_m \alpha''(\omega)\sin \omega t \qquad (C.10)$$

where

$$\alpha'(\omega) = \int_0^\infty \frac{da(x)}{dx}\cos \omega x \, dx , \qquad (C.11)$$

$$\alpha''(\omega) = \int_0^\infty \frac{da(x)}{dx}\sin \omega x \, dx \qquad (C.12)$$

both being real functions of ω. We now define the complex polarizability $\alpha(\omega)$ by

$$\alpha(\omega) = \alpha'(\omega) - i\alpha''(\omega) \qquad (C.13)$$

and we see that

$$\alpha(-\omega) = \alpha'(\omega) + i\alpha''(\omega) \qquad (C.14)$$

We may also write

$$\alpha(\omega) = \int_0^\infty \frac{da(x)}{dx} e^{-i\omega x} \, dx = \int_0^\infty \frac{da(t)}{dt} e^{-i\omega t} \, dt = -\int_0^\infty \frac{db(t)}{dt} e^{-i\omega t} \, dt \qquad (C.15)$$

which on integrating by parts becomes

$$\alpha(\omega) = -b(t)e^{-i\omega t}\Big|_0^\infty - i\omega \int_0^\infty b(t)e^{-i\omega t} \, dt \qquad (C.16)$$

$$= a(\infty) - a(0) - i\omega \int_0^\infty b(t)e^{-i\omega t} \, dt \qquad (C.17)$$

But

$$a(\infty) - a(0) = \int_0^\infty \frac{da(x)}{dx} \, dx = \alpha'(0) = \alpha(0) \qquad (C.18)$$

Setting $\omega = 0$ we see that $\alpha'(0)$ is the static polarizability due to the field **E** so we denote it by α_s and we express the previous equation as

$$\alpha(\omega) = \alpha_s - i\omega \int_0^\infty b(t)e^{-i\omega t}\, dt \qquad (C.19)$$

or

$$\frac{\alpha(\omega)}{\alpha'(0)} = 1 - i\omega \int_0^\infty R(t)e^{-i\omega t}\, dt \quad \left(R(t) = \frac{b(t)}{b(0)} \right) \qquad (C.20)$$

This equation effectively connects the alternating and after-effect solutions provided the response is linear. We may now make use of the Kramers–Kronig dispersion relations [66, 67] to rewrite equations (C.11) and (C.12) as

$$\alpha'(\omega) = \frac{2}{\pi} \mathfrak{B} \int_0^\infty \frac{\alpha''(\mu)\mu\, d\mu}{\mu^2 - \omega^2} \qquad (C.21)$$

$$\alpha''(\omega) = \frac{2}{\pi} \mathfrak{B} \int_0^\infty \frac{\alpha'(\mu)\omega\, d\mu}{\omega^2 - \mu^2} \qquad (C.22)$$

where the $\mathfrak{B}$ indicates that the Cauchy principal value (see Jeffreys and Jeffreys [68], Art. 12.02) of the integral is to be taken. Let us put $\omega = 0$ in Eq. (C.21) and since ω and μ may be interchanged, we must have

$$\alpha'(0) = \alpha_s = \frac{2}{\pi} \int_0^\infty \frac{\alpha''(\omega)\, d\omega}{\omega} \qquad (C.23)$$

This is a most interesting and important relation since it connects the *equilibrium* polarizability α_s with the *dissipative* part of the *frequency-dependent* polarizability; in other words, it provides a link between the equilibrium and the nonequilibrium properties of the body. We now go on to describe a most important theorem known as the fluctuation-dissipation theorem, of which Eq. (C.23) is one form.

The method we now describe follows closely that of Scaife [66, 67]). It is known that the polarizability of a dielectric body may be written

$$\alpha_s = \frac{\langle M^2 \rangle_0}{3kT} \qquad (C.24)$$

where $\langle M^2 \rangle_0 = \langle \mathbf{M}.\mathbf{M} \rangle_0$ is the ensemble average of the square of the fluctuating dipole moment **M** of the body in the absence of an external field. By the ergodic hypothesis we must have

$$\langle M^2 \rangle_0 = \overline{\mathbf{M} \cdot \mathbf{M}} = \overline{M^2} = \lim_{T \to \infty} \frac{1}{T} \int_{-T/2}^{T/2} \mathbf{M}(t) \cdot \mathbf{M}(t) \, dt \qquad \text{(C.25)}$$

Let us now write the Fourier transform pair:

$$\tilde{\mathbf{M}}(\omega) = \int_{-T/2}^{T/2} \mathbf{M}(t) e^{-i\omega t} \, dt \quad (T \to \infty) \qquad \text{(C.26)}$$

$$\mathbf{M}(t) = \frac{1}{2\pi} \int_{-\infty}^{\infty} \tilde{\mathbf{M}}(\omega) e^{i\omega t} \, d\omega \qquad \text{(C.27)}$$

Inserting Eq. (C.26) into Eq. (C.25) then leads after a short calculation (or immediately from Parseval's theorem; [69]) to

$$\langle M^2 \rangle_0 = \overline{M^2} = \frac{1}{2\pi} \int_{-\infty}^{\infty} \lim_{T \to \infty} \frac{|\tilde{\mathbf{M}}(\omega) \cdot \tilde{\mathbf{M}}^*(-\omega)|}{T} \, d\omega \qquad \text{(C.28)}$$

where

$$\mathfrak{M}(\omega) = \lim_{T \to \infty} \frac{|\tilde{\mathbf{M}}(\omega) \cdot \tilde{\mathbf{M}}^*(-\omega)|}{T} = \lim_{t \to \infty} \frac{|\tilde{\mathbf{M}}(\omega)|^2}{T} \qquad \text{(C.29)}$$

is the *spectral density* of the fluctuations in $\mathbf{M}(t)$. Then we have as a *direct* consequence of the Parseval theorem and the ergodic hypothesis

$$\langle M^2 \rangle_0 = \overline{M^2} = \frac{1}{2\pi} \int_{-\infty}^{\infty} \mathfrak{M}(\omega) \, d\omega \qquad \text{(C.30)}$$

With the aid of Eq. (C.23)

$$\alpha_s = \frac{2}{\pi} \int_0^{\infty} \frac{\alpha''(\omega)}{\omega} \, d\omega = \frac{1}{3kT} \frac{1}{2\pi} \int_{-\infty}^{\infty} \mathfrak{M}(\omega) \, d\omega$$

$$= \frac{1}{3kT} \frac{1}{\pi} \int_0^{\infty} \mathfrak{M}(\omega) \, d\omega \qquad \text{(C.31)}$$

since $\mathfrak{M}(\omega)$ is an even function ω, from which it follows that

$$6kT\alpha''(\omega) = \omega \mathfrak{M}(\omega) \qquad \text{(C.32)}$$

We have related the *dissipative* part of the *frequency-dependent* complex polarizability to the *spectral density* of the spontaneous fluctuations in the dipole moment at *equilibrium* of the body. This is the fluctuation-dissipation theorem [70, 71]. In identifying the integrals in Eq. (C.31), we have asserted that *macroscopic fluctuations* decay according to *macroscopic* laws. Let us now introduce the concept of the autocorrelation function (a.c.f.), which in this case is the time average of $\mathbf{M}(t)$ with $\mathbf{M}(t + t')$ or $\mathbf{M}(t - t')$, that is

$$C_M(t) = \overline{\mathbf{M}(t') \cdot \mathbf{M}(t + t')} = \overline{\mathbf{M}(t - t') \cdot \mathbf{M}(t')}$$

The reader will recall, from Appendix A, the Wiener–Khinchin theorem; namely, the a.c.f. and the spectral density are each other's Fourier cosine transform. Thus

$$C_M(t) = \frac{1}{\pi} \int_0^\infty \mathfrak{M}(\omega) \cos \omega t \, d\omega \tag{C.33}$$

which with Eq. (C.32) gives

$$C_M(t) = \frac{6kT}{\pi} \int_\pi^\infty \frac{\alpha''(\omega)}{\omega} \cos \omega t \, d\omega \tag{C.34}$$

This on inversion gives

$$\alpha''(\omega) = \omega \left(\frac{\pi}{6kT} \right) \left(\frac{2}{\pi} \right) \int_0^\infty C_M(t) \cos \omega t \, dt$$

$$= \omega \int_0^\infty b(t) \cos \omega t \, dt \tag{C.35}$$

from Eq. (C.19) and thus

$$3b(t) = \frac{1}{kT} C_M(t) \tag{C.36}$$

Hence

$$\alpha(\omega) = \alpha_s - \frac{i\omega}{kT} \int_0^\infty C_M(t) e^{-i\omega t} \, dt \tag{C.37}$$

or

$$\alpha(\omega) = \frac{1}{3kT} \left(\overline{\mathbf{M} \cdot \mathbf{M}} - i\omega \int_0^\infty \overline{\mathbf{M}(t') \cdot \mathbf{M}(t' + t)} e^{-i\omega t} \, dt \right)$$

$$= \frac{1}{3kT} \left[\langle \mathbf{M} \cdot \mathbf{M} \rangle_0 - i\omega \int_0^\infty \langle \mathbf{M}(t') \cdot \mathbf{M}(t' + t) \rangle_0 e^{-i\omega t} \, dt \right] \tag{C.38}$$

This is the Kubo relation, and is the generalization of the Fröhlich [29] relation,

$$\alpha_s = \frac{\langle M^2 \rangle_0}{3kT} \tag{C.39}$$

to cover the dynamical behavior of the dielectric. Either the Kubo relation or the fluctuation-dissipation theorem [67] may be used to calculate $\alpha(\omega)$. Similar considerations apply to the linear magnetic response.

In order to use this result consider the two-dimensional rotator model, here

$$R(t) = e^{-t/\tau_D} \tag{C.40}$$

which with the connection formula yields

$$\alpha(\omega)/\alpha'(0) = (1 + i\omega\tau_D)^{-1} \tag{C.41}$$

Consider now the AC solution of Eq. (C.1) with $\mathbf{E}(t) = \mathbf{E}_m e^{i\omega t}$

$$\zeta \frac{\partial f(\theta, t)}{\partial t} = \frac{\partial}{\partial \theta} \left[kT \frac{\partial f}{\partial \theta} + \mu \mathbf{E}_m e^{i\omega t} \sin \theta f \right] \tag{C.42}$$

We readily find that the mean dipole moment is in the linear case

$$\langle \mu \cos \theta \rangle = \frac{\mu^2}{2kT} \mathbf{E}_m e^{i\omega t} (1 + i\omega\tau_D)^{-1} \tag{C.43}$$

so that again Eq. (C.41) is obtained.

We have illustrated the concept of linear response using the simple rotational diffusion model of Debye. We will now illustrate how the concept applies for a rotator in an external potential. For convenience we will describe the procedure for the itinerant oscillator model given by Coffey [72].

We begin with the equations of motion of an itinerant oscillator model under the influence of an external electric field $\mathbf{E}(t)$ which are [72]

$$I_1 \ddot{\phi}_1 + \zeta_1 \dot{\phi}_1 + V'(\phi_1 - \phi_2) + \mu_1 E \sin \phi_1 = \lambda_1(t) \tag{C.44}$$

$$I_2 \ddot{\phi}_2 + \zeta_2 \dot{\phi}_2 - V'(\phi_1 - \phi_2) + \mu_2 E \sin \phi_2 = \lambda_2(t) \tag{C.45}$$

where

$$V(\phi_1 - \phi_2) = -2V_0 \cos(\phi_1 - \phi_2) \tag{C.46}$$

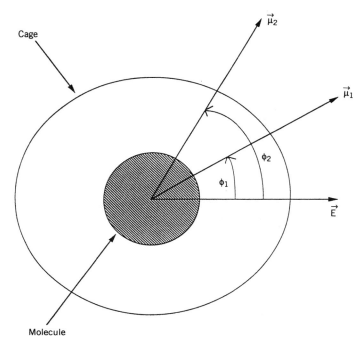

Figure 8. The two dimensional itinerant oscillator model.

so that

$$V'(\phi_1 - \phi_2) = 2V_0 \sin(\phi_1 - \phi_2) \tag{C.47}$$

The quantities subscripted by 1 refer to the inner dipole while those subscripted by 2 refer to the outer cage, I_i, μ_i, $i = 1, 2$, are the corresponding moments of inertia and dipole moments, $\zeta_i \dot{\phi}_i$ and λ_i are the stochastic torques and ϕ_i is the angular position. $\lambda_i(t)$ is as usual a white noise driving torque with correlation function

$$\langle \lambda_i(0)\lambda_j(t) \rangle = 2kT\delta_{ij}\beta_i I_i \delta(t) \tag{C.48}$$

where δ_{ij} and $\delta(t)$ are the Kronecker and Dirac delta functions, respectively. The direction of $\mathbf{E}(t)$ is taken as the initial line.

It should be noted that Eqs. (C.44) and (C.45) represent an approximate form of the itinerant oscillator model which has the virtue that it allows one to use the *same* transformations, namely

$$\chi = \frac{I_1\phi_1 + I_2\phi_2}{I_1 + I_2}, \qquad \eta = \frac{\phi_1 - \phi_2}{2}$$

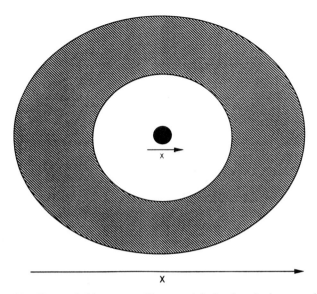

Figure 9. Van Kampen's itinerant oscillator model, that is, a body moves in a fluid and contains in its interior a damped oscillator. The displacement of the body is X relative to the fluid and the displacement of the oscillator x.

for uncoupling the damped equations as used for the undamped ones. The exact equations of the damped itinerant oscillator have been given by van Kampen [58, p.240]. These take account of the fact that the *mutual* damping (braking) torque on the two dipoles is proportional to the *difference* of their angular velocities. Thus the exact equations of motion, taking account of Newton's third law, are

$$I_1 \ddot{\phi}_1 + \zeta_1(\dot{\phi}_1 - \dot{\phi}_2) + V'(\phi_1 - \phi_2) = \lambda_1$$

$$I_2 \ddot{\phi}_2 + \zeta_1(\dot{\phi}_2 - \dot{\phi}_1) + \zeta_2 \dot{\phi}_2 - V'(\phi_1 - \phi_2) = -\lambda_1 + \lambda_2$$

Now the cage is supposed to be massive in comparison to the inner dipole so that $I_2 \gg I_1$ and $\dot{\phi}_2 \ll \dot{\phi}_1$, $\zeta_1(\dot{\phi}_2 - \dot{\phi}_1) \ll \zeta_2 \dot{\phi}_2$. If these conditions are satisfied, the above equations reduce to (C.44), (C.45), recalling that λ_1 must be neglected in the second equation if $\zeta_1(\dot{\phi}_2 - \dot{\phi}_1)$ is neglected. The three-dimensional form of the exact itinerant oscillator equations also coincides with the egg model of Shliomis and Stepanov [80] if the inertia of the egg is neglected and if the fluid surrounding the egg instead of being at rest is supposed to have a local angular velocity $\mathbf{\Omega}$. Thus the equations of motion of the egg model are

$$\zeta_1(\omega_1 - \omega_2) - T_1 = 0$$

$$\zeta_2(\omega_2 - \Omega) + \zeta_1(\omega_2 - \omega_1) - T_2 = 0$$

ω_1 is the angular velocity of the yolk, ω_2 that of the egg-shell, $\zeta_1 = 6\mu v$, $\zeta_2 = 6\eta V'$ in the notation of Section VIII. T_1 is the sum of the magnetic

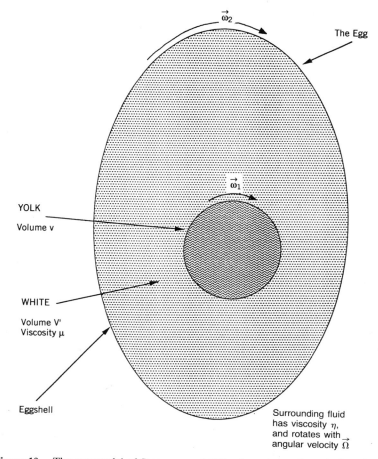

$\vec{\omega}_2$

The Egg

YOLK

Volume v

$\vec{\omega}_1$

WHITE

Volume V'
Viscosity μ

Eggshell

Surrounding fluid
has viscosity η,
and rotates with
angular velocity $\vec{\Omega}$

Figure 10. The egg model of Stepanov and Shliomis ω_1 is angular velocity of yolk. ω_2 is angular velocity of eggshell. Ω is local angular velocity of the surrounding fluid. μ = magnetic viscosity represented by the viscosity of the white. η is the viscosity of the surrounding fluid. v is the volume of the yolk. V' is the hydrodynamic volume.

and random torques acting on the yolk of volume v representing the magnet of moment $\mathbf{M}v$. $\mathbf{T}_2$ is the sum of the magnetic and random torques acting on the egg-shell (The magnetic torques are transmitted from the yolk via the white to the egg-shell). In the absence of an external field the sum of the *magnetic* torques acting on yolk and eggshell will vanish just as the $V'(\phi_1 - \phi_2)$ term vanishes in the sum of (C.44) and (C.45) and the equations decouple in $\boldsymbol{\omega}_2$ and $\boldsymbol{\omega}_2 - \boldsymbol{\omega}_1$ if $\boldsymbol{\Omega} = 0$.

The equations of motion, (C.44) and (C.45), are the starting point of the derivation of the Kramers equation for our problem. By applying the standard methods of constructing the Kramers equation from the Langevin equation, described in Appendix B one finds that the Kramers equation here is

$$\partial W / \partial t = [L + L_{\text{ext}}(t)]W \qquad (C.49)$$

where the joint distribution function is

$$W = W(\phi_1, \phi_2, \dot{\phi}_1, \dot{\phi}_2, t) \equiv W(\phi_1, \phi_2, \dot{\phi}_1, \dot{\phi}_2, t; \phi_1, \phi_2, \dot{\phi}_1, \dot{\phi}_2, t_0) \qquad (C.50)$$

Here L is termed the Kramers operator [31] and is

$$L = -\sum_{i=1}^{2} \left[\dot{\phi}_i \frac{\partial}{\partial \phi_i} - \frac{1}{I_i} \frac{\partial}{\partial \dot{\phi}_i} \frac{\partial V}{\partial \phi_i} - \frac{\zeta_i}{I_i} \frac{\partial}{\partial \dot{\phi}_i} \left(\dot{\phi}_i + \frac{kT}{I_i} \frac{\partial^2}{\partial \dot{\phi}_i^2} \right) \right] \qquad (C.51)$$

and $L_{\text{ext}}(t)$ the external field operator is

$$E(t) \sum_{i=1}^{2} \frac{\mu_i \sin \phi_i}{I_i} \frac{\partial}{\partial \dot{\phi}_i} \qquad (C.52)$$

We wish to calculate from the Kramers equation above the linear response of the system to an alternating electric field

$$\mathbf{E}(t) = \text{Re } \mathbf{E}_m \exp(i\omega t) \qquad (C.53)$$

This is carried out most easily by finding the after-effect solution of the Kramers equation.

This is the solution following the removal of a unidirectional electric field of unit magnitude at $t = 0$.

In general, the solution of the Kramers equation will consist of a term not involving the external field, a term linear in, a term quadratic in that field and so on. The solution to the first order in the field strength is the linear response solution which is the one we wish to find. Let us denote by $W^{(0)}$ the distribution in the absence of the external field $\mathbf{E}(t)$ and let us write in the Kramers equation

$$W = W^{(0)} + W^{(1)} + W^{(2)} + \cdots \qquad (C.54)$$

where $W^{(1)}$ denotes the portion of W linear in E, $W^{(2)}$ that quadratic in E and so on again. Then to the first order in E

$$\partial W^{(1)}/\partial t = LW^{(1)} + L_{\text{ext}}(t)W^{(0)} \qquad (C.55)$$

because

$$LW^{(0)} = 0 \qquad (C.56)$$

by definition. Recall that $W^{(0)}$ is the stationary solution in the absence of the external field $\mathbf{E}(t)$.

We wish to solve the linear Kramers equation above on $0 < t < \infty$ where t is the time at which $\mathbf{E}(t)$ is switched off; L_{ext} thus vanishes for $t > 0$. Equation (C.55) now becomes

$$\partial W^{(1)}/\partial t = LW^{(1)} \qquad (C.57)$$

The formal solution of this is

$$W_t^{(1)} = (\exp Lt)W_1^{(0)} \equiv e^{Lt}W_1^{(0)} \qquad (C.58)$$

Thus to terms linear in E,

$$W_t = W^{(0)} + e^{Lt}W_1^{(0)} \qquad (C.59)$$

The distribution function before the field is switched off is

$$A \exp\{-[I_1\dot\phi_1^2 + I_2\dot\phi_2^2 + V(\phi_1 - \phi_2)]/$$
$$2kT\} \exp\{(\mu_1 E \cos\phi_1 + \mu_2 E \cos\phi_2)/kT\} \qquad (C.60)$$

where A is a constant that normalizes the distribution function. To first order in E, this expression is

$$(1 + (\mu_1 E/kT) \cos\phi_1 + (\mu_2 E/kT) \cos\phi_2)W^{(0)} \qquad (C.61)$$

which is

$$W^{(0)} + (E/kT)(\mu_1 \cos \phi_1 + \mu_2 \cos \phi_2)W^{(0)} = W^{(0)} + W^{(1)}$$

(C.62)

by definition. Thus

$$W_t = W^{(0)} + (E/kT)e^{Lt}(\mu_1 \cos \phi_1 + \mu_2 \cos \phi_2)W^{(0)}$$ (C.63)

The mean dipole moment is

$$M(t) = \langle \mathbf{m} \cdot \mathbf{e} \rangle = C_m(t) = \langle \mu_1 \cos \phi_1 + \mu_2 \cos \phi_2 \rangle$$

$$= \int_0^{2\pi} \int_0^{2\pi} \int_{-\infty}^{\infty} \int_{-\infty}^{\infty} (\mu_1 \cos \phi_1 + \mu_2 \cos \phi_2)W_t \, d\phi_1 \, d\phi_2 \, d\dot{\phi}_1 \, d\dot{\phi}_2$$

$$= \int_0^{2\pi} \int_0^{2\pi} \int_{-\infty}^{\infty} \int_{-\infty}^{\infty} (\mu_1 \cos \phi_1 + \mu_2 \cos \phi_2)e^{Lt}$$

$$\times [(\mu_1 \cos \phi_1 + \mu_2 \cos \phi_2)W^{(0)}] \, d\phi_1 \, d\phi_2 \, d\dot{\phi}_1 \, d\dot{\phi}_2$$

$$= \langle (\mathbf{m}(0) \cdot \mathbf{e})(\mathbf{m}(t) \cdot \mathbf{e}) \rangle_0$$

(C.64)

because the first term $W^{(0)}$ in the previous equation vanishes when integrated over $\cos \phi_i$.

The above equation is the correlation function of the total dipole moment. This result was derived in a slightly different manner by Schröer [73], Coffey et al. [74], and Risken and Vollmer [75] and is implicit in the work of Budó [76].

Budó [76] originally addressed himself to this problem when a weak alternating field is applied. We have on the other hand discussed the problem by using the after-effect solution and the linear response theory. We now reconcile the two methods. When an alternating field $\mathbf{E}(t)$ is applied at time $t = -\infty$ the appropriate form of the Kramers equation is

$$\partial W/\partial t = (L + L_{\text{ext}}(t))$$ (C.65)

where L and L_{ext} are defined in Eqs. (C.51, C.52) with $\mathbf{E}(t) = \mathbf{E}_m e^{i\omega t}$. As before we write

$$W = W^{(0)} + W^{(1)} + W^{(2)} + \cdots$$ (C.66)

where $W^{(0)}$ is the steady state distribution function in the absence of the AC field and so on as before with

$$LW^{(0)} = 0 \qquad (C.67)$$

and $W^{(1)}$, satisfying

$$\partial W^{(1)}/\partial t = LW^{(1)} + L_{\text{ext}}(t)W^{(0)} \qquad (C.68)$$

Because the AC field was applied at $t = -\infty$ we require the stationary solution of the above Kramers equation which is

$$W_t^{(1)} = \int_{-\infty}^{t} e^{L(t-t')} L_{\text{ext}}(t')W^{(0)} \, dt' \,, \quad -\infty < t' < t \qquad (C.69)$$

Now

$$L_{\text{ext}}(t) = \frac{1}{I} E_m e^{i\omega t} \sum_{i=1}^{2} \mu_i \sin \phi_i \frac{\partial}{\partial \dot\phi_i} \qquad (C.70)$$

Further the mean dipole moment is

$$\langle \mathbf{m} \cdot \mathbf{e} \rangle_{E_m} = \langle \mu_1 \cos \phi_1 + \mu_2 \cos \phi_2 \rangle$$

$$= \int_0^{2\pi} \int_0^{2\pi} \int_{-\infty}^{\infty} \int_{-\infty}^{\infty} (\mu_1 \cos \phi_1 + \mu_2 \cos \phi_2)$$

$$\times \int_{-\infty}^{t} e^{L(t-t')} L_{\text{ext}}(t')W^{(0)} \, dt' \, d\phi_1 \, d\phi_2 \, d\dot\phi_1 \, d\dot\phi_2 \qquad (C.71)$$

Returning to Eq. (C.70) we note that

$$W^{(0)} \sin \phi_i \, \partial/\partial \dot\phi_i = -(I/kT)L(\cos \phi_i W^{(0)}) \qquad (C.72)$$

and therefore

$$L_{\text{ext}}W^{(0)} = -(E_m/kT)e^{i\omega t}L(\mu_i \cos \phi_i W^{(0)}) \qquad (C.73)$$

Thus Eq. (C.71) is reduced to

$$-\int_0^{2\pi} \int_0^{2\pi} \int_{-\infty}^{\infty} \int_{-\infty}^{\infty} (\mu_1 \cos \phi_1 + \mu_2 \cos \phi_2) \int_{-\infty}^{t} e^{L(t-t')} \frac{E_m}{kT}$$

$$\times e^{i\omega t'} L[(\mu_1 \cos \phi_1 + \mu_2 \cos \phi_2)W^{(0)}] \, d\phi_1 \, d\phi_2 \, d\dot\phi_1 \, d\dot\phi_2 \, dt' \qquad (C.74)$$

Let us now write $t' = t - \tau$. This then becomes

$$-\frac{1}{kT} \int_0^{2\pi} \int_0^{2\pi} \int_{-\infty}^{\infty} \int_{-\infty}^{\infty} (\mu_1 \cos \phi_1 + \mu_2 \cos \phi_2) \int_0^{\infty} e^{L\tau} E_m e^{i\omega(t-\tau)}$$

$$\times L[(\mu_1 \cos \phi_1 + \mu_2 \cos \phi_2) W^{(0)}] \, d\phi_1 \, d\phi_2 \, d\dot\phi_1 \, d\dot\phi_2 \, d\tau$$

$$= -\frac{1}{kT} \int_0^{\infty} d\tau E_m e^{i\omega(t-\tau)} \Bigg[\int_0^{2\pi} \int_0^{2\pi} \int_{-\infty}^{\infty} \int_{-\infty}^{\infty} (\mu_1 \cos \phi_1$$

$$+ \mu_2 \cos \phi_2) e^{L\tau} L(\mu_1 \cos \phi_1 + \mu_2 \cos \phi_2) W^{(0)} \, d\phi_1 \, d\phi_2 \, d\dot\phi_1 \, d\dot\phi_2 \Bigg]$$

$$= -\frac{1}{kT} \int_0^{\infty} d\tau E_m e^{i\omega(t-\tau)} \frac{dC_m(\tau)}{d\tau} \tag{C.75}$$

where C_m is given by (C.64) with $E = 1$, that is, the after-effect solution following the removal of a steady field of unit magnitude. Thus the original work of Budó [76] is reconciled with the present solution. This is a useful illustration of the concept of linear response in the presence of a potential. In conclusion it should be noted that in the itinerant oscillator as applied to ferrofluids, the quantities which are to be evaluated are $\langle \cos(\phi_1 - \phi_2) \rangle$ and $\langle \sin(\phi_1 - \phi_2) \rangle$ corresponding to fields applied. $\parallel$ and $\perp$ to the cage axis.

APPENDIX D: THE CALCULATION OF DRIFT AND DIFFUSION COEFFICIENTS IN CURVILINEAR COORDINATES

D.1. The Gilbert–Langevin Equation in Spherical Polar Coordinates

Gilbert's equation when augmented by a random field term as the Gilbert–Langevin equation is

$$\frac{d\mathbf{M}}{dt} = \gamma \mathbf{M} \times \left(\mathbf{H} - \eta \, \frac{d\mathbf{M}}{dt} + \mathbf{h}(t) \right) \tag{D.1}$$

It is possible to rewrite this equation in the spherical polar coordinates (r, ϑ, ϕ) given by Fig. 11. The magnitude M_s of $\mathbf{M}$ is constant therefore only the angular coordinates ϑ and ϕ will vary. The Gilbert–Langevin equation in spherical polar coordinates may therefore be transformed into two equations in functions of ϑ and ϕ. In order to accomplish this transformation we start with Brown's form of Gilbert's equation which we derived in Section I, that is,

$$\frac{d\mathbf{M}}{dt} = M_s g' \mathbf{M} \times \mathbf{H} + h'(\mathbf{M} \times \mathbf{H}) \times \mathbf{M} \tag{D.2}$$

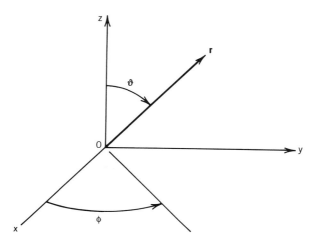

Figure 11. The vector **r** in spherical polar coordinates.

The quantity $\mathbf{M} \times \mathbf{H}$ is

$$\mathbf{M} \times \mathbf{H} = M_s \begin{vmatrix} \mathbf{e}_r & \mathbf{e}_\vartheta & \mathbf{e}_\phi \\ 1 & 0 & 0 \\ H_r & H_\vartheta & H_\phi \end{vmatrix} = (-H_\phi \mathbf{e}_\vartheta + H_\vartheta \mathbf{e}_\phi) M_s \qquad (D.3)$$

and $(\mathbf{M} \times \mathbf{H}) \times \mathbf{M}$ is

$$(\mathbf{M} \times \mathbf{H}) \times \mathbf{M} = M_s^2 \begin{vmatrix} \mathbf{e}_r & \mathbf{e}_\vartheta & \mathbf{e}_\phi \\ 0 & -H_\phi & H_\vartheta \\ 1 & 0 & 0 \end{vmatrix} = (H_\vartheta \mathbf{e}_\vartheta + H_\phi \mathbf{e}_\phi) M_s^2 \qquad (D.4)$$

Substitution of these into Eq. (D.2) will give

$$\frac{d\mathbf{M}}{dt} = M_s \left(\frac{d\vartheta}{dt} \mathbf{e}_\vartheta + \sin \vartheta \frac{d\phi}{dt} \mathbf{e}_\phi \right) = M_s^2 g' \left(-H_\phi \mathbf{e}_\vartheta + H_\vartheta \mathbf{e}_\phi \right)$$
$$+ M_s^2 h' (H_\vartheta \mathbf{e}_\vartheta + H_\phi \mathbf{e}_\phi) \qquad (D.5)$$

The separation of the two angular components leads to the two equations

$$\frac{d\vartheta}{dt} = M_s \left(h' H_\vartheta - g' H_\phi \right) \qquad (D.6)$$

$$\sin \vartheta \frac{d\phi}{dt} = M_s (h' H_\phi + g' H_\vartheta) \qquad (D.7)$$

Equations (D.6) and (D.7) are Gilbert's equation in spherical polar coordinates. To obtain the Gilbert–Langevin equation in such coordinates we augment the field components H_ϑ and H_ϕ with random field terms h_ϑ and h_ϕ. By graphical comparison of the Cartesian and spherical polar coordinate systems, we find these to be

$$h_\vartheta = h_1 \cos \vartheta \cos \phi + h_2 \cos \vartheta \sin \phi - h_3 \sin \vartheta \qquad \text{(D.8)}$$

$$h_\phi = -h_1 \sin \phi + h_2 \cos \phi \qquad \text{(D.9)}$$

The Gilbert–Langevin equations in spherical polar coordinates are therefore

$$\frac{d\vartheta}{dt} = M_s(h'H_\vartheta - g'H_\phi) + M_s(h' \cos \vartheta \cos \phi + g' \sin \phi)h_1$$
$$+ M_s(h' \cos \vartheta \sin \phi - g' \cos \phi)h_2$$
$$+ M_s(-h' \sin \vartheta)h_3 \qquad \text{(D.10)}$$

$$\sin \vartheta \, \frac{d\phi}{dt} = M_s(h'H_\phi + g'H_\vartheta) + M_s(-h' \sin \phi + g' \cos \vartheta \cos \phi)h_1$$
$$+ M_s(h' \cos \phi + g' \cos \vartheta \sin \phi]h_2$$
$$+ M_s(-g' \sin \vartheta)h_3 \qquad \text{(D.11)}$$

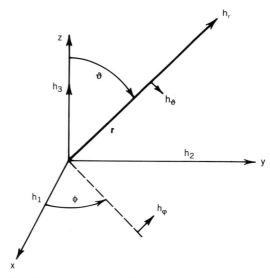

Figure 12. The random field $h(t)$ in spherical polar coordinates.

These equations are of the form

$$\frac{d\vartheta}{dt} = F_\vartheta(\vartheta, \phi, t) + G_{\vartheta 1}(\vartheta, \phi)h_1(t)$$
$$+ G_{\vartheta 2}(\vartheta, \phi)h_2(t)$$
$$+ G_{\vartheta 3}(\vartheta, \phi)h_3(t) \qquad \text{(D.12)}$$

and

$$\frac{d\phi}{dt} = F_\phi(\vartheta, \phi, t) + G_{\phi 1}(\vartheta, \phi)h_1(t)$$
$$+ G_{\phi 2}(\vartheta, \phi)h_2(t)$$
$$+ G_{\phi 3}(\phi, \vartheta)h_3(t) \qquad \text{(D.13)}$$

D.2. The Kramers–Moyal Expansion Coefficients for Nonlinear Langevin Equations

More generally the equations of the last section have the form

$$\frac{dx_i}{dt} = F_i(x_1, x_2, t) + G_{ij}(x_1, x_2, t)h_j(t) \qquad \text{(D.14)}$$

where

$$\vartheta = x_1 \quad \text{and} \quad \phi = x_2$$

and summation over $j = 1, 2, 3$ is implied. When (as in Eqs. (D.12) and (D.13)) the functions G_{ij} are dependent on x_i, the noise term is said to be multiplicative, whereas with constant G_{ij} the noise term is additive. Such a distinction between additive and multiplicative noise terms is not of importance for a single variable Langevin equation since the "offending" multiplicative function can be removed by normalization to give

$$\frac{1}{G(x)} \frac{dx}{dt} = \frac{F(x, t)}{G(x)} + h(t) \qquad \text{(D.15)}$$

However in the multivariable case, such a normalization is not possible and it will be seen that a "spurious" or "noise induced" drift will arise as a result of the multiplicative noise term. This drift term will occur in the drift or first order coefficients of the Kramers–Moyal [31] expansion which are as follows:

$$D_i^{(1)}(x_1, x_2, t) = \lim_{\tau \to 0} \frac{1}{\tau} \left\langle [x_i(t + \tau) - x_i] \right\rangle \Big|_{x_i = x_i(t)} \qquad \text{(D.16)}$$

$$D_{ij}^{(2)}(x_1, x_2, t) = \lim_{\tau \to 0} \frac{1}{2\tau} \langle [(x_i(t + \tau) - x_i)(x_j(t + \tau) - x_j)] \rangle \big|_{x_i = x_i(t)}$$

(D.17)

$$D^{(n)} = 0, \quad n \geq 3$$

(D.18)

where $x_i(t + \tau)$, $\tau > 0$ is a solution to the Langevin equation where $x_i(t)$ has a definite or sharp value at time t.

It is possible to set up from these coefficients, the Fokker–Planck equation which describes the probability density of the orientation of **M**.

Stratonovich [12] finds the Kramers–Moyal [31] coefficients for an equation of this form to be

$$D_i^{(1)}(x_1, x_2, t) = F_i(x_1, x_2, t) + G_{kj}(x_1, x_2, t) \frac{\partial}{\partial x_k} G_{ij}(x_1, x_2, t)$$

(D.19)

and

$$D_i^{(2)}(x_1, x_2, t) = G_{ik}(x_1, x_2, t) G_{jk}(x_1, x_2, t)$$

(D.20)

For the two variable case, the last term in the drift coefficient is the term due to multiplicative noise, which arises because during changes in $\mathbf{h}(t)$ changes in $\mathbf{M}(t)$ also occur and consequently the correlation

$$\langle G_{ij}(\mathbf{M}(t))\mathbf{h}(t) \rangle \neq 0$$

(D.21)

Itô [51] obtains different expressions for the Kramers–Moyal co-efficients in which the spurious drift term is absent. However use of Itô coefficients involves new rules for calculus and so Stratonovich's method will be used here since it is also in agreement with the original method of Brown [8] and is the correct definition to use in the case of a physical noise which always has a finite correlation time [58] (see B.2).

D.3.   Evaluation of the Kramers–Moyal Coefficients for the Gilbert–Langevin Equation

We next evaluate the Kramers–Moyal coefficients for the Gilbert–Langevin equation using the Stratonovich method. Here we define $G_{ik} = G_{\vartheta k}$, $G_{jk} = G_{\phi k}$.

$$D_1^{(1)}(\vartheta, t) = F_\vartheta + G_{\vartheta 1} \frac{\partial}{\partial \vartheta} G_{\vartheta 1} + G_{\vartheta 2} \frac{\partial}{\partial \vartheta} G_{\vartheta 2} + G_{\vartheta 3} \frac{\partial}{\partial \vartheta} G_{\vartheta 3}$$

$$+ G_{\phi 1} \frac{\partial}{\partial \phi} G_{\vartheta 1} + G_{\phi 2} \frac{\partial}{\partial \phi} G_{\vartheta 2} + G_{\phi 3} \frac{\partial}{\partial \phi} G_{\vartheta 3}$$

$$= F_\vartheta + M_s^2 \Big\{ (h' \cos \vartheta \cos \phi + g' \sin \phi)(-h' \sin \vartheta \cos \phi)$$

$$+ (h' \cos \vartheta \sin \phi - g' \cos \phi)(-h' \sin \vartheta \sin \phi)$$

$$+ h'^2 \cos \vartheta \sin \vartheta$$

$$+ \left(-h' \frac{\sin \phi}{\sin \vartheta} + g' \frac{\cos \vartheta \cos \phi}{\sin \vartheta} \right)(-h' \cos \vartheta \sin \phi + g' \cos \phi)$$

$$+ \left(h' \frac{\cos \phi}{\sin \vartheta} + g' \frac{\cos \vartheta \sin \phi}{\sin \vartheta} \right)(h' \cos \vartheta \cos \phi + g' \sin \phi) \Big\}$$

$$= F_\vartheta + M_s^2 \Big\{ -h'^2 [\cos^2\phi \sin \vartheta \cos \vartheta + \sin^2\phi \sin \vartheta \cos \vartheta - \sin \vartheta \cos \vartheta]$$

$$+ g'h'[-\sin \phi \cos \phi \sin \vartheta + \sin \vartheta \cos \phi \sin \phi]$$

$$+ h'^2 \left(\sin^2\phi \frac{\cos \vartheta}{\sin \vartheta} + \cos^2\phi \frac{\cos \vartheta}{\sin \vartheta} \right)$$

$$+ g'^2 \left(\cos^2\phi \frac{\cos \vartheta}{\sin \vartheta} + \sin^2\phi \frac{\cos \vartheta}{\sin \vartheta} \right)$$

$$+ h'g' \left(\frac{\cos \vartheta}{\sin \vartheta} \cos \phi \cos \vartheta \sin \phi - \frac{\cos \vartheta}{\sin \vartheta} \cos \phi \cos \vartheta \sin \phi \right) \Big\}$$

$$= F_\vartheta + M_s^2 (h'^2 + g'^2) \frac{\cos \vartheta}{\sin \vartheta} \tag{D.22}$$

$$D_2^{(1)}(\phi, t) = F_\phi + G_{\phi 1} \frac{\partial}{\partial \phi} G_{\phi 1} + G_{\phi 2} \frac{\partial}{\partial \phi} G_{\phi 2} + G_{\phi 3} \frac{\partial}{\partial \phi} G_{\phi 3}$$

$$+ G_{\vartheta 1} \frac{\partial}{\partial \vartheta} G_{\phi 1} + G_{\vartheta 2} \frac{\partial}{\partial \vartheta} G_{\phi 2} + G_{\vartheta 3} \frac{\partial}{\partial \vartheta} G_{\phi 3}$$

$$= F_\phi + M_s^2 \Big\{ \frac{1}{\sin^2\vartheta} [(-h' \sin \phi + g' \cos \vartheta \cos \phi)(-h \cos \phi$$

$$- g' \cos \vartheta \sin \phi)$$

$$+ (h' \cos \phi + g' \cos \vartheta \sin \phi)(-h' \sin \phi + g' \cos \vartheta \cos \phi)]$$

$$+ (h' \cos \vartheta \cos \phi + g' \sin \phi) \left(h' \frac{\cos \vartheta}{\sin^2\vartheta} \sin \phi - g' \csc^2\vartheta \cos \phi \right)$$

$$+ (h' \cos \vartheta \sin \phi - g' \cos \phi) \left(-h' \frac{\cos \vartheta}{\sin^2\vartheta} \cos \phi - g' \csc^2\vartheta \sin \phi \right) \Big\}$$

$$= F_\phi + M_s^2 \left\{ h'^2 \left(\frac{\sin \phi \cos \phi}{\sin^2 \vartheta} - \frac{\sin \phi \cos \phi}{\sin^2 \vartheta} \right. \right.$$

$$+ \cos \phi \sin \phi \frac{\cos^2 \vartheta}{\sin^2 \vartheta} - \cos \phi \sin \phi \left. \frac{\cos^2 \vartheta}{\sin^2 \vartheta} \right)$$

$$+ g'^2 \left(-\frac{\cos^2 \vartheta}{\sin^2 \vartheta} \cos \phi \sin \phi + \frac{\cos^2 \vartheta}{\sin^2 \vartheta} \cos \phi \sin \phi \right.$$

$$+ \sin \phi \cos \phi \frac{\cos \vartheta}{\sin \vartheta} - \sin \vartheta \cos \phi \left. \frac{\cos \vartheta}{\sin \vartheta} \right)$$

$$+ h'g' \left(\frac{\cos \vartheta \sin^2 \phi - \cos \vartheta \cos^2 \phi}{\sin^2 \vartheta} + \frac{\cos \vartheta \cos^2 \phi}{\sin^2 \vartheta} \right.$$

$$- \frac{\sin^2 \phi \cos \vartheta}{\sin^2 \vartheta} - \frac{\cos \vartheta}{\sin^2 \vartheta} \cos^2 \phi$$

$$\left. \left. + \frac{\cos \vartheta}{\sin^2 \vartheta} \sin^2 \phi - \frac{\cos \vartheta}{\sin^2 \vartheta} \sin^2 \phi + \frac{\cos^2 \phi}{\sin^2 \vartheta} \cos \vartheta \right) \right\}$$

$$= F_\phi \tag{D.23}$$

The second order Kramers–Moyal coefficients or diffusion coefficients are

$$D_{11}^{(2)}(\vartheta, t) = G_{\vartheta 1}^2 + G_{\vartheta 2}^2 + G_{\vartheta 3}^2 \tag{D.24}$$

$$= M_s^2 [(h' \cos \vartheta \cos \phi + g' \sin \phi)^2$$

$$+ (h' \cos \vartheta \sin \phi - g' \cos \phi)^2 + h'^2 \sin^2 \vartheta]$$

$$= M_s^2 [h'^2 (\cos^2 \vartheta \cos^2 \phi + \cos^2 \vartheta \sin^2 \phi + \sin^2 \vartheta)$$

$$+ g'^2 (\sin^2 \phi + \cos^2 \phi)$$

$$+ h'g' (2 \sin \phi \cos \vartheta \cos \phi - 2 \sin \phi \cos \vartheta \cos \phi)]$$

$$= M_s^2 (h'^2 + g'^2) \tag{D.25}$$

$$D_{22}^{(2)}(\phi, t) = G_{\phi 1}^2 + G_{\phi 2}^2 + G_{\phi 3}^2$$

$$= \frac{M_s^2}{\sin^2 \vartheta} [(-h' \sin \phi + g' \cos \vartheta \cos \phi)^2$$

$$+ (h' \cos \phi + g' \cos \vartheta \sin \phi)^2 + g'^2 \sin^2 \vartheta]$$

$$= \frac{M_s^2}{\sin^2 \vartheta} [h'^2 (\sin^2 \phi + \cos^2 \phi) + g'^2 (\cos^2 \vartheta + \sin^2 \vartheta)$$

$$+ h'g' (-2 \sin \phi \cos \phi \cos \vartheta + 2 \sin \phi \cos \phi \cos \vartheta)]$$

$$= \frac{M_s^2 (h'^2 + g'^2)}{\sin^2 \vartheta} \tag{D.26}$$

$$D_{12}^{(2)}(\vartheta, t) = D_{21}(\vartheta, t)$$

$$= G_{\vartheta 1}G_{\phi 1} + G_{\vartheta 2}G_{\phi 2} + G_{\vartheta 3}G_{\phi 3}$$

$$= \frac{M_s^2}{\sin \vartheta} [(h' \cos \vartheta \cos \phi + g' \sin \phi)$$
$$\times (-h' \sin \phi + g' \cos \vartheta \cos \phi)$$
$$+ (h' \cos \vartheta \sin \phi - g' \cos \phi)$$
$$\times (h' \cos \phi + g' \cos \vartheta \sin \phi)$$
$$+ g'h' \sin^2 \vartheta]$$

$$= \frac{M_s^2}{\sin \vartheta} [h'^2(-\cos \vartheta \cos \phi \sin \phi + \cos \phi \cos \vartheta \sin \phi)$$
$$+ g'^2(\cos \vartheta \sin \phi \cos \phi - \cos \phi \cos \vartheta \sin \phi)$$
$$\times h'g'(-\sin^2\phi + \cos^2\vartheta \cos^2\phi - \cos^2\phi$$
$$+ \cos^2\vartheta \sin \phi + \sin^2\vartheta)]$$

$$= 0 \tag{D.27}$$

D.4. Formulation of the Fokker–Planck Equation from the Kramers–Moyal Expansion Coefficients

The complete Kramers–Moyal coefficients are thus

$$D_1 = M_s(h'H_\vartheta - g'H_\phi) + M_s^2(h'^2 + g'^2) \cot \vartheta \tag{D.28}$$

$$D_2 = \frac{M_s}{\sin \vartheta} (h'H_\phi + g'H_\vartheta) \tag{D.29}$$

$$D_{11} = M_s^2(h'^2 + g'^2) \tag{D.30}$$

$$D_{22} = M_s^2(h'^2 + g'^2) \csc^2\vartheta \tag{D.31}$$

D.5. Derivation of Brown's Equation

The Fokker–Planck equation for two variables is [31]

$$\frac{\partial P}{\partial t} + \frac{\partial S_i}{\partial x_i} = 0, \quad \text{for } i = 1, 2 \tag{D.32}$$

where $P(x_1, x_2, t) \, dx_1 \, dx_2$ is the probability of a value in $dx_1 dx_2$ at time t and where the probability current is defined by

$$S_i = D_i P - \frac{\partial}{\partial x_i} D_{ij} P \tag{D.33}$$

where D_i and D_{ij} are the diffusion and drift coefficients. The diffusion current components from equations (D.28)–(D.31) are therefore

$$S_1 = M_s(h'H_\vartheta - g'H_\phi)P + M_s^2(h'^2 + g'^2)\cot\vartheta P - \frac{\partial}{\partial x_1} M_s^2(h'^2 + g'^2)P$$

$$S_2 = M_s \frac{1}{\sin\vartheta}(h'H_\phi + g'H_\vartheta)P - \frac{\partial}{\partial x_2} M_s^2(h'^2 + g'^2)\csc^2(\vartheta)P \qquad (D.34)$$

The Fokker–Planck equation in this case is therefore

$$\frac{\partial P}{\partial t} = -\frac{\partial}{\partial x_1}\Big\{M_s(h'H_\vartheta - g'H_\phi)P + M_s^2(h'^2 + g'^2)\,P\cot\vartheta$$

$$-\frac{\partial}{\partial x_1} M_s^2(h'^2 + g'^2)P\Big\}$$

$$-\frac{\partial}{\partial x_2}\Big\{M_s \frac{1}{\sin\vartheta}(h'H_\phi + g'H_\vartheta)P - \frac{\partial}{\partial x_2} M_s^2(h'^2 + g'^2)\frac{1}{\sin^2\vartheta}P\Big\}$$
$$(D.35)$$

Substitution of $\vartheta = x_1$ and $\phi = x_2$ yields

$$\frac{\partial P}{\partial t} = -\frac{\partial}{\partial\vartheta}\Big\{M_s(h'H_\vartheta - g'H_\phi)P$$

$$+ M_s^2(h'^2 + g'^2)\Big[\cot\vartheta P - \frac{\partial P}{\partial\vartheta}\Big]\Big\}$$

$$-\frac{\partial}{\partial\phi}\Big\{M_s \frac{1}{\sin\vartheta}(h'H_\phi + g'H_\vartheta)P - M_s^2(h'^2 + g'^2)\frac{1}{\sin^2\vartheta}\frac{\partial P}{\partial\phi}\Big\}$$
$$(D.36)$$

In order to write this equation as a function of the distribution $W(\vartheta, \phi, t)\,d\Omega = W(\vartheta, \phi, t)\sin\vartheta\,d\vartheta\,d\phi$ which is the probability of finding a value within the solid angle $d\Omega$ at time t we need to make the substitution

$$P = W\sin\vartheta \qquad (D.37)$$

On making this substitution, we have

$$\frac{\partial W}{\partial t} = -\frac{1}{\sin\vartheta}\frac{\partial}{\partial\vartheta}\Big\{\sin\vartheta\Big[M_s(h'H_\vartheta - g'H_\phi)W$$

$$- M_s^2(h'^2 + g'^2)\frac{\partial W}{\partial\vartheta}\Big]\Big\}$$

$$
-\frac{\partial}{\partial \phi} \left\{ M_s \frac{1}{\sin \vartheta} (h'H_\phi + g'H_\vartheta)W \right.
$$

$$
\left. -\frac{\partial}{\partial \phi} M_s^2 (h'^2 + g'^2) \frac{1}{\sin^2 \vartheta} W \right\} \tag{D.38}
$$

From equations (1.32) and (1.33) we have

$$
h' = \frac{\gamma}{M_s} \frac{\alpha}{1 + \alpha^2} \quad \text{and} \quad g' = \frac{\gamma}{M_s} \frac{1}{1 + \alpha^2}
$$

The quantity $M_s^2(h'^2 + g'^2)$ is therefore

$$
M_s^2(h'^2 + g'^2) = M_s^2 \left(\frac{\gamma}{M_s} \right)^2 \frac{1}{1 + \alpha^2}
$$

$$
= M_s \gamma g' = M_s \gamma \frac{h'}{\alpha}
$$

$$
= h'\eta \tag{D.39}
$$

from $\alpha = \eta \gamma M_s$. The Fokker–Planck equation is therefore

$$
\frac{\partial W}{\partial t} = -\frac{1}{\sin \vartheta} \frac{\partial}{\partial \vartheta} \left\{ \sin \vartheta \left[M_s(h'H_\vartheta - g'H_\phi)W - \eta h' \frac{\partial W}{\partial \vartheta} \right] \right\}
$$

$$
-\frac{\partial}{\partial \phi} \left\{ M_s \frac{1}{\sin \vartheta} (h'H_\phi + g'H_\vartheta)W - \frac{\partial}{\partial \phi} h' \frac{1}{\sin^2 \vartheta} W \right\} \tag{D.40}
$$

Furthermore the field components H_ϑ and H_ϕ from $\mathbf{H} = -\partial V/\partial \mathbf{M}$ are

$$
H_\vartheta = -\frac{1}{M_s} \frac{\partial V}{\partial \vartheta} \tag{D.41}
$$

and

$$
H_\phi = -\frac{1}{M_s} \frac{1}{\sin \vartheta} \frac{\partial V}{\partial \phi} \tag{D.42}
$$

Substitution of these field components leads to

$$
\frac{\partial W}{\partial t} = \frac{1}{\sin \vartheta} \frac{\partial}{\partial \vartheta} \left\{ \sin \vartheta \left[\left(h' \frac{\partial V}{\partial \vartheta} - g' \frac{\partial V}{\partial \phi} \right)W + \eta h' \frac{\partial W}{\partial \vartheta} \right] \right\}
$$

$$
+ \frac{1}{\sin \vartheta} \frac{\partial}{\partial \phi} \left\{ \left(\frac{h'}{\sin \vartheta} \frac{\partial V}{\partial \phi} + g' \frac{\partial V}{\partial \vartheta} \right)W + \eta h' \frac{\partial W}{\partial \phi} \right\} \tag{D.43}
$$

Imposition of equilibrium conditions as in Section II leads to

$$\frac{\partial W}{\partial t} = 0 \quad \text{only if} \quad \frac{vh'}{kT} = h'\eta$$

and so we have for Gilbert's damping factor η, the condition that

$$\frac{kT\eta}{v} = 1 \tag{D.44}$$

It is important to note that the generalized form of the equations used in this analysis, Eqs. (D.12) and (D.13) is such that the noise term is normalized to

$$\langle h_i(t)h_j(t')\rangle = 2\delta_{ij}\delta(t - t') \tag{D.45}$$

Brown [8] used an arbitrary noise correlation such that (see Section II)

$$\langle h_i(t)h_j(t')\rangle = \mu\delta_{ij}\delta(t - t') \tag{D.46}$$

Use of this arbitrary noise correlation will mean that equilibrium will now impose the condition

$$\frac{kT\eta}{v} = \frac{\mu}{2} \tag{D.47}$$

This however will not affect Brown's equation which by writing

$$\eta h' = k' \tag{D.48}$$

is

$$\frac{\partial W}{\partial t} = \frac{1}{\sin \vartheta} \frac{\partial}{\partial \vartheta} \left\{ \sin \vartheta \left[\left(h' \frac{\partial V}{\partial \vartheta} - \frac{g'}{\sin \vartheta} \frac{\partial V}{\partial \phi} \right) W + k' \frac{\partial W}{\partial \vartheta} \right] \right\}$$

$$+ \frac{1}{\sin \vartheta} \frac{\partial}{\partial \phi} \left\{ \left[\left(\frac{h'}{\sin \vartheta} \frac{\partial V}{\partial \phi} + g' \frac{\partial V}{\partial \vartheta} \right) W + \frac{k'}{\sin \vartheta} \frac{\partial W}{\partial \phi} \right] \right\} \tag{D.49}$$

This is Brown's equation in the form it appeared in his 1963 paper [8].

APPENDIX E: THE NONINERTIAL LANGEVIN EQUATION

We study, following Lewis et al. [77], the rotational Brownian movement of a sphere which is supposed homogeneous, the motion being due entirely to random couples that have no preferential direction. The sphere contains a rigid electric dipole $\boldsymbol{\mu}$. We take through the center of the sphere a unit vector $\mathbf{u}(t)$ in the direction of $\boldsymbol{\mu}$. Then the rate of change of $\boldsymbol{\mu}(t)$ is [46]

$$\frac{d\boldsymbol{\mu}(t)}{dt} = \boldsymbol{\omega}(t) \times \boldsymbol{\mu}(t) \quad \left(\mathbf{u}(t) = \frac{\boldsymbol{\mu}(t)}{|\boldsymbol{\mu}|} \right) \tag{E.1}$$

where $\boldsymbol{\omega}(t)$ is the angular velocity of the body. It should be noted that Eq. (E.1) is a purely kinematic relation with no particular reference either to the Brownian movement or to the shape of the body. We specialize it to the rotational Brownian motion of a sphere by supposing that $\boldsymbol{\omega}$ obeys the Euler–Langevin equation

$$I \frac{d\boldsymbol{\omega}(t)}{dt} + \zeta \boldsymbol{\omega}(t) = \boldsymbol{\lambda}(t) + \boldsymbol{\mu}(t) \times \mathbf{F}(t) \tag{E.2}$$

In Eq. (E.2), I is the moment of inertia of the sphere, $\zeta \boldsymbol{\omega}$ is the damping torque due to Brownian movement and $\boldsymbol{\lambda}(t)$ is the white noise driving torque, again due to Brownian movement so that $\boldsymbol{\lambda}(t)$ has the following properties

$$\langle \lambda_i(t) \rangle = 0 \tag{E.3}$$

$$\langle \lambda_i(t) \lambda_j(t') \rangle = 2kT\zeta \delta_{ij} \delta(t - t') \tag{E.4}$$

where δ_{ij} is Kronecker's delta, $i, j = 1, 2, 3$, which correspond to the Cartesian axes, x, y, z, fixed in the sphere. $\delta(t)$ is the Dirac delta function. The term $\boldsymbol{\mu} \times \mathbf{F}(t)$, in Eq. (E.2), is the torque due to an externally applied field.

Equation (E.2) includes the inertia of the sphere. The noninertial response is the response when I tends to zero or when ζ, the friction coefficient, becomes very large. In this limit the angular velocity vector may be immediately written down from Eq. (E.2) as

$$\boldsymbol{\omega}(t) = \frac{\boldsymbol{\lambda}(t)}{\zeta} + \frac{\boldsymbol{\mu} \times \mathbf{F}(t)}{\zeta} \tag{E.5}$$

We can now combine this with the kinematic relation (Eq. (E.1) to obtain

$$\frac{d\boldsymbol{\mu}(t)}{dt} = \frac{\boldsymbol{\lambda}(t)}{\zeta} \times \boldsymbol{\mu}(t) + \frac{\{\boldsymbol{\mu}(t) \times \mathbf{F}(t)\} \times \boldsymbol{\mu}(t)}{\zeta} \tag{E.6}$$

which, with the properties of the triple vector product, becomes

$$\frac{d\boldsymbol{\mu}}{dt} = \frac{\boldsymbol{\lambda}(t)}{\zeta} \times \boldsymbol{\mu} + \frac{\mu^2 \mathbf{F}(t)}{\zeta} - \frac{\boldsymbol{\mu}\{\boldsymbol{\mu} \cdot \mathbf{F}(t)\}}{\zeta} \tag{E.7}$$

This is the Langevin equation for the motion of $\boldsymbol{\mu}$ in the noninertial limit.

Let us suppose that a strong DC field F_0 has been applied along the polar or z axis for a long time and let us suppose that at $t = 0$ (F_0 having been applied at $t = -\infty$) a small field $F_1 U(t)$, where $U(t)$ is the unit step function is also applied along the z axis, so that

$$\mathbf{F}(t) = F_0 \mathbf{k} + F_1 U(t)\mathbf{k} \tag{E.8}$$

Equation (E.2) then becomes, with the aid of Eq. (E.8),

$$\dot{\mu}_x = \frac{1}{\zeta}(\lambda_y \mu_z - \lambda_z \mu_y) - \frac{\mu_x \mu_z}{\zeta}[F_0 + F_1 U(t)] \tag{E.9}$$

$$\dot{\mu}_y = \frac{1}{\zeta}(\lambda_z \mu_x - \lambda_x \mu_z) - \frac{\mu_y \mu_z}{\zeta}[F_0 + F_1 U(t)] \tag{E.10}$$

$$\dot{\mu}_z = \frac{1}{\zeta}(\lambda_x \mu_y - \lambda_y \mu_x) + \frac{(\mu_x^2 + \mu_y^2)}{\zeta}[F_0 + F_1 U(t)] \tag{E.11}$$

The quantity of interest to us is the average behavior of the component of the dipole moment in the field direction, namely $\langle \mu_z \rangle$ (we postulate an assembly of such rotating spheres). Since

$$\mu^2 = \mu_x^2 + \mu_y^2 + \mu_z^2 \tag{E.12}$$

Eq. (E.11) becomes

$$\dot{\mu}_z = \frac{1}{\zeta}(\lambda_x \mu_y - \lambda_y \mu_x) + \frac{(\mu^2 - \mu_z^2)}{\zeta}[F_0 + F_1 U(t)] \tag{E.13}$$

Equation (E.13) contains multiplicative noise terms; $\lambda_x \mu_y$ and $\lambda_y \mu_x$. This poses an interpretation problem as discussed by Risken [31]. Risken has shown, taking the Langevin equation for N stochastic variables $\{\xi\} = \{\xi_1, \xi_2, \xi_3, \dots, \xi_N\}$ as

$$\dot{\xi}_i = h_i(\{\xi\}, t) + g_{ij}(\{\xi\}, t)\Gamma_j(t) \tag{E.14}$$

with

$$\overline{\Gamma_j(t)} = 0 \tag{E.15}$$

$$\overline{\Gamma_i(t)\Gamma_j(t')} = 2\delta_{ij}\delta(t - t') \tag{E.16}$$

and interpreting it as a Stratonovich [12] equation, that the drift coefficient is

$$D_i(\{\xi\}, t) = \dot{x}_i = \lim_{\tau \to 0} \frac{1}{\tau} \{\overline{(\xi_i(t + \tau) - x_i)}\}|_{\xi_k(t) = x_k}$$

$$= h_i(\{x\}, t) + g_{kj}(\{x\}, t) \frac{\partial}{\partial x_k} g_{ij}(\{x\}, t)$$

$$k = 1, 2, \ldots, N \tag{E.17}$$

The last term in Eq. (E.17) is as we have seen called the noise-induced or spurious drift [31].

In Eq. (E.17), $\xi_i(t + \tau)$ is a solution of (E.14) which at time t has the *sharp* value $\xi_k(t) = x_k$ for $k = 1, 2, \ldots, N$. The quantities x_k are themselves random variables with p.d.f. $W(\{x\}, t)$ such that $W dx_k$ is the probability of finding x_k in x_k to $x_k + dx_k$. We now use this theorem to evaluate the average of the multiplicative noise terms in Eq. (E.13). We have nine tensor components, namely

$$g_{11} = 0, \qquad g_{12} = \frac{\mu_z}{\zeta}, \qquad g_{13} = \frac{-\mu_y}{\zeta}$$

$$g_{21} = \frac{-\mu_z}{\zeta}, \qquad g_{22} = 0, \qquad g_{23} = \frac{\mu_x}{\zeta}$$

$$g_{31} = \frac{\mu_y}{\zeta}, \qquad g_{32} = \frac{-\mu_x}{\zeta}, \qquad g_{33} = 0 \tag{E.18}$$

whence with the aid of Eq. (E.17),

$$D_3(\{x\}, t) = kT\zeta \left[g_{11} \frac{\partial}{\partial x_1} g_{31} + g_{12} \frac{\partial}{\partial x_1} g_{32} + g_{13} \frac{\partial}{\partial x_1} g_{33} \right.$$

$$+ g_{21} \frac{\partial}{\partial x_2} g_{31} + g_{22} \frac{\partial}{\partial x_2} g_{32} + g_{23} \frac{\partial}{\partial x_2} g_{33}$$

$$+ g_{31} \frac{\partial}{\partial x_3} g_{31} + g_{32} \frac{\partial}{\partial x_3} g_{32} + g_{33} \frac{\partial}{\partial x_3} g_{33} \left. \right]$$

$$= kT\zeta \left[\frac{\mu_z}{\zeta} \frac{\partial}{\partial \mu_x} \left(\frac{-\mu_x}{\zeta} \right) + \left(\frac{-\mu_z}{\zeta} \right) \frac{\partial}{\partial \mu_y} \left(\frac{\mu_y}{\zeta} \right) \right.$$

$$\left. + \frac{\mu_z}{\zeta} \frac{\partial}{\partial \mu_z} \left(\frac{\mu_y}{\zeta} \right) + \left(\frac{-\mu_x}{\zeta} \right) \frac{\partial}{\partial \mu_z} \left(\frac{-\mu_x}{\zeta} \right) \right] \qquad \text{(E.19)}$$

and implicitly taking a second average over the distribution functions $W(\{\mu\}, t)$ of μ at time t so that

$$\langle f(\{\mu\}) \rangle = \int f(\{\mu\}) W(\{\mu\}, t) \, d\mu$$

$$\langle D_3(\{\mu\}, t) \rangle = \frac{-2kT}{\zeta} \langle \mu_z \rangle \qquad \text{(E.20)}$$

which is the noise-induced drift. The deterministic drift, from Eq. (E.13) is

$$h_3(\{\mu\}, t) = \frac{\mu^2 - \mu_z^2}{\zeta} \qquad \text{(E.21)}$$

Thus the averaged equation of motion of the dipole component μ_z is

$$\langle \dot{\mu}_z \rangle + \frac{2kT}{\zeta} \langle \mu_z \rangle = \frac{\mu^2 [F_0 + F_1 U(t)]}{\zeta} \langle 1 - u_z^2 \rangle \qquad \text{(E.22)}$$

Note that since

$$u_z = \cos \vartheta \qquad \text{(E.23)}$$

Eq. (E.22) is

$$\frac{d}{dt} \langle P_1(\cos \vartheta) \rangle + \frac{2kT}{\zeta} \langle P_1(\cos \vartheta) \rangle$$

$$= \frac{2\mu [F_0 + F_1 U(t)]}{3\zeta} [1 - \langle P_2(\cos \vartheta) \rangle] \qquad \text{(E.24)}$$

$$P_1(u_z) = u_z \qquad \text{(E.25)}$$

$$P_2(u_z) = \frac{1}{2} (3u_z^2 - 1) \qquad \text{(E.26)}$$

are the Legendre polynomials of order 1 and 2.

We consider the same problem as above but this time the step change in the field is applied parallel to the x axis so that we need to determine the behavior of $\langle \mu_x \rangle$. By using the formula for the noise-induced drift in exactly the same manner as before we find that the x component of the dipole moment satisfies

$$\langle \dot{\mu}_x \rangle + \frac{2kT}{\zeta} \langle \mu_x \rangle = \frac{\mu^2}{\zeta} F_1 U(t)$$

$$- \frac{\langle \mu_x^2 \rangle}{\zeta} F_1 U(t) - \frac{\langle \mu_z \mu_x \rangle}{\zeta} F_0 \qquad \text{(E.27)}$$

Now

$$\mu_x = \mu \sin \vartheta \cos \phi = \mu X_{11}(\vartheta, \phi) \qquad \text{(E.28)}$$

where X_{11} is the spherical harmonic of order $(1, 1)$ [37]. Thus in terms of spherical harmonics Eq. (E.27) becomes

$$\mu \langle \dot{X}_{11} \rangle + \frac{2kT}{\zeta} \mu \langle X_{11} \rangle = -\frac{\mu^2 F_0}{3\zeta} \langle X_{21} \rangle$$

$$+ \frac{\mu^2 F_1 U(t)}{3\zeta} [2 - \tfrac{1}{2} \langle X_{22} \rangle + \langle P_2 \rangle] \qquad \text{(E.29)}$$

$$X_{21} = 3 \sin \vartheta \cos \vartheta \cos \phi \qquad \text{(E.30)}$$

$$X_{22} = 3 \sin^2 \vartheta \cos 2\phi \qquad \text{(E.31)}$$

Now we illustrate how the differential-difference equation for $\langle P_2 \rangle$ may be obtained by averaging the noninertial Langevin equation. We recall that for a longitudinal field F_0,

$$\dot{\mu}_x = \frac{1}{\zeta} (\lambda_y \mu_z - \lambda_z \mu_y) - \frac{\mu_x \mu_z}{\zeta} F_0 \qquad \text{(E.32)}$$

$$\dot{\mu}_y = \frac{1}{\zeta} (\lambda_z \mu_x - \lambda_x \mu_z) - \frac{\mu_y \mu_z}{\zeta} F_0 \qquad \text{(E.33)}$$

$$\dot{\mu}_z = \frac{1}{\zeta} (\lambda_x \mu_y - \lambda_y \mu_x) + \frac{\mu^2 - \mu_z^2}{\zeta} F_0 \qquad \text{(E.34)}$$

We now make the transformation

$$\mu_x \rightarrow x \,, \tag{E.35}$$

$$\mu_y \rightarrow y \tag{E.36}$$

$$\mu_z^2 \rightarrow z \tag{E.37}$$

so that

$$\dot{x} = \frac{1}{\zeta} (\lambda_y \sqrt{z} - \lambda_z y) - \frac{x\sqrt{z}}{\zeta} F_0 \tag{E.38}$$

$$\dot{y} = \frac{1}{\zeta} (\lambda_z x - \lambda_x \sqrt{z}) - \frac{y\sqrt{z}}{\zeta} F_0 \tag{E.39}$$

$$\dot{z} = \frac{1}{\zeta} (2\lambda_x y\sqrt{z} - 2\lambda_y x\sqrt{z}) + \frac{2\sqrt{z}}{\zeta} (\mu^2 - z)F_0 \tag{E.40}$$

We now apply the formula for the noise-induced drift terms in exactly the same manner as for the dipole moments. The relevant tensor components are

$$g_{11} = 0 \,, \qquad g_{12} = \frac{\sqrt{z}}{\zeta} \,, \qquad g_{13} = \frac{-y}{\zeta}$$

$$g_{21} = -\frac{\sqrt{z}}{\zeta} \,, \qquad g_{22} = 0 \,, \qquad g_{23} = \frac{x}{\zeta}$$

$$g_{31} = \frac{2y\sqrt{z}}{\zeta} \,, \qquad g_{32} = \frac{-2x\sqrt{z}}{\zeta} \,, \qquad g_{33} = 0 \tag{E.41}$$

hence

$$\begin{aligned}
D_3 &= \zeta kT \left[\frac{\sqrt{z}}{\zeta} \frac{\partial}{\partial x} \left(\frac{-2x\sqrt{z}}{\zeta} \right) + \left(-\frac{\sqrt{z}}{\zeta} \right) \frac{\partial}{\partial y} \left(\frac{2y\sqrt{z}}{\zeta} \right) \right. \\
&\quad \left. + \frac{2y\sqrt{z}}{\zeta} \frac{\partial}{\partial z} \left(\frac{2y\sqrt{z}}{\zeta} \right) + \left(\frac{-2x\sqrt{z}}{\zeta} \right) \frac{\partial}{\partial z} \left(\frac{-2x\sqrt{z}}{\zeta} \right) \right] \\
&= \zeta kT \left[-\frac{2z}{\zeta^2} - \frac{2z}{\zeta^2} + \frac{2y\sqrt{z}}{\zeta^2} \frac{2y}{2\sqrt{z}} + \frac{2x\sqrt{z}}{\zeta^2} \frac{2x}{2\sqrt{z}} \right] \\
&= \frac{2kT}{\zeta} (y^2 + x^2 - 2z) \\
&= \frac{2kT}{\zeta} (\mu^2 - 3z) \tag{E.42}
\end{aligned}$$

This is the noise-induced part of $\dot{z}$. The deterministic contribution to $\dot{z}$ is

$$+ \frac{2\sqrt{z}}{\zeta} (\mu^2 - z) F_0 \qquad \text{(E.43)}$$

Thus

$$\langle \dot{z} \rangle = \frac{2kT}{\zeta} \langle \mu^2 - 3z \rangle + \frac{2\sqrt{z}}{\zeta} \langle \mu^2 - z \rangle \qquad \text{(E.44)}$$

We note that

$$P_2(u_z) = \frac{1}{2} (3u_z^2 - 1) \qquad \text{(E.45)}$$

so that

$$\langle \dot{P}_2 \rangle = \frac{3}{2\mu^2} \langle \dot{z} \rangle \qquad \text{(E.46)}$$

Therefore Eq. (E.44) becomes

$$\frac{d}{dt} \langle P_2 \rangle = - \frac{6kT}{\zeta} \langle P_2 \rangle + \frac{6}{5} [\langle P_1 \rangle - \langle P_3 \rangle] \frac{\mu^2 F_0}{\zeta} \qquad \text{(E.47)}$$

That is,

$$\frac{d}{dt} \langle P_2 \rangle + \frac{3}{\tau_D} \langle P_2 \rangle = \frac{3}{5} \frac{\xi}{\tau_D} [\langle P_1 \rangle - \langle P_3 \rangle] \qquad \text{(E.48)}$$

The calculation thus shows how the hierarchy of equations is formed.

References

1. C. Kittel, *Phys. Rev.* **70**, 965 (1946).
2. R. Carey and E. D. Isaac, *Magnetic Domains*, The English Universities Press, 1966.
3. J. Frenkel and J. Dorfman, *Nature* **126**, 274 (1930).
4. C. P. Bean and J. D. Livingston, *J. Appl. Phys.* **30**, 120S (1959).
5. L. D. Landau and E. M. Lifshitz, *Phys. Z.*, *Sowjetunion* **8**, 153 (1935), reprinted in Collected Works of Landau, Pergamon Press, London, 1965, No. 18.
6. T. L. Gilbert, *Phys. Rev.* **100**, 1243 (1955) (Abstract only; full report, Armour Research Foundation project No. A059, Supplementary report, May 1, 1956).
7. T. L. Gilbert and J. M. Kelly, *Proc. Conf. on Magnetism and Magnetic Materials*, Pittsburg, USA (AIEE Special Publication T-78) 253 (1955).
8. W. F. Brown Jr., *Phys. Rev.* **130**, 1677 (1963).

9. P. Langevin, *Comptes Rendus* **146**, 530 (1908).

10. M. C. Wang and G. E. Uhlenbeck, *Rev. Mod. Phys.* **17**, 323 (1945).

11. N. Wax, Editor, *Selected Papers on Noise and Stochastic Processes*, Dover, New York, 1954.

12. R. L. Stratonovich, *Conditional Markov Processes and their Application to the Theory of Optimal Control*, Elsevier, New York, 1968.

13. A. Einstein, in R. Fürth, ed., *Investigations on the Theory of the Brownian Movement*, Dover, New York, 1954.

14. P. Debye, *Polar Molecules*, Chemical Catalog (1929); reprinted by Dover Publications, New York.

15. H. A. Kramers, *Physica* **7**, 284 (1940); S. Chandrasekhar, in [36], pp. 63–70.

16. M. A. Martsenyuk, Yu. L. Raĭkher, and M. I. Shliomis, *Sov. Phys. JETP* **38**, 413 (1974).

17. Yu. L. Raĭkher and M. I. Shliomis, *Sov. Phys. JETP* **40**, 526 (1974).

18. M. I. Shliomis, *Sov. Phys. Usp.* **17**, 153 (1974).

19. M. I. Shliomis and Yu. L. Raĭkher, *IEEE Trans. Magn.* **16**, 237 (1980).

20. W. T. Coffey and Yu. P. Kalmykov, *J. Mol. Liq.* **49**, 79 (1991).

21. G. T. Rado and H. Suhl, *Magnetism* Vol. 1, Academic Press, New York, 1963, Chap. 10.

22. T. H. D'Dell, *Ferromagnetodynamics*, Macmillan, London, 1981, Chap. 2.

23. R. Kikuchi, *J. Appl. Phys.* **27** 1352 (1956).

24. H. E. Burke, *Handbook of Magnetic Phenomena*, Van Nostrand Rheinhold, New York, 1986.

25. L. Néel, *Ann. Geophys.* **5**, 99 (1949); *Comptes Rendus* **228**, 664 (1949).

26. B. K. P. Scaife, *J. Phys. D.* **19**, L195 (1986).

27. P. J. Cregg, D. S. F. Crothers, K. P. Quinn, and C. N. Scully, *J. Mol. Liq.* **49**, 169 (1991).

28. D. S. F. Crothers, *Adv. Phys.* **20**, 405 (1971).

29. H. Fröhlich, *Theory of Dielectrics*, 2nd edition, Oxford University Press, 1958.

30. J. L. Doob, Ann. Math. **43**, 351 (1942).

31. H. Risken, *The Fokker–Planck Equation*, Springer-Verlag, Berlin, 1984.

32. B. Gnedenko, *The Theory of Probability*, Mir, Moscow, 1966.

33. W. T. Coffey, *J. Chem. Phys.* **93**, 724 (1990).

34. D. G. Frood and P. Lal, unpublished, 1975.

35. W. T. Coffey and S. G. McGoldrick, *Chem. Phys.* **120**, 1 (1988).

36. S. Chandrasekhar, *Rev. Mod. Phys.* **15**, 1 (1943).

37. H. Bateman, *Partial Differential Equations*, Cambridge University Press, 1932.

38. A. Aharoni, *Phys. Rev.* **177**, 793 (1969).

39. B. K. P. Scaife, *Principles of Dielectrics*, Oxford University Press, Oxford 1989.

40. W. T. Coffey and B. V. Paranjape, *Proc. R. Ir. Acad.* **78A**, 17 (1978).

41. H. Block and E. F. Hayes, *Trans. Faraday Soc.* **66**, 2512 (1970).

42. R. Ullman, *J. Chem. Phys.* **56**, 1869 (1972).

43. P. C. Fannin, B. K. P. Scaife, and S. W. Charles, *J. Phys. D: Appl. Phys.* **21**, 533 (1988).

44. P. C. Fannin and S. W. Charles, *J. Phys. E: Sci. Instrum.* **22**, 412 (1989).

45. P. C. Fannin, B. K. P. Scaife, and S. W. Charles, *J. Magn. Magn. Mater* **85**, 54 (1990).

46. E. A. Milne, *Vectorial Mechanics*, Methuen, London, 1948.

47. H. Goldstein, *Classical Mechanics*, Addison Wesley, Reading, MA, 1950.

48. W. T. Coffey, J. L. Déjardin, Yu. P. Kalmykov, and K. P. Quinn, *Chem. Phys.*, in press (1992).

49. W. T. Coffey, *J. Chem. Phys.* **95**, 2029 (1991).

50. L. D. Landau and E. M. Lifshitz, *Mechanics, A Course of Theoretical Physics*, Vol. 1, 3rd edition, Pergamon, Oxford, 1976.

51. K. Itô, *Proc. Imp. Acad.* **20**, 519 (1944).

52. W. T. Coffey, Yu. P. Kalmykov, and K. P. Quinn, *J. Chem. Phys.*, **96**, 5471 (1992).

53. M. Abramowitz and E. A. Stégun, eds., *Mathematical Handbook*, Dover, New York, 1964.

54. Yu. P. Kalmykov, *Sov. J. Chem. Phys.* **6**, 1099 (1990).

55. Yu. P. Kalmykov, *Phys. Rev. A*, **45**, 7184 (1992).

56. A. Morita, *J. Phys. D.* **11**, 1357 (1978).

57. J. R. McConnell, *Rotational Brownian Motion and Dielectric Theory*, Academic Press, London, 1980.

58. N. G. van Kampen, *Stochastic Processes in Physics and Chemistry*, North-Holland, Amsterdam, 1981.

59. L. Van Hove, *Phys. Rev.* **95**, 249 (1954).

60. M. W. Evans, G. J. Evans, W. T. Coffey, P. Grigolini, *Molecular Dynamics*, Wiley, New York, 1982.

61. M. Morse and H. Feschbach, *Methods of Theoretical Physics*, McGraw-Hill, New York, 1953.

62. R. L. Stratonovich, *Topics in the Theory of Random Noise*, Vol. 1, Gordon and Breach, New York, 1963.

63. M. Mörsch, H. Risken and H. D. Vollmer, *Z. Phys. B* **32**, 245 (1979).

64. E. T. Whittaker and G. N. Watson, *Modern Analysis*, 4th edition, Cambridge University Press, Cambridge, 1927.

65. H. Bateman, *Differential Equations*, Chelsea, New York, 1956.

66. B. K. P. Scaife, ERA Report L/T 392, Electrical Research Association, Leatherhead, Surrey, 1959.

67. B. K. P. Scaife, *Complex Permittivity*, English Universities Press, London, 1971.

68. H. Jeffreys and B. S. Jeffreys, *Mathematical Physics*, Cambridge University Press, Cambridge, 1950.

69. E. C. Titchmarsh, *Introduction to the Theory of Fourier Integrals*, Oxford University Press, Oxford, 1937.

70. H. B. Callen and J. A. Welton, *Phys. Rev.* **83**, 34 (1951).

71. R. F. Green and H. B. Callen, *Phys. Rev.* **83**, 1231 (1951).

72. W. T. Coffey, J. K. Vij, and P. M. Corcoran, *Proc. R. Soc. London Ser. A* **425**, 169 (1989).

73. W. Schröer, private communication, 1982.

74. W. T. Coffey, C. Rybarsch, and W. Schröer, *Chem. Phys. Lett.* **92**, 247 (1982).

75. H. Risken and H. D. Vollmer, *Mol. Phys* **46**, 1073 (1982).

76. A. Budó, *J. Chem. Phys.* **17**, 686 (1949).

77. J. T. Lewis, J. R. McConnell, and B. K. P. Scaife, *Proc. R. Ir. Acad.* **76A**, 43 (1976).

78. M. San Miguel, L. Pesquara, M. A. Rodriquez, and A. Hernández-Machado, *Phys. Rev. A* **35**, 208 (1987).

79. M. I. Stepanov and M. I. Shliomis, *Izv. Acad. Sci.* **55** 1042 (1991).

80. M. I. Shliomis and V. I. Stepanov, private communication, 1991.

81. A. F. Pshenichnikov and A. V. Lebedev, *Sov. Phys. JEPT* **68**, 498 (1989).

82. P. C. Fannin, B. K. P. Scaife, and S. W. Charles, *J. Magn. Magn. Mater.* **72**, 95 (1988).

83. P. C. Fannin and S. W. Charles, *J. Phys. D: Appl. Phys.* **24**, 76 (1991).

84. M. M. Maiorov, *Magnetohydrodynamics* **2**, 21 (1979) (cover to cover translation of *Magnitnaia Hidrodinamika*).

85. P. C. Fannin and S. W. Charles, *J. Phys. D: Appl. Phys.* **22**, 187 (1989).

86. C. N. Scully, P. J. Cregg, and D. S. F. Crothers, *Phys. Rev. B* **45** 474 (1992).

THE ALGEBRA OF EFFECTIVE HAMILTONIANS AND OPERATORS: EXACT OPERATORS

VINCENT HURTUBISE AND KARL F. FREED

*The James Franck Institute and the Department of Chemistry
The University of Chicago, Chicago, Illinois*

CONTENTS

Advances in Chemical Physics, Volume LXXXIII, Edited by I. Prigogine and Stuart A. Rice.
ISBN 0-471-54018-8 © 1993 John Wiley & Sons, Inc.

ABSTRACT

A classification is provided for all transformations of a time-independent Hamiltonian H into an effective Hamiltonian **h** that generates exact eigenvalues of H while acting within a model subspace of the full Hilbert space. This classification is used to determine how the Hermiticity of **h** depends on norm and scalar-product conservation by the transformation operators. Some literature assertions on this topic are disproved by the existence of new transformations found here. Each category of transformation is applied to an arbitrary time-independent operator A to provide all possible definitions of the corresponding effective operator **a** that, while acting only within the model space, provides exact diagonal and off-diagonal matrix elements of A for the states corresponding to the eigenfunctions of **h**. Several important properties of the effective operators and the transformations are studied. No effective operator definition is found to conserve the commutation relation between two arbitrary operators, but many definitions preserve the commutation relations between an arbitrary operator and H and/or that between an arbitrary operator and any constant of the motion. The latter choices of effective operators enable the definition of a complete set of commuting effective observables corresponding to a complete set of commuting observables. These results have a bearing on the conservation of symmetries in effective Hamiltonian calculations, a subject analyzed further in a future paper. Length and dipole velocity transition moments are proven equivalent when computed with some effective operator definitions. Some commutation relations that are used in semi-empirical theories of chemical bonding are examined based on exact effective operator commutation relations. Some formal and computational advantages and drawbacks of the possible effective operator definitions are discussed, and a review is provided of the varied forms that have appeared in the literature. This review includes Hellmann–Feynman theorem based "effective operators" that only provide diagonal matrix elements of A in special cases. Norm-preserving transformations are found to yield a simpler effective operator formalism from both formal and computational viewpoints. Further comparisons are presented in a future paper where finite order approximations are studied.

I.   INTRODUCTION

Three widely used methods to obtain accurate solutions of the time independent Schrödinger equation are large basis set variational calculations, Rayleigh–Schrödinger (RSPT) and Brillouin–Wigner (BWPT) perturbation theories, and "exact" (nonperturbative) reformulations, for example, coupled-cluster (CC) theory. RSPT, its exact reformulations, and BWPT have been applied quite successfully to the calculation of energy levels in many systems for which a reasonable zeroth order approximation is provided by a single reference function. Examples of such applications to ordinary (non-many-body) systems are well known and can be found in textbooks. For many-body systems, the single function is a configurational or determinantal function, for example, the ground state of infinite nuclear matter and of doubly magic nuclei, closed shell atomic states, and many near equilibrium geometry ground states of closed shell molecules.

On the other hand, many systems, such as finite nuclei with particles beyond a doubly magic core, open shell atomic states, transition states on potential surfaces, and excited states of molecular systems, have degenerate or quasi-degenerate zeroth order states which must be described with multireference determinantal wave functions. Similarly, the zeroth order states of many other types of systems also involve a linear superposition of several functions, for example, states encountered in problems of vibration-rotation coupling and of vibrational relaxation. Generalizations of perturbation theory to these cases are known as degenerate perturbation theory (DPT) and quasi-degenerate perturbation theory (QDPT). Some theories [1–5] develop perturbation expansions for the eigenfunctions and eigenvalues of the Hamiltonian H, and others [6–10] give ones for an effective Hamiltonian $\mathbf{h}$ [11]. Only the second type of theory is considered in this paper. Although the dimension d of the space Ω_0 in which $\mathbf{h}$ is defined is finite and usually small, the d eigenvalues of $\mathbf{h}$ are identical to d eigenvalues of H. Effective Hamiltonian calculations thus offer an interesting alternative to variational calculations because they avoid the problem of diagonalizing large matrices and because d energies are obtained simultaneously from a single calculation. Since order by order expansions may not always converge well and sometimes may even diverge, effective Hamiltonian methods have also been formulated using methods that are, in principle, exact, for example, iterative schemes [10, 12–14] and CC theories [10, 15–21]. Both perturbative and exact formulations are considered in this and a subsequent paper [22], hereafter referred to as paper II.

Similar transformations may also be applied to an arbitrary time-independent operator A, producing in the model space Ω_0 an effective operator a which gives exact expectation values and transition moments of A between the eigenfunctions of H corresponding to those of h. Effective operators thus permit the calculation of properties other than the energy for all the states corresponding to the eigenvalues of h. These effective operators provide useful computation tools and a complementary assessment of the quality of the eigenfunctions of h beyond that given by the energies.

Both perturbative and exact effective Hamiltonian formulations usually choose the model space by selecting d eigenfunctions of a zeroth order Hamiltonian H_0, where the choice is often made solely to provide good convergence, for example, in ab initio calculations of atomic energy levels and molecular potential surfaces [23–25]. The zeroth order H_0 can also be chosen to correspond to a well defined physical model; then perturbation expansions, or their exact equivalent, offer a systematic way to deduce corrections to the model. There are also cases where both criteria are (or at least are hoped to be) met simultaneously. For example, England and co-workers develop zeroth order pairing Hamiltonians corresponding to the Bardeen–Cooper–Schrieffer and alternate molecular orbital ansatzes for eventual (field theoretic) RSPT calculation of molecular potential surfaces [26].

Effective Hamiltonians and effective operators are used to provide a theoretical justification and, when necessary, corrections to the semi-empirical Hamiltonians and operators of many fields. In such applications, H_0 may, but does not necessarily, correspond to a well defined model. For example, Freed and co-workers utilize ab initio DPT and QDPT calculations to study some semi-empirical theories of chemical bonding [27–29] and the Slater–Condon parameters of atomic physics [30]. Lindgren and his school employ a special case of DPT to analyze atomic hyperfine interaction model operators [31]. Ellis and Osnes [32] review the extensive body of work on the derivation of the nuclear shell model. Applications to other problems of nuclear physics, to solid state, and to statistical physics are given in reviews by Brandow [33, 34], while those to vibration and rotation-vibration problems, microwave spectroscopy of molecules with small and large amplitude motions, Stark effect, and quadrupole coupling can be found in the work of Jørgensen and co-workers [35, 36] and the references therein. Several workers [24, 35, 37, 38] use effective Hamiltonians to transform the four-component Dirac equation to two-component Pauli-like equations [39], some (but not all) of which are suitable for variational calculations. General pseudopotential theory may also be formulated in terms of effective Hamiltonians where the model space may either be finite or infinite [40]. The present derivations assume finite dimensional model spaces, although some results may extend to infinite spaces.

When effective Hamiltonians are applied to many-body problems, the zeroth order H_0 is usually represented as a sum of one-particle operators that are defined in terms of a set of one-particle functions. These functions are called spin-orbitals in atomic and molecular physics and orbitals in nuclear physics. We use the latter term for simplicity. The orbitals are divided into core, valence or active, and excited orbitals. The model space configurations are characterized by having all core orbitals occupied and all excited orbitals empty. If these model space configurations include all possible ways of distributing the remaining particles into the valence orbitals, the model space is called "complete". Proofs have been given for the existence of a fully linked perturbation expansion for some types of effective Hamiltonians in a complete model space (see also Section V) [8, 41, 42]. Some included model space configurations can, however, cause a wide spread in the eigenvalues of H_0, potentially leading to serious convergence problems. In many cases the number of configurations in a complete model space can be prohibitively large. To alleviate these problems, Hose and Kaldor have suggested the use of an "incomplete" model space defined by retaining only the "important" configurations of the complete model space [43]. Much work with incomplete model spaces is progressing, and there are now several different schemes for selecting the configurations retained. Many of these alternatives are computationally convenient [44]. So far, there have been no incomplete model space applications to the calculation of properties other than the energy, and when this is contemplated, results derived here and in subsequent papers will be useful in choosing between various different possible formalisms.

Incomplete model spaces generate disconnected diagrams not only for the h used by Hose and Kaldor, but for other perturbative and CC formulations [20, 45, 46]. The effects of such diagrams on the size extensivity of the calculated energies depend both on the particular h and on the kind of model space used [21, 47–52]. Most of the recent work on incomplete model spaces uses Fock space effective Hamiltonian formalisms similar to those introduced earlier by Kutzelnigg and co-workers for complete model spaces [53–55]. Such formulations determine **h** by projection of a Fock space effective Hamiltonian.

Given a choice of the model space, there are many ways of transforming H into an effective Hamiltonian [37]. The different possible transformations generate as many equally valid effective Hamiltonians, all having the same set of d exact eigenvalues. Section II shows that if these transformations are employed to convert an arbitrary time-independent operator A into an effective operator **a**, an even greater array of possible definitions results. Some effective operator definitions are shown here to have substantially more convenient general properties. Only brief considerations are made of computational features as these are best addressed in conjunction with approximation methods, as will be done in paper II. One main goal of this paper is to exploit the freedom in defining **h** and **a** to determine the kinds of transformations producing effective operators that are computationally convenient and that satisfy some important physical constraints as we now discuss. In this context, we note that recent ab initio computations [56–58] of effective dipole operators have chosen a representation based upon the general considerations derived here.

A fundamental aspect of semi-empirical chemical bonding theories is their requirement that the model operators be state independent [56]. This property is, of course, not required of effective operators if only the numerical values are desired for the matrix elements of operators. Indeed, some semi-empirical theories, used in other areas of physics, do not impose the requirement of state independence. For instance, LS-dependent parameters are employed in describing the hyperfine coupling of two-electron atoms [31]. However, whenever effective operators themselves are the quantities of interest, as when studying semi-empirical theories of chemical bonding, state independence of effective operators becomes a necessity. This paper thus examines conditions leading to the generation of state-independent effective operators.

Previous work has not investigated if commutation relations are conserved upon transformation to effective operators. Many important consequences emerge from particular commutation relations, for example, the equivalence between the dipole length and dipole velocity forms for transition moments follows from the commutation relation between the position and Hamiltonian operators. Hence, it is of interest to determine if these consequences also apply to effective operators. In particular, commutation relations involving constants of the motion are of central importance since these operators are associated with fundamental symmetries of the system. Effective operator definitions are especially useful if the effective Hamiltonian **h** commutes with the effective constants of the motion because this condition implies that the eigenfunctions of **h** retain, or, if degenerate, can be chosen to retain, these fundamental symmetries. In turn, choosing the degenerate eigenfunctions of **h** this way ensures that the corresponding eigenfunctions of the full H also have the desired symmetries since corresponding matrix elements of constants of the motion and of their effective counterparts are equal to each other. The eigenfunctions of a few types of **h** have been shown to possess the desired symmetries when H_0 commutes with the constants of the motion [32, 35, 36, 41, 59]. In such cases, however, the corresponding eigenvectors of H, if degenerate, must also be proven to have these symmetries, a step sometimes neglected [35, 36]. Moreover, as discussed in paper II, it is not always possible or convenient to select H_0 this way, in which case the preservation of symmetries in **h** calculations depends on the conservation of commutation relations involving constants of the motion. Thus, we study here the conditions on the transformation to effective Hamiltonians and operators that preserve commutation relations and basic symmetries of the system for both state dependent and independent effective operators.

Some semi-empirical theories of chemical bonding assume without proof that the commutation relations between particular model operators are the same as the ones between the corresponding original operators [60–62]. Criteria for the validity of this assumption are investigated here for the first time based on our results for conservation of commutation relations by effective operators.

The above mentioned size consistency considerations in the definition of and calculations with effective Hamiltonians in incomplete model spaces also apply to effective operators but are not addressed here. The topics we treat are studied in complete generality. Our results apply equally well to all types of (finite) model spaces, complete or incomplete, degenerate or quasi-degenerate. (The derivation of a diagrammatic representation of the various choices of **a** is an interesting problem that should be studied in the future.) Our analyses utilize mapping operators [37] that transform a given Hilbert space Hamiltonian H into an effective Hamiltonian **h**. These transformations are not directly used in all literature effective Hamiltonian derivations, but they are implicit in the final results. As we show here, the properties of the mapping operators determine the correct form of **a**, how degenerate eigenfunctions of **h** must be chosen in order to correspond to orthogonal eigenfunctions of H, and whether commutation relations and symmetries are conserved. Thus, the mapping operators provide a convenient method for characterizing the properties of effective Hamiltonians and operators.

Section II reviews the general theory of effective Hamiltonians and effective operators. Although many possible mapping operators can, in principle, be used to transform any arbitrary time-independent operator A into an effective one **a**, we demonstrate that most of these transformations produce effective operators that have formal and computational drawbacks. This is shown to stem from the lack of scalar product conservation by many mappings. Scalar product preservation and its relation to the Hermiticity of **h** are explored in Section III, where we prove the existence of new types of transformations and disprove literature claims [63, 64] that a Hermitian **h** can only be generated by transformations that conserve the scalar product. Section IV studies commutation relations between effective operators and derives conditions on the mapping operators which conserve the commutation relations for particular operators including constants of motion. A review, commentary and classification is given in Sections V and VI of **h** and **a** definitions, respectively, that have appeared in the literature. The review in Section V demonstrates that most of the mappings of Section II are considered in previous work and that our analysis of the various possible **a** definitions will prove useful in choosing among the many possibilities. Section VI also reviews Hellmann–Feynman theorem based "effective operators", which provide only diagonal matrix elements of A when A does not lift the degeneracy of the states of interest. Section VII applies the results of Section IV to the conservation of symmetries in **h** calculations and to the study of the validity of some commutation relations between model operators of semi-empirical chemical bonding theories. Paper II considers other formal and computational aspects of the **a** definitions presented in this paper, in particular those associated with calculations of **h** and **a** to finite orders of perturbation theory.

II. EFFECTIVE HAMILTONIANS AND EFFECTIVE OPERATORS

A. Effective Hamiltonians and Mapping Operators

The eigenfunctions $|\phi_\alpha\rangle$ of H are called *true* eigenfunctions to distinguish them from those of an effective Hamiltonian **h**. The latter is defined in a finite subspace, called the model space, of the full Hilbert space. Its eigenfunctions are called *model* eigenfunctions and are denoted by $|\phi_\alpha\rangle_0$. Other full Hilbert space operators A and their model space counterparts **a** are called, respectively, true and effective operators. Let Ω_0 be the d-dimensional model space, and let P_0 and $Q_0 = \mathbf{1} - P_0$ be the projection operators onto Ω_0 and its orthogonal complement $\Omega_0^\perp$, respectively. An effective Hamiltonian **h** is defined by requiring that its eigenvalues be identical to d of the eigenvalues of H, that is,

$$\mathbf{h}|\phi_\alpha\rangle_0 = E_\alpha|\phi_\alpha\rangle_0 , \quad \alpha \in \{d\} \tag{2.1}$$

where E_α is the exact energy of the αth eigenfunction $|\phi_\alpha\rangle$ of H and where the model eigenfunction index is that of the corresponding true eigenfunction. As discussed further below, the definition (2.1) does not fully determine $\mathbf{h}$, and, in fact, for a given model space there exists an infinity of effective Hamiltonians having the same eigenvalues but different eigenvectors.

Define k as the (linear) operator, called the wave operator, transforming the model eigenfunctions into the corresponding true ones,

$$|\phi_\alpha\rangle = k|\phi_\alpha\rangle_0 , \quad \alpha \in \{d\} . \tag{2.2}$$

Designate the space defined by the d true eigenfunctions as Ω, and let P and $Q = 1 - P$ be the projection operators onto Ω and its orthogonal complement $\Omega^\perp$, respectively. Note that k operates only on functions in Ω_0 and transforms them to functions only in Ω. Thus, we have the conditions on k that

$$kP_0 = k = Pk \tag{2.3}$$

$$kQ_0 = 0 = Qk \tag{2.4}$$

As noted by Jørgensen [37], since $k : \Omega_0 \rightarrow \Omega$ maps a basis set in Ω_0 into one in Ω, there exists a linear operator $l : \Omega \rightarrow \Omega_0$ effecting the inverse mapping,

$$|\phi_\alpha\rangle_0 = l|\phi_\alpha\rangle , \quad \alpha \in \{d\} \tag{2.5}$$

The operator l acts only on functions in Ω and transforms them to functions only in Ω_0. This introduces similar conditions on l as in (2.3),

$$lP = l = P_0 l \tag{2.6}$$

$$lQ = 0 = Q_0 l \tag{2.7}$$

Left-multiplying the full Hilbert space Schrödinger equation,

$$H|\phi_\alpha\rangle = E_\alpha|\phi_\alpha\rangle \tag{2.8}$$

by l and using (2.2) and (2.5) yields

$$lHk|\phi_\alpha\rangle_0 = E_\alpha|\phi_\alpha\rangle_0 , \quad \alpha \in \{d\} \tag{2.9}$$

Comparing with Eq. (2.1) implies that Eq. (2.9) provides one expression for the effective Hamiltonian $\mathbf{h}$ as

$$\mathbf{h} = lHk = lPHPk \qquad (2.10)$$

The last equality is obtained from (2.3) and (2.6) and emphasizes that $\mathbf{h}$ gives eigenvalues of H only for the states in Ω. If $k^\dagger$ is used instead of l in the derivation of (2.9), the generalized eigenvalue equation

$$k^\dagger Hk|\phi_\alpha\rangle_0 = E_\alpha k^\dagger k|\phi_\alpha\rangle_0, \quad \alpha \in \{d\} \qquad (2.11)$$

is produced in which the operator $k^\dagger k = P_0 k^\dagger k P_0$ is called the overlap [65] or metric [33] associated with the Hermitian effective Hamiltonian $k^\dagger Hk$. Section III shows that this metric is unity (in Ω_0) iff $k^\dagger = l$, in which case (2.9) and (2.11) are identical (see Eq. (2.12)). If $k^\dagger \neq l$, then the use of Eq. (2.11) involves more computational labor than (2.9) because of the additional calculation of the metric. In fact, Eq. (2.11) has only been used in some formal work [6] but, to our knowledge, not in any numerical applications.

The space Ω_0 is usually chosen by selecting d eigenfunctions of a zeroth order Hamiltonian H_0, and k, l and $\mathbf{h}$ are obtained using perturbation theory or one of its formally exact reformulations, for example, an iterative scheme. In general, only these operators are perturbatively expanded and not the model or the true eigenfunctions. By multiplying (2.2) on the left by l and (2.5) on the left by k, it follows that l and k are related to one another by [37]

$$lk = P_0 \qquad (2.12)$$

$$kl = P \qquad (2.13)$$

Equations (2.12) and (2.13) show that $l(k)$ is the left *and* the right inverse mapping of $k(l)$. Hence, these relations generalize the equivalence of left and right inverses of finite space operators to mapping operators between two finite spaces. Not surprisingly, it may be shown that Eqs. (2.12) and (2.13) are indeed equivalent to each other [66]. There are an infinite number of operator pairs (k, l) that satisfy the relations (2.12) and (2.13) [67]. Any such operator pair (k, l) can be used to generate an effective Hamiltonian $\mathbf{h}$. The many different possible transformations produce as many, equally valid, effective Hamiltonians, although some of them might be more convenient for particular calculations. The remainder of this section classifies the possible categories of operator pairs (k, l), their properties, and the effective operators generated by these transformations.

B. Effective Operators and Classification of Mapping Operators

Consider a time-independent operator A whose matrix elements, $A_{\alpha\beta}$, α, $\beta \in \{d\}$ (both expectation values and transition moments), in the space Ω we wish to compute. This goal is to be achieved by transforming the calculation from Ω into one in Ω_0, resulting in an effective operator **a** whose matrix elements, taken between appropriate model eigenfunctions of an effective Hamiltonian **h**, are the desired $A_{\alpha\beta}$. As we now discuss, numerous possible definitions of **a** arise depending on the type of mapping operators that are used to produce **h** and on the choice of model eigenfunctions.

In accord with previous works, the overlaps, between degenerate true eigenfunctions are assumed to be null. This is, of course, useful for computing the $A_{\alpha\beta}$, but it often requires additional calculations when mappings do not conserve overlaps, as discussed in Sections II.D and II.E. Henceforth, *all* greek indices lie in the set $\{d\}$, unless otherwise indicated.

Without loss of generality, the right model eigenfunctions of **h** are taken to be unity normalized. If k is norm preserving, then the corresponding true eigenfunctions are also unity normalized. Thus, matrix elements of A are then given by

$$A_{\alpha\beta} = \langle \phi_\alpha | A | \phi_\beta \rangle \qquad (2.14)$$

where

$$\langle \phi_\alpha | \phi_\alpha \rangle = \delta_{\alpha\beta} \qquad (2.15)$$

If, on the other hand, k does not conserve norms, then, in general, matrix elements of A must be computed from the relation

$$A_{\alpha\beta} = \langle \phi'_\alpha | A | \phi'_\beta \rangle / n_\alpha n_\beta \qquad (2.16)$$

where the n are the norms of the true eigenvectors,

$$n_\alpha \equiv \langle \phi'_\alpha | \phi'_\alpha \rangle^{1/2} \qquad (2.17)$$

and where primes designate unnormalized true eigenfunctions corresponding to unity normed model eigenfunctions by a non-norm-preserving mapping k. Various **a** formulations differ in how they treat the norms

n_α, so a consideration of how changing normalization converts between **a** classes is helpful in understanding relations between slightly differing possible formulations.

Equations (2.14) and (2.16) show that the form of $A_{\alpha\beta}$ to be transformed into a matrix element in Ω_0 depends on whether or not k is norm-preserving. It also depends on how the n_α are treated because expression (2.14) applies if the norms are incorporated into either new model eigenfunctions, normalized so their true counterparts are unity normed, or into new mappings that relate unity normed model eigenfunctions to true eigenfunctions also unity normalized [68]. Furthermore, when **h** is non-Hermitian, as is Bloch's effective Hamiltonian [6–9] discussed in Section V, different transformed expressions emerge if the effective operator is defined to act between right, left, or both right and left model eigenvectors. Hence, there exists a multiplicity of possible effective operator definitions whose form is determined to a large extent by the Hermiticity of **h** and the norm conservation property of the mapping operators.

Section III demonstrates that k conserves norms iff

$$k^\dagger = l \tag{2.18}$$

which in turn occurs iff k conserves the scalar product. Condition (2.18) is also shown as equivalent to norm preservation and to scalar product conservation by l. Mappings satisfying the condition (2.18) produce a Hermitian effective Hamiltonian, as follows by substituting (2.18) into (2.10). An example of such mappings is provided by the des Cloizeaux (also called "canonical") formalism [7, 37, 65] which is discussed in Section V. Despite contrary claims [63, 64], Eq. (2.18) is *not* a necessary condition for the Hermiticity of **h**: Section III demonstrates the existence of mapping operators that violate (2.18) but that generate Hermitian effective Hamiltonians.

We are thus led to classify mapping operators into the following three general categories:

1. Norm-preserving mappings; these necessarily generate a Hermitian effective Hamiltonian;
2. Non-norm-preserving mappings that produce a non-Hermitian effective Hamiltonian;
3. Non-norm-preserving mappings that generate a Hermitian effective Hamiltonian.

The next three subsections introduce effective operator definitions for each category and study some of their properties. Effective operators from non-norm-preserving mappings are shown to have some formal and computational drawbacks relative to those generated by norm-preserving mappings.

C. Effective Operators Generated by Norm-Preserving Mappings

Norm-preserving mappings are denoted by $(\hat{K}, \hat{K}^{\dagger})$ and, as discussed in Section II.B, generate a Hermitian effective Hamiltonian $\hat{K}^{\dagger}H\hat{K} \equiv \hat{\mathbf{H}}$. The orthonormalized model eigenfunctions of $\hat{\mathbf{H}}$ are written as $|\hat{\alpha}\rangle_0$ and the corresponding true eigenfunctions are designated by $|\hat{\Psi}_{\alpha}\rangle$. Thus, Eq. (2.2) specializes to

$$|\hat{\Psi}_{\alpha}\rangle = \hat{K}|\hat{\alpha}\rangle_0 \qquad (2.19)$$

Since $(\hat{K}, \hat{K}^{\dagger})$ conserves scalar products, orthonormalizing the model eigenvectors ensures that the corresponding true eigenvectors also form an orthonormal set,

$$\langle \hat{\Psi}_{\alpha}|\hat{\Psi}_{\beta}\rangle = {}_0\langle \hat{\alpha}|\hat{\beta}\rangle_0 = \delta_{\alpha,\beta} \qquad (2.20)$$

Formal expressions for P_0, $\hat{K}$, and $\hat{\mathbf{H}}$ are thus

$$P_0 = \sum_{\beta} |\hat{\beta}\rangle_0 \, {}_0\langle \hat{\beta}| \qquad (2.21)$$

$$\hat{K} = \sum_{\beta} |\hat{\Psi}_{\beta}\rangle \, {}_0\langle \hat{\beta}| \qquad (2.22)$$

$$\hat{\mathbf{H}} = \sum_{\beta} |\hat{\beta}\rangle_0 \, E_{\beta} \, {}_0\langle \hat{B}| \qquad (2.23)$$

Replacing $|\phi_{\alpha}\rangle$ by $|\hat{\Psi}_{\alpha}\rangle$ into (2.14) and substituting (2.19) gives

$$A_{\alpha\beta} = {}_0\langle \hat{\alpha}|\hat{K}^{\dagger}A\hat{K}|\hat{\beta}\rangle_0 \qquad (2.24)$$

The operator $\hat{K}^{\dagger}A\hat{K} \equiv \hat{\mathbf{A}}$ in (2.24) does not depend on the indices α and β. Thus, this case produces the state-independent effective operator definition $\hat{\mathbf{A}}$ of Table I. Equation (2.3) implies that $\hat{\mathbf{A}}$ may be rewritten as $\hat{K}^{\dagger}PAP\hat{K}$, a form which emphasizes that $\hat{\mathbf{A}}$ gives matrix elements of A only for the states in Ω (a property of all effective operator definitions).

D. Effective Operators Generated by Non-Norm-Preserving Mappings which Produce a Non-Hermitian Effective Hamiltonian

As Section II.B explains, expression (2.14) or (2.16) for matrix elements $A_{\alpha\beta}$ applies here depending on which model eigenfunctions are selected

TABLE I

Effective Hamiltonian and Effective Operator Definitions Corresponding to a Time-Independent Operator A

Mappings	Norms Conserved[a]	Effective Hamiltonian — Hermitian	Effective Hamiltonian — Eigenfunctions	Effective operator definition	Bra	Ket	Effective Hamiltonian is the $A = H$ Case	State-Independent
$(\hat{K}, \hat{K}^\dagger)$	Yes	Yes — $\hat{H} \equiv \hat{K}^\dagger H\hat{K}$	$|\hat{\alpha}\rangle_0{}^b$	$\hat{A} = \hat{K}^\dagger A\hat{K}$	$_0\langle\hat{\alpha}'|$	$|\hat{\beta}\rangle_0$	Yes	Yes
$(K, L)^c$	No	No — $H = LHK$	$|\alpha'\rangle_0{}^d,\;\;_0\langle\alpha'|{}^e$; $|\alpha\rangle_0 = N_\alpha^{-1}|\alpha'\rangle_0$; $_0\langle\bar{\alpha}|{}^{f,g} = N_\alpha\,_0\langle\bar{\alpha}'|$	$A_{\alpha\beta}^{I} = K^\dagger AK/N_\alpha N_\beta$	$_0\langle\alpha'|$	$|\beta'\rangle_0$	No	No
				$A_{\alpha\beta}^{II} = LAKN_\alpha N_\beta^{-1}$	$_0\langle\bar{\alpha}|$	$|\beta'\rangle_0$	Yes	No
				$A_{\alpha\beta}^{III} = K^\dagger AL^\dagger N_\alpha^{-1} N_\beta$	$_0\langle\alpha'|$	$|\bar{\beta}'\rangle_0$	No	No
				$A^{I} = K^\dagger AK$	$_0\langle\alpha|$	$|\beta\rangle_0$	No	Yes
				$A^{II} = LAK$	$_0\langle\bar{\alpha}|$	$|\beta\rangle_0$	Yes	Yes
				$A^{III} = K^\dagger AL^\dagger$	$_0\langle\alpha|$	$|\bar{\beta}\rangle_0$	No	Yes
				$\bar{A}^{I} = \bar{K}^\dagger A\bar{K}$	$_0\langle\alpha'|$	$|\beta'\rangle_0$	No	Yes
				$\bar{A}^{II} = \bar{L}A\bar{K}$	$_0\langle\bar{\alpha}'|$	$|\beta'\rangle_0$	Yes	Yes
				$\bar{A}^{III} = \bar{K}^\dagger A\bar{L}^\dagger$	$_0\langle\alpha'|$	$|\bar{\beta}'\rangle_0$	No	Yes
$(\tilde{K}, \tilde{L})^h$	No	Yes — $\tilde{H} \equiv \tilde{L}H\tilde{K}$	$|\tilde{\alpha}'\rangle_0{}^{b,i}$; $|\tilde{\alpha}\rangle_0{}^f = \tilde{N}_\alpha^{-1}|\tilde{\alpha}'\rangle_0$; $_0\langle\tilde{\bar{\alpha}}| = \tilde{N}_\alpha\,_0\langle\tilde{\bar{\alpha}}'|$	$\tilde{A}_{\alpha\beta}^{I} = \tilde{K}^\dagger A\tilde{K}/\tilde{N}_\alpha \tilde{N}_\beta$	$_0\langle\tilde{\alpha}'|$	$|\tilde{\beta}'\rangle_0$	No	No
				$\tilde{A}_{\alpha\beta}^{II} = \tilde{L}A\tilde{K}\tilde{N}_\alpha \tilde{N}_\beta^{-1}$	$_0\langle\tilde{\alpha}'|$	$|\tilde{\beta}'\rangle_0$	Yes	No
				$\tilde{A}_{\alpha\beta}^{III} = \tilde{K}^\dagger A\tilde{L}^\dagger \tilde{N}_\alpha^{-1} \tilde{N}_\beta$	$_0\langle\tilde{\alpha}'|$	$|\tilde{\beta}'\rangle_0$	Yes	No
				$\tilde{A}_{\alpha\beta}^{IV} = \tilde{L}A\tilde{L}^\dagger \tilde{N}_\alpha \tilde{N}_\beta$	$_0\langle\tilde{\alpha}'|$	$|\tilde{\beta}'\rangle_0$	No	No
				$\tilde{A}^{I} = \tilde{K}^\dagger A\tilde{K}$	$_0\langle\tilde{\alpha}|$	$|\tilde{\beta}\rangle_0$	No	Yes
				$\tilde{A}^{II} = \tilde{L}A\tilde{K}$	$_0\langle\tilde{\bar{\alpha}}|$	$|\tilde{\beta}\rangle_0$	Yes	Yes
				$\tilde{A}^{III} = \tilde{K}^\dagger A\tilde{L}^\dagger$	$_0\langle\tilde{\alpha}|$	$|\tilde{\bar{\beta}}\rangle_0$	Yes	Yes
				$\tilde{A}^{IV} = \tilde{L}A\tilde{L}^\dagger$	$_0\langle\tilde{\bar{\alpha}}|$	$|\tilde{\bar{\beta}}\rangle_0$	No	Yes

[a] Equivalent to scalar product conservation.

[b] Taken to be orthonormalized.

[c] Transforming these mappings so as to absorb the normalization factors of Eq. (2.16) gives the mappings $(\bar{K}, \bar{L})$; see text.

[d] Normed to unity; N_α is the norm of the corresponding true eigenvector.

[e] The $_0\langle\bar{\alpha}'|$ are biorthonormalized to the $|\alpha'\rangle_0$; see text.

[f] Corresponds to a unity normed true eigenfunction.

[g] The $_0\langle\bar{\alpha}|$ are biorthonormalized to the $|\alpha\rangle_0$; see text.

[h] Transforming these mappings so as to absorb the normalization factors of Eq. (2.16) gives mappings of the category $(\tilde{K}, \tilde{K}^\dagger)$. Other non-norm-preserving mappings (see Table II) can produce a Hermitian effective Hamiltonian and require different effective operator definitions; see text.

[i] $\tilde{N}$ is the norm of the corresponding true eigenvector.

and on how the normalization factors are treated. Analysis of the various resulting effective operator definitions is facilitated by first introducing two sets of model bras and kets in terms of which the mapping operators are expressed.

Section III demonstrates how mapping operators that generate a non-Hermitian **h** can be classified into the four subcategories 2a–2d of Table II based on the conservation of the angles between (arbitrary) degenerate eigenfunctions and that of the norms of a set of eigenfunctions. For the purpose of defining effective operators, however, it turns out to be unnecessary to distinguish mappings based on the former conservation (see below). Hence, mappings of categories 2a and 2b are denoted by (K, L) and those of categories 2c and 2d by $(\bar{K}, \bar{L})$. The mappings $(\bar{K}, \bar{L})$ are proven below to result from transforming mappings (K, L) so as to absorb the normalization factors of Eq. (2.16).

TABLE II

Classification of all Mapping Operators

Conserve				
Overlaps Between Eigenvectors		Norms of		
Nondegenerate[b]	Degenerate	Eigenfunctions	Category[a]	Symbol
Yes	Yes	Yes	1[c]	$(\hat{K}, \hat{K}^\dagger)$
	Yes	No[d]	3a	$(\tilde{K}, \tilde{L})$
	No	No	3b	$(\tilde{K}, \tilde{L})^e$
	No	Yes	3c	$(\tilde{k}, \tilde{l})$ $(\tilde{k}, \tilde{l})^f$
No	Yes	No[d]	2a	(K, L)
	No	No	2b	
	Yes	Yes	2c[h]	$(\bar{K}, \bar{L})^g$
	No	Yes	2d[i]	

[a]Category numbers are those introduced in Section II.B.

[b]Equivalent to the Hermiticity of the effective Hamiltonian, see text.

[c]Only mappings of this category conserve norms or, equivalently, scalar products; see Appendix B.

[d]Conserve the norms of the degenerate eigenvectors.

[e]Conserve the overlaps between orthogonal degenerate eigenfunctions.

[f]Result from transforming $(\tilde{k}, \tilde{l})$ so as to absorb the normalization factors of Eq. (2.16) and generate the same effective Hamiltonian as $(\tilde{k}, \tilde{l})$; see Appendix B.2.

[g]Result from transforming (K, L) so as to absorb the normalization factors of Eq. (2.16) and generate the same effective Hamiltonian as (K, L); see Section II.D.3.

[h]Result from transforming category 2a mappings and those mappings of category 2b that conserve the angles between orthogonal degenerate eigenvectors.

[i]Result from transforming the other category 2b mappings.

1. *Eigenvectors, Norms and Mappings* (K, L)

The unity normalized right eigenvectors of the non-Hermitian effective Hamiltonian $LHK \equiv \mathbf{H}$ are denoted by $|\alpha'\rangle_0$. Because condition (2.18) is not fulfilled, the corresponding true eigenvectors $|\Psi'_\alpha\rangle$ are not unity normed. As is customary [6–9], the $|\Psi'_\alpha\rangle$ are taken to be mutually orthogonal, a generally desirable property. Following Bloch [6], $|\alpha\rangle_0$ designates the unnormalized right model eigenvectors corresponding to orthonormal true eigenvectors $|\Psi_\alpha\rangle$. Primed and unprimed right model and true eigenfunctions differ only in their norm and yield two specializations of Eqs. (2.2) and (2.5):

$$|\Psi'_\alpha\rangle = K|\alpha'\rangle_0 \qquad (2.25)$$

$$|\alpha'\rangle_0 = L|\Psi'_\alpha\rangle \qquad (2.26)$$

and

$$|\Psi_\alpha\rangle = K|\alpha\rangle_0 \qquad (2.27)$$

$$|\alpha\rangle_0 = L|\Psi_\alpha\rangle \qquad (2.28)$$

With category 2a mappings, orthogonalization of the degenerate model eigenfunctions ensures that of their true counterparts. However, if (K, L) belong to category 2b, which as Section III shows is generally the case, other, more complicated, procedures must [69] be used. These procedures are nonetheless preferable to orthogonalizing the model eigenvectors [70]. Hence, the true eigenvectors are taken to be orthogonal with mappings of both category 2a and 2b, which, consequently, need not be distinguished from one another.

The left and right eigenvectors of $\mathbf{H}$ form a bi-orthogonal set, that is, any left (right) eigenvector with eigenvalue E_α is orthogonal to all right (left) eigenvectors with eigenvalue $E_\beta \neq E_\alpha$. It may be shown [71] that a degenerate left (right) eigenvector can always be chosen to be orthogonal to all but one of the right (left) eigenvectors with the same eigenvalue. Setting this unique nonzero overlap to unity produces the bi-orthonormality conditions

$$_0\langle \bar{\alpha} \mid \beta \rangle_0 = {}_0\langle \alpha \mid \bar{\beta} \rangle_0 = \delta_{\alpha,\beta} \qquad (2.29)$$

$$_0\langle \bar{\alpha}' \mid \beta' \rangle_0 = {}_0\langle \alpha' \mid \bar{\beta}' \rangle_0 = \delta_{\alpha,b} \qquad (2.30)$$

where bars denote left eigenvectors. Thus, given one set of eigenvectors, its bi-orthonormal complement is unique and can be determined without another effective Hamiltonian diagonalization [72]. Equation (2.30) implies that P_0 may be written as the alternative forms

$$P_0 = \sum_\beta |\beta'\rangle_0 \, _0\langle \bar{\beta}'| = \sum_\beta |\bar{\beta}'\rangle_0 \, _0\langle \beta'| \tag{2.31}$$

Primed and unprimed true eigenfunctions are related by

$$|\Psi'_\alpha\rangle = N_\alpha |\Psi_\alpha\rangle \tag{2.32}$$

where N_α is the norm of $|\Psi'_\alpha\rangle$. Left-multiplying (2.32) by L and using (2.26) and (2.28) gives

$$|\alpha'\rangle_0 = N_\alpha |\alpha\rangle_0 \tag{2.33}$$

which upon substitution into (2.30) and use of (2.29) yields

$$_0\langle \bar{\alpha}'| = N_\alpha^{-1} \, _0\langle \bar{\alpha}| \tag{2.34}$$

Formal expressions for K and L in terms of these functions are

$$K = \sum_\beta |\Psi_\beta\rangle \, _0\langle \bar{\beta}| = \sum_\beta |\Psi'_\beta\rangle \, _0\langle \bar{\beta}'| \tag{2.35}$$

$$L = \sum_\beta |\beta\rangle_0 \, \langle \Psi_\beta| = \sum_\beta |\beta'\rangle_0 \, N_\beta^{-2}\langle \Psi'_\beta| \tag{2.36}$$

Substituting (2.35) and (2.36) into the definition of **H** produces

$$\mathbf{H} = \sum_\beta |\beta'\rangle_0 \, E_\beta \, _0\langle \bar{\beta}'| = \sum_\beta |\beta\rangle_0 \, E_\beta \, _0\langle \bar{\beta}| \tag{2.37}$$

The left model and true eigenvectors are related as follows. Using (2.36) in conjunction with (2.29) and (2.30) gives, respectively,

$$_0\langle \bar{\alpha}|L = \langle \Psi_\alpha| \tag{2.38}$$

and

$$_0\langle \bar{\alpha}'|L = N_\alpha^{-2}\langle \Psi'_\alpha| \tag{2.39}$$

Likewise, from (2.35), (2.15) and (2.17) we obtain

$$\langle \Psi_\alpha|K = \, _0\langle \bar{\alpha}| \tag{2.40}$$

and

$$\langle \Psi_\alpha' | K = N_\alpha^2 \,_0\langle \bar{\alpha}' | \tag{2.41}$$

Equations (2.38)–(2.41) imply that $L^\dagger$ is the wave operator for $|\bar{\alpha}\rangle_0$ and $|\bar{\alpha}'\rangle_0$ and that $K^\dagger$ provides the inverse mapping. This is fully consistent with (2.2)–(2.10) since these vectors are right eigenvectors of $\mathbf{H}^\dagger = K^\dagger H L^\dagger$. Notice however that while $_0\langle \bar{\alpha} |$ and $\langle \Psi_\alpha |$ are simply related via K and L, this is not true of $_0\langle \bar{\alpha}' |$ and $\langle \Psi_\alpha' |$. Thus, both $|\alpha\rangle_0$ and $_0\langle \bar{\alpha} |$ are mapped into unity normed true eigenvectors, but neither are $|\alpha'\rangle_0$ or $_0\langle \bar{\alpha}' |$.

2. Effective Operators with Mappings (K, L)

We now turn to the effective operator definitions generated by the non-norm-preserving mappings (K, L). As mentioned in Section II.B, expression (2.16), in which $|\phi_\alpha'\rangle$ is replaced by $|\Psi_\alpha'\rangle$, applies if the $|\alpha'\rangle_0$ are used, while expression (2.14), with $|\phi_\alpha\rangle$ replaced by $|\Psi_\alpha\rangle$, applies if the N_α are incorporated into new model eigenfunctions. We first discuss the effective operator expressions that may be obtained from (2.16).

1. Using right eigenvectors only, the substitution of (2.25) into (2.16) and (2.17) gives

$$A_{\alpha\beta} = \,_0\langle \alpha' | K^\dagger A K | \beta' \rangle_0 / N_\alpha N_\beta \tag{2.42}$$

   with

$$N_\alpha = \,_0\langle \alpha' | K^\dagger K | \alpha' \rangle_0^{1/2} \tag{2.43}$$

2. Using both right and left eigenvectors the insertion of (2.25) and (2.39) into (2.16) yields

$$A_{\alpha\beta} = \,_0\langle \bar{\alpha}' | L A K | \beta' \rangle_0 \, N_\alpha N_\beta^{-1} \tag{2.44}$$

   and

$$A_{\alpha\beta} = \,_0\langle \alpha' | K^\dagger A L^\dagger | \bar{\beta}' \rangle_0 \, N_\alpha^{-1} N_\beta \tag{2.45}$$

However, substituting (2.25) and (2.39) into (2.17) produces only the identity $N_\alpha = N_\alpha$. Thus, the N_α must be computed with either only right (Eq. (2.43)) or only left eigenvectors (Eq. (2.47)).

3. Using left eigenvectors only the substitution of (2.39) into (2.16) and (2.17) gives

$$A_{\alpha\beta} = {}_0\langle \bar{\alpha}' | LAL^\dagger | \bar{\beta}' \rangle_0 \, N_\alpha N_\beta \qquad (2.46)$$

with

$$N_\alpha = {}_0\langle \bar{\alpha}' | LL^\dagger | \bar{\alpha}' \rangle_0^{-1/2} \qquad (2.47)$$

The ${}_0\langle \bar{\alpha}' |$ are not unity normed but are normalized according to (2.30) which implies that the $|\alpha'\rangle_0$ must first be evaluated. Hence, Eq. (2.46) provides a somewhat cumbersome expression for $A_{\alpha\beta}$ in terms of left eigenvectors only. An expression with unity normed left eigenvectors is simpler. Since these vectors are the unity normed right eigenvectors of $\mathbf{H}^\dagger$, the latter expression reduces to case 1 (with mappings $(L^\dagger, K^\dagger)$; see remark below Eq. (2.41)). Thus, only expressions (2.42), (2.44) and (2.45) need to be considered. These produce the effective operator definitions $\mathbf{A}^I_{\alpha\beta}$, $\mathbf{A}^{II}_{\alpha\beta}$ and $\mathbf{A}^{III}_{\alpha\beta}$ of Table I, each of which is state dependent, that is, depends on the particular α and β through the presence of N_α and N_β.

We now turn to the effective operator definitions produced by (2.14) with model eigenfunctions that incorporate the normalization factors of (2.16) so their true counterparts are unity normed. Equations (2.27) and (2.38) show these model eigenfunctions to be the $|\alpha\rangle_0$ and ${}_0\langle\bar{\alpha}|$ that are defined in (2.33) and (2.34). Substituting Eqs. (2.27) and (2.38) into (2.14) and proceeding as in the derivation of the forms $\mathbf{A}^i_{\alpha\beta}$, $i = I$–III, yields the state-independent definitions $\mathbf{A}^I$, $\mathbf{A}^{II}$ and $\mathbf{A}^{III}$ of Table I. Notice that the effective Hamiltonian $\mathbf{H}$ is identically produced upon taking $A = H$ in the effective operator $\mathbf{A}^{II}$. Table I indicates that this convenient property is not shared by all the effective operator definitions.

The effective operator $\mathbf{A}^i$ is the state-independent part of the definition $\mathbf{A}^i_{\alpha\beta}$, $i = I$–III. The operator $\mathbf{A}^i_{\alpha\beta}$ can thus be obtained by combining the perturbation expansions of its normalization factors and of $\mathbf{A}^i$ into a single expression [73] or by computing these normalization factors and $\mathbf{A}^i$ separately. These combined and noncombined forms of $\mathbf{A}^i_{\alpha\beta}$ may differ when computed approximately (see Section VI and paper II). The calculation of $A_{\alpha\beta}$ with the noncombined form is the same as with $\mathbf{A}^i$ since the model eigenvectors used with $\mathbf{A}^i$ are obtained by multiplying those utilized with $\mathbf{A}^i_{\alpha\beta}$ by the above normalization factors. The operators $\mathbf{A}^i_{\alpha\beta}$ and $\mathbf{A}^i$ are nevertheless different and, thus, do not have necessarily the same properties, for example, the conservation of commutation relations studied in Section IV.

3. Mappings $(\bar{K}, \bar{L})$ and Associated Effective Operators

Section II.B explains that expression (2.14) for $A_{\alpha\beta}$, with $|\phi_\alpha\rangle$ replaced by $|\Psi_\alpha\rangle$, may also be employed if the normalization factors present in (2.16) are incorporated into new mapping operators. In order to do so, let $\mathbf{N}$ designate the following operator of Ω_0:

$$\mathbf{N}|\alpha'\rangle_0 \equiv N_\alpha |\alpha'\rangle_0 \tag{2.48}$$

Equation (2.30) implies that a formal expression for $\mathbf{N}$ is

$$\mathbf{N} = \sum_\beta |\beta'\rangle_0 \ N_\beta \ _0\langle \bar{\beta}'| \tag{2.49}$$

As can be verified using (2.30) and (2.31), the inverse of $\mathbf{N}$ is given by

$$\mathbf{N}^{-1} = \sum_\beta |\beta'\rangle_0 \ N_\beta^{-1} \ _0\langle \bar{\beta}'| \tag{2.50}$$

and satisfies

$$\mathbf{N}^{-1}|\alpha'\rangle_0 = N_\alpha^{-1}|\alpha'\rangle_0 \tag{2.51}$$

Introducing these operators into Eqs. (2.35) and (2.36) produces the new mapping operators [74]

$$\bar{K} \equiv K\mathbf{N}^{-1} = \sum_\beta |\Psi'_\beta\rangle N_\beta^{-1} \ _0\langle \bar{\beta}'| = \sum_\beta |\Psi_\beta\rangle \ _0\langle \bar{\beta}'| \tag{2.52}$$

and

$$\bar{L} \equiv \mathbf{N}L = \sum_\beta |\beta'\rangle_0 \ N_\beta^{-1}\langle \Psi'_\beta| = \sum_\beta |\beta'\rangle_0 \ \langle \Psi_\beta| \tag{2.53}$$

which, as desired, obey the relations

$$\bar{K}|\alpha'\rangle_0 = |\Psi_\alpha\rangle, \qquad _0\langle \bar{\alpha}'|\bar{L} = \langle \Psi_\alpha| \tag{2.54}$$

Substituting Eqs. (2.54) into (2.14), and proceeding as in the derivation of the definitions $\mathbf{A}^i$, $i = \text{I–III}$, yields the state-independent effective operator definitions $\bar{\mathbf{A}}^{\text{I}}$, $\bar{\mathbf{A}}^{\text{II}}$ and $\bar{\mathbf{A}}^{\text{III}}$ of Table I.

The left relation in (2.54) implies that $|\alpha'\rangle_0$ and its true counterpart have the same length irrespective of how the former is normed. Hence, the mappings $(\bar{K}, \bar{L})$ actually conserve the norms of a set of right eigenvectors. On the other hand, the right equation in (2.54) shows that these mappings do not preserve the norms of left eigenvectors since $_0\langle\bar{\alpha}'|$ is not unity normed but, instead, normalized according to (2.30) [75]. Since the mappings $(\bar{K}, \bar{L})$ relate true eigenfunctions to $|\alpha'\rangle_0$ and $_0\langle\bar{\alpha}'|$, they generate the same effective Hamiltonian **H** as (K, L). Indeed, the last expressions in (2.52) and (2.53) for, respectively, $\bar{K}$ and $\bar{L}$ transform $\bar{L}H\bar{K}$ into the first expression of (2.37) for **H**.

All mapping pairs (K, L) that differ *only* in the values of the norms N_α yield the *same* effective Hamiltonian **H** and transformed mappings $(\bar{K}, \bar{L})$. This follows from, respectively, the first part of (2.37) and the last equalities in (2.52) and (2.53) because $|\alpha'\rangle_0$ and $_0\langle\bar{\alpha}'|$ are independent of N_α. The mappings $(\bar{K}, \bar{L})$ can generally be obtained only by explicitly calculating, or expanding, the N_α and combining them with the mappings (K, L). Hence, use of the mappings $(\bar{K}, \bar{L})$ does not obviate the calculation of the N_α in evaluating effective operators. Appendix A shows that all non-norm-preserving mappings that produce a non-Hermitian **h** and conserve the norms of a set of eigenvectors are related to mappings of the category (K, L) by equations analogous to (2.52), (2.53).

E. Effective Operators Generated by Non-Norm-Preserving Mappings which Produce a Hermitian Effective Hamiltonian

We demonstrate in Section III the existence of non-norm-preserving mappings that produce a Hermitian effective Hamiltonian and study some of their properties. These mappings preserve the overlaps between nondegenerate eigenvectors and, possibly, either those between arbitrary degenerate eigenvectors or the norms of a set of eigenfunctions, thus yielding the subcategories 3a, 3b and 3c of Table II.

A special but important case of subcategory 3b involve mappings that preserve the angles between degenerate eigenvectors (only) if they are orthogonal. These mappings and those of subcategory 3a are denoted by the same symbols $(\tilde{K}, \tilde{L})$ because of their similarity and because their difference does not affect the effective operator definitions they produce. In both cases, orthonormalized model eigenfunctions have true counterparts that are also orthogonal but not (all) unity normed. Hence, the formal expressions for the mappings and the various choices for effective operator definitions are the same in both cases. These definitions are given in Table I and are derived in Appendix B.3 using an analysis similar to that of Section II.D. Appendix B.3 also demonstrates that combining $(\tilde{K}, \tilde{L})$ with the normalization factors needed when unity normed model eigenfunctions are used, produces norm-preserving mappings and, thus, does not yield new effective operator definitions. Neither the other subcategory 3b mappings $(\tilde{k}, \tilde{l})$ nor the subcategory 3c mappings $(\bar{k}, \bar{l})$

conserve the angles between degenerate eigenfunctions, and this yields two choices as follows: The model eigenfunctions can be chosen to make their true counterparts orthogonal, or they can be orthogonalized. Contrary to the analogous situation with the mappings (K, L) in Section II.D, both choices are convenient here because the effective Hamiltonian is Hermitian [76]. The effective operator definitions produced by these two choices differ from each other and from those obtainable with the mappings $(\tilde{K}, \tilde{L})$ [71]. To be concise only the latter definitions are presented here, but all mappings are considered in our comparison of the relative merits of the various choices.

F. Summary

Mappings (k, l) that satisfy the condition $k^\dagger = l$ are norm-preserving and produce a Hermitian effective Hamiltonian. For any operator A, they generate a state-independent effective operator whose matrix elements are computed using orthonormalized model eigenfunctions. Mappings violating this condition do not conserve norms and can produce either a Hermitian (case 1) or a non-Hermitian (case 2) effective Hamiltonian. Both cases generate state-dependent effective operators if the model eigenfunctions are unity normalized and state-independent ones if the model eigenfunctions are normalized such that the corresponding true eigenfunctions are unit normed. Except for one type of case 1 transformations, namely the mappings $(\tilde{K}, \tilde{L})$, state-independent effective operators can also be produced by incorporating into new mapping operators the normalization factors needed with unity normed model eigenfunctions. With the mappings $(\tilde{K}, \tilde{L})$ this last option produces norm-preserving mappings and, thus, does not yield new effective operator definitions. Table I summarizes all possible effective operator definitions with norm-preserving, case 2, and $(\tilde{K}, \tilde{L})$ mappings. Several advantages of various effective operator definitions become apparent in Section IV which considers conservation of commutation relations and symmetries.

III. CONSERVATION OF SCALAR PRODUCT AND HERMITICITY OF EFFECTIVE HAMILTONIANS

Section II describes how the state independence of effective operators is closely related to the normalization of the true eigenfunctions. Because the model eigenfunctions can always be taken to be normalized (but not necessarily mutually orthogonal), state independence of effective operators is assured if the transformations k and l preserve lengths. When k and l are not norm-preserving, effective operators may still be state independent if the model eigenfunctions are normalized such that the corresponding true eigenfunctions are unity normalized or if the norms are absorbed into new mapping operators. Both of these cases require determination of the requisite norms by a separate expansion. This section presents a basic theorem concerning scalar product and norm-preserving transformations, along with a study of the relation between norm conservation and the Hermiticity of the effective Hamiltonian **h**. This leads to a second theorem proving the existence of non-norm-preserving mappings that produce a Hermitian **h**, providing their classification, and relating a subcategory of these mappings to norm-preserving ones. A third theorem classifies all mappings.

Let $|\gamma\rangle_0$ and $|\delta\rangle_0$ be two arbitrary functions in Ω_0, and let $|\gamma\rangle$ and $|\delta\rangle$ be their respective corresponding true counterparts, namely,

$$|\gamma\rangle \equiv k|\gamma\rangle_0 \tag{3.1}$$

$$|\delta\rangle \equiv k|\delta\rangle_0 \tag{3.2}$$

Define $|\Lambda\rangle$ and $|\Sigma\rangle$ to be two arbitrary functions of Ω with $|\Lambda\rangle_0$ and $|\Sigma\rangle_0$ their respective model counterparts,

$$|\Lambda\rangle_0 \equiv l|\Lambda\rangle \tag{3.3}$$

$$|\Sigma\rangle_0 \equiv l|\Sigma\rangle \tag{3.4}$$

where (k, l) are general mapping operators. Theorem I shows that *the following nine properties are all equivalent to each other*:

(1)
$$k^\dagger = l \tag{3.5}$$

(2)
$$kk^\dagger = P \tag{3.6}$$

(3)
$$ll^\dagger = P_0 \tag{3.7}$$

(4)
$$k^\dagger k = P_0 \tag{3.8}$$

(5)
$$l^\dagger l = P \tag{3.9}$$

(6)
$$\langle\gamma|\delta\rangle = {}_0\langle\gamma|\delta\rangle_0 \text{ for all } |\gamma\rangle_0 \text{ and } |\delta\rangle_0 \tag{3.10}$$

(7)
$${}_0\langle\Lambda|\Sigma\rangle_0 = \langle\Lambda|\Sigma\rangle \text{ for all } |\Lambda\rangle \text{ and } |\Sigma\rangle \tag{3.11}$$

(8)
$$\||\gamma\rangle\| = \||\gamma\rangle_0\| \text{ for all } |\gamma\rangle_0 \tag{3.12}$$

(9)
$$\||\Lambda\rangle_0\| = \||\Lambda\rangle\| \text{ for all } |\Lambda\rangle \tag{3.13}$$

A proof of Theorem I is given in Appendix C.

The properties in (3.5)–(3.13) are a generalization of the usual conditions on transformations of functions within the same subspace. Jørgensen [37, 64] states that the scalar product is conserved by mappings between two spaces iff $k^\dagger = l$, which in our notation is (1) iff (6) and (1) iff (7). He then derives conditions (4) and (5) from condition (1) and indicates that transformations which conserve the scalar product and, thus, norms are known as partial isometries. Kvaniscka [9] mentions that mappings satisfying condition (4) are partial isometries.

The other conditions listed in (3.5)–(3.13) have important consequences. Conditions (2) and (5) are shown in the next section and in Appendix E to be related to the conservation of commutation relations. Condition (4) implies that the metric or overlap operator $k^\dagger k$ of Eq. (2.11) is unity (in Ω_0) iff condition (1) holds, a result used when discussing (2.11) in Section II.A. If instead of right eigenvectors, the left ones are utilized in deriving Eq. (2.11), then an analogous relation is obtained in which **h** and its associated overlap operator are, respectively, $lHl^\dagger$ and $ll^\dagger$ [71]. Hence, condition (3) is analogous to condition (4) and, by Theorem I, is equivalent to (4) iff the mappings are norm-preserving. Conditions (6)–(9) imply that norm-preserving mappings conserve angles as well. Since nondegenerate true eigenfunctions are orthogonal to each other, so are the corresponding model ones with such mappings. Hence, norm-preserving mappings must produce a Hermitian **h**. This is indeed the case since substitution of (3.4) into (2.10) gives the effective Hamiltonian $\hat{\mathbf{H}}$ defined in Table I.

As just shown, (general) conservation of lengths, or, equivalently of angles, implies that **h** is Hermitian. However, the converse is not true, that is, the Hermiticity of the effective Hamiltonian does not imply the conservation of lengths or of angles. Since statements to the contrary exist [63, 64], we prove this point by demonstrating in Appendix B the existence of non-norm-preserving mappings which nonetheless produce a Hermitian effective Hamiltonian. One subcategory of these mappings, herein denoted by $(\tilde{K}, \tilde{L})$, is closely related to norm-preserving mappings. Theorem II.a of Appendix B.1 specifies this relation and, in particular, shows that any mapping pair $(\tilde{K}, \tilde{L})$ generates the *same* **h** as some norm-preserving mapping pairs. Other types of non-norm-preserving mappings can also produce a Hermitian **h**. These are not closely related to norm-preserving mappings.

Theorem II.b presents the following classification: *all non-norm-preserving mappings that produce a Hermitian* **h** *fall into the subcategories 3a, 3b, and 3c of Table II.* A proof of Theorem II.b is given in Appendix B.

Theorem II.b can be generalized to Theorem III as follows: *all mappings can be classified in the various categories presented in Table II.*

The many mapping types given in Table II arise as follows. Appendix B.2 shows that the mappings (k, l) produce a Hermitian effective Hamiltonian $\mathbf{h}$ iff its *nondegenerate* eigenfunctions are mutually orthogonal. Thus, the Hermiticity of $\mathbf{h}$ is equivalent to the conservation of the overlaps between nondegenerate eigenfunctions by the mappings (k, l), but it is independent of angle conservation between degenerate eigenvectors or of norm conservation for eigenfunctions. Appendix B.2 further demonstrates that the second and third conservations are independent of one another, and can thus be used to distinguish four different kinds of mappings producing a Hermitian $\mathbf{h}$. Likewise, four different types of mappings generate a non-Hermitian $\mathbf{h}$, but, as explained in Section II.D, the same symbol is used irrespective of whether the second conservation is obeyed. The relations between categories 2a–2d are established using some of the results demonstrated in Appendix B [71].

Since mapping operators need only satisfy (2.12), (2.13), those fulfilling any of these three conservations are less numerous than the ones that do not. Hence, the (K, L) mappings are the most numerous and, as mentioned in Section II.D, most of them do not preserve the angles between degenerate eigenvectors. Norm-preserving mappings are the least numerous since they obey all three conservations.

Some mapping categories of Table II have somewhat counterintuitive properties. Mappings $(\tilde{K}, \tilde{L})$ conserve the overlaps between orthogonal eigenvectors but are not angle-preserving. Appendix B shows that this arises because the conservation of basis vector overlaps is equivalent to angle preservation only if the basis vector norms are also conserved. Mappings $(\bar{K}, \bar{L})$ and $(\bar{k}, \bar{l})$ conserve the eigenfunction norms but are not norm-preserving. This follows because norm preservation is equivalent to the conservation of basis vector norms only if the overlaps between these vectors are also preserved. Appendix A demonstrates this for the mappings $(\bar{K}, \bar{L})$; the reasoning for the mappings $(\bar{k}, \bar{l})$ is very similar [71].

Section II explains how mappings that conserve the norms of eigenfunctions lead to a simpler effective operator formalism. Table II shows that the non-norm-preserving mapping categories $(\bar{K}, \bar{L})$ and $(\bar{k}, \bar{l})$ satisfy this condition but are related to other non-norm-preserving mappings by a transformation that incorporates the normalization factors needed with the latter. The former mappings can in general be obtained only through such a transformation, and, thus, their use in effective operators involves the calculation or perturbation expansion of the norms of *all d* true eigenfunctions.

In summary, Theorem I pertains to scalar product and norm-preserving mapping operators. Such transformations generate a Hermitian effective Hamiltonian **h** and ensure that the true and model eigenfunctions have the same norm and the same angles. Theorems II.a and II.b demonstrate that a Hermitian **h** can be produced by three kinds of *non*-norm-preserving mappings, one of which is closely related to norm-preserving transformation. Theorem III classifies all mapping operators. Two types of non-norm-preserving mappings, one producing a Hermitian **h** and the other a non-Hermitian **h**, are shown to conserve the eigenfunction norms. These mappings ensure that the corresponding true eigenfunctions are unity normalized if the model ones are, a desirable property for defining effective operators. Both types can generally be obtained only by combining other non-norm-preserving mappings with normalization factors, and, thus, do not obviate the need to deal explicitly with norms in effective operator calculations. Effective operators produced by the various kinds of mappings are considered in Section II, and some of their important properties are studied in the next section.

IV. COMMUTATION RELATIONS

This section studies the previously unaddressed problem of commutation relation conservation upon transformation to effective operators [77]. State-independent effective operators are treated first.

A. State-Independent Operators

Section II and Table I show that state-independent effective operators can be obtained with norm-preserving mappings $(\hat{K}, \hat{K}^\dagger)$ or with any of the three kinds of non-norm-preserving mappings (K, L), $(\bar{K}, \bar{L})$, and $(\tilde{K}, \tilde{L})$. This section first proves that the commutation relations between two arbitrary operators cannot generally be conserved upon transformation to any of these state-independent effective operators. A determination is then made of operators whose commutation relations are preserved by at least some state-independent effective operator definitions, and a few applications are then presented. Particular interest is focused on operators which commute with H, including constants of the motion.

1. Nonconservation of Commutation Relations Between Two Arbitrary Operators

Theorem IV pertains to commutation relations for all possible state-independent effective operators: *Let A and B be two arbitrary operators and let F be their commutator $F \equiv [A, B]$. The commutation relation between A and B is, in general, not conserved upon transformation to state-independent effective operators.* This theorem is first demonstrated for state-independent effective operators obtained with norm-preserving mappings, in which case it is equivalent to

$$[\hat{\mathbf{A}}, \hat{\mathbf{B}}] \neq \hat{\mathbf{F}} \tag{4.1}$$

The proof is then modified to apply to the other state-independent effective operator definitions. To prove (4.1) it is first shown that

$$[\hat{\mathbf{A}}, \hat{\mathbf{B}}] = \hat{\mathbf{F}} \tag{4.2}$$

is true iff

$$\hat{K}\hat{K}^{\dagger} = \mathbf{1} \tag{4.3}$$

where $\mathbf{1}$ is the unit operator of the full Hilbert space. Equation (4.3) is next proven to be an impossibility.

Applying the definition $\hat{\mathbf{A}}$ in Table I to the left hand and right hand sides of (4.1) transforms them, respectively, to

$$[\hat{A}, \hat{B}] = \hat{K}^{\dagger}(A\hat{K}\hat{K}^{\dagger}B - B\hat{K}\hat{K}^{\dagger}A)\hat{K} \tag{4.4}$$

and

$$\hat{\mathbf{F}} = \hat{K}^{\dagger}[A, B]\hat{K} \tag{4.5}$$

Equation (4.3) implies (4.2) as follows from (4.4) and (4.5). Equation (4.3) is also necessary for (4.2) since if (4.3) is violated, the equality of (4.4) and (4.5) requires a relation between one or both of the mapping operators ($\hat{K}$ and $\hat{K}^{\dagger}$) and A, B, or both, contrary to the hypothesis of A and B being *arbitrary* operators. Equation (4.3), however, is not valid because Eqs. (2.3) and (2.4) imply that $\hat{K}\hat{K}^{\dagger}$ acts only in the subspace Ω of the whole Hilbert space [78]. Thus, Theorem IV follows for state-independent effective operators obtained with norm-preserving mappings.

The reasoning needed for the other possible state-independent effective operator definitions is simply obtained by replacing $\hat{K}^{\dagger}$ and $\hat{K}$ in the above proof by, respectively, the operators left- and right-multiplying A in the chosen definition. This process shows that the commutation relation between two arbitrary operators is conserved iff Eq. (4.3) is satisfied with the above mapping replacements. In all cases the product analogous to $\hat{K}\hat{K}^{\dagger}$, appearing in the left hand side of the analog of (4.3), is defined solely in Ω, leading to the same contradiction as demonstrated for norm-preserving mappings. Therefore, Theorem IV follows for all state-independent effective operator definitions.

2. Conservation of Commutation Relations of Special Operators and Applications

We now determine particular classes of commutation relations that are, indeed, conserved upon transformation to state-independent effective operators. The proof of (4.1) demonstrates that the preservation of $[A, B]$ by definition $\hat{A}$ requires the existence of a relation between $\hat{K}$, $\hat{K}^{\dagger}$, or both and one or both of the true operators A or B. Likewise, there must be a relation between the appropriate wave operator, the inverse mapping operator, or both, and A, B, or both for other state-independent effective operator definitions to conserve $[A, B]$. All mapping operators depend on the spaces Ω_0 and Ω. Although the model space is often specified by selecting eigenfunctions of a zeroth order Hamiltonian, it may, in principle, be arbitrarily defined. On the other hand, the space Ω necessarily depends on H. Therefore, the existence of a relation between mapping operators and A, B, or both, implies a relation between H and A, B, or both.

Consider first norm-preserving mappings, whereupon condition (2) of Theorem I implies that

$$\hat{K}\hat{K}^{\dagger} = P \tag{4.6}$$

Since $\hat{H} = \hat{K}^{\dagger}H\hat{K}$ (see Table I), taking $A = H$ into (4.4) gives an expression for $[\hat{H}, \hat{B}]$ and, thus, provides the simplest example of the above mentioned relation between H and A, B, or both. Introducing (4.6) into this expression yields

$$[\hat{H}, \hat{B}] = \hat{K}^{\dagger}(HPB - BPH)\hat{K} = \hat{K}^{\dagger}[H, B]\hat{K} \tag{4.7}$$

where P and H are commuted and then (2.3) and its Hermitian conjugate are used. Hence, the commutation relation between an arbitrary operator and H is preserved upon transformation to state-independent effective operators generated with norm-preserving mappings.

The key step in deriving (4.7) is the commutation of P with H. Clearly, a similar reasoning applies when replacing H with any operator that commutes with H because such an operator also commutes with P. Therefore, this leads to Theorem V as follows: *state-independent effective operators produced by norm-preserving mappings conserve the commutation relations between H and an arbitrary operator B and between B and any operator that commutes with H.* Given particular choices of P, additional operators may commute with P. The preceding theorem may thus be generalized into Theorem VI as follows: *if the mapping operators are norm-preserving, then the commutation relation between any arbitrary operator and any operator that commutes with P is conserved by the corresponding state-independent effective operators.*

Other preserved commutation relations can be obtained from the following general necessary and sufficient condition on A and B for the conservation of their commutator: Multiplying Eq. (4.5) on the left and on the right, respectively, by $\hat{K}$ and $\hat{K}^\dagger$ and using (4.6) yields

$$\hat{K}\hat{\mathbf{F}}\hat{K}^\dagger = P[A, B]P \qquad (4.8)$$

Conversely, left- and right-multiplying Eq. (4.8), respectively, by $\hat{K}^\dagger$ and $\hat{K}$ and using the norm-preserving case of Eq. (2.12) produces Eq. (4.5). Similarly, Eq. (4.4) is equivalent to

$$\hat{K}[\hat{\mathbf{A}}, \hat{\mathbf{B}}]\hat{K}^\dagger = P(APB - BPA)P \qquad (4.9)$$

Hence, the commutation relation between A and B is conserved iff the right hand sides of Eqs. (4.8) and (4.9) are equal to each other, thereby leading to Theorem VII as follows: *the commutation relation between two operators A and B is preserved upon transformation to state-independent effective operators obtained with norm-preserving mappings iff A and B satisfy*

$$P(AQB - BQA)P = 0_\Omega \qquad (4.10)$$

where 0_Ω is the null operator in the space Ω.

Notice that the condition $[A, P] = 0$ in Theorems V and VI is indeed sufficient for (4.10). There clearly exist many other sufficient conditions on A and B causing (4.10) to be fulfilled, for example, $PAQ = 0 = PBQ$, $PA = 0 = PB$, $PA = 0 = BQ$, etcetera. All such conditions necessarily imply a relation between A, B, or both, and P, Q, or both. Consequently, although more general, Theorems VI and VII have fewer applications than Theorem V because all the sufficient conditions obtainable from (4.10), except that of Theorem V, cannot be verified exactly. This is because the d true eigenfunctions that define P are not known a priori and can be determined a posteriori only approximately. Except for H and operators commuting with H, it is thus in general impossible to determine if a particular operator commutes with P. Also, conditions that involve Q explicitly, for example, $PAQ = 0 = PBQ$ are not readily established.

Theorem V, on the other hand, has important consequences. First,

(Corollary V.1) *to any complete set of commuting observables (CSCO) there corresponds a complete set of commuting effective observables (CSCEO) if the mappings are norm-preserving.* This corollary is proven using the definition of a complete commuting set.

$$[H, O_i] = [O_i, O_j] = 0 , \quad i, j = 1, 2, \ldots n$$

which by Theorem V implies that

$$[\hat{\mathbf{H}}, \hat{\mathbf{O}}_i] = [\hat{\mathbf{O}}_i, \hat{\mathbf{O}}_j] = 0 , \quad i, j = 1, 2, \ldots n$$

where $\hat{\mathbf{O}}_1, \hat{\mathbf{O}}_2, \ldots, \hat{\mathbf{O}}_n$ are the effective observables corresponding, respectively, to $O_1, O_2, \ldots, O_n$. Section VII applies Corollary V.1 to the preservation of symmetries in effective Hamiltonian calculations.

Another important application of Theorem V is that (Corollary V.2) *the dipole length and dipole velocity transition moments are equivalent when computed with state-independent effective operators obtained with norm-preserving mappings.* According to definition $\hat{\mathbf{A}}$ (see Table I), these computations evaluate ${}_0\langle \hat{\alpha} | \hat{\hat{\mathbf{p}}} | \hat{\beta} \rangle_0$, and ${}_0\langle \hat{\alpha} | \hat{\hat{\mathbf{r}}} | \hat{\beta} \rangle_0$ with

$$\hat{\hat{\mathbf{p}}} \equiv \hat{K}^\dagger \vec{p} \hat{K} , \quad \hat{\hat{\mathbf{r}}} \equiv \hat{K}^\dagger \vec{r} \hat{K} \tag{4.11}$$

$|\hat{\alpha}\rangle_0$ and $|\hat{\beta}\rangle_0$ orthonormalized eigenvectors of $\hat{\mathbf{H}}$, and H, $\vec{p}$ and $\vec{r}$ the total Hamiltonian, position and momentum operators, respectively, for a system of n identical particles of mass m. Corollary V.2 follows because Theorem V implies that the commutation relation

$$\frac{i\hbar}{m} \vec{p} = [\vec{r}, H] \tag{4.12}$$

is also obeyed by the corresponding effective operators,

$$\frac{i\hbar}{m} \hat{\hat{\mathbf{p}}} = [\hat{\hat{\mathbf{r}}}, \hat{\mathbf{H}}] \tag{4.13}$$

Left- and right-multiplying (4.13) by ${}_0\langle \hat{\alpha} |$ and $|\hat{\beta}\rangle_0$, respectively, and using the eigenvalue equation (2.1) with $\mathbf{h}$ replaced by $\hat{\mathbf{H}}$ yields

$${}_0\langle \hat{\alpha} | \hat{\hat{\mathbf{p}}} | \hat{\beta} \rangle_0 = i m \omega_{\alpha\beta} \; {}_0\langle \hat{\alpha} | \hat{\hat{\mathbf{r}}} | \hat{\beta} \rangle_0 \tag{4.14}$$

where the exact transition frequency is

$$\omega_{\alpha\beta} \equiv (E_\alpha - E_\beta)/\hbar \tag{4.15}$$

thereby proving Corollary V.2.

Equation (4.12) implies an equivalence similar to (4.14) for true operators and eigenfunctions which may be used to derive the Thomas–Reiche–Kuhn sum rule. Equation (4.14), however, does not produce a sum rule for transition moments as shown in Appendix D. A general study of sum rules for effective operators will be presented elsewhere [79].

Table III summarizes Theorems V–VII and their corollaries along with similar results for the other state-independent effective operator definitions. Appendix E demonstrates the analogs of Theorems V–VII, except the conservation by definitions $\mathbf{A}^{III}$ and $\bar{\mathbf{A}}^{III}$ of $[H, C]$ for C a constant of the motion which commutes separately with H_0 and V. This last point is proven in paper II. The analogs of Corollaries V.1 and V.2 are obtained similarly to, respectively, Corollaries V.1 and V.2. Just as with Corollary V.2, none of the equivalences between the dipole length and dipole velocity transition moments for definitions $\mathbf{A}^{II}$, $\bar{\mathbf{A}}^{II}$, or $\tilde{\mathbf{A}}^i$, $i = I$–IV, produces a sum rule for transition moments (see Appendix D).

B. State-Dependent Operators

This section studies the commutation properties of the state-dependent effective operators generated by the (K, L) and $(\tilde{K}, \tilde{L})$ mappings (see Table I). As in Section IV.A, let A and B designate two arbitrary operators and F be their commutator $F \equiv [A, B]$. A state-dependent definition $\mathbf{A}^j_{\alpha\beta}$ conserves F iff

$$[\mathbf{A}^j_{\alpha\beta}, \mathbf{B}^j_{\alpha\beta}] = \mathbf{F}^j_{\alpha\beta} \qquad (4.16)$$

Complications arise here that are absent with state-independent effective operators. The effective operators in (4.16) cannot merely be replaced by their definitions from Table I since the latter may not be applied directly to any vector of the space Ω_0 because of their normalization factors. These factors are associated with the model eigenvectors on which the operators act to produce the matrix elements $A_{\alpha\beta}$. Consequently, arbitrary bras and kets of Ω_0 must first be expanded in the basis of these eigenvectors before a state-dependent definition can be used with them. This represents a serious drawback with the use of state-dependent effective operators.

The possible validity of Eq. (4.16) can be checked by comparing the matrix elements of each of its sides taken between such expansions (procedure 1). Alternatively (procedure 2), a new operator expression can be constructed by inserting to the left and to the right of the effective operator a resolution of the unit operator P_0 expressed in terms of the above model eigenbras and eigenkets [80]. This new operator form may act on arbitrary vectors and, thus, be substituted in (4.16). Procedures 1

TABLE III

Conservation of Commutation Relations and Some Consequences; A and B are Arbitrary Operators and C is a Constant of the Motion

Mappings	Effective operator definition $\hat{A}$	$[A, B]$ conserved iff	$[A, B]$ conserved if $[A, P] = 0$	$[H, B]$ conserved	$[H, C]$ conserved	$[C, B]$ conserved	CSCO has corresponding CSCEO	$\langle \tilde{p}_{\text{eff}} \rangle_{\alpha\beta}$ equivalent to $\langle \tilde{r}_{\text{eff}} \rangle_{\alpha\beta}$
$(\tilde{K}, \tilde{K}^\dagger)$	$\hat{A}$	$X = 0_\Omega$ [a]	Yes	Yes	Yes	Yes	Yes	Yes
(K, L)	$A^I_{\alpha\beta}$	$X - P(A\bar{Y}B - B\bar{Y}A)P = 0_\Omega$ [b]	No	No	No	No	No	No
	$A^{II}_{\alpha\beta}$	$X = 0_\Omega$	Yes	Yes	Yes	Yes	Yes	No
	$A^{III}_{\alpha\beta}$	$X = 0_\Omega$	Yes	No	No [c]	Yes	No [d]	No
	A^I	$X - P(AYB - BYA)P = 0_\Omega$ [e]	No	No	No	No	No	No
	A^{II}	$X = 0_\Omega$	Yes	Yes	Yes	Yes	Yes	Yes
	A^{III}	$X = 0_\Omega$	Yes	No	No [c]	Yes	No [d]	No
	$\bar{A}^I$	$X - P(A\bar{Y}B - B\bar{Y}A)P = 0_\Omega$	No	No	No	No	No	No
	$\bar{A}^{II}$	$X = 0_\Omega$	Yes	Yes	Yes	Yes	Yes	Yes
	$\bar{A}^{III}$	$X = 0_\Omega$	Yes	No	No [c]	Yes	No [d]	No
$(\tilde{K}, \tilde{L})$	$\tilde{A}^I_{\alpha\beta}$	$X = 0_\Omega$	Yes	Yes	Yes	Yes	Yes	Yes
	$\tilde{A}^{II}_{\alpha\beta}$	$X = 0_\Omega$	Yes	Yes	Yes	Yes	Yes	Yes
	$\tilde{A}^{III}_{\alpha\beta}$	$X = 0_\Omega$	Yes	Yes	Yes	Yes	Yes	Yes
	$\tilde{A}^{IV}_{\alpha\beta}$	$X = 0_\Omega$	Yes	Yes	Yes	Yes	Yes	Yes
	$\tilde{A}^I$	$X - P(A\tilde{Y}B - B\tilde{Y}A)P = 0_\Omega$ [f]	No	Yes	Yes	No	No	Yes
	$\tilde{A}^{II}$	$X = 0_\Omega$	Yes	Yes	Yes	Yes	Yes	Yes
	$\tilde{A}^{III}$	$X = 0_\Omega$	Yes	Yes	Yes	Yes	Yes	Yes
	$\tilde{A}^{IV}$	$X - P(A\tilde{Z}B - B\tilde{Z}A)P = 0_\Omega$ [g]	No	Yes	Yes	No	No	Yes

[a] $X \equiv P(AQB - BQA)P$.

[b] $\bar{Y} \equiv \bar{K}\bar{K}^\dagger - P$.

[c] Yes if $[C, H_0] = 0 = [C, V]$, a result demonstrated in paper II.

[d] Yes if H_0 commutes with all the observables.

[e] $Y \equiv KK^\dagger - P$.

[f] $\tilde{Y} \equiv \tilde{K}\tilde{K}^\dagger - P$.

[g] $\tilde{Z} \equiv \tilde{r}^\dagger\tilde{r} - P$

496

and 2 are equivalent, but the latter turns out to be simpler and is thus applied here.

We now illustrate procedure 2 with the "noncombined" form (see Section II.D.2) of definitions $A_{\alpha\beta}^{I}$ and $A_{\alpha\beta}^{II}$ [81]. Left- and right-multiplying $A_{\alpha\beta}^{II}$ by the first expression in (2.31) for P_0 yields

$$A_{\alpha\beta}^{II} = \sum_{\alpha,\beta} |\alpha'\rangle_0 \, N_\alpha \, {}_0\langle \bar{\alpha}'|A^{II}|\beta'\rangle_0 \, N_\beta^{-1} \, {}_0\langle \bar{\beta}'| \tag{4.17}$$

Using (2.49) and (2.50) converts (4.17) to

$$A_{\alpha\beta}^{II} = NA^{II}N^{-1} \tag{4.18}$$

Applying (4.18) to each operator of (4.16) for definition $A_{\alpha\beta}^{II}$ produces

$$[A_{\alpha\beta}^{II}, B_{\alpha\beta}^{II}] - F_{\alpha\beta}^{II} = N([A^{II}, B^{II}] - F^{II})N^{-1} \tag{4.19}$$

which implies that $[A, B]$ is conserved by definition $A_{\alpha\beta}^{II}$ iff it is conserved by definition A^{II}. Furthermore, as Table I shows, H is the $A = H$ case of both of these definitions. Consequently, *all* commutation relation conservations for the latter definition also apply to the former.

Applying the same procedure to definition $A_{\alpha\beta}^{I}$, both forms of P_0 in Eq. (2.31) are used to re-express $A_{\alpha\beta}^{I}$ as

$$A_{\alpha\beta}^{I} = \sum_{\alpha,\beta} |\bar{\alpha}'\rangle_0 \, N_\alpha^{-1} \, {}_0\langle \alpha'|A^{I}|\beta'\rangle_0 \, N_\beta^{-1} \, {}_0\langle \bar{\beta}'| \tag{4.20}$$

Substituting (2.50) into (4.20) yields

$$A_{\alpha\beta}^{I} = (N^{-1})^{\dagger}A^{I}N^{-1} \tag{4.21}$$

Applying (4.21) to all operators of (4.16) for definition $A_{\alpha\beta}^{I}$ gives

$$[A_{\alpha\beta}^{I}, B_{\alpha\beta}^{I}] - F_{\alpha\beta}^{I} = (N^{-1})^{\dagger}(A^{I}N^{-1}(N^{-1})^{\dagger}B^{I} - B^{I}N^{-1}(N^{-1})^{\dagger}A^{I} - F^{I})N^{-1} \tag{4.22}$$

Thus, $[A, B]$ is conserved iff

$$A^{I}N^{-1}(N^{-1})^{\dagger}B^{I} - B^{I}N^{-1}(N^{-1})^{\dagger}A^{I} = F^{I} \tag{4.23}$$

Left- and right-operating on (4.23) with, respectively, $L^{\dagger}$ and L and using the identity (2.13) with $k = K$ and $l = L$ produces

$$P[AKN^{-1}(N^{-1})^{\dagger}K^{\dagger}B - BKN^{-1}(N^{-1})^{\dagger}K^{\dagger}A]P = PFP \tag{4.24}$$

Equation (4.24) can be transformed back to (4.23) by left- and right-operating with, respectively, $K^\dagger$ and K and by using the second part of identity (2.3) with k replaced by K. Hence, Eqs. (4.23) and (4.24) are equivalent. Equation (2.52) can be used to show that Eq. (4.24) is the same as the condition for the conservation of $[A, B]$ by the definition $\bar{\mathbf{A}}^{\mathrm{I}}$ in Table III. Thus, $[A, B]$ is conserved by definition $\mathbf{A}^{\mathrm{I}}_{\alpha\beta}$ iff it is preserved by definition $\bar{\mathbf{A}}^{\mathrm{I}}$.

Since the effective Hamiltonian $\mathbf{H}$ is not the $A = H$ case of definition $\mathbf{A}^{\mathrm{I}}_{\alpha\beta}$, the result just demonstrated cannot be used to deduce anything about the conservation of $[H, B]$. The analysis of this commutator is similar to that of $[A, B]$ and uses the fact that $\mathbf{H}$ commutes with $\mathbf{N}^{-1}$. It is found that $[H, B]$ is conserved by definition $\mathbf{A}^{\mathrm{I}}_{\alpha\beta}$ iff it is preserved by definition $\bar{\mathbf{A}}^{\mathrm{I}}$.

The above results for definitions $\mathbf{A}^{\mathrm{I}}_{\alpha\beta}$ and $\mathbf{A}^{\mathrm{II}}_{\alpha\beta}$ are summarized in Table III, which also displays the commutation properties for all other state-dependent effective operator definitions. These are derived similarly to those for definitions $\mathbf{A}^{\mathrm{I}}_{\alpha\beta}$ and $\mathbf{A}^{\mathrm{II}}_{\alpha\beta}$, except for the preservation by definition $\mathbf{A}^{\mathrm{III}}_{\alpha\beta}$ of $[H, C]$ for C a constant of the motion that commutes with H_0 and V separately. This last result is demonstrated in paper II.

Table III shows that definitions $\mathbf{A}^{i}_{\alpha\beta}$, $i = \mathrm{I}\text{–}\mathrm{III}$, and $\tilde{\mathbf{A}}^{j}_{\alpha\beta}$, $j = \mathrm{I}\text{–}\mathrm{IV}$, have the *same* commutation properties as definitions $\bar{\mathbf{A}}^{i}$ and $\hat{\mathbf{A}}$, respectively. This arises because the operators $\mathbf{A}^{i}_{\alpha\beta}$, $i = \mathrm{I}\text{–}\mathrm{III}$, and $\tilde{\mathbf{A}}^{j}_{\alpha\beta}$, $j = \mathrm{I}\text{–}\mathrm{IV}$, have the same matrix representation as, respectively, $\bar{\mathbf{A}}^{i}$ and one particular operator $\hat{\mathbf{A}}$ and, thus, are formally equivalent to the latter (although computationally the procedures are different). These equivalences are most simply established in the basis sets of the eigenvectors used to produce the $A_{\alpha\beta}$. Table I shows that ${}_0\langle \bar{\beta}' | \mathbf{A}^{\mathrm{II}}_{\alpha\beta} | \alpha'\rangle_0 = A_{\alpha\beta} = {}_0\langle \bar{\beta}' | \bar{\mathbf{A}}^{\mathrm{II}} | \alpha'\rangle_0$, which implies that the matrix representations of $\mathbf{A}^{\mathrm{II}}_{\alpha\beta}$ and $\bar{\mathbf{A}}^{\mathrm{II}}$ are the same since the (β, α) matrix element of any operator M of Ω_0 in the nonorthogonal set $\{|\alpha'\rangle_0\}$ is computed as

$$ {}_0\langle \bar{\beta}' | M | \alpha'\rangle_0 \tag{4.25} $$

The matrix representations of $\mathbf{A}^{\mathrm{III}}_{\alpha\beta}$ and $\bar{\mathbf{A}}^{\mathrm{III}}$ in the set $\{|\bar{\alpha}'\rangle_0\}$ are similarly proven identical, as are the matrix representations in the set $\{|\tilde{\alpha}'\rangle_0\}$ of $\tilde{\mathbf{A}}^{j}_{\alpha\beta}$, $j = \mathrm{I}\text{–}\mathrm{IV}$, and of the operator $\hat{\mathbf{A}}$ that is generated by the (unique) "reference" norm-preserving mappings for the pair $(\tilde{K}, \tilde{L})$ (see Appendix B). Finally, $\mathbf{A}^{\mathrm{I}}_{\alpha\beta}$ and $\bar{\mathbf{A}}^{\mathrm{I}}$ have the same matrix representation in $\{|\alpha'\rangle_0\}$ as follows from taking $M = P_0 \mathbf{A}^{\mathrm{I}}_{\alpha\beta}$ and $M = P_0 \bar{\mathbf{A}}^{\mathrm{I}}$ into (4.25), substituting for P_0 the second expression in Eq. (2.31), and using ${}_0\langle \beta' | \mathbf{A}^{\mathrm{I}}_{\alpha\beta} | \epsilon'\rangle_0 = A_{\beta\epsilon} = {}_0\langle \beta' | \bar{\mathbf{A}}^{\mathrm{I}} | \epsilon'\rangle_0$.

C. Summary

The commutation relation between two arbitrary operators is not conserved upon transformation to effective operators by any of the definitions. Many state-independent effective operator definitions preserve the commutation relations involving H or a constant of the motion, as well as those involving operators which are related to P in a special way, for example, A with $[P, A] = 0$. Many state-dependent definitions also conserve these special commutation relations. However, state-dependent definitions are not as convenient for formal and possibly computational reasons. The most important preserved commutation relations are those involving observables, since, as discussed in Section VII, they ensure that the basic symmetries of the system are conserved in effective Hamiltonian calculations.

V. REVIEW OF EFFECTIVE HAMILTONIAN FORMALISMS

Section II provides a wide array of different possible categories of effective Hamiltonians $\mathbf{h}$ and operators $\mathbf{a}$. This section briefly reviews several of the previously used effective Hamiltonian formalisms and shows that many of the mapping categories have been proposed in various contexts. This literature review also serves to elucidate several important properties of the different choices. A few articles [33, 37, 47, 65, 82, 83] review and interrelate many of the known effective Hamiltonians, and interested readers may consult them for more details. Here, we summarize the salient points and provide an update on recent formalisms. Several methods are not discussed because they are not amenable to the type of analysis presented in Section IV. For example, Padé approximants [32, 44, 84–86] provide a purely numerical procedure for improving the eigenvalues but neither alter the mapping operators nor $\mathbf{h}$. For our purpose, it is most convenient to classify the various $\mathbf{h}$'s by their defining mapping operators. Since Bloch's effective Hamiltonian $\mathbf{H_B}$ has been applied most and since many of the other $\mathbf{h}$'s can be related to $\mathbf{H_B}$, a brief summary of Bloch's [6] effective Hamiltonian leads naturally into a discussion of many other methods.

A. Bloch's Effective Hamiltonian $\mathbf{H_B}$

Bloch defines the right model eigenfunctions $\{|\alpha_B\rangle_0\}$ of $\mathbf{H_B}$ to be the projections onto Ω_0 of d orthonormal true eigenfunctions. Thus, Bloch's inverse mapping operator is $P_0 P$ and does not conserve norms. The resulting effective Hamiltonian $\mathbf{H_B}$ is generally not Hermitian since the projections of orthogonal functions need not be orthogonal. Therefore, the mapping operators used by Bloch are of the (K, L) category,

$$\mathbf{H}_B = L_B H K_B , \qquad L_B \equiv P_0 P \qquad (5.1)$$

Following the notation of Section II.D.1, the right model eigenfunctions $\{|\alpha_B\rangle_0\}$ of $\mathbf{H}_B$ correspond to orthonormalized true eigenfunctions $\{|\Psi_\alpha\rangle\}$. (The notation $|\alpha_B\rangle_0$ just appends the subscript B to that used by Bloch.) The form of L_B implies upon use of (2.28) that

$$_0\langle \alpha_B | \alpha_B \rangle_0 = {}_0\langle \alpha_B | \Psi_\alpha \rangle \qquad (5.2)$$

The relation (5.2) applies for any normalization of $|\alpha_B\rangle_0$, provided $|\Psi_\alpha\rangle$ is its corresponding true eigenfunction, since right model and true eigenfunctions scale together. In particular, introducing the definition of Section II.D.1 for unity normed model eigenfunctions converts Eq. (5.2) into

$$_0\langle \alpha_B' | \alpha_B' \rangle_0 = {}_0\langle \alpha_B' | \Psi_\alpha' \rangle = 1 \qquad (5.3)$$

Equation (5.3) is often called the "intermediate normalization" (IN) condition. Substituting the definition (5.1) of L_B into Eq. (2.12) applied to (K_B, L_B) and using the second part of (2.3) produces

$$P_0 K_B = P_0 \qquad (5.4)$$

Conversely, the second equation in (5.1) can be derived from either (5.2), (5.3), or (5.4). Hence, these four relations are all equivalent. Inserting the second equation in (5.1) into the first, commuting P with H, and applying the second equality in (2.3) gives

$$\mathbf{H}_B = P_0 H K_B = P_0 H_0 + P_0 V K_B \qquad (5.5)$$

where the last equality follows from Eq. (5.4) (and $[H_0, P_0] = 0$) and provides the usual form in which $\mathbf{H}_B$ is written.

The bi-orthonormal complement of the right eigenvectors $\{|\alpha_B\rangle_0$ and $\{|\alpha_B'\rangle_0\}$ are denoted, respectively, by $\{_0\langle \bar{\alpha}_B|\}$ and $\{_0\langle \bar{\alpha}_B'|\}$. Bloch [6] shows that the unity normed true eigenfunction $|\Psi_\alpha\rangle$ is also the projection onto Ω of $|\bar{\alpha}_B\rangle_0$,

$$|\Psi_\alpha\rangle = PP_0|\bar{\alpha}_B\rangle_0 = L_B^\dagger|\bar{\alpha}_B\rangle_0 \qquad (5.6)$$

Equation (5.6) illustrates the general fact (see Eq. (2.38)) that $l^\dagger$ is a wave operator for left model eigenvectors, but this relation is *not* valid when $|\bar{\alpha}_B\rangle_0$ and $|\Psi_\alpha\rangle$ are replaced by, respectively, $|\bar{\alpha}'_B\rangle_0$ and $|\Psi'_\alpha\rangle$ (see Eq. (2.39)) [87].

Bloch provides Rayleigh–Schrödinger (RS) perturbation expansions of K_B and H_B for degenerate Ω_0, whereupon the first term on the right hand side of (5.5) reduces to $E_0 P_0$. Kvanička [9] extends Bloch's work to a quasi-degenerate model space. des Cloizeaux [7] derives H_B starting from a degenerate Brillouin–Wigner (BW) perturbation theory by expanding about E_0. Following des Cloizeaux's method, Brandow [8] derives rules for the diagrams representing contributions to the BW and RS perturbation expansions of H_B. Brandow proves that if Ω_0 is a *complete* model space, the unlinked diagrams of any order in the RS perturbation expansion cancel each other exactly. Brandow designates this property of the H_B expansion as being "fully linked". A different proof of the "fully linked" nature of the RS expansion of H_B is given by Lindgren [19, 41]. Diagrammatic analyses of H_B based on time-dependent perturbation theory are provided by Oberlechner et al. [88] for a complete, degenerate model space with results equivalent to those of Brandow. Kuo et al. [89] also use a time-dependent approach in their "Q-box" method, which involves an infinite sum of diagrams (some disconnected) with identical "structures". The complete degenerate model space expansion of Kuo et al. is not "fully linked" but is in a form suited to infinite order summations [59]. Other formal work using time-dependent perturbation theory for H_B is described in [65] and [33].

Recent developments include exact [12–14, 44, 90, 91] and approximate [14, 90, 92–94] iterative schemes to determine H_B, the intermediate Hamiltonian method [21, 24, 95], the use of *incomplete* model spaces [43, 44] and some multireference open-shell coupled-cluster (CC) formalisms [16–20, 96, 97]. Only *some* eigenvalues of the intermediate Hamiltonian H_I are also eigenvalues of H. The corresponding model eigenvectors of H_I are related to their true counterparts as in Bloch's theory. Provided effective operators **a** are restricted to act solely between *these* model eigenvectors, the possible **a** definitions from Bloch's formalism (see Section VI.A) can be used.

B.   des Cloizeaux's or Canonical Effective Hamiltonian $\hat{H}_C$

The simple projection relation between the right model eigenfunctions of H_B and their true counterparts is an appealing aspect of Bloch's formalism. However, the non-Hermiticity of the resulting effective Hamiltonian represents a strong drawback, as discussed in Section VII. This has led many, beginning with des Cloizeaux [7], to derive Hermitian effective Hamiltonians. des Cloizeaux's method transforms the $\{|\alpha_B\rangle_0\}$ (*not* the

$\{|\alpha'_{\rm B}\rangle_0\}$) into an orthonormal set $\{|\hat{\alpha}_{\rm C}\rangle_0\}$) which are the eigenfunctions of the Hermitian effective Hamiltonian

$$\hat{\mathbf{H}}_{\rm C} = (P_0 P P_0)^{(-1/2)} \mathbf{H}_{\rm B} (P_0 P P_0)^{(1/2)} = \hat{K}^\dagger_{\rm C} H \hat{K}_{\rm C} , \qquad (5.7)$$

where the wave operator mapping $|\hat{\alpha}_{\rm c}\rangle_0$ into $|\Psi_\alpha\rangle$ is [98]

$$\hat{K}_{\rm C} \equiv K_{\rm B}(P_0 P P_0)^{(1/2)} = L^\dagger_{\rm B}(P_0 P P_0)^{(-1/2)} . \qquad (5.8)$$

The mappings $(\hat{K}_{\rm C}, \hat{K}^\dagger_{\rm C})$ are, thus, norm-preserving.

Klein [65] notes that the des Cloizeaux transformation is a different formulation of the "symmetric" transformation of Löwdin [99] and uses this fact to prove that $\{|\hat{\alpha}_{\rm c}\rangle_0\}$ is the set of *orthonormal* functions in Ω_0 which differs minimally (in a least squares sense) from the d true eigenfunctions $|\Psi_\alpha\rangle$ of Ω. This is often referred to as a "maximum similarity" property.

Jørgensen [37] demonstrates that $\hat{K}_{\rm C}$ is the only norm-preserving wave operator satisfying the condition

$$(P_0 \hat{K})^\dagger = P_0 \hat{K} \qquad (5.9)$$

which he uses [37, 82] to determine whether other choices [100] of Hermitian effective Hamiltonians are identical to $\hat{\mathbf{H}}_{\rm C}$. One equivalent of $\hat{\mathbf{H}}_{\rm C}$ arises in the one-transformation version of the generalized van Vleck formalism [9, 37, 101, 102] in which the wave operator is expressed in terms of a unitary transformation of the whole Hilbert space,

$$\hat{K}_{\rm C} = \exp(-G) P_0 , \qquad G^\dagger = -G \qquad (5.10)$$

that satisfies the "Kemble condition",

$$P_0 G P_0 = 0 = Q_0 G Q_0 \qquad (5.11)$$

des Cloizeaux [7] provides explicit RS expressions for the low (≤ 3) orders of $\hat{\mathbf{H}}_{\rm C}$ for Ω_0 degenerate, and Klein [65] gives RS expressions through sixth and fifth order for, respectively, $\hat{\mathbf{H}}_{\rm C}$ and the "canonical unitary" transformation operator U which is defined as follows. The whole Hilbert space is decomposed into a direct sum of degenerate subspaces i,

$$\sum_i P_0(i) = \mathbf{1} \qquad (5.12)$$

The operator

$$U \equiv \sum_i \hat{K}^\dagger_C(i) \qquad (5.13)$$

then generates canonical effective Hamiltonians on *all* degenerate subspaces,

$$UHU^\dagger = \sum_i \hat{\mathbf{H}}_C(i) . \qquad (5.14)$$

Kvaniščka [9, 101] uses the one-transformation van Vleck formalism to obtain low expressions for $\hat{\mathbf{H}}_C$ for Ω_0 quasi-degenerate. Hurtubise [66] provides a much simpler derivation. Brandow [8] and Westhaus [42, 103] prove that the RS perturbation expansion of $\hat{\mathbf{H}}_C$ is fully linked for a degenerate, or quasi-degenerate, complete space Ω_0. Westhaus' proof actually applies to other Hermitian effective Hamiltonians as well [104]. Explicit expressions for all $\hat{\mathbf{H}}_C$ diagrams through third order are provided by Sheppard and Freed [105]. Kvaniščka [106] formulates $\hat{\mathbf{H}}_C$ into a CC scheme for Ω_0 a complete model space. Redmon and Bartlett [45] give the second order diagrams of $\hat{\mathbf{H}}_C$ for an incomplete Ω_0 consisting of a Hartree–Fock determinant plus selected singly and doubly excited determinants. We are not aware of any applications of their formalism.

C. Other Effective Hamiltonians

Bloch demonstrates that the left eigenvectors of $\mathbf{H}_B$ emerge as the eigenvectors of the generalized eigenvalue equation obtained by setting $k = PP_0$ in Eq. (2.11). The effective Hamiltonian in this generalized eigenvalue equation is the projection onto Ω_0 of the operator PH that is considered by Kato in his classic study of degenerate perturbation theory [107]. We are unaware of any numerical implementations of this formalism. Nevertheless, because the eigenvectors of this generalized eigenvalue equation are the left eigenvectors of $\mathbf{H}_B$, extension of this formalism to effective operators would necessarily produce one of the **a** definitions obtainable from Bloch's formalism. These are discussed in the next section.

The effective Hamiltonian of Kirtman and of Certain and Hirschfelder [108–112] emerges from a *different* [102] generalization of van Vleck's transformation than that of the canonical formalism. As in the latter, the mapping operators are norm-preserving and are thus of the $(\hat{K}, \hat{K}^\dagger)$ category. The resulting **h** is Hermitian and satisfies a generalized $(2n + 1)$ rule, but has been shown to contain unlinked diagrams begining in fifth order for a complete model space [102, 109]. If van Vleck's transformation method is generalized using *multiple* transformations, yet another norm-preserving mapping formalism is produced with a Hermitian **h** [35, 37]. Similar to Klein's "canonical unitary" transformation, the exponential transformation of Primas [113, 114] simultaneously produces a Hermitian **h** on each degenerate subspace of the whole Hilbert space. The resulting mappings are norm-preserving but differ [37, 82] from the canonical ones and from the other norm-preserving formalisms mentioned above.

The Hermitian **h**, utilized by Westhaus and co-workers in some [103–115] of their numerical computations, is not really canonical as conditions other than (5.11) are applied to fully specify their exponential transformation operator. Suzuki [116] obtains a recursive scheme for a Hermitian **h** which he claims to be $\hat{\mathbf{H}}_C$. This is not the case, however, because his method is based on multiple transformations, which as mentioned above, produce a Hermitian **h** that differs [37] from $\hat{\mathbf{H}}_C$.

Mukherjee and co-workers [15, 117] use an open-shell CC formalism based on a non-Hermitian **h** that does not coincide with Bloch's. The corresponding mappings are of the (K, L) category. They also suggest [117] a Hermitian **h** whose mappings are of the type (5.10) and, thus, are norm-preserving. It is not clear to us whether these mappings are canonical [118]. Both **h**'s are apparently fully linked in complete and some incomplete model spaces. Further studies by Mukherjee [21, 49] of incomplete model space Ω_0 open-shell CC formalisms involve other kinds of **h**'s, which are rederived by Kutzelnigg et al. [50, 51] in their general Fock space study of effective Hamiltonian connectedness in an incomplete Ω_0. The effective Hamiltonian and the mapping operators are represented in terms of Fock space transformation operators W by

$$\mathbf{h} = P_0 \mathbf{1} W^{-1} H W \mathbf{1} P_0 = P_0 W^{-1} H W P_0 \tag{5.15}$$

$$k = W P_0 , \qquad l = P_0 W^{-1} \tag{5.16}$$

where **1** is the unit operator of the full (fixed particle number) Hilbert space to which the model space Ω_0 belongs. Four different types of Fock space transformation operators W are shown to yield an **h** free of disconnected diagrams. Two types of W, the exponential and the normal-order exponential ansatzes, produce a non-Hermitian **h**, and, thus, the corresponding mappings are of the (K, L) category. Because IN is not applied, these mappings and the **h**'s they generate are different from, respectively, (K_B, L_B) and Bloch's effective Hamiltonian $\mathbf{H}_B$, as discussed after (5.3). A third kind of W is unitary, and, consequently, the corresponding mapping operators satisfy $k^\dagger = l$, which is property (1) of

Theorem I of Section III. Hence, this third category is of the norm-preserving variety $(\hat{K}, \hat{K}^{\dagger})$. The last kind of W is not unitary but satisfies $P_0 W^{\dagger} W P_0 = P_0$, which is property (2) of Theorem I. Hence, the corresponding mappings are also norm-preserving.

Other studies of Fock space $\mathbf{h}$ formulations by Kutzelnigg and Koch consider two different definitions of the "diagonal" part of a Fock space operator. Fock space transformations yield different $\mathbf{h}$'s from those produced by the analogous Hilbert space transformations with their second definition [54], but apparently not with their first one [53, 119]. Their numerous W transformation variants yield $\mathbf{h}$'s which can be classified using the same method utilized above with the four transformations that Kutzelnigg et al. find to yield connected $\mathbf{h}$'s. Table IV presents this classification.

The norm-preserving mappings discussed in this subsection and $(\hat{K}_C, \hat{K}_C^{\dagger})$ may all be expressed [37, 65, 102] in the form

$$\hat{K} = U P_0 \tag{5.17}$$

where U is a *unitary* operator of the whole Hilbert space which transforms H into a "block-diagonal" form having no off-diagonal elements between Ω_0 and $\Omega_0^{\perp}$. Of course, U is different for each different $\hat{K}$. (It is easily seen that Eq. (5.17) implies that condition (4) of Theorem I is indeed satisfied.)

It is not clear to us whether the norm-preserving operators from the time-dependent formalism of Johnson and Barranger [121, 122] can necessarily be expressed as in Eq. (5.17). These authors prove that if the "time-base" of their folded diagrams is chosen to preserve the symmetry between past and future, then the resulting $\mathbf{h}$ is Hermitian and the norms of arbitrary functions in Ω_0 are preserved. This last property implies by Theorem I, property (8) that the corresponding mappings are norm-preserving. An infinite number of possible time-base choices conserve the above symmetry, and, thus, lead to as many different norm-preserving mapping formalisms. Disconnected diagrams never appear in their complete model space formalism.

Banerjee et al. [123, 124] follow a method similar to that of Johnson and Barranger to obtain an apparently Hermitian $\mathbf{h}$. Because no relation between model and true eigenfunctions is provided, it is not clear to us if the mappings are of the category $(\hat{K}, \hat{K}^{\dagger})$, or of any subcategories, for example, $(\tilde{K}, \tilde{L})$, of non-norm-preserving mappings that produce a Hermitian $\mathbf{h}$. Extension of this work to an incomplete Ω_0 is provided by Haque and Mukherjee [46] who find that this $\mathbf{h}$ contains disconnected diagrams.

TABLE IV

Classification of Mapping Operators Corresponding to the Fock Space Transformation W Variants of Kutzelnigg and Koch

Fock Space "Diagonal" Definition	Variant	W Unitary	Effective Hamiltonian **h** Hermitian	Corresponding Mapping Category
1^a	a	No	No	$(K, L)^b$
	b	Yes	Yes	$(\hat{K}, \hat{K}^{\dagger})$
	b'^c	Yes	Yes	$(\hat{K}, \hat{K}^{\dagger})$
	c	Yes	Yes	$(\hat{K}, \hat{K}^{\dagger})$
	c'^c	Yes	Yes	$(\hat{K}, \hat{K}^{\dagger})$
2^d	a	No	No	(K, L)
	$\tilde{a}^e$	No	No	$(K, L)^f$
	b	Yes	Yes	$(\hat{K}, \hat{K}^{\dagger})$
	c	Yes	Yes	$(\hat{K}, \hat{K}^{\dagger})$
	c'^c	Yes	Yes	$(\hat{K}, \hat{K}^{\dagger})$
	d^g	No	Yes	$(\hat{K}, \hat{K}^{\dagger})^{h,i}$
	d'^c	No	Yes	$(\hat{K}, \hat{K}^{\dagger})^h$

^{a}[53].

^{b}The IN is apparently used implying the mappings are (K_B, L_B).

^{c}Modification of the "normalization" of the preceeding variant to obtain a $(2n + 1)$ rule. The resulting **h** is different but the unitarity of W and, thus, the corresponding mapping category are apparently unaffected.

^{d}[54].

^{e}Produces the same Fock space effective Hamiltonian and, thus, Hilbert space **h** as variant a.

^{f}It is not clear to us whether these mappings are the same as those for variant a [120].

^{g}Produces the same Fock space effective Hamiltonian and, thus, Hilbert space **h** as variant b.

^{h}The relations $\mathbf{h} = P_0 W^{\dagger} H W P_0$, $k = W P_0$, and $l = P_0 W^{\dagger}$ hold here instead of (5.9), (5.10). Hence, $k^{\dagger} = l$, which by Theorem I, property (1) implies that (k, l) are norm-preserving.

^{i}Similarly to variant $\tilde{a}$, it is not clear to us whether these mappings are the same as those for variant b.

VI.　REVIEW OF EFFECTIVE OPERATOR FORMALISMS AND CALCULATIONS

Some expressions that appear in the literature for effective operators **a** are quite complicated because of the particular perturbation expansion or recursive scheme used and/or because of the introduction of a core-valence separation. Section II presents each **a** definition only in its compact formal form. These simpler expressions have the advantage of enabling the **a** structures to be studied more readily. Some literature **a** definitions are in a Brillouin–Wigner (BW) form and therefore depend on the (exact) energies of the model eigenfunctions. More precisely, the associated mapping operators are state-dependent. A given d-dimensional model space Ω_0 yields d such BW mapping operators, which, strictly speaking, do not correspond to the kind of mappings considered in Section II. However, upon removal of the energy dependence through Rayleigh–Schrödinger (RS) type expansions, etcetera, these d mappings become identical, and the **a** definitions are transformed into the types considered in Section II. Clearly, the RS and BW types of **a** are closely related as they operate on the same model functions and produce the same results if computed exactly. Consequently, they may be considered to be different versions of the same operator, just as H_B and the BW effective Hamiltonian used as the starting point in Brandow's derivation [8] of H_B (see Section V.A). Some recursive schemes mentioned in Section V are based on a BW formalism, and, thus, their extension to definitions of **a** yield BW type operators. It is thus convenient to classify together BW and RS versions of the same **a**, although they generally have different computational advantages and drawbacks.

We first examine the **a** definitions for Bloch's formalism. The only **a** definition possible with the canonical formalism is considered next, followed by **a** definitions based on other norm-preserving mappings which have been suggested. Considerably fewer calculations exist for **a** than for **h**. Ellis and Osnes [32, 125] review **a** calculations made in nuclear physics, which, as we discuss below, are all effectively decomposed into one-dimensional calculations. We also discuss the few **a** calculations performed in the context of atomic and molecular physics.

A.   Effective Operators Based on H_B

Brandow [8] studies the effective operator

$$K_B^\dagger A K_B / N_\alpha^B N_\beta^B \equiv A_{\alpha\beta}^{IB} , \quad {}_0\langle\alpha_B'|, |\beta_B'\rangle_0 \tag{6.1}$$

where the norm of the true eigenfunction corresponding to the unity normalized $|\delta_B'\rangle_0$ is

$$N_\delta^B = {}_0\langle\delta_B'|K_B^\dagger K_B|\delta_B'\rangle_0^{1/2} = (1 + x_\delta)^{1/2} , \quad \chi_\delta = {}_0\langle\delta_B'|K_B^\dagger K_B - P_0|\delta_B'\rangle_0 \tag{6.2}$$

and where (6.1) indicates the model functions used. Equation (6.1) defines a state-dependent effective operator of the $A_{\alpha\beta}^I$ category. Brandow's derivation is similar to the one he uses for H_B. It shows that the RS diagrammatic expansion of $A_{\alpha\beta}^{IB}$ is not "fully linked". Brandow's analysis implies that unlinked terms also appear in some other **a** definitions, as we shall show elsewhere [22]. Brandow's final expression appears quite different from (6.1) and (6.2) because he introduces a core-valence separation. A similar remark holds for the $\mathbf{A}_{\alpha\beta}^{IB}$ expression of Krenciglowa and Kuo [126]. These authors use essentially the same formalism as Kuo et al. [89] for $\mathbf{H}_B$ (see Section V.A), and, consequently, their $\mathbf{A}_{\alpha\beta}^{IB}$ expression is not "fully linked". Oberlechner et al. [88] derive $\mathbf{A}_{\alpha\beta}^{IB}$ with the same time-dependent formalism they apply to obtain $\mathbf{H}_B$ (see Section V.A), and their discussion of its properties is merely a summary of Brandow's.

Leeinas and Kuo [127] use both the BW and the RS forms in a model calculation in which Ω_0 is one-dimensional, whereupon Eqs. (6.1) and (6.4) reduce to

$$K_B^\dagger A K_B (N^B)^{-2} = \mathbf{A}_{ii}^{IB} , \quad {}_0\langle i|, |i\rangle_0 \tag{6.3}$$

$$(N^B)^2 = {}_0\langle i|K_B^\dagger K_B|i\rangle_0 \tag{6.4}$$

where $|i\rangle_0$ is the (unity normed) basis ket defining Ω_0. These are merely the equations of ordinary, that is, nondegenerate, perturbation theory and lead to a fully linked RS expansion [128] for the combined form (see Section II.D.2) of $\mathbf{A}_{ii}^{IB}$. Leeinas and Kuo find that the combined and non-combined forms of $\mathbf{A}_{ii}^{IB}$ yield different results and that the latter converges better [129]. Harvey and Khanna [130] begin with the BW form of $\mathbf{A}_{\alpha\beta}^{IB}$ in a nonrigorous analysis of many-body effective operators, in particular, of the effective electromagnetic operator of nuclear physics. They apply [131] their theory to a semi-empirical study of the "effective charge" (essentially the quadrupole moment operator) of protons and neutrons for nuclei in the ^{16}O region.

Ellis and Osnes [32, 125] review **a** calculations in nuclear physics. The latter have apparently been limited to mass 17 nuclei effective charge calculations that use RS expansions for a very special form of $\mathbf{A}_{\alpha\beta}^{IB}$. Mass 17 nuclei have one particle outside a closed shell, and, consequently, the d-dimensional model space Ω_0 reduces to the direct sum of d separate one-dimensional model spaces. This reduction occurs because H_0 has enough exact symmetries of the full Hamiltonian that $\mathbf{H}_B$ is diagonal in the one-particle basis set [59]. Each model subspace Ω_0^α thus yields a different wave operator K_β^α. Diagonal matrix elements of A are computed using $\mathbf{A}_{ii}^{IB}$ of Eq. (6.3) in which K_B is replaced by K_B^α. The off-diagonal matrix elements of A are evaluated using model eigenfunctions in different model subspaces and the operator $K_B^{\alpha\dagger} A K_B^\beta / N_\alpha^B N_\beta^B$. As Ellis and Osnes discuss, the RS expansion of this special $\mathbf{A}_{\alpha\beta}^{IB}$ form is fully linked because it is essentially one-dimensional.

The $\mathbf{A}^{II}$ category definition

$$L_B A K_B \equiv \mathbf{A}_B^{II}, \quad {}_0\langle \bar{\alpha}_B|, |\beta_B\rangle_0 \qquad (6.5)$$

is suggested by Schucan and Weidenmüller [132] in a formal study of intruder state problems. No attention is paid there to obtaining the necessary model eigenfunctions, which, as discussed in Section II.D, necessitates computing the norms N_α^B. Subsequent work [63] by Weidenmüller, however, notes this problem. Leeinas and Kuo [127] use the RS forms of $\mathbf{A}_B^{II}$ in the one-dimensional model calculation mentioned above [129]. Notice that the one-dimensional case reduces $\mathbf{A}_B^{II}$ identically to the $\mathbf{A}_{\alpha\beta}^{II}$ category definition

$$L_B A K_B N_\alpha^B (N_\beta^B)^{-1} \equiv \mathbf{A}_{\alpha\beta}^{IIB}, \quad {}_0\langle \bar{\alpha}_B'|, |\beta_B'\rangle_0 \qquad (6.6)$$

because the normalization factors in (6.6) cancel each other. Consequently, a one-dimensional $\mathbf{A}_B^{II}$ calculation is performed using the (unity normed) model eigenfunctions $|\alpha_B'\rangle_0 = |i\rangle_0 = |\bar{\alpha}_B'\rangle_0$.

B. Effective Operators Based on Norm-Preserving Transformations

Section II.C demonstrates that only one **a** definition emerges from any formalism that involves norm-preserving mappings. Many such formalisms are mentioned in Section V, and most of the corresponding **a** definitions have been suggested in the literature. All these definitions belong to category $\hat{\mathbf{A}}$ of Table I.

Brandow [8] considers the effective operator

$$\hat{K}_C^\dagger A \hat{K}_C \equiv \hat{\mathbf{A}}_C, \quad {}_0\langle \hat{\alpha}_C|, |\hat{\beta}_C\rangle_0 \qquad (6.7)$$

where $\hat{K}_C$ and $\{|\hat{\alpha}_C\rangle_0\}$ are, respectively, the canonical wave operator and the orthonormalized eigenfunctions of $\hat{\mathbf{H}}_C$ discussed in Section V.B. Brandow claims to have proven that the RS expansion of $\hat{\mathbf{A}}_C$ is fully linked for a complete, degenerate or quasi-degenerate model space Ω_0. This same **a** definition has also been suggested by Klein [65] and by Kvanisčka [106] in a coupled cluster (CC) reformulation of the canonical formalism (see Section V.B). Hurtubise and Freed [133] show that the RS expansion is fully linked through second order in the perturbation V for a complete model space, and they obtain all the diagrams through that order. Sun and co-workers [56–58] use these results to calculate dipole and transition moments for many states of CH, CH^+, and OH by evaluating the effective dipole operator through first order in V. To our knowledge, these are the *first* **a** computations that do *not* reduce to one-dimensional calculations.

Jørgensen [35] introduces the general definition

$$P_0 U^\dagger A U P_0 \equiv \hat{\mathbf{A}}_J \qquad (6.8)$$

where U is the *unitary* operator appearing in the various norm-preserving mappings of Eq. (5.17). The kets and bras operated upon by $\hat{\mathbf{A}}_J$ are, respectively, the orthonormalized right eigenfunctions of the effective Hamiltonian $P_0 U^\dagger H U P_0$ and the corresponding bras. Although Jørgensen's suggestion appears in a paper on the many-transformation version of generalized van Vleck theory, the definition applies for any operator U satisfying the above conditions and, thus, to all **h**'s obtained with norm-preserving mappings expressible as in Eq. (5.17). This includes the canonical mappings, as shown by Jørgensen [37], and the other norm-preserving mappings mentioned in Section V.C with possible exceptions [46, 121–124] noted there.

The **a** definition of Harris [134] represents a special case of (6.8) because it arises from a van Vleck formalism. Similar considerations apply to that of Westhaus et al. [103] since, as mentioned in Section V.C, they use mappings which are *not* canonical, but which can be written as in Eq. (5.17). Westhaus et al. introduce approximate mapping operators to obtain the one- and two-body parts of their **a** when A is a one-body operator. Finally, Johnson and Barranger [121, 122] suggest an **a** definition of the form $\hat{\mathbf{A}}$ for any one of the norm-preserving mappings possible from their formalism (see Section V.C).

C.   Special Form of a for Expectation Values

Kuo et al. [89, 126] and Lindgren and co-workers [31, 135] introduce an "effective operator" that is based on Bloch's effective Hamiltonian $\mathbf{H}_B$ and that may be used only for the calculation of *diagonal* matrix elements, that is, expectation values. The latter authors apply their formalism to hyperfine interactions in atoms containing one valence electron beyond a closed shell. A similar definition may actually be derived for any **h** formulation [71]. These diagonal "effective operator" definitions have serious limitations not shared by the more general **a** definitions of Table I. Besides being applicable only to diagonal matrix elements of a true operator A, the operator A is *not* permitted to lift the degeneracy of the true states of interest. This last essential restriction, although not generally mentioned, is implicitly assumed in the literature derivations. In order to better understand these limitations, we now review and clarify the derivation of the diagonal "effective operator". This process also determines the correct model eigenbra that must be used with this diagonal "effective operator".

We assume A to be Hermitian since a non-Hermitian A may always be decomposed into a pair of Hermitian $X_1 \equiv (A + A^\dagger)/2$ and $X_2 \equiv i(A - A^\dagger)/2$ which may be considered separately. The auxiliary Hamiltonian

$$H_\mu \equiv H + \mu A \tag{6.9}$$

is Hermitian and has the set of eigenfunctions

$$H_\mu | \phi_\alpha^\mu \rangle = E_\alpha^\mu | \phi_\alpha^\mu \rangle \tag{6.10}$$

It is convenient for later use not to require that the eigenfunctions of (6.10) be unity normalized. Consequently, the eigenvalues of (6.10) are written as

$$\langle \phi_\alpha^\mu | H_\mu | \phi_\alpha^\mu \rangle = E_\alpha^\mu \langle \phi_\alpha^\mu | \phi_\alpha^\mu \rangle \tag{6.11}$$

Taking the derivative with respect to μ on both sides of (6.11) and using (6.10) gives

$$\langle \phi_\alpha^\mu | \partial H_\mu / \partial \mu | \phi_\alpha^\mu \rangle = \langle \phi_\alpha^\mu | \phi_\alpha^\mu \rangle \, \partial E_\alpha^\mu / \partial \mu \tag{6.12}$$

which is just the Hellmann–Feynman theorem. The limit $\mu \rightarrow 0$ of Eq. (6.12) and (6.9) may be combined to yield

$$\langle \phi_\alpha^0 | A | \phi_\alpha^0 \rangle / \langle \phi_\alpha^0 | \phi_\alpha^0 \rangle = \lim_{\mu \to 0} \partial E_\alpha^\mu / \partial \mu \tag{6.13}$$

where

$$| \phi_\alpha^0 \rangle \equiv \lim_{\mu \to 0} | \phi_\alpha^\mu \rangle \tag{6.14}$$

The left hand side of (6.13) is the desired diagonal matrix element of A only if $| \phi_\alpha^0 \rangle$ is the true eigenfunction of interest with eigenvalue E_α,

$$| \phi_\alpha^0 \rangle = | \phi_\alpha \rangle \tag{6.15}$$

Equations (6.9) and (6.10) imply that $| \phi_\alpha^0 \rangle$ is an eigenfunction of H with eigenvalue $\lim_{\mu \to 0} E_\alpha^\mu$. If this eigenvalue is degenerate, then it is well known [136] that $| \phi_\alpha^0 \rangle$ is one of the eigenvectors obtained by diagonalizing A in the degenerate subspace. Thus, in order to transform (6.13) into

$$\langle \phi_\alpha | A | \phi_\alpha \rangle / \langle \phi_\alpha | \phi_\alpha \rangle = \lim_{\mu \to 0} \partial E_\alpha^\mu / \partial \mu \tag{6.16}$$

it is necessary to assume that A cannot lift the degeneracy of the true state $|\phi_\alpha\rangle$.

The next step involves transforming the right hand side of (6.16) into a calculation in the model space Ω_0. Given a transformation (k, l) from H to $\mathbf{h}$, the modified mappings (k_μ, l_μ), generated by replacing $V \rightarrow V + \mu A$ in (k, l), transform H_μ into an effective Hamiltonian with the same eigenvalues E_α^μ. Substituting this effective Hamiltonian into the right hand side of (6.16) produces an operator of Ω_0 that is bracketed between two model eigenfunctions, thus leading to an "effective operator" definition for diagonal matrix elements of A. It may be shown that this is possible for any choice of $\mathbf{h}$ [71]. Here the discussion is limited to the non-Hermitian $\mathbf{h}$ produced by the mappings (K, L), as this encompasses the effective Hamiltonian $\mathbf{H}_B$ used most frequently in the literature.

Section II.D shows how the mappings (K, L) generate the non-Hermitian effective Hamiltonian $\mathbf{H} = LHK$ corresponding to H,

$$\mathbf{H}|\alpha'\rangle_0 = E_\alpha|\alpha'\rangle_0 \,, \qquad {}_0\langle\bar{\alpha}'|\mathbf{H} = {}_0\langle\bar{\alpha}'|E_\alpha \tag{6.17}$$

where the normalizations are specified in (2.30). Replacing V by $V + \mu A$ transforms H into H_μ of Eq. (6.9) and, thus, $\mathbf{H}$ into the effective Hamiltonian

$$\mathbf{H}_\mu \equiv K_\mu H_\mu L_\mu \tag{6.18}$$

corresponding to H_μ,

$$\mathbf{H}_\mu|\alpha'_\mu\rangle_0 = E_\alpha^\mu|\alpha'_\mu\rangle_0 \tag{6.19}$$

$$_0\langle\bar{\alpha}'_\mu|\mathbf{H}_\mu = {}_0\langle\bar{\alpha}'_\mu|E_\alpha^\mu \tag{6.20}$$

The $K_\mu(L_\mu)$ in (6.18)–(6.20) designate $K(L)$ with the replacement $V \rightarrow V + \mu A$, and the normalizations are

$$_0\langle\alpha'_\mu|\alpha'_\mu\rangle_0 = 1 \,, \qquad {}_0\langle\bar{\alpha}'_\mu|\beta'_\mu\rangle_0 = \delta_{\alpha\beta} \tag{6.21}$$

Expanding K_μ and L_μ in powers of μ converts Eq. (6.18) into

$$\mathbf{H}_\mu \equiv \mathbf{H} + \mu\mathbf{H}^{\{1\}} + \mu^2\mathbf{H}^{\{2\}} + \cdots \tag{6.22}$$

where $\mathbf{H}^{\{j\}}$ results from replacing j operators V in each term of a perturbation expansion for $\mathbf{H}$ (of order $\geq j$) by A in all possible ways.

The eigenvalues E_α^μ can be written as

$$_0\langle \bar{\alpha}_\mu' | \mathbf{H}_\mu | \alpha_\mu' \rangle_0 = E_\alpha^\mu \tag{6.23}$$

Taking the derivative with respect to μ in (6.23) and using Eqs. (6.19), (6.20), and (6.21) gives

$$\partial E_\alpha^\mu / \partial \mu = {}_0\langle \bar{\alpha}_\mu' | \partial \mathbf{H}_\mu / \partial \mu | \alpha_\mu' \rangle_0 \tag{6.24}$$

which is the Hellmann–Feynman theorem for $\mathbf{H}$ with model eigenfunctions normed according to (6.21). Because the norm of the true eigenfunction $|\phi_\alpha^\mu\rangle$ has been left arbitrary, it can now be chosen to be that prescribed by the non-norm-preserving (K_μ, L_μ), that is, for these mappings Eqs. (2.26) and (2.39) read

$$|\alpha_\mu'\rangle_0 = L_\mu |\phi_\alpha^\mu\rangle \tag{6.25}$$

$$_0\langle \bar{\alpha}_\mu' | L_\mu = (N_\alpha^\mu)^{-2} \langle \phi_\alpha^\mu | \tag{6.26}$$

where $N_\alpha^\mu = {}_0\langle \alpha_\mu' | K_\mu^\dagger K_\mu | \alpha_\mu' \rangle_0$ is the norm of $|\phi_\alpha^\mu\rangle$. The limit $\mu \to 0$ of Eqs. (6.25) and (6.26) nay be combined with the assumption (6.12) and the $\lim_{\mu \to 0} L_\mu = L$ to produce

$$\lim_{\mu \to 0} |\alpha_\mu'\rangle_0 = |\alpha'\rangle_0 \ , \qquad \lim_{\mu \to 0} |\bar{\alpha}_\mu'\rangle_0 = |\bar{\alpha}'\rangle_0 \tag{6.27}$$

Consequently, Eqs. (6.24) and (6.22) yield

$$\lim_{\mu \to 0} \partial E_\alpha^\mu / \partial \mu = {}_0\langle \bar{\alpha}' | \mathbf{H}^{\{1\}} | \alpha' \rangle_0 \tag{6.28}$$

Comparing (6.28) with (6.16) gives the Hellmann–Feynman "effective operator"

$$\mathbf{H}^{\{1\}} = \mathbf{A}^{\text{HF}} , \quad {}_0\langle \alpha' |, |\alpha'\rangle \tag{6.29}$$

valid only for diagonal matrix elements of A.

Notice that $\mathbf{A}^{\text{HF}}$ operates on the unity-normed right eigenkets and on their bi-orthonormal complement. Consequently, the operator $\mathbf{A}_B^{\text{HF}}$ within Bloch's formalism operates on $_0\langle \bar{\alpha}_B' |$ and $|\alpha_B'\rangle_0$. The first expression of Kuo et al. [89] for $\mathbf{A}_B^{\text{HF}}$ is therefore incorrect because it is bracketed between $_0\langle \alpha_B' |$ and $|\alpha_B'\rangle_0$. However, this mistake is corrected in a later publication [126]. The $\mathbf{A}_B^{\text{HF}}$ expression of Lindgren and co-workers [31,

135] is the same as the first expression of Kuo et al., and, hence, is also incorrect. However, the former authors perform calculations in a d-dimensional model space Ω_0 which reduces by symmetry to the direct sum of d one-dimensional model spaces, just as with the model spaces used in the nuclear physics a calculations reviewed by Ellis and Osnes. As noted above (see (6.6)), the one-dimensional Ω_0 has its model eigenfunctions simply given by $|\alpha_B'\rangle = |i\rangle_0 = |\bar{\alpha}_B'\rangle_0$. Thus, the A_B^{HF} of Lindgren and co-workers is indeed correct for their calculations. Nevertheless, such an expression is definitely incorrect for a general d-dimensional Ω_0 [137].

We are not aware of any simply useable literature formula that relates *off-diagonal* matrix elements of A to the derivative of the energy as in Eq. (6.16). It is easy to show that the "effective operator" in Eq. (6.29) does not provide off-diagonal matrix elements. If it did, then it would be the same as $\mathbf{A}^{II}$ of Table I and, in particular, for Bloch's formalism we would have

$$\mathbf{A}_B^{II} = \mathbf{H}_B^{\{1\}} \qquad (6.30)$$

where $\mathbf{A}_B^{II}$ is given in (6.5). However, the perturbation expansions of both sides of Eq. (6.30) differ starting in second order in V for a degenerate model space. Similar differences are found [71] between the perturbation expansions of $\hat{\mathbf{A}}_C$ of (6.7) and the Hellmann–Feynman "effective operator" $\hat{\mathbf{H}}_C^{\{1\}}$ of the canonical formalism. Consequently, Hellman–Feynman "effective operators" generally differ from the a that acts on the same eigenbras and eigenkets, and, thus, cannot provide off-diagonal matrix elements of A.

VII. DISCUSSION

Effective operators a are defined by requiring that they produce *exact* matrix elements $A_{\alpha\beta}$ of a time independent full Hilbert space operator A when acting between the appropriate eigenbras and eigenkets of an effective Hamiltonian h. Most possible definitions of h can generate several possible a definitions depending on various choices for the eigenvectors of h and on whether the mappings between these eigenvectors and those of H, respectively called model and true eigenvectors, preserve the eigenvector norms. This paper analyzes whether the different a definitions are independent of the states α and β, conserve commutation relations and symmetries, and maintain the Hermiticity of operators. These are desirable properties since it is generally preferable for effective operators to be as similar to their true counterparts as possible. State independence is actually necessary for certain applications, for example, when studying model or semi-empirical operators that are defined as state-independent. Conservation of symmetries is also important computationally since it permits block diagonalization of the effective Hamiltonian. Some of these properties may be sacrificed for numerical convenience when effective Hamiltonians and operators are used strictly as computational tools. For example, many workers employ Bloch's non-Hermitian effective Hamiltonian [138]. However, when approximating non-Hermitian $\mathbf{h}$'s there is no guarantee that the eigenvalues remain real. The non-Hermiticity of $\mathbf{h}$ thus represents a serious drawback. Some computational aspects of the various $\mathbf{a}$ definitions are also discussed.

State-independent effective operators can be applied directly to arbitrary bras and kets of the model space Ω_0. However, when used with state-dependent effective operators these vectors must be expanded in the basis of the model eigenbras and eigenkets, respectively, that are utilized to produce the $A_{\alpha\beta}$ with the latter operators. Consequently, as shown in Section IV.B, general expressions for state-dependent effective operators are more complicated than for state-independent ones. State dependence of effective operators is thus a conceptual drawback.

We demonstrate that commutation relations between arbitrary operators are not generally preserved by effective Hamiltonians and operators, but that some effective operator definitions conserve particular commutation relations. The most important preserved commutation relations are those involving H or constants of the motion. Table III shows that each type of mapping [139] between model and true eigenfunctions produces at least one effective operator definition that conserves the commutators $[H, C]$ and $[C, D]$, where C and D are constants of the motion. Hence, nondegenerate model eigenvectors automatically have the symmetries associated with constants of the motion, and degenerate ones can be chosen to have these symmetries. In turn, choosing the degenerate model eigenfunctions this way ensures that their true counterparts indeed have these symmetries. If a complete set of observables exists, the degenerate true eigenvectors are then mutually orthogonal. This is particularly useful with non-norm-preserving mappings since they generally do not conserve the overlaps between degenerate eigenvectors [140]. If a complete set of observables is lacking, the degenerate model eigenfunctions must then be chosen by a more complicated calculation which corresponds to an orthogonalization of their true counterparts.

Paper II shows how the preservation of symmetries through the conservation of $[H, C]$ is greatly simplified when the true constant of the motion C commutes with H_0 and V separately. This case is the only one previously considered in the literature. Our more general theorems must be used when it is not convenient or possible to choose H_0 as commuting with every constant of the motion of interest.

Section IV proves that the conservation of the commutation relation (4.12) between H and the position operator $\vec{r}$ leads to the equivalence of the dipole length and dipole velocity transition moments computed with certain effective operator definitions. Contrary to the similar equivalence for transition moments computed with true operators, however, this does not yield a sum rule. Many other sum rules follow from commutation relations between true operators. In view of the many useful applications of sum rules [141, 142] the existence of sum rules for quantities computed using effective operators is of interest and will be studied elsewhere [79]. A potential application lies in determining the amount, or proportion, of transition strengths carried by a particular state or group of states [142, 143].

Some semi-empirical theories of chemical bonding [60–62] use the commutation relation (4.12) with the true operators replaced by model ones to obtain additional relations between the parameters of the theories. This yields a reduction in the number of these parameters and, often a simpler determination of their values. It is also hoped [62] that this improves the theory by building in the "right" physics. This belief that "useful" theorems retain their validity upon replacement of nonempirical quantities by semi-empirical ones is commonly held by semi-empiricists [144] but remains, in fact, an assumption. Our analysis of the conservation of this commutation relation in Section IV is useful in this context.

Semi-empirical Hamiltonians and operators are taken to be state independent [56] and have the same Hermiticity as their true counterparts. Consequently, the valence shell effective Hamiltonians and operators they mimic must also have these two properties. Table I shows that the effective Hamiltonian and operator definitions $\hat{H}$ and $\hat{A}$, as well as $\tilde{H}$ and either $\tilde{A}^I$ or $\tilde{A}^{IV}$ fulfill these criteria. Thus, these definition pairs may be used to derive the valence shell effective Hamiltonians and operators mimicked by the semi-empirical methods. Table III indicates that the commutation relation (4.12) is preserved by all three definition pairs. Hence, the validity of the relations derived from the semi-empirical version of (4.12) depends on the extent to which the semi-empirical Hamiltonians and operators actually mimic, respectively, *exact* valence shell effective Hamiltonians and operators. In particular, the latter Hamiltonians and operators contain higher-body terms which are neglected, or ignored, in semi-empirical theories. These nonclassical higher body interactions have been shown to be nonnegligible for the valence shell Hamiltonians of many atoms and molecules [27, 145–149] and for the dipole moment operators of some small molecules [56–58]. There is no a priori reason to believe that Eq. (4.12) is still conserved by effective operators if higher body interactions are neglected or are averaged into the fewer body operators. In fact, the neglect of higher body terms may destroy the conservation of the commutation relation between $\vec{r}$ and H. Semiempirical Hamiltonians effectively average higher body interactions into the traditional ones, [27, 149, 150] and the preservation of Eq. (4.12) by effective operators may provide useful constraints on the nature of this averaging for both the semi-empirical Hamiltonians and operators.

It is of interest, for both computational and theoretical considerations, to compare the various effective operator definitions. Such a comparison should include the relative computational labor involved by each definition. This paper is concerned more with formal than computational aspects. Hence, we limit ourselves here to a few remarks and present further comparisons in paper II. Non-norm-preserving mappings are demonstrated in this paper to produce formal and computational problems that are absent with norm-preserving ones. They necessitate the calculation of normalization factors in one form or another. These factors cause the effective operator to be state-dependent if they are computed separately from the "rest" of the operator, for example, $K^{\dagger}AK$ for definition $A_{\alpha\beta}^{I}$, or combined with it. Alternatively, they can be incorporated into either new mapping operators or into new model eigenfunctions. Both of the latter choices produce state-independent effective operators that are desirable, especially for certain formal or theoretical work. Non-norm-preserving mappings also generally require special calculations in order to ensure the orthogonality of the degenerate true eigenvectors. In contrast, norm-preserving mappings generate state-independent effective operators which yield the desired matrix elements with orthogonalized model eigenfunctions.

Non-Hermitian effective Hamiltonians are proven to be generated only by non-norm-preserving mappings, but Hermitian effective Hamiltonian can be produced by both norm-preserving mappings and non-norm-mappings. Consequently, literature derivations of effective Hamiltonians must be examined in order to determine which type of mapping is actually used. This is not necessarily a simple task since some formalisms do not directly use mappings operators. However, the properties of the mapping operators determine the correct effective operator definitions and procedures for orthogonalizing degenerate true eigenvectors. Section V classifies most of the literature effective Hamiltonian formalisms according to Tables I and II. While previous applications of effective Hamiltonians are not always represented in terms of mapping operators, these mappings provide a useful taxonomical scheme that automatically yields a wealth of properties for the effective operators.

Here we examined only properties of *exact* effective operators. Actual calculations require approximations. Although state dependence and independence are not modified by approximations, conservation of commutation relations may be affected. This feature is somewhat analogous to the use of the semi-empirical version of Eq. (4.12) discussed above. Paper II addresses the conservation of commutation relations and how symmetries can be preserved when perturbation expansions for effective Hamiltonians and operators are truncated at a finite order.

APPENDIX A: FURTHER STUDY OF MAPPINGS OF THE CATEGORY $(\bar{K}, \bar{L})$

The first part demonstrates that all mappings of the category $(\bar{K}, \bar{L})$, that is, which produce a non-Hermitian **h** and conserve the norms for a set of eigenvectors, can be obtained by transforming the mappings (K, L) that generate the same **h** but do not conserve these norms. The transformation incorporates the normalization factors needed with (K, L) into the mappings $(\bar{K}, \bar{L})$. A particular mapping pair $(\bar{K}, \bar{L})$ is related in this process to a nondenumerable infinity of (K, L) mappings. Part 2 shows that although the $(\bar{K}, \bar{L})$ mappings conserve the norms of eigenvectors, they are not norm-preserving because they do not conserve the overlaps between eigenvectors. Proofs are presented for mappings $(\bar{K}, \bar{L})$ conserving norms of right eigenvectors, but the treatment for left eigenvector norm preservation is similar [71].

1.   Relation Between Mappings $(\bar{K}, \bar{L})$ and (K, L)

Let $(\bar{s}, \bar{t})$ be an arbitrary $(\bar{K}, \bar{L})$ mapping pair. Right eigenvectors $|\alpha'\rangle_0$ of the effective Hamiltonian

$$\bar{\mathbf{h}} \equiv \bar{t} H \bar{s} \tag{A.1}$$

produced by these mappings, have true counterparts $|\Psi_\alpha\rangle$,

$$|\Psi_\alpha\rangle = \bar{s}|\alpha'\rangle_0 \ , \qquad |\alpha'\rangle_0 = \bar{t}|\Psi_\alpha\rangle \ . \tag{A.2}$$

By hypothesis, $|\Psi_\alpha\rangle$ and $|\alpha'\rangle_0$ have the same norm, which is chosen without any loss of generality as unity.

Because $\bar{\mathbf{h}}$ is non-Hermitian, the mappings $(\bar{s}, \bar{t})$ do not conserve angles. As described in Section II.D, degenerate model eigenfunctions may be taken such that their true counterparts are mutually orthogonal. Thus, the true eigenvectors form an orthonormal set,

$$\langle \Psi_\alpha | \Psi_\beta \rangle = \delta_{\alpha,\beta} \tag{A.3}$$

but the $|\alpha'\rangle_0$ do not. Their bi-orthonormal complements $_0\langle\bar{\beta}'|$ obey (see (2.30))

$$_0\langle\bar{\beta}'|\alpha'\rangle_0 = \delta_{\alpha,\beta} \tag{A.4}$$

Equations (A.2)–(A.4) generate the formal expressions for $\bar{s}$ and $\bar{t}$

$$\bar{s} = \sum_\beta |\Psi_\beta\rangle \,_0\langle\bar{\beta}'|, \qquad \bar{t} = \sum_\beta |\beta'\rangle_0 \,\langle\Psi_\beta| \tag{A.5}$$

Substituting Eqs. (A.5) into (A.1) produces

$$\bar{\mathbf{h}} = \sum_\beta |\beta'\rangle_0 \, E_\beta \,_0\langle\bar{\beta}'| \tag{A.6}$$

Now introduce a set of new model eigenfunctions $|\alpha\rangle_0$ through the scaling transformation

$$|\alpha\rangle_0 \equiv c_\alpha^{-1}|\alpha'\rangle_0 \tag{A.7}$$

where c_α is an arbitrary nonzero real number and $c_\alpha \neq 1$ for at least one α. Their bi-orthonormal complements $_0\langle\bar{\beta}|$ are related to those of $|\alpha'\rangle_0$ by

$$|\bar{\alpha}\rangle_0 = c_\alpha|\bar{\alpha}'\rangle_0 \tag{A.8}$$

Substituting (A.7) and (A.8) into (A.6) yields

$$\bar{\mathbf{h}} = tHs \tag{A.9}$$

which is written in terms of the transformed mapping operators

$$s \equiv \sum_\beta |\Psi_\beta\rangle \,_0\langle\bar{\beta}| = |\Psi'_\beta\rangle \,_0\langle\bar{\beta}'| \tag{A.10}$$

$$t \equiv \sum_\beta |\beta\rangle_0 \,\langle\Psi_\beta| = |\beta'\rangle_0 \, c_\beta^{-2}\langle\Psi'_\beta| \tag{A.11}$$

and the true eigenfunctions

$$|\Psi'_\alpha\rangle \equiv c_\alpha|\Psi_\alpha\rangle \tag{A.12}$$

where $|\Psi'_\alpha\rangle$ has a nonunity norm. The (s, t) mappings belong to the category (K, L) because they generate a non-Hermitian effective Hamiltonian ($\bar{\mathbf{h}}$) and relate eigenvectors having different norms, for example, $|\alpha'\rangle_0$ to $|\Psi'_\alpha\rangle$, $|\alpha\rangle_0$ to $|\Psi_\alpha\rangle$ and $_0\langle\bar\alpha|$ to $\langle\Psi_\alpha|$. Indeed, (A.10) and (A.11) are isomorphic to expressions (2.35) and (2.36) for the mappings (K, L) because Eq. (A.12) displays c_α as the norm of the true eigenfunction $|\Psi'_\alpha\rangle$. Consequently, applying the transformation $(K, L) \rightarrow (\bar K, \bar L)$ of Section II.D to (s, t) produces the mappings $(\bar s, \bar t)$ (compare Eqs. (A.5) to (2.52), (2.53)). Finally, because the c_α can be chosen arbitrarily, the mappings $(\bar s, \bar t)$ are related to an infinite number of mappings (s, t).

2.   The Mappings $(\bar K, \bar L)$ Are Not Norm-Preserving

Mappings $(\bar K, \bar L)$ conserve the norms of the right eigenfunctions. In particular, they relate by (A.2) the unity normalized eigenvectors $|\alpha'\rangle_0$ and $|\Psi_\alpha\rangle$. This, however, does not imply that they conserve the norm of an *arbitrary* vector because angles between basis vectors differ in these two sets of eigenfunctions. Consider an arbitrary vector $|\gamma\rangle_0$ of Ω_0 and its true counterpart $|\gamma\rangle \equiv \bar K |\gamma\rangle_0$. Left-operating on the former with the first expression for P_0 in (2.31) produces the expansion in the set $\{|\alpha'\rangle_0\}$

$$|\gamma\rangle_0 = \sum_\alpha f_\alpha |\alpha'\rangle_0 \, , \qquad f_\alpha \equiv {}_0\langle\bar\alpha'|\gamma\rangle_0 \qquad (A.13)$$

from which it follows that

$$_0\langle\gamma|\gamma\rangle_0 = \sum_{\alpha,\beta} f_\alpha^* f_\beta \; _0\langle\alpha'|\beta'\rangle_0 \qquad (A.14)$$

Left-multiplying Eq. (A.13) by $\bar K$ and using the left equation in (2.54) gives $|\gamma\rangle = \sum_\alpha f_\alpha |\Psi_\alpha\rangle$ and hence

$$\langle\gamma|\gamma\rangle = \sum_\alpha |f_\alpha|^2 \qquad (A.15)$$

The expressions in the right hand sides of (A.14) and (A.15) are in general different because the $|\alpha'\rangle_0$ do not form an orthogonal set. Similar reasoning applies for an arbitrary vector $|\Lambda\rangle$ of Ω and its model counterpart $|\Lambda\rangle_0 \equiv \bar L |\Lambda\rangle$. These two examples demonstrate that conservation of the lengths of basis vectors is equivalent to norm preservation only if the angles between the basis vectors are also conserved.

APPENDIX B: NON-NORM-PRESERVING MAPPINGS THAT PRODUCE A HERMITIAN EFFECTIVE HAMILTONIAN

We first present and prove Theorem II.a relating one subset of this group of mappings to norm-preserving transformations. Next, Theorem II.b of

Section III is proven to classify this group of mappings, and then the effective operator definitions for the subset of Theorem II.a are derived.

1.a   Theorem II.a

Theorem II.a relates one subgroup, herein denoted by $(\tilde{K}, \tilde{L})$, of non-norm-preserving mappings that generate a Hermitian effective Hamiltonian to norm-preserving mappings as follows: For any norm-preserving mapping pair $(\hat{K}, \hat{K}^{\dagger})$ there exists a non-denumerable infinity of non-norm-preserving mappings $(\tilde{K}, \tilde{L})$ that produce the same effective Hamiltonian as $(\hat{K}, \hat{K}^{\dagger})$ but that relate the model eigenfunctions to true eigenfunctions with different norms and the same or different phases from those related through $(\hat{K}, \hat{K}^{\dagger})$. Conversely, for any pair of non-norm-preserving mappings $(\tilde{K}, \tilde{L})$ that produces a Hermitian effective Hamiltonian $\mathbf{H}$ *and* that conserves the angles between orthogonal degenerate model eigenfunctions, there exist one "reference" and a non-denumerable infinity of other norm-preserving mapping pairs that also yield $\mathbf{H}$ but that relate the orthonormalized model eigenfunctions to orthonormalized true eigenfunctions with, respectively, the same and different phases than those of the true eigenfunctions generated by $(\tilde{K}, \tilde{L})$.

1.b   Proof and Discussion of Theorem II.a

Consider the set of orthonormalized model eigenfunctions $\{|\hat{\alpha}\rangle_0\}$ of the effective Hamiltonian $\hat{\mathbf{H}}$ defined in Table I. As explained in section II.C, the corresponding true eigenfunctions $|\hat{\Psi}_\alpha\rangle$ also form an orthonormalized set. Multiplying $|\hat{\Psi}_\alpha\rangle$ by an arbitrary complex number c_α, with $|c_\alpha| \neq 1$ for at least one α, generates another set of orthogonal true eigenfunctions $\{|\tilde{\Pi}'_\alpha\rangle\}$,

$$|\tilde{\Pi}'_\alpha\rangle \equiv c_\alpha |\hat{\Psi}_\alpha\rangle \tag{B.1}$$

Let $\tilde{u}$ be the wave operator transforming $|\hat{\alpha}\rangle_0$ into $|\tilde{\Pi}'_\alpha\rangle$ and let $\tilde{v}$ be the inverse mapping operator, $\tilde{u}|\hat{\alpha}\rangle_0 = |\tilde{\Pi}'_\alpha\rangle$, $\tilde{v}|\tilde{\Pi}'_\alpha\rangle = |\hat{\alpha}\rangle_0$. Formal expressions for $\tilde{u}$ and $\tilde{v}$ are

$$\tilde{u} = \sum_\beta |\tilde{\Pi}'_\beta\rangle \, _0\langle \hat{\beta}|, \qquad \tilde{v} = \sum_\beta |\hat{\beta}\rangle_0 \, \tilde{s}_\beta^{-2} \langle \tilde{\Pi}'_\beta| \tag{B.2}$$

where $\tilde{s}_\beta$ is the norm of $|\tilde{\Pi}'_\beta\rangle$,

$$\tilde{s}_\beta \equiv \langle \tilde{\Pi}'_\beta | \tilde{\Pi}'_\beta \rangle^{1/2} = |c_\beta| \tag{B.3}$$

Equations (B.2) demonstrate that $\tilde{u}$ and $\tilde{v}$ violate condition (1) of Theorem I given in Eq. (3.5), and, thus, are not norm-preserving. However, since they relate the $\{|\tilde{\Pi}'_\alpha\}$ to the $\{|\hat{\alpha}\rangle_0\}$, the effective Hamiltonian $\tilde{\mathbf{h}}$ which they produce should be identical to $\hat{\mathbf{H}}$. The general definition (2.10) indeed implies that $\tilde{\mathbf{h}}$ is given by

$$\tilde{\mathbf{h}} = \tilde{v}H\tilde{u} \tag{B.4}$$

Substituting Eqs. (B.2) into (B.4), and using, in order, (B.1), (2.8) with $|\phi_\alpha\rangle$ replaced by $|\hat{\Psi}_\alpha\rangle$, (2.20) and (B.3) yields

$$\tilde{\mathbf{h}} = \sum_\alpha |\hat{\alpha}\rangle_0\; E_\alpha\; {}_0\langle\hat{\alpha}| = \hat{\mathbf{H}} \tag{B.5}$$

where the last equality follows from Eq. (2.23), thereby completing the proof that the non-norm-preserving mappings $(\tilde{u}, \tilde{v})$ produce $\hat{\mathbf{H}}$. The first part of Theorem II.a now follows by noting that an infinity of different pairs of non-norm-preserving mappings $(\tilde{u}, \tilde{v})$ give $\hat{\mathbf{H}}$ and may be produced by varying the c_α of Eq. (B.1).

It might be counterintuitive that the mappings $(\tilde{u}, \tilde{v})$ do not conserve angles since overlaps are preserved for the above eigenfunctions, ${}_0\langle\hat{\alpha}|\hat{\beta}\rangle_0 = \langle\tilde{\Pi}'_\alpha|\tilde{\Pi}'_\beta\rangle$ for $\alpha \neq \beta$. This relation, however, does not imply the conservation of angles between arbitrary vectors. Expanding two arbitrary vectors $|\Lambda\rangle_0$ and $|\Sigma\rangle_0$ of Ω_0 in the set $\{|\hat{\alpha}\rangle_0\}$ and their true counterparts $|\Lambda\rangle \equiv \tilde{u}|\Lambda\rangle_0$ and $|\Sigma\rangle \equiv \tilde{u}|\Sigma\rangle_0$ in the $|\tilde{\Pi}'_\alpha\rangle$'s shows that ${}_0\langle\Lambda|\Sigma\rangle_0 \neq \langle\Lambda|\Sigma\rangle$ because ${}_0\langle\hat{\alpha}|\hat{\alpha}\rangle_0 \neq \langle\tilde{\Pi}'_\alpha|\tilde{\Pi}'_\alpha\rangle$. Similarly, the overlap between two arbitrary vectors of Ω differs from that for their model counterparts obtained using $\tilde{v}$. Hence, the preservation of the angles between basis vectors is equivalent to angle conservation only if the lengths of the basis vectors are also preserved.

The non-norm-preserving mappings $(\tilde{u}, \tilde{v})$ are, by construction, closely related to the norm-preserving mappings $(\hat{K}, \hat{K}^\dagger)$. We now show that *all* non-norm-preserving mappings that generate a Hermitian effective Hamiltonian *and* preserve the angles between orthogonal eigenfunctions are related in a similar fashion to norm-preserving mappings. This result then leads naturally to the second part of Theorem II.a.

Let $\{|\tilde{\alpha}'\rangle_0\}$ be the set of orthonormalized eigenfunctions of $\tilde{\mathbf{H}}$. The corresponding true eigenfunctions are orthogonal to one another but are not necessarily normalized to unity. Denote by $\tilde{N}_\alpha$ the norm of the αth corresponding true eigenfunction $|\tilde{\Psi}'_\alpha\rangle$,

$$\tilde{N}_\alpha \equiv \langle\tilde{\Psi}'_\alpha|\tilde{\Psi}'_\alpha\rangle^{1/2} \tag{B.6}$$

where $\tilde{N}_\alpha \neq 1$ for at least one α. We have

$$\tilde{K} = \sum_\beta |\tilde{\Psi}'_\beta\rangle\ _0\langle\tilde{\beta}'| \tag{B.7}$$

$$\tilde{L} = \sum_\beta |\tilde{\beta}'\rangle_0\ \tilde{N}_\beta^{-2}\langle\tilde{\Psi}'_\beta| \tag{B.8}$$

$$\tilde{\mathbf{H}} = \tilde{L}H\tilde{K} = \sum_\beta |\tilde{\beta}'\rangle_0\ E_\beta\ _0\langle\tilde{\beta}'| \tag{B.9}$$

Introduce orthonormal true eigenfunctions $\{|\Delta_\alpha\rangle\}$ by normalizing the $|\tilde{\Psi}'_\alpha\rangle$ to unity but leaving their phases arbitrary. The $|\Delta_\alpha\rangle$ are related to the $|\tilde{\alpha}'\rangle_0$ by the operators $\Sigma_\beta |\Delta_\beta\rangle\ _0\langle\tilde{\beta}'| \equiv \hat{k}$ and $\Sigma_\beta |\tilde{\beta}'\rangle_0\langle\Delta_\beta| \equiv \hat{l}$. The mappings $\hat{k}$ and $\hat{l}$ are norm-preserving since they satisfy Eq. (3.5). The effective Hamiltonian $\hat{l}H\hat{k} = \hat{k}^\dagger H\hat{k}$ which they produce is identical to $\tilde{\mathbf{H}}$ as can easily be verified by using (B.12) and the orthonormality property of the $\{|\Delta_\alpha\rangle\}$. This derivation implies that $\tilde{\mathbf{H}}$ can be generated by an infinity of norm-preserving mapping pairs since the phases of the true eigenfunctions $|\Delta_\alpha\rangle$ are arbitrary. Taking these phases to be the same as those of the true eigenfunctions $|\tilde{\Psi}_\alpha\rangle$ uniquely specifies one of the possible norm-preserving mapping pairs, termed "reference" pair. This completes the proof for the second part of Theorem II.a.

2. Proof of Theorem II.b

The Hermiticity of an effective Hamiltonian is first shown to be equivalent to the orthogonality of its *nondegenerate* eigenfunctions. This does not impose any conditions on the conservation of true eigenvector norms or of angles between degenerate model eigenfunctions, which are next demonstrated to be independent of one another. Theorem II.b then follows.

We first require Theorem 1. Let X be an operator defined in a finite space. X is Hermitian iff (1) its eigenvalues x_α are real, and (2) its nondegenerate eigenvectors are mutually orthogonal. It is well known [151] that if X is Hermitian then (1) and (2) follow, and that neither (1) nor (2) is a sufficient condition for the Hermiticity of X. However, properties (1) and (2) *together* imply that X is Hermitian [152].

Since the eigenvectors of a general effective Hamiltonian $\mathbf{h} = lHk$ are real, Theorem 1 implies that $\mathbf{h}$ is Hermitian iff its *nondegenerate* eigenvectors are mutually orthogonal [153]. Hence, $\mathbf{h}$ is Hermitian iff (k, l) conserve the overlaps between *nondegenerate* eigenfunctions (conservation A). Conservation A can be used to broadly categorize the mapping operators. The preservation of the angles between arbitrarily chosen degenerate true eigenfunctions (conservation B) and that of the lengths of a set of eigenvectors (conservation C) can be utilized to further classify the mapping operators. This is because conservations B and C are independent of the Hermiticity of $\mathbf{h}$ and, as we now show, of one another.

Let $|\phi_1\rangle_0, \ldots, |\phi_n\rangle_0$ be arbitrarily chosen degenerate eigenfunctions of $\mathbf{h}$ with true counterparts $|\phi_i\rangle \equiv k|\phi_i\rangle_0$. Applying an arbitrary linear transformation to the $|\phi\rangle_0$ produces other arbitrary model eigenfunctions $|\phi_i'\rangle_0 \equiv \Sigma_m c_{mi}|\phi_m\rangle_0$ with overlaps

$$_0\langle \phi_i'|\phi_j'\rangle_0 = \sum_{m,n} c_{mi}^* c_{nj} \, _0\langle \phi_m|\phi_n\rangle_0 \qquad (B.10)$$

The angles between the corresponding true eigenfunctions $|\phi_i'\rangle \equiv k|\phi_i'\rangle_0$ are given by

$$\langle \phi_i'|\phi_j'\rangle = \sum_{m,n} c_{mi}^* c_{nj} \langle \phi_m|\phi_n\rangle \qquad (B.11)$$

Conservation B holds iff the right hand sides of (B.10) and (B.11) are identical, that is, iff

$$\langle \phi_m|\phi_n\rangle = \, _0\langle \phi_m|\phi_n\rangle_0 \quad \text{for all } m \text{ and } n \qquad (B.12)$$

Hence, mappings satisfying conservation B preserve norms of arbitrarily chosen degenerate eigenvectors [154], but not necessarily those of non-degenerate eigenfunctions. Thus, conservation B does not imply conservation C. The converse is also true since the condition $\langle \phi_i|\phi_i\rangle = \, _0\langle \phi_i|\phi_i\rangle_0$, implied by conservation C for a set of model eigenfunctions that includes the $|\phi_i\rangle_0$, differs from Eq. (B.12).

Conservations A–C yield the four mapping categories 1, 3a–3c of Table III that all generate a Hermitian effective Hamiltonian.

Category 1 mappings are norm-preserving. If (and only if) all three conservations A–C are fulfilled, the model eigenvectors $|\phi_\alpha\rangle_0$ can be orthonormalized *and* correspond to orthonormalized true eigenvectors $|\phi_\alpha\rangle$. Thus, formal expressions for category 1 mappings are $k = \Sigma_\beta |\phi_\beta\rangle \, _0\langle \phi_\beta|$, $l = \Sigma_\beta |\phi_\beta\rangle_0 \langle \phi_\beta|$, which thus satisfy condition 1 of Theorem I (see Eq. (3.5)).

Conservation A and Eq. (B.12), which is equivalent to conservation B, imply that formal expressions for category 3a mappings are the right hand sides of (B.7) and (B.8) in which $\tilde{N}_\alpha = 1$ for each degenerate eigenvector. Hence, category 3a consists of the subset of mappings $(\tilde{K}, \tilde{L})$ which satisfy this last condition. The other $(\tilde{K}, \tilde{L})$ mappings do not fulfill conservation B and, thus, only preserve the angles between *orthogonal* degenerate eigenfunctions, as shown in Appendix A.1. They form a special subcase of category 3b. Other category 3b mappings $(\bar{k}, \bar{l})$ do not conserve these overlaps. Transforming the mappings $(\bar{k}, \bar{l})$ in order to absorb the normalization factors of Eq. (2.16) produces the category 3c mappings $(\bar{k}, \bar{l})$ [71]. Each $(\bar{k}, \bar{l})$ pair is generated by an infinity of $(\bar{k}, \bar{l})$ pairs that yield the same effective Hamiltonian, but differ in the norms of the true counterparts of the unity normed model eigenvectors, and produces the same effective Hamiltonian as the latter pairs [71]. This completes the proof of Theorem II.B.

3. Effective Operator Definitions Generated by Mappings of the Type $(\tilde{K}, \tilde{L})$

Consider the orthonormalized model eigenfunctions $|\tilde{\alpha}'\rangle_0$ and their true counterparts $|\tilde{\Psi}'_\alpha\rangle$ in terms of which the mappings $(\tilde{K}, \tilde{L})$ are formally expressed in Eqs. (B.7) and (B.8). A set of orthonormalized true eigenfunctions $|\tilde{\Psi}_\alpha\rangle$ may be obtained by normalizing $|\tilde{\Psi}'_\alpha\rangle$ to unity,

$$|\tilde{\Psi}'_\alpha\rangle = \tilde{N}_\alpha |\tilde{\Psi}_\alpha\rangle \qquad (B.13)$$

Since the mappings $(\tilde{K}, \tilde{L})$ do not conserve norms, Eq. (2.16) with $|\phi'_\alpha\rangle$ replaced by $|\tilde{\Psi}'_\alpha\rangle$ that is,

$$A_{\alpha\beta} = \langle \tilde{\Psi}'_\alpha | A | \tilde{\Psi}'_\beta \rangle / \tilde{N}_\alpha \tilde{N}_\beta \qquad (B.14)$$

applies here, unless the normalization factors present in (B.14) are incorporated into new model eigenfunctions or new mapping operators, whereupon Eq. (2.14) with $|\phi_\alpha\rangle$ replaced by $|\tilde{\Psi}_\alpha\rangle$ implies that

$$A_{\alpha\beta} = \langle \tilde{\Psi}_\alpha | A | \tilde{\Psi}_\beta \rangle \qquad (B.15)$$

Because the effective Hamiltonian $\tilde{\mathbf{H}}$ is Hermitian, there is only one set of model eigenfunctions. More than one Ω_0 expression for (B.14) can nonetheless be obtained because Eqs. (B.7) and (B.8) yield, respectively,

$$\tilde{K}|\tilde{\alpha}'\rangle_0 = |\tilde{\Psi}'_\alpha\rangle \qquad (B.16)$$

and

$$\tilde{L}^\dagger|\tilde{\alpha}'\rangle_0 = \tilde{N}_\alpha^{-2}|\tilde{\Psi}'_\alpha\rangle \qquad (B.17)$$

both of which may be used for $|\tilde{\Psi}'_\alpha\rangle$. Introducing these replacements produces the state-dependent effective operator definitions $\tilde{\mathbf{A}}^i_{\alpha\beta}$, $i = \text{I–IV}$, of Table I. The same replacements in (B.6) produce the two expressions $_0\langle \tilde{\alpha}' | \tilde{K}^\dagger \tilde{K} | \tilde{\alpha}' \rangle_0^{1/2}$ and $_0\langle \tilde{\alpha}' | \tilde{L}\tilde{L}^\dagger | \tilde{\alpha}' \rangle_0^{-1/2}$ for $\tilde{N}_\alpha$ which are in fact equivalent because $\tilde{K}^\dagger \tilde{K}$ and $\tilde{L}\tilde{L}^\dagger$ are mutual inverses and are diagonal in the set $\{|\tilde{\alpha}'\rangle_0\}$.

In order to use Eq. (B.15), the normalization factors in (B.14) are incorporated (see Table I) into the new model eigenfunctions $|\tilde{\alpha}\rangle_0$ and $|\bar{\tilde{\alpha}}\rangle_0$, which are related to $|\tilde{\Psi}_\alpha\rangle$ by $(\tilde{K}, \tilde{L})$. Substituting (B.13) into (B.16) and using the definition of $|\tilde{\alpha}\rangle_0$ gives

$$\tilde{K}|\tilde{\alpha}\rangle_0 = |\tilde{\Psi}_\alpha\rangle \tag{B.18}$$

and, likewise, Eqs. (B.13) and (B.17) imply

$$\tilde{L}^\dagger|\bar{\tilde{\alpha}}\rangle_0 = |\tilde{\Psi}_\alpha\rangle \tag{B.19}$$

Inserting (B.18) and (B.19) into (B.15) and proceeding as in the derivation of the definitions $\tilde{\mathbf{A}}^i_{\alpha\beta}$, $i = \text{I–IV}$, gives the state-independent effective operator definitions $\tilde{\mathbf{A}}^i$, $i = \text{I–IV}$, of Table I.

Equation (B.15) also applies if the normalization factors are absorbed into new mapping operators. Proceeding similarly as in Section II.D, the definition

$$\tilde{\mathbf{N}}|\tilde{\alpha}'\rangle_0 \equiv \tilde{N}_\alpha|\tilde{\alpha}'\rangle_0 \tag{B.20}$$

has the formal solution

$$\tilde{\mathbf{N}} = \sum_\beta |\tilde{\beta}'\rangle_0 \; \tilde{N}_\beta \; _0\langle \tilde{\beta}'| \tag{B.21}$$

and is Hermitian. Its inverse is also Hermitian, obeys the equation

$$\tilde{\mathbf{N}}^{-1}|\tilde{\alpha}'\rangle_0 = \tilde{N}_\alpha^{-1}|\tilde{\alpha}'\rangle_0 \tag{B.22}$$

and has the formal representation

$$\tilde{\mathbf{N}}^{-1} = \sum_\beta |\tilde{\beta}'\rangle_0 \; \tilde{N}_\beta^{-1} \; _0\langle \tilde{\beta}'| \tag{B.23}$$

The transformed mappings

$$\tilde{K}\tilde{\mathbf{N}}^{-1} = \sum_\beta |\tilde{\Psi}_\beta\rangle \; _0\langle \tilde{\beta}'| , \qquad \tilde{\mathbf{N}}\tilde{L} = \sum_\beta |\tilde{\beta}'\rangle_0 \; \langle \tilde{\Psi}_\beta| \tag{B.24}$$

relate $|\tilde{a}'\rangle_0$ and $|\tilde{\Psi}_\alpha\rangle$ to each other, as desired. These new mappings satisfy Eq. (2.18) and are thus norm-preserving. Consequently, they yield the state-independent effective operator definition $\hat{A}$ of Section II.C. Since $|\tilde{\Psi}_\alpha\rangle$ has the same phase as $|\tilde{\Psi}'_\alpha\rangle$, the new mappings $(\tilde{K}\tilde{N}^{-1}, \tilde{N}\tilde{L})$ actually are the "reference" pair (see Appendix A.1) of the mappings $(\tilde{K}, \tilde{L})$.

APPENDIX C: PROOF OF THEOREM I

We establish the implications $(1) \Rightarrow (2) \Rightarrow (4) \Rightarrow (6) \Rightarrow (8) \Rightarrow (4) \Rightarrow (1)$ and $(1) \Rightarrow (3) \Rightarrow (5) \Rightarrow (7) \Rightarrow (9) \Rightarrow (5) \Rightarrow (1)$ of (3.5)–(3.13). This demonstrates that any of the nine conditions can be reached from any of the other eight, thereby establishing their logical equivalence.

$(1) \Rightarrow (2)$: Substituting Eq. (3.5) into Eq. (2.13) gives Eq. (3.6).

$(2) \Rightarrow (4)$: Multiplying Eq. (3.6) from the right by k and substituting the second part of (2.3) in the right hand side of the resulting equation we have

$$kk^\dagger k = k \qquad (C.1)$$

Left-multiplying (C.1) by l and substituting Eq. (2.12) in the resulting relation gives

$$P_0 \ k^\dagger k = P_0 \qquad (C.2)$$

Using the Hermitian conjugate of the first part of (2.3) in (C.2) produces Eq. (3.8).

$(4) \Rightarrow (6)$: Forming scalar products from (3.1) and (3.2) we have

$$\langle \gamma | \delta \rangle = {}_0\langle \gamma | k^\dagger k | \delta \rangle_0 \qquad (C.3)$$

Then substituting (3.8) into (C.3) yields (3.10).

$(6) \Rightarrow (8)$: Eq. (3.12) follows from taking $\gamma = \delta$ in (3.10).

$(8) \Rightarrow (4)$: Consider the norm of $k|\gamma\rangle_0$,

$$\| k|\gamma\rangle_0 \|^2 = {}_0\langle k\gamma | k\gamma \rangle_0 = {}_0\langle \gamma | k^\dagger k\gamma \rangle_0 \qquad (C.4)$$

Equation (3.12) implies that

$$\| k|\gamma\rangle_0 \|^2 = \| |\gamma\rangle_0 \|^2 = {}_0\langle \gamma | \gamma \rangle_0 \qquad (C.5)$$

Therefore, comparing (C.4) and (C.5) gives

$$_0\langle\gamma|(k^\dagger k - P_0)\gamma\rangle_0 = 0 \text{ for all } |\gamma\rangle_0 \qquad (C.6)$$

We now require Theorem a [155]. If T is a self-adjoint linear transformation in an inner-product space (real or complex), then $T = 0$ iff $\langle x|Tx\rangle = 0$ for all $|x\rangle$. Section II shows that $(k^\dagger k)$ is an operator only in Ω_0. Therefore, $(k^\dagger k - P_0)$ is also defined only in Ω_0. Hence, Eq. (3.8) follows from applying Theorem a to (C.6).

(4) $\Rightarrow$ (1): Right-multiplying Eq. (3.8) by l gives

$$k^\dagger k l = P_0\ l \qquad (C.7)$$

Using (2.13) in the left hand side and the second part of (2.6) in the right hand side of (C.7), respectively, produces

$$k^\dagger P = l \qquad (C.8)$$

Using the Hermitian conjugate of the second part of (2.3) in (C.8) yields (3.5).

The establishment of (1) $\Rightarrow$ (3) $\Rightarrow$ (5) $\Rightarrow$ (7) $\Rightarrow$ (9) $\Rightarrow$ (5) $\Rightarrow$ (1) is accomplished similarly as follows:

(1) $\Rightarrow$ (3): Substitution of the Hermitian conjugate of (3.5) into Eq. (2.12) gives Eq. (3.7).

(3) $\Rightarrow$ (5): Multiplying (3.7) from the right by l and substituting the second part of (2.6) in the right hand side of the resulting equation produces

$$ll^\dagger l = l \qquad (C.9)$$

Left-multiplying (C.9) by k and substituting (2.13) in the resulting relation gives

$$Pl^\dagger l = P \qquad (C.10)$$

Substituting the Hermitian conjugate of the first part of (2.6) in (C.10) yields (3.9).

(5) $\Rightarrow$ (7): The scalar product of (3.3) with (3.4) gives

$$_0\langle\Lambda|\Sigma\rangle_0 = \langle\Lambda|l^\dagger l|\Sigma\rangle \qquad (C.11)$$

Substituting (3.9) into (C.11) produces (3.11).

(7) $\Rightarrow$ (9): This result follows from setting $\Lambda = \Sigma$ in (3.11).

(9) $\Rightarrow$ (5): Here we have

$$\|l|\Lambda\rangle\|^2 = \langle l\Lambda|l\Lambda\rangle = \langle \Lambda|l^\dagger l\Lambda\rangle \tag{C.12}$$

Equation (3.13) implies that

$$\|l|\Lambda\rangle\|^2 = \||\Lambda\rangle\| = \langle \Lambda|\Lambda\rangle \tag{C.13}$$

Subtracting (C.13) from (C.12) gives

$$\langle \Lambda|(l^\dagger l - P)\Lambda\rangle = 0 \text{ for all } |\Lambda\rangle \tag{C.14}$$

The first part of Eqs. (2.6) and that of (2.7) imply that $l^\dagger l$ is defined only in Ω. Therefore, $(l^\dagger l - P)$ is also defined only in Ω, so application of Theorem a to (C.14) yields Eq. (3.9).

(5) $\Rightarrow$ (1): Right-multiplying Eq. (3.9) by k produces

$$l^\dagger l k = Pk \tag{C.15}$$

Using (2.12) in the left hand side and the second part of (2.3) in the right hand side of (C.15) we get

$$l^\dagger P_0 = k \tag{C.16}$$

Introducing the Hermitian conjugate of the second part of (2.6) into (C.16) yields the Hermitian conjugate of (3.5).

APPENDIX D: LACK OF SUM RULE FOR TRANSITION MOMENTS COMPUTED WITH EFFECTIVE OPERATORS

This appendix demonstrates that the equivalence between the dipole length and dipole velocity transition moments, computed using effective operators, does not produce a sum rule for these moments. The proof is first provided for effective operator definition $\hat{A}$, and then modifications required for definitions $A_{\alpha\beta}^{II}$, A^{II}, $\bar{A}^{II}$, $\tilde{A}^i$, and $\tilde{A}_{\alpha\beta}^i$, $i = $ I–IV are described.

Equation (4.14) provides the equivalence between the dipole length and dipole velocity transition moments for a system of n identical particles of mass m with state-independent effective operator definition $\hat{A}$. To see that this equivalence does not produce a sum rule, consider first the usual derivation of the Thomas–Reiche–Kuhn sum rule for the true operators. Left- and right-multiplying equation (4.12) by $\langle \Psi_\alpha|$ and $|\Psi_\beta\rangle$, respectively, the z component yields

$$\langle \Psi_\alpha|p_z|\Psi_\beta\rangle = im\omega_{\alpha\beta}\langle \Psi_\alpha|z|\Psi_\beta\rangle \tag{D.1}$$

where $\omega_{\alpha\beta}$ is defined in (4.15). Equation (D.1) implies that

$$(-2)im\,\omega_{\alpha\beta}|\langle\Psi_\alpha|z|\Psi_\beta\rangle|^2 = \langle\Psi_\beta|p_z|\Psi_\alpha\rangle\langle\Psi_\alpha|z|\Psi_\beta\rangle - \langle\Psi_\beta|z|\Psi_\alpha\rangle$$
$$\times\langle\Psi_\alpha|p_z|\Psi_\beta\rangle \tag{D.2}$$

The indices α and β in (D.1) and (D.2) are not limited to indices of states in Ω. Therefore, summing over a complete set of states α in (D.2) gives

$$(-2)im\sum_\alpha\omega_{\alpha\beta}|\langle\Psi_\alpha|z|\Psi_\beta\rangle|^2 = \langle\Psi_\beta|[p_z,z]|\Psi_\beta\rangle \tag{D.3}$$

Substituting the commutation relation

$$[z,p_z] = i\hbar n\mathbf{1} \tag{D.4}$$

in Eq. (D.3) produces the well-known Thomas–Reiche–Kuhn sum rule,

$$\sum_\alpha f^z_{\alpha\beta} = n \tag{D.5}$$

where $f^z_{\alpha\beta}$ is the oscillator strength for a z-polarized transition $|\Psi_\beta\rangle\to|\Psi_\alpha\rangle$,

$$f^z_{\alpha\beta} \equiv 2m\hbar^{-1}\omega_{\alpha\beta}|\langle\Psi_\alpha|z|\Psi_\beta\rangle|^2 \tag{D.6}$$

We now attempt to follow this derivation with the z components

$$\hat{\mathbf{p}}_z \equiv \hat{K}^\dagger p_z\hat{K}, \qquad \hat{\mathbf{z}} \equiv \hat{K}^\dagger z\hat{K} \tag{D.7}$$

of the effective operators $\hat{\tilde{\mathbf{p}}}$ and $\hat{\tilde{\mathbf{r}}}$ defined in (4.11). Equation (D.2) implies that

$$(-2)im\,\omega_{\alpha\beta}|_0\langle\hat{\alpha}|\hat{\mathbf{z}}|\hat{\beta}\rangle_0|^2 = {}_0\langle\hat{\beta}|\hat{\mathbf{p}}_z|\hat{\alpha}\rangle_0\,{}_0\langle\hat{\alpha}|\hat{\mathbf{z}}|\hat{\beta}\rangle_0$$
$$- {}_0\langle\hat{\beta}|\hat{\mathbf{z}}|\hat{\alpha}\rangle_0\,{}_0\langle\hat{\alpha}|\hat{\mathbf{p}}_z|\hat{\beta}\rangle_0\,, \quad \alpha,\beta\in\{d\} \tag{D.8}$$

Summing over all model states α and using Eq. (2.21) produces

$$(-2)im\sum_\alpha\omega_{\alpha\beta}|_0\langle\hat{\alpha}|\hat{\mathbf{z}}|\hat{\beta}\rangle_0|^2 = {}_0\langle\hat{\beta}|[\hat{\mathbf{p}}_z,\hat{\mathbf{z}}]|\hat{\beta}\rangle_0\,, \quad \alpha,\beta\in\{d\} \tag{D.9}$$

The further transformation of the right hand side of (D.9) requires a general expression for $[\hat{\mathbf{p}}_z, \hat{z}]$. However, such an expression is unavailable from (D.4) and Theorem V because neither z nor p_z commutes with H. In fact, Theorem VII implies that $[z, p_z]$ is not conserved except for very special and unlikely true Hamiltonians such that $P(zQp_z - p_zQz)P = 0_\Omega$. The proof of Theorem IV and Eq. (4.6) imply that $[z, p_z]$ would generally be conserved if Ω and, thus, Ω_0 were the whole Hilbert space. Hence, only if the dimension of Ω_0 were infinite could $[\hat{\mathbf{p}}_z, \hat{z}] = i\hbar n P_0$. This fact also follows from the well-known result that in a *finite* space no two operators have their commutator equal to a multiple of the unit operator in that space [156].

Using any other effective operator definition (mentioned in the first paragraph) and proceeding similarly yields a relation analogous to (D.9) whose right hand side contains a matrix element of the commutator between the effective operators corresponding to z and p_z. This commutator cannot be transformed any further for the same reason that applies to the commutator in (D.9).

APPENDIX E: PRESERVATION OF COMMUTATION RELATIONS BY STATE-INDEPENDENT EFFECTIVE OPERATOR DEFINITIONS OTHER THAN Â

Theorems V and VI of Section IV show that definition Â conserves $[H, B]$ and $[A, B]$, with B an arbitrary operator and A an operator commuting with H or, more generally, with P. Theorem VII provides the necessary and sufficient condition for the conservation of a general commutator by definition Â. Analogous results for the other state-independent effective operator definitions are presented in columns 3–7 of Table III and are derived in this appendix by modifying the proofs of Theorems V–VII.

The derivations of Theorems V and VI for the preservation of $[A, B]$ consist of (1) the replacement by P of the product $\hat{K}\hat{K}^\dagger$ that is present between A and B in $[\hat{\mathbf{A}}, \hat{\mathbf{B}}]$, (2) the commutation of A with P, and (3) the incorporation of P into $\hat{K}$ and $\hat{K}^\dagger$. Step (1) uses Eq. (4.6), which is valid *only* for norm-preserving mappings. With other definitions, however, the products analogous to $\hat{K}\hat{K}^\dagger$ may be replaced by P if (criterion 1) they satisfy the fundamental relation (2.13). Step (2) clearly applies to any effective operator definitions. So does step (3) since P can be combined with any mapping operators using Eqs. (2.3) and (2.6). Hence, all effective operator definitions that fulfill criterion 1 conserve $[A, B]$ if $[A, P] = 0$. When these definitions, like Â, produce the associated effective Hamiltonian upon taking $A = H$ (criterion 2), the derivation for $[H, B]$ just replaces A by H in the one for $[A, B]$ and, thus, the preservation of $[H, B]$ follows from that of $[A, B]$. New analyses for these two commutators, however, are needed with definitions that violate criterion 1, 2 or both.

In Theorem VII, $F \equiv [A, B]$ is the commutator of two arbitrary operators A and B. The proof of this theorem first reversibly transforms $[\hat{A}, \hat{B}]$ and $\hat{F}$ into Ω-space operators. Both expressions are then equated to obtain necessary and sufficient conditions for the conservation of F. The first step can be accomplished with *any* other effective operator definition. The general form of an effective operator is $\mathbf{a} = sAt$, with mapping operators $s = P_0 sP$ and $t = PtP_0$. Left- and right-multiplying $\mathbf{a}$ by, respectively, the mappings u and v that satisfy $us = P = tv$ (see Eq. (2.13)) gives the operator PAP. Left- and right-operating on PAP with s and t, respectively, yields $\mathbf{a}$. However, the second step of the derivation with other definitions does not necessarily produce the same condition as in Theorem VII. This is because step (1) in the derivation of Theorems V and VI is used to simplify the condition of Theorem VII. The above discussion of step (1) implies that only the effective operator definitions that satisfy criterion 1 can yield the same condition as that of Theorem VII.

1.   *Effective Operator Definitions* $\mathbf{A}^\mathrm{I}$, $\mathbf{A}^\mathrm{II}$ *and* $\mathbf{A}^\mathrm{III}$

Criteria 1 and 2 are now applied to the three state-independent effective operator definitions possible with the mappings (K, L) and summarized in Table I. The product analogous to $\hat{K}\hat{K}^\dagger$ for definition $\mathbf{A}^\mathrm{II}$ is KL since

$$[\mathbf{A}^\mathrm{II}, \mathbf{B}^\mathrm{II}] = L(AKLB - BKLA)K \qquad (\mathrm{E}.1)$$

Equation (2.13) implies that KL is equal to P. Hence, criterion 1 is satisfied. So is criterion 2 because Table I shows that $\mathbf{H}$ is the $A = H$ case of definition $\mathbf{A}^\mathrm{II}$. Therefore, the results of Theorems V–VII ensue for this definition.

Definition $\mathbf{A}^\mathrm{I}$ leads to

$$[\mathbf{A}^\mathrm{I}, \mathbf{B}^\mathrm{I}] = K^\dagger(AKK^\dagger B - BKK^\dagger A)K \qquad (\mathrm{E}.2)$$

but condition (2) of Theorem 1 implies $KK^\dagger \neq P$ because $K^\dagger \neq L$. Since criterion 1 is violated, the equivalent of step (1) of Theorems V–VI does not follow here. Definition $\mathbf{A}^\mathrm{I}$ is now proven *not* to conserve the commutation relation between an operator A that commutes with P and an arbitrary operator B. As shown in Section II.A, $KK^\dagger$ is defined solely in Ω, and consequently

$$KK^\dagger = P + Y \tag{E.3}$$

where $Y = PYP$ is a nonzero operator in Ω. Substituting (E.3) into (E.2) and using $[A, P] = 0$ and (2.3) yields

$$[\mathbf{A}^I, \mathbf{B}^I] = K^\dagger[A, B]K + K^\dagger(AYB - BYA)K \tag{E.4}$$

The presence of the second term in the right hand side of (E.4) renders impossible the preservation of $[A, B]$. Conditions on A and/or B causing this term to be null may, of course, be found, for example, $AP = 0$, $PB = 0 = BP$, etcetera, but such conditions are restrictions that general A and B do not fulfill.

The commutator corresponding to $[H, B]$ cannot be derived by setting $A = H$ into (E.4) because criterion 2 is not obeyed by definition $\mathbf{A}^I$ (see Table I). This definition and that of $\mathbf{H}$ given in Table I imply that

$$[\mathbf{H}, \mathbf{B}^I] = LHKK^\dagger BK - K^\dagger BKLHK = LHBK - K^\dagger BHK + LHYBK \tag{E.5}$$

where (2.3) and (2.13) for mappings (K, L), (E.3) and $[P, H] = 0$ are used to obtain the last equality. Equation (E.5) clearly shows the lack of conservation of $[H, B]$ in this case.

Following our discussion of the derivation of Theorem VII, Eq. (E.4) is left- and right-multiplied by, respectively, $L^\dagger$ and L, Eqs. (2.13) for mappings (K, L) and (E.3) are used, and the resulting relation is subtracted from $L^\dagger \mathbf{F}^I L = P[A, B]P$. This produces

$$P(AQB - BQA)P - P(AYB - BYA)P = 0_\Omega \tag{E.6}$$

as the necessary and sufficient condition for the conservation of the commutation relation between two arbitrary operators A and B. This condition differs from Eq. (4.10) of Theorem VII. Notice that (E.6) requires a relation between either one or both of the operators A and B, and P, Q, and/or Y. The operator Y depends on the mapping operators and, thus, on the model space and the space Ω. Hence, many of the sufficient conditions for the conservation of $[A, B]$ obtainable from Eq. (E.6) are more complicated and more restrictive than those produced by Eq. (4.10) of Theorem VII, and, as the latter, cannot be verified a priori.

Definition $\mathbf{A}^{III}$ yields the commutation relation

$$[\mathbf{A}^{III}, \mathbf{B}^{III}] = K^\dagger(AL^\dagger K^\dagger B - BL^\dagger K^\dagger A)L^\dagger \qquad (E.7)$$

Since, from (2.13), $L^\dagger K^\dagger = P$, criterion 1 is fulfilled. Hence, except for $A = H$, Theorems V–VII are also valid with this definition. The restriction $A \neq H$ is necessary here because Table I shows that criterion 2 is not obeyed by definition $\mathbf{A}^{III}$. The latter and the definition of $\mathbf{H}$ lead to

$$[\mathbf{H}, \mathbf{B}^{III}] = LHKK^\dagger BL^\dagger - K^\dagger BL^\dagger LHK \qquad (E.8)$$

Condition (5) of Theorem 1 implies $L^\dagger L \neq P$ since $K^\dagger \neq L$, so similarly to (E.3) we may write

$$L^\dagger L = P + Z \qquad (E.9)$$

where $Z = PZP$ is a nonzero operator in Ω. Substituting (E.3) and (E.9) into (E.8) and using (2.3) and (2.6) for mappings (K, L) and $[P, H] = 0$ yields

$$[\mathbf{H}, \mathbf{B}^{III}] = LHBL^\dagger - K^\dagger BHK + LHYBL^\dagger - K^\dagger BZHK \qquad (E.10)$$

Equation (E.10) implies that $[H, B]$ is not conserved. The complicated expression in the right hand side of (E.10) results from the fact that criterion 2 does not hold with this definition. Indeed, $\mathbf{H}^\dagger$ is the $A = H$ case of definition $\mathbf{A}^{III}$, and it is easy to verify that

$$[\mathbf{H}^\dagger, \mathbf{B}^{III}] = K^\dagger[H, B]L^\dagger \qquad (E.11)$$

implying that $[H, B]$ is conserved by $\mathbf{A}^{III}$ effective operators if $\mathbf{H}^\dagger$ is used instead of $\mathbf{H}$.

2. Other State-Independent Effective Operator Definitions

For the other effective operator definitions, advantage is taken of "isomorphisms" with definitions $\mathbf{A}^i$, $i = $ I–III. Table I shows that definition $\bar{\mathbf{A}}^i$ can formally be obtained from definition $\mathbf{A}^i$, $i = $ I–III, by replacing (K, L) by $(\bar{K}, \bar{L})$. Since the products analogous to $\hat{K}\hat{K}^\dagger$ for $\mathbf{A}^i$, $i = $ I–III, are, respectively, $KK^\dagger$, KL, and $L^\dagger K^\dagger$ (see above), those for definitions $\bar{\mathbf{A}}^i$ are thus $\bar{K}\bar{K}^\dagger$, $\bar{K}\bar{L}$, and $\bar{L}^\dagger \bar{K}^\dagger$, respectively. As the product for $\bar{\mathbf{A}}^i$ fulfills (2.13) or not (criterion 1), just as the one for $\mathbf{A}^i$, $i = $ I–III, we have an "isomorphism" between definitions $\mathbf{A}^i$ and $\bar{\mathbf{A}}^i$, $i = $ I–III. Thus, all definition $\bar{\mathbf{A}}^i$ results about the conservation of $[A, B]$, for $A \neq H$, and $B \neq H$

follow by applying the above replacement to those of definition $\mathbf{A}^i$, $i = $ I–III. The definition $\bar{\mathbf{A}}^i$ results for the preservation of $[H, B]$ may similarly be derived from those of definition $\mathbf{A}^i$, $i = $ I–III, since the effective Hamiltonian $\mathbf{H} = LHK$ can also be written as $\bar{L}H\bar{K}$ (see Section II.D.3).

The definition $\tilde{\mathbf{A}}^i$, $i = $ I–IV, results for the conservation of $[A, B]$, with $A \neq H$ and $B \neq H$, follow from similar "isomorphisms" between $\mathbf{A}^i$ and $\tilde{\mathbf{A}}^i$, $i = $ I–III $[(K, L) \rightarrow (\tilde{K}, \tilde{L})]$, and between $\mathbf{A}^I$ and $\tilde{\mathbf{A}}^{IV}$ $[K \rightarrow \tilde{L}^\dagger, K^\dagger \rightarrow \tilde{L}]$. Comparing the definitions of $\mathbf{H}$ and $\tilde{\mathbf{H}}$ in Table I shows that the above "isomorphism" between $\mathbf{A}^{II}$ and $\tilde{\mathbf{A}}^{II}$ is also valid for $A = H$. Hence, the conservation of $[H, B]$ demonstrated for the former definition also holds for the latter. Since $\tilde{\mathbf{H}}$ is Hermitian, it can be written (see Table I) as $\tilde{L}H\tilde{K}$ (first form), or as $\tilde{K}^\dagger H\tilde{L}^\dagger$ (second form). The second form of $\tilde{\mathbf{H}}$ can formally be obtained from $\mathbf{H}^\dagger$ by the replacement $(K, L) \rightarrow (\tilde{K}, \tilde{L})$. This result in conjunction with the above "isomorphism" between $\mathbf{A}^{III}$ and $\tilde{\mathbf{A}}^{III}$ and with Eq. (E.11) implies that $[H, B]$ is preserved by definition $\tilde{\mathbf{A}}^{III}$. This commutator is also conserved by definitions $\tilde{\mathbf{A}}^I$ and $\tilde{\mathbf{A}}^{IV}$ because both forms of $\tilde{\mathbf{H}}$ can be used. Choosing the first form in $\tilde{\mathbf{B}}^I\tilde{\mathbf{H}}$ and the second form in $\tilde{\mathbf{H}}\tilde{\mathbf{B}}^I$ gives

$$[\tilde{\mathbf{H}}, \tilde{\mathbf{B}}^I] = \tilde{K}^\dagger(H\tilde{L}^\dagger\tilde{K}^\dagger B - B\tilde{K}\tilde{L}H)\tilde{K} \tag{E.12}$$

Since both products $\tilde{K}\tilde{L}$ and $\tilde{L}^\dagger\tilde{K}^\dagger$ satisfy (2.13), Eq. (E.12) yields the desired result. The proof for definition $\tilde{\mathbf{A}}^{IV}$ proceeds similarly, with the first form chosen for $\tilde{\mathbf{H}}\tilde{\mathbf{B}}^{IV}$ and the second for $\tilde{\mathbf{B}}^{IV}\tilde{\mathbf{H}}$.

Acknowledgments

This research is supported, in part, by NSF Grant CHE89-13123. V. H. is grateful to the NSERCC (Canada) for a graduate scholarship during the begining of this work, to Professor Bill Reinhardt for his hospitality during its completion, and to Dr. Reinhold Blümel and Professor Michel Vallières for helpful discussions.

References

1. A. Messiah, *Quantum Mechanics* (Wiley, New York, 1962), Vol. II, pp. 698–700.
2. J. O. Hirschfelder, *Int. J. Quantum Chem.* **3**, 731 (1969).
3. H. J. Silverstone, *J. Chem. Phys.* **54**, 2325 (1971).
4. J. O. Hirschfelder and P. R. Certain, *J. Chem. Phys.* **60**, 1118 (1974).
5. H. J. Silverstone and R. K. Moats, *Phys. Rev. A* **23**, 1645 (1981).
6. C. Bloch, *Nucl. Phys.* **6**, 329 (1958).
7. J. des Cloizeaux, *Nucl. Phys.* **20**, 321 (1960).
8. B. H. Brandow, *Rev. Mod. Phys.* **39**, 771 (1967).
9. V. Kvanisčka, *Czech. J. Phys.* **B24**, 605 (1974).

10. A more complete list of references can be found in Section V.

11. RSPT and BWPT are special cases of effective Hamiltonian theory where the dimension of the model space is equal to unity; see [7, 8].

12. S. Y. Lee and K. Suzuki, *Phys. Lett.* **91B**, 173 (1980).

13. K. Suzuki and S. Y. Lee, *Prog. Theory. Phys.* **64**, 2091 (1980).

14. Ph. Durand, *J. Phys. Lett.* **43**, L461 (1982).

15. D. Mukherjee, R. K. Moitra, and A. Mukhopadhyay, *Mol. Phys.* **30**, 1861 (1975).

16. R. Offerman, W. Ey, and H. Kümmel, *Nucl. Phys.* **A273**, 349 (1976).

17. R. Offerman, *Nucl. Phys.* **A273**, 368 (1976).

18. W. Ey, *Nucl. Phys.* **A296**, 189 (1978).

19. I. Lindgren, *Int. J. Quantum Chem.* **S12**, 33 (1978).

20. B. Jeziorski and H. J. Monkhorst, *Phys. Rev. A* **24**, 1688 (1981).

21. D. Mukherjee, *Int. J. Quantum Chem.* **S20**, 409 (1986) and refs. therein.

22. V. Hurtubise and K. F. Freed, manuscript in preparation.

23. See [21] and U. Kaldor, *Phys. Rev. A* **38**, 6013 (1988) and refs. therein for examples of such applications.

24. Ph. Durand and J. P. Malrieu, *Adv. Chem. Phys.* **67**, 321 (1987) and refs. therein.

25. G. Jolicard and M. Y. Perrin, *Chem. Phys.* **123**, 249 (1988).

26. W. B. England, *J. Phys. Chem.* **86**, 1204 (1982); *Int. J. Quantum Chem.* **23**, 905 (1983); *Int. J. Quantum Chem.* **S17**, 357 (1983); W. B. England, D. M. Silver and E. Otto Steinborn, *J. Chem. Phys.* **81**, 4546 (1984); W. B. England and D. M. Silver, *J. Chem. Phys.* **85**, 5847 (1986); W. B. England, T. E. Sorensen, and D. M. Silver, *Int. J. Quantum Chem.* **S20**, 81 (1986).

27. K. F. Freed, in: *Many-Body Methods in Quantum Chemistry*, U. Kaldor (ed.) (Springer-Verlag, Berlin, 1989) p.1.

28. X. C. Wang and K. F. Freed, *J. Chem. Phys.* **91**, 1142 (1989); Erratum **94**, 5253 (1991); **91**, 1151 (1989); **91**, 3002 (1989).

29. A. W. Kanzler and K. F. Freed, *J. Chem. Phys.* **94**, 3778 (1991).

30. J. J. Oleksik and K. F. Freed, *J. Chem. Phys.* **79**, 1396 (1983).

31. I. Lindgren, *Rep. Prog. Phys.* **47**, 345 (1984).

32. P. J. Ellis and E. Osnes, *Rev. Mod. Phys.* **49**, 777 (1977).

33. B. H. Brandow, in: *Effective Interactions and Operators in Nuclei*, B. R. Barret (ed.) (Springer-Verlag, Berlin, 1975), p.1.

34. B. H. Brandow, *Adv. Quantum Chem.* **10**, 187 (1977).

35. F. Jørgensen, and T. Pedersen, *Mol. Phys.* **27**, 33 (1974); **27**, 959 (1974); **28**, 599 (1974).

36. F. Jørgensen, T. Pedersen, and A. Chedin, *Mol. Phys.* **30**, 1377 (1975).

37. F. Jørgensen, *Mol. Phys.* **29**, 1137 (1975).

38. Ph. Durand, *C. R. Acad. Sci. Paris, II* **303**, 119 (1986).

39. L. L. Foldy and S. A. Wouthuysen, *Phys. Rev.* **78**, 29 (1950) and [1, pp. 940–948].

40. J. D. Weeks, A. Hazi, and S. A. Rice, *Adv. Chem. Phys.* **16**, 283 (1969); K. F. Freed, Chem. Phys. Lett. **29**, 143 (1974).

41. I. Lindgren, *J. Phys. B* **7**, 2441 (1974).

42. P. Westhaus, *Int. J. Quantum Chem.* **S7**, 463 (1973).

43. G. Hose and U. Kaldor, *J. Phys. B* **12**, 3827 (1979); *Phys. Scripta* **21**, 357 (1980); *Chem. Phys.* **62**, 469 (1981); *J. Phys. Chem.* **86**, 2133 (1982).

44. G. Hose, in: *Many-Body Methods in Quantum Chemistry*, U. Kaldor (ed.) (Springer-Verlag, Berlin, 1989) p. 43.

45. L. T. Redmon and R. J. Bartlett, *J. Chem. Phys.* **76**, 1938 (1982).

46. M. A. Haque and D. Mukherjee, *Pramana* **23**, 651 (1984).

47. B. H. Brandow, in: *New Horizons of Quantum Chemistry*, P. O. Löwdin and B. Pullmann (eds.) (Reidel, Dordrecht, 1983), p. 15.

48. M. G. Sheppard, *J. Chem. Phys.* **80**, 1225 (1984).

49. D. Mukherjee, *Chem. Phys. Lett.* **125**, 207 (1986).

50. W. Kutzelnigg, D. Mukherjee, and S. Koch, *J. Chem. Phys.* **87**, 5902 (1987).

51. D. Mukherjee, W. Kutzelnigg, and S. Koch, *J. Chem. Phys.* **87**, 5911 (1987).

52. R. Chowdhuri, D. Mukherjee, and M. D. Prasad, in: *Aspects of Many-Body effects in Molecules and Extended Systems*, D. Mukherjee, (ed.) (Springer-Verlag, Berlin, 1989) p. 3; R. Chowdhuri, D. Mukhopadhyay, and D. Muhkerjee, p. 165; other papers in the same volume.

53. W. Kutzelnigg, *J. Chem. Phys.* **77**, 3081 (1982).

54. W. Kutzelnigg and S. Koch, *J. Chem. Phys.* **79**, 4315 (1983).

55. W. Kutzelnigg, *J. Chem. Phys.* **80**, 822 (1984).

56. H. Sun and K. F. Freed, *J. Chem. Phys.* **88**, 2659 1988).

57. H. Sun, Y. S. Lee, and K. F. Freed, *Chem. Phys. Lett.* **150**, 529 (1988).

58. A. W. Kanzler, H. Sun, and K. F. Freed, *Int. J. Quantum Chem.* **39**, 269 (1991).

59. K. F. Ratcliff, in: *Effective Interactions and Operators in Nuclei*, B. R. Barret (ed.) (Springer-Verlag, Berlin, 1975) p. 42.

60. J. Lindberg, *Chem. Phys. Lett.* **1**, 39 (1967).

61. K. Sakamoto, T. Hayashi, and Y. J. I'Haya, *Int. J. Quantum Chem.* **S7**, 337 (1973).

62. M. C. Zerner and R. G. Parr, *J. Chem. Phys.* **69**, 3858 (1978).

63. H. A. Weidenmüller, in: *Effective Interactions and Operators in Nuclei*, B. R. Barret (ed.) (Springer-Verlag, Berlin, 1975) p. 152.

64. F. Jørgensen, *Int. J. Quantum Chem.* **12**, 397 (1977).

65. D. J. Klein, *J. Chem. Phys.* **61**, 786 (1974).

66. V. Hurtubise, (unpublished).

67. The only unknowns in (2.12), (2.13) are k and l because even though P is defined in terms of unknown true eigenfunctions, its perturbation expansion is known (see [6, 107]) and is independent of k and l. Therefore, if either k or l is chosen, then the other is uniquely determined by (2.12) or (2.13). Although no previous work has taken advantage of this simplification, we will show elsewhere [66] that Eq. (2.12) or (2.13) can indeed be used to simplify the derivation of the operators k and l for many of the known different forms of effective Hamiltonians.

68. The latter possibility does not necessarily produce norm-preserving mappings because, as shown in Section III, conservation of the norms of the model eigenvectors is *not* equivalent to norm preservation for arbitrary vectors.

69. A "small" subset of category 2b mappings conserves (only) the angles between orthogonal degenerate eigenvectors (similar to the category 3b subset $(\tilde{K}, \tilde{L})$) [71].

70. Section VII discusses such a procedure that applies if a complete set of commuting

observables exists. If this set is lacking, it is necessary to use a more complicated calculation equivalent to an orthogonalization, for example, Gram–Schmidt, of the true model eigenfunctions. This process requires evaluating the overlaps $\langle \Psi_\alpha' | \Psi_\beta' \rangle$ or $\langle \Psi_\alpha | \Psi_\beta \rangle$ for the degenerate states. If the true eigenfunctions are not required to be orthogonal, these overlaps are generally still needed for a meaningful calculation of the $A_{\alpha\beta}$. However, orthogonal true eigenvectors can then be obtained at little extra cost since the lengthiest part of the orthogonalization process is the calculation of these overlaps. Thus, orthogonalizing the degenerate model eigenvectors involves only slightly fewer computations than the most complicated procedure for orthogonalizing their true counterparts, but it loses the advantages of the latter option.

71. V. Hurtubise, (unpublished).

72. If the $|\alpha\rangle_0$ ($|\alpha'\rangle_0$) are known, then Eq. (2.29) (Eq. (2.30)) gives for each β a uniquely solvable inhomogeneous set of d equations in the d unknown components of $_0\langle \bar\beta |$ ($_0\langle \bar\beta' |$) (and vice versa).

73. For an example, see [8] where definition $\mathbf{A}_{\alpha\beta}^I$ is used with K as Bloch's wave operator [6].

74. Just as with mappings (K, L) of categories 2a and 2b, it is unnecessary to distinguish between mappings $(\bar K, \bar L)$ of categories 2a and 2d. Table II specifies the relation between the two sets of categories.

75. Mappings that conserve the norm of left, but not right, eigenvectors can be obtained by first choosing the $_0\langle \bar\alpha' |$ to be unity normed, followed by normalizing the $|\alpha'\rangle_0$ according to (2.30), and then proceeding in a manner analogous to Eqs. (2.48)–(2.53) [71]. This yields mappings that relate $_0\langle \bar\alpha' |$ to a true left eigenfunction that is also unity normed. The effective operator definitions produced by these mappings reduce to the ones considered above because the $_0\langle \bar\alpha' |$ are then the unity normed right eigenvectors of $\mathbf{H}^\dagger$ (just as those of case (c) above with the mappings (K, L)).

76. This Hermiticity supports the procedure of orthogonalizing the model eigenvectors.

77. Ellis and Osnes [32], Jørgensen and co-workers [35, 36], Lindgren [41], and Ratcliff [59] have shown that if a constant of the motion C commutes with the zeroth order Hamiltonian H_0, then it commutes with particular effective Hamiltonians, truncated at any order, whose model eigenfunctions, therefore, have the symmetry due to C. These authors do not consider this symmetry preservation in terms of commutation relation conservation, and paper II shows that the former can be obtained as a special case of the latter.

78. Alternatively, considered as an operator over the whole Hilbert space $\hat K$ has a nontrivial kernel (see Eq. (2.4)) and, thus, has no inverse.

79. V. Hurtubise and K. F. Freed, (unpublished).

80. Procedure 1 also requires resolutions of P_0 to evaluate the effective operator products in the left hand side of (4.16).

81. Both the combined and noncombined forms yield the same results since they produce the same effective operator when computed exactly (see Section II.D.2). The latter is selected because its closer relation to state-independent operators leads to simpler derivations. When approximate calculations are performed, however, the two forms are not generally equivalent and may thus yield different results.

82. F. Jørgensen, J. Chem. Phys. 68, 3952 (1978).

83. B. H. Brandow, Int. J. Quantum Chem. 15, 207 (1979).

84. T. H. Schucan, in: *Effective Interactions and Operators in Nuclei*, B. R. Barret (ed.) (Springer-Verlag, Berlin, 1975) p. 228.

85. U. Kaldor, *J. Chem. Phys.* **81**, 2406 (1984).

86. G. Hose, *J. Chem. Phys.* **84**, 4505 (1986).

87. Krenciglowa and Kuo [126] show that upon adiabatically turning on the perturbation, $|\bar{\alpha}'_B\rangle_0$ becomes $|\Psi'_\alpha\rangle$ when Ω_0 is degenerate.

88. G. Oberlechner, F. Owono-N'-Guema, and J. Richert, *Nuovo Cimento* **B68**, 23 (1970).

89. T. T. S. Kuo, S. Y. Lee, and K. F. Ratcliff, *Nucl. Phys.* **A176**, 65 (1971).

90. Ph. Durand, *Phys. Rev. A* **28**, 3184 (1983).

91. G. Jolicard, *Chem. Phys.* **115**, 57 (1987).

92. D. Mayneau, Ph. Durand, J. P. Daudey, and J. P. Malrieu, *Phys. Rev. A* **28**, 3193 (1983).

93. D. Hegarty and M. A. Robb, *Mol. Phys.* **37**, 1455 (1979).

94. H. Baker, M. A. Robb, and Z. Slattery, *Mol. Phys.* **44**, 1035 (1981).

95. J. P. Malrieu, Ph. Durand, and J. P. Daudey, *J. Phys. A* **18**, 809 (1985).

96. B. Jeziorski and J. Paldus, *J. Chem. Phys.* **88**, 5673 (1988).

97. M. A. Haque and D. Mukherjee, *J. Chem. Phys.* **80**, 5058 (1984).

98. The last equality in (5.8) is obtained by applying the identities $k = l^\dagger k^\dagger k$ and $k^\dagger k = (ll^\dagger)^{-1}$ to Bloch's formalism. The former identity follows from (2.13) and the second part of (2.3), and the latter is demonstrated in [37, p. 1157].

99. P. O. Löwdin, *J. Chem. Phys.* **18**, 365 (1965).

100. See [65, 102] for other **h**'s not considered by Jørgensen.

101. V. Kvaniscka and A. Holubec, *Chem. Phys. Lett.* **32**, 489 (1975).

102. I. Shavitt and L. T. Redmon, *J. Chem. Phys.* **73**, 5711 (1980).

103. P. Westhaus, E. G. Bradford, and D. Hall, *J. Chem. Phys.* **62**, 1607 (1975).

104. Westhaus' proof uses the exponential formulation (5.10) but does not depend on the form of G. Indeed, their operator G is not fully specified and may thus satisfy (5.11) or other conditions.

105. M. G. Sheppard and K. F. Freed, *J. Chem. Phys.* **75**, 4507 (1981); M. G. Sheppard, *Brandow Diagrams and Algebraic Formulas* (unpublished handbook).

106. V. Kvaniscka, *Chem. Phys. Lett.* **79**, 89 (1981).

107. T. Kato, *Prog. Theor. Phys.* **4**, 514 (1949).

108. B. Kirtman, *J. Chem. Phys.* **49**, 3890 (1968).

109. B. Kirtman, *J. Chem. Phys.* **75**, 798 (1981).

110. P. R. Certain and J. O. Hirschfelder, *J. Chem. Phys.* **52**, 5977 (1970).

111. P. R. Certain and J. O. Hirschfelder, *J. Chem. Phys.* **53**, 2992 (1970).

112. J. O. Hirschfelder, *Chem. Phys. Lett.* **54**, 1 (1978).

113. H. Primas, *Helv. Phys. Acta* **34**, 331 (1961).

114. H. Primas, *Rev. Mod. Phys.* **35**, 710 (1963).

115. P. Westhaus and E. G. Bradford, *J. Chem. Phys.* **63**, 5416; **64**, 4276 (1976).

116. K. Suzuki, *Prog. Theor. Phys.* **58**, 1064 (1977).

117. D. Mukherjee, R. K. Moitra, and A. Mukhopadhyay, *Mol. Phys.* **33**, 955 (1977).

118. The operator G for these mappings is given as $G = T^{\dagger} - T$, with $P_0 TP_0 = 0$. It thus follows from (5.11) that the mappings are canonical iff $Q_0 TQ_0 = 0$. This relation appears to hold in the example provided, but there is no mention concerning its generality.

119. Some properties of the Fock space transformations W and effective Hamiltonians $\mathbf{h}^F$ and, thus, of the resulting $\mathbf{h}$, appear to differ from those obtained by Hilbert space transformations. For example, their "canonical unitary" W is not "separable" and yields an $\mathbf{h}^F$ and, thus, an $\mathbf{h}$ with disconnected diagrams on each degenerate subspace. However, the analogous U of Eq. (5.13) may be shown to be "separable" [71], and the resulting $\hat{\mathbf{H}}_C$ on each complete subspace Ω_0 is "fully linked", as proven by Brandow [8]. These differences are not explained.

120. As discussed in Section III, mappings for variants a and $\tilde{a}$ can differ in the true eigenfunction norms and/or degenerate true eigenvector angles and still generate the same $\mathbf{h}$. Since variant $\tilde{a}$ sets the "irrelevant" part W_B of variant a to be null, Eq. (5.17) implies that the mappings for both variants are the same iff $W_B P_0 = 0$.

121. M. B. Johnson and M. Barranger, *Ann. Phys.* (*N.Y.*) **62**, 172 (1971).

122. M. B. Johnson, in: *Effective Interactions and Operators in Nuclei*, B. R. Barret (ed.) (Springer-Verlag, Berlin, 1975) p. 25.

123. A. Banerjee, D. Mukherjee, and J. Simons, *J. Chem. Phys.* **76**, 1979 (1982).

124. A. Banerjee, D. Mukherjee, and J. Simons, *J. Chem. Phys.* **76**, 1995 (1982).

125. P. J. Ellis, in: *Effective Interactions and Operators in Nuclei*, B. R. Barret (ed.) (Springer-Verlag, Berlin, 1975) p. 296.

126. E. M. Krenciglowa and T. T. S. Kuo, *Nucl. Phys.* **A240**, 195 (1975).

127. J. M. Leeinas and T. T. S. Kuo, *Ann. Phys.* (N.Y.) **98**, 177 (1976).

128. See [8, p. 786] for a review of the various proofs.

129. Leeinas and Kuo actually present their calculations in the noncombined form of $\mathbf{A}_{ii}^{IB}$ and the state independent $\mathbf{A}_B^{II}$ of Eq. (6.5). When applied to a one-dimensional model space, however, the perturbation expansion of the latter may be proven [71] to be identical to that of the combined $\mathbf{A}_{ii}^{IB}$ form (or of any other combined $\mathbf{a}$ definition) since they operate on the same functions.

130. M. Harvey and F. C. Khanna, *Nucl. Phys.* **A152**, 588 (1970).

131. M. Harvey and F. C. Khanna, *Nucl. Phys.* **A155**, 337 (1970).

132. T. H. Schucan and H. A. Weidenmüller, *Ann. Phys.* (*N.Y.*) **76**, 483 (1973).

133. V. Hurtubise and K. F. Freed, (unpublished).

134. R. A. Harris, *J. Chem. Phys.* **47**, 3967 (1967).

135. S. Garpman, I. Lindgren, J. Lindgren, and M. Morrison, *Phys. Rev. A* **11**, 758 (1975).

136. See, for example, [1, p. 699].

137. It might be thought that an effective operator acting on $_0\langle \alpha'_\mu |$ and $| \alpha'_\mu \rangle_0$ can be obtained using $E_\alpha^\mu = {}_0\langle \alpha'_\mu | \mathbf{H}_\mu | \alpha'_\mu \rangle_0$ instead of (6.23) in the above derivation. However, this gives $\lim_{\mu \to 0} \partial E_\alpha^\mu / \partial \mu$ as the sum of $_0\langle \alpha' | \mathbf{H}^{(1)} | \alpha' \rangle_0$ and of other terms which depend on $\lim_{\mu \to 0} \partial | \alpha'_\mu \rangle_0 / \partial \mu$.

138. See, for example, [12, 25, 32] and references therein and [43, 44, 92].

139. This is also true for the non-norm-preserving mappings $(\tilde{k}, \tilde{l})$ and $(\bar{k}, \bar{l})$ of Table II which yield effective operators that are not presented here [71].

140. The exceptions are the $(\tilde{K}, \tilde{L})$ mappings, the categories 2a and 2c mappings, and those of category 2b that preserve the angles between degenerate orthogonal eigenvectors. Each of these mapping sets is "small".

141. H. A. Bethe and R. Jackiw, *Intermediate Quantum Mechanics*, 2nd edition (Benjamin, 1968) Chapters 11 and 17.

142. T. Suzuki, *Ann. Phys. Fr.* **9**, 535 (1984).

143. C. C. Chang, in: *Topics in Nuclear Physics II*, T. T. S. Kuo (ed.) (Springer-Verlag, Berlin, 1981) p. 889.

144. J. Sadlej, *Semiempirical Methods of Quantum Chemistry* (Wiley, New York, 1985) p. 44.

145. H. Sun, K. F. Freed, M. F. Herman, and D. L. Yeager, *J. Chem. Phys.* **72**, 4158 (1980).

146. D. L. Yeager, M. G. Sheppard, and K. F. Freed, *J. Am. Chem. Soc.* **102**, 1270 (1980).

147. M. G. Sheppard and K. F. Freed, *J. Chem. Phys.* **75**, 4525 (1981).

148. Y. S. Lee and K. F. Freed, *J. Chem. Phys.* **79**, 839 (1983)

149. Y. S. Lee, K. F. Freed, H. Sun, and D. L. Yeager, *J. Chem. Phys.* **79**, 3865 (1983).

150. K. F. Freed and H. Sun, *Israel J. Chem.* **19**, 99 (1980).

151. See, for example, F. W. Byron Jr., and R. W. Fuller, *Mathematics of Classical and Quantum Physics* (Addison-Wesley, Reading, 1969), Vol. 1, pp. 158–171.

152. As we have not found a proof, this is now demonstrated. Property (2) implies that the eigenvectors $|a\rangle$ of X form an orthonormal set if the nondegenerate ones are unity normed and if the Gram–Schmidt procedure is applied to the degenerate ones. The formal representation of X in this set is then $X = \Sigma_a |a\rangle x_a \langle a|$. This relation and property (1) imply that $X^\dagger = X$, which completes our proof.

153. This result can also be obtained [71] from the condition, translated into our notation, $[l^\dagger l, PHP] = 0_\Omega$ given by Schucan and Weidenmüller, *Ann. Phys.* (N.Y.) **73**, 108 (1972) as equivalent to the Hermiticity of **h**.

154. Mappings satisfying conservation B actually preserve norms and scalar products in the transformations between corresponding degenerate subspaces of Ω and Ω_0. This follows from substituting (B.12) in the expression for the overlap of two arbitrary vectors of the degenerate subspace of Ω_0 (Ω) spanned by the $|\phi_i\rangle_0$ ($|\phi_i'\rangle$). This result is useful in establishing the relations between mappings of categories 2a–2d [71].

155. [151, p. 154].

156. C. Cohen-Tannoudji, B. Diu, and F. Laloë, *Quantum Mechanics* (Wiley, New York, 1977) Vol. I. p. 188.

MELTING AND LIQUID STRUCTURE IN TWO DIMENSIONS

MATTHEW A. GLASER and NOEL A. CLARK

Condensed Matter Laboratory, Department of Physics, University of Colorado, Boulder, Colorado

CONTENTS

Advances in Chemical Physics, Volume LXXXIII, Edited by I. Prigogine and Stuart A. Rice.
ISBN 0-471-54018-8 © 1993 John Wiley & Sons, Inc.

I.　INTRODUCTION

Melting and freezing are among the most common yet fascinating physical phenomena encountered in everyday life. The abrupt changes in the properties of a substance when it is heated through its melting point, such as the dramatic transition from solidity to fluidity achieved with very little change in density, are both intriguing and mysterious. Not surprisingly, the melting transition has been studied since the dawn of experiment. Since the rise of atomic and statistical physics in the last century, many attempts have been made to formulate a general theory of melting. Despite this effort, no generally accepted understanding of melting has emerged.

The problematic nature of the melting transition can be illustrated by comparison with other well-known first-order phase transitions, for instance the normal metal–(low T_C) superconductor transition. The normal metal–superconductor and melting transitions have similar symptomatic definitions, the former being a loss of resistance to current flow, and the latter being a loss of resistance to shear. However, superconductivity can also be neatly described as a phonon-mediated (Cooper) pairing of electrons and condensation of Cooper pairs into a coherent ground state wave function. This mechanistic description of the normal metal–superconductor transition has required considerable theoretical effort for its development, but nevertheless boils down to a simple statement, indicating an ultimate understanding of the *mechanism* of superconductivity and of the *essence* of the change that occurs at the transition. In contrast, the mechanism for melting has remained elusive, and it is still not possible to make a simple statement about the essence of the change that occurs upon melting, although there have been many interesting proposals (see Section II).

One might argue, following Landau, that a straightforward distinction between liquid and crystal can be made on the basis of order and symmetry: the solid structure is periodic and invariant only under a discrete set of translations and rotations, whereas the liquid has only short-range order and invariance under any translation or rotation. This view leads to a phenomenological description of melting in which the order parameters characterize this symmetry change, giving a measure of the localization of atoms in the crystal. However, this description suffers the same limitation as its companion Landau–Ginsberg description of superconductivity: it sheds little light on the underlying mechanism of the phase transition and is therefore unable to describe important aspects of its physics.

We believe that the lack of a mechanistic description of melting reflects a fundamental lack of understanding of liquid structure, which has traditionally been (and remains) one of the most intractable problems in theoretical physics. This may seem a surprising and extreme statement to those familiar with the modern theory of liquids. In fact, the modern integral equation approach to dense liquid structure, although approximate, allows the pair correlation functions and thermodynamic properties of dense simple liquids to be calculated with a high degree of quantitative success [1]. However, modern liquid theory does not provide an intuitive picture of the microscopic structure of dense liquids, nor does it yield a fundamental understanding of their entropy. What is needed, in our opinion, is a more direct approach to understanding the entropy of the dense liquid state and the entropy difference between the solid and liquid phases.

In fact, such an approach has been pioneered by J. D. Bernal, who laid the groundwork for the *statistical geometry*-based analysis of liquid structure in the late 1950s and early 1960s [2–5]. Working from the hypothesis that dense liquid structure is primarily determined by excluded volume effects, Bernal attempted to relate the problem of liquid structure to the essentially geometrical problem of the packing of hard spheres. By studying mechanically produced dense random packings of hard spheres, he attempted to identify the important geometrical degrees of freedom in dense liquids, or, conversely, to quantify the constraints imposed on local particle arrangements by excluded volume effects. Bernal's ultimate goal was to compute the entropy of the liquid state in terms of local geometrical degrees of freedom. Unfortunately, his goal was never realized, although he elucidated most of the important features of the statistical geometry of dense liquids. His approach was primarily descriptive, and although it provided insight into dense liquid structure, it did not lend itself to the quantitative calculation of liquid state properties. As a result, this approach was subsequently all but abandoned (although it has never quite died) in favor of the integral-equation approach. However, we believe that Bernal's work (described in detail in Section IV.A) represents an important milestone in the development of a general theory of melting, because it opens the door to an intuitive understanding of liquid properties.

Whereas Bernal had to rely primarily on mechanically produced random sphere packings in his work on liquid structure, the molecular dynamics and Monte Carlo computer simulation techniques have, during the past three decades, provided researchers with new and powerful experimental tools enabling a much closer look into the structure of the liquid—one has available the trajectory of every atom. Despite this, computer simulation has been used principally to calculate liquid-state correlation functions. This situation, to quote Lumsden and Wilson [6], "... appears to stem in part from a peculiar and fundamental relation that has always existed between experiment and theory in science: the importance of experimental data is judged by the theory to which it is applied. As the physicist Arthur Eddington said half seriously, no fact should be accepted as true until it has been confirmed by theory. Unless an attractive theory exists that decrees certain kinds of information to be important, few scientists will set out to acquire the information." Thus, it is only infrequently that computer simulations have been used to characterize liquid structure in ways other than those dictated by the prevailing liquid theory.

The subject of two-dimensional (2D) melting provides a classic example of a case in which progress in basic science has been driven almost entirely by theory. In this instance, the development of a novel theory of 2D melting by Kosterlitz, Thouless, Halperin, Nelson, and Young (the KTHNY theory) in the mid-to-late 1970s (motivated in part by earlier theoretical work proving the absence of true long-range order in 2D solids) led to a great deal of interest in 2D melting, as evidenced by the large number of experimental and theoretical studies of 2D melting that have been performed in the past 15 years. It is only a slight exaggeration to maintain that the subject of 2D melting did not exist before the development of the KTHNY theory. Furthermore, because of its novel and detailed predictions, the KTHNY theory has largely dictated the types of measurements that have been made in experimental and computer simulation studies of 2D melting. However, no strong evidence for the KTHNY melting mechanism has emerged from these studies, and the results of most simulation studies indicate that some other melting mechanism preempts the KTHNY mechanism. In our opinion, the success of the KTHNY theory is limited by its unrealistic treatment of 2D liquid structure. Thus, we believe that a fundamental understanding of the 2D melting transition requires the development of new ways of characterizing the structure of 2D liquids, independent of any particular theoretical scheme.

In this paper we present a computer simulation study of the melting transition of a model two-dimensional system (the 2D Weeks–Chandler–Andersen (WCA) system, which consists of particles interacting via a short-ranged, purely repulsive, circularly symmetric pair potential). This work is driven by the notions that the most important requirement for a melting model is to exhibit the mechanism of melting, and that the key to doing so is to understand and describe the structure of the 2D liquid phase in more depth than has been done to date. Based on our observations, we have identified a mechanism for 2D melting and have formulated a model for 2D melting that incorporates the essential features of our proposed mechanism.

The mechanistic picture of two-dimensional melting suggested by our work can be concisely stated as follows: *two-dimensional melting is a condensation of localized, thermally generated geometrical defects.* Condensation (e.g., Bose condensation or liquid condensation from the vapor) is a phase transition resulting from attractive interactions between the excitations or particles of interest. Thus, our observations lead to the identification of the geometrical excitations relevant to 2D melting and provide evidence that their effective attractive interactions produce a condensation transition that we identify with melting.

The statement of the previous paragraph, although concise, is necessarily somewhat obscure. Our basic idea, however, is really quite simple, and can be easily verified by the sceptical reader by, for example, pouring marbles into a box with rough or irregular sides. If identical disks are packed densely but haphazardly in the plane (to create a so-called *dense random packing*), one finds that the number of distinct local arrangements of disks is strictly limited. In particular, a limited set of "holes" (rings of disks arranged so that all disks touch their neighbors) are found in such packings, namely the equilateral triangle, the rhombus or square, the pentagon, and the hexagon (larger holes are comparatively rare). These basic packings units are shown in Fig. 1a. The structure of a dense random packing of disks can be uniquely described as a space-filling array

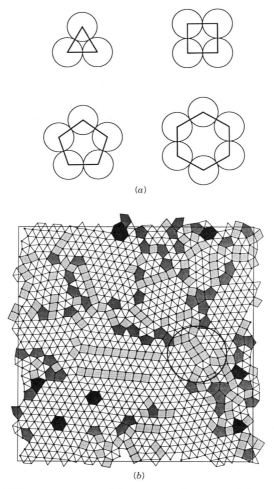

(a)

(b)

Figure 1. (a) Elementary polygonal holes in two dimensions. (b) Polygon construction for a representative dense random packing of hard disks at a packing fraction of 0.8334. Polygons having more than three sides are shaded according to the number of sides, and particle positions are marked by bullets.

of such elementary holes. On the other hand, a perfect triangular lattice corresponds to a periodic array of equilateral triangles. In the work presented here, we regard polygons with more than three sides, representing localized, easily identifiable deviations from triangular packing, as *geometrical defects*.

Figure 1*b* provides an example of a decomposition of a computer-generated dense random packing of hard disks into elementary holes (this decomposition, which we will subsequently refer to as the *polygon construction*, is described in more detail in Section III.D). Perhaps the most striking feature of such dense random packings is their *inhomogeneity*: they contain large regions of nearly perfect triangular ordering (consisting of triangular packing units), separated by regions containing numerous geometrical defects (polygons with more than three sides). This inhomogeneity is directly related to another prominent feature of dense random packings, namely the high degree of *correlation* between geometrical defects, which leads to an aggregation of geometrical defects into broad "domain boundaries" separating nearly defect-free regions, that is, to the formation of a *defect condensate*. This high degree of correlation between defects is manifested by the appearance of a limited number of characteristic local arrangements of polygons, and implies that strong, effective *attractive interactions* between geometrical defects are present. Such interactions essentially result from packing constraints—in a dense packing of hard disks, it is impossible to produce a large-amplitude local deviation from triangular packing without disrupting the triangular ordering in the vicinity of the defect. Because geometrical defects represent a local dilation, or increase in volume, geometrical defects are elements of "free volume," or voids, and thus the inhomogeneity in the spatial distribution of defects is associated with a corresponding inhomogeneity in the particle density. Finally, we point out that the description of the structure of dense random packings of hard disks in terms of elementary polyhedral packing units is an extremely sensitive diagnostic of disorder—essentially all of the structural disorder evident in Fig. 1*b* is directly associated with polygons having four or more sides.

Our "polygon" description of the structure of 2D dense random packings of hard disks parallels Bernal's description of three-dimensional (3D) dense random packings of hard spheres as space-filling arrays of elementary polyhedral units ("Bernal holes," or "canonical polyhedra") [2–5]. Bernal's approach to 3D liquid structure is discussed in more detail in Section IV.A.

These ideas can be carried over directly to the description of 2D solid and liquid structure. As for dense random packings of hard disks, we identify local arrangements of particles that deviate significantly from triangular ordering as geometrical defects. The resulting decompositions into elementary polyhedral packing units for typical configurations of the 2D WCA solid near melting and the 2D WCA liquid near freezing are shown in Figs. 2*a* and 2*b*, respectively. Geometrical defects are relatively scarce in the WCA solid near melting (Fig. 2*a*), generally appearing as isolated four-sided polygons or in short linear aggregates. The dense

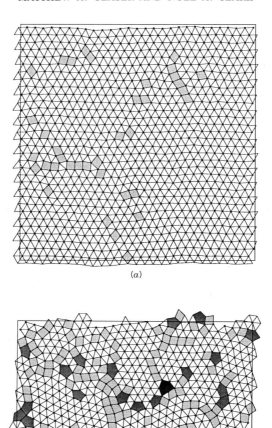

Figure 2. (*a*) Polygon construction for a representative configuration of the 896-particle WCA solid near melting: $\rho = 0.88$. (*b*) Polygon construction for a representative configuration of the 896-particle WCA liquid near freezing: $\rho = 0.85$. (*c*) Low defect density configuration of the topologically constrained generalized tiling model: $p_{max} = 6$, $r = 4$, $t = 0.8$. (*d*) High defect density configuration of the topologically constrained generalized tiling model: $p_{max} = 6$, $r = 4$, $t = 1.2$.

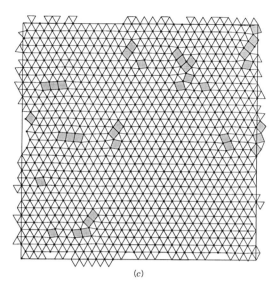

(c)

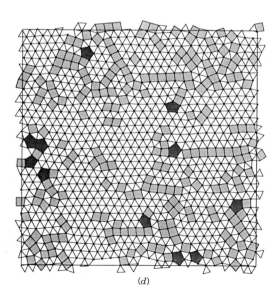

(d)

Figure 2. (*Continued*)

WCA liquid near freezing (Fig. 2*b*), on the other hand, bears a strong qualitative resemblance to dense random packings of hard disks (Fig. 1*b*). In particular, the spatial distribution of geometrical defects in the dense WCA liquid is quite inhomogeneous, and extensive regions of solid-like triangular ordering are evident. The defects show a clear tendency to aggregate, indicative of strong effective interactions between defects, and form a defect condensate similar in appearance to that seen in dense random packings of hard disks, with a few characteristic types of local arrangements of polygons prevalent. At the freezing point, the defect aggregates nearly percolate across the system (see Fig. 2*b*). As for dense random packings of hard disks, geometrical defects in the WCA system represent local elements of free volume, and, as shown in Section III.D, most of the volume expansion that occurs upon melting can be traced to the creation of polygons with four or more sides.

These observations form the basis for our view of 2D melting as a geometrical defect condensation transition. The geometrical defects are thermally generated excitations with strong attractive interactions. The attractive interactions between geometrical defects lead to defect aggregation and can, if sufficiently strong, give rise to a first-order defect condensation transition, which we associate with the 2D melting transition. The solid phase is thus viewed as a *defect gas*, with a low density of geometrical defects, while the liquid phase is viewed as a *defect liquid* or *defect condensate*, with a high density of strongly correlated geometrical defects. Thus, the number density of geometrical defects is the order parameter for the 2D melting transition. The 2D melting transition is an entropy-driven transition—creation of geometrical defects is energetically costly, but a finite density of defects contributes a finite entropy due to a multiplicity of possible arrangements of polygons. At a particular temperature the (negative) entropic contribution to the free energy exceeds the (positive) energy cost for defect creation, and the system melts. *The entropy of melting is essentially the configurational entropy of geometrical defects.*

To formulate a melting model based on these qualitative ideas, we need a way of incorporating defect interactions into our model. Some insight into the form of the defect–defect interactions can be gained by studying the characteristic defect aggregates that are observed in the WCA solid and liquid and in dense random packings of hard disks, and, in particular, by noting that the polygon network obtained from the polygon construction qualitatively resembles a *tiling* (see Figs. 2*a* and 2*b*). For example, three- and four-sided polygons tend to adopt local arrangements typical of plane tilings composed of squares and equilateral triangles, an example of which is shown in Fig. 3. This is illustrated by the

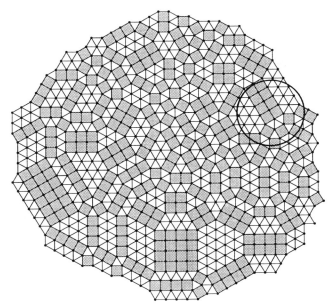

Figure 3. Typical random ST tiling grown using an algorithm described in Section V.B.

circled region of the dense random packing shown in Fig. 1*b*, which bears
a striking resemblance to the circled portion of the tiling shown in Fig. 3.
In general, polygons appear to adopt relative arrangements that involve a
minimal distortion of polygons from their "ideal" shapes (the ideal shape
is that of a regular polygon). This suggests that we can regard the polygon
network as a *generalized tiling*, that is, a tiling composed of deformable
tiles, in contrast to an ideal tiling, which consists of tiles of fixed shape.
Generalized tilings can contain *tiling faults*, local arrangements of tiles
that would not be allowed in a tiling composed of rigid tiles. Such tiling
faults involve some distortion of tiles from their ideal shape, to which we
associate an energy cost. Associating an energy cost with tiling faults
gives rise to defect–defect interactions, because thermally generated
geometrical defects will tend to adopt relative arrangements that minimize the strength of tiling faults. The resulting effective interactions are
anisotropic (only particular relative arrangements of defects are favorable) and *attractive* (defect aggregation can lower the energy of a system
of defects). The Hamiltonian for our *generalized tiling model* (GTM) for
2D melting thus contains two terms: an energy cost for defect creation
(the defect chemical potential), and an energy cost for tiling faults (which
generates the defect interactions).

We have obtained the thermodynamic, statistical, and structural properties of the GTM from Monte Carlo simulations. To assess the role of topological defects in the 2D melting process, we have studied both *topologically constrained* and *topologically unconstrained* versions of this model. We should point out at the outset that the loss of long-range order upon melting is a direct consequence of the proliferation of topological defects. In other words, topological defects are responsible for the symmetry change associated with the 2D melting transition. However, contrary to prevailing notions about 2D melting, this does not imply that 2D melting and 2D liquid structure can be understood entirely in terms of the interactions and statistical mechanics of topological defects. In fact, it is possible that, thermodynamically, their appearance is a side effect of a different process. Thus, symmetry alone tells us nothing about the relative contributions of topological and geometrical fluctuations to the entropy of melting. By studying topologically constrained and unconstrained versions of the GTM, we can begin to address this issue in a well-defined way.

One of our principal findings is that both the topologically constrained and unconstrained versions of the GTM display first-order phase transitions for sufficiently strong interactions between geometrical defects. In both cases, the high-temperature phase exhibits local structural features characteristic of 2D liquids. The phase transition in the topologically constrained model is a pure geometrical defect condensation transition, whereas the phase transition in the topologically unconstrained model involves both geometrical defect condensation and topological defect proliferation. A phase transition is observed in the topologically constrained model only for very strong defect interactions, in the limit in which the "tiles" are quite rigid, approaching the ideal tiling limit. In particular, the interactions required to produce a geometrical defect condensation transition are much stronger than those present in real 2D liquids. This implies that both geometrical and topological fluctuations are important sources of entropy in the 2D melting transition. This conclusion is supported by our studies of the topologically unconstrained GTM, which exhibits a first-order phase transition even for very weak geometrical defect interactions. For defect interactions approximately equal to those present in 2D liquids, the topologically unconstrained model exhibits a phase transition with a transition entropy comparable to the melting entropy measured in computer simulations of 2D melting. As discussed in Section V, however, this version of the model appears to severely overestimate the contribution of topological defects to the entropy of the 2D liquid, and so it is not possible to calculate the relative contributions of geometrical and topological degrees of freedom to the entropy of melting.

The qualitative behavior of the topologically constrained GTM is illustrated by Figs. 2c and 2d, which show typical low and high temperature configurations of this model (the method used to generate these configurations is described in Section V). The defect interactions are comparable to those present in the 2D WCA system. Because of the constraint on topology, this version of the model possesses long-range order even at high temperatures (it appears to have quasi-long-range translational order and long-range bond orientational order in general), and so does not realistically reproduce the large-scale structure of the liquid. Nevertheless, the *local* geometrical structure of the model resembles that of the WCA system, in that the same local arrangements of polygons occur with comparable probabilities (compare with Figs. 2a and 2b).

Our work suggests that the 2D melting transition is a defect condensation transition involving both geometrical and topological defects. The appropriate order parameter for 2D melting is the (scalar) number density of geometrical defects rather than the symmetry-derived order parameter(s) (Fourier components of the particle density). Thus, the global symmetry difference between the 2D solid and liquid phases provides little guidance in identifying the basic mechanism of the 2D melting transition. In fact, the configurational degrees of freedom of the GTM are essentially *lattice-gas* and *connectivity* degrees of freedom (see Section V), with no reference to real spatial variables or spatial symmetries. Melting, in our view, is a first-order transition because it is a *condensation* transition, and not because of the particular symmetry groups of the solid and liquid phases. Thus, the expectation that 2D melting should be qualitatively different from 3D melting, based on the absence of true long-range translational order in the 2D solid, probably has little basis. In fact, Bernal's work on 3D liquid structure strongly suggests that the mechanism for 3D melting is similar to the 2D melting mechanism we have identified.

Our melting model, as we have described it, is purely phenomenological. We have simply attempted to empirically determine the important features of the 2D melting process from MD simulation studies of 2D melting, and have formulated a melting model that contains these essential features. We are currently unable to explain why short wavelength geometrical fluctuations of the type that we have identified are the important configurational degrees of freedom in the melting process, nor can we derive the geometrical defect interactions from first principles. Clearly, however, these features of the melting process derive from excluded volume effects, because such excluded volume effects are expected to strongly condition the short wavelength fluctuation spectrum in solids and dense liquids and lead to strong correlations between

large-amplitude geometrical fluctuations. We believe that we have elucidated the basic mechanism for 2D melting, but much work remains to be
done to construct a complete theory of 2D melting based on the
qualitative ideas described here.

The remainder of this article is organized as follows: in Section II we
give a review of previous theories of 2D melting. Section III contains a
description of our molecular dynamics simulation study of melting in the
2D WCA system, and serves to motivate our model for 2D melting. In
Section IV we describe our study of the structure of dense random
packings of hard disks, which gives some insight into the role of excluded
volume effects in determining the statistical geometry of dense 2D
liquids. In Section V we describe our generalized tiling model for 2D
melting, and present the results of Monte Carlo simulation studies of this
model. Finally, in Section VI we conclude with a summary of our results
and a prognosis for the future.

II. THEORIES OF TWO-DIMENSIONAL MELTING

A. Introduction

In general, theories of 2D melting cannot be considered separately from
theories of three-dimensional (3D) melting, because many of the same
basic approaches have been tried in both cases. Thus, the discussion of
this section, which focuses on 2D melting, provides a thumbnail sketch of
some of the more promising approaches to both 2D and 3D melting. No
comprehensive review of melting theory exists, although there are a
number of articles and books that review various aspects of the problem
[7–20]. In this section, we will simply point out some of the more
interesting theoretical developments in the area of 2D melting.

The overall purpose of this section is to provide a background for
discussion of our own theoretical work, to be described in Section V.
Thus, our aim is to elucidate the *mechanism* for melting underlying each
of the theories described below. The concept of a melting mechanism
deserves to be explained in more detail. Intuitively, it seems that it should
be possible to describe the melting transition in terms of a relatively small
number of *active* or *relevant* degrees of freedom rather than the nominal
$2N$ (in 2D) or $3N$ (in 3D) configurational degrees of freedom. This is
plausible because the density of a liquid at the freezing point is high,
comparable to solid densities. Coupled with a hard core repulsion between molecules, this means that the set of possible particle configurations in a liquid near freezing is highly constrained, and that the liquid

must possess considerable short-range order, as is, in fact, evident from scattering experiments and computer simulations. These *packing* constraints might be expected to greatly reduce the number of effective degrees of freedom, just as constraints reduce the number of independent variables in a mechanics problem. In other words, only a limited set of collective coordinates are needed to characterize the difference between the solid and the liquid. This is the intuitive idea underlying most melting theories. Where melting theories differ is in their choice of relevant degrees of freedom, that is, in their identification of a melting mechanism. In defect-mediated melting theories, for example, the defect coordinates are the relevant degrees of freedom, whereas in density-functional theories, the relevant degrees of freedom are Fourier components of the density. Once a set of relevant degrees of freedom has been chosen, an effective partition function or free energy expression is usually constructed in terms of these collective coordinates.

There is one feature of 2D systems that suggests that 2D melting may be fundamentally different than 3D melting, namely the absence of true long-range order order in 2D crystalline solids, a feature that differentiates 2D solids from 3D solids. The type of order present in 2D solids (and in other 2D systems) has come to be known as *quasi-long-range order*. This theoretical background is described in Section II.B.

In spite of this fundamental difference between 2D and 3D solids, there is considerable empirical evidence that the melting transition in 2D is similar to that in 3D. We review this evidence in Section II.C, where regularities in the melting process (so-called empirical melting criteria) are discussed.

As mentioned in Section I, the Kosterlitz–Thouless–Halperin–Nelson–Young (KTHNY) theory has stimulated considerable interest in the subject of 2D melting. Consequently, no discussion of 2D melting would be complete without mention of this seminal theory, which provides a standard against which competing theoretical approaches (including ours) must be compared. The KTHNY theory has been extensively reviewed elsewhere [21–23]. However, for the sake of completeness, we have provided a brief summary of the KTHNY theory in Section II.D.

The KTHNY theory is a specific example of a *defect-mediated* melting theory, of which there are numerous other examples. In Section II.E we discuss some of the other 2D defect-mediated melting theories, including computer simulations of systems of interacting defects.

Finally, in Section II.F we discuss density-functional theories (DFTs) of melting. At present, DFT is perhaps the most popular approach to 3D melting, and has also been applied to 2D melting.

B.   Ordering in Two Dimensions

It has long been recognized that the degree of ordering present in a system is strongly dependent on its spatial dimensionality d and on the the dimensionality, n, of its order parameter. In general, the disruption of ordering due to thermal fluctuations becomes more pronounced for decreasing d and increasing n. The modern theory of phase transitions, in particular the renormalization-group approach, has succeeded in making these qualitative ideas more concrete [24–34]. For models that exhibit a second-order phase transition, thermal fluctuations cause the critical temperature T_C (below which the order parameter has a nonzero expectation value) to be lower than the mean-field theory estimate for all d, and lead to critical exponents that differ from the mean-field theory predictions for $d < 4$. The deviations from mean-field behavior become increasingly pronounced for decreasing d. Indeed, for small enough d, it is possible for ordering to be completely eliminated in the sense that the expectation value of the order parameter vanishes for $T > 0$ (T_C is suppressed to zero). The largest value of d for which this occurs is called the *lower critical dimension*, and depends on n.

The systems of most interest here are crystalline solids, which correspond to $n = 2$ because the symmetry-based order parameters of a crystalline solid are the Fourier components of the density,

$$\langle \rho_{\mathbf{G}} \rangle = \left\langle \sum_{i=1}^{N} e^{i\mathbf{G} \cdot \mathbf{r}_i} \right\rangle \tag{2.1}$$

which are two-component objects. Here $\mathbf{G}$ is a reciprocal lattice vector, $\mathbf{r}_i$ is the position of the ith particle, and the sum ranges over all N particles in the system. For $n = 2$ systems, there is considerable evidence from the study of specific models that the lower critical dimension is $d = 2$. Arguments for the absence of long-range order in $d = 2$, $n = 2$ systems date back to the work of Bloch [35], Peierls [36, 37] and Landau [38, 39], who presented nonrigorous arguments suggesting that long wavelength excitations (spin waves or phonons) are sufficient to destroy long-range order in two-dimensional magnets and crystals. Later, Hohenberg, Mermin, and Wagner presented rigorous proofs (based on the Bogoliubov inequality) of the absence of long-range order in $d = 2$ superfluids [40] (which correspond to $n = 2$ because the order parameter is the complex superfluid wave function), classical spin models with $n > 1$ [41], and two-dimensional crystals [42]. In general, ordered phases of $d = 2$, $n = 2$ systems are unstable against the long wavelength Goldstone modes which exist as a result of spontaneously broken continuous symmetry [32, 33].

Mermin's proof [42] of the lack of long-range order in 2D crystals is independent of the harmonic approximation, but only applies to particular forms of the interparticle potential, for example, potentials of the Lennard–Jones type and power law potentials $v(r) = \epsilon(\sigma/r)^m$ with $m > 2$. This proof does not apply to the hard disk system or to systems with long-range interactions (e.g., power law potentials with $m \leq 2$). For hard disks and for power law potentials with $m = 1$, however, there is evidence from computer simulations [43–45] that true long-range translational order is absent. Thus, although no completely general proof of the absence of long-range order in $d = 2$ crystals exists, it is likely that the Hohenberg–Mermin–Wagner theorem is generally applicable.

The bound on $\langle \rho_G \rangle$ found by Mermin for large but finite 2D solids is quite weak, namely

$$\langle \rho_G \rangle \lesssim \frac{1}{(\ln N)^{1/2}} \qquad (2.2)$$

Although $\langle \rho_G \rangle$ clearly goes to zero in the thermodynamic limit, the weakness of this bound led Mermin to suggest that some sort of ordering might still be present. For a 2D harmonic lattice, Mermin showed that, although the displacement-displacement correlation function diverges,

$$\langle |\mathbf{u}(\mathbf{R}) - \mathbf{u}(\mathbf{0})|^2 \rangle \sim \ln R , \quad R \to \infty \qquad (2.3)$$

reflecting the absence of long-range translational order, long-range *bond-orientational* order is present, in the sense that

$$\langle [\mathbf{r}(\mathbf{R} + \mathbf{a}) - \mathbf{r}(\mathbf{R})] \cdot [\mathbf{r}(\mathbf{a}) - \mathbf{r}(\mathbf{0})] \rangle \to a^2 , \quad R \to \infty \qquad (2.4)$$

as for a perfect triangular lattice. Here $\mathbf{R}$ is a direct lattice vector, $\mathbf{r}(\mathbf{R}) = \mathbf{R} + \mathbf{u}(\mathbf{R})$ is the instantaneous position of the particle associated with the lattice site at $\mathbf{R}$, and $\mathbf{a}$ is a basis vector of the direct lattice.

As Mermin's work suggested, and subsequent work has verified, the absence of long-range order in $d = 2$, $n = 2$ systems does not preclude the existence of low temperature phases with unusual properties. For example, Jancovici showed that the susceptibility of the 2D harmonic lattice,

$$\chi_G = \frac{1}{Nk_B T} (\langle |\rho_G|^2 \rangle - |\langle \rho_G \rangle|^2) \qquad (2.5)$$

which measures the response of the lattice to a periodic external potential with wave vector G, diverges as a particular finite temperature T_G is approached from above,

$$\chi_{\mathbf{G}} \sim C(T - T_G)^{-1} \tag{2.6}$$

where C is a constant, and the "critical" temperature T_G depends on the magnitude G of the reciprocal lattice vector [46]. This is somewhat surprising, because a divergent susceptibility is usually associated with a second-order phase transition, and the harmonic model exhibits no phase transitions. In this case, the divergence of the susceptibility is not associated with any thermal singularities. In the thermodynamic limit (in which $\langle \rho_{\mathbf{G}} \rangle \to 0$), the structure factor is proportional to the susceptibility, so that

$$S(\mathbf{G}) = \frac{1}{N} \langle |\rho_{\mathbf{G}}|^2 \rangle = k_B T \chi_{\mathbf{G}} \tag{2.7}$$

diverges for $T \leq T_G$. In other words, Jancovici's result implies that the 2D harmonic lattice exhibits pseudo-Bragg peaks at low temperatures, in spite of the lack of long-range translational order. However, unlike the structure factor of the 3D harmonic solid, which exhibits delta-function Bragg singularities,

$$S(\mathbf{k}) \sim \delta(\mathbf{k} - \mathbf{G}) \tag{2.8}$$

power law singularities are observed in the structure factor of the 2D harmonic lattice [47, 48],

$$S(\mathbf{k}) \sim |\mathbf{k} - \mathbf{G}|^{-2 + \eta_G(T)} \tag{2.9}$$

The power law singularities in $S(\mathbf{k})$ reflect the power law (algebraic) decay of the Debye–Waller correlation function $C_{\mathbf{G}}(\mathbf{R})$ for large R [47, 48],

$$C_{\mathbf{G}}(\mathbf{R}) = \langle e^{i\mathbf{G}\cdot[\mathbf{u}(\mathbf{R}) - \mathbf{u}(\mathbf{0})]} \rangle \sim R^{-\eta_G(T)}, \quad R \to \infty \tag{2.10}$$

of which $S(\mathbf{k})$ is the Fourier transform. This differs qualitatively from the exponential decay of correlations expected for a liquid, and also differs from an ordinary 3D solid, for which $C_{\mathbf{G}}(\mathbf{R})$ approaches a nonzero constant for large R. The dependence of the power law exponent $\eta_G(T)$ on G and T can be derived within the framework of the Kosterlitz–Thouless–Halperin–Nelson–Young theory of 2D melting, described in Section II.D.

The lack of long-range translational order in the 2D harmonic lattice is reflected in the mean-square displacement of a particle from its lattice site, which diverges logarithmically with increasing system size [47],

$$\langle u^2 \rangle \sim \ln N \,, \quad N \to \infty \qquad (2.11)$$

in contrast to the 3D harmonic lattice, for which $\langle u^2 \rangle$ is finite in the thermodynamic limit. This behavior is also quite different from that of a 2D fluid, for which

$$\langle u^2 \rangle \sim N \,, \quad N \to \infty \qquad (2.12)$$

Even though the mean-square displacement diverges logarithmically, the nearest-neighbor separation has small fluctuations, and has a well-defined and finite average value in the thermodynamic limit. This is why the harmonic approximation is expected to apply to the low temperature 2D solid, even though the mean-square displacement diverges.

The 2D XY model and 2D superfluid also possess low temperature phases characterized by power law decay of correlations. Power law decay of correlations is characteristic of a system at an ordinary critical point, and in this sense these low temperature phases resemble *phases of critical points* with a temperature-dependent critical exponent $\eta(T)$. The term *quasi-long-range order* (QLRO) has been coined to describe the ordering of such low temperature phases. Indeed, the loss of order in these systems is very weak. We have seen that the mean-square fluctuation in the separation of two particles in a 2D crystal grows only logarithmically with separation. Thus, the ordering in such systems can, with some justification, be regarded as long-range, although qualitatively different than the true long-range order (LRO) present in a 3D crystal. It can also be shown that a 2D crystal possesses a nonvanishing shear modulus, and so resembles a solid in this respect as well.

It is clear that the order present in a 2D crystal is qualitatively different than the short-range order (SRO) present in a liquid. Thus, if a 2D crystal is stable at finite temperatures, it is possible to have a phase transition from a low temperature phase exhibiting QLRO to a high temperature phase having only SRO. It is interesting to consider whether a phase transition is strictly required between a phase having QLRO and one having only SRO. Because the symmetry group of both phases is the same in the thermodynamic limit (QLRO does not result in a macroscopic broken symmetry), it seems that a phase transition is not necessary. However, as we have seen, the loss of order in the low temperature phase is very weak, and samples of macroscopic but finite size are expected to exhibit broken symmetry, so that for all practical purposes there is a symmetry difference between the two phases, and we would expect a phase transition. Perhaps a better way to look at the situation is in terms of the correlation length, which is infinite in the low temperature phase, due to algebraic decay of correlations. Consequently, there must exist a singular point where the correlation length diverges. Because a diverging correlation length implies a nonanalytic free energy, a phase transition must intervene between the high and low temperature phases.

The first evidence for such a phase transition in $d = 2$, $n = 2$ systems came from computer simulations of the hard disk system [49], where a melting transition was observed, and from high temperature expansions for the 2D XY model [50], which suggested that the susceptibility diverges as a particular finite temperature is approached from above, indicative of a phase transition. These two pieces of evidence, among others, led Kosterlitz and Thouless [51, 52] to propose a theory of phase transitions in $d = 2$, $n = 2$ systems. Their work was later extended by Halperin, Nelson, and Young [53–55], who formulated a complete theory of two-dimensional melting. This melting theory is described in Section II.D.

C. Empirical Melting Criteria in Two Dimensions

There are a number of empirical regularities associated with the melting transition in both two and three dimensions. Empirical melting criteria in 3D have been reviewed in [7, 13]. Frenkel and McTague [13] have also reviewed some of the 2D work. The body of "experimental" data on 2D melting is necessarily more limited than that on 3D melting, coming primarily from computer simulations. Nevertheless, distinct trends are evident in the properties of the 2D melting transition, which we briefly review here.

We should mention at the outset that the majority of computer simulation studies (including our own; see Section III) find that 2D melting is a first-order transition. However, there are a number of simulation studies that are inconclusive as to the order of the melting transition. We will not discuss these studies in this section, focusing instead of those simulations in which a first-order melting transition is clearly indicated. We provide a more balanced survey of the 2D simulation work in Section III.A.

Perhaps the most suggestive regularity observed in simulation studies of 2D melting is in the entropy change upon melting, $\Delta s = \Delta S/N$. In a large number of computer simulation studies, an apparent first-order transition is observed, with $\Delta s \sim 0.3 - 0.5 k_B$. This constitutes a 2D version of *Richard's rule*, which states that $\Delta s \sim k_B$ for 3D melting. A transition entropy in this range is observed for hard disks [49, 56, 57], the Weeks–Chandler–Andersen (WCA) model [58], the Gaussian core model [59], the Yukawa system [60], the Lennard–Jones (LJ) system [61–64], systems with power law pair potentials (r^{-1} [65, 66], r^{-3} [66], r^{-6}

[67], and r^{-12} [62, 68, 69]) and systems with a logarithmic pair potential [70–72]. There is a clear correlation between the entropy change Δs and the fractional volume change on melting, $\Delta v / v_s$, where $\Delta v = \Delta V / N$ is the volume change per particle and v_s is the volume per particle in the solid at melting. For example, the hard disk and LJ systems have the largest entropy change ($\Delta s \sim 0.5 k_B$) and the largest volume increase on melting ($\Delta v / v_s \sim 0.05$), while systems with r^{-1} and logarithmic pair potentials, for which the volume change on melting is rigorously zero [73], have transition entropies near the low end of the quoted range ($\Delta s \sim 0.3 k_B$). In spite of these variations, the "quasi-universal" value of the entropy of melting for widely different systems strongly suggests some sort of universal mechanism for two-dimensional melting.

Perhaps the best known empirical melting rule for 3D melting is the *Lindemann melting criterion*, which states that a solid will melt when the root-mean-square displacement of particles from their lattice sites reaches a certain critical value, $\delta \equiv \langle u^2 \rangle^{1/2} / a \sim 0.1$ (a more extensive discussion of the Lindemann criterion can be found in [14] and references therein). The Lindemann criterion is not strictly applicable to 2D melting because, as discussed above, the mean-square displacement $\langle u^2 \rangle$ diverges as $\ln N$ for 2D crystals, and so the Lindemann parameter δ is also divergent. This behavior has in fact been verified for hard disks [43, 44], the WCA system [58], and systems with power law potentials (r^{-1} [45], r^{-12} [44]). Nevertheless, for a given system size, δ is found to be nearly constant along the melting line, and nearly the same for different systems. For example, for $N = 256$, $\delta \approx 0.17$ for hark disks [43] and $\delta \approx 0.15$ for the LJ system [74], while for $N = 100$ and a r^{-1} pair potential, $\delta \approx 0.16$ [45], and for $N = 504$ and a r^{-3} pair potential, $\delta \approx 0.18$ [75]. A more general approach is that of Bedanov et al. [76], who propose the use of the modified Lindemann parameter $\delta' = \langle |\mathbf{u}(\mathbf{R} + \mathbf{a}) - \mathbf{u}(\mathbf{R})|^2 \rangle^{1/2} / a$, where $\mathbf{a}$ is a basis vector of the direct lattice. In contrast to the usual Lindemann parameter, δ' is finite in the thermodynamic limit. For the r^{-1}, r^{-3}, and LJ potentials, Bedanov et al. find $\delta' \approx 0.3$ at melting [76, 77].

The *Hansen–Verlet freezing criterion* states that a 3D fluid will freeze when the height of the first peak in the structure factor reaches a critical value ($S(k_m) \sim 2.85$, where k_m is the wave vector corresponding to the primary maximum in $S(k)$). There are relatively few tests of the Hansen–Verlet freezing criterion in 2D. Caillol et al. compared the structure factors of hard disk and logarithmic systems and found that $S(k_m) \approx 4.4$ at freezing for both systems, although the oscillations in the structure factor for $k > k_m$ are much more strongly damped in the logarithmic system than in the hard disk system [71]. On the other hand, Broughton et al. found $S(k_m) \approx 5.3$ for the r^{-12} system at freezing [68], while Vashishta and Kalia obtained a value of $S(k_m) \approx 5.5$ for the r^{-1} system at freezing [66]. Although these values show considerable variation, the value of $S(k_m)$ at freezing in 2D is clearly quite a bit larger than the corresponding quantity in 3D.

Finally, *Ross' melting rule* states that the excess Helmholtz free energy of a 3D solid at the melting point is $F_{ex} \approx 6Nk_BT$. There have been several tests of Ross' melting rule in 2D. The excess Helmholtz free energy of the 2D solid at melting was found to be $F_{ex}/Nk_BT \approx 3.7$ for the LJ system [74] and for the r^{-12} system [62], while $F_{ex}/Nk_BT \approx 3.9$ for hard disks at melting [79]. This suggests that Ross' melting rule is at least approximately obeyed in 2D (notice that the excess free energy at melting in 2D is smaller than the corresponding 3D value).

Overall, the situation in 2D appears to be very similar to that in 3D, although more extensive tests of these empirical melting criteria are clearly needed. The available evidence suggests that the 2D melting mechanism may be qualitatively similar to the 3D melting mechanism, in spite of the peculiar character of the 2D solid.

D. The Kosterlitz–Thouless–Halperin–Nelson–Young Theory of Two-Dimensional Melting

1. Introduction

In our discussion of ordering in two dimensions, we pointed out that $d = 2$, $n = 2$ systems have unusual low temperature phases with *quasi-long-range order* (characterized by power law decay of correlations). Because these systems exhibit short-range order (exponential decay of correlations) at high temperatures, the possibility of a novel type of phase transition (from QLRO to SRO) arises. This type of phase transition might be expected to differ qualitatively from more familiar order-disorder transitions, in which the more ordered phase possesses true long-range order. Based on these considerations, Kosterlitz and Thouless proposed a theory of phase transitions in $d = 2$, $n = 2$ systems, in which the phase transition is driven by the *unbinding* or *ionization* of neutral pairs of topological defects [51, 52]. The relevant topological defects are dislocations in the case of the 2D solid and vortices in the 2D XY model and 2D superfluid. The Kosterlitz–Thouless theory predicts a continuous phase transition with several novel features, including a universal jump discontinuity in the relevant coupling constant at the phase transition, an exponential divergence of the correlation length as the transition is approached from above, and a (probably unobservable) essential singularity in the specific heat at the transition. Because of these unusual characteristics, such phase transitions have come to be known as Kosterlitz–Thouless phase transitions.

Kosterlitz and Thouless exploited the analogy between a 2D system of dislocations and a 2D *Coulomb gas* to obtain the properties of the 2D melting transition. However, the Coulomb gas consists of a system of *scalar* charges, whereas dislocations in a 2D solid have a *vector* character, with corresponding orientation-dependent interaction terms [80, 81]. In addition, the unbinding of dislocations alone is not sufficient to completely destroy long-range order. As shown by Nelson and Halperin [54], a lattice containing a finite density of isolated dislocations still possesses quasi-long-range *bond-orientational* order. To correct these shortcomings, Halperin and Nelson [53, 54] and Young [55] have extended the Kosterlitz–Thouless melting theory.

The Kosterlitz–Thouless–Halperin–Nelson–Young (KTHNY) theory is similar in spirit to other defect-mediated melting theories. However, the assumptions and predictions of the KTHNY theory differ markedly from those of other defect-mediated melting theories. In particular, the KTHNY theory predicts that 2D melting can occur in two stages, via two continuous phase transitions, and posits the existence of a novel bond-orientationally ordered *hexatic* phase. Because of its unique predictions, the KTHNY theory of 2D melting has been seminal, stimulating a great deal of recent interest in 2D melting. In light of its importance, we will briefly review the physical content and predictions of the KTHNY theory.

Because topological defects (dislocations and disclinations) play a key role as a disordering mechanism in two dimensions in general, and in the KTHNY theory in particular, we will start with a short discussion of these defects. In two dimensions, an isolated dislocation is formed by inserting an extra half row of particles into a triangular lattice (Fig. 4b). Similarly, an isolated disclination is formed by inserting (removing) a 60° wedge of material into (from) a triangular lattice, to form a $+1$ (-1) disclination. A $+1$ disclination corresponds to a point having sevenfold symmetry, while a -1 disclination corresponds to a point of fivefold symmetry (Fig. 4a).

The strength of a dislocation is uniquely defined by the Burgers' vector, defined as the amount by which a "Burgers' circuit" (a circuit that would close in a perfect triangular lattice) around the defect fails to close, as shown in Fig. 4b. We will usually measure the strength of a dislocation in units of a lattice spacing. For example, the dislocation shown in Fig. 4b has unit strength because the Burgers' circuit fails to close by one lattice spacing. Similarly, the strength of an isolated disclination can be defined as the amount by which a vector aligned along one of the lattice directions is rotated upon "parallel transport" along a closed path around the disclination, in units of 60° (Fig. 4a). The sign of the disclination determines whether the rotation is in the same sense, or opposite to, the direction of parallel transport. Clearly, disclinations have a profound

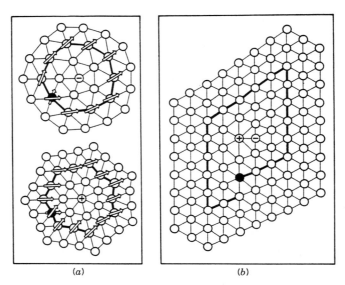

Figure 4. Isolated topological defects in a triangular lattice. (*a*) Isolated −1 and +1 disclinations. A vector aligned along a local lattice direction is rotated by 60° upon "parallel transport" around a unit strength disclination. (*b*) An isolated dislocation. The heavy line represents a Burgers' circuit around the dislocation, and the Burgers' vector of the dislocation is the amount by which the circuit fails to close. The core of the dislocation is a tightly bound pair of +1 and −1 disclinations (Reproduced from [78] by permission of Oxford University Press.)

effect on the orientation of the lattice, and so can be considered *orientational* defects, while dislocations primarily disrupt translational order, and so can be considered *translational* defects. Finally, as shown in Fig. 4*b*, a dislocation can be thought of as a tightly bound pair of disclinations, namely, a particle with sevenfold symmetry (a +1 disclination) separated by a lattice spacing from a particle with fivefold symmetry (a −1 disclination).

Dislocations and disclinations are considered *topological* defects because they cannot be eliminated from the lattice without a *global* rearrangement of particles (inserting or removing a half-row or a 60° wedge of particles). Alternatively, we can say that dislocations and disclinations have topological stability because their removal requires global changes in the topology, or connectivity, of the nearest-neighbor bond network (the so-called Delaunay triangulation, discussed in more detail in Section III.D), in that a macroscopic number of bonds must be rearranged.

In formulating their theory of 2D melting [51, 52], Kosterlitz and Thouless emphasized the ability of a 2D solid to support a shear stress as the key property distinguishing a solid from a liquid. This difference they attributed to the presence of "free" dislocations in the liquid near freezing, which move to the surface in response to an applied shear stress, giving rise to a finite shear viscosity but a vanishing shear elastic constant. In the solid state, on the other hand, there are no free dislocations (any dislocations that are present occur in tightly bound pairs), and so the system is able to support a shear stress, and the shear elastic constant is finite. This view led Kosterlitz and Thouless to introduce the concept of *topological long-range order* to characterize the order present in a 2D solid. To assess the topological order of a system (i.e., detect free dislocations), one makes a large Burgers' circuit, enclosing many lattice sites, and measures the amount by which the circuit fails to close. If free dislocations are present, the number of free dislocations enclosed by the circuit will be proportional to the area of the loop ($A \sim L^2$, where L is the length of the circuit), so that the path will fail to close by an amount proportional to L (the Burgers' vectors of the dislocations enclosed by the loop are randomly oriented along the three crystal axes, and so the net Burgers' vector corresponds to the end-to-end displacement in a random walk process). On the other hand, if only bound pairs of dislocations are present, only pairs cut by the Burgers' circuit will contribute to the net Burgers' vector. The number of such pairs is proportional to Ld, where d is the average pair separation, and so the path will fail to close by an amount proportional to $(Ld)^{1/2}$. Thus, the dependence of the total Burgers' vector on L can be used to assess the presence or absence of topological long-range order.

In the Kosterlitz–Thouless theory, the melting transition is associated with a dislocation-unbinding transition, in which the system loses its resistance to shear stress. To estimate the transition temperature at which dislocation pairs become unbound, Kosterlitz and Thouless presented a simple argument based on the free energy of a single dislocation [51, 52]. Because an isolated dislocation produces a strain that falls off at large distances as r^{-1}, the energy of an isolated dislocation depends logarithmically on the size of the system. The energy of an isolated dislocation can be computed in the framework of linear continuum elastic theory, which yields [81]

$$U = \frac{a^2 \mu(\mu + \lambda)}{4\pi(2\mu + \lambda)} \ln\left(\frac{A}{a_0^2}\right) = \frac{K}{16\pi} \ln\left(\frac{A}{a_0^2}\right) \tag{2.13}$$

where μ and λ are the Lamé elastic constants of the 2D solid [82], a is the lattice constant, A is the area of the system, a_0 is a distance on the order of the lattice constant which represents the diameter of the dislocation "core" (the linear extent of the region around the dislocation within which the continuum elastic description fails), and the coupling constant K is given by

$$K = \frac{4a^2\mu(\mu + \lambda)}{(2\mu + \lambda)} \qquad (2.14)$$

Strictly speaking, we should include a core energy term E_C in the expression for the dislocation energy, but this term is negligible in the limit $A \to \infty$, so it can be omitted for the purposes of this discussion. Because there are approximately A/a_0^2 possible locations for the dislocation, the entropy (to within terms of order unity) is

$$S = k_B \ln\left(\frac{A}{a_0^2}\right) \qquad (2.15)$$

The Helmholtz free energy of a single dislocation is therefore

$$F = U - TS = \left(\frac{K}{16\pi} - k_B T\right) \ln\left(\frac{A}{a_0^2}\right) \qquad (2.16)$$

At low temperatures, the energy term in the free energy dominates the entropy term, and the density of dislocations is small, with dislocations occurring in bound pairs. The density of bound dislocations increases with increasing temperature, until, at a critical temperature T_{KT} given by

$$k_B T_{KT} = \frac{K}{16\pi} \qquad (2.17)$$

the free energy for an isolated dislocation becomes negative, free dislocations proliferate, and the system loses its resistance to shear stress. This estimate of the melting temperature represents the simplest possible *instability* theory involving dislocations in 2D. Amazingly, the expression for the melting temperature obtained from this crude argument is *identical* to that obtained from the complete KTHNY theory, with the difference that K is replaced by a "renormalized" coupling constant K_R.

The full KTHNY theory improves upon the simple instability theory presented above in several respects. Kosterlitz and Thouless recognized that the elastic constants of a solid containing thermally excited dislocation pairs are *renormalized* by the presence of such excitations, so that the coupling constant that appears in the above formula for the transition temperature, Eq. (2.17), should be replaced by its renormalized (softened) counterpart, K_R. In practice, the interaction between a pair of dislocations with a separation r is screened by dislocation pairs with smaller separations. This results in a lengthscale-dependent coupling between dislocations, and a renormalization-group formalism is required to obtain the macroscopic elastic constants. At small relative separations, dislocations interact via the "bare" elastic constants, and the coupling constants are renormalized downward as the separation increases. The melting temperature is determined by the infinite-separation coupling constant, K_R. Kosterlitz and Thouless carried out the renormalization-group calculation of K_R for *scalar* dislocations.

Halperin and Nelson [53, 54] and Young [55] recognized that the vector character of dislocations must be taken into account in calculating the melting temperature, and also recognized that the dislocation-unbinding transition results in a sixfold bond orientationally ordered fluid phase, the *hexatic* phase, and that a second, *disclination*-unbinding transition is required to obtain an isotropic fluid.

In the remainder of this subsection, we describe the major assumptions and key results of the KTHNY theory.

2.   The Solid Phase and the Dislocation-Unbinding Transition

The KTHNY theory is based on linear continuum elastic theory [82], so that the statistical properties of the system are determined by the reduced elastic Hamiltonian

$$\bar{H} = -\frac{H}{k_B T} = -\frac{1}{2} \int \frac{d\mathbf{r}}{a^2} [2\bar{\mu} u_{ij}^2 + \bar{\lambda} u_{kk}^2] \tag{2.18}$$

where $\bar{\mu} = \mu a^2/k_B T$ and $\bar{\lambda} = \lambda a^2/k_B T$ are the reduced Lamé elastic constants, a is the lattice constant of the underlying lattice, and the symmetric strain tensor u_{ij} is related to the displacement field $\mathbf{u}(\mathbf{r})$,

$$u_{ij}(\mathbf{r}) = \frac{1}{2} \left[\frac{\partial u_i(\mathbf{r})}{\partial r_j} + \frac{\partial u_j(\mathbf{r})}{\partial r_i} \right] \tag{2.19}$$

The Einstein summation convention has been used in Eq. (2.18) and will be employed in the remainder of this section. If the system has periodic boundary conditions, Eq. (2.18) can be integrated by parts to obtain a description in terms of the displacement field,

$$\bar{H} = -\frac{1}{2} \int \frac{d\mathbf{r}}{a^2} [2\bar{\mu}(\nabla \mathbf{u})^2 + (\bar{\lambda} + \bar{\mu})(\nabla \cdot \mathbf{u})^2] \tag{2.20}$$

Equilibrium averages can be calculated from Eq. (2.18) or Eq. (2.20) by integrating over all configurations of the displacement field, provided a short distance cutoff that simulates the effect of the underlying lattice is imposed to eliminate ultraviolet divergences.

It is convenient to separate the displacement field $\mathbf{u}(\mathbf{r})$ into a smoothly varying part $\mathbf{u}_0(\mathbf{r})$ and a singular part $\mathbf{u}_D(\mathbf{r})$ due to dislocations,

$$\mathbf{u}(\mathbf{r}) = \mathbf{u}_0(\mathbf{r}) + \mathbf{u}_D(\mathbf{r}) \tag{2.21}$$

The smooth part of the displacement field satisfies the requirement

$$\oint_C d\mathbf{u}_0 = 0 \tag{2.22}$$

for all closed contours C, while

$$\oint_C d\mathbf{u}_D = a \sum_\alpha \mathbf{b}_\alpha \tag{2.23}$$

where the sum ranges over all dislocations within the contour C, and $\mathbf{b}_\alpha$ is the Burgers' vector of the αth dislocation. When the decomposition, Eq. (2.21), is inserted into the elastic Hamiltonian, the Hamiltonian splits into two pieces,

$$\bar{H} = \bar{H}_0 + \bar{H}_D \tag{2.24}$$

where $\bar{H}_0$ is the purely harmonic part,

$$\bar{H}_0 = -\frac{1}{2} \int \frac{d\mathbf{r}}{a^2} \left[2\bar{\mu} u_{0ij}^2 + \bar{\lambda} u_{0kk}^2 \right] \tag{2.25}$$

and the dislocation part is given by

$$\bar{H}_D = -\frac{\bar{K}}{8\pi} \sum_{\alpha \neq \beta} \left[\mathbf{b}_\alpha \cdot \mathbf{b}_\beta \ln\left(\frac{r_{\alpha\beta}}{a_0}\right) - \frac{(\mathbf{b}_\alpha \cdot \mathbf{r}_{\alpha\beta})(\mathbf{b}_\beta \cdot \mathbf{r}_{\alpha\beta})}{r_{\alpha\beta}^2} \right]$$
$$+ \frac{E_C}{k_B T} \sum_\alpha |\mathbf{b}_\alpha|^2 \tag{2.26}$$

Here u_{0ij} is the part of the strain tensor derived from $\mathbf{u}_0$, $\mathbf{r}_{\alpha\beta} = \mathbf{r}_\alpha - \mathbf{r}_\beta$ is the separation of dislocations α and β, the reduced coupling constant is defined as

$$\bar{K} = \frac{K}{k_B T} \tag{2.27}$$

where K is given by Eq. (2.14), a_0 is the dislocation core diameter introduced earlier, and

$$E_C = \frac{(C+1)\bar{K}k_B T}{8\pi} \tag{2.28}$$

is the "core energy," or chemical potential, of a dislocation. C is a measure of the ratio of the dislocation core diameter to the lattice spacing. The dislocation Hamiltonian is subject to the "charge neutrality" constraint

$$\sum_\alpha \mathbf{b}_\alpha = 0 \tag{2.29}$$

For a derivation of Eq. (2.26) see [53, 83].

In the KTHNY theory, the properties of the 2D solid are calculated by considering how the elastic constants in Eqs. (2.25) are (2.26) are renormalized by the dislocation degrees of freedom. The underlying physical idea is that of a *lengthscale-dependent coupling*, which results from *screening* of the interactions between a a pair of dislocations with separation r by dislocation pairs with smaller separations. This causes the effective coupling to become weaker as the separation between dislocation increases. In the limit $r \to 0$ the coupling constant $\bar{K}$ reduces to its bare, or unrenormalized, value, while in the opposite limit ($r \to \infty$), the coupling constant takes on its macroscopic (renormalized) value $\bar{K}_R$ (the reduced Lamé elastic constants $\bar{\mu}$ and $\bar{\lambda}$ behave similarly). As mentioned above, the calculation of the renormalized coupling constant as a function of temperature requires the apparatus of the renormalization group. An analysis based on the resulting renormalization group equations yields a variety of properties of the solid phase and the dislocation-unbinding transition, which we will simply list here (for details see [54] and references therein):

• The coupling constant $\bar{K}_R(T)$ approaches the universal value 16π as $T \to T_m^-$. Since $\bar{K}_R(T)$ vanishes for $T > T_m$, the coupling constant has a universal jump at the phase transition. The behavior of $\bar{K}_R(T)$ just below the melting point is governed by an exponent ν,

$$\bar{K}_R = \frac{16\pi}{(1 - C|t|^\nu)} \tag{2.30}$$

where $\nu = 0.369635\ldots$ (an analytic expression for ν can be found in [54]), $t = (T - T_m)/T_m$, and C is a positive, nonuniversal constant.

- In the solid phase, the Debye–Waller correlation function decays algebraically to zero for large r,

$$C_G(\mathbf{r}) = \langle e^{i\mathbf{G}\cdot[\mathbf{u}(\mathbf{r})-\mathbf{u}(\mathbf{0})]}\rangle \sim r^{-\eta_G(T)} \tag{2.31}$$

with corresponding power law singularities in the structure factor,

$$S(\mathbf{k}) \sim |\mathbf{k} - \mathbf{G}|^{-2+\eta_G(T)} \tag{2.32}$$

The exponent $\eta_G(T)$ is related to the renormalized elastic constants,

$$\eta_G(T) = \frac{G^2 a^2 (3\bar{\mu}_R + \bar{\lambda}_R)}{4\pi\bar{\mu}_R(2\bar{\mu}_R + \bar{\lambda}_R)} \tag{2.33}$$

Unlike the coupling constant $\bar{K}_R(T)$, the Lamé constants $\bar{\mu}_R(T)$ and $\bar{\lambda}_R(T)$ and the exponents $\eta_G(T)$ do not approach universal values as $T \to T_m^-$. However, the universal value of $\bar{K}_R(T)$ for $T \to T_m^-$ represents a universal relation between $\bar{\mu}_R(T)$ and $\bar{\lambda}_R(T)$ right at the melting point, which leads to

$$\eta_G^* \equiv \lim_{T \to T_m^-} \eta_G(T)$$

$$= \frac{G^2 a^2}{64\pi^2} \lim_{T \to T_m^-} [1 + \sigma_R(T)][3 - \sigma_R(T)] \tag{2.34}$$

where $\sigma_R(T) = \bar{\lambda}_R/(2\bar{\mu}_R + \bar{\lambda}_R)$ is the two-dimensional Poisson's ratio. Because $\sigma_R(T)$ cannot exceed unity [82], the above relation places bounds on the exponents η_G^*. For example, the exponent for the first Bragg peak of a triangular lattice at the melting point, $\eta_{G_1}^*$, cannot exceed $1/3$.

- Above T_m there is a finite density of free dislocations, which leads to exponentially decaying translational order,

$$C_G(\mathbf{r}) \sim e^{-r/\xi(T)} \tag{2.35}$$

where the correlation length ξ diverges exponentially as $T \to T_m^+$,

$$\xi(T) \sim \exp(C|t|^{-\nu}) \tag{2.36}$$

Here ν is the same exponent that appears in Eq. (2.30) and C is a positive constant. This is very different from the power law divergence of the correlation length associated with an ordinary second-order transition. An exponential divergence of the correlation length is typical of a system at its lower critical dimension [84]. The structure factor is free of singularities above T_m, but develops singularities at Bragg positions as $T \to T_m^+$,

$$S(\mathbf{G}) \sim \xi^{2-\eta_G^*} \tag{2.37}$$

where η_G^* is defined in Eq. (2.34).

• The density of free dislocations above T_m is simply related to the translational correlation length,

$$n_f \sim \xi^{-2} \tag{2.38}$$

which approaches zero as $T \to T_m^+$.

• The specific heat exhibits an unobservable essential singularity at T_m,

$$C_P \sim \xi^{-2} \tag{2.39}$$

However, there is a small bump in the specific heat above T_m [85], representing the entropy associated with unbinding dislocation pairs. This feature may be unobservably small, however, because the density of such pairs is assumed to be small.

3. The Hexatic Phase and the Disclination-Unbinding Transition

Several years after Kosterlitz and Thouless proposed their theory of 2D melting [51, 52], Halperin and Nelson [53, 54] pointed out that the dislocation-unbinding transition does not completely destroy long-range order, because the fluid phase formed by dislocation unbinding possesses quasi-long-range sixfold bond orientational order. Halperin and Nelson named the resulting phase (having short-range translational order and quasi-long-range bond orientational order) the *hexatic* phase, and argued that a second (disclination-unbinding) transition is necessary to transform the hexatic phase into an ordinary fluid phase (having short-range translational and bond orientational order).

With some appropriate definition of nearest neighbors (for example, the Voronoi construction, described in Section III.D, provides a unique definition of nearest neighbors), a bond angle field θ can be defined, where θ is the orientation of a "bond" between nearest neighbors with respect to some reference direction. In the continuum limit, the bond angle field is related to the displacement field by

$$\theta(\mathbf{r}) = \frac{1}{2} \left[\frac{\partial u_y(\mathbf{r})}{\partial x} - \frac{\partial u_x(\mathbf{r})}{\partial y} \right] \tag{2.40}$$

Once the bond angle field is defined, a sixfold bond orientational order parameter field can be constructed,

$$\psi(\mathbf{r}) = e^{6i\theta(\mathbf{r})} \tag{2.41}$$

Variations (modulo 60°) in bond orientation produce variations in the phase of ψ, which causes the spatial correlation function

$$C_6(\mathbf{r}) = \langle \psi^*(\mathbf{r})\psi(\mathbf{0}) \rangle \tag{2.42}$$

to decrease with increasing $\mathbf{r}$. In the solid phase, $C_6(\mathbf{r})$ decays to a nonzero constant [42], so there is long-range bond orientational order. In the hexatic phase, on the other hand, $C_6(\mathbf{r})$ decays algebraically to zero for large r [54],

$$C_6(\mathbf{r}) \sim r^{-\eta_6(T)} \tag{2.43}$$

with a temperature-dependent exponent $\eta_6(T)$. As shown by Nelson and Halperin [54], the properties of the hexatic phase are governed by a Hamiltonian of the form

$$H_A = \frac{1}{2} K_A(T) \int d\mathbf{r} |\nabla\theta(\mathbf{r})|^2 \tag{2.44}$$

where $K_A(T)$ is analogous to a Frank elastic constant of a liquid crystal. With a Hamiltonian of this form it can be shown [86] that

$$\eta_6(T) = \frac{18}{\pi \bar{K}_A(T)} \tag{2.45}$$

where the reduced Frank constant $\bar{K}_A(T) = K_A(T)/k_B T$ is related to the correlation length derived above,

$$\bar{K}_A(T) \sim \xi^2(T) \tag{2.46}$$

The Frank constant thus diverges as $T \to T_m^+$ and $\eta_6(T)$ tends to zero in the same limit.

The quasi-long-range bond orientational order present in the hexatic phase is destroyed at a second, disclination-unbinding, transition, which occurs at a temperature $T_i > T_m$. In this transition the tightly bound disclination pairs that make up a dislocation unbind in a way quite analogous to the dislocation-unbinding transition described above. The bond angle field can be decomposed into a smooth part θ_0 and a singular part θ_D, which, as before, results in a decomposition of the Hamiltonian, Eq. (2.44), into two parts,

$$
\begin{aligned}
H_A = \frac{1}{2} K_A &\int d\mathbf{r} |\nabla \theta_0(\mathbf{r})|^2 \\
&- \frac{\pi K_A}{36} \sum_{\alpha \neq \beta} s_\alpha s_\beta \ln \left(\frac{r_{\alpha\beta}}{a_{0A}} \right) \\
&+ E_{CA} \sum_\alpha s_\alpha^2
\end{aligned}
\tag{2.47}
$$

where $s_\alpha = 0, \pm 1, \pm 2, \ldots$ is an integer measure of the strength of disclination α, $r_{\alpha\beta} = |\mathbf{r}_\alpha - \mathbf{r}_\beta|$ is the separation of disclinations α and β, and a_{0A} and E_{CA} are the disclination core diameter and core energy, respectively. This Hamiltonian (involving scalar charges) is isomorphic to Kosterlitz' model [87] of the 2D superfluid and the 2D XY model, and his results are immediately applicable to this case. We will merely list the important results of Kosterlitz' analysis, which is analogous to the analysis of the dislocation-unbinding transition discussed above:

- The renormalized Frank constant approaches a universal value as $T \to T_i^-$,

$$
\lim_{T \to T_i^-} \bar{K}_{AR} = \frac{72}{\pi}
\tag{2.48}
$$

and vanishes for $T > T_i$. Consequently, the exponent $\eta_6(T)$ also approaches a universal value at the melting temperature,

$$
\lim_{T \to T_i^-} \eta_6(T) = \frac{1}{4}
\tag{2.49}
$$

- Above T_i, the sixfold bond orientational correlation function decays exponentially,

$$
C_6(\mathbf{r}) \sim e^{-r/\xi_6(T)}
\tag{2.50}
$$

with a correlation length that diverges exponentially as $T \to T_i^+$,

$$
\xi_6(T) \sim \exp[C(T - T_i)^{-1/2}]
\tag{2.51}
$$

where C is a positive constant.

• As before, the specific heat has only an essential singularity,

$$C_P \sim \xi_6^{-2} \qquad (2.52)$$

4. Discussion

The most interesting prediction of the KTHNY theory is that melting in 2D can occur via two continuous phase transitions, associated with the successive unbinding of dislocations and disclinations. The KTHNY theory also predicts a novel intermediate phase, the *hexatic* phase, having short-range translational order and quasi-long-range bond orientational order. This phase has been found in a variety of liquid crystal systems [88]. In addition, this theory makes a number of detailed predictions about the behavior of translational and orientational correlation functions and elastic constants near the two phase transitions, predictions that can be tested in suitably designed experiments. On the other hand, verification of a particular prediction of the KTHNY theory cannot be regarded as verification of the theory as a whole. For example, observation of a hexatic phase does not imply that the solid–hexatic and hexatic–liquid phase transitions are of the type envisioned by KTHNY.

The novel predictions of the KTHNY theory have led to a large number of experimental and simulation studies of the 2D melting transition in recent years. A review of these studies is outside the scope of this article, although we will present a brief survey of the simulation results in Section III.A. Reasonably up to date surveys of the experimental and simulation results can be found in [21–23]. Here we will confine ourselves to a few general remarks. The experimental situation is mixed, with some studies apparently verifying specific predictions of the KTHNY theory, and others finding melting behavior at variance with the KTHNY predictions. On the other hand, simulation studies generally find that the 2D melting transition is first order, with a defect density that increases dramatically upon melting. These studies suggest that some other melting mechanism supercedes the KTHNY melting mechanism.

With this in mind, it is worthwhile to review the major assumptions and approximations of the KTHNY theory:

1. The KTHNY theory is based on linear continuum elastic theory. This is probably inadequate for describing the short-range interactions between dislocations and disclinations, in the regime where defect cores overlap. If the defect density is high, or if defects show a strong tendency to aggregate, the form of the short-range interaction potential could be important. If the 2D melting transition is second order, as predicted by KTHNY theory, then short wavelength fluctuations can with some justification be neglected, and the form of the short-range interaction potential may be irrelevant (if the defects have a strong tendency to aggregate, however, then short-range interactions will strongly condition the long wavelength fluctuation spectrum). If, on the other hand, the transition is first-order, short wavelength fluctuations may dominate the thermodynamics of the phase transition, and the form of the interaction potential for small defect separations will be important in determining the short wavelength fluctuation spectrum.

Numerical calculations of dislocation pair interactions have been carried out for systems of particles with r^{-1} [89] and LJ [90] potentials. For the r^{-1} potential, Fisher et al. [89] find that the elastic dislocation interaction potential is accurate for dislocation separations as small as ~3 lattice spacings, while Joos and Duesbery [90] find that separations of ~30 lattice spacings are necessary to reach the asymptotic elastic limit. The adequacy of the continuum elastic approximation in describing the short-range interactions between defects is thus still something of an open question, and may depend on the range of the interparticle potential.

Because of its use of linear continuum elastic theory, the KTHNY theory would be expected to describe the solid phase much better than the liquid phase. Also, the KTHNY theory neglects anharmonic effects other than those due to topological defects. Nontopological anharmonic excitations may make a significant contribution to the properties of the solid and liquid phases, and may be important in determining the nature of the 2D melting transition.

2. *The KTHNY theory assumes that the density of topological defects is small.* The KTHNY theory is based on a perturbation expansion in the fugacity, which converges only if the fugacity is small (or the defect core energy is large). Although this may be a reasonable assumption for the solid phase, the soundness of this assumption in the liquid phase is questionable, because most simulation studies (see Section III) and studies of analog systems (colloidal suspensions and mechanical analog systems) [91–97] indicate that the defect density is high in the liquid phase. Thus, although the KTHNY theory may be an accurate theory of the solid phase, its treatment of the liquid phase is probably unrealistic. In this sense the KTHNY theory resembles an instability theory in that it seems to predict the temperature at which the 2D solid becomes unstable to dislocation unbinding rather than the thermodynamic melting temperature.

E. Other Defect-Mediated Melting Theories

1. Introduction

The various defect-mediated melting theories that have been proposed differ primarily in the type of defects that are assumed to be important in the melting transition. In 2D there are two general classes of defects, namely point defects (e.g., dislocations and vacancies) and *line* defects (domain walls, or grain boundaries). In general, defect-mediated melting theories can be further classified as *instability* theories, which predict the point at which the solid phase becomes unstable to defect proliferation, *condensation* theories, which regard the melting transition as a defect condensation transition and (in principle) calculate the free energy of both the solid and liquid phases to determine the thermodynamic melting point, and *ionization* theories, in which the transition is driven by the ionization of neutral defect pairs, as in the KTHNY theory described above (this mechanism can lead to either a first- or second-order phase transition). In 2D defect condensation theories based on point defects, the point defects can condense into linear defect structures upon melting. In this case the distinction between point defect-mediated melting theories and line defect-mediated melting theories blurs.

2. Vacancy-Mediated Melting Theories

The idea that melting may be associated with the proliferation of vacancies is a natural one considering that most materials expand upon melting. As far as we know, the first vacancy-mediated melting theory was proposed by Frenkel [98]. A more recent vacancy-mediated melting theory has been developed by O'Reilly [99]. Although these theories were developed to describe 3D melting, the extension to 2D melting is straightforward.

The earliest vacancy-mediated melting theories were essentially instability theories, which predicted the temperature at which the free energy of formation of vacancies becomes negative (an example of such arguments can be found in the section on KTHNY theory above). More recent theories (such as O'Reilly's) view melting as a vacancy condensation transition. Vacancy-mediated melting theories have never enjoyed wide acceptance because the concept of a vacancy is ill-defined in the liquid phase, and because such theories seem to be inapplicable to substances that contract upon melting.

3. Dislocation- and Disclination-Mediated Melting Theories

Dislocation-mediated melting theories are by far the most popular of the defect-based melting theories, with a correspondingly large literature. The literature on dislocation-mediated melting in 3D up to about 1980 has been reviewed by Cotterill [9, 10]. More recently, Edwards and Warner [100] used techniques from polymer physics to treat the statistics of an ensemble of dislocation loops, and found that the interactions between dislocations are screened by a finite density of dislocations, leading to a first-order transition. Nelson and Toner showed that a 3D solid with a finite concentration of unbound dislocation loops possesses long-range bond orientational order and a resistance to torsion not present in an ordinary liquid [101], analogous to the hexatic phase in 2D. Thus, the proliferation of dislocation loops by itself is not sufficient to completely melt a 3D solid. Kleinert, in an impressive series of papers (see [23] and references therein), has formulated a gauge theory of dislocation and disclination lines, exploiting an analogy between the statistical mechanics of defect lines and the Landau–Ginsberg theory of superconductivity. In Kleinert's view, the inability of a shear stress to penetrate into a fluid is analogous to the Meissner effect in superconductors. Kleinert's theory yields a first-order transition with a transition entropy of $\Delta s \approx 1.2 k_B$, consistent with Richard's rule. River and Duffy have argued that a different sort of line defect, the "odd line," is the relevant configurational degree of freedom that distinguishes a liquid from a solid [102]. Based on rather simple arguments, these authors derive a universal value of $\Delta s = k_B \ln 2$ for the melting entropy, which is close to the measured transition entropy for simple metals in the limit of zero volume change [103, 104].

Kleinert has extended his defect theory of melting to two dimensions, and finds that a single first-order disclination-unbinding transition with a transition entropy of $\Delta s \approx 0.3 k_B$ preempts the successive dislocation-unbinding and disclination-unbinding transitions predicted by the KTHNY theory [23, 105]. This value of the transition entropy is comparable to that measured in atomistic simulations of 2D melting (see Section II.C). Kleinert has also criticized the KTHNY theory, arguing that the defect core energy term in Eq. (2.26) is unphysical and leads to erroneous predictions (specifically, this term prevents dislocation "pileup," which Kleinert claims is important in disclination-mediated melting). More recently, Janke and Kleinert have generalized Kleinert's theory by including a rotational stiffness term in the elastic Hamiltonian [106]. They find a crossover from a single first-order melting transition to two successive Kosterlitz–Thouless transitions as the lengthscale of rotational stiffness increases. Their theory may account for the fact that a hexatic phase is observed in liquid crystal thin films (for which the lengthscale of rotational stiffness is large), but is not observed in systems composed of spherically symmetric objects (for which the lengthscale of rotational stiffness is small).

4. Grain Boundary-Mediated Melting Theories

Grain boundary-mediated melting theories date back to the work of Mott and Gurney [107, 108], who envisioned the liquid as a very fine grained polycrystal, and formulated a simple melting theory based on grain boundary proliferation. In 2D, computer simulation studies (see Section III) and experiments on analog systems [91–97] demonstrate that the density of topological defects increases precipitously upon melting, and that topological defects tend to aggregate into "strings" or "clumps" that qualitatively resemble grain boundaries, both in the liquid and in the transition region. These observations led Fisher et al. [89] and Chui [109–111] to consider grain boundary formation as a possible mechanism for 2D melting.

In the melting theories described here, grain boundaries are regarded as lines of dislocations with equal Burgers' vectors [112, 113], and the energetics are calculated within the framework of linear continuum elastic theory. In this approximation, the lowest energy configuration consists of equally spaced dislocations with Burgers' vectors perpendicular to the grain boundary [89, 109–111]. The adequacy of continuum elastic theory for describing the energetics and fluctuations of grain boundaries depends somewhat on the type of grain boundary considered. For small-angle grain boundaries, the constituent dislocations are widely spaced, and it is likely that elastic theory is a good approximation. For large-angle grain boundaries, on the other hand, in which the constituent dislocations are closely spaced, the continuum elastic approximation may be seriously in error, due to the breakdown of this approximation for small dislocation separations. Indeed, from numerical calculations for a system of particles with r^{-1} pair interactions, Fisher et al. [89] found energies per unit length for large-angle grain boundaries ($22° < \theta < 30°$, where θ is the angle of the grain boundary) roughly 20% lower than the elastic theory estimates. Thus, the predictions of grain boundary-mediated melting theories based on elastic theory may be unreliable for large-angle grain boundaries.

Fisher et al. [89] have calculated the temperature at which the 2D solid becomes unstable to grain boundary formation. These authors discuss two types of fluctuation in the shape of grain boundaries, namely "vibration," in which the Burgers' vectors of the dislocation remain parallel but fluctuate in position, and "meandering," in which the orientation of the grain boundary and of the Burgers' vectors of its constituent dislocations vary by 60° along the boundary. For small-angle grain boundaries, they find that vibration gives the dominant contribution to the entropy, and, for this case, they obtain an instability temperature identical to the Kosterlitz–Thouless instability temperature, T_{KT} (see Eq. (2.17)). They therefore conclude that the small-angle grain boundary formation instability is essentially a dislocation-formation instability in disguise. For large-angle grain boundaries, Fisher et al. assume that meandering gives the dominant contribution to the entropy, and perform a crude calculation for a meandering 30° grain boundary. The resulting estimate of the transition temperature is a factor of two larger than T_{KT}, indicating that melting via large-angle grain boundary formation is unfavorable relative to dislocation formation. However, this crude calculation cannot be considered definitive.

A more detailed theory of grain boundary melting based on linear elastic theory has been formulated by Chui [109–111]. In addition to taking into account the fluctuations of isolated grain boundaries, Chui considers how such fluctuations are modified by interactions between parallel grain boundaries, and includes contributions to the overall free energy from the grain boundary crossing energy, coupling between grain boundaries and bound dislocation pairs, and coupling of grain boundaries to a finite density change. Taking all of these factors into account, Chui obtains a Landau free energy expansion in the grain boundary number density, which yields a strong first-order melting transition for small dislocation core energies ($E_C < 2.84 k_B T_{KT}$), with the transition becoming much more weakly first-order for $E_C > 2.84 k_B T_{KT}$. Chui also showed that the proliferation of grain boundaries completely destroys long-range orientational order, so that no hexatic phase occurs in his theory.

Chui's theory is interesting because it is an example of a defect *condensation* melting mechanism (albeit a very restrictive one), and demonstrates how such a mechanism can give rise to a first-order phase transition.

5. Simulations of Defect Hamiltonians

Several researchers have carried out simulations of systems of dislocations and disclinations in order to test the predictions of the KTHNY theory directly. Saito carried out Monte Carlo simulations of systems of dislocations on a triangular lattice, with interactions given by the Hamiltonian of Eq. (2.26) [114–117]. He found that the melting transition is first-order for small dislocation core energies and continuous for large core energies, with a crossover between $E_C = 2.28 k_B T_{KT}$ and $E_C = 3.28 k_B T_{KT}$. This is interesting in light of Chui's grain boundary-mediated melting theory, which predicts that the melting transition crosses over from weakly first-order to strongly first-order at $E_C = 2.85 k_B T_{KT}$. Saito also found that melting is mediated by dislocation unbinding for large core energies, and is caused by the formation of grain boundary loops for large core energies.

Systems of disclinations interacting via the Hamiltonian of Eq. (2.47) have been studied in Monte Carlo simulations of the Laplacian roughening model, a model for interface roughening which is dual to the disclination system [118]. In simulations of this model on a triangular lattice, Strandburg et al. found two Kosterlitz–Thouless transitions for large disclination core energies, crossing over to a single phase transition for small core energies, with the crossover point (bicritical point) near $E_C = 2.7 k_B T_{KT}$ [119, 120]. Again, this value is remarkably close to the crossover point predicted by Chui, although it is not clear how the disclination core energy is related to the dislocation core energy. On the other hand, Janke and co-workers simulated another model related by duality to the Laplacian roughening model, and found a first-order phase transition at a value of E_C for which Strandburg et al. found two Kosterlitz–Thouless transitions [121, 122]. More work is clearly needed to clarify the situation with regard to the melting transition of the disclination system.

The results of these simulation studies demonstrate that, even within the framework of linear elastic theory, the melting behavior of dislocation and disclination systems is more complex than that suggested by the KTHNY theory, and depends critically on the defect core energies. For small core energies, the behavior of these systems is more suggestive of a defect condensation melting mechanism, whereas the behavior for large core energies appears to be consistent with the KTHNY melting mechanism.

F. Density-Functional Theories of Melting

1. Overview

Density-functional theories (DFTs) of melting date back to the work of Kirkwood and Monroe [123], who derived an integral equation for the single-particle density based on the concept of a local free energy density, and showed that periodic solutions characteristic of a crystalline phase develop for certain values of temperature and density. More recently, a number of approximate theories for the solid–liquid transition have developed out of the density-functional theory of classical inhomogeneous systems [1, 124]. The density-functional formalism is based on the density-functional theorems of classical statistical mechanics [125], which state that thermodynamic quantities such as the Helmholtz free energy F and the grand potential Ω are unique functionals of the single-particle density $\rho(\mathbf{r})$, e.g., $F = F[\rho(\mathbf{r})]$. In practice, the functional dependence of F on $\rho(\mathbf{r})$ is unknown, and a functional relationship must be guessed at.

All of the approximate DFTs of melting assume that the properties of the uniform (liquid) phase are well characterized, and, in particular, that the free energy and pair correlation function of the liquid are known. The crystalline phase is viewed as a highly inhomogeneous fluid with a spatially varying density $\rho(\mathbf{r})$ having the symmetry of the crystalline lattice. A simple parametrization of the density $\rho(\mathbf{r})$ is chosen, and then the appropriate free energy is minimized as a function of the variational parameters for each set of state variables (e.g., the grand potential Ω is minimized for each μ, V, T). The applicability of this approach is not limited to the solid–liquid phase transition, but can be extended to a variety of phenomena involving inhomogeneous systems, including solid–solid phase transitions, quasicrystal formation, glass formation, the structure of solid–liquid interfaces, etcetera. It is also possible to study metastable phases and to investigate the relative stability of various competing phases under the same thermodynamic conditions.

DFT relies heavily on recent developments in the theory of classical liquids, which allow the equilibrium properties of simple dense classical liquids to be calculated with considerable accuracy. Because DFT (in principle, if not in practice) treats the liquid state exactly, whereas the solid state is described approximately, it is often stated that DFT is a theory of freezing rather than melting. However, DFT is a theory of solid–liquid coexistence, and consequently describes the melting and freezing processes equally well (or poorly). The accurate treatment of the liquid phase is an attractive feature of DFT, because many theories of melting treat the liquid phase quite unrealistically (e.g., as an ideal gas of dislocations or vacancies), and any correct theory of solid–liquid coexistence must describe both coexisting phases accurately. Whether or not DFT provides an adequate structural and thermodynamic description of the solid phase remains an open question, however. There is a competing approach to the melting problem that employs liquid-state theory to calculate the free energy of the liquid and some other theory (e.g., self-consistent phonon theory or cell theory) to calculate the free energy of the solid. This approach gives fairly good results in specific cases (see, e.g., [64, 126]). However, the density-functional formalism has the aesthetic virtue of treating both phases within the same theoretical framework, and also has the advantage of computational simplicity. Further, DFT shows promise of explaining some of the quasi-universal features of melting in a straightforward way.

The DFT approach to ordering and phase transitions in classical systems is currently an extremely active area of research, with new developments occurring rapidly. A variety of different formulations DFT have been used, and so it is difficult to give a comprehensive overview of this field. We will merely try to sketch the major trends in DFT.

The first density-functional theory of melting to make use of the modern direct correlation function-based approach to liquid theory was formulated by Ramakrishnan and Yussouff [127], who treated crystalline order as a perturbation on liquid order. This *perturbative* approach to melting was subsequently taken up by a number of other researchers (for reviews see [17, 18]), and was found to give reasonably good results in a number of cases. In such perturbative DFTs, the excess Helmholtz free energy is assumed to be an analytic functional of $\rho(\mathbf{r})$, so that it can be expanded in a functional Taylor series about a uniform state with density ρ_l in powers of the density difference $\Delta\rho(\mathbf{r}) = \rho_s(\mathbf{r}) - \rho_l$. The Taylor series is typically truncated at second order, and the solid density $\rho_s(\mathbf{r})$ is usually parametrized as a sum of Gaussians centered at lattice sites (in which case the Gaussian width is the basic variational parameter in the theory) or, more generally, using a Fourier expansion of the solid density (so that the Fourier components of density are the variational parameters).

Because the excess free energy is assumed to be an *analytic* functional of the single-particle density, the perturbative approach is a mean-field approach, which, in the case of a continuous transition, would lead to mean-field critical exponents. In other words, order parameter fluctuations are neglected in this approach. The perturbative DFTs are, in fact, closely related to various Landau theories of melting that have been proposed [38, 128, 129] (the Landau theory of melting is essentially a simplified version of perturbative DFT). Because the symmetry of the crystalline phase leads to the appearance of third-order terms in the Landau free energy expansion, the Landau theory predicts that melting should be a first-order transition independent of spatial dimensionality. Further, Alexander and McTague have argued that because of the third-order terms the body centered cubic (BCC) lattice should be favored over other possible lattice types near the melting transition. This prediction has considerable empirical support in that all metallic elements (except Mg) on the left side of the periodic table, and most lanthanides and actinides, have a BCC structure near melting. The relation between Landau theory and DFT has recently been discussed by Laird et al., who argue that the analysis of Alexander and McTague is qualitatively and quantitatively incorrect, due to its neglect of higher-order Fourier components of the density and of the density change associated with melting. They argue that the relative stability of various lattice types represent the net effect of competing contributions from many Fourier components of density, and show explicitly that a one-order-parameter Landau theory does not exhibit a freezing transition.

Recently, the approximations underlying the perturbative DFTs have been called into question [130–132]. The solid density $\rho_s(\mathbf{r})$ predicted by this theory and measured in computer simulations is very sharply peaked about lattice sites, as can be inferred from the relative smallness of the Lindemann parameter near melting. Thus, the expansion parameter $\rho_s(\mathbf{r}) - \rho_l$ is by no means small, so that the truncation of the functional

Taylor series at second order is a questionable approximation. This approximation is justified if the third- and higher-order direct correlation functions of the liquid phase are negligibly small, but this does not appear to be the case. In particular, Curtin [130] and Cerjan et al. [132] have studied the effect of including third-order terms in the perturbation series, and have found that the agreement with computer simulations is significantly worse than for the second-order theory. This clearly shows that third-order (and perhaps higher-order) terms are important, and that the convergence properties of the perturbation series are poor.

The perceived weaknesses of the perturbative approach have led a number of researchers to formulate *nonperturbative* DFTs (for a review see [133]), which are based on the idea of a *mapping* of the solid onto an *effective liquid*. The various formulations differ primarily in the specific mapping used to determine the density, $\hat{\rho}$, of the effective liquid. Formally, we can write $\hat{\rho}$ as a functional of $\rho_s(\mathbf{r})$,

$$\hat{\rho} = \hat{\rho}[\rho_s(\mathbf{r})] \tag{2.53}$$

(in the most general case the effective liquid density may be spatially varying, that is, $\hat{\rho}(\mathbf{r}) = \hat{\rho}[\rho_s(\mathbf{r})]$).

The most recent nonperturbative DFTs [133, 134] give extremely good agreement with the results of computer simulations in particular cases. This is surprising, because the nonperturbative DFTs make rather severe, uncontrolled approximations, for which the physical motivation is unclear. Thus, it is something of a mystery why these theories work so well. Recently, Beale and Holger [135] have found that one of the most successful nonperturbative DFTs (the Generalized Effective Liquid Approximation, or GELA [133, 134]) produces unphysical solutions when the functional form of the peaks in $\rho_s(\mathbf{r})$ is allowed to vary (Lutsko and Baus [133, 134] assume a Gaussian form for the solid peaks). This observation calls the approximations underlying the GELA into question. Nevertheless, the successes of the nonperturbative DFTs are striking enough to warrant the current interest in these theories. In contrast to the perturbative DFTs, the nonperturbative theories make no assumptions as to the analyticity of the free energy, and so are not mean-field theories.

Another class of DFTs, which are close in spirit to the original Kirkwood–Monroe theory, are the bifurcation theories. It is usually stated that these are theories of solid phase instability. However, Bagchi et al. argue that the bifurcation analysis, if carried through exactly, predicts the equilibrium melting point [136]. We will not attempt to review these theories here, but simply refer the reader to the recent work of Bagchi and co-workers [132, 136–139], and to a recent review article that discusses bifurcation theories [17].

A qualitative discussion of the mechanism for freezing (or melting) in density-functional theory has been presented by Baus [16]. His argument is particularly simple for the freezing transition of hard spheres, in which case the transition is driven entirely by entropy. Roughly speaking, there are two contributions to the entropy difference between a hard sphere solid and a hard sphere liquid at the same density. The *localization* entropy represents the entropy cost of localizing particles around lattice sites (i.e., the entropy loss associated with setting up a nonuniform density distribution). This contribution to the entropy difference is negative (the solid phase has a smaller localization entropy than the liquid). The *correlation* entropy represents the entropy loss due to interparticle correlations. Generally speaking, the most probable nearest-neighbor distance in a hard sphere solid is larger than in a hard sphere liquid at the same density (the particles are less "jammed"), and so particles have more available volume within which to execute short-time vibrational motions. Thus, the correlation contribution to the entropy difference is positive (the solid phase has a larger correlation entropy than the liquid). The correlation entropy difference increases more rapidly with increasing density than does the magnitude of the localization entropy difference, so that at some critical density the total entropy difference becomes positive, and the system freezes.

2. Density-Functional Theories of Two-Dimensional Melting

A number of density-functional theories of 2D melting have been proposed, beginning with the pioneering work of Ramakrishnan [140]. Using perturbative DFT, Ramakrishnan found an entropy change on melting of $\Delta s \approx 0.3 k_B$, and showed that the height of the primary peak in the structure factor is $S(k_m) \approx 5.0$ at the freezing point. As discussed above in the section on empirical melting criteria, both of these findings are consistent with the results of computer simulations. What is particularly interesting about Ramakrishnan's analysis is that his results are *independent of the form of the interatomic potential*, at least to the degree of approximation implied by his approach. Thus, his work seems to provide a natural explanation of the quasi-universal features of melting discussed in Section II.C. However, he is able to formulate a potential-independent theory only by assuming an infinite compressibility for the fluid phase and by considering theories with only one or two order parameters (Fourier components of density). As mentioned above, such a drastic truncation of the Fourier expansion is likely to produce erroneous results. Nevertheless, Ramakrishnan's analysis is quite compelling, as it suggests that quasi-universal features of melting are observed because fluids with quite different interatomic potentials can have very similar direct correlation functions (the interatomic potential enters into DFT only implicitly through the direct correlation functions of the fluid phase).

More recent 2D DFTs have been formulated for specific interatomic potentials. Ballone et al. [141] studied the melting transition of a 2D electron system (r^{-1} pair potential) using perturbative DFT, and found a transition at a coupling strength of $\Gamma_C = 149$ (in reasonable agreement with both experiment and computer simulation), with an entropy change of $\Delta s = 1.2 k_B$ (a factor of four larger than the value found in computer simulations), and found that the primary peak in the structure factor has a value of $S(k_m) = 4.73$ at the freezing point (in reasonable agreement with simulation).

Similar results were obtained by Radloff et al. [142], who studied the 2D one-component plasma (logarithmic pair potential), using both the perturbative and bifurcation approaches. They find reasonable agreement between the coupling constants at melting obtained using the two approaches, finding $\Gamma_C = 133$ (perturbative analysis) and $\Gamma_C = 124$ (bifurcation analysis), values which are in reasonable agreement with computer simulations. The entropy change upon melting obtained from their perturbative DFT is $\Delta s \approx 0.97 k_B$ (a factor of three larger than the value obtained from simulation), and $S(k_m) = 4.1$ at freezing (in fair agreement with simulation).

The freezing transition of hard disks has been studied using both perturbative [143] and nonperturbative [144–146] DFTs. These results are summarized in [143]. The nonperturbative theory of Colot and Baus [145] and the perturbative theory of Laird et al. [143] are the most successful in predicting the melting and freezing densities of hard disks, although there are still substantial deviations from the values measured in computer simulations. Colot and Baus find that $S(k_m) = 5.25$ at freezing, which is in reasonable agreement with computer simulation. None of the DFTs for hard disks predicts the entropy change on melting.

The studies described above do not take into account the quasi-long-range translational order of the 2D solid phase, using instead a parametrization of the solid density that assumes true long-range translational order. It would be interesting to investigate whether the results obtained are strongly dependent on the assumed translational order of the solid (QLRO versus LRO). Also, a Fourier-space DFT treatment of the hexatic phase should be possible in principle. This could be useful for identifying systems in which a hexatic phase is likely to occur.

III. SIMULATION STUDIES OF LIQUID STRUCTURE AND MELTING IN TWO DIMENSIONS

A. Introduction

Beginning with Alder and Wainwright's pioneering molecular dynamics simulations of the melting transition of hard disks in the late 1950s and early 1960s [49, 147], a large number of computer simulation studies of 2D melting have been carried out. The majority of this activity has taken place in the past 15 years, in direct response to the development of the KTHNY theory. 2D systems with a variety of pair interaction potentials have been investigated, including hard disks [43, 44, 49, 56, 57, 79, 147–151], systems with power law (r^{-12} [44, 62, 68, 69, 152], r^{-6} [67, 153], r^{-5} [154, 155], r^{-3} [66, 75–77, 156, 157], and r^{-1} [45, 65, 66, 76, 77, 158–161]), logarithmic [70–72, 162], Yukawa [60], Lennard–Jones (LJ) [61–64, 74, 76, 77, 150, 163–177], and Weeks–Chandler–Andersen (WCA) [58, 178, 179] pair potentials, the Gaussian core model [59, 180], and a system with a piecewise-linear force law [181, 182] (we restrict our discussion to simulation studies of monatomic systems in the absence of a substrate potential). We have already discussed some of this work in the previous section, with regard to empirical melting criteria for 2D melting. Much of the simulation work on 2D melting has been recently reviewed [21–23]. Our discussion here will be limited to a few general comments.

Although the situation is still controversial, it is safe to say that no really strong evidence for the KTHNY melting scenario has emerged from the simulation work. The majority of simulation studies find that 2D melting occurs via a single first-order phase transition, rather than the two continuous phase transitions predicted by the KTHNY theory, although there have been several reports of an orientationally ordered intermediate phase between the solid and liquid phases [77, 153, 160, 165, 166, 176]. The most recent and thorough of the studies that report an orientationally ordered intermediate phase [176], in fact, finds a *first-order* melting transition, but also finds that the two-phase coexistence region has unusual properties, exhibiting algebraic decay of bond orientational correlations out to at least 100σ, where σ is the Lennard–Jones length parameter. Further, many of the simulation studies that clearly indicate a first-order melting transition (including our own work) find that bond orientational correlations are significantly longer ranged than translational correlations in the coexistence region. Thus, the nature of solid–liquid coexistence in 2D may be qualitatively different from 3D solid–liquid coexistence [176, 177].

In addition, a large number of studies find a high density of topological defects in the 2D liquid phase, which are observed to aggregate into linear structures separating nearly defect-free regions. In other words, the spatial distribution of defects in the 2D liquid is quite inhomogeneous. These observations suggest that 2D melting occurs via a different mechanism than that proposed by KTHNY, and, in particular, suggest that melting is a sort of *defect condensation* transition rather than the *defect ionization* transition envisioned by KTHNY. In fact, Weber and Stillinger [180] have proposed that the first-order transition observed in their molecular dynamics simulation study of the 2D Gaussian core model can be described as a type of disclination condensation transition, and a number of researchers have suggested a grain boundary melting mechanism on the basis of their simulation studies.

On the other hand, some simulation studies suggest that the observed melting point is close to that predicted by the KTHNY theory, in the sense that the value of the elastic coupling constant $\bar{K}_R$ at the melting point is nearly equal to the universal value (16π) predicted by the KTHNY theory [67, 68, 154, 159, 173, 174, 182]. Abraham has argued that $\bar{K}_R$ approaches 16π only in superheated solids, and has shown that the value of $\bar{K}_R$ at the thermodynamic melting point is significantly larger than 16π for the 2D LJ system [170]. However, more recent work on the phase diagram of the 2D LJ system suggests that the estimate of the thermodynamic melting temperature used by Abraham is seriously in error, and implies that the thermodynamic melting point is closer to the dislocation-unbinding instability temperature than was previously thought [177]. Overall, it appears that the dislocation-unbinding transition is preempted by a first-order transition produced by some other mechanism, but that the thermodynamic melting temperature is close to the KTHNY melting temperature.

The simulation work described above is certainly not free from criticism, because finite computing resources impose limits on the size of the systems that can be simulated, and also limit the duration of such computer experiments. The majority of simulations of 2D melting have been performed with systems of ~1000 particles or less, and total run times on the order of nanoseconds (in units appropriate for atomic systems). It has been argued that the simulated systems are too small to reveal the behavior in the thermodynamic limit, and that the simulated systems are out of equilibrium due to the short run times. In fact, Toxvaerd [172] has shown that the density of disclinations in the LJ solid near melting depends logarithmically on system size, and Zollweg et al. [44] have found that the disclination density, shear elastic constant, and global bond orientational order parameter have logarithmic size dependencies in the hard disk and LJ solids near melting. On the basis of their findings, Zollweg et al. conclude that reliable results cannot be obtained from simulations of systems containing less than ~16,000 particles. However, Zollweg et al. did not study the size dependence of the melting transition itself (i.e., the size dependence of the solid and liquid free energies), and there is considerable empirical evidence that the characteristics of the 2D melting transition do not vary strongly as the system size is varied. Nevertheless, an adequate study of finite-size effects in 2D melting is still lacking.

Finite-time effects in simulation studies of 2D melting are not well understood. However, there is little evidence that the simulated systems are far from equilibrium. The most common nonequilibrium effect that is observed is superheating (or superexpansion) of small systems beyond the thermodynamic melting point. This effect becomes rapidly smaller for increasing system sizes, and all but disappears for 2D systems containing more than ~4000 particles. In general, it appears that the severity of finite-time effects in the melting region decreases rapidly with increasing system size. In particular, the extremely slow fluctuations in thermodynamic quantities and in bond orientational order observed by Novaco and Shea [154] in the transition region of the r^{-5} system appear to be due, at least in part, to the small size ($N = 256$) of the system they studied.

On balance, then, we believe that the evidence from computer simulations suggests a first-order 2D melting transition, at least for the systems that have been simulated. If one accepts this view, then it is obviously of interest to investigate melting mechanisms other than that proposed by KTHNY. Therefore, in our simulation work, described below, we have refrained from carrying out extensive tests of the KTHNY theory (although we do present considerable new evidence for a first-order melting transition), and have instead focused on identifying the actual melting mechanism. To this end, we have analyzed the statistical geometry of the 2D liquid in an attempt to identify the configurational degrees of freedom that drive the 2D melting transition. This work has led us to view 2D melting as a *geometrical defect condensation* transition. Most of the remainder of this section is devoted to a description of our MD simulation studies of the statistical geometry of a model 2D system (the Weeks–Chandler–Andersen, or WCA, system). The primary purpose of this section is to provide the "experimental" data that motivates our model for 2D melting, which is presented in Section V.

B.   Thermodynamics of the Weeks–Chandler–Andersen System

The results described in the remainder of this section were obtained from standard microcanonical molecular dynamics (MD) simulations of the two-dimensional Weeks–Chandler–Andersen (WCA) system [183]. The WCA potential represents the repulsive part of the Lennard–Jones (LJ) pair potential, and is obtained by truncating the LJ potential at its minimum and shifting it upward by an amount ϵ. As a result, both the potential and its first derivative are continuous at the cutoff distance $(r_c = 2^{1/6}\sigma)$. The pair potential is given by

$$v(r) = \begin{cases} 4\epsilon[(\sigma/r)^{12} - (\sigma/r)^6] + \epsilon, & \text{for } r < r_c \\ 0, & \text{for } r \geq r_c \end{cases} \qquad (3.1)$$

The WCA and LJ potentials are shown in Fig. 5. We will consistently use reduced units in the discussion to follow, in which σ is the unit of length, ϵ is the unit of energy, and the particle mass m is the unit of mass. Temperature is expressed in units of ϵ/k_B. Once the units of length, energy, and mass are specified, reduced units for other quantities can be derived. For instance, the unit of time is $\tau = (m\sigma^2/\epsilon)^{1/2}$. To make contact with real atomic systems, we note that the LJ parameters appropriate for argon are $\sigma = 3.405$ Å, $\epsilon/k_B = 119.8$ K, and $m = 6.63 \times 10^{-23}$ g [1]. This leads to a unit of time of $\tau = 2.16 \times 10^{-12}$ s. The WCA system is similar to the hard disk system in that the potential is short-ranged and purely repulsive, but the potential is continuous so that MD simulations are convenient to perform. In addition, simulation of the WCA system requires less computational effort than is required for most other potentials, due to the fact that the number of pair interactions is minimized.

We carried out simulations of 896-, 3584-, and 8064-particle systems for a wide range of densities at a reduced temperature of $T = 0.6$, and measured a variety of thermodynamic, structural, and dynamic properties. A preliminary account of our results has been given elsewhere

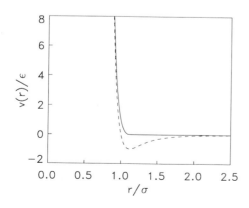

Figure 5. Weeks–Chandler–Andersen pair potential (solid line). The Lennard–Jones pair potential is shown for comparison (dashed line).

[178, 179]. Six series of simulations (density scans) were performed: the 896-particle system was simulated for decreasing densities in the range $0.05 \leq \rho \leq 0.91$ (run 1), for decreasing densities in the range $0.91 < \rho \leq 1.00$ (run 2), and for increasing densities in the range $0.80 \leq r \leq 0.91$ (run 3), the 3584-particle system was simulated for decreasing densities in the range $0.70 \leq \rho \leq 0.91$ (run 4) and for increasing densities in the range $0.80 \leq \rho \leq 0.91$ (run 5), and the 8064-particle system was simulated for decreasing densities in the range $0.80 \leq \rho \leq 0.91$ (run 6). We used the Verlet algorithm [1, 184] to integrate the equations of motion, using an integration timestep of 0.005 (in reduced units) for all of our MD simulations. At each density, the system was thermalized for 50,000 timesteps, starting with a thermalized configuration from a slightly higher density, suitably rescaled. Following thermalization, the runs were continued for 20,000–100,000 timesteps and a variety of properties were measured.

The equations of state obtained for runs 1, 4, and 6 in the region of the melting transition are shown in Fig. 6. The equations of state for all three runs exhibit a plateau as would be expected for a first-order melting transition (the density is a discontinuous function of pressure). This plateau is more distinct for the larger system sizes, but there is no evidence for a strong system-size dependence of the coexistence pressure. Smaller systems generally exhibit metastable extensions of the solid and/or liquid branches of the equation of state, due to the fact that the free energy required to create a solid–liquid interface is comparable to

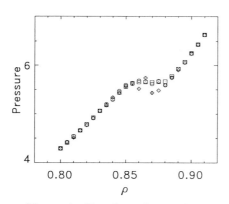

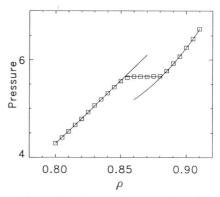

Figure 6. Equation of state for the WCA system in the melting region, for $N = 8064$ (squares), $N = 3584$ (circles), and $N = 896$ (diamonds).

Figure 7. Construction to determine the limits of coexistence for the WCA system. The data points are for the 8064-particle system, and the solid lines represent fits (described in the text).

the bulk free energy. As the system size increases, the interface free energy becomes a negligible fraction of the bulk free energy, and metastability is less commonly observed. The 8064-particle system, in fact, shows no evidence of a metastable (superexpanded) solid branch.

The flatness of the plateau in the equation of state for the 8064-particle system makes it possible to accurately determine the limits of coexistence for the WCA system. We accomplished this by fitting the solid and liquid branches of the equation of state to polynomials and fitting the equation of state in the coexistence region to a zero-slope straight line, and determining the points at which the linear fit intersects the polynomial fits to the solid and liquid branches. The resulting fits are shown in Fig. 7. This procedure yields a melting density of $\rho_s = 0.881$, a freezing density of $\rho_l = 0.853$, and a coexistence pressure of $P_{sl} = 5.66$. The fractional change in the specific volume upon melting is therefore $\Delta v / v_s = 0.033$, where v_s is the specific volume of the solid at melting. We can estimate the entropy change on melting from the condition for chemical equilibrium, $\Delta g = 0 = \Delta u - T \Delta s + P \Delta v$, where g is the Gibbs free energy per particle, u is the internal energy per particle and s is the entropy per particle. This leads to

$$\Delta s = \frac{1}{T} (\Delta u + P \Delta v) \tag{3.2}$$

Inserting the measured values into Eq. (3.2) yields $\Delta s = 0.39 k_B$. As discussed in Section II.C, this value of the melting entropy is similar to the values measured for other simulated 2D systems.

The potential energy per particle for runs 1, 4, and 6 (Fig. 8) also shows behavior in the transition region indicative of a first-order transition, in that it evidently interpolates linearly between its solid and liquid values. This is the expected behavior for quantities which represent averages of microscopic observables over all particles. In the coexistence region, as the relative proportions of solid and liquid vary, the average values of such quantities interpolate linearly between the coexisting solid and liquid values. Similarly, *distribution functions* of microscopic observables are, in the coexistence region, equal to the weighted sum of the corresponding distributions for the solid and liquid at coexistence. We will point out numerous examples of such behavior in the discussion to follow. All of the evidence, taken together, strongly suggests a first-order phase transition with a well-defined coexistence region.

In Figs. 9 and 10 we have shown in the pressure and potential energy, respectively, for the 3584-particle system, obtained on expansion and compression (runs 4 and 5). The only hysteresis that is observed is in the solid phase, in which the values of pressure and potential energy obtained

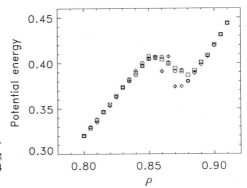

Figure 8. Potential energy per particle for the WCA system in the melting region, for $N = 8064$ (squares), $N = 3584$ (circles), and $N = 896$ (diamonds).

on compression are consistently higher than those obtained on expansion. Closer inspection of the solid configurations obtained on compression reveals that the system crystallizes with crystal axes that are misaligned with the computational cell, which freezes in a certain number of defects. The presence of such frozen-in defects increases the potential energy and pressure over that of a perfect crystal at the same density. This effect should become less pronounced as the system size increases (it should scale as $N^{-1/2}$), and in fact we see significantly less hysteresis in the 3584-particle system than in the 896-particle system.

There is no evidence for a metastable extension of the liquid phase into the coexistence region, in marked contrast to the situation in 3D, in which a metastable liquid branch is all too easily obtained. A possible explanation of this difference may lie in the fact that 3D systems are inherently frustrated, in the sense that the most efficient *local* packing of

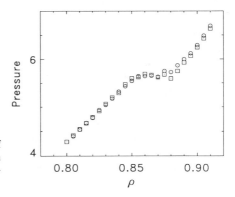

Figure 9. Hysteresis in the equation of state for the 3584-particle WCA system in the melting region. Squares, expansion; circles, compression.

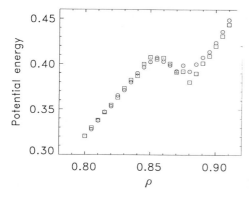

Figure 10. Hysteresis in the potential energy per particle for the 3584-particle WCA system in the melting region. Squares, expansion; circles, compression.

particles (tetrahedral packing) cannot be replicated to form any space-filling structure. This feature of 3D systems inhibits crystal nucleation. In 2D, on the other hand, the most efficient local packing (triangular) also corresponds to the preferred crystal structure (a triangular lattice), so that nucleation is not inhibited. In fact, as we show below, the dense 2D liquid contains numerous embryonic crystal nuclei.

The specific heat at constant volume (C_V) for the 896-particle system (runs 1 and 2) is shown in Fig. 11. The specific heat approaches the ideal gas value ($C_V/Nk_B = 1$) in the low density limit, increasing to $C_V/Nk_B \sim 1.7$ near freezing. The specific heat on the solid side of the transition is $C_V/Nk_B \sim 1.65$, approximately 17% smaller than the harmonic solid value ($C_V/Nk_B = 2$), indicating considerable anharmonicity in the WCA solid near melting. The specific heat approaches the harmonic solid value for increasing density in the solid phase, increasing to

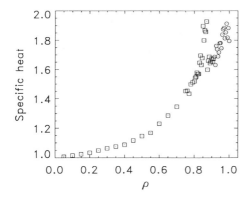

Figure 11. Specific heat per particle for the 896-particle WCA system, for run 1 (squares) and run 2 (circles).

$C_V / N k_B \sim 1.9$ at $\rho = 1.0$. The behavior of the 896-particle system in the melting region is typical of that observed for larger systems, with a maximum appearing near $\rho = 0.865$, roughly in the middle of the coexistence region. The increased specific heat in this region can be attributed to fluctuations of the interface between coexisting solid and liquid phases.

C. Inherent Structure of Two-Dimensional Liquids

It is generally recognized that there are at least two distinct characteristic timescales for the microscopic dynamics of dense liquids: a short timescale typical of single-particle vibrational motion, and a longer timescale associated with collective structural changes [185]. In dense 2D liquids, the longer timescale appears to be associated with the formation and dissolution of local solid-like fluctuations. In fact, we have found that the solid-like regions present in the dense WCA liquid have a broad distribution of sizes (see Section III.D), suggesting that there is a correspondingly broad distribution of structural relaxation times. Single-particle vibrational motion is responsible for oscillatory behavior in the velocity autocorrelation function at short times, while slower structural relaxation processes produce long-time tails in various time correlation functions, for instance in the stress and sixfold bond orientational order parameter autocorrelation functions. The short timescale is almost independent of density in the WCA liquid near freezing, while the long timescale increases steeply with increasing density near freezing. This increase is associated with an increase in the mean size of solid-like regions near freezing.

This conceptual division of the dynamics into vibrational and structural components has led a number of researchers to attempt to remove the vibrational component of the dynamics in order to better visualize the "inherent" structure of the liquid and study its dynamics. For this purpose, we have used a simple (although nonunique) procedure for suppressing vibrational fluctuations, namely, time-averaging of instantaneous configurations over a time window comparable to the period of single-particle vibrational motion, but short compared to the mean lifetime of solid-like regions in the dense WCA liquid. We consistently used an averaging time of 1 LJ unit, which is ~ 2 vibrational periods (as estimated from the velocity autocorrelation function). This procedure amounts to a time-domain filtering of the particle trajectories. The nonuniqueness of the procedure enters in the choice of an averaging time, but we feel that this is not a serious drawback because it is possible to make a physically motivated choice. Most of the results discussed below were obtained from time-averaged configurations. However, we will indicate whether time averaging was performed on a case by case basis.

An alternative approach is that used by Stillinger and Weber [186], who developed a unique procedure for mapping a finite-temperature liquid configuration onto a zero-temperature configuration by means of a steepest descents "quench" that carries the system to the nearest local minimum in the potential energy hypersurface (this procedure is simply a conjugate gradient minimization of the potential energy). For 2D systems with long-range pair potentials, this procedure produces configurations that are extremely well ordered compared to those obtained by simple time averaging [186]. It is not clear, however, that such highly ordered configurations provide a useful reference system for the relatively disordered liquid state. In addition, the method of Stillinger and Weber is inapplicable to systems with discontinuous pair potentials (such as hard spheres).

For the short-ranged WCA potential, we have found that steepest descents quenches produce configurations that are much more disordered than those obtained by Stillinger and Weber for a long-ranged potential [186], and that in fact strongly resemble the configurations obtained by simple time averaging. In the case of the WCA potential (and other short-ranged potentials) the potential energy minima are not distinct points in configuration space, but are flat-bottomed valleys corresponding to zero potential energy (the topography of the potential energy hypersurface resembles the mountains and playas of Nevada rather than the rolling hills of Pennsylvania). Consequently, for a short-ranged potential, a steepest descents quench takes the system to the edge of a potential energy "playa" and stops there. For a long-ranged potential, on the other hand, the condition of zero net force on each particle results in much greater ordering than is present in time-averaged configurations.

D. Statistical Geometry of the Weeks–Chandler–Andersen System

1. Overview

In this section we give an overview of the statistical geometry of the WCA system, both to introduce the reader to the types of analysis we have used and to describe the most important ideas that have emerged from this work. A more detailed discussion of our work has been relegated to the following sections, with an emphasis on results that lend further support to the key findings presented in this section. Most of the results described below were obtained from time-averaged configurations of the WCA system.

Our techniques for probing local geometrical and topological structure in the liquid and solid are largely based on the *Voronoi* construction [187], which is perhaps the most frequently used method for characterizing local topological order in 2D particle systems. The Voronoi construction assigns a cellular domain to each particle, namely, that region of the plane that is closer to the given particle than to any other particle. In general, the Voronoi cell so defined is an irregular polygon having three or more sides. For a triangular lattice, the Voronoi construction yields a regular array of hexagonal cells. The Voronoi construction also provides a unique definition of nearest neighbors. Any pairs of particles whose Voronoi cells share an edge are considered to be nearest neighbors. The number of nearest neighbors (the *coordination number*) of each particle characterizes the local topology, or connectivity, of the nearest-neighbor bond network, and makes it possible to identify topological defects (dislocations and disclinations) in 2D systems by finding particles with a coordination number different than six (the triangular lattice value). It can be shown that the average coordination number in the 2D Voronoi construction is 6 [188]. Thus, in 2D we have a topological conservation law, in contrast to 3D, where there is no such constraint on the average coordination number.

For the particle configurations we will be considering, vertices in the Voronoi construction represent the intersection of three Voronoi cells. This is not a completely general result—in a perfect square lattice, for example, the vertices in the Voronoi construction are points where four cells intersect. However, in thermally disordered configurations such four-coordinated (and higher-coordinated) vertices occur with essentially zero probability. In the absence of four- or higher-coordinated vertices, the definitions of nearest neighbors (and of local topology) provided by the Voronoi construction is unique, which in large part explains its popularity.

Disclinations in 2D systems are simply identified as particles having a coordination number different from 6. A five-coordinated particle is a -1 disclination, while a seven-coordinated particle is a $+1$ disclination (these can be thought of as resulting from the removal or insertion of a 60° wedge of material on a microscopic level). As shown in Fig. 4a, an isolated disclination results in a 60° rotation of a vector oriented along one of the lattice directions upon "parallel transport" around a closed circuit enclosing the disclination. An isolated dislocation is identified as a nearest-neighbor pair of oppositely charged (unit charge) disclinations, that is, a 5–7 pair (see Fig. 4b), and has a nonzero Burger's vector (defined as the amount by which a "Burger's circuit" around the dislocation fails to close). The concept of a dislocation is ill-defined in the liquid phase, because there is no reference lattice upon which to construct a Burger's circuit.

In Figs. 12a–17a we have shown the Voronoi construction for typical time-averaged configurations of the 896-particle WCA system for several densities spanning the melting region. We have chosen configurations representative of the solid near melting ($\rho = 0.89$), the solid at the melting point ($\rho = 0.88$), the system in the solid–liquid coexistence region ($\rho = 0.865$), the liquid at the freezing point ($\rho = 0.85$), and the dense liquid near freezing ($\rho = 0.83$ and $\rho = 0.80$). Topological defects are rare in the time-averaged WCA solid, occur-

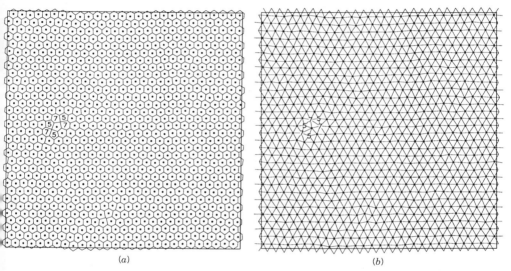

(a)   (b)

Figure 12. Various types of structural analysis for a representative configuration of the time-averaged 896-particle WCA solid at $\rho = 0.89$. (a) Voronoi construction, showing Voronoi cell boundaries (solid lines), with particle positions marked by bullets (for six-coordinated particles) or numbers (indicating the coordination number of non-six-coordinated particles). (b) Delaunay construction, showing nearest-neighbor bonds (solid lines), with particle positions marked as in (a). (c) Ordered regions, showing Voronoi cell boundaries for ordered particles (solid lines), and boundaries of ordered regions (heavy solid lines). Particle positions are marked by bullets. (d) Polygon construction. Polygons having more than three sides are shaded according to the number of sides, and intact nearest-neighbor bonds are indicated by solid lines. Particle positions are marked by bullets. (e) Vertex classification. Each vertex type is identified by letter except for vertices of types A (bullets), B (circles), C (squares), D (diamonds), E (triangles), and F (inverted triangles). Intact nearest-neighbor bonds are indicated by solid lines. (f) Tiling charges. Each vertex is marked by its tiling charge (in tenths), except for charge zero vertices, which are marked by bullets (type A vertices) or circles (other charge zero vertices). Negative charges are indicated by overbars, and intact nearest-neighbor bonds are indicated by solid lines.

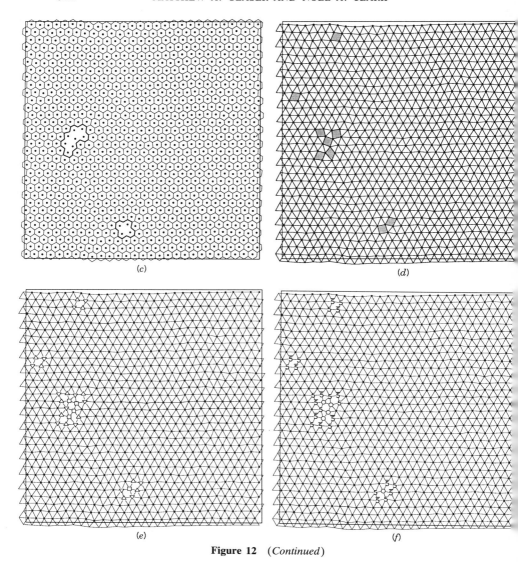

Figure 12 (*Continued*)

ring most frequently in neutral, closely bound groups of four disclinations ("disclination quadrupoles") associated with particles that have locally adopted a nearly square arrangement (see Fig. 13a), or in neutral "rings" composed of six disclinations (see Fig. 12a). This ring structure is associated with two groups of particles that have adopted a nearly square arrangement, such that the "squares" share a corner. In Figs. 12*a* and 13*a* there are also several examples of large amplitude, localized geometrical distortions in which the particles have adopted a nearly square arrangement, but which are not associated with disclinations. The signature of these "square" fluctuations is a short Voronoi cell side. When the

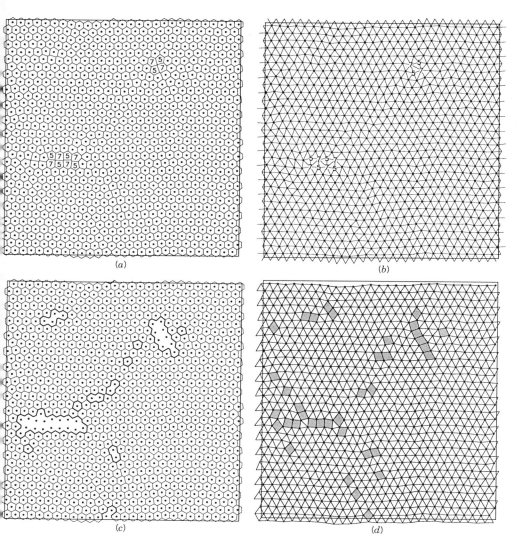

Figure 13. Various types of structural analysis for a representative configuration of the time-averaged 896 particle WCA solid at $\rho = 0.88$ (see the caption for Fig. 12).

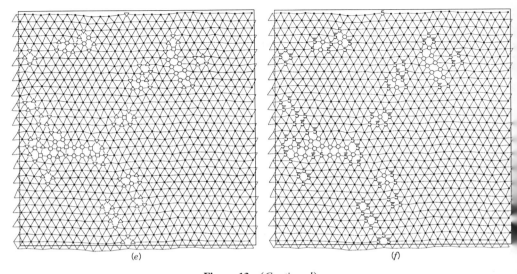

(e)   (f)

Figure 13   (*Continued*)

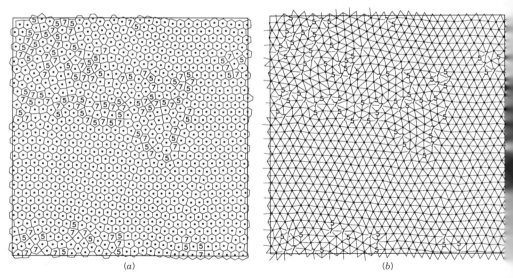

(a)   (b)

Figure 14.   Various types of structural analysis for a representative configuration of the time-averaged 896-particle WCA system in the coexistence region at $\rho = 0.865$ (see the caption for Fig. 12).

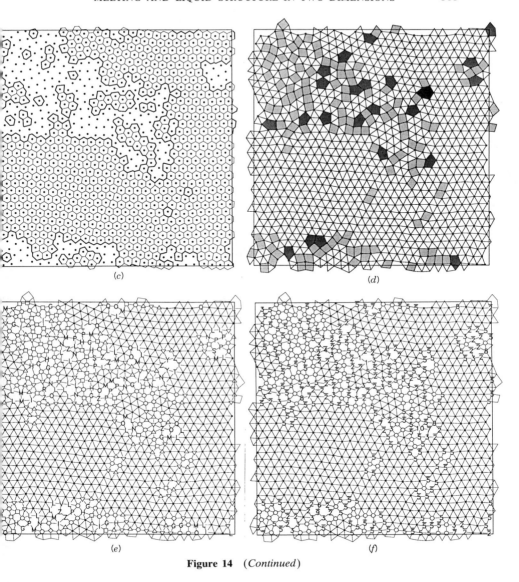

Figure 14   (*Continued*)

cell sidelength goes to zero, a disclination quadrupole appears, as shown in Fig. 18. Because the Voronoi construction fails to identify "squares" in a large number of cases, it appears that another approach is needed to identify such localized structural defects. We will have much more to say about this later.

As the solid melts, the number of topological defects increases dramatically. In the WCA liquid (see Figs. 15a–17a), the topological defects are highly correlated, typically aggregating into chains or clumps of disclinations of alternating sign. The net disclination charge of such aggregates is generally small. Local, nearly square arrangements of

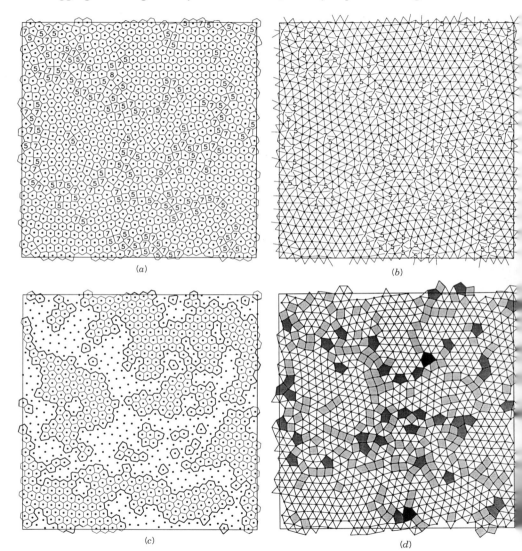

(a)   (b)

(c)   (d)

Figure 15. Various types of structural analysis for a representative configuration of the time-averaged 896-particle WCA liquid at $\rho = 0.85$ (see the caption for Fig. 12).

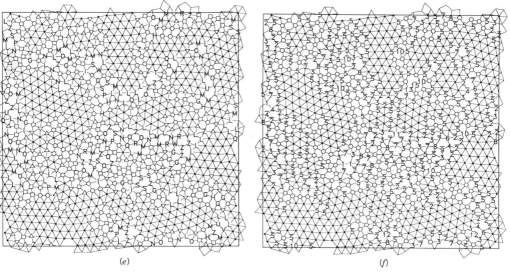

(e)   (f)

Figure 15   (*Continued*)

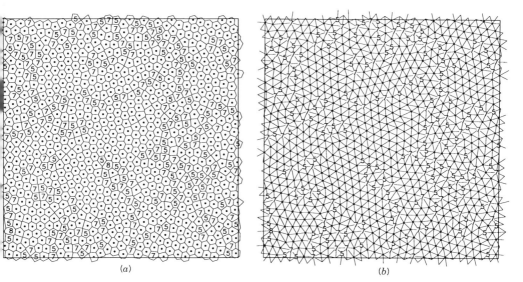

(a)   (b)

Figure 16.   Various types of structural analysis for a representative configuration of the time-averaged 896-particle WCA liquid at $\rho = 0.83$ (see the caption for Fig. 12).

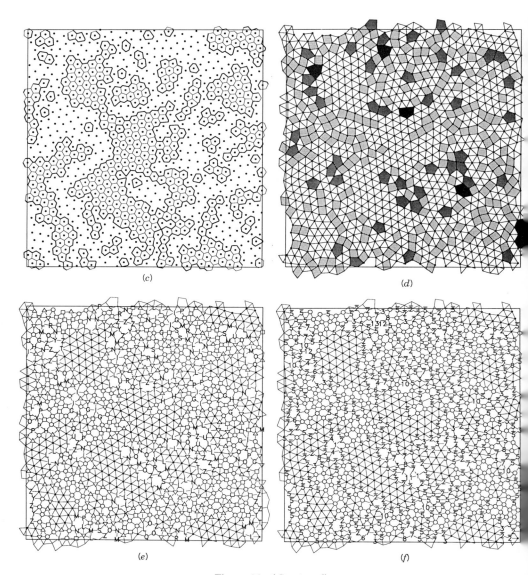

Figure 16   (*Continued*)

particles, as well as chains of "squares," are also ubiquitous in the dense WCA liquid, and in many cases are not associated with topological defects. Although the disclination aggregates are quite extended, they do not appear to percolate across the system, even at the lowest density investigated ($\rho = 0.70$). For all of the liquid states discussed here, six-coordinated particles are in the majority. Isolated dislocations and disclinations (separated by more than one or two nearest-neighbor spacings from other topological defects) are comparatively rare in the dense WCA liquid. We will discuss the statistics of topological defects in more detail below.

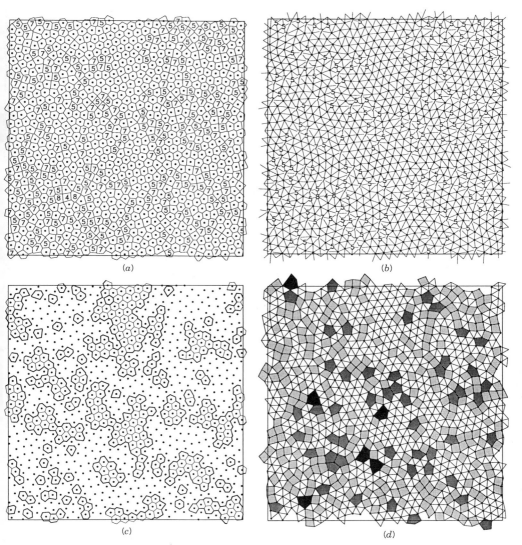

(a)   (b)

(c)   (d)

Figure 17. Various types of structural analysis for a representative configuration of the time-averaged 896-particle WCA liquid at $\rho = 0.80$ (see the caption for Fig. 12).

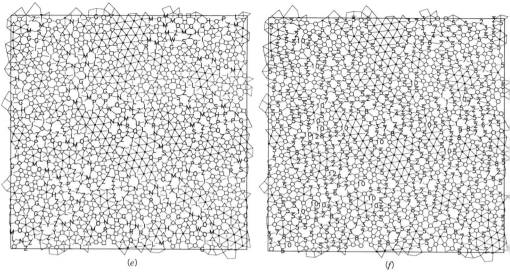

(e)   (f)

Figure 17   (*Continued*)

The spatial aggregation of topological defects evident in Figs. 15*a*–17*a* is directly related to the other prominent feature of the dense WCA liquid: extensive solid-like regions, which appear as rafts of nearly hexagonal Voronoi cells. As we will show, this dramatic spatial in-homogeneity is a consequence of the defect condensation transition that produces the liquid phase. Thus, it is the key qualitative feature of liquid structure that is required to understand 2D melting.

The dual of the Voronoi construction, the so-called *Delaunay* construc-

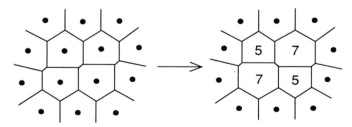

Figure 18.   Voronoi construction for two nearly square arrangements of points, showing how a small displacement of the points can create disclination quadrupole (right) from a configuration with no disclinations (left).

tion, is the planar graph obtained by connecting all nearest neighbor pairs by "bonds." The Delaunay construction thus represents the nearest-neighbor bond network. The Delaunay construction *triangulates* the plane, that is, it partitions the plane into (in general, irregular) triangular regions, with the particles residing on the vertices of the triangles. This triangulation property is a consequence of the fact that vertices in the Voronoi construction represent the intersection of three cells.

The Delaunay constructions for the configurations of Figs. 12a–17a are shown in Figs. 12b–17b. The Delaunay construction does not contain any information not contained in the Voronoi construction, but it does give a somewhat different picture of local structure. The general impression of polycrystalline disorder in the dense WCA liquid is more striking in the Delaunay construction than in the Voronoi construction, and a considerable variation in nearest-neighbor bond lengths and bond angles is evident in the Delaunay construction. The concept of a nearest-neighbor bond network will figure prominently in the subsequent discussion, and we will have occasion to refer back to the Delaunay construction.

In order to probe local solid-like fluctuations in the dense 2D WCA liquid, we measured the local sixfold bond-orientational order parameter

$$\psi_6 \equiv \psi_{6i} = (1/n_i) \sum_{j=1}^{n_i} e^{i6\theta_{ij}} \qquad (3.3)$$

where n_i is the number of nearest neighbors of particle i and θ_{ij} is the orientation of the nearest-neighbor bond between particle i and its jth neighbor relative to the an arbitrary axis. For most of the subsequent discussion we will drop the subscript i and refer to this order parameter simply as ψ_6.

For a perfect triangular lattice, the magnitude of ψ_6 is 1, and the phase of ψ_6 depends on the orientation of the lattice. To identify solid-like (or sixfold ordered) regions in the WCA liquid, we selected particles for which the magnitude of the sixfold order parameter was large. In practice we used the criterion $|\psi_6| > 0.75$ to identify sixfold bond orientationally ordered particles (we provide a partial justification of this criterion below). Sixfold ordered regions were then constructed by searching for continuous nearest-neighbor clusters of sixfold ordered particles. In the subsequent discussion, we will refer to sixfold ordered regions and non-sixfold ordered regions simply as "ordered" and "disordered" regions, respectively. In Figs. 12c–17c we have shown the ordered regions in the configurations of Figs. 12a–17a.

Figures 15c–17c reveal extensive regions of sixfold order in the dense

WCA liquid. The typical size of these regions increases rapidly near the freezing density. At the lowest density studied ($\rho = 0.70$), only relatively small ordered regions occur. The orientation of the local crystalline axes can vary considerably across the larger ordered regions, as is apparent from Figs. 15c–17c. The statistics of ordered regions in the WCA system will be discussed in detail below.

An examination of Figs. 12c and 13c reveals that, in addition to serving as a means of identifying local sixfold order in the dense liquid, $|\psi_6|$ can be used to identify regions of local geometrical disorder in the WCA solid near melting, in particular regions of local "square" ordering. In this regard $|\psi_6|$ is a much more sensitive indicator of disorder than coordination number. This is reflected in the fact that the number of sixfold ordered particles is significantly smaller than the number of six-coordinated particles in both the solid and liquid phases, as can been seen from a comparison of Figs. 12a–17a with Figs. 12c–17c. This is not unexpected, because $|\psi_6|$ measures deviations of nearest-neighbor bond angles from $60°$, and so is sensitive to local geometrical distortions that may not have a topological signature. In general, to each topology there corresponds a *region* of configuration space that represents non-topology changing, local geometrical distortions of the nearest-neighbor bond network. In any case, it is clear that there is considerable structural disorder present in both the solid and liquid phases that is not revealed by simply identifying particles whose coordination number differs from 6. Most of the remainder of this section is devoted to a discussion of methods for characterizing this structural disorder.

As mentioned previously, there is considerable variation in nearest-neighbor bond lengths and bond angles in the nearest-neighbor bond network of the dense WCA liquid. This suggests that it may be useful to characterize structural disorder in the WCA system in terms of the geometry of the nearest-neighbor bond network. To this end, we have constructed the nearest-neighbor bond length and bond angle probability distributions for the WCA system at various densities. Further, we have tabulated the contributions to these distributions from ordered and disordered regions separately. The resulting bond length and bond angle distributions for the WCA liquid at $\rho = 0.83$ are shown in Figs. 19 and 20, respectively.

The bond length and bond angle distribution functions for ordered regions in the WCA liquid have roughly the form expected for a triangular lattice: a single asymmetric peak in the bond length distribution function, peaked near the most probable nearest-neighbor separation (the position of the maximum in the overall distribution), and a single, nearly symmetric peak in the bond angle distribution function, centered

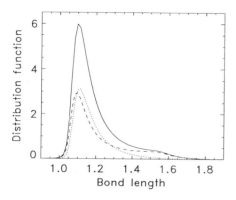

Figure 19. Nearest-neighbor bond length distribution functions for the time-averaged 3584-particle WCA liquid at $\rho = 0.83$. The figure shows the contributions from ordered regions (dotted line), disordered regions (dashed line), and the overall distribution (solid line).

near 60°. On the other hand, the *distributions for disordered regions are multipeaked*. The bond length distribution for disordered regions has peaks near 1.10 and 1.55, and the corresponding bond angle distribution exhibits three peaks, located near 48°, 58°, and 82°. These features can be qualitatively explained by assuming that the *disordered regions contain numerous local groups of particles with a nearly square local arrangement*. In the Delaunay triangulation the shortest diagonal of a "square" is a nearest-neighbor bond, so that each such "square" contains one bond with a length roughly equal to $\sqrt{2}$ times the most probable nearest-neighbor spacing. With a most probable nearest-neighbor distance of 1.10 (the observed value at $\rho = 0.83$), we would expect to see a secondary peak at 1.56, close to the location of the observed secondary peak. In addition, each such "square" contains four 45° bond angles and two 90° bond angles, and so we would expect to see an additional peak in the

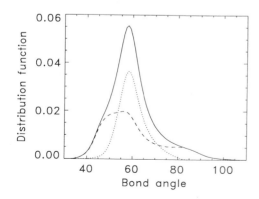

Figure 20. Nearest-neighbor bond angle distribution functions for the time-averaged 3584-particle WCA liquid at $\rho = 0.83$. The figure shows the contributions from ordered regions (dotted line), disordered regions (dashed line), and the overall distribution (solid line). The horizontal scale is in degrees.

bond angle distribution near 45° and another peak with half the weight near 90°. This is in qualitative accord with the observed features of the bond angle distribution function. The peak near 58° in the bond angle distribution for disordered regions indicates that disordered regions also contain groups of particles with local (equilateral) triangular order. Although the presence of "squares" explains the general features of the bond length and bond angle distribution functions reasonably well, the geometry of the WCA liquid turns out to be a bit more complicated than this simple picture suggests.

The form of the bond length and bond angle distributions demonstrates that nearest-neighbor bonds in the WCA liquid can be naturally divided into two groups: bonds whose length is comparable to the most probable nearest-neighbor bond length (which contribute to the primary peaks in the bond length and bond angle distributions), and bonds that are significantly longer (which contribute to the secondary peaks in these distributions). This in turn suggests a novel method for characterizing the local geometrical structure of the WCA liquid. We *dilute* the nearest-neighbor bond network (Delaunay construction), removing bonds that are significantly longer than the most probable nearest-neighbor separation. We will refer to bonds that have been removed as "broken" or "missing" bonds, and to the remaining bonds as "intact" bonds. Because the starting point for the bond dilution procedure is the Delaunay construction, which triangulates the plane, the resulting diluted bond network partitions the plane into polygonal areas having three or more sides, with the particles residing on the vertices of the polygons. For this reason we refer to this construction as the *polygon construction*. For example, for groups of particles having a nearly square arrangement, the bond dilution procedure removes the shortest diagonal of the "square," leaving a four-sided polygon in the place of two triangles. Thus "squares" show up as four-sided polygons in the polygon construction.

In practice, we chose to dilute the nearest-neighbor bond network using a bond angle criterion, removing bonds opposite bond angles exceeding 75°. This criterion gives a result very similar to that obtained by removing bonds more than ~20% longer than the most probable nearest-neighbor distance, and is much easier to apply in practice, because the same criterion can be applied at all densities (and to different systems) without an a priori knowledge of the most probable nearest-neighbor distance. The resulting polygon constructions for the configurations of Figs. 12a–17a are shown in Figs. 12d–17d.

Comparison of Figs. 12d–17d with Figs. 12c–17c reveals that the polygon picture is complementary to the picture of ordered regions in that there is a very close correspondence between disordered regions and regions containing polygons having more than three sides. It is also clear that the polygon construction gives a much more detailed characterization of structural disorder than a characterization in terms of coordination number (compare Figs. 12*d*–17*d* with Figs. 12*a*–17*a*). The polygon construction, which specifies the connectivity of the polygon network, is a more *primitive* characterization of local structure than the Voronoi construction, which specifies the connectivity of the nearest-neighbor bond network. Both constructions contain essentially the same information about the topology of the nearest-neighbor bond network, but the polygon construction contains additional information about local geometry (bond lengths and bond angles). This additional information is contained in the new variable that specifies the state (broken or intact) of each nearest-neighbor bond. In fact, the polygon construction strongly resembles an Ising or lattice-gas model defined on the bonds of a "random" triangulation. This is an analogy we will exploit fully in Section V.

Examination of Figs. 12*d*–17*d* reveals that polygons having four or more sides are associated with local regions of relatively low density, and therefore have something of the character of voids. This is not surprising, because the method we use to identify non-three-sided polygons (removal of bonds significantly longer than the most probable nearest-neighbor distance) naturally identifies regions of relatively loose packing. Four-sided polygons are the smallest (and most numerous) type of void, and polygons with more than four sides are associated with larger voids. For example, vacancies (holes large enough to accommodate an additional particle) typically appear as polygons having six or more sides (see Figs. 14*d*–17*d*). As we show below, most of the volume increase upon melting is due to the creation of such "geometrical voids" rather than an increase in the most probable nearest-neighbor spacing (the most probable nearest-neighbor distance actually decreases slightly upon melting!). Thus, the density distribution in the dense WCA liquid is quite inhomogeneous. Because polygons having four or more sides are localized, easily identifiable geometrical deviations from triangular ordering, we consider such polygons to be particle-like excitations, or *geometrical defects*.

As is evident from Figs. 12*d*–17*d*, the geometrical defects are highly correlated, and in the WCA liquid aggregate into a variety of characteristic structures. Among the most common structures are chains, or "ladders," of four-sided polygons. In general, however, the geometrical defect aggregates that appear in the dense WCA liquid are rather complex. One way of characterizing geometrical defect *correlations* is by means of a *vertex classification*, in which vertices in the polygon construction are classified according to the sequence of polygons around the vertex. Our classification scheme is shown in Fig. 21. A letter is assigned to each vertex type, roughly in the order of frequency of occurrence in the WCA liquid at $\rho = 0.85$. All sequences obtained by cyclic permutation or inversion of a given sequence are considered equivalent. Any vertices that are not of types A–Y are classified as type Z.

The vertex classifications for the configurations of Figs. 12a–17a are shown in Figs. 12e–17e. It is interesting to note that in plane tilings composed of squares and equilateral triangles (ST tilings), there are only four possible types of vertex (types A, B, C and D in Fig. 21). An example of a "random" ST tiling is shown in Fig. 3 (this tiling was grown using an algorithm described in Section V.B). What is remarkable is that type B and C vertices are also quite common in the dense WCA liquid (see Figs. 12d–17d). In this sense, the local arrangements of three- and four-sided polygons in the dense WCA liquid appear to be conditioned by *tiling rules*, with a clear tendency for three- and four-sided polygons to form local structures characteristic of ST tilings. Of course, many violations of the tiling rules are also observed. Vertex types E and F are the most common types of "tiling fault."

Based on these observations, we can regard the polygon construction as a type of *generalized tiling*, composed of deformable tiles having three or more sides. Because the tiles are deformable, tiling faults are possible. We can assign a strength to tiling faults by considering how much "ideal" tiles (i.e., *regular* polygons) would have to be deformed in order to fit around a given vertex. For example, the sum of *ideal* bond angles for the polygons that make up a type E vertex is only 330°, 30° shy of the 360° required for the tiles to fit together without gaps. As a result, the bond angles of the polygons must be increased slightly. We use 60° as the unit of charge for tiling faults, to make contact with the usual definition of disclination charge. For instance, vertex types J and K, which are isolated disclinations in the Voronoi construction and have disclination charges of -1 and $+1$, respectively, also have "tiling charges" of -1 and $+1$. Vertex types E and F have charges of $-1/2$ and $+1/2$, respectively. The general formula for computing the tiling charge c_α of vertex α is

$$c_\alpha = 6\left[\sum_{j=1}^{n_\alpha} \left(\frac{1}{2} - \frac{1}{p_j}\right) - 1\right] \tag{3.4}$$

where the sum ranges over the n_α polygons surrounding vertex α, and p_j is the number of sides of the jth polygon. It is easy to verify that the "quantum" of tiling charge is $1/10$ in generalized tilings consisting of tiles

Vertex type	Sequence	Diagram	Charge (in tenths)		Vertex type	Sequence	Diagram	Charge (in tenths)
A	333333		0		L	53333		-2
B	44333		0		M	54333		3
C	43433		0		N	53433		3
D	4444		0		O	5433		-7
E	43333		-5		P	5443		-2
F	433333		5		Q	533333		8
G	4443		-5		R	5343		-7
H	44433		5		S	4543		-2
I	44343		5		T	63333		0
J	33333		-10		U	5533		-4
K	3333333		10		V	55333		6
					W	54433		8
					X	53533		6
					Y	6433		-5

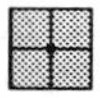

Figure 21. Classification scheme for vertices in the polygon construction. The fourth column gives the tiling charge, as defined in the text.

having six or fewer sides. For this reason, we will usually express tiling charges in tenths. The tiling charges of the most common vertex types are listed in Fig. 21.

The tiling charges of vertices in the configuration of Figs. 12a–17a are shown in Figs. 12f–17f. The charges are expressed in tenths, and negative charges are indicated by an overbar. The most common vertex charges in the WCA solid and dense WCA liquid are 0 and ±1/2. A number of other types of low charge vertices associated with five-sided polygons are also apparent (1/10, −2/10, 3/10, −7/10, 8/10, etc.). Vertices of charge 1 or greater are comparatively rare. The fact that the tiling charges of most vertices in the dense WCA liquid are zero or small suggests that the local arrangements of polygons are strongly conditioned by tiling rules. *There is a clear tendency for polygons to aggregate into structures that minimize the tiling charges on surrounding vertices.* This leads to the possibility of formulating the statistical mechanics of the polygon network in a simple way, as we will show in Section V.

Based on this overview of the statistical geometry of the 2D WCA system, we propose the following mechanism for 2D melting: the 2D melting transition is mediated by *geometrical defects* (polygons having four or more sides), which are localized excitations of the triangular lattice ground state, with a finite creation energy. These geometrical defects have attractive, anisotropic interactions ("shape" interactions) that cause them to form aggregates in which the local arrangement of polygons is governed by a tendency to minimize the tiling charge on surrounding vertices. The melting transition results from the simultaneous proliferation and condensation of geometrical defects into grain-boundary-like structures, and so can be regarded as a *geometrical defect condensation* transition. *The number density of geometrical defects (a scalar) is the order parameter of the 2D melting transition* (alternatively, the fraction of broken bonds could be taken to be the order parameter). We will present a detailed phenomenological model based on these ideas in Section V. In Fig. 22 we have shown the fraction of broken bonds (ϕ) for the 3584-particle WCA system in the transition region, and in Figs. 23 and 24 we have shown the polygon order parameters $\zeta_p = N_p/2N$, where N_p is the number of p-sided polygons and N is the number of particles. ζ_3 shows the behavior expected of an order parameter, while ϕ, ζ_4, ζ_5, and ζ_6 are "disorder" parameters.

It is possible to relate the order parameters ζ_p to the volume change associated with melting in a straightforward way. If we assume that the average length of intact nearest-neighbor bonds does not change through the melting transition (as we show below, the *most probable* nearest-neighbor distance actually *decreases* upon melting), and assume that a

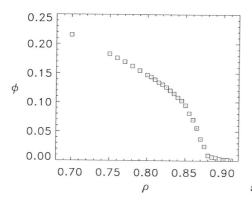

Figure 22. ϕ versus ρ for the time-averaged 3584-particle WCA system.

p-sided polygon has its "ideal" area (the area of a p-sided regular polygon of sidelength a),

$$A_p = \left(\frac{a^2 p}{4}\right) \tan\left[\pi\left(\frac{1}{2} - \frac{1}{p}\right)\right] \qquad (3.5)$$

then the volume per particle is

$$v = 2 \sum_p \zeta_p A_p \qquad (3.6)$$

The "theoretical" volume expansion upon melting can be obtained by calculating the volumes of the solid and liquid at coexistence from Eq. (3.6), using the measured ζ_p. For the time-averaged WCA system, we obtain a theoretical fractional volume expansion of

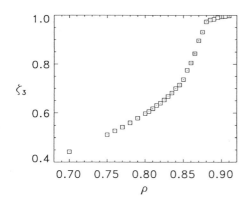

Figure 23. ζ_3 versus ρ for the time-averaged 3584-particle WCA system.

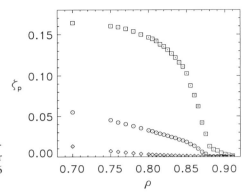

Figure 24. ζ_p versus ρ for the time-averaged 3584-particle WCA system, for $p = 4$ (squares), $p = 5$ (circles), and $p = 6$ (diamonds).

$$\frac{\Delta v}{v_s} = \frac{v_l - v_s}{v_s} = 0.047 \qquad (3.7)$$

where v_s and v_l are the specific volumes of the solid and liquid at coexistence, respectively. For comparison, the actual fractional volume change on melting is

$$\frac{\Delta v}{v_s} = 0.033 \qquad (3.8)$$

Considering the crude nature of our calculation, the agreement is not bad. It is likely that our simple argument overestimates the volume change because polygons generally have areas smaller than their ideal values, and because we have neglected the possibility that the most probable nearest-neighbor separation decreases upon melting (see below). All things considered, it appears that *geometrical defect creation* can account relatively straightforwardly for the volume change of the WCA system upon melting.

The vertex type statistics and tiling charge statistics for the time-averaged 3584-particle WCA system for several densities spanning the melting region are listed in Tables I and II. A number of interesting regularities are apparent in these statistics. Not surprisingly, the number of tiling faults increases precipitously upon melting (the fraction of charge 0 vertices decreases from 0.9319 at $\rho = 0.88$ to 0.567 at $\rho = 0.85$). However, the tiling faults present in the dense liquid tend to be low strength tiling faults, so that, for instance, 95% of the vertices have tiling charges of magnitude 1/2 or less at $\rho = 0.85$ (this number decreases to 78% at $\rho = 0.70$). Most of the vertices in the dense WCA liquid are of a

TABLE I

Probabilities of Occurrence of Various Vertex Types for the Time-Averaged 3584-Particle WCA System in the Solid Near Melting ($\rho = 0.88$), the Liquid Near Freezing ($\rho = 0.85$), and the Dense Liquid ($\rho = 0.83$ and $\rho = 0.80$)

Vertex type	WCA, $\rho = 0.88$	WCA, $\rho = 0.85$	WCA, $\rho = 0.83$	WCA, $\rho = 0.80$
A	0.915 ± 0.002	0.381 ± 0.004	0.2549 ± 0.0016	0.1582 ± 0.0013
B	0.0120 ± 0.0005	0.1055 ± 0.0013	0.1214 ± 0.0004	0.1230 ± 0.0010
C	$(4.9 \pm 0.2) \times 10^{-3}$	0.0733 ± 0.0007	0.0942 ± 0.0007	0.1056 ± 0.0005
D	$(4.4 \pm 1.1) \times 10^{-5}$	$(2.73 \pm 0.05) \times 10^{-3}$	$(4.17 \pm 0.06) \times 10^{-3}$	$(5.39 \pm 0.09) \times 10^{-3}$
E	0.0318 ± 0.0005	0.1111 ± 0.0002	0.1178 ± 0.0003	0.1115 ± 0.0004
F	0.0316 ± 0.0005	0.1090 ± 0.0004	0.1131 ± 0.0003	0.1020 ± 0.0003
G	$(3.6 \pm 0.5) \times 10^{-4}$	0.0146 ± 0.0003	0.0208 ± 0.0002	0.0273 ± 0.0004
H	$(2.4 \pm 0.2) \times 10^{-4}$	$(9.2 \pm 0.2) \times 10^{-3}$	0.01309 ± 0.00015	0.01735 ± 0.00015
I	$(1.49 \pm 0.15) \times 10^{-4}$	$(7.75 \pm 0.09) \times 10^{-3}$	0.01106 ± 0.00015	0.01543 ± 0.00017
J	$(7 \pm 7) \times 10^{-7}$	$(7.3 \pm 0.5) \times 10^{-5}$	$(9.5 \pm 0.9) \times 10^{-5}$	$(1.13 \pm 0.04) \times 10^{-4}$
K	$(1.7 \pm 0.9) \times 10^{-5}$	$(1.54 \pm 0.05) \times 10^{-3}$	$(1.86 \pm 0.05) \times 10^{-3}$	$(1.76 \pm 0.03) \times 10^{-3}$
L	$(1.28 \pm 0.18) \times 10^{-3}$	0.0559 ± 0.0006	0.0655 ± 0.0006	0.0684 ± 0.0006
M	$(6.6 \pm 1.0) \times 10^{-4}$	0.02865 ± 0.00017	0.03635 ± 0.00018	0.0455 ± 0.0003
N	$(3.1 \pm 0.6) \times 10^{-4}$	0.0217 ± 0.0003	0.0304 ± 0.0003	0.0394 ± 0.0003
O	$(4.1 \pm 0.6) \times 10^{-4}$	0.01494 ± 0.00015	0.01972 ± 0.00007	0.02552 ± 0.00018
P	$(9 \pm 2) \times 10^{-5}$	$(9.93 \pm 0.14) \times 10^{-3}$	0.01521 ± 0.00014	0.02323 ± 0.00019
Q	$(4.4 \pm 0.4) \times 10^{-4}$	$(7.20 \pm 0.08) \times 10^{-3}$	$(8.41 \pm 0.12) \times 10^{-3}$	$(9.03 \pm 0.12) \times 10^{-3}$
R	$(1.2 \pm 0.2) \times 10^{-4}$	$(5.43 \pm 0.05) \times 10^{-3}$	$(8.11 \pm 0.10) \times 10^{-3}$	0.01149 ± 0.00007
S	$(9 \pm 2) \times 10^{-5}$	$(5.19 \pm 0.08) \times 10^{-3}$	$(7.55 \pm 0.10) \times 10^{-3}$	0.01170 ± 0.00014
T	$(1.5 \pm 0.8) \times 10^{-4}$	$(3.3 \pm 0.02) \times 10^{-3}$	$(4.24 \pm 0.09) \times 10^{-3}$	$(6.17 \pm 0.12) \times 10^{-3}$
U	$(2.9 \pm 1.2) \times 10^{-5}$	$(2.49 \pm 0.04) \times 10^{-3}$	$(3.81 \pm 0.08) \times 10^{-3}$	$(6.41 \pm 0.16) \times 10^{-3}$
V	$(2.2 \pm 1.1) \times 10^{-5}$	$(2.24 \pm 0.08) \times 10^{-3}$	$(2.97 \pm 0.06) \times 10^{-3}$	$(4.15 \pm 0.07) \times 10^{-3}$
W	$(2.2 \pm 0.8) \times 10^{-5}$	$(1.90 \pm 0.04) \times 10^{-3}$	$(2.80 \pm 0.07) \times 10^{-3}$	$(4.19 \pm 0.03) \times 10^{-3}$
X	$(9 \pm 3) \times 10^{-6}$	$(1.57 \pm 0.04) \times 10^{-3}$	$(2.27 \pm 0.04) \times 10^{-3}$	$(3.25 \pm 0.08) \times 10^{-3}$
Y	$(2.9 \pm 1.3) \times 10^{-5}$	$(1.56 \pm 0.07) \times 10^{-3}$	$(2.72 \pm 0.11) \times 10^{-3}$	$(4.96 \pm 0.09) \times 10^{-3}$

TABLE II

Probabilities of Occurrence of Various Values of the Tiling Charge for the Time-Averaged 3584-Particle WCA System in the Solid Near Melting ($\rho = 0.88$), the Liquid Near Freezing ($\rho = 0.85$), and the Dense Liquid ($\rho = 0.83$ and $\rho = 0.80$)

Tiling charge (in tenths)	WCA, $\rho = 0.88$	WCA, $\rho = 0.85$	WCA, $\rho = 0.83$	WCA, $\rho = 0.80$
-10	$(5.1 \pm 0.7) \times 10^{-5}$	$(2.45 \pm 0.03 \times 10^{-3}$	$(3.70 \pm 0.06) \times 10^{-3}$	$(5.67 \pm 0.09) \times 10^{-3}$
-9	$(9 \pm 5) \times 10^{-6}$	$(1.01 \pm 0.03) \times 10^{-3}$	$(1.38 \pm 0.04) \times 10^{-3}$	$(2.25 \pm 0.07) \times 10^{-3}$
-7	$(5.3 \pm 0.8) \times 10^{-4}$	0.02046 ± 0.00017	0.02811 ± 0.00013	0.0378 ± 0.0002
-6	0	$(8.1 \pm 1.2) \times 10^{-5}$	$(1.32 \pm 0.16) \times 10^{-4}$	$(3.13 \pm 0.17) \times 10^{-4}$
-5	0.0322 ± 0.0006	0.1279 ± 0.0005	0.1425 ± 0.0004	0.1462 ± 0.0006
-4	$(4.8 \pm 1.5) \times 10^{-5}$	$(3.49 \pm 0.05) \times 10^{-3}$	$(5.51 \pm 0.11) \times 10^{-3}$	$(9.6 \pm 0.2) \times 10^{-3}$
-2	$(1.5 \pm 0.2) \times 10^{-3}$	0.0715 ± 0.0007	0.0893 ± 0.0007	0.1056 ± 0.0007
0	0.9319 ± 0.0015	0.567 ± 0.002	0.4804 ± 0.0013	0.4017 ± 0.0012
1	$(1.2 \pm 0.6) \times 10^{-5}$	$(2.20 \pm 0.04) \times 10^{-3}$	$(4.20 \pm 0.11) \times 10^{-3}$	$(8.26 \pm 0.13) \times 10^{-3}$
3	$(9.8 \pm 1.6) \times 10^{-4}$	0.0517 ± 0.0004	0.0694 ± 0.0004	0.0901 ± 0.0005
4	$(3 \pm 3) \times 10^{-7}$	$(1.05 \pm 0.10) \times 10^{-4}$	$(2.16 \pm 0.18) \times 10^{-4}$	$(5.7 \pm 0.3) \times 10^{-4}$
5	0.0320 ± 0.0005	0.1281 ± 0.0006	0.1408 ± 0.0004	0.1413 ± 0.0005
6	$(3.2 \pm 1.4) \times 10^{-5}$	$(4.06 \pm 0.08) \times 10^{-3}$	$(5.72 \pm 0.09) \times 10^{-3}$	$(8.53 \pm 0.10) \times 10^{-3}$
8	$(5.1 \pm 0.5) \times 10^{-4}$	0.01266 ± 0.00013	0.01671 ± 0.00011	0.02243 ± 0.00015
9	0	$(1.8 \pm 0.5) \times 10^{-5}$	$(3.4 \pm 0.4) \times 10^{-5}$	$(1.16 \pm 0.14) \times 10^{-4}$
10	$(1.5 \pm 0.2) \times 10^{-4}$	$(5.34 \pm 0.09) \times 10^{-3}$	$(7.88 \pm 0.07) \times 10^{-3}$	0.01077 ± 0.00011

few common types. For example, at $\rho = 0.85$, 6 vertex types (A, B, C, E, F, and L) account for 84% of the vertices. Type E and F vertices are the most common types of tiling faults for all densities except the lowest ($\rho = 0.70$), and type E vertices are more probable than type F vertices for all densities. Similarly, there are significantly more type B vertices than type C vertices for all densities.

2. Translational and Bond Orientational Correlations

Perhaps the most widely used measure of liquid and solid structure is the *pair correlation function*,

$$g^{(2)}(\mathbf{r}_1, \mathbf{r}_2) = \frac{1}{\rho} [G(\mathbf{r}_1, \mathbf{r}_2) - \delta(\mathbf{r}_1 - \mathbf{r}_2)] \tag{3.9}$$

where ρ is the average particle density, $G(\mathbf{r}_1, \mathbf{r}_2)$ is the density–density correlation function,

$$G(\mathbf{r}_1, \mathbf{r}_2) = \frac{1}{N} \langle \rho(\mathbf{r}_1)\rho(\mathbf{r}_2) \rangle \tag{3.10}$$

and

$$\rho(\mathbf{r}) = \sum_{i=1}^{N} \delta(\mathbf{r} - \mathbf{r}_i) \tag{3.11}$$

is the instantaneous particle density. The subtraction of the delta function in Eq. (3.9) removes the self-correlation from the overall correlation function $G(\mathbf{r}_1, \mathbf{r}_2)$. $g^{(2)}(\mathbf{r}_1, \mathbf{r}_2)$ is the second member of a hierarchy of n-particle correlation functions (see, for example [1, 189, 190]), of which the first member is

$$g^{(1)}(\mathbf{r}) = \frac{1}{\rho} \langle \rho(\mathbf{r}) \rangle \tag{3.12}$$

Both $g^{(1)}$ and $g^{(2)}$ have simple physical interpretations. $\rho g^{(1)}(\mathbf{r}) \, d\mathbf{r}$ is the probability of finding a particle in the volume element $d\mathbf{r}$ at $\mathbf{r}$, and $\rho^2 g^{(2)}(\mathbf{r}_1, \mathbf{r}_2)$, $d\mathbf{r}_1 \, d\mathbf{r}_2$ is the probability of simultaneously finding a particle in $d\mathbf{r}_1$ at $\mathbf{r}_1$ and another particle in $d\mathbf{r}_2$ at $\mathbf{r}_2$.

The pair correlation function is important for various reasons. For one thing, a knowledge of the pair correlation function is sufficient to calculate various thermodynamic quantities (particularly the potential energy and the pressure), assuming that the total interaction potential is a sum of pairwise interactions [1, 189, 190]. For another, the *structure factor*, which is the Fourier transform of the pair correlation function,

$$S(\mathbf{k}) = 1 + \frac{\rho^2}{N} \int \int \exp[-i\mathbf{k} \cdot (\mathbf{r}_1 - \mathbf{r}_2)] g^{(2)}(\mathbf{r}_1, \mathbf{r}_2) \, d\mathbf{r}_1 \, d\mathbf{r}_2 \quad (3.13)$$

is the quantity that is measured (more or less) directly in scattering experiments. Thus, for both theoretical and experimental reasons, the pair correlation function is quite important. In addition, such correlation functions figure prominently in various melting theories, because $\rho^{(1)}(\mathbf{r}) = \rho g^{(1)}(\mathbf{r})$ is the order parameter for the solid–liquid phase transition, in the sense that it characterizes the symmetry difference between the solid and liquid phases. In particular, density functional theories of melting are essentially correlation function-based theories in which the correlation functions of the liquid phase are used as input. Also, in the context of two-dimensional melting, the KTHNY theory makes specific predictions concerning the behavior of the translational (density–density) and bond orientational correlation functions in the melting region, and so measuring the correlation functions provides a means of verifying the KTHNY theory.

In a homogeneous, isotropic liquid, $g^{(2)}(\mathbf{r}_1, \mathbf{r}_2)$ only depends on the magnitude of the separation vector, $r_{12} = |\mathbf{r}_1 - \mathbf{r}_2|$, that is, $g^{(2)}(\mathbf{r}_1, \mathbf{r}_2) = g^{(2)}(r_{12})$. From now on we will drop the superscript and refer to the pair correlation function simply as $g(r)$, where r is the pair separation. In this case $S(\mathbf{k})$ depends only on $|\mathbf{k}|$, and in the two-dimensional case we obtain

$$S(k) = 1 + 2\pi\rho \int_0^\infty r[g(r) - 1] J_0(kr) \, dr \quad (3.14)$$

where J_0 is a zero-order Bessel function, and the delta function part of $S(k)$ (corresponding to forward scattering) has been explicitly removed ($g(r)$ has been replaced by $g(r) - 1$ in the integral). In the inhomogeneous, anisotropic solid phase, one can average $g^{(2)}(\mathbf{r}_1, \mathbf{r}_2)$ over one of the particle coordinates (say $\mathbf{r}_1$) to obtain a quantity that only depends on the difference vector $\mathbf{r}_{12} = \mathbf{r}_1 - \mathbf{r}_2$, and then perform an angular average over all orientations of the difference vector to obtain a quantity that depends only on the magnitude of the difference vector $r_{12} = |\mathbf{r}_{12}|$. We will refer to the resulting correlation function as $g(r)$, in analogy to the liquid case, and will refer to the corresponding angular-averaged structure factor as $S(k)$.

Finally, in analogy to the pair correlation function, we can define a two-point sixfold bond orientational correlation function $g_6^{(2)}(\mathbf{r}_1, \mathbf{r}_2)$,

$$g_6^{(2)}(\mathbf{r}_1, \mathbf{r}_2) = \frac{1}{\rho} \left[\frac{G_6(\mathbf{r}_1, \mathbf{r}_2) - \langle |\psi_6|^2 \rangle \delta(\mathbf{r}_1 - \mathbf{r}_2)}{g^{(2)}(\mathbf{r}_1, \mathbf{r}_2)} \right] \quad (3.15)$$

where $G_6(\mathbf{r}_1, \mathbf{r}_2)$ is the sixfold bond orientational order parameter density correlation function,

$$G_6(\mathbf{r}_1, \mathbf{r}_2) = \frac{1}{N} \langle \rho_6(\mathbf{r}_1) \rho_6^*(\mathbf{r}_2) \rangle \tag{3.16}$$

$\rho_6(\mathbf{r})$ is the sixfold bond orientational order parameter density,

$$\rho_6(\mathbf{r}) = \sum_{i=1}^{N} \psi_{6i} \delta(\mathbf{r} - \mathbf{r}_i) \tag{3.17}$$

and

$$\langle |\psi_6|^2 \rangle = \frac{1}{N} \left\langle \sum_{i=1}^{N} |\psi_{6i}|^2 \right\rangle \tag{3.18}$$

is the mean-square local sixfold bond orientational order parameter.

The local sixfold bond orientational order parameter ψ_{6i} is defined in Eq. (3.3). $g^{(2)}(\mathbf{r}_1, \mathbf{r}_2)$ is divided out of Eq. (3.15) in order to remove translational correlations from the bond orientational correlation function. In the homogeneous and isotropic liquid phase $g_6^{(2)}(\mathbf{r}_1, \mathbf{r}_2)$ reduces to a function of r_{12} only, which we will denote by $g_6(r)$, and a corresponding translation- and rotation-invariant quantity can be defined for the solid phase by performing suitable averages.

We have calculated $g(r)$, $g_6(r)$, and $S(k)$ for the WCA system over a range of densities spanning the melting region. In addition, we have performed least-squares fits of $g(r)$ and $g_6(r)$ to obtain translational and bond orientational correlation lengths (ξ and ξ_6, respectively) in the liquid phase and power law decay exponents in the solid phase, and have examined the behavior of the correlation functions in the coexistence region. Our results are consistent with those found in other simulation studies, but are ambiguous with regard to verification of falsification of the KTHNY predictions. In particular, finite-size effects appear to have a severe effect on the asymptotic behavior of $g_6(r)$ in the coexistence region, so that it is difficult to rule out the existence of a hexatic phase on the basis of $g_6(r)$. Our fitting procedure and results have been described elsewhere [191]. We simply note that the translational and bond orientational correlation lengths are of comparable magnitude in the liquid phase, increasing rapidly with increasing density near freezing, and attaining values of $\xi = 3.38 \pm 0.02$ and $\xi_6 = 4.22 \pm 0.03$, respectively, for the non-time-averaged WCA liquid near freezing ($\rho = 0.85$). These results make qualitative sense if we view a 2D liquid as a sort of fluctuating polycrystal, in which case ξ and ξ_6 should be approximately equal and comparable to the characteristic linear dimension of crystallites.

The density dependence of the position of the primary maximum in $g(r)$, r_m, provides a dramatic illustration of the inhomogeneity of the volume expansion that occurs upon melting. r_m defines the most probable nearest-neighbor separation. In Fig. 25 we have plotted r_m versus ρ for the time-averaged 3584-particle WCA system. This figure shows that the most probable nearest-neighbor separation *decreases* by nearly 2% upon melting, even though the system as a whole undergoes a volume *expansion* of 3.3%. A homogeneous expansion would necessarily involve some increase in the most probable nearest-neighbor separation, and so this result implies that the volume increase upon melting is inhomogeneous, involving the creation of pockets of "free volume" with no increase in the most probable nearest-neighbor spacing. This supports our previous finding that most of the volume increase upon melting goes into the formation of small voids (polygons having more than three sides in the polygon construction).

This point is further illustrated by Fig. 26, which shows the variation of the position of the primary maximum in $S(k)$, k_m, with density. This figure also shows how the magnitude of the first reciprocal lattice vector of a perfect triangular lattice varies with density (solid line), illustrating the density dependence of k_m that would be expected for a homogeneous system. The data follows the solid line faithfully in the solid phase, indicating that the solid is reasonably homogeneous, but then exhibits a positive deviation from the solid line as the system melts, showing that the predominant "lattice periodicity" in the liquid is significantly shorter than that predicted by the homogeneous model. The decrease in the most probable nearest-neighbor separation upon melting should be even more

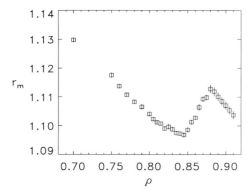

Figure 25. r_m versus ρ for the time-averaged 3584-particle WCA system.

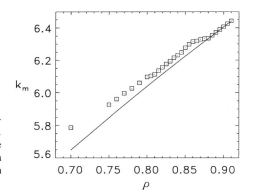

Figure 26. k_m versus ρ for the time-averaged 3584-particle WCA system. The solid line shows how the magnitude of the first reciprocal lattice vector of a perfect triangular lattice varies with density.

dramatic in systems with long-ranged pair potentials, which have a much smaller volume change upon melting than the WCA system.

In passing, we should mention that we find $S(k_m) = 5.016 \pm 0.009$ for the (non-time-averaged) 3584-particle WCA system near freezing ($\rho = 0.85$). This value is similar to the values found for other 2D systems (see Section II.C), and provides further evidence that a version of the Hansen–Verlet criterion is applicable to 2D melting.

3. Voronoi Analysis

Local topological order in 2D systems can be characterized by the fraction of c-coordinated particles, f_c. Figures 27–29 show f_c for $c = 4$–8 as a function of ρ for the time-averaged 3584-particle WCA system. f_6, the fraction of six-coordinated particles, serves as a sort of order parameter, being nearly unity in the time-averaged WCA solid ($f_6 = 0.9961 \pm$

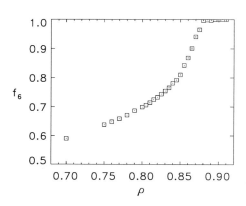

Figure 27. f_6 versus ρ for the time-averaged 3584-particle WCA system.

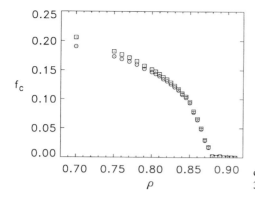

Figure 28. f_5 (squares) and f_7 (circles) versus ρ for the time-averaged 3584-particle WCA system.

0.0006 at $\rho = 0.88$) and significantly smaller in the liquid (ranging from $f_6 = 0.8102 \pm 0.0018$ at $\rho = 0.85$ to $f_6 = 0.5900 \pm 0.0010$ at $\rho = 0.70$). f_5 (the fraction of -1 disclinations) is larger than f_7 (the fraction of -1 disclinations) for all liquid densities. This asymmetry is reflected in a corresponding asymmetry in the probabilities of 4- and 8-coordinated particles (f_8 is larger than f_4 for all liquid densities), because the average coordination number for 2D systems is 6 [188], and there are very few particles with a coordination number less than 4 or greater than 8. In fact, this asymmetry may have little significance, as it is also observed in the coordination number statistics of a random set of points in 2D [97]. We should point out, however, that 8-coordinated particles are often associated with vacancies (or large voids), so that f_8 may be larger than f_4 in the WCA liquid because vacancies have a larger probability of occurrence than interstitials in hard core systems.

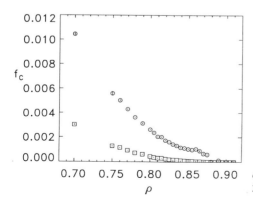

Figure 29. f_4 (squares) and f_8 (circles) versus ρ for the time-averaged 3584-particle WCA system.

The Voronoi cell area distribution functions for the time-averaged 3584-particle WCA system are shown in Fig. 30. These distributions consist of a single asymmetric peak that broadens and moves outward as the density decreases. As shown in Fig. 30*b*, these distributions exhibit an *isosbestic* point (a single point at which all the distributions cross) in the coexistence region (in fact, there are *two* isosbestic points, but the isosbestic point near 1.08 is less well defined than the one near 1.17). This suggests that the distributions in this region are really the sum of two

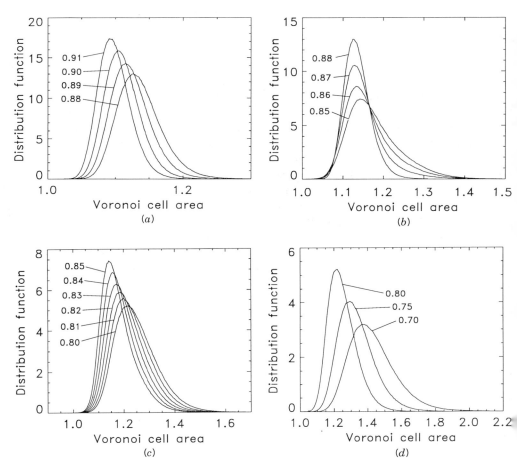

Figure 30. Voronoi cell area distribution functions for the time-averaged 3584-particle WCA system, with each curve labeled by the corresponding value of ρ: (*a*) solid region; (*b*) coexistence region; (*c*) liquid near freezing; (*d*) liquid away from freezing.

contributions, and that varying the density in the coexistence region merely changes the relative weight of the two contributions. As discussed above, this observation provides additional support for the assumption of two-phase coexistence, because microscopic distribution functions in the coexistence region should be a weighted sum of solid and liquid contributions. As we show below, a number of other distribution functions show similar behavior.

We have also separated the Voronoi cell area distributions into contributions from sixfold ordered and disordered regions, as was done above for the nearest-neighbor bond length and bond angle distribution functions. The result is shown in Fig. 31, for a liquid density of $\rho = 0.83$. Not surprisingly, the distribution for disordered regions is broader and centered at a larger value of the cell area than the distribution for ordered regions. This is to be expected because disordered regions are more loosely packed (they contain numerous voids in the form of polygons having more than three sides), and so the mean Voronoi cell area (which defines a local inverse number density) is larger in these regions.

The Voronoi cell sidelength distribution functions for the time-averaged 3584-particle WCA system are shown in Fig. 32. These distributions have a more complex shape than the Voronoi cell area distributions, and provide an interesting diagnostic for disorder. As the system melts, this distribution develops a "tail" at small values of the cell sidelength. In fact, this feature of the Voronoi cell sidelength distribution function has been used extensively by Fraser et al. to locate the solid–liquid transition in hard disk mixtures [192]. As mentioned above (and illustrated in Fig. 18), a group of particles with a nearly square arrangement is associated with a short Voronoi cell side. In general, short sides in the Voronoi construction are associated with long bonds in the dual Delaunay con-

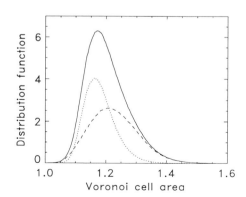

Figure 31. Voronoi cell area distribution functions for the time-averaged 3584-particle WCA liquid at $\rho = 0.83$. The figure shows the contributions from ordered regions (dotted line), disordered regions (dashed line), and the overall distribution (solid line).

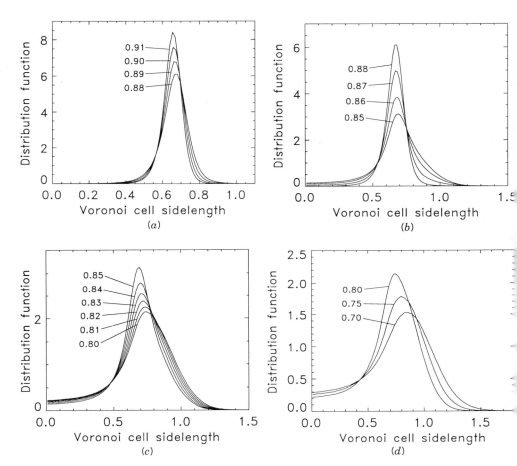

Figure 32. Voronoi cell sidelength distribution functions for the time-averaged 3584-particle WCA system, with each curve labeled by the corresponding value of ρ: (a) solid region (b) coexistence region; (c) liquid near freezing; (d) liquid away from freezing.

struction, or, alternatively, with polygons having more than three sides in the polygon construction. Thus, the tail in this distribution is a consequence of the type of geometrical disorder we have already identified.

In the coexistence region, the Voronoi cell sidelength distributions exhibit two isosbestic points (Fig. 32b), consistent with two-phase coexistence. As before, we have separated the Voronoi cell sidelength distribution for the WCA liquid at $\rho = 0.83$ into contributions from ordered and disordered regions (Fig. 33). Most of the tail in the overall distribution comes from disordered regions. This is to be expected, because

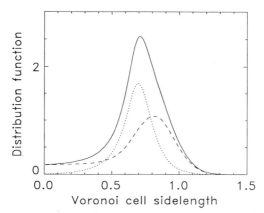

Figure 33. Voronoi cell sidelength distribution functions for the time-averaged 3584-particle WCA liquid at $\rho = 0.83$. The figure shows the contributions from ordered regions (dotted line), disordered regions (dashed line), and the overall distribution (solid line).

short cell sides are associated with nearest-neighbor bond angles that deviate significantly from 60°, and consequently with regions with low sixfold bond orientational order.

What is more interesting is that the Voronoi cell sidelength distributions in the dense liquid exhibit isosbestic points, near 0.5 and 0.8, suggesting that the distributions in the dense liquid region are the sum of two distinct (density invariant) contributions. A further implication of this observation is that the dense liquid can be considered to be composed of two types of local environments (roughly corresponding to the ordered and disordered regions discussed above), each with its own characteristic geometrical structure and hence its own characteristic signature in the Voronoi cell sidelength distributions (and in other measures of local geometry; see below). In the dense liquid, while the relative proportions of these two types of regions vary with density, the characteristic geometrical structure of each type of region remains nearly invariant, and so isosbestic points are observed in distribution functions that probe the local geometry of the liquid. We will refer to this picture of dense 2D liquid structure as a "two-fluid" model. This view emphasizes the *inhomogeneity* of dense 2D liquids, a feature of liquid structure that is largely ignored in more traditional treatments of liquid structure [1].

4. Bond Orientational Order

As discussed above, the local sixfold bond orientational order parameter ψ_6 is a sensitive probe of the local geometry of 2D systems. This is illustrated by Fig. 34, which shows that $|\psi_6|$ distribution functions for the time-averaged 3584-particle WCA system at various densities. In the solid

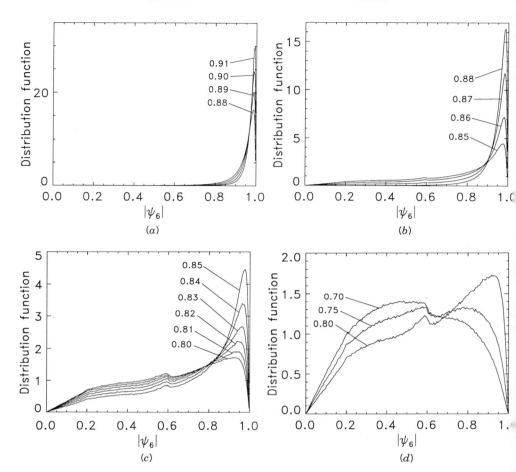

Figure 34. $|\psi_6|$ distribution functions for the time-averaged 3584-particle WCA system, with each curve labeled by the corresponding value of ρ: (*a*) solid region; (*b*) coexistence region; (*c*) liquid near freezing; (*d*) liquid away from freezing.

(Fig. 34*a*), the $|\psi_6|$ distributions are strongly peaked near $|\psi_6| = 1$, indicative of a high degree of local triangular order. As the system melts (Fig. 34*b*), the $|\psi_6|$ distributions develop a tail at small values of $|\psi_6|$. As we have seen for other microscopic distribution functions, the $|\psi_6|$ distributions display an isosbestic point in the coexistence region (Fig. 34*b*), at $|\psi_6| \approx 0.9$, consistent with the behavior expected for solid–liquid coexistence.

The $|\psi_6|$ distributions for the liquid near freezing (Fig. 34c) show a number of interesting features. First, there is a distinct peak near $|\psi_6| = 1$, reflecting considerable solid-like ordering in the liquid near freezing. As the density decreases from freezing, the solid-like peak decreases in amplitude, while the small $|\psi_6|$ part of the distribution rises. In addition, the $|\psi_6|$ distributions in the dense liquid region exhibit an isosbestic point, near $|\psi_6| = 0.8$, providing further support of the "two-fluid" picture of dense liquid structure discussed above. Finally, the $|\psi_6|$ distributions for the dense liquid (Fig. 34c) exhibit small cusps near $|\psi_6| = 0.6$ and $|\psi_6| = 0.2$. We will discuss the origin of these features in a moment.

The discussion of the previous paragraph provides a partial justification for our method of partitioning of the WCA liquid into sixfold ordered and disordered regions on the basis of the magnitude of the local sixfold order parameter, $|\psi_6|$. Examination of Fig. 34c shows that the criterion we used for selecting ordered particles ($|\psi_6| > 0.75$) is reasonable, in that the solid-like peak observed in these distributions falls in the range $0.75 \lesssim |\psi_6| \lesssim 1.0$.

We now return to a discussion of the small features that were observed in the $|\psi_6|$ distributions for the dense time-averaged WCA liquid (Fig. 34c) at $|\psi_6| \approx 0.6$ and $|\psi_6| \approx 0.2$. The small "bump" at $|\psi_6| \approx 0.6$ appears to be associated with type B vertices, while the "bend" at $|\psi_6| \approx 0.2$ seems to be associated with type C vertices. This is illustrated by Table III, which lists the values of $|\psi_6|$ corresponding to the "ideal" vertices shown in Fig. 21. Table III also lists values of the fourfold and twelvefold order parameters, which we will discuss in a moment. As this table shows, ideal type B and C vertices have $|\psi_6| = 0.6$ and $|\psi_6| = 0.2$, respectively, so it is at least plausible that the features we observe in the $|\psi_6|$ distributions are associated with these vertex types. If this is indeed the case, then these features should be even more pronounced in $|\psi_6|$ distribution functions calculated including only "intact" bonds in the sum in Eq. (3.3), because including "broken" bonds (such as the diagonal bonds of four sided polygons) causes the value of $|\psi_6|$ for a given vertex type to deviate from its ideal value. The $|\psi_6|$ distributions calculated using only intact bonds are shown in Fig. 35, for the time-averaged 3584-particle system. As expected, the distributions in the dense liquid region (Fig. 35c) have cusp-like features at $|\psi_6| \approx 0.6$ and $|\psi_6| \approx 0.2$ that are more pronounced than in the distributions calculated using both intact and broken bonds (Fig. 34c).

The bond orientational order of 2D liquids can be characterized in more detail by considering other types of order besides sixfold order. In general, we can define a *bond angle density* for particle i,

TABLE III

Estimated Magnitudes of the Fourfold, Sixfold, and Twelvefold Bond Orientational Order Parameters for the Vertex Types Shown in Fig. 21

Vertex type	Coordination number	$|\psi_4|$	$|\psi_6|$	$|\psi_{12}|$
A	6	0.0	1.0	1.0
B	5	0.4	0.6	1.0
C	5	0.2	0.2	1.0
D	4	1.0	0.0	1.0
E	5	~ 0.24	~ 0.70	~ 0.10
F	6	~ 0.15	~ 0.69	~ 0.09
G	4	~ 0.81	~ 0.27	~ 0.09
H	5	~ 0.49	~ 0.41	~ 0.13
I	5	~ 0.17	~ 0.24	~ 0.05
J	5	~ 0.00	~ 0.00	~ 0.00
K	7	~ 0.00	~ 0.00	~ 0.00
L	5	~ 0.24	~ 0.95	~ 0.82
M	5	~ 0.40	~ 0.65	~ 0.62
N	5	~ 0.22	~ 0.46	~ 0.58
O	4	~ 0.48	~ 0.74	~ 0.22
P	4	~ 0.74	~ 0.46	~ 0.80
Q	6	~ 0.17	~ 0.37	~ 0.24
R	4	~ 0.67	~ 0.79	~ 0.33
S	4	~ 0.64	~ 0.04	~ 0.81
T	5	0.2	1.0	1.0
U	4	~ 0.39	~ 0.95	~ 0.81
V	5	~ 0.41	~ 0.45	~ 0.48
W	5	~ 0.50	~ 0.39	~ 0.31
X	5	~ 0.25	~ 0.15	~ 0.70
Y	4	~ 0.36	~ 0.68	~ 0.23

$$\rho_{bi}(\theta) = \frac{2\pi}{n_i} \sum_{j=1}^{n_i} \delta(\theta - \theta_{ij}) \qquad (3.19)$$

where, as in Eq. (3.3), the sum runs over the n_i nearest neighbors of particle i, and θ_{ij} is the angle of the bond between particle i and its jth neighbor with respect to a reference direction. Because $\rho_{bi}(\theta)$ has a periodicity of 2π, we can write a Fourier decomposition,

$$\rho_{bi}(\theta) = \sum_{m=0}^{\infty} \psi_{mi} e^{-im\theta} \qquad (3.20)$$

where the complex Fourier coefficients ψ_m given by

$$\psi_m \equiv \psi_{mi} = \frac{1}{n_i} \sum_{j=1}^{n_i} e^{im\theta_{ij}} \qquad (3.21)$$

play the role of m-fold bond orientational order parameters. We will concentrate on ψ_4 and ψ_{12} here, because $|\psi_4|$ is sensitive to local square

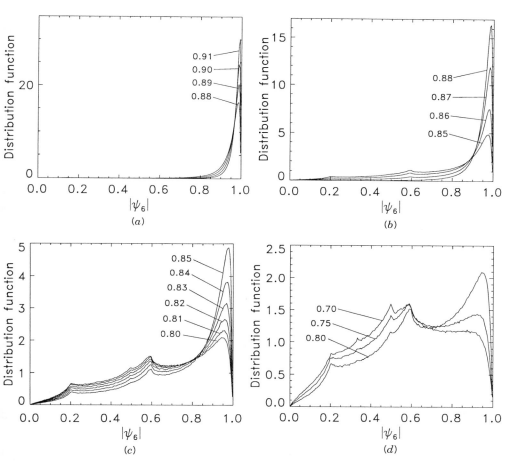

Figure 35. $|\psi_6|$ distribution functions (calculated using only intact bonds) for the time-averaged 3584-particle WCA system, with each curve labeled by the corresponding value of ρ: (a) solid region; (b) coexistence region; (c) liquid near freezing; (d) liquid away from freezing.

ordering, and $|\psi_{12}|$ is unity for the four types (A–D) of vertex that appear in ST tilings, as shown in Table III. In particular, $|\psi_{12}|$ can be used to detect type B and C vertices, which, as we have seen, are prevalent in the dense WCA liquid. In Fig. 36 we have shown the $|\psi_4|$ distribution functions for the time-averaged 3584-particle WCA system at various densities, calculated with only intact bonds included in the sum in Eq. (3.21). Similarly, Fig. 37 shows the $|\psi_{12}|$ distribution functions for the

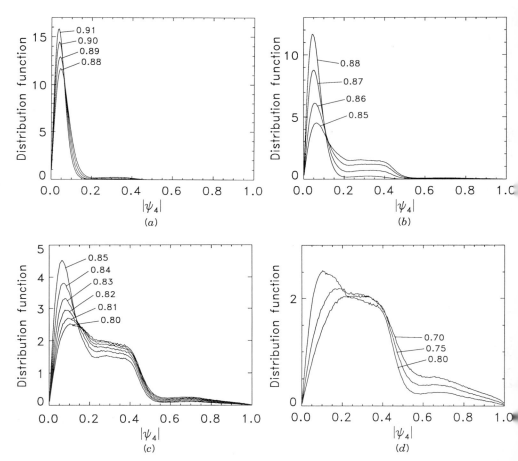

Figure 36. $|\psi_4|$ distribution functions (calculated using only intact bonds) for the time-averaged 3584-particle WCA system, with each curve labeled by the corresponding value of ρ: (a) solid region; (b) coexistence region; (c) liquid near freezing; (d) liquid away from freezing.

time-averaged 3584-particle WCA system at various densities, calculated using only intact bonds. These distributions show features that can be identified with particular vertex types, together with features that are not so easily explained. Like the $|\psi_6|$ distributions, the $|\psi_4|$ and $|\psi_{12}|$ distributions exhibit isosbestic points in the coexistence and dense liquid regions, which provides additional support for the conclusions drawn above.

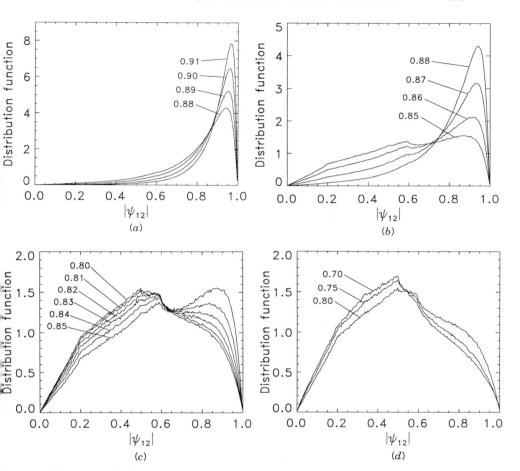

Figure 37. $|\psi_{12}|$ distribution functions (calculated using only intact bonds) for the time-averaged 3584-particle WCA system, with each curve labeled by the corresponding value of ρ: (*a*) solid region; (*b*) coexistence region; (*c*) liquid near freezing; (*d*) liquid away from freezing.

An even more detailed picture of bond orientational order in 2D systems can be gained by studying the *joint* probability distributions $P(|\psi_4|, |\psi_6|)$, $P(|\psi_4|, |\psi_{12}|)$, and $P(|\psi_6|, |\psi_{12}|)$. Contour plots of these distributions for the time-averaged 3584-particle WCA liquid at $\rho = 0.83$, calculated using only intact bonds, are shown in Figs. 38*a*–40*a*. In Figs. 38*a*–40*a* we have also marked the approximate locations of the various vertex types in order-parameter space (from Table III) with the corre-

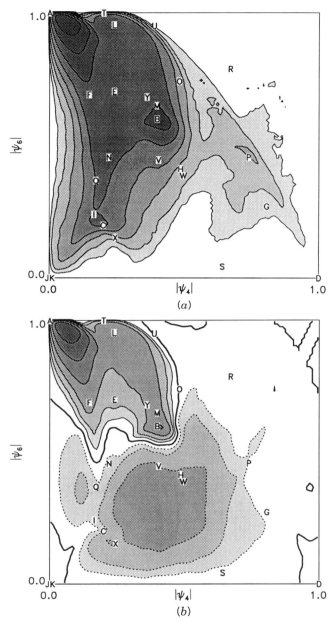

Figure 38. Contour plot of $P(|\psi_4|, |\psi_6|)$ (calculated using only intact bonds) for the time-averaged 3584-particle WCA liquid at $\rho = 0.83$. The letters indicate the approximate locations of the corresponding vertex types in order-parameter space. (a) Raw distribution. The contour levels are 0.125, 0.25, 0.5, 1, 2, 4, 8, 16, 32, 64, and 128. (b) Rationalized distribution (see text). The contour levels are -16, -8, -4, -2, -1, -0.5, 0, 0.5, 1, 2, 4, 8, and 16. Positive contours are indicated by solid lines, negative contours by dotted lines, and the zero contour by a heavy solid line.

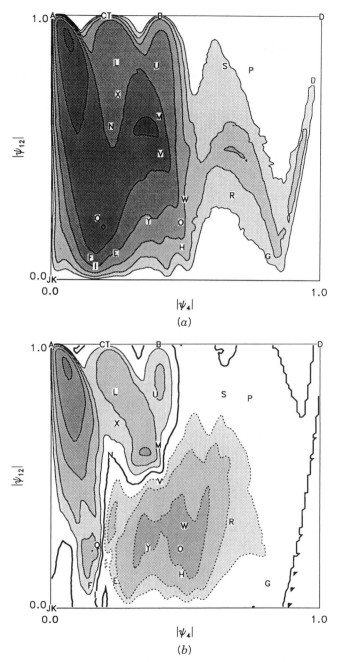

Figure 39. Contour plot of $P(|\psi_4|, |\psi_{12}|)$ (calculated using only intact bonds) for the time-averaged 3584-particle WCA liquid at $\rho = 0.83$. The letters indicate the approximate locations of the corresponding vertex types in order-parameter space. The contour levels are the same as for Fig. 38. (*a*) Raw distribution. (*b*) Rationalized distribution (see text).

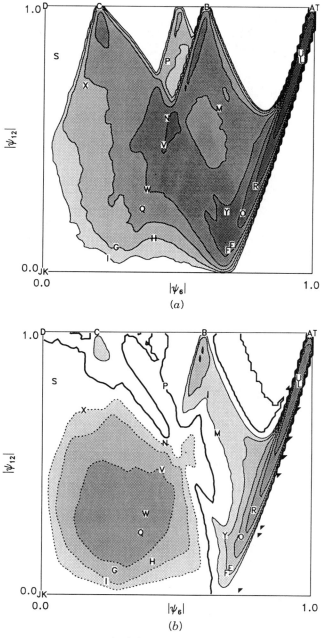

Figure 40. Contour plot of $P(|\psi_6|, |\psi_{12}|)$ (calculated using only intact bonds) for the time-averaged 3584-particle WCA liquid at $\rho = 0.83$. The letters indicate the approximate locations of the corresponding vertex types in order-parameter space. The contour levels are the same as for Fig. 38. (*a*) Raw distribution. (*b*) Rationalized distribution (see text).

sponding letters. These distributions are highly structured, in part due to the fact that certain regions of order parameter space are simply inaccessible (for example, the region around $|\psi_6| = 1$, $|\psi_{12}| = 0$ is inaccessible; it is impossible to have nearly perfect sixfold order without also having a high degree of twelvefold order). To remove this underlying structure, we computed "random" distribution functions using randomly chosen bond angles and the same (intact bond) coordination number statistics as the WCA liquid, and subtracted these random distributions from the distributions for the WCA liquid. The resulting "rationalized" distributions, $P_R(|\psi_4|, |\psi_6|)$, $P_R(|\psi_4|, |\psi_{12}|)$, and $P_R(|\psi_6|, |\psi_{12}|)$, are shown in Figs. 38b–40b. As before, the approximate locations of various vertex types in order parameter space are indicated by letters. These distributions have prominent features associated with specific vertex types. In particular, the distributions exhibit distinct peaks associated with type A, B, and C vertices, and less distinct features associated with other vertex types (in particular vertex types E, F, L, and M). This is compelling evidence for the tiling picture of 2D liquid structure described above.

Fluctuations in local m-fold bond orientational order are characterized by the local m-fold bond orientational susceptibility,

$$\chi_m = \langle |\psi_m|^2 \rangle - |\langle \psi_m \rangle|^2 \qquad (3.22)$$

Figures 41–43 show χ_4, χ_6, and χ_{12} as a function of ρ for the time-averaged 3584-particle WCA system (the susceptibilities calculated using only intact bonds behave similarly). The steep increase in χ_6 and χ_{12} for increasing densities in the liquid near freezing is presumably related to the increasing size of solid-like fluctuations. The decrease in χ_4 for increasing densities in the liquid near freezing is not surprising in light of

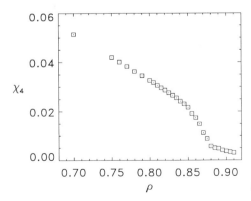

Figure 41. χ_4 versus ρ for the time-averaged 3584-particle WCA system.

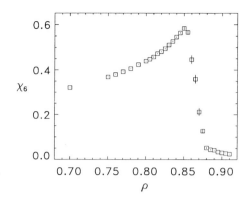

Figure 42. χ_6 versus ρ for the time-averaged 3584-particle WCA system.

the fact that the degree of local sixfold ordering is increasing in this region (fourfold order is incompatible with sixfold order in the sense that $|\psi_4| = 0$ in a perfect triangular lattice). Finally, as one would expect, fluctuations in local bond orientational order are much smaller in the solid phase than in the liquid phase, so that χ_4, χ_6 and χ_{12} are all much smaller in the solid than in the liquid. All three susceptibilities increase for decreasing solid densities, however.

A quantity of interest in the KTHNY theory of 2D melting is the so-called Halperin–Nelson order parameter,

$$\Psi_6 = \frac{1}{N} \sum_{i=1}^{N} \psi_{6i} \tag{3.23}$$

which measures the *global* sixfold bond orientational order present in a

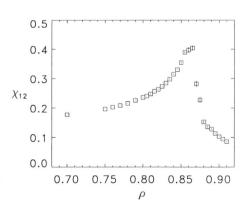

Figure 43. χ_{12} versus ρ for the time-averaged 3584-particle WCA system.

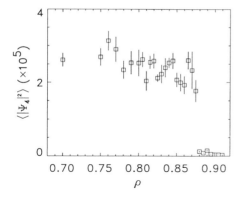

system, and is expected to have a nonzero expectation value in the solid phase. Global fourfold and twelvefold order parameters, Ψ_4 and Ψ_{12}, can be defined similarly. In Figs. 44–46 we have shown $\langle|\Psi_4|^2\rangle$, $\langle|\Psi_6|^2\rangle$, and $\langle|\Psi_{12}|^2\rangle$ as a function of ρ for the time-averaged 3584-particle WCA system (again, the quantities calculated using only intact bonds are qualitatively similar). $\langle|\Psi_6|^2\rangle$ and $\langle|\Psi_{12}|^2\rangle$ are $O(1)$ in the solid phase and $O(1/N)$ (as expected on the basis of a simple random walk model) in the liquid phase. $\langle|\Psi_4|^2\rangle$, on the other hand, is nearly zero in the solid phase ($\langle|\Psi_4|^2\rangle$ is exactly zero for a perfect triangular lattice), and *increases* to $O(\epsilon/N)$ in the liquid phase, where ϵ is a small number (the value of $\langle|\Psi_4|^2\rangle$ in the liquid phase is still described by a random walk model, but the effective step size $\epsilon = \langle|\psi_4|\rangle$ is small).

In the melting model to be presented in Section V, the liquid phase is produced by a condensation of defects. This gives rise to pronounced

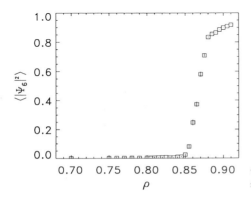

Figure 45. $\langle|\Psi_6|^2\rangle$ versus ρ for the time-averaged 3584-particle WCA system.

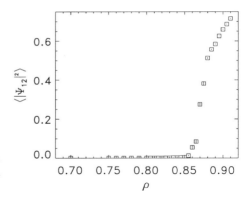

Figure 46. $\langle|\Psi_{12}|^2\rangle$ versus ρ for the time-averaged 3584-particle WCA system.

spatial inhomogeneity in the liquid, with solid-like clusters separated by defect aggregates, as is observed in the WCA liquid near freezing. Thus, the correct prediction of the characteristics of the spatial inhomogeneity of the liquid phase is an important test for our model (or for any other melting model). Also, the extensive solid-like fluctuations that we observe in the dense WCA liquid are interesting in their own right, both as an essential feature of liquid structure and because the transport properties of the dense liquid are strongly conditioned by such fluctuations. As far as we know, however, contemporary liquid-state theory is incapable of predicting the detailed characteristics of such solid-like fluctuations.

To characterize solid-like fluctuations in the WCA liquid we measured the size distribution of sixfold-ordered clusters,

$$n_s = \frac{N_s}{N} \tag{3.24}$$

where N_s is the average number of ordered clusters containing s particles, and N is the number of particles. Because of this normalization, we have

$$\sum_s n_s = \frac{N_C}{N} \tag{3.25}$$

where N_C is the average total number of ordered clusters. This is the usual normalization used in percolation theory [193, 194]. In Fig. 47 we show an example of the small s portion of n_s for the time-averaged 3584-particle WCA liquid at $\rho = 0.83$. The distribution is quite broad in the dense liquid, and apparently has a power law dependence on cluster size for small s. In fact, we have found that these distributions are described quite well by the functional form

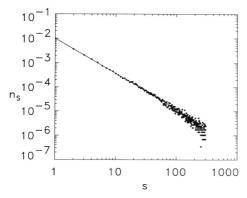

Figure 47. Small s part of n_s for the time-averaged 3584-particle WCA liquid at $\rho = 0.83$ (bullets). The solid line represents a fit to Eq. (3.26).

$$n_s = As^{-\tau_s}e^{-s/\xi_s} \tag{3.26}$$

This is roughly the form that is predicted by the Fisher droplet model of condensation [195], and represents a special case of the scaling ansatz used in percolation theory [193, 194]. This functional form has power law behavior for small s, and crosses over an exponential dependence for large s. The solid line in Fig. 47 represents a fit to Eq. (3.26). The fits obtained for the dense WCA liquid are generally quite good, although some systematic deviations are seen for large s. The power-law exponents τ_s and correlation sizes ξ_s obtained from such fits for the time-averaged WCA liquid are shown in Figs. 48 and 49.

For the time-averaged liquid, τ_s varies from $\tau_s \approx 1.2$ to $\tau_s \approx 1.5$ as the density increases from $\rho = 0.70$ to $\rho = 0.85$. This is in contrast to simple percolation models, for which τ_s is constant for varying site or bond

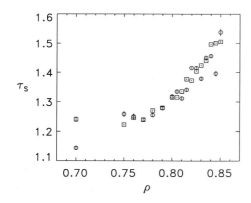

Figure 48. τ_s versus ρ for the time-averaged WCA liquid, for $N = 3584$ (squares) and $N = 896$ (circles).

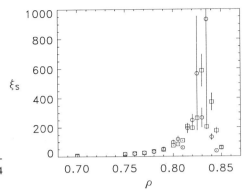

Figure 49. ξ_s versus ρ for the time-averaged WCA liquid, for $N = 3584$ (squares) and $N = 896$ (circles).

occupation probabilities. For site percolation on a triangular lattice, for example, $\tau_s = 187/91$, independent of the site occupancy. The physical interpretation of the exponent τ_s is still somewhat obscure, although we believe that it is related to the geometry of the boundaries between ordered regions, and therefore to the "tiling rules" that govern dense liquid structure. We will return to this point in Section V.

ξ_s, which characterizes the typical size of ordered clusters, increases strongly near freezing, as might have been expected from the qualitative appearance of Figs. 15c–17c. In fact, for the time-averaged system, ξ_s has an apparent cusp or divergence near $\rho = 0.83$. Although it is tempting to interpret this feature as the signature of a percolation transition, we believe that this feature is a finite-size effect. This can be seen by comparing the 3584- and 896-particle systems at $\rho = 0.85$ (Figs. 50a and 50b). The 3584-particle exhibits a distinct peak in n_s at $s \approx 2000$, while the peak in n_s for the 896-particle system is at a much smaller value of s, $s \approx 500$. The location of this peak is roughly proportional to the size of the system. We believe that this peak represents very large ordered clusters that in an infinite system would be distributed over a wide range of s, but in a finite system are "bunched up" at $s \sim N/2$ because they span the system. For a finite system, therefore, n_s consists of two parts: a size-independent portion at small s; and a size-dependent part (associated with spanning clusters) that develops at large s near the freezing density. Near $\rho = 0.83$, this size-dependent portion contributes to the tail of the small s distribution, producing a very large effective ξ_s. As $\rho \rightarrow 0.85$, however, the size-dependent part moves to larger s and becomes decoupled from the small s part of the distribution, and ξ_s decreases.

In Fig. 51 we have shown the average fraction of ordered particles for the time-averaged 3584-particle WCA system. The fraction of ordered

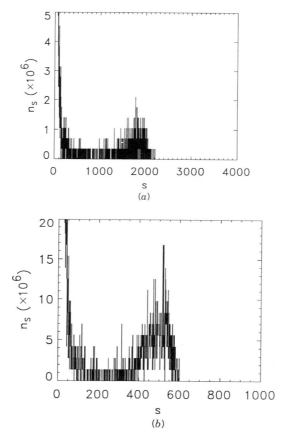

Figure 50. Large s part of n_s for the time-averaged WCA liquid at $\rho = 0.85$: (a) $N = 3584$; (b) $N = 896$.

particles is an order parameter-like quantity analogous to the fraction of six-coordinated particles, f_6. However, the fraction of ordered particles is always significantly smaller than f_6 (compare Figs. 27 and 51), reflecting the fact that $|\psi_6|$ is more sensitive to local geometrical disorder than the coordination number.

In Figs. 52 and 53 we have shown the average size of ordered clusters ($\langle s \rangle$) and the (normalized) average number of ordered clusters (N_C/N) for the time-averaged 3584-particle WCA liquid. In the dense liquid region, $\langle s \rangle$ increases steeply with increasing ρ, reflecting the behavior of the translational and bond orientational correlation lengths in the same

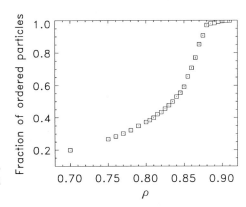

Figure 51. Fraction of ordered particles versus ρ for the time-averaged 3584-particle WCA system.

region. The dependence of N_C/N on ρ in the liquid region is not straightforward to explain, as it represents the interplay between two competing trends (the fraction of ordered particles and the average size of ordered clusters both decrease with decreasing ρ).

5. Bond Length and Bond Angle Distribution Functions

The nearest-neighbor bond length distribution functions for the time-averaged 3584-particle WCA system are shown in Fig. 54, while the corresponding bond angle distributions are shown in Fig. 55. The characteristics of these distributions in the dense liquid phase have been discussed briefly above. Here we simply point out that these distributions exhibit isosbestic points in the coexistence and dense liquid regions, features that were observed in other microscopic distribution functions. The observation of isosbestic points in the coexistence region provides additional evidence for a first-order transition, while the appearance of isosbestic points in the dense liquid region is supporting evidence for the "two-fluid" picture of dense liquid structure presented above.

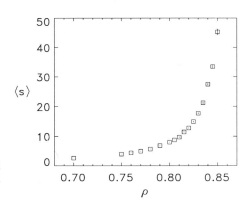

Figure 52. Average size of ordered clusters versus ρ for the time-averaged 3584-particle WCA liquid.

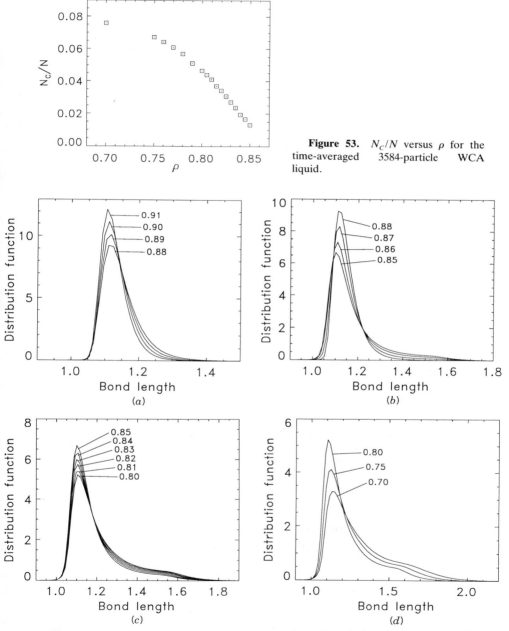

Figure 53. N_C/N versus ρ for the time-averaged 3584-particle WCA liquid.

Figure 54. Bond length distribution functions for the time-averaged 3584-particle WCA system, with each curve labeled by the corresponding value of ρ: (*a*) solid region; (*b*) coexistence region; (*c*) liquid near freezing; (*d*) liquid away from freezing.

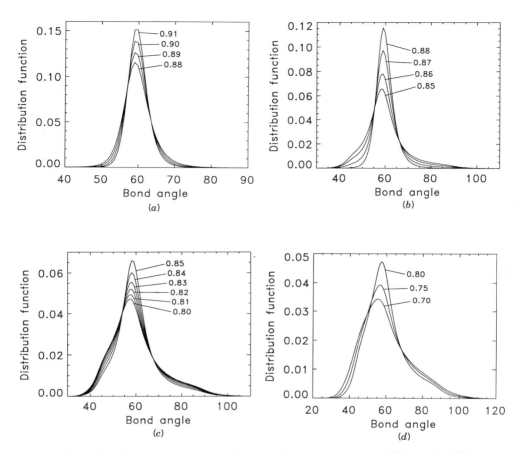

Figure 55. Bond angle distribution functions for the time-averaged 3584-particle WCA system, with each curve labeled by the corresponding value of ρ: (*a*) solid region; (*b*) coexistence region; (*c*) liquid near freezing; (*d*) liquid away from freezing.

IV. STRUCTURAL PROPERTIES OF DENSE RANDOM PACKINGS OF HARD DISKS

A. Introduction

It is generally believed that the structural properties of dense simple liquids are primarily determined by the harsh, repulsive part of the pair potential. This accounts for the success of perturbation theories that treat the attractive part of the pair potential as a perturbation on the steep, repulsive part of the potential [1, 183, 196]. Thus, it seems likely that the characteristic geometrical structure of the dense 2D WCA liquid is a result of excluded volume effects, or packing constraints. To test this assumption, we generated dense random packings of hard disks (DRPs) and analyzed their structure using the same types of structural analysis that were used to characterize the WCA system. Our approach parallels that of Bernal, who proposed in 1937 that the problem of liquid structure could be related to the essentially geometrical problem of the packing of hard spheres [197, 198]. Because Bernal's work is closely related to our own, we will discuss the findings of Bernal and his co-workers briefly. A more detailed discussion of this work can be found in Bernal's eloquent reviews [2–5] and in two more recent reviews of the geometrical approach to liquid structure [199] and melting [188].

The guiding principle of Bernal's theory of liquids is the view that liquids are [2] "... *homogeneous, coherent* and *essentially irregular* assemblages of molecules containing no crystalline regions or holes large enough to admit another molecule." Bernal recognized that, although a liquid is disordered, this disorder is strongly conditioned by excluded volume effects, which severely limit the possible arrangements of molecules in a dense liquid. Deducing the nature of the constraints imposed on possible molecular arrangements in a dense liquid by excluded volume effects represents a formidable (and analytically intractable) problem in statistical geometry. Rather than attacking this mathematical problem head on, Bernal and his co-workers took an empirical approach, attempting to characterize the statistical geometry of dense liquids by studying dense random packings (DRPs) of hard spheres (and other models), using both mechanical models and computer-generated packings [2–5, 200–202]. A central idea of the Bernal theory of liquids is that the large entropy (relative to a crystalline solid) of the liquid state is due to the irregularity of local liquid structure, and that, in principle, the entropy of a liquid can be calculated from a knowledge of its statistical geometry.

The most basic characteristic of DRPs is their packing fraction (the fraction of the total volume occupied by spheres). A number of studies find a packing fraction for 3D DRPs of $\eta \simeq 0.637$ [188]. Investigations by Bernal and others have also identified a "loose random packing" (LRP) limit near $\eta \simeq 0.60$, which appears to be the stability limit for random packings in 3D [188]. For regular close packing (CP) in 3D (fcc of hcp) the packing fraction is $\eta = 0.740$. It appears that the various methods for generating random packings rarely result in packing fractions between 0.740 and 0.637. On the basis of considerable experience, Bernal states [2],

I have always found it impossible to construct irregular arrangements of any intermediate density. We believe this is a necessary consequence of the irregularity, and that there is *an absolute impossibility of forming a homogeneous assembly of points of volume intermediate between those of long-range order and closest packed disorder.* Regular and irregular close packing are evidently arrangements of quite different local co-ordination, and a transition between them is necessarily as abrupt as that between two regular phases of different structure such as occurs in the α-(body-centred) and γ-(face-centred) phases of iron.

From this observation Bernal infers that the melting transition must be first-order. It is interesting to note that the volume increase on passing from close packing to dense random packing is 16%, close to the volume expansion upon melting for argon [188]. Perhaps a more relevant comparison is with the melting transition for hard spheres in 3D, which melt at a packing fraction of 0.545 and freeze at a packing fraction of 0.494 [79], corresponding to a 10% volume increase on melting. This is at least of the same order of magnitude as the volume difference between the DRP and CP limits. Bernal's statement must be tempered somewhat, since it is easy to devise ways of constructing packings having densities intermediate between the DRP and close-packed limits, for instance by inserting numerous defects into a close-packed crystal. Bernal's statement has to do with the *improbability* of generating such structures in a stochastic process. The number of possible packings (or the entropy) as a function of packing fraction evidently decreases drastically for $\eta > 0.637$, with the result that few configuration-space pathways lead to packings having intermediate densities.

Bernal's studies were able to establish relatively close correspondence between 3D DRP structure (as characterized by the pair correlation function, $g(r)$) and the structure of real liquids and glasses. As described below, $g(r)$ for 2D DRPs exhibits large and extremely sharp peaks, much sharper and higher than those observed in the 2D liquid. Because we would expect to see similar features in $g(r)$ for 3D DRPs (we would, for example, expect to see a δ function contribution to $g(r)$ at $r = d$, where d is the hard sphere diameter), it appears that the close match between the pair correlation functions of 3D DRPs and real liquids found in these early studies is largely fortuitous, probably resulting from the use of rather wide radial bins in constructing $g(r)$, which would tend to smooth out sharp features such as those we observe in 2D dense random packings. Nevertheless, the level of agreement found by Bernal et al. is quite suggestive, even if their correlation functions have been inadvertantly smoothed, because it shows that DRPs contain approximately the right number of particles in each coordination shell.

To characterize the statistical geometry of DRPs, Bernal and his co-workers carried out a Voronoi analysis of DRP structure, similar to the analysis of the 2D WCA system described in Section III.D, and compiled the statistics of Voronoi polyhedra. The most striking feature observed in these studies was the predominance of pentagonal faces for Voronoi polyhedra. Bernal recognized that this was an important feature of liquid structure, because of the impossibility of forming space filling arrangements of objects having fivefold symmetry. Thus, the appearance of pentagonal faces is intimately connected to the lack of long-range order in 3D liquids. Rivier and Duffy have recently formulated a 3D melting theory derived from this basic idea [102].

One of the most interesting measures of the statistical geometry of DRPs employed by Bernal is that derived from the so-called "reduced simplicial graph." The term "simplicial graph" is used by Bernal to describe the dual of the Voronoi construction, namely the bond network formed by connecting every particle to all of its geometrical neighbors (particles with whom it shares a Voronoi cell face). The simplicial graph is what we have referred to as the Delaunay construction in Section III.D, and in 3D, the simplicial graph partitions space into (distorted) tetrahedral volumes, much as its partitions the plane into triangular areas in 2D. The *reduced* simplicial graph is obtained by removing bonds from the simplicial graph on a *metrical* basis, for instance removing bonds that are significantly longer than the hard sphere diameter. This partitions space into polyhedral volumes (tetrahedra, octahedra, and larger polyhedra), and is seen to be the 3D analog of the polygon construction described in Section III.D. According to Bernal, there are only five "deltahedra" (convex polyhedra with triangular faces and edges of equal length) small enough that another sphere cannot be inserted into the center without the edge length exceeding the hard sphere diameter by more than 10%: the tetrahedron (with 4 vertices); the octahedron (6 vertices); the tetragonal dodecahedron (8 vertices); the capped trigonal prism (9 vertices); and the Archimedean antiprism (10 vertices). One large DRP studied by Bernal et al. was found to consist of 73.0% (by number) tetrahedra, 20.3% half octahedra, 3.1% tetragonal dodecahedra, 3.2% trigonal prisms, and 0.4% Archimedean antiprisms (accounting, respectively, for 48.4%, 26.9%, 14.8%, 7.8%, and 2.1% of the total volume) [4]. More recently, Ashby et al. [203] have argued that the list of so-called *canonical* polyhedra should be expanded to include three additional polyhedra having 7, 11, and 12 vertices.

Although Bernal found no evidence of crystalline regions in DRPs, he did find evidence of so-called "pseudonuclei," which are exceptionally dense local arrangements of spheres consisting of helical chains of tetrahedra [4]. Such polytetrahedral aggregates can branch but cannot fill space (for a recent review of polytetrahedral order in condensed matter see [208]). Bernal also observed numerous *collineations* in DRPs, that is, lines of spheres in close contact, consisting of as many as eight spheres [5]. Such collineations would be expected to produce extremely sharp peaks in the pair distribution function.

The structure of a dense 3D liquid can, in Bernal's view, be described as a packing, or mosaic, of canonical polyhedra. Bernal and his co-workers showed that the five canonical polyhedra can be packed in at least 197 different ways around a point, allowing an ~5% variation in edge lengths, and that every point on the surface of such a cluster can be the center of a new cluster, so that space filling mosaic of polyhedra can be formed [4]. It is easily seen that the close-packed crystal structures (fcc and hcp) can be regarded as periodic packings of tetrahedra and octahedra, and that a variety of other common crystal structures (for instance, bcc) can be regarded as periodic arrays of slightly distorted tetrahedra and/or octahedra. Because the arrangement of tetrahedra and octahedra in a crystal is regular, the configurational entropy of the polyhedra in a perfect crystal is zero, whereas the liquid state has a relatively large configurational entropy, due to the many possible arrangements of poly-hedra. This suggests a model for 3D melting very similar to the model we have developed for 2D melting, in which the canonical polyhedra are the "relevant" configurational degrees of freedom for melting, and the number densities of various types of polyhedra can be regarded as order parameters for the melting transition. The work described in this article is primarily concerned with the application of this idea to 2D melting. However, we believe that the basic mechanism for melting is the same in both 2D and 3D. What is missing from the work of Bernal is a way of taking into account the interactions or correlations between the various types of polyhedra, which is essential for a calculation of the entropy and for development of a model for the melting transition. In Section V we describe our approach to this problem, which is based on generalized tiling models.

It is worth noting that the characterization of liquid structure in terms of canonical polyhedra has never gained wide acceptance, primarily because it is difficult to define the reduced simplicial graph in a unique manner. Typically the reduced simplicial graph is obtained from the full simplicial graph by removing bonds exceeding some critical length. Because the choice of critical length is necessarily somewhat arbitrary, an element of arbitrariness enters into the mapping of a particle configura-tion into a configuration of canonical polyhedra. Because of this difficul-ty, most researchers have preferred to characterize the statistical geome-try of disordered systems in terms of the coordination numbers, which can be specified in a unique way from the Voronoi construction, with no dependence on specific metrical parameters. We believe, however, that this approach omits some important physics in the quest for formal precision, since configurations of particles described by the same set of coordination numbers can have significantly different local geometries, and, conversely, small changes in local geometry can result in markedly different coordination numbers. The neglect of metrical information (information about bond lengths and bond angles) implicit in the coordination number description makes it an insufficiently sensitive measure of local geometry, and consequently any estimate of the entropy of a disordered state based on this description will be too small.

B.   Previous Work on Hard Disk Packings

To the best of our knowledge, the earliest study of dense random packings of hard spheres in 2D is that of Nowick and Mader [204], who built an apparatus that allowed either slow or rapid deposition of random mixtures of spheres of two different sizes. Diffraction patterns of the resulting two-dimensional arrays of spheres were then obtained by preparing photographic transparencies of the arrays and then producing Fraunhofer diffraction patterns of the transparencies. Their experiment also permitted vibrational annealing of the as-deposited two-dimensional sphere packings. The single component sphere packings they obtain for moderate rates of deposition contain numerous crystalline regions (regions of nearly perfect packing) with different orientations, separated by broad disordered regions. These packings are quite similar to those obtained in our study of dense random packings (described below), and so we omit a detailed discussion here.

More recently, Rubinstein and Nelson [205] studied computer-generated binary arrays of hard disks. Starting with an initial seed, successive disks are brought into contact with the growing cluster subject to the condition that they be as close as possible to the center of the seed. Of particular interest is an array consisting of 3200 disks packed around a smaller one (Fig. 1 in [205]), which appears to be *isomorphic* to a tiling composed of squares and equilateral triangles (ST tiling), with disks situated on the vertices of tiles. *This result clearly shows that packing constraints lead, in certain cases, to highly specific tiling rules.* This conclusion finds support in the results described in Section III and in our studies of dense random packings of hard disks, described below. Of course, the deterministic growth algorithm of Rubinstein and Nelson leads to a variety of more general structures, depending on the size mismatch and the particular seed used, but the resemblance of some of their disk arrays to tilings is quite compelling.

Equally compelling is the work of Onoda and Toner [206], who studied computer-generated deterministic packings of disks around rhomboidal seeds, using an algorithm similar to that of Rubinstein and Nelson [205]. In particular, disks are added sequentially to the perimeter of the growing cluster at the site closest to the center of the seed. *The resulting arrays of disks are, in all cases, isomorphic to ST tilings.* In fact, the disks in such packings reside on the vertices of a perfect tiling composed of rhombi and equilateral triangles. The tiling patterns formed depend on the acute angle of the rhombi and the configuration of rhombi used as a seed. Onoda and Toner found that the network of rhombi present in such packings is fractal, and calculated the corresponding fractal dimension as a function of the acute angle of the rhombi both analytically and numerically. However, we are particularly interested in this work because it provides another example of a disk packing problem that can be translated into a tiling problem.

In the remainder of this section, we present the results of our detailed study of dense random packings of hard disks. As we will show, the structure of such dense random packings strongly resembles that of the dense WCA liquid (see Section III), and illustrates even more dramatically the way in which packing constraints are translated into tiling rules.

C. Computational Method

Dense random packings were generated using the algorithm of Finney [207]. First, a low density, random configuration of disks is generated, and the system is compressed in stages. At each stage of compression, a search for disk overlaps is made, and overlaps are removed by moving disks apart along the line joining their centers. The search for overlaps is iterated until all overlaps have been removed at a given packing fraction before the next stage of compression is initiated. If all the overlaps are not removed after 5000 sweeps through the system, then the system is considered to have reached DRP limit, and the process is terminated.

Using this algorithm, we generated 167 DRPs, each consisting of 896 hard disks, with periodic boundary conditions. The distribution of ultimate packing fractions so generated is shown in Fig. 56. The average packing fraction for DRPs was found to be 0.8271 ± 0.0004. This is ~9% lower than the packing fraction of the close-packed triangular lattice, $\pi/2\sqrt{3} \cong 0.9069$, and ~14% higher than the packing fraction of the hard disk system at the melting point, 0.723 [57, 79]. The fact that the DRP packing fraction is significantly higher than the packing fraction of hard disks at melting is interesting because it suggests that the crystalline hard disk solid is nonergodic. In order words, there exist glassy "pockets" in configuration space, corresponding to DRP configurations, that will never

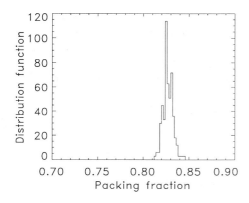

Figure 56. Packing fraction distribution for dense random packings of hard disks generated using the algorithm described in the text.

be explored by a system that starts in a crystalline configuration, so that time averages, which never sample such glassy configurations, are not equivalent to ensemble averages, which sample the entire accessible region of configuration space. Conversely, this implies that it might be possible to form a *thermally* stable one-component 2D glass (DRPs appear to be *mechanically* stable), even though 2D systems lack the inherent frustration that aids amorphization in 3D [208]. It would be interesting to determine whether 2D DRPs are stable at finite temperature, or whether they anneal to form large crystalline regions.

D. Statistical Geometry of Dense Random Packings of Hard Disks

1. Overview

We have investigated the statistical geometry of DRPs using the same methods that we used to characterize the 2D WCA system. To facilitate comparison of the two systems, we present our results on DRPs in a sequence that roughly parallels the discussion of the WCA system.

In Fig. 57a we show the Voronoi construction for a typical DRP. Figure 57b shows the corresponding Delaunay construction. The general appearance is similar to that of the dense WCA liquid (compare with Figs. 15a–17a and Figs. 15b–17b). Topological defects occur in grain boundary-like aggregates consisting of chains or clumps of disclinations of alternating sign, and there are well-defined solid-like regions. As in the WCA system, there is considerable local structural disorder that lacks a topological signature, in particular local groups of particles with a nearly square arrangement. Extended chains of "squares" are observed in some cases (in crystallography, such structural defects are called stacking faults or antiphase boundaries). The impression of polycrystalline order is even more striking in DRPs than in the dense WCA liquid. As in the WCA liquid, the majority of particles are six-coordinated, although the Delaunay construction reveals considerable variation in bond lengths and bond angles, even for six-coordinated particles.

Figure 57c show the sixfold-ordered regions in the DRP of Fig. 57a.

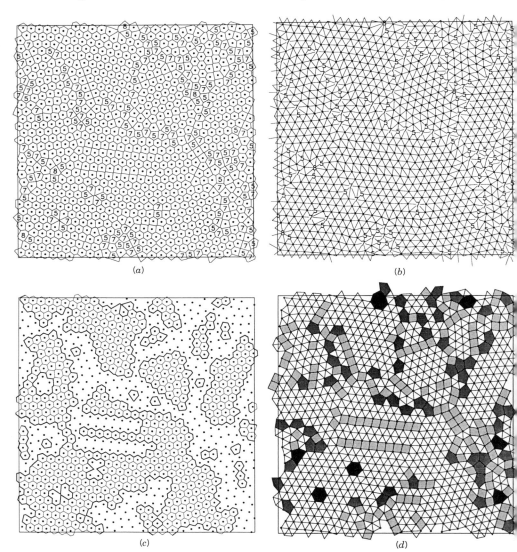

(a)

(b)

(c)

(d)

Figure 57. Various types of structural analysis for a representative dense random packing of hard disks at a packing fraction of 0.8334 (see the caption for Fig. 12).

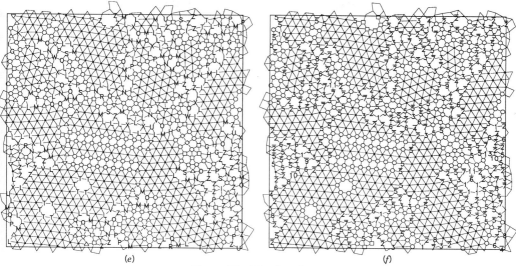

(e)

(f)

Figure 57   (*Continued*)

As in the WCA liquid, extensive ordered regions are apparent, reinforcing the general impression of polycrystalline order, and there are significantly fewer sixfold-ordered particles than six-coordinated particles.

The nearest-neighbor bond length and bond angle distribution functions for DRPs are shown in Figs. 58 and 59, respectively (here, and in the subsequent discussion, the disk diameter is taken to be the unit of length). The bond length distribution exhibits an extremely sharp peak near 1, resulting from the large number of disk contacts present in DRPs. The bond angle distribution also exhibits a sharp peak near 60°, coming

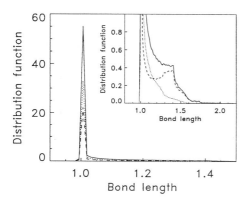

Figure 58. Nearest-neighbor bond length distribution functions for dense random packings of hard disks. The figure shows the contributions from ordered regions (dotted line), disordered regions (dashed line), and the overall distribution (solid line). The inset shows an expanded view.

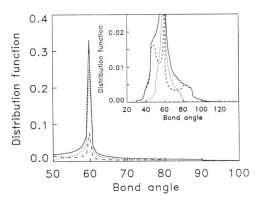

Figure 59. Nearest-neighbor bond angle distribution functions for dense random packings of hard disks. The figure shows the contributions from ordered regions (dotted line), disordered regions (dashed line), and the overall distribution (solid line). The horizontal scale is in degrees. The inset shows an expanded view.

from the extensive regions of near perfect triangular packing that are present in DRPs. The fine structure of these distributions (insets of Figs. 58 and 59) is particularly interesting. The overall bond length distribution has a sharp edge near $\sqrt{2}$, and the distribution for disordered particles exhibits two peaks, centered near 1 and 1.35. The overall bond angle distribution has sharp edges near 45° and 90°, and the distribution for disordered particles has three distinct peaks, centered near 47°, 60°, and 85°. These features are clear signatures of local "square" ordering in DRPs. The sharp edges observed in both distributions correspond to the maximum bond lengths and bond angles that can appear in a rhombus of edgelength 1, namely those that appear in a perfect square.

The polygon construction for the DRP of Fig. 57a is shown in Fig. 57d. The impression of a tiling is even more striking in DRPs than in the dense WCA liquid, and many of the same local arrangements of "tiles" are observed. As in the dense WCA liquid, four-sided polygons are the most prevalent non-three-sided polygons, with decreasing numbers of five- and more-sided polygons. However, relatively more large polygons (polygons with six or more sides) are observed in DRPs than in the WCA liquid, and a number of "perfect" vacancies are observed (these appear as hexagonal six-sided polygons).

The vertex classification for the DRP of Fig. 57a is shown in Fig. 57e, and the tiling charges for the DRP are shown in Fig. 57f. Again, the appearance of the vertex and tiling charge plots is quite similar to those of the dense WCA liquid, with a similar tendency for polygons to form local

TABLE IV

Polygon Order Parameters ζ_p for Dense Random Packings of Hard Disks (DRP), the Dense WCA Liquid, the Topologically Constrained Generalized Tiling Model with $p_{max} = 6$, $r = 4$ (GTM1), and the Topologically Unconstrained Generalized Tiling Model with $p_{max} = 6$, $r = 4$ (GTM2); the Fraction of Broken Bonds is Approximately the Same for All Four Systems ($\phi \approx 0.117$)

p	DRP	WCA, $\rho = 0.83$	GTM1, $t = 1.4$	GTM2, $t = 0.685$
3	0.694 ± 0.002	0.6671 ± 0.0008	0.6589 ± 0.0003	0.6602 ± 0.0006
4	0.0917 ± 0.0009	0.1268 ± 0.0004	0.15643 ± 0.00015	0.1537 ± 0.0003
5	0.0338 ± 0.0004	0.02430 ± 0.00018	$(9.17 \pm 0.04) \times 10^{-3}$	0.01058 ± 0.00005
6	$(4.28 \pm 0.13) \times 10^{-3}$	$(1.52 \pm 0.04) \times 10^{-3}$	$(1.86 \pm 0.09) \times 10^{-4}$	$(1.77 \pm 0.06) \times 10^{-4}$

arrangements that minimize the tiling charge. In Tables IV–VIII we have compared the polygon, coordination number, vertex and tiling charge statistics of the DRP system with those of the WCA system at $\rho = 0.83$, the density at which the order parameter ϕ for the WCA system matches that of the DRP system ($\phi = 0.1169 \pm 0.0009$). The polygon statistics of the two systems are similar, but the WCA system has more four-sided polygons, while the DRP system has more five- and six-sided polygons. The DRP system has considerably more six-coordinated particles than the WCA liquid, but, like the WCA system, has fewer seven-coordinated particles than five-coordinated particles. The DRP system has significantly more vertices of types A, L, M, and T than the WCA system, and significantly fewer of types C, E, and F. Also, the asymmetry between type B and C vertices is much more pronounced in the DRP system than in the WCA system. The DRP system has significantly more charge 0, $-2/10$, and $+3/10$ vertices than the WCA system, but the total fraction of vertices with $|c_\alpha| \le \frac{1}{2}$ is nearly the same for the two systems (0.92 for the DRP system and 0.94 for the WCA system). Overall, the statistics for

TABLE V

Probabilities of Occurrence of Various Values of the Coordination Number (All Bonds) for Dense Random Packings of Hard Disks (DRP), the Dense WCA Liquid, the Topologically Constrained Generalized Tiling Model with $p_{max} = 6$, $r = 4$ (GTM1), and the Topologically Unconstrained Generalized Tiling Model with $p_{max} = 6$, $r = 4$ (GTM2); the Fraction of Broken Bonds is Approximately the Same for all Four Systems ($\phi \approx 0.117$)

Coordination number	DRP	WCA, $\rho = 0.83$	GTM1, $t = 1.4$	GTM2, $t = 0.685$
3	0	0	0	0
4	$(4.1 \pm 0.6) \times 10^{-4}$	$(1.68 \pm 0.13) \times 10^{-4}$	0	$(5.71 \pm 0.04) \times 10^{-3}$
5	0.1086 ± 0.0009	0.1234 ± 0.0006	0	0.1973 ± 0.0003
6	0.7861 ± 0.0018	0.7542 ± 0.0011	1	0.5988 ± 0.0006
7	0.1003 ± 0.0009	0.1209 ± 0.0005	0	0.1876 ± 0.0003
8	$(4.48 \pm 0.18) \times 10^{-3}$	$(1.41 \pm 0.05) \times 10^{-3}$	0	0.01034 ± 0.00005
9	$(4.7 \pm 1.7) \times 10^{-5}$	$(2.4 \pm 1.9) \times 10^{-6}$	0	$(1.56 \pm 0.05) \times 10^{-4}$

TABLE VI

Probabilities of Occurrence of Various Values of the Coordination Number (Intact Bonds Only) for Dense Random Packings of Hard Disks (DRP), the Dense WCA Liquid, the Topologically Constrained Generalized Tiling Model with $p_{max} = 6$, $r = 4$ (GTM1), and the Topologically Unconstrained Generalized Tiling Model with $p_{max} = 6$, $r = 4$ (GTM2); the Fraction of Broken Bonds is Approximately the Same for All Four Systems ($\phi \approx 0.117$)

Coordination number	DRP	WCA, $\rho = 0.83$	GTM1, $t = 1.4$	GTM2, $t = 0.685$
3	$(8.4 \pm 0.3) \times 10^{-3}$	$(3.82 \pm 0.04) \times 10^{-3}$	$(2.15 \pm 0.06) \times 10^{-4}$	$(1.4 \pm 1.0) \times 10^{-5}$
4	0.1088 ± 0.0016	0.0990 ± 0.0004	0.0770 ± 0.0003	0.0517 ± 0.0002
5	0.458 ± 0.003	0.5133 ± 0.0012	0.5466 ± 0.0006	0.5986 ± 0.0010
6	0.424 ± 0.004	0.3820 ± 0.0015	0.3761 ± 0.0005	0.3492 ± 0.0012
7	$(3.7 \pm 0.5) \times 10^{-4}$	$(1.90 \pm 0.05) \times 10^{-3}$	0	$(5.41 \pm 0.09) \times 10^{-4}$
8	0	0	0	0
9	0	0	0	0

the two systems show similar trends. The most important conclusion to be drawn from our study of DRPs is that the tendency for particles in the dense WCA liquid to adopt local arrangements characteristic of tilings, as if conditioned by particular tiling rules, is the result of *packing constraints*, because dense random packings exhibit a very similar tendency.

TABLE VII

Probabilities of Occurrence of Various Vertex Types for Dense Random Packings of Hard Disks (DRP), the Dense WCA Liquid, the Topologically Constrained Generalized Tiling Model with $p_{max} = 6$, $r = 4$ (GTM1), and the Topologically Unconstrained Generalized Tiling Model with $p_{max} = 6$, $r = 4$ (GTM2); the Fraction of Broken Bonds is Approximately the Same for All Four Systems ($\phi \approx 0.117$)

Vertex type	DRP	WCA, $\rho = 0.83$	GTM1, $t = 1.4$	GTM2, $t = 0.685$
A	0.369 ± 0.004	0.2549 ± 0.0016	0.2407 ± 0.0005	0.2391 ± 0.0012
B	0.1144 ± 0.0016	0.1214 ± 0.0004	0.1856 ± 0.0003	0.1926 ± 0.0004
C	0.0536 ± 0.0011	0.0942 ± 0.0007	0.1226 ± 0.0005	0.1726 ± 0.0006
D	$(3.8 \pm 0.2) \times 10^{-3}$	$(4.17 \pm 0.06) \times 10^{-3}$	$(9.99 \pm 0.13) \times 10^{-3}$	$(8.37 \pm 0.06) \times 10^{-3}$
E	0.0481 ± 0.0006	0.1178 ± 0.0003	0.1470 ± 0.0002	0.1208 ± 0.0002
F	0.0492 ± 0.0006	0.1131 ± 0.0003	0.12762 ± 0.00016	0.10785 ± 0.00017
G	$(8.4 \pm 0.3) \times 10^{-3}$	0.0208 ± 0.0002	0.02769 ± 0.00011	0.01879 ± 0.00008
H	$(5.5 \pm 0.2) \times 10^{-3}$	0.01309 ± 0.00015	0.02280 ± 0.00008	0.01647 ± 0.00009
I	$(4.60 \pm 0.18) \times 10^{-3}$	0.01106 ± 0.00015	0.01937 ± 0.00007	0.01497 ± 0.00006
J	$(7 \pm 7) \times 10^{-6}$	$(9.5 \pm 0.9) \times 10^{-5}$	0	$(9.41 \pm 0.10) \times 10^{-4}$
K	$(3.7 \pm 0.5) \times 10^{-4}$	$(1.86 \pm 0.05) \times 10^{-3}$	0	$(5.41 \pm 0.10) \times 10^{-4}$
L	0.0969 ± 0.0011	0.0655 ± 0.0006	0.01446 ± 0.00008	0.0394 ± 0.0002
M	0.0455 ± 0.0006	0.03635 ± 0.00018	0.01955 ± 0.00005	0.01926 ± 0.00010
N	0.0335 ± 0.0006	0.0304 ± 0.0003	0.01109 ± 0.00008	0.01898 ± 0.00010
O	0.0181 ± 0.0004	0.01972 ± 0.00007	$(6.79 \pm 0.05) \times 10^{-3}$	$(2.08 \pm 0.02) \times 10^{-3}$
P	0.0176 ± 0.0005	0.01521 ± 0.00014	$(8.89 \pm 0.09) \times 10^{-3}$	0.01081 ± 0.00007
Q	$(4.99 \pm 0.17) \times 10^{-3}$	$(8.41 \pm 0.12) \times 10^{-3}$	$(2.52 \pm 0.03) \times 10^{-3}$	$(1.044 \pm 0.016) \times 10^{-3}$
R	$(5.0 \pm 0.2) \times 10^{-3}$	$(8.11 \pm 0.10) \times 10^{-3}$	$(2.23 \pm 0.03) \times 10^{-3}$	$(1.016 \pm 0.017) \times 10^{-3}$
S	0.0109 ± 0.0003	$(7.55 \pm 0.10) \times 10^{-3}$	$(8.28 \pm 0.07) \times 10^{-3}$	$(4.85 \pm 0.04) \times 10^{-3}$
T	0.0200 ± 0.0008	$(4.24 \pm 0.09) \times 10^{-3}$	$(2.64 \pm 0.11) \times 10^{-4}$	$(1.03 \pm 0.04) \times 10^{-3}$
U	$(8.4 \pm 0.3) \times 10^{-3}$	$(3.81 \pm 0.08) \times 10^{-3}$	$(2.19 \pm 0.03) \times 10^{-3}$	$(6.36 \pm 0.17) \times 10^{-4}$
V	$(8.0 \pm 0.2) \times 10^{-3}$	$(2.97 \pm 0.06) \times 10^{-3}$	$(2.64 \pm 0.09) \times 10^{-4}$	$(1.79 \pm 0.06) \times 10^{-4}$
W	$(2.75 \pm 0.13) \times 10^{-3}$	$(2.80 \pm 0.07) \times 10^{-3}$	$(9.99 \pm 0.09) \times 10^{-4}$	$(2.82 \pm 0.07) \times 10^{-4}$
X	$(7.0 \pm 0.2) \times 10^{-3}$	$(2.27 \pm 0.04) \times 10^{-3}$	$(1.35 \pm 0.08) \times 10^{-4}$	$(1.80 \pm 0.06) \times 10^{-4}$
Y	$(5.2 \pm 0.2) \times 10^{-3}$	$(2.72 \pm 0.11) \times 10^{-3}$	$(3.58 \pm 0.12) \times 10^{-4}$	$(1.79 \pm 0.10) \times 10^{-4}$

TABLE VIII

Probabilities of Occurrence of Various Values of the Tiling Charge for Dense Random Packings of Hard Disks (DRP), the Dense WCA Liquid, the Topologically Constrained Generalized Tiling Model with $p_{max} = 6$, $r = 4$ (GTM1), and the Topologically Unconstrained Generalized Tiling Model with $p_{max} = 6$, $r = 4$ (GTM2); the Fraction of Broken Bonds is Approximately the Same for All Four Systems ($\phi \approx 0.117$)

Tiling charge (in tenths)	DRP	WCA, $\rho = 0.83$	GTM1, $t = 1.4$	GTM2, $t = 0.685$
-10	$(1.80 \pm 0.11) \times 10^{-3}$	$(3.70 \pm 0.06) \times 10^{-3}$	$(3.68 \pm 0.05) \times 10^{-3}$	$(1.863 \pm 0.019) \times 10^{-3}$
-9	$(3.17 \pm 0.14) \times 10^{-3}$	$(1.38 \pm 0.04) \times 10^{-3}$	$(7.1 \pm 0.4) \times 10^{-5}$	0
-7	0.0240 ± 0.0004	0.02811 ± 0.00013	$(9.05 \pm 0.06) \times 10^{-3}$	$(3.10 \pm 0.03) \times 10^{-3}$
-6	$(5.3 \pm 0.6) \times 10^{-4}$	$(1.32 \pm 0.16) \times 10^{-4}$	$(1.30 \pm 0.10) \times 10^{-5}$	0
-5	0.0632 ± 0.0007	0.1425 ± 0.0004	0.17517 ± 0.00013	0.1399 ± 0.0002
-4	0.0105 ± 0.0003	$(5.51 \pm 0.11) \times 10^{-3}$	$(2.57 \pm 0.04) \times 10^{-3}$	$(9.91 \pm 0.14) \times 10^{-4}$
-2	0.1291 ± 0.0014	0.0893 ± 0.0007	0.03184 ± 0.00014	0.0551 ± 0.0003
0	0.564 ± 0.003	0.4804 ± 0.0013	0.5596 ± 0.0004	0.6141 ± 0.0007
1	$(7.3 \pm 0.3) \times 10^{-3}$	$(4.20 \pm 0.11) \times 10^{-3}$	$(2.68 \pm 0.03) \times 10^{-3}$	$(1.54 \pm 0.02) \times 10^{-3}$
3	0.0836 ± 0.0010	0.0694 ± 0.0004	0.03310 ± 0.00012	0.03976 ± 0.00017
4	$(7.4 \pm 0.8) \times 10^{-4}$	$(2.16 \pm 0.18) \times 10^{-4}$	$(8.4 \pm 0.4) \times 10^{-5}$	$(1.8 \pm 0.2) \times 10^{-5}$
5	0.0663 ± 0.0008	0.1408 ± 0.0004	0.17029 ± 0.00007	0.1396 ± 0.0002
6	0.0161 ± 0.0004	$(5.72 \pm 0.09) \times 10^{-3}$	$(5.24 \pm 0.13) \times 10^{-3}$	$(3.95 \pm 0.08) \times 10^{-4}$
8	0.0138 ± 0.0003	0.01671 ± 0.00011	$(5.14 \pm 0.02) \times 10^{-3}$	$(1.843 \pm 0.019) \times 10^{-3}$
9	$(1.5 \pm 0.3) \times 10^{-4}$	$(3.4 \pm 0.4) \times 10^{-5}$	0	0
10	$(1.93 \pm 0.12) \times 10^{-3}$	$(7.88 \pm 0.07) \times 10^{-3}$	$(5.61 \pm 0.04) \times 10^{-3}$	$(1.80 \pm 0.02) \times 10^{-3}$

Just how packing constraints get translated into tiling rules is not entirely clear. This is a question we will return to in Section VI.

In analogy to the procedure we used for the WCA system, we can use Eq. (3.6) to calculate the "theoretical" volume deviation of DRPs from close packing from the order parameters ζ_p, assuming that the polygons have their "ideal" areas (Eq. (3.5)). In this way we obtain a theoretical fractional volume expansion from close packing of

$$\frac{\Delta v}{v_{CP}} = \frac{v_{DRP} - v_{CP}}{v_{CP}} = 0.072 \qquad (4.1)$$

where v_{DRP} and v_{CP} are the specific volumes of DRPs and of the close-packed triangular lattice, respectively. For comparison, the actual volume deviation from close packing is

$$\frac{\Delta v}{v_{CP}} = 0.096 \qquad (4.2)$$

The agreement is reasonably good, which indicates that identifying non-three-sided polygons is a good way of identifying the excess volume in DRPs.

2. Translational and Bond Orientational Correlations

The pair correlation function $g(r)$ and the bond orientational correlation function $g_6(r)$ for DRPs are shown in Figs. 60 and 61, respectively. These

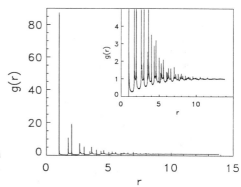

Figure 60. Pair correlation function for dense random packings of hard disks. The inset shows an expanded view.

correlation functions (in particular $g(r)$) exhibit extremely sharp peaks, at locations generally corresponding to the direct lattice vectors of a perfect triangular lattice having a lattice spacing of 1. These very sharp peaks are apparently associated with the extensive regions of nearly perfect local triangular order evident in Fig. 57c. In fact, the height of the first few peaks in $g(r)$ seems to be limited only by the width of the radial bins chosen for constructing the correlation functions. However, the *relative* heights of the peaks are expected to be significant. Because $g(r)$ and $g_6(r)$ are so highly structured, exhibiting many orders of direct lattice coordination shells, the methods used to extract the correlation lengths of the WCA liquid are inapplicable [191]. Instead, we fit the heights of peaks corresponding to direct lattice vectors in both correlation functions to exponentials. Using this fitting method, we obtain a translational correlation length of $\xi = 2.1 \pm 0.2$, and a somewhat larger orientational correla-

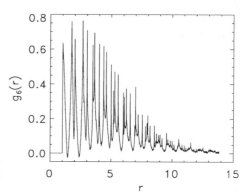

Figure 61. Sixfold bond orientational correlation function for dense random packings of hard disks.

tion length of $\xi_6 = 2.34 \pm 0.10$. These values are roughly comparable to those of the time-averaged WCA liquid at $\rho = 0.82$. It is remarkable that the correlation lengths are similar to those of the dense WCA liquid, even though the correlation functions of DRPs are much more highly structured.

3.   Voronoi Analysis

The Voronoi cell area distribution function for DRPs is shown in Fig. 62 (solid line), together with the separate contributions from ordered and disordered regions (dotted and dashed lines, respectively). In addition to a sharp peak at a cell area of 0.866 (the Voronoi cell area for ideal type A vertices), this distribution has a distinct peak near 0.933, which is the Voronoi cell area for ideal type B and C vertices. This is rather remarkable evidence that the particles in DRPs tend to adopt local arrangements characteristic of ST tilings. Similar features are evident in the Voronoi cell sidelength distribution (Fig. 63), which has a sharp peak at a sidelength of 0.577 (from type A, B, and C vertices), an abrupt "bend" near 0.789 (from type B and C vertices), and a small peak near 1.0 (from type B and D vertices), while the Voronoi cell sidelength distribution for disordered regions (dashed line in Fig. 63) exhibits a distinct peak near 0.75, evidently due to type B and C vertices. Even more striking is the distribution of triangle areas in the Delaunay triangulation, shown in Fig. 64. This distribution has distinct peaks at 0.433 (the area of an equilateral triangle of unit sidelength) and 0.5 (the area of a right triangle). The peak at 0.5 is convincing evidence of a tendency for particles to adopt a local square arrangement. Altogether, then, these distributions provide strong support for the tiling picture of DRP and liquid structure discussed above.

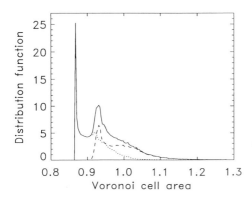

Figure 62. Voronoi cell area distribution functions for the DRP system. The figure shows the contributions from ordered regions (dotted line), disordered regions (dashed line), and the overall distribution (solid line).

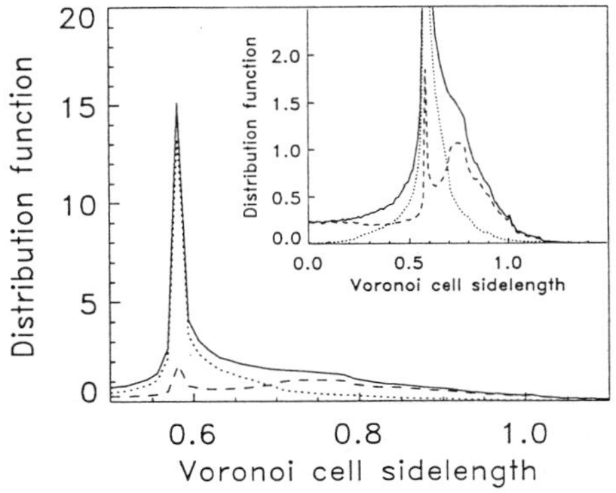

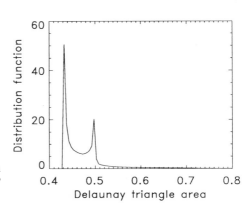

Figure 63. Voronoi cell sidelength distribution functions for the DRP system. The figure shows the contributions from ordered regions (dotted line), disordered regions (dashed line), and the overall distribution (solid line). The inset shows an expanded view.

4. Bond Orientational Order

In Figs. 65–67 we have shown the fourfold, sixfold, and twelvefold bond orientational order parameter distribution functions for DRPs, calculated using only intact bonds. For comparison, we have also shown the corresponding "random" distributions, calculated using randomly chosen bond angles and the same (intact bond) coordination number statistics as the DRP system. As for the WCA system, the distributions for the DRP system have sharp peaks corresponding to particular vertex types (see Table III), features which do not appear in the random distributions. These observations support the tiling description we have developed.

The joint order-parameter distributions $P(|\psi_4|, |\psi_6|)$, $P(|\psi_4|, |\psi_{12}|)$, and $P(|\psi_6|, |\psi_{12}|)$ for the DRP system, calculated using only intact bonds, are shown in Figs. 68a, 69a, and 70a. As for the WCA system, we have

Figure 64. Delaunay triangle area distribution function for the DRP system.

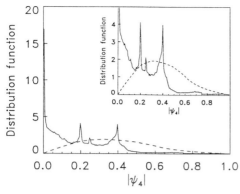

Figure 65. $|\psi_4|$ distribution function (calculated using only intact bonds) for the DRP system. The dashed curve shows the $|\psi_4|$ distribution calculated using randomly chosen bond angles and the same coordination number statistics as the WCA system. The inset shows an expanded view.

subtracted the corresponding random distributions from these distributions to obtain the "rationalized" joint order-parameter distribution functions $P_R(|\psi_4|, |\psi_6|)$, $P_R(|\psi_4|, |\psi_{12}|)$, and $P_R(|\psi_6|, |\psi_{12}|)$, shown in Figs. 68b, 69b, and 70b. As we saw for the WCA system, there are distinct features in these distributions associated with particular vertex types, most notably vertices of types A, B, C, E, F, L, M, and T.

The size distribution of sixfold-ordered regions, n_s, also resembles that of the 2D WCA liquid, being well described by the Fisher droplet model. We fit n_s to Eq. (3.26) and obtained a power law exponent of $\tau_s = 1.35 \pm 0.02$ and a correlation size of $\xi_s = (3 \pm 6) \times 10^2$ (the large uncertainty in ξ_s is due to poor statistics in the tail of n_s). This value of τ_s is similar to the values obtained for the dense time-averaged WCA liquid. The small s part of n_s is shown in Fig. 71, together with the fit to Eq. (3.26). A extended tail (out to $s \sim 600$) is observed in the large s

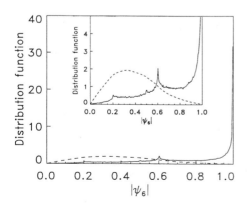

Figure 66. $|\psi_6|$ distribution function (calculated using only intact bonds) for the DRP system. The dashed curve shows the $|\psi_6|$ distribution calculated using randomly chosen bond angles and the same coordination number statistics as the WCA system. The inset shows an expanded view.

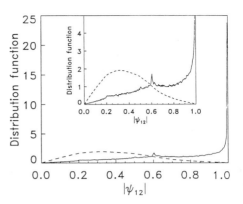

Figure 67. $|\psi_{12}|$ distribution function (calculated using only intact bonds) for the DRP system. The dashed curve shows the $|\psi_{12}|$ distribution calculated using randomly chosen bond angles and the same coordination number statistics as the WCA system. The inset shows an expanded view.

distribution. The statistics of our data are not sufficient to determine whether the DRP system exhibits the type of finite-size effect discussed in Section III.D, however.

As for the WCA system, the fraction of sixfold-ordered particles in DRPs (0.547 ± 0.004) is significantly less than the fraction of six-coordinated particles ($f_6 = 0.7861 \pm 0.0018$), confirming that the sixfold bond orientational order parameter is a more sensitive indicator of local geometrical disorder than is the coordination number. The average size of ordered clusters in DRPs is $\langle s \rangle = 30.4 \pm 0.8$, and the normalized average number of ordered clusters is $N_C/N = 0.0180 \pm 0.0005$. These values are comparable to those measured in the dense WCA liquid near freezing (see Figs. 52 and 53).

V. GEOMETRICAL DEFECT CONDENSATION MODELS FOR TWO-DIMENSIONAL MELTING

A. Introduction

In Section III we described our attempts to characterize the geometrical disorder that is introduced in the melting of a 2D solid in order to identify the mechanism for 2D melting. Perhaps the key conclusion that arose from our simulation studies was that the local structures observed in the dense 2D WCA liquid (and in dense random packings of hard disks) can be usefully described as *polygon packings* or *imperfect tilings*, and that the effect of packing constraints on the local geometry of these dense systems can be embodied in particular tiling rules that condition the local arrangements of polygons. Polygons having four or more sides represent local groups of particles that adopt a nearly square (pentagonal, hexagonal, etc.) arrangement. The crucial structural change associated with 2D

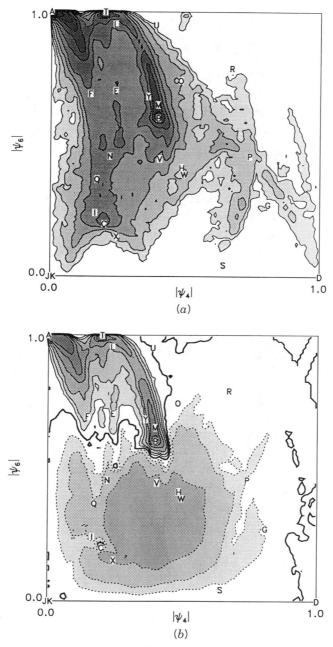

Figure 68. Contour plot of $P(|\psi_4|, |\psi_6|)$ (calculated using only intact bonds) for the DRP system. The letters indicate the approximate locations of the corresponding vertex types in order-parameter space. The contour levels are the same as for Fig. 38. (*a*) Raw distribution. (*b*) Rationalized distribution (see text).

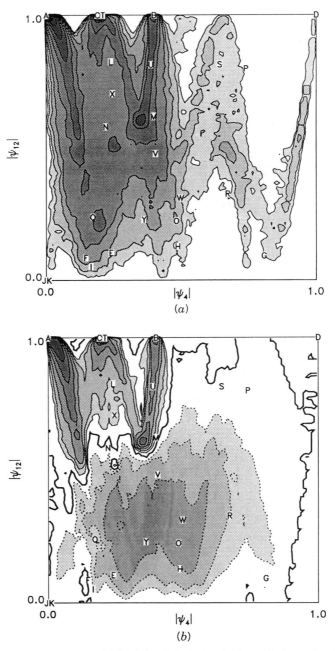

Figure 69. Contour plot of $P(|\psi_4|, |\psi_{12}|)$ (calculated using only intact bonds) for the DRP system. The letters indicate the approximate locations of the corresponding vertex types in order-parameter space. The contour levels are the same as for Fig. 38. (*a*) Raw distribution. (*b*) Rationalized distribution (see text).

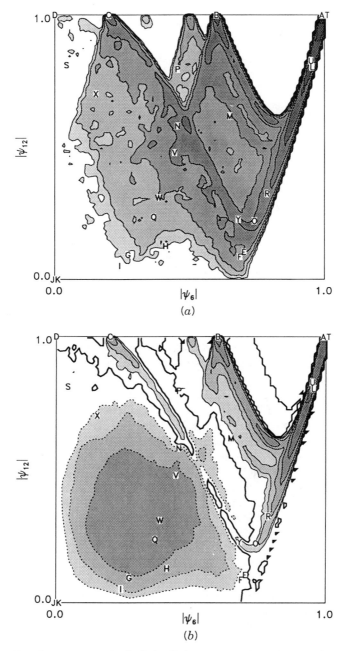

Figure 70. Contour plot of $P(|\psi_6|, |\psi_{12}|)$ (calculated using only intact bonds) for the DRP system. The letters indicate the approximate locations of the corresponding vertex types in order-parameter space. The contour levels are the same as for Fig. 38. (*a*) Raw distribution. (*b*) Rationalized distribution (see text).

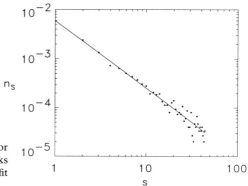

Figure 71. Small s part of n_s for dense random packings of hard disks (bullets). The solid line represents a fit to Eq. (3.26).

melting appears to be the proliferation and condensation of non-three-sided polygons into characteristic tiling-like structures. The volume increase of the WCA system upon melting is directly associated with the creation of polygons having more than three sides.

These observations have led us to propose the following general 2D melting mechanism. We regard large amplitude, local geometrical distortions of the triangular lattice (associated with polygons having four or more sides in our polygon construction) as particle-like excitations, or localized *geometrical defects*. A geometrical defect is essentially a deviation from local triangular packing, and the energy cost of defect creation is the energy cost of disrupting the triangular lattice structure. For hard disks, for example, this energy cost is roughly the volume increase associated with defect creation multiplied by the pressure. In addition, geometrical defects have anisotropic, attractive interactions that cause defects to aggregate into characteristic structures in order to minimize the distortion of polygons from their "ideal" shapes—such "shape" interactions arise from the tendency of polygons to arrange themselves in order to produce minimal long-range distortion of the triangular lattice. In certain relative arrangements, two neighboring geometrical defects disrupt the packing less than two widely separated defects, so there is a net attractive interaction. Such attractive interactions, if sufficiently strong, can lead to a first-order condensation transition from a low temperature phase containing few geometrical defects (solid phase) to a high temperature phase containing a high density of defects (liquid phase). In this transition, which we identify with the melting transition, defects condense into the characteristic defect aggregates observed in the 2D WCA liquid. We will refer to this general mechanism for 2D melting as *geometrical defect condensation*.

The melting model described in this section represents a specific realization of the geometrical defect condensation mechanism, in which the polygon network is regarded as an imperfect tiling, namely a tiling composed of deformable tiles in which tiling faults can occur. In this model the interactions between geometrical defects are conditioned by tiling rules, which are enforced in practice by assigning an energy cost to tiling faults (the strength of a tiling fault, or the "tiling charge," represents the amount by which "ideal" tiles must be deformed in order to fit around a vertex). This causes defects to aggregate into characteristic local structures in order to minimize the tiling charge on surrounding vertices, and thus leads to effective short-range, attractive, anisotropic interactions between defects that make a defect condensation transition possible. We will refer to this model as a *generalized tiling model*. In general, one would also expect long-range interactions between geometrical defects; such long-range interactions are neglected in the generalized tiling model.

In this model, the 2D melting transition is driven by the condensation of geometrical defects. The proliferation of topological defects is a secondary effect of geometrical defect proliferation, and the creation energy of topological defects and their effective interactions are entirely derived from the creation energy and interactions of geometrical defects. Topological defects are quite important, however, as they are responsible for the loss of long-range order upon melting. Also, the increased configuration space volume associated with a dynamic topology makes an important contribution to the configurational entropy of the liquid state. As we shall show, topologically constrained and unconstrained versions of our model have quite different phase diagrams, and it appears that the contribution of topological fluctuations to the thermodynamics of melting cannot be neglected.

In our view, then, the entropy of melting is essentially the configurational entropy of the geometrical defects; roughly speaking, the logarithm of the number of possible tilings at the freezing density, minus the corresponding quantity for the solid at melting. For systems composed of hard particles, our model accounts for the volume change upon melting in a natural way. Because the creation of a defect involves a local dilation, or volume increase, the proliferation of defects associated with the condensation transition results in a finite volume change.

Not surprisingly, similar ideas have been advanced by other researchers, going back at least to the pioneering work of Bernal on liquid structure (see Section IV.A). Inspired by Bernal's work, Collins developed a 2D melting model [209] based on ideal tilings composed of squares and equilateral triangles (ST tilings). We will refer to this model as the Collins model for 2D melting. Collins' original mean-field treatment of this model did not exhibit a phase transition, except under rather special (and somewhat artificial) conditions. More recently, other researchers have extended the Collins model [210–214] and have found evidence for a first-order phase transition, although most of this work involves approximations of questionable validity.

Most of the properties of ST tilings that we will discuss have come from the literature on quasicrystals. As has recently been discussed [215], ST tilings form the basis for random tiling models for quasicrystals. In fact, an experimental example of a 2D quasicrystal based on ST tilings has recently been discovered [216]. This fact explains much of the recent interest in such tilings, but also illustrates clearly why models based on ideal ST tilings cannot describe the 2D liquid accurately. It has been argued [217] that random tilings possess long-range orientational order and quasi-long-range translational order, giving rise to sharp (power law) diffraction peaks having noncrystallographic (in this case twelvefold, or dodecagonal) symmetry. For this reason random tiling models provide a poor description of 2D liquid structure.

Our generalized tiling model, described in detail in Section V.B, can be considered a generalization of the Collins model. We have considered two variants of this generalized tiling model: a *topologically constrained* version based on nonideal tilings (tilings containing tiling faults) in which the topology of the triangular lattice is retained, and a *topologically unconstrained* version in which the topology of the underlying lattice is allowed to vary. We have also studied the contribution of polygons having more than four sides to the properties of the model. The most general form of our model has unconstrained topology and includes polygons with up to six sides.

By studying several versions of our model subject to various constraints, we have attempted to separately assess the contributions of geometrical and topological degrees of freedom to the configurational entropy and phase diagram of this tiling model. As we will show, the most general version of our tiling model is capable of quantitatively reproducing the local geometrical changes observed in 2D melting, and exhibits a phase transition with an entropy comparable to the quasi-universal value $\Delta s \sim 0.3 - 0.5 k_B$. The results are sufficiently encouraging that we believe that we have elucidated a general mechanism for 2D melting, although much work clearly remains to be done. At the very least, we have demonstrated that no serious effort at understanding melting and liquid structure can be made without incorporating local geometrical degrees of freedom.

It is interesting to contrast this melting model with the KTHNY theory. In the KTHNY theory, 2D melting is associated with the unbinding of a dilute gas of thermally generated, bound dislocation pairs, leading to a loss of long-range translational order, followed by an unbinding of the disclination pairs that make up the dislocations, resulting in a loss of long-range orientational order. 2D melting in the KTHNY theory thus occurs via two successive defect-unbinding transitions, with the density of defects remaining nearly constant (and small) through the transition. In our model, 2D melting is viewed as a defect *condensation* process, in which defects simultaneously and suddenly proliferate and aggregate into clusters separating relatively defect-free areas. This scenario seems to better describe the behavior that is observed in computer simulations and analog systems, in which the density of defects increases dramatically near melting, producing a liquid characterized by pronounced spatial inhomogeneity.

There are several choices for the type of defect that is considered to condense in the 2D melting transition. One fairly natural choice would be topological defects (dislocations and/or disclinations). In fact, a topological defect condensation mechanism has been proposed by Weber and Stillinger [180], and simulations of dislocation and disclination systems suggest a topological defect condensation transition (see Section II.E). However, the defect interactions in these simulations are based upon continuum elastic theory, which might be expected to fail at high defect densities, particularly in describing the energetics of extended defect clusters. Also, there has been (as far as we know) no attempt to relate the structure of the disordered phase in these models to liquid structure. The grain boundary melting model of Chui [109–111] can also be regarded as a sort of dislocation condensation model. Other melting models are based on the mechanism of vacancy condensation [99].

Our model differs from the models mentioned above in that we consider *metrical* degrees of freedom (i.e., nearest-neighbor bond angles and bond lengths) to be as important as *topological* degrees of freedom. As discussed in Section III.D, large amplitude geometrical fluctuations from the local triangular lattice "vacuum state" often have no topological signature, but may make a significant contribution to the configurational entropy of the liquid state. Our description of 2D solid and liquid structure in terms of *tilings* or *polygon packings* is a more *primitive* description than a description in terms of dislocations and disclinations, because it contains metrical information in addition to topological information and, hence, gives a more complete picture of the local structure (and, we believe, a better estimate of the entropy of the dense liquid).

B.   General Properties of Random Square-Triangle Tilings

For purposes of illustration, a representative random tiling composed of squares and equilateral triangles (ST tiling) is shown in Fig. 3. This tiling was grown using a nondeterministic growth algorithm, which we describe below. As mentioned in Section III.D, these tilings are composed of only four types of vertex, Types A, B, C, and D in Fig. 21, which embody the tiling "rules" for ST tilings. The local constraints on the arrangements of tiles represented by these tiling rules give rise to extended correlations, which are apparent in the characteristic texture of the ST tiling shown in Fig. 3. Such constraints are also responsible for the long-range orientational and quasi-long-range translational order that has been conjectured to be present in such tilings [217].

ST tilings can be characterized by the ratio of the number of squares to the number of equilateral triangles, $R = N_s/N_t$. It is clear that the entropy (i.e., the logarithm of the number of possible arrangements) of ST tilings vanishes in the limits $R = 0$ (all triangles) and $R = \infty$ (all squares). The entropy therefore has a maximum at some intermediate value of R. Kawamura found from his transfer-matrix calculations that the entropy is maximized at $R \simeq 0.45$ [211]. More recently, Leung et al. have shown that the entropy is maximized at $R = \sqrt{3}/4$ [215]. Further, Kawamura estimated that the maximum value of the entropy is $S/N \approx 0.22 k_B$ [211], where N is the number of vertices.

To study the statistics of ST tilings in more detail, we developed an algorithm for growing ST tilings from a seed (either a triangle or a square). At each stage of the growth process, we search the perimeter of the tiling for "forced" locations (positions with exterior angles of 60°, 90°, or 120°, which can only accommodate an equilateral triangle, a square, or two equilateral triangles, respectively), and add tiles at forced locations until no more forced moves are found. We then add a tile at an unforced location on the perimeter. Roughly speaking, we search for the most concave portion of the perimeter, and add a tile there, choosing a square or a triangle with a relative probability governed by the desired R. After an unforced tile has been added, we repeat the search for forced locations.

Using this algorithm, we grew 500 tilings, each consisting of 100,000 tiles, with a target R of $R = \sqrt{3}/4 = 0.433013\ldots$. The algorithm described above is not foolproof, and we encountered dead ends (30° exterior angles) $\sim 2\%$ of the time before we reached 100,000 tiles. The average R obtained was slightly lower than the target value, namely $R = 0.43247 \pm 0.00011$. We measured the probabilities of the four types of allowed vertices, and found $f_A = 0.1694 \pm 0.0002$, $f_B = 0.4685 \pm 0.0002$, $f_C = 0.2633 \pm 0.0005$ and $f_D = 0.0988 \pm 0.0002$. Interestingly, the probabilities of type B and C vertices show the same asymmetry that was observed in the WCA and DRP systems (see Table VII), with type B vertices almost twice as numerous as type C vertices.

C. The Collins Model for Two-Dimensional Melting

Inspired by the work of Bernal (see Section IV.A), Collins in 1964 proposed a simple model for 2D melting, in which the 2D liquid is regarded as a random tiling of equilateral triangles and squares, and the 2D solid is regarded as a periodic array of equilateral triangles. In this model, the creation of a square has an associated energy cost; however, the entropy also increases with the number of squares, due to the multiplicity of tilings that are possible with a finite ratio of squares to triangles, and so there exists the possibility of an entropy-driven phase transition at a sufficiently high temperature.

In his mean-field treatment of this model, Collins found no evidence for a phase transition except in certain special cases. However, many years later, Kawamura [210, 211], Jing et al. [212], and Yi and Guo [213, 214] reexamined the Collins model and found evidence for a first-order phase transition. More recently, ST tilings have been the subject of renewed interest in the context of random tiling models for dodecagonal quasicrystals [215, 217]. The fact that ST tilings are used as models for quasicrystals illustrates a major weakness of melting models based on ideal ST tilings. As discussed above, even random ST tilings probably possess long-range orientational and quasi-long-range translational order [215, 217], manifested by the appearance of Bragg peaks in scattering experiments. A liquid, on the other hand, has short-ranged orientational and translational order. Thus, ideal ST tilings are inadequate models for the liquid state, and the Collins model does not describe the structure of the liquid state realistically. Nevertheless, the properties of this model can provide guidance in the formulation of more realistic tiling models for melting.

We will restrict the present discussion to the results of Kawamura [210, 211], who has given the most rigorous treatment of the Collins model. Kawamura used the transfer-matrix approach to obtain the thermodynamic properties of the Collins model, finding a first-order phase transition with an associated volume change of $\Delta v / v_s \approx 0.07$ and an entropy change of $\Delta S / N \approx 0.22 k_B$. Although this model provides an unrealistic description of liquid structure, Kawamura's results are intriguing in that they show how the degree of cooperativity implicit in the tiling rules can give rise to a first-order phase transition. As Kawamura put it [210]:

> Strong correlation effect due to the short-range repulsion gives the topo-logical disorder highly *cooperative nature* and leads to the *first-order transition* from the crystalline state to the highly random state. In other words, topological disorder appears neither gradually nor individually, but quite abruptly and collectively.

As we show below, the generalized tiling model we have developed exhibits very similar cooperative features that in general lead to a first-order transition.

D. Formulation of the Generalized Tiling Model for Two-Dimensional Melting

The tiling models discussed in the previous section suffer from a number of shortcomings, the most serious of which are related to their unrealistic treatment of liquid structure. In an attempt to improve on the models based on ideal ST tilings, we have formulated a generalized tiling model in which tiling faults are allowed. This introduces another energy into the model (the energy required to create a tiling fault). Another generalization that we have considered is the inclusion of tiles having more than four sides. We have also studied variants of the model having both constrained and unconstrained topologies. As Kawamura has shown [210, 211], ideal ST tilings can be mapped onto a triangular lattice, and thus have the topology of a triangular lattice. In our discussion of generalized tiling models, we will retain the concept of an underlying bond network, and will distinguish between topologically constrained models, for which the underlying bond network is that of a triangular lattice, and topologically unconstrained models, for which the topology of the underlying bond network is allowed to vary. As we shall see, the removal of topological constraints is crucial for reproducing the thermodynamics of the 2D melting transition.

The generalized model we have developed is described by the following Hamiltonian:

$$H = E_c \sum_{i=1}^{N_{pol}} \left(\frac{\Delta A_{p i}}{\Delta A_4} \right) + E_d \sum_{\alpha=1}^{N} c_\alpha^2 \qquad (5.1)$$

where N_{pol} is the number of polygons, p_i is the number of sides of the ith polygon, N is the number of vertices, and

$$\Delta A_p = A_p - (p - 2) A_3 \qquad (5.2)$$

is the excess area of a regular p-sided polygon of unit sidelength relative to $p - 2$ equilateral triangles of unit sidelength. The first term in the Hamiltonian represents the energy cost of creating a regular p-sided polygon from $p - 2$ equilateral triangles in the presence of an external pressure. Note that

$$A_p = \frac{p}{4} \tan\left[\pi \left(\frac{1}{2} - \frac{1}{p} \right) \right] \qquad (5.3)$$

The tiling charge c_α of vertex α is defined as in Section III.D,

$$c_\alpha = 6 \left[\sum_{j=1}^{n_\alpha} \left(\frac{1}{2} - \frac{1}{p_j} \right) - 1 \right] \qquad (5.4)$$

where the sum runs over the n_α polygons which impinge on vertex α. The tiling charge simply measures the deviation of the sum of ideal polygon angles around the vertex from 360°, in units of 60°. Effectively this is a measure of how much the polygons must be distorted in order to fit around the vertex. In the work described here, we have only considered tilings composed of polygons having six or fewer sides. In this case, the smallest increment, or "quantum," of tiling charge is $1/10$ (or 6°).

In reality there are additional terms in the Hamiltonian corresponding to various constraints that we place on the model: for instance, we assign an infinite energy to configurations containing tiles having more than the maximum allowed number of sides (in our work, the maximum number of sides, p_{max}, ranges from 4 to 6). In the case of constrained topology, we assign infinite energy to those configurations of tiles which cannot be mapped onto a triangular lattice. Even in the case of "unconstrained" topology we have certain constraints: a bond cannot connect a given vertex to itself, and two vertices cannot be connected by more than one bond (these constraints arise from the requirement that the underlying bond network be a triangulation).

The mapping of a given arrangement of polygons onto a bond network is, in principle, straightforward: the bond network divides the plane into triangular areas (analogous to the Delaunay construction described in Section III.D), with some bonds being "broken" and others intact. Bonds internal to polygons having four or more sides are considered to be broken, while the bonds that form the edges of the polygons are considered intact. Since each bond has two states, we can assign a bond spin variable σ to each bond, which takes on the value 0 for intact bonds and 1 for broken bonds. The model thus bears a generic resemblance to Ising, lattice gas, or bond dilution models. We will have more to say about these analogies in the discussion to follow. In the topologically unconstrained case, there is an additional dynamical variable that specifies the connectivity of the bond network, namely the *connectivity matrix* $G_{\alpha\beta}$, which is 1 if vertices α and β are connected by a bond and 0 if not. The most general version of the model is thus a type of lattice-gas model defined on the bonds of a dynamically triangulated lattice.

For polygons having four or more sides, there are several ways of placing the broken bonds, which are shown in Fig. 72. Requiring the bond network to have the topology of a triangular lattice forces a

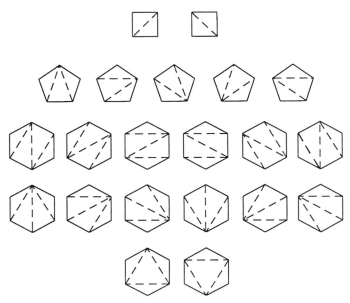

Figure 72. Possible arrangements of internal bonds for four-, five-, and six-sided polygons.

particular choice for the placement of the broken bonds. Of course, there are many configurations of polygons that cannot be mapped onto a triangular lattice. These configurations are not in the accessible region of phase space for topologically constrained versions of our model. As a consequence, the entropy of the topologically constrained model is always less than or equal to that of the topologically unconstrained model.

There are a variety of quantities that can be taken as order parameters (or disorder parameters) in our model. We view the melting transition as a condensation of geometrical defects (polygons having more than three sides). Thus, it is appropriate to take the populations of the various types of polygons as order parameters:

$$\zeta_p = \frac{1}{2N} \left\langle \sum_{i=1}^{N_{\text{pol}}} \delta_{p,p_i} \right\rangle \tag{5.5}$$

where δ is a Kroneger delta function, and $\langle \ldots \rangle$ denotes an ensemble average. This normalization is chosen so that $\zeta_3 = 1$ for a perfectly ordered system (a triangular lattice with no broken bonds). A closely related (dis)order parameter is the fraction of broken bonds,

$$\phi = \frac{1}{3N} \left\langle \sum_{j=1}^{3N} \sigma_k \right\rangle = \frac{2}{3} \sum_{p=4}^{p_{max}} (p - 3)\zeta_p \qquad (5.6)$$

Other measures of order (or disorder) in this model include the fraction of charge c vertices and the probabilities of various types of vertex, which provide a sensitive probe of the local geometry of the model. The topology of the bond network is characterized by the statistics of the vertex coordination numbers.

In the case of constrained topology, with $p_{max} = 4$, the Hamiltonian for our model can be written

$$H = E_c \sum_{j=1}^{3N} \sigma_k + E_d \sum_{\alpha=1}^{N} c_\alpha^2 + E_r \sum_{(k,l)} \sigma_k \sigma_l \qquad (5.7)$$

where the first sum ranges over the $3N$ bonds in the lattice and the third sum runs over nearest-neighbor pairs of bonds. The bond spin variables σ_k reside on a Kagomé lattice (the lattice formed by bond centers in a triangular lattice). The vertex tiling charge c_α is related to the bond spin variables by

$$c_\alpha = \frac{1}{2} \left(-\sum_{m=1}^{6} \sigma_m + \sum_{n=1}^{6} \sigma_n \right) \qquad (5.8)$$

where the two sums range over the first and second shells of bonds surrounding vertex α, respectively. The third term in the Hamiltonian, which runs over nearest-neighbor pairs of bonds, enforces the constraint $p_{max} = 4$ if we make the choice $E_r = +\infty$. This term prevents the breaking of any pair of neighboring bonds. Although the Hamiltonian has a simple form when written in terms of vertex changes, the effective bond spin interactions contained in the second term in the Hamiltonian are extended. The energy of a given bond depends on the state of 28 neighboring bonds, with a complicated template of ferromagnetic and antiferromagnetic interactions. This model is effectively a very complicated Ising or lattice-gas model on a Kagomé lattice, with up to sixth-neighbor interactions. A preliminary account of the properties of this variant of our model has appeared elsewhere [218]. In the limit $E_d \to \infty$, this model reduces to the ideal ST tiling model of Kawamura [210, 211].

The properties of the generalized tiling model depend on two parameters, a dimensionless temperature $t = k_B T/E_c$, and a dimensionless tiling fault energy $r = E_d/E_c$. We have obtained the statistical and thermodynamic properties of our model as a function of t and r from Metropolis Monte Carlo (MC) simulations. The basic variables used in the MC simulations are the bond spin variables, which can take on the values 0 (intact) or 1 (broken), and connectivity variables, which keep track of the connectivity of the bond (or polygon) network. A MC move consists of an attempt to change the state of a bond, with the transition probability governed by the Hamiltonian of Eq. (5.7). A MC sweep consists of one such "bond-altering" move per bond. Several examples of possible bond-altering moves are shown in Fig. 73 (move (e) is not allowed for $p_{max} \leq 6$). In the topologically unconstrained case, the MC sweep also contains a "bond-flipping" sweep, in which randomly chosen broken

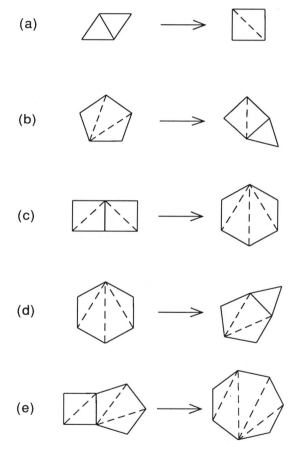

Figure 73. Examples of possible bond altering moves in the Monte Carlo procedure discussed in the text. The solid lines represent intact bonds and the dotted lines broken bonds. Move (e) would not be allowed for $p_{max} \leq 6$.

bonds have their connectivity changed. Several examples of such bond-flipping moves are shown in Fig. 74. Since a bond flip results in no energy change, a randomly chosen broken bond is flipped with unit probability, and a sweep consists of one such flip per broken bond. It can be shown that any triangulation can be transformed into any other by a finite sequence of bond flips [219], so that all possible bond network topologies are accessible. Various constraints were enforced in the MC simulations by simply rejecting attempted moves that would produce configurations violating the constraints.

We used Monte Carlo simulations to obtain the statistical and thermodynamic properties of topologically constrained and unconstrained versions of our model, for $p_{max} = 4$ and $p_{max} = 6$. For each of these versions of the generalized tiling model, we carried out heating runs (and a more

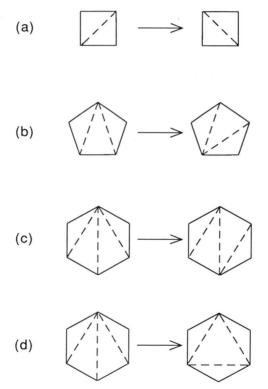

Figure 74. Examples of possible bond flipping moves in the Monte Carlo procedure discussed in the text. The solid lines represent intact bonds and the dotted lines broken bonds.

limited number of cooling runs) for various values of r. Each heating (cooling) run consisted of a series of simulations for increasing (decreasing) t, with each simulation (except the first simulation for heating runs) started from a well-thermalized configuration from the previous value of t. Each simulation of the topologically constrained systems consisted of 20,000 MC sweeps, of which the first 10,000 were discarded and the second 10,000 were used in computing averages. In the topologically unconstrained case, we found that much longer simulations were required to obtain well-converged results. This indicates that the characteristic timescale for topology-changing fluctuations of the bond network is considerably longer than that associated with fluctuations in the bond spin variables. In this case, each simulation consisted of 100,000 MC sweeps, of which the first 50,000 were discarded and the second 50,000 used in computing averages. For the topologically constrained model with $p_{max} = 4$, we studied systems containing between 56 and 8064 vertices. In the other three cases, only 896-vertex systems were simulated. Periodic boundary conditions were used in all simulations. The results of these simulations are described in the following two sections.

E. Statistical and Thermodynamic Properties of the Generalized Tiling Model

In Figs. 75 and 76 we have shown the order parameter (ϕ) and internal energy per vertex $(u = \langle H \rangle / N)$ measured in MC simulations of the topologically constrained generalized tiling model for $p_{max} = 4$, $N = 3584$.

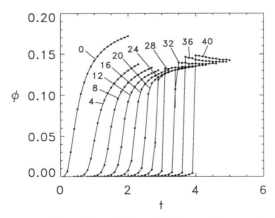

Figure 75. ϕ versus t from Monte Carlo simulations of the topologically constrained generalized tiling model, for $p_{max} = 4$, $N = 3584$. Each series of simulations is labeled by its corresponding r value.

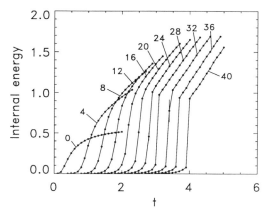

Figure 76. Internal energy versus t from Monte Carlo simulations of the topologically constrained generalized tiling model, for $p_{max} = 4$, $N = 3584$. Each series of simulations is labeled by its corresponding r-value.

Each curve represents a series of simulations for increasing t at a given r, and is labeled by the corresponding value of r. For $r < 20$, the curves are continuous, with no evidence of a phase transition, while for $r > 20$, both ϕ and the internal energy exhibit clear discontinuities indicative of a first-order transition. This version of the model evidently has a critical point at $r_C \approx 20$. This conclusion is strengthened by a comparison of the internal energy measured in heating and cooling runs for various values of r. We observe no hysteresis for $r = 20$, but see an increasingly large amount of hysteresis for larger r, indicative of a first-order transition (in fact, for $r \geq 32$, the system never returns to the ordered state upon cooling). Of course, hysteresis can also be associated with a second-order phase transition (due to critical slowing down), but the appearance of a tight hysteresis loop with clear discontinuities both on heating and cooling (Fig. 77) seems to rule out a second-order phase transition.

In Figs. 78 and 79 we have shown the fraction of broken bonds and internal energy of the topologically constrained version of our model with $p_{max} = 6$, for an 896-vertex system. The overall behavior is quite similar to that of the topologically constrained model with $p_{max} = 4$. The critical point appears at a somewhat smaller value of r ($r_C \approx 16$) and the transition temperatures are shifted downward slightly. This is the behavior we expect; the additional configurational degrees of freedom contributed by five- and six-sided polygons increase the entropy of the disordered (high temperature) phase, and so the transition takes place at a lower temperature (roughly speaking, the U and TS terms in the free

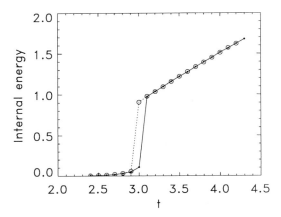

Figure 77. Internal energy versus t from Monte Carlo simulations of the topologically constrained generalized tiling model, for $p_{max} = 4$, $r = 28$, $N = 3584$, obtained on heating (bullets) and cooling (circles).

energy become equal at a lower temperature). Based on our observations, we have constructed a (somewhat schematic) phase diagram for the topologically constrained tiling model, shown in Fig. 80a.

The order parameter and internal energy of the topologically unconstrained tiling model for $p_{max} = 4$ and $N = 896$ are shown in Figs. 81 and 82, respectively. These data are noisier than those for the topologically

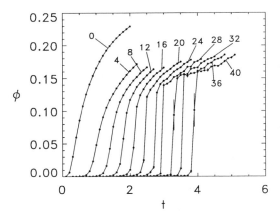

Figure 78. ϕ versus t from Monte Carlo simulations of the topologically constrained generalized tiling model, for $p_{max} = 6$, $N = 896$. Each series of simulations is labeled by its corresponding r-value.

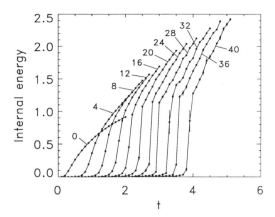

Figure 79. Internal energy versus t from Monte Carlo simulations of the topologically constrained generalized tiling model, for $p_{max} = 6$, $N = 896$. Each series of simulations is labeled by its corresponding r-value.

constrained system, but there is good evidence (particularly in the internal energy data) for a first-order transition for $r \geq 0.3$. Notice that a transition occurs at much smaller values of r than in the topologically constrained case. This is presumably due to the fact that topological fluctuations greatly increase the configurational entropy of the disordered phase. We see no evidence for a phase transition for $r = 0$, so there is a critical point between $r = 0$ and $r = 0.3$.

Finally, in Figs. 83 and 84 we have shown the order parameter and internal energy for the topologically unconstrained tiling model for $p_{max} = 6$ and $N = 896$. These data are qualitatively similar to those for $p_{max} = 4$, with a critical point between $r = 0$ and $r = 0.3$, and first-order transitions for $r \geq 0.3$. Comparison of the internal energy measured in heating and cooling runs for various values of r (e.g., Fig. 85) provides additional evidence for a first-order transition. No hysteresis is observed for $r = 0$, consistent with our conclusion that there is no phase transition at this value of r. For all other values of r, however, there is pronounced hysteresis, and system never returns to the ordered state upon cooling. As for the constrained system, the transition temperatures are slightly lower for $p_{max} = 6$ than for $p_{max} = 4$. A schematic phase diagram for the topologically unconstrained tiling model is shown in Fig. 80b.

The appearance of a critical point in the phase diagram of our model would, at first glance, appear to be a disaster for our proposed melting model. The fact that the solid and liquid phases have different symmetries (even in 2D) excludes the possibility of a critical point for melting

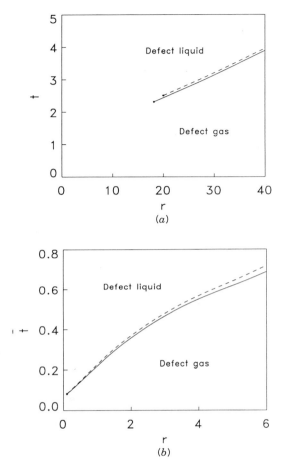

Figure 80. Phase diagram for the generalized timing model, showing the first-order transition lines for $p_{max} = 4$ (dashed line) and $p_{max} = 6$ (solid line), and the approximate locations of the critical points (bullets). (*a*) Topologically constrained model. (*b*) Topologically unconstrained model.

(various types of multicritical point are not excluded, however). On the other hand, our model can have a critical point, because this model is formulated in an abstract configuration space defined by *connectivity* variables specifying the topology of the bond network, and *lattice-gas* variables which describe the state (intact or broken) of each bond. The model makes no reference to actual spatial configurations of particles,

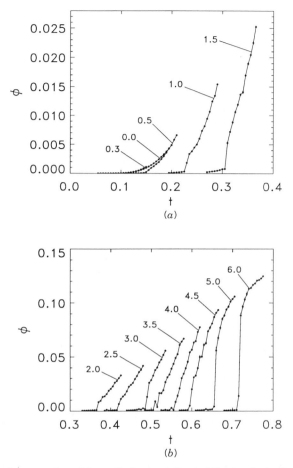

Figure 81. ϕ versus t from Monte Carlo simulations of the topologically unconstrained generalized tiling model, for $p_{max} = 4$, $N = 896$. Each series of simulations is labeled by its corresponding r-value. (a) Small r. (b) Large r.

and so contains no information about spatial symmetries. In the generalized tiling model, there is no symmetry difference between the high and low temperature phases, and so a critical point is allowed. A more realistic model would have to be formulated in real space. Such a real-space geometrical defect condensation model would, in the topologically unconstrained case, exhibit a symmetry change between the low and high temperature phases, because of the proliferation of topological

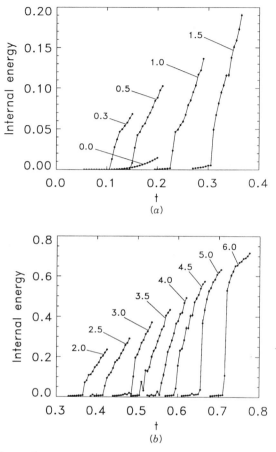

Figure 82. Internal energy versus t from Monte Carlo simulations of the topologically unconstrained generalized tiling model, for $p_{max} = 4$, $N = 896$. Each series of simulations is labeled by its corresponding r-value. (a) Small r. (b) Large r.

defects, and so a critical point would never be observed. The generalized tiling model can be considered an approximation to such a hypothetical real-space defect condensation model. The basic question is whether the generalized tiling model treats the defect interactions accurately, so that defect configurations are weighted properly. Clearly, the *long-range* interactions between defects are treated quite unrealistically in the generalized tiling model (this model contains only short-range interactions between defects). As a result, the entropy of the liquid state is

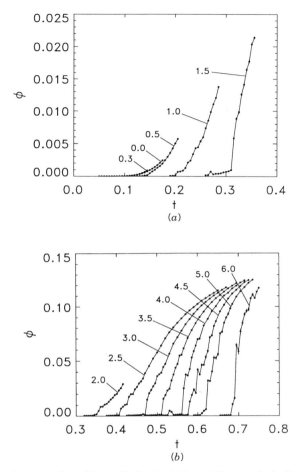

Figure 83. ϕ versus t from Monte Carlo simulations of the topologically unconstrained generalized tiling model, for $p_{max} = 6$, $N = 896$. Each series of simulations is labeled by its corresponding r-value. (a) Small r. (b) Large r.

overestimated by the topologically unconstrained generalized tiling model, whereas the topologically constrained model severely underestimates the entropy of the liquid state because of the exclusion of topological degrees of freedom. Nevertheless, that fact that a phase transition is observed in both versions of the model indicates that the defect condensation mechanism is robust, and that the qualitative features of the 2D melting transition are reasonably well reproduced by our model.

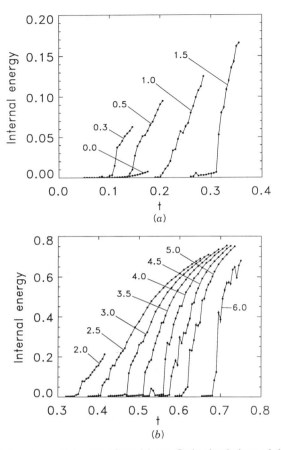

Figure 84. Internal energy versus t from Monte Carlo simulations of the topologically unconstrained generalized tiling model, for $p_{max} = 6$, $N = 896$. Each series of simulations is labeled by its corresponding r-value. (a) Small r. (b) Large r.

We have made a detailed comparison of the statistical geometry (order parameters, coordination numbers, vertex type and tiling charge statistics) of the generalized tiling model for various r with that of the WCA and DRP systems. The best overall agreement with the WCA system is found in the range $2 \leq r \leq 6$. The topologically constrained tiling model has no phase transition in this range of r, implying that this model severely underestimates the configurational entropy of the liquid phase. The topologically unconstrained tiling model, on the other hand, exhibits a first-order phase transition in this range, and so appears to provide a better estimate of the entropy of the 2D liquid.

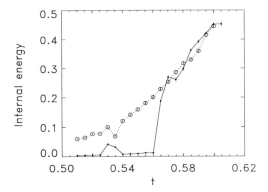

Figure 85. Internal energy versus t from Monte Carlo simulations of the topologically unconstrained generalized tiling model, for $p_{max} = 6$, $r = 4$, $N = 896$, obtained on heating (bullets) and cooling (circles).

In Tables IV–VIII we have compared the statistics of the topologically constrained and unconstrained generalized tiling models for $r = 4$, $p_{max} = 6$ (denoted by GTM1 and GTM2, respectively) with those of the WCA and DRP systems. The density of the WCA system and the temperatures of the two variants of the generalized tiling model have been chosen so that the order parameters of all three systems are nearly equal to the order parameter of the DRP system ($\phi = 0.1169 \pm 0.0009$). We have chosen to show the statistics for $r = 4$ because this is the value of r at which the statistics of the generalized tiling model are in the best overall agreement with those of the WCA system.

Examination of these tables reveals that, while there is certainly not perfect agreement between the two versions of the generalized tiling model and the WCA and DRP systems, many of the same trends are observed. Both versions of the model predict significantly (~20%) more four-sided polygons and significantly fewer five- and six-sided polygons than are observed in the WCA system, even though the predicted number of three-sided polygons is about right. The deviation of the polygon statistics of the two versions of the model from those of the DRP system are even more pronounced. The topologically unconstrained model exhibits significantly fewer six-coordinated particles (counting all bonds) than either the WCA or DRP systems, although the model correctly predicts that five-coordinated particles are more prevalent than seven-coordinated particles. *The intact-bond* coordination number statistics for the two versions of the tiling model are in better agreement with those of the WCA and DRP systems than are the overall coordination number statistics. In this case, the topologically constrained model shows somewhat better agreement with the WCA and DRP systems than does the topologically unconstrained model.

Because the tiling model exhibits significantly fewer five- and six-sided polygons than the WCA and DRP systems, there are correspondingly fewer vertices of types L–Y in the model. However, most of the trends observed in the vertex statistics of the WCA and DRP systems are also seen in the tiling model. For example, type B vertices are more probable than type C vertices in the WCA and DRP systems, an asymmetry that is reflected in both versions of the tiling model (as noted above, random ST tilings exhibit the same asymmetry); however, both vertex types are much more prevalent in the tiling model than in either the WCA or DRP systems, due to the larger number of four-sided polygons. The fractions of vertex types E–I predicted by the topologically unconstrained tiling model are in very good agreement with the corresponding values for the WCA system. Although this level of agreement is somewhat fortuitous, it at least shows that the relative probabilities of these vertex types are predicted correctly, and in particular that type E vertices are slightly more prevalent than type F vertices in both the tiling model and the WCA system. As mentioned before, the tiling model exhibits systematically lower probabilities of vertex types L–Y than do the WCA and DRP systems. However, the topologically unconstrained model does, generally speaking, predict the correct *sequence* of probabilities, and the relative probabilities of type L, M, N, and P vertices predicted by the topologically unconstrained tiling model are in good agreement with those of the WCA and DRP systems. Finally, the generalized tiling model exhibits significantly more charge zero vertices than does the WCA system, although the fractions of charge $\pm 1/2$ vertices for the topologically constrained tiling model are nearly the same as for the WCA system. As expected, the tiling model exhibits a lower incidence of those values of the tiling charge that are associated with five-sided polygons ($-2/10$, $+3/10$, etc.) than does the WCA system, but the relative probabilities show similar trends. Although there are many detailed differences between the generalized tiling model and the WCA and DRP systems, the overall agreement is encouraging, and suggests that we have correctly taken into account the factors that govern the local geometry of dense 2D liquids.

It is perhaps more important to compare the properties of the phase transition in the tiling model with those of the melting transition in the WCA system. The WCA system exhibits an order parameter discontinuity upon melting of $\Delta\phi \approx 0.085$, and an entropy change per particle of $\Delta s \approx 0.39 k_B$. On the other hand, for the topologically unconstrained generalized tiling model with $p_{\max} = 6$ and $r = 4$, the order parameter discontinuity associated with the phase transition is $\Delta \phi \approx 0.03$, and the entropy change per vertex is $\Delta s = \Delta u / t_C \approx 0.35 k_B$. Although the entropy change for the tiling model is comparable to that for the WCA system, the order parameter discontinuity in the tiling model is smaller (by more than a factor of two) than that of the WCA system. There are various possible explanations for this discrepancy, but we think the most likely one is that the topology of the bond network is underconstrained, which leads to an overestimation of the entropy of the disordered phase. This is apparently due to the neglect of long-range interactions between defects in our model (the Hamiltonian of Eq. (5.7) contains only short-range interactions). We will give more evidence for this hypothesis in the next section.

F. Structural Properties of the Generalized Tiling Model

As discussed above, the generalized tiling model is formulated in an abstract configuration space that contains only lattice-gas and connectivity variables; the Hamiltonian for this model does not depend on spatial variables, nor do the Monte Carlo simulations described above yield spatial configurations of particles or tiles. In order to make structural comparisons between our model and real liquids and solids, we developed a procedure for mapping configurations of our model from the abstract configuration space of the model into real two-dimensional space. The basic idea is to treat polygons as elastic tiles having a preferred shape and sidelength, and to assign an elastic potential energy to the polygon network,

$$U(\{r_{\alpha\beta}\}, \{\theta_{\alpha\beta\gamma}\}) = \frac{1}{2} K_r \sum_{(\alpha,\beta)} (r_{\alpha\beta} - a)^2 + \frac{1}{2} K_\theta \sum_{(\alpha,\beta,\gamma)} (\theta_{\alpha\beta\gamma} - \theta_p)^2$$

(5.9)

where the first sum ranges over all intact bonds and the second sum ranges over all polygon angles. $r_{\alpha\beta}$ is the separation of vertices α and β, a is the equilibrium bond length, and $\theta_{\alpha\beta\gamma}$ is the polygon angle defined by vertices α, β, and γ. The "ideal" polygon angle θ_p is given by

$$\theta_p = \pi \left(1 - \frac{2}{p}\right)$$

(5.10)

where p is the number of polygon sides.

Once we have assigned an elastic energy to the polygon network, we can allow the vertices to move in order to minimize the elastic energy. For the topologically constrained model, we can use a perfect triangular lattice as the starting configuration. The elastic energy (Eq. (5.9)) is then minimized with respect to the vertex coordinates $\mathbf{r}_\alpha$ using a conjugate gradient minimization routine. We will refer to this as a "relaxation" of the elastic network.

The topologically unconstrained case is more problematic, because the triangular lattice is not a good starting point for the relaxation procedure in the later stages of a MC simulation, when the topology of the bond network may be quite different than that of a triangular lattice. In this case we must start at the very beginning of a MC run, at which point the bond network still has the topology of a triangular lattice, and periodically relax the bond network over the course of the MC run, using the previous relaxed configuration as the starting point for each successive relaxation. Even if this is done, however, there is no guarantee that the process will produce meaningful results in the later stages of a simulation.

Assuming that such a relaxation procedure has been successfully carried out, we can analyze the resulting spatial structures using the same types of analysis that were used in Section III and IV. We have in fact done this for the topologically constrained model. Some typical results are shown in Fig. 86, namely the size distribution of sixfold-ordered clusters, n_s, for a 3584-vertex system at $t = 1.4$, $r = 4$, for $p_{max} = 6$. This distribution represents an average over 100 configurations from a 10,000-sweep MC run. The fraction of broken bonds at this value of r and t is $\phi = 0.11690 \pm 0.00011$, which is comparable to the order parameter of the time-averaged WCA system at $\rho = 0.83$ and to that of the DRP system. The solid line represents a fit to Eq. (3.26), which yields a power law

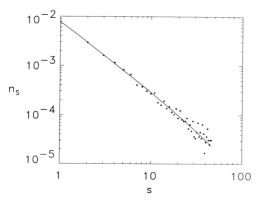

Figure 86. Small s part of n_s calculated from relaxed configurations of the topologically constrained generalized tiling model for $p_{max} = 6$, $r = 4$, $t = 1.4$ (bullets). The solid line represents a fit to Eq. (3.26).

exponent $\tau_s = 1.394 \pm 0.013$ and a correlation size of $\xi_s = (1.0 \pm 0.4) \times 10^2$. This value of τ_s is comparable to the values obtained for the WCA liquid near freezing (see Fig. 48) and for the DRP system. This supports our conjecture that the form of n_s is primarily determined by the geometry of the boundaries between sixfold-ordered domains, and is not sensitive to the topology of the underlying bond network.

Not surprisingly, we have found that, like ideal ST tilings, relaxed configurations of the topologically constrained generalized tiling model possess long-range bond orientational order and quasi-long-range translational order. Clearly, then, this version of our model does not properly describe the structure of a 2D liquid. Thus, we have applied the relaxation procedure to the topologically unconstrained version of the model and have studied the resulting configurations.

In Fig. 87 we have shown the result of our mapping procedure for the topologically unconstrained model with $p_{max} = 6$ at $r = 4$, $t = 0.635$ (this value of t was chosen so that the order parameter ϕ was nearly equal to that of the WCA system at $\rho = 0.85$), showing the dynamic evolution out to 12,000 Monte Carlo sweeps.

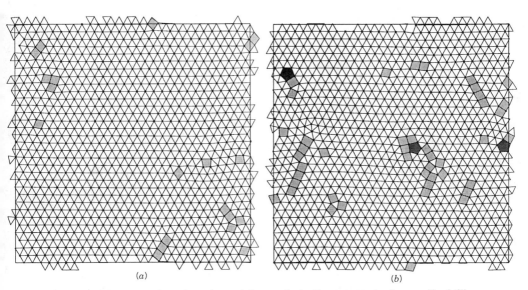

(a)   (b)

Figure 87. Relaxed configurations of the topologically unconstrained generalized tiling model for $p_{max} = 6$, $r = 4$, $t = 0.635$: (a) sweep 500; (b) sweep 1000; (c) sweep 1500; (d) sweep 2000; (e) sweep 5000; (f) sweep 12,000.

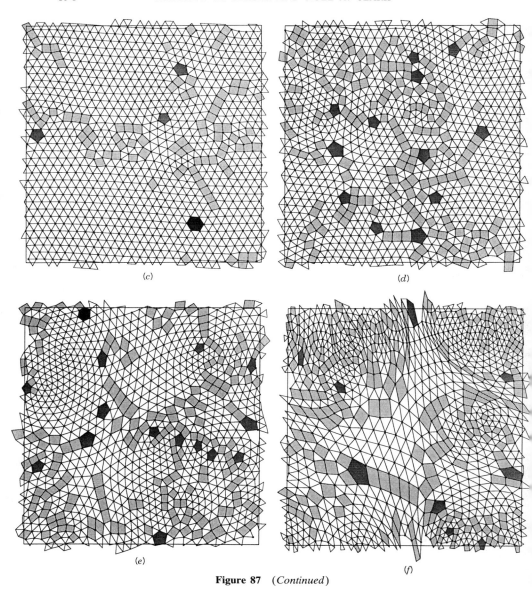

Figure 87 (*Continued*)

In the early stages of the evolution, geometrical defects proliferate, forming structures reminiscent of the dense WCA liquid. The subsequent evolution appears to be dominated by large-scale changes in topology. By sweep 12,000, the system of polygons has a highly deformed topology accompanied by large density fluctuations, features that bear little resemblance to the WCA liquid. This unphysical situation appears to be related to the formation of isolated disclinations (two such disclinations can be seen in the upper left-hand corner of Fig. 87e). This indicates that the long-range interactions between the topological defects are not properly taken into account in our model, and supports our earlier statement that the topology of the model is underconstrained. Further, the elastic energy of the network as a function of the number of sweeps (Fig. 88) shows no signs of leveling off, so that the system does not appear to be converging toward a steady-state structure. As a result, it is not possible to measure equilibrium structural properties for this version of our model.

Quite different behavior is exhibited by the topologically constrained model, as illustrated by Fig. 89, which shows the dynamic evolution of the topologically constrained model out to 12,000 Monte Carlo sweeps, for $p_{max} = 6$, $r = 4$, $t = 1.2$ (at this temperature, the order parameter ϕ is nearly equal to that of the WCA system at $\rho = 0.85$). A steady-state structure develops rather quickly in this version of the model, and large-amplitude density fluctuations such as those observed in the topologically unconstrained model are absent.

To properly estimate the entropy and calculate the structural properties of the liquid phase, we must incorporate long-range interactions between defects into our model. We are currently studying a melting model formulated in real space in an attempt to take into account the long-range interactions, and will report on the results of these studies in a future publication.

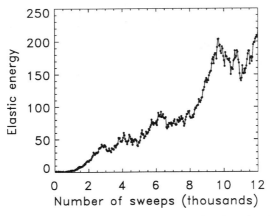

Figure 88. Elastic potential energy versus sweep number for relaxed configurations of the topologically unconstrained generalized tiling model for $p_{max} = 6$, $r = 4$, $t = 0.635$.

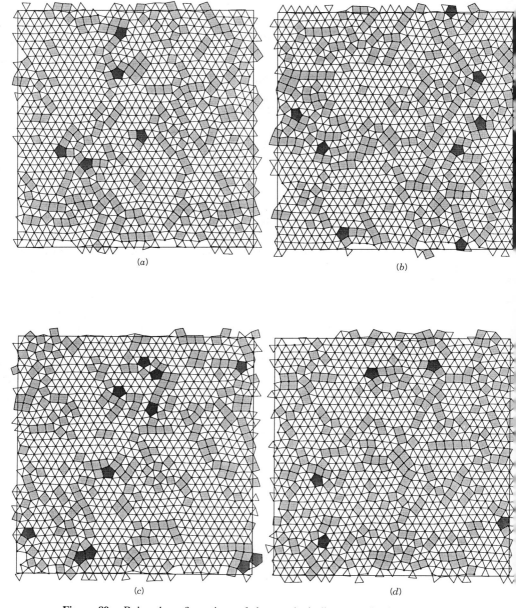

Figure 89. Relaxed configurations of the topologically constrained generalized tiling model for $p_{max} = 6$, $r = 4$, $t = 1.2$: (*a*) sweep 500; (*b*) sweep 1000; (*c*) sweep 1500; (*d*) sweep 2000; (*e*) sweep 5000; (*f*) sweep 12,000.

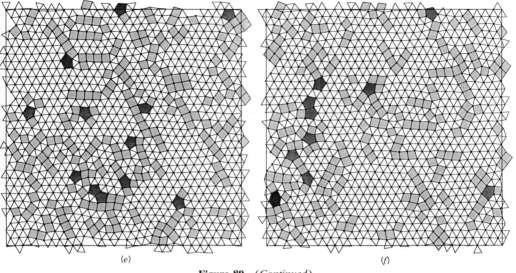

(e)   (f)

Figure 89   (*Continued*)

VI.   CONCLUSION

The primary objective of our research was to identify the mechanism for 2D melting. To this end, we carried out a series of molecular dynamics simulation studies of 2D liquid structure and the 2D melting transition. Based on these studies, we identified *geometrical defect condensation* as the likely mechanism for 2D melting. To investigate whether this is a plausible mechanism for melting, we formulated a *generalized tiling model* that incorporates the essential features of this melting mechanism, and obtained the thermodynamic and statistical properties of this model from Monte Carlo simulations. It is appropriate at this point to summarize what we have learned from our investigations of this model.

The most general (topologically *unconstrained*, with $p_{max} = 6$) version of our generalized tiling model exhibits a first-order phase transition that bears a strong qualitative resemblance to the 2D melting transition, reproducing both the statistical geometry (polygon and vertex statistics) of the 2D liquid and the thermodynamics of the phase transition for a particular choice of parameters in the Hamiltonian. This agreement is certainly not perfect. Although the transition entropy per particle in the model is $\Delta s \approx 0.35 k_B$, which is comparable to the transition entropy of the melting transition in the WCA system, the order-parameter discontinuity upon melting for the tiling model is more than a factor of two smaller than that of the WCA system. Evidently, the topologically unconstrained version of our model is *underconstrained*, so that it overestimates the entropy of the disordered phase. In addition , the statistical geometry of the model differs in detail from that of the WCA system, although the overall agreement is quite encouraging. It is worth pointing out that the tiling model we have formulated is a *specific* realization of the defect condensation melting mechanism (in fact the simplest conceivable realization incorporating the essential physics). Other realizations of this mechanism, with different forms for the Hamiltonian, may give an improved description of the melting transition. Overall, however, we feel that the tiling model provides strong support for the melting mechanism we have proposed.

On the other hand, the topologically *constrained* version of our model appears to severely underestimate the entropy of the disordered phase. As a result, this version of the model exhibits no phase transition in the region of parameter space in which the statistical geometry of the model most closely matches that of the WCA system. However, the topologically constrained model does exhibit a phase transition for relatively large values of the tiling fault energy parameter r, so that a geometrical defect condensation transition does occur in this variant of our model as well, although in a less physically interesting region of parameter space. The topologically constrained version of the model also reproduces the statistical geometry (polygon and vertex statistics) of the WCA and DRP systems reasonably well, and produces a size distribution of ordered clusters that is quite similar to the corresponding distributions for the WCA and DRP systems. Thus, it appears that certain characteristics of dense 2D liquid structure are relatively insensitive to the topology of the underlying bond network.

In our view, 2D melting is a defect condensation transition involving both geometrical and topological defects. The (scalar) order parameter of the transition (the number density of geometrical defects) is unrelated to the symmetry-derived order parameter(s) (Fourier components of the particle density), and thus the global symmetry change associated with the 2D melting transition provides little insight into the underlying mechanism of the transition. In a sense, the global symmetry change is a *side effect* of the defect condensation mechanism that drives the 2D melting transition.

An analogous situation is encountered with the 2D q-state Potts model, which exhibits a first-order transition for $q \geq 3$. On the basis of renormalization group studies of the Potts *lattice-gas* [220–221], the mechanism producing a first-order phase transition for $q \geq 3$ has been identified as *condensation of effective vacancies*, and so the relevant (scalar) order parameter for this transition is the density of effective vacancies, and not the magnetization (which is the order parameter that symmetry would indicate). It is possible that this is a very common situation with regard to first-order phase transitions; symmetry can mislead, as well as enlighten.

Topological degrees of freedom appear to play an essential role in the melting transition. The topologically constrained version of our tiling model (for which the accessible region of configuration space contains only those tilings that can be mapped onto a triangular lattice) fails to reproduce the thermodynamics of the 2D melting transition. Removal of the constraint on topology leads to a substantial increase in the entropy of the disordered phase, and produces a first-order transition in the physically interesting region of parameter space, where the topologically constrained model has no phase transition. However, it is clear that the topologically unconstrained version of our tiling model does not treat long-range interactions between topological (or geometrical) defects correctly. As discussed in Section V, this version of the model appears to be topologically underconstrained, and hence overestimates the entropy of the disordered phase. This is why the entropy of the phase transition in this model is comparable to that measured for the WCA system, even though the order parameter discontinuity is much smaller. Although the local geometry of the 2D liquid is reproduced rather well by our model, the global topology of the model is rather unphysical (see Fig. 87*f*) due to the presence of isolated disclinations. In a real 2D liquid, isolated disclinations are screened by other defects, and one does not observe the exotic topologies exhibited by our model. We believe that a more realistic treatment of the long-range interactions between geometrical defects is critical in improving the agreement between our model and the WCA system, and in enabling the calculation of the pair correlation function of the 2D liquid (and other properties that depend on long-range ordering) from our model. To this end we have begun studying a model which is essentially a real-space formulation of the geometrical defect condensation mechanism.

Packing constraints appear to be responsible for the tiling structure apparent in the dense 2D WCA liquid. This conclusion follows from our observation that dense random packings of hard disks exhibit a very similar type of ordering. The mechanism by which packing constraints get translated into tiling rules is obscure, but may be related to *mechanical stability* (ability to support a compressive stress). Dense random packings of hard disks appear to be mechanically stable in this sense. It is possible that local tiling structures of the type we have identified are the only possible mechanically stable local arrangements of particles. This may also explain why such structures are relatively long-lived in the dense WCA liquid. In any case, this is a possibility that should be explored further.

Clearly, there is a need for further work in a number of areas, especially considering the provocative nature of our conclusions. Ultimately, we must be able to derive the parameters in our phenomenological melting model from the true microscopic Hamiltonian of a 2D system in order to make this model plausible. One possible way of doing this would be to directly calculate the energies of various geometrical defect configurations in systems of particles with realistic pair potentials, using rigid bond constraints to put in geometrical defects by hand. Also, it is obviously quite important to understand how packing constraints get translated into tiling rules. In the short term, the influence of mechanical stability on local structure in dense random packings of hard disks should be explored. In addition, we have not considered the effect of *vibrational* degrees of freedom in our treatment of the melting transition. Because there is a volume change associated with melting in most cases, the mean amplitude of vibrational motion might be expected to change in the transition, and there will be a corresponding contribution to the entropy of melting. This effect will change the phase diagram of our melting model.

We believe that the mechanism for three-dimensional melting is qualitatively similar to the mechanism we have proposed for 2D melting. Indeed, if global symmetry is irrelevant in determining the characteristics of the melting transition, then there is no compelling reason that 3D melting should be qualitatively different from 2D melting.

Acknowledgements

We would like to thank Allen Armstrong, Rainer Malzbender, Paul Beale, Michael Fisher, and Tom Lubensky for their invaluable suggestions and comments. This work was supported by the National Science Foundation under Grant Nos. DMR-8807443 and DMR-9003431, and was partially supported by the National Center for Supercomputing Applications under Grant No. DMR-900011N.

References

1. J. P. Hansen and I. R. McDonald, *Theory of Simple Liquids* (Academic Press, London, 1986).

2. J. D. Bernal, *Nature* **183**, 141 (1959).

3. J. D. Bernal, *Nature* **185**, 68 (1960).

4. J. D. Bernal, *Proc. R. Soc. A* **280**, 299 (1964).

5. J. D. Bernal, in *Liquids: Structure, Properties, Solid Interactions*, T. J. Hughel, Ed. (Elsevier, Amsterdam, 1965) p. 25.

6. C. Lumsden and E. O. Wilson, *Promethioam Fire: Reflections on the Origin of the Mind* (Harvard University Press, Cambridge, 1983).

7. A. R. Ubbelohde, *The Molten State of Matter* (Wiley, Chichester, 1978).

8. R. M. J. Cotterill, E. J. Jensen, and W. D. Kristensen, in *Anharmonic Lattices, Structural Transitions and Melting*, T. Riste, Ed. (Noordhoff, Leiden, 1974) p. 405.

9. R. M. J. Cotterill, *J. Cryst. Growth* **48**, 582 (1980).

10. R. M. J. Cotterill, in *Ordering in Strongly Fluctuating Condensed Matter Systems*, T. Riste, Ed. (Plenum Press, New York, 1980) p. 261.

11. R. M. J. Cotterill, in *The Physics of Superionic Conductors and Electrode Materials*, J. W. Perram, Ed. (Plenum Press, New York, 1983) p. 79.

12. S. Simozar, *Theory of Melting*, Ph.D. thesis, University of Pennsylvania, 1982.

13. D. Frenkel and J. P. McTague, *Annu. Rev. Phys. Chem.* **31**, 491 (1980).

14. L. L. Boyer, *Phase Transitions* **5**, 1 (1985).

15. W. G. Hoover and M. Ross, *Contemp. Phys.* **12**, 339 (1971).

16. M. Baus, *J. Stat. Phys.* **48**, 1129 (1987).

17. A. D. J. Haymet, *Annu. Rev. Phys. Chem.* **38**, 89 (1987).

18. A. D. J. Haymet, *Science* **236**, 1076 (1987).

19. T. V. Ramakrishnan, *Mater. Sci. Forum* **3**, 57 (1985).

20. T. V. Ramakrishnan, *Pramāna* **22**, 365 (1984).

21. G. Venkataraman and D. Sahoo, *Pramāna* **24**, 317 (1985).

22. K. J. Strandburg, *Rev. Mod. Phys.* **60**, 161 (1988).

23. H. Kleinert, *Gauge Fields in Condensed Matter* (World Scientific, Singapore, 1989).

24. S. K. Ma, *Modern Theory of Critical Phenomena* (Benjamin/Cummings, London, 1976).

25. M. N. Barber, *Phys. Rep.* **29**, 1 (1977).

26. D. J. Wallace and R. K. P. Zia, *Rep. Prog. Phys.* **41**, 1 (1978).

27. L. E. Reichl, *A Modern Course in Statistical Physics* (University of Texas Press, Austin, 1980).

28. M. Toda, R. Kubo, and N. Saito, *Statistical Physics I* (Springer-Verlag, Berlin, 1983).

29. P. Pfeuty and G. Toulouse, *Introduction to the Renormalization Group and to Critical Phenomena* (Wiley, London, 1983).

30. M. E. Fisher, in *Critical Phenomena*, F. J. W. Hahne, Ed. (Springer-Verlag, Berlin, 1983) p. 1.

31. D. J. Amit, *Field Theory, the Renormalization Group, and Critical Phenomena* (World Scientific, Singapore, 1984).

32. K. Huang, *Statistical Mechanics* (Wiley, New York, 1987).

33. G. Parisi, *Statistical Field Theory* (Addison-Wesley, Redwood City, 1988).

34. C. Itzykson and J.-M. Drouffe, *Statistical Field Theory* (Cambridge University Press, Cambridge, 1989).

35. F. Bloch, *Z. Phys.* **61**, 206 (1930).

36. R. Peierls, *Helv. Phys. Acta.* **7**, 81 (1934).

37. R. Peierls, *Ann. Inst. Henri Poincaré* **5**, 177 (1935).

38. L. D. Landau, *Phys. Z. Sowjetunion* **11**, 26 (1937).

39. L. D. Landau and E. M. Lifshitz, *Statistical Physics* (Pergamon Press, Oxford, 1986).

40. P. C. Hohenberg, *Phys. Rev.* **158**, 383 (1967).
41. N. D. Mermin and H. Wagner, *Phys. Rev. Lett.* **17**, 1133 (1966).
42. N. D. Mermin, *Phys. Rev.* **176**, 250 (1968).
43. D. A. Young and B. J. Alder, *J. Chem. Phys.* **60**, 1254 (1974).
44. J. A. Zollweg, G. V. Chester, and P. W. Leung, *Phys. Rev. B* **39**, 9518 (1989).
45. R. C. Gann, S. Chakravarty, and G. V. Chester, *Phys. Rev. B* **20**, 326 (1979).
46. B. Jancovici, *Phys. Rev. Lett.* **19**, 20 (1967).
47. Y. Imry and L. Gunther, *Phys. Rev. B* **3**, 3939 (1971).
48. H. Mikeska and H. Schmidt, *J. Low Temp. Phys.* **2**, 371 (1970).
49. B. J. Alder and T. E. Wainright, *Phys. Rev.* **127**, 359 (1962).
50. H. E. Stanley and T. A. Kaplan, *Phys. Rev. Lett.* **17**, 913 (1966).
51. J. M. Kosterlitz and D. J. Thouless, *J. Phys. C* **5**, L124 (1972).
52. J. M. Kosterlitz and D. J. Thouless, *J. Phys. C* **6**, 1181 (1973).
53. B. I. Halperin and D. R. Nelson, *Phys. Rev. Lett.* **41**, 121 (1978).
54. D. R. Nelson and B. I. Halperin, *Phys. Rev. B* **19**, 2457 (1979).
55. A. P. Young, *Phys. Rev. B* **19**, 1855 (1979).
56. W. G. Hoover and B. J. Alder, *J. Chem. Phys.* **46**, 686 (1967).
57. F. H. Ree, in *Physical Chemistry, An Advanced Treatise, Volume VI–IIA*, D. Henderson, Ed. (Academic Press, New York, 1971) p. 157.
58. S. Toxvaerd, *Phys. Rev. Lett.* **51**, 1971 (1983).
59. F. H. Stillinger and T. A. Weber, *J. Chem. Phys.* **74**, 4015 (1981).
60. H. Cheng, P. Dutta, D. E. Ellis, and R. Kalia, *J. Chem. Phys.* **85**, 2232 (1986).
61. F. Tsien and J. P. Valleau, *Mol. Phys.* **27**, 177 (1974).
62. L. V. Woodcock, F. van Swol, and J. N. Cape, *J. Chem. Phys.* **73**, 913 (1980).
63. F. F. Abraham, *Phys. Rev. Lett.* **44**, 463 (1980).
64. J. A. Barker, D. Henderson, and F. F. Abraham, *Physica A* **106**, 226 (1981).
65. R. K. Kalia, P. Vashishta, and S. W. de Leeuw, *Phys. Rev. B* **23**, 4794 (1981).
66. P. Vashishta and R. K. Kalia, in *Melting, Localization, and Chaos*, R. K. Kalia and P. Vashishta, Eds. (North-Holland, Amsterdam, 1982) p. 43.
67. M. P. Allen, D. Frenkel, W. Gignac, and J. P. McTague, *J. Chem. Phys.* **78**, 4206 (1983).
68. J. Q. Broughton, G. H. Gilmer, and J. D. Weeks, *Phys. Rev. B* **25**, 4651 (1982).
69. D. J. Evans, *Phys. Lett. A* **88**, 48 (1982).
70. S. W. de Leeuw and J. W. Perram, *Physica A* **113**, 546 (1982).
71. J. M. Caillol, D. Levesque, J. J. Weis, and J. P. Hansen, *J. Stat. Phys.* **28**, 325 (1982).
72. Ph. Choquard and J. Clerouin, *Phys. Rev. Lett.* **50**, 2086 (1983).
73. J. D. Weeks, *Phys. Rev. B* **24**, 1530 (1981).
74. S. Toxvaerd, *J. Chem. Phys.* **69**, 4750 (1978).
75. V. M. Bedanov, G. V. Gadiyak, and Y. E. Lozovik, *Sov. Phys. Solid State* **25**, 113 (1983).
76. V. M. Bedanov, G. V. Gadiyak, and Y. E. Lozovik, *Phys. Lett. A* **109**, 289 (1985).
77. V. M. Bedanov, G. V. Gadiyak, and Y. E. Lozovik, *Sov. Phys. JETP* **61**, 967 (1985).
78. M. P. Allen and D. J. Tildesley, *Computer Simulation of Liquids* (Oxford University Press, Oxford, 1989).

79. W. G. Hoover and F. H. Ree, *J. Chem. Phys.* **49**, 3608 (1968).

80. F. R. N. Nabarro, *Theory of Crystal Dislocations* (Dover Publications, New York, 1987).

81. J. Friedel, *Dislocations* (Pergamon Press, London, 1964).

82. L. D. Landau and E. M. Lifshitz, *Theory of Elasticity* (Pergamon Press, Oxford, 1986).

83. D. R. Nelson, *Phys. Rev. B* **18**, 2318 (1978).

84. Michael N. Barber, *Phys. Rep.* **59**, 375 (1980).

85. A. N. Berker and D. R. Nelson, *Phys. Rev. B* **19**, 2488 (1979).

86. F. Wegner, *Z. Phys.* **206**, 465 (1967).

87. J. M. Kosterlitz, *J. Phys. C* **7**, 1046 (1974).

88. P. S. Pershan, *Structure of Liquid Crystal Phases* (World Scientific, Singapore, 1988).

89. D. S. Fisher, B. I. Halperin, and R. Morf, *Phys. Rev. B* **20**, 4692 (1979).

90. B. Joos and M. S. Duesbery, *Phys. Rev. Lett.* **55**, 1997 (1985).

91. C. A. Murray and D. H. Van Winkle, *Phys. Rev. Lett.* **58**, 1200 (1987).

92. C. A. Murray and R. A. Wenk, *Phys. Rev. Lett.* **62**, 1643 (1989).

93. A. Armstrong, *Defect Mediated Melting in Colloidal Monolayers*, Ph.D. thesis, University of Colorado, Boulder, 1988.

94. A. J. Armstrong, R. C. Mockler, and W. J. O'Sullivan, *J. Phys.: Condens. Matter* **1**, 1707 (1989).

95. Y. Tang, *Study of Two Dimensional Melting in Colloidal Monolayers*, Ph.D. thesis, University of Colorado, Boulder, 1989.

96. Y. Tang, A. J. Armstrong, R. C. Mockler, and W. J. O'Sullivan, *Phys. Rev. Lett.* **62**, 2401 (1990).

97. R. M. Malzbender, *Freezing Flatland: Analog Simulation of a Phase Transition in a Classical Two-Dimensional System*, Ph.D. thesis, University of Colorado, Boulder, 1990.

98. J. Frenkel, *Kinetic Theory of Liquids* (Dover, New York, 1955).

99. D. E. O'Reilly, *Phys. Rev. A* **15**, 1198 (1977).

100. S. F. Edwards and M. Warner, *Philos. Mag. A* **40**, 257 (1979).

101. D. R. Nelson and J. Toner, *Phys. Rev. B* **24**, 363 (1981).

102. N. Rivier and D. M. Duffy, *J. Phys. C* **15**, 2867 (1982).

103. S. M. Stishov, I. N. Makarenko, V. A. Ivanov, and A. M. Nikolaenko, *Phys. Lett. A* **45**, 18 (1973).

104. M. Lasocka, *Phys. Lett. A* **51**, 137 (1975).

105. W. Janke and H. Kleinert, *Phys. Lett. A* **105**, 134 (1984).

106. W. Janke and H. Kleinert, *Phys. Rev. Lett.* **61**, 2344 (1988).

107. R. W. Gurney and N. F. Mott, *Rep. Prog. Phys.* **5**, 46 (1938).

108. R. W. Gurney and N. F. Mott, *Trans. Faraday Soc.* **35**, 364 (1939).

109. S. T. Chui, *Phys. Rev. Lett.* **48**, 933 (1982).

110. S. T. Chui, *Phys. Rev. B* **28**, 178 (1983).

111. S. T. Chui, in *Melting, Localization, and Chaos*, R. K. Kalia and P. Vashishta, Eds. (North-Holland, Amsterdam, 1982) p. 29.

112. W. T. Read and W. Shockley, *Phys. Rev.* **78**, 275 (1950).

113. W. Shockley, in *L'État Solide*, R. Stoops, Ed. (Coudenberg, Bruxelles, 1952) p. 431.

114. Y. Saito, *Phys. Rev. B* **26**, 6239 (1982).

115. Y. Saito, *Phys. Rev. Lett.* **48**, 1114 (1982).

116. Y. Saito, *Surf. Sci.* **125**, 285 (1983).

117. H. Muller-Krumbhaar and Y. Saito, *Surf. Sci.* **144**, 84 (1984).

118. D. R. Nelson, *Phys. Rev. B* **26**, 269 (1982).

119. K. J. Strandburg, S. A. Solla, and G. V. Chester, *Phys. Rev. B* **28**, 2717 (1983).

120. K. J. Strandburg, *Phys. Rev. B* **34**, 3536 (1986).

121. W. Janke and H. Kleinert, *Phys. Lett. A* **114**, 255 (1986).

122. W. Janke and D. Toussaint, *Phys. Lett. A* **116**, 387 (1986).

123. J. G. Kirkwood and E. Monroe, *J. Chem. Phys.* **9**, 514 (1941).

124. R. Evans, *Adv. Phys.* **28**, 143 (1979).

125. N. D. Mermin, *Phys. Rev. A* **137**, 1441 (1965).

126. D. Stroud and N. W. Ashcroft, *Phys. Rev. B* **5**, 371 (1972).

127. T. V. Ramakrishnan and M. Yussouff, *Phys. Rev. B* **19**, 2775 (1979).

128. S. Alexander and J. McTague, *Phys. Rev. Lett.* **41**, 702 (1978).

129. T. Lubensky and P. Chaikin, *Soft Condensed Matter Physics*, unpublished (1991).

130. W. A. Curtin, *J. Chem. Phys.* **88**, 7050 (1988).

131. M. Baus and J. L. Colot, *Mol. Phys.* **55**, 653 (1985).

132. C. Cerjan, B. Bagchi, and S. A. Rice, *J. Chem. Phys.* **83**, 2376 (1985).

133. J. F. Lutsko and M. Baus, *Phys. Rev. A* **41**, 6647 (1990).

134. J. F. Lutsko and M. Baus, *Phys. Rev. Lett.* **64**, 761 (1990).

135. P. D. Beale and D. Holger, private communication (1991).

136. B. Bagchi, C. Cerjan, and S. A. Rice, *J. Chem. Phys.* **79**, 6222 (1983).

137. B. Bagchi, C. Cerjan, and S. A. Rice, *J. Chem. Phys.* **79**, 5595 (1983).

138. C. Cerjan and B. Bagchi, *Phys. Rev. A* **31**, 1647 (1985).

139. B. Bagchi, C. Cerjan, and S. A. Rice, *Phys. Rev. B* **29**, 2857 (1984).

140. T. V. Ramakrishnan, *Phys. Rev. Lett.* **48**, 541 (1982).

141. P. Ballone, G. Pastore, M. Rovere, and M. P. Tosi, *J. Phys. C* **18**, 4011 (1985).

142. P. L. Radloff, B. Bagchi, C. Cerjan, and S. A. Rice, *J. Chem. Phys.* **81**, 1406 (1984).

143. B. B. Laird, J. D. McCoy, and A. D. J. Haymet, *J. Chem. Phys.* **88**, 3900 (1988).

144. P. Tarazona, *Mol. Phys.* **52**, 81 (1984).

145. J.-L. Colot and M. Baus, *Phys. Lett. A* **119**, 135 (1986).

146. L. Mederos, P. Tarazona, and G. Navascués, *Phys. Rev. B* **35**, 3376 (1987).

147. T. Wainwright and B. J. Alder, *Suppl. Nuovo Cimento* **9**, 116 (1958).

148. B. J. Alder, W. G. Hoover, and D. A. Young, *J. Chem. Phys.* **49**, 3688 (1968).

149. J. A. Zollweg, in *Ordering in Two Dimensions*, S. K. Sinha, Ed. (North-Holland, Amsterdam, 1980) p. 331.

150. K. J. Strandburg, J. A. Zollweg, and G. V. Chester, *Phys. Rev. B* **30**, 2755 (1984).

151. M. A. A. da Silva, A. Caliri, and B. J. Mokross, *Phys. Rev. Lett.* **58**, 2312 (1987).

152. J. Q. Broughton, G. H. Gilmer, and J. D. Weeks, *J. Chem. Phys.* **75**, 5128 (1981).

153. J. P. McTague, D. Frenkel, and M. P. Allen, in *Ordering in Two Dimensions*, S. K. Sinha, Ed. (North-Holland, Amsterdam, 1980) p. 147.

154. A. D. Novaco and P. A. Shea, *Phys. Rev. B* **26**, 284 (1982).

155. A. D. Novaco, *Phys. Rev. B* **35**, 8621 (1987).

156. R. K. Kalia and P. Vashishta, *J. Phys. C* **14**, L643 (1981).

157. V. M. Bedanov, G. V. Gadiyak, and Y. E. Lozovik, *Phys. Lett. A* **92**, 400 (1982).

158. R. W. Hockney and T. R. Brown, *J. Phys. C* **8**, 1813 (1975).

159. R. H. Morf, *Phys. Rev. Lett.* **43**, 931 (1979).

160. R. Morf, in *Physics of Intercalation Compounds*, L. Pietronero and E. Tosatti, Ed. (Springer-Verlag, Berlin, 1981) p. 252.

161. R. K. Kalia and P. Vashishta, in *Physics of Intercalation Compounds*, L. Pietronero and E. Tosatti, Eds. (Springer-Verlag, Berlin, 1981), p. 244.

162. A. T. Fiory, *Phys. Rev. B* **28**, 236 (1983).

163. F. W. de Wette, R. E. Allen, D. S. Hughes, and A. Rahman, *Phys. Lett. A* **29**, 548 (1969).

164. R. M. J. Cotterill and L. B. Pedersen, *Solid State Commun.* **10**, 439 (1972).

165. D. Frenkel and J. P. McTague, *Phys. Rev. Lett.* **42**, 1632 (1979).

166. D. Frenkel, F. E. Hanson, and J. P. McTague, in *Ordering in Strongly Fluctuating Condensed Matter Systems*, T. Riste, Ed. (Plenum Press, New York, 1980) p. 285.

167. F. F. Abraham, in *Ordering in Two Dimensions*, S. K. Sinha, Ed. (North-Holland, Amsterdam, 1980) p. 155.

168. F. F. Abraham, *Phys. Rev. B* **23**, 6145 (1981).

169. F. F. Abraham, in *Melting, Localization, and Chaos*, R. K. Kalia and P. Vashishta, Eds. (North-Holland, Amsterdam, 1982) p. 75.

170. F. F. Abraham, *Phys. Rep.* **80**, 339 (1981).

171. S. Toxvaerd, *Phys. Rev. Lett.* **44**, 1002 (1980).

172. S. Toxvaerd, *Phys. Rev. A* **24**, 2735 (1981).

173. J. Tobochnik and G. V. Chester, in *Ordering in Two Dimensions*, S. K. Sinha, Ed. (North-Holland, Amsterdam, 1980) p. 339.

174. J. Tobochnik and G. V. Chester, *Phys. Rev. B* **25**, 6778 (1982).

175. A. F. Bakker, C. Bruin, and H. J. Hilhorst, *Phys. Rev. Lett.* **52**, 449 (1984).

176. C. Udink and J. van der Elsken, *Phys. Rev. B* **35**, 279 (1987).

177. C. Udink and D. Frenkel, *Phys. Rev. B* **35**, 6933 (1987).

178. M. A. Glaser and N. A. Clark, *Phys. Rev. A* **41**, 4585 (1990).

179. M. A. Glaser and N. A. Clark, in *Geometry and Thermodynamics*, J.-C. Tolédano, Ed. (Plenum Press, New York, 1990) p. 193.

180 T. A. Weber and F. H. Stillinger, *J. Chem. Phys.* **74**, 4020 (1981).

181. J. A. Combs, *Phys. Rev. Lett.* **61**, 714 (1988).

182. J. A. Combs, *Phys. Rev. B* **38**, 6751 (1988).

183. J. D. Weeks, D. Chandler, and H. C. Andersen, *J. Chem. Phys.* **54**, 5237 (1971).

184. G. Ciccotti and W. G. Hoover, Eds. *Molecular Dynamics Simulation of Statistical-Mechanical Systems*. (North-Holland, Amsterdam, 1986).

185. B. J. Alder, in *Molecular Dynamics Simulation of Statistical-Mechanical Systems*, G. Ciccotti and W. G. Hoover, Eds. (North-Holland, Amsterdam, 1986) p. 66.

186. F. H. Stillinger and T. A. Weber, *Phys. Rev. A* **25**, 978 (1982).

187. G. F. Voronoi, *J. Reine, Angew. Math.* **134**, 198 (1908).

188. R. Collins, in *Phase Transitions and Critical Phenomena*, C. Domb and M. S. Green, Eds. (Academic Press, New York, 1972) p. 271.

189. S. A. Rice and P. Gray, *The Statistical Mechanics of Simple Liquids* (Wiley Interscience, New York, 1965).

190. D. A. McQuarrie, *Statistical Mechanics* (Harper & Row, New York, 1976).

191. M. A. Glaser, *Melting as Defect Condensation: A Tiling Model for Two-Dimensional Melting*, Ph.D. thesis, University of Colorado, Boulder, 1991.

192. D. P. Fraser, M. J. Zuckermann, and O. G. Mouritsen, *Phys. Rev. A* **43**, 6642 (1991).

193. D. Stuaffer, *Phys. Rep.* **54**, 1 (1979).

194. D. Stauffer, *Introduction to Percolation Theory* (Taylor & Francis, London, 1985).

195. M. E. Fisher, *Physics* **3**, 255 (1967).

196. J. A. Barker and D. Henderson, *Rev. Mod. Phys.* **48**, 587 (1976).

197. J. D. Bernal, *Proc. R. Soc. A* **163**, 320 (1937).

198. J. D. Bernal, *Trans. Farad. Soc.* **33**, 27 (1937).

199. J. M. Ziman, *Models of Disorder* (Cambridge University Press, Cambridge, 1979).

200. J. D. Bernal and J. Mason, *Nature* **188**, 910 (1960).

201. J. D. Bernal and S. V. King, *Discuss. Farad. Soc.* **43**, 60 (1967).

202. J. D. Bernal and J. L. Finney, *Discuss. Farad. Soc.* **43**, 62 (1967).

203. M. F. Ashby, F. Spaepen, and S. Williams, *Acta Metall.* **26**, 1647 (1978).

204. A. S. Nowick and S. R. Mader, *IBM J.* **September–November**, 358 (1965).

205. M. Rubinstein and D. R. Nelson, *Phys. Rev. B* **26**, 6254 (1982).

206. G. Y. Onoda and J. Toner, *IBM Res. Rep.* **RC 11827** (Log #53135) (1986).

207. J. L. Finney, *Mater. Sci. Eng.* **23**, 207 (1976).

208. D. R. Nelson and F. Spaepen, *Solid State Phys.* **42**, 1 (1989).

209. R. Collins, *Proc. Phys. Soc.* **83**, 553 (1964).

210. H. Kawamura, in *Topological Disorder in Condensed Matter*, F. Yonezawa and T. Ninomiya, Eds. (Springer-Verlag, Berlin, 1983) p. 181.

211. H. Kawamura, *Prog. Theor. Phys.* **70**, 352 (1983).

212. D. Yi-Jing, C. Li-Rong, and Y. Tzu-Tung, *J. Phys. C* **15**, 3059 (1982).

213. Y. M. Yi and Z. C. Guo, *Commun. Theor. Phys.* **8**, 17 (1987).

214. Y. M. Yi and Z. C. Guo, *J. Phys.: Condens. Matter* **1**, 1731 (1989).

215. P. W. Leung, C. L. Henley, and G. V. Chester, *Phys. Rev. B* **39**, 446 (1989).

216. D. X. Li, H. Chen, and K. H. Kuo, *Phys. Rev. Lett.* **60**, 1645 (1988).

217. C. L. Henley, *J. Phys. A* **21**, 1649 (1988).

218. M. A. Glaser and N. A. Clark, in *Dynamics and Patterns in Complex Fluids*, A. Onuki and K. Kawasaki, Eds. (Springer-Verlag, Berlin, 1990) p. 141.

219. D. V. Boulatov, V. A. Kazakov, I. K. Kostov, and A. A. Migdal, *Nucl. Phys. B* **275**, 641 (1986).

220. A. N. Berker and D. Anderman, *J. Appl. Phys.* **53**, 7923 (1982).

221. B. Nienhuis, A. N. Berker, E. K. Riedel, and M. Schick, *Phys. Rev. Lett.* **43**, 737 (1979).

NOTE ADDED IN PROOF

Subsequent to the writing of this chapter we became aware of several interesting recent developments, as well as some relevant earlier work that we had overlooked.

Quite recently, Lee and Strandburg have studied the finite size scaling behavior of the bulk free energy barrier between the solid and liquid states in the hard disk system, and have obtained the first unambiguous evidence for a first-order melting transition [1].

Janke and Kleinert have recently reported a high-statistics Monte Carlo study of the Laplacian roughening model [2]. Based on a finite-size scaling analysis, they conclude that this model exhibits a single first-order melting transition, in agreement with their earlier work, rather than the two Kosterlitz–Thouless transitions found by Strandburg et al. (see Section II.E.5). Janke and Kleinert have also recently published a more detailed account of their proposed model for defect melting [3] (see Section II.E.4).

Two earlier studies of dense random packings of hard disks have been brought to our attention [4, 5]. Both studies obtain a packing fraction in the range 0.82–0.83 for dense random packings of hard disks, in general agreement with our findings. However, Quickenden and Tan [5] also find that 2D dense random packings are unstable to coordinate contraction, slowly and asymptotically approaching the close-packed limit with time. The possible nonexistence of stable 2D dense random packings certainly deserves further investigation.

Finally, we would like to direct the reader's attention to a recent monograph on bond orientational order in condensed matter [6], which contains some discussion of bond orientational order in 2D systems.

References

1. J. Lee and K. J. Strandburg, submitted to *Phys. Rev. Lett.* (1992).
2. W. Janke and H. Kleinert, *Phys. Lett. A* **140**, 513 (1989).
3. W. Janke and H. Kleinert, *Phys. Rev. B* **41**, 6848 (1990).
4. H. H. Kausch, D. G. Fesko, and N. W. Tschoegl, *J. Colloid Interface Sci.* **37**, 603 (1971).
5. T. I. Quickenden and G. K. Tan, *J. Colloid Interface Sci.* **48**, 382 (1974).
6. *Bond Orientational Order in Condensed Matter Systems*, K. J. Strandburg, ed. (Springer-Verlag, New York, 1992).

AUTHOR INDEX

Numbers in parentheses are reference numbers and indicate that the author's work is referred to although his name is not mentioned in the text. Numbers in *italic* show the pages on which the complete references are listed.

711

SUBJECT INDEX